PREFACE TO THE FIFTH EDITION

This book is intended for a comprehensive first course in physical chemistry. It emphasizes the fundamentals that provide a basis for the understanding of chemistry. The objective of the book over the years has been to make physical ideas about chemistry accessible to students. As more physical chemistry has been taught in lower level courses, it has been possible to eliminate certain elementary material and add more advanced material.

The fifth edition has the same four main sections as the fourth edition, but the order has been changed so that Quantum Chemistry comes before Chemical Dynamics. This makes it possible to use quantum and statistical mechanical ideas in discussing chemical kinetics. It also allows photochemistry to be discussed in the context of kinetics. Since gas-phase kinetics and liquid-phase kinetics are so different, they are treated in separate chapters in this edition. The number of chapters has been kept to 20 by combining the fourth-edition chapters on One-Component Systems and Phase Equilibria.

In this edition the treatment of the second law of thermodynamics has been expanded by the inclusion of the Carnot cycle for heat engines, refrigerators, and heat pumps. Although it is true that the chemical applications of entropy can be developed without the Carnot cycle, as they were in the fourth edition, the idea of cyclic processes is important in chemistry and chemical engineering. In addition, the chapter on the second law has been expanded to give a more complete treatment of the thermodynamics of perfect gases and their mixtures. Two major thermodynamic tables have been added to the Appendix to replace four tables that were scattered through the text in previous editions. This has the advantage of showing the close interrelationships of the various thermodynamic quantities.

The treatments of quantum mechanics and statistical mechanics have been expanded in this edition because of their growing importance in chemistry. For example, statistical mechanics is used to calculate the thermodynamic properties of atomic and molecular hydrogen at 3000 K. In this way it is possible to calculate the degree of dissociation of hydrogen at 3000 K using only information on hydrogen atoms and hydrogen molecules.

The movement toward SI units continues in this edition, but several non-SI units are retained, primarily the calorie, atmosphere, and electron volt. The Angstrom unit has been eliminated in favor of the nanometer (1 nm = 10^{-9} m) and the picometer (1 pm = 10^{-12} m). According to the SI system, the calorie should be replaced by the joule, but since the major reference tables (NBS Technical Note 270 and JANAF Thermochemical Tables, see Tables A.1 and A.2 of this book) are not yet available in joules, it did not seem practical to make the change in this edition.

Because the number of credits in physical chemistry courses, and therefore the need for more advanced material, varies at different universities and colleges, more topics have been included in this edition than can be covered in some courses. Some of the more advanced material has been set in smaller type to indicate that it might be avoided in a first course.

At the end of each chapter there are two parallel problem sets. The answers are given for the first set. The *Solutions Manual for Physical Chemistry*, 2nd ed., John Wiley, New York, 1979, contains worked-out solutions for all of the different types of problems in the first set and answers for the second set. This edition contains 161 new problems.

Outlines of Theoretical Chemistry, as it was then entitled, was written in 1913 by Dr. Frederick Getman, who carried it through 1927 in four editions. The next four editions were written by Dr. Farrington Daniels. In 1955 I joined Dr. Daniels in the first edition of this book. We worked together on three editions and started working on the fourth before Dr. Daniels died on June 23, 1972. His good judgment and wise counsel have been greatly missed. It is remarkable that the present edition traces its origins back 66 years.

Many individuals made helpful suggestions in the preparation of this edition. I want to especially acknowledge the suggestions provided by Malcolm W. Chase, John M. Deutch, Robert W. Field, Carl W. Garland, Edward L. King, James L. Kinsey, Irwin Oppenheim, Martin A. Paul, John Ross, Clark C. Stephenson, Paul R. Schimmel, Robert J. Silbey, Ralph H. Staley, Jeffrey I. Steinfeld, John S. Waugh, and Mark S. Wrighton.

Philip A. Lyons, Mark A. Ratner, Philip H. Rieger, and Peter E. Yankwich read the entire manuscript and made many useful suggestions. Philip H. Rieger reread the entire manuscript after it had been completed and found further ways in which the book could be improved.

I especially thank Lillian Alberty for the difficult job of typing the manuscript and for encouraging me in the preparation of this new edition.

Cambridge, Massachusetts, 1978 Robert A. Alberty

A NOTE TO THE STUDENT

A problems book consisting of selected problems from the first set of problems, including worked-out solutions, and answers for the second set of problems, is available as a companion to this text. Please ask for *Solutions Manual for Physical Chemistry*, SI Version, by Robert A. Alberty.

R. A. A.

CONTENTS

PHYSICAL QUANTITIES AND UNITS

The measurement of any physical quantity consists of a comparison with a standard amount of that quantity, which is referred to as a unit. Thus the statement of a physical measurement consists of two parts: (1) a number that represents the number of times the unit has to be used to give the physical quantity, and (2) the unit itself. The SI system of units is founded on the seven base units listed in the following table. (SI stands for Systeme International d'Unites.) The definitions of these units are given in the Appendix. All other physical quantities may be expressed in terms of combinations of these base units. Combinations of units are referred to as derived units. Some derived units have their own special symbols. For example, the unit of work in the SI system is the joule, which is represented by J and is defined in terms of base units by kg m^2 s^{-2}. The symbols for other derived units are given in the Appendix. In the book we will use SI units except for some units, such as the atmosphere and electron volt which, however, may be expressed in terms of SI units.

Basic Physical Quantity	Symbol for Quantity	Name of SI Unit	Symbol for SI Unit
Length	l	meter	m
Mass	m	kilogram	kg
Time	t	second	s
Electric current	I	ampere	A
Thermodynamic temperature	T	kelvin	K
Amount of substance	n	mole	mol
Luminous intensity	I_v	candela	cd

(Note that symbols for quantities are always printed in italic type, and symbols for units are always printed in roman type.)

A physical quantity may be expressed in various units. For example, a certain quantity of heat q may be expressed as $q = 1$ cal $= 4.184$ J. Note that it is perfectly correct to use equal signs in this way.

When you look at an equation relating physical quantities, it is important to realize that the symbols represent numbers and associated units. In carrying out algebra with physical quantities the mathematical processes of addition, subtraction, and equating can be applied only to quantities of the same kind. If the quantities are of the same kind, they can be expressed in the same unit.* When the terms to be added, subtracted, or equated are of the same kind, the equation is said to be dimensionally homogeneous.

It is also important to remember that x in e^x, $\ln x$, and $\sin x$ can only be a pure number and cannot have a unit.

If units are chosen arbitrarily, additional numerical factors may appear in equations relating different physical quantities. In practice it is more convenient to choose a system of units so that the equations between physical quantities have exactly the same form as the corresponding equations between pure numbers. A system of units that has this property is said to be coherent. The SI system is coherent. This means that if all quantities in a calculation are expressed in SI base units, the result will be expressed in SI base units without including any numerical factors. It is, however, a good habit to check a calculation to see that units cancel to yield the correct units for the final result.

A physical quantity may be converted from one unit to another by multiplying by a conversion factor. To find a conversion factor, it is necessary to express one unit in terms of another. For example, one calorie is equal to 4.184 joules; that is, 1 cal = 4.184 J. Dividing both sides of the equation by 1 cal yields 1 = 4.184 J cal^{-1}. To convert the change in enthalpy of a reaction from calories to joules, simply multiply the change in enthalpy in calories by 4.184 J cal^{-1}; note that this is equivalent to multiplying by one. A number of useful conversion factors are listed inside the front cover.

In order to make a table or graph of a physical quantity, it is convenient to divide the physical quantity by the unit to obtain a pure number. For example, $\Delta H^\circ = 100\ \text{kJ mol}^{-1}$ may be written $\Delta H^\circ/(\text{kJ mol}^{-1}) = 100$. This system is used in this textbook. A physical quantity may be divided by a power of 10 as well as by a unit to obtain a convenient numerical value for plotting or tabulating. For example, $k = 1.53 \times 10^{10}\ \text{m}^3\ \text{s}^{-1}$ or $k/(10^{10}\ \text{m}^3\ \text{s}^{-1}) = 1.53$.

In the preceding edition bars were placed over thermodynamic quantities *per mole*. That practice has been discontinued in this edition, except for partial molar quantities; chemists primarily use thermodynamic quantities per mole, so that bars would have to be used almost all the time. However, chemists do *not* regularly use bars or other ways of indicating thermodynamic quantities per mole. The use of the mole as a unit, as required by the SI system, is a constant reminder of the size of the system being considered.

---

* Note that the inverse of this statement is not necessarily true. For example, the entropy and heat capacity may both be expressed in $\text{J K}^{-1}\ \text{mol}^{-1}$, but this does not mean that they can be added, subtracted, or equated.

PART ONE

THERMODYNAMICS

Thermodynamics deals with relationships between properties of systems at equilibrium and with differences in properties between various equilibrium states. It has nothing to do with time. Even so, it is one of the most powerful tools of physical chemistry; because of its importance, the first part of this book is devoted to it. The first law of thermodynamics deals with the amount of work that can be done by a chemical or physical process and the amount of heat that is absorbed or evolved. On the basis of the first law it is possible to build up tables of enthalpies of formation that may be used to calculate enthalpy changes for reactions that have not yet been studied. With information on heat capacities of reactants and products also available, it is possible to calculate the heat of a reaction at a temperature where it has not previously been studied.

The second law of thermodynamics deals with the natural direction of processes and the question of whether a given chemical reaction can occur by itself. The second law was formulated initially in terms of the efficiencies of heat engines, but it also leads to the definition of entropy, which is important in determining the direction of chemical change. The second law provides the basis for the definition of the equilibrium constant for a chemical reaction. It provides an answer to the question, "To what extent will this particular reaction go before equilibrium is reached?" It also provides the basis for reliable predictions of the effects of temperature, pressure, and concentration on chemical and physical equilibrium. The third law provides the basis for calculating equilibrium constants from calorimetric measurements only. This is an illustration of the way in which thermodynamics interrelates apparently unrelated measurements on systems at equilibrium.

Here, the ideas of thermodynamics are first applied to equilibria between different phases. This provides the basis for the quantitative treatment of fractional distillation and for the interpretation of the phase changes in mixtures of solids. The ideas of thermodynamics are next applied to chemical reactions. The use of thermodynamic tables to calculate equilibrium compositions is discussed in detail. These methods are also applied to the reactions of electrochemical cells, which determine the maximum electromotive force that a cell can produce.

Finally, the concepts and relationships of thermodynamics are applied to the equilibrium properties of surfaces. These include surface tension, surface pressure, adsorption, adhesion, and the effect of curvature of a surface on vapor pressure.

Equilibrium conditions are independent of mechanism; the great strength (and weakness) of thermodynamics is that it is not concerned with mechanisms or models (as, for example, molecules). In Part Two we will see how various thermodynamic quantities may be calculated from information about individual molecules using statistical mechanics. The relations between the various thermodynamic quantities derived classically also apply in statistical mechanics. Statistical mechanics provides insight into thermodynamics, but it is difficult to apply statistical mechanics to very complicated systems and to highly interactive systems.

CHAPTER 1

FIRST LAW OF THERMODYNAMICS

The quantitative concepts of temperature, work, internal energy, and heat play an important role in the understanding of chemical phenomena. These concepts will be developed in this chapter, and the relationship between heat and work as forms of energy will be emphasized. The chapter opens with a discussion of the thermodynamic concept of temperature. The principle involved in defining temperature was not recognized until after the establishment of the first and second laws of thermodynamics, and therefore it is referred to as the "zeroth" law.

The first law is often called the law of conservation of energy. This concept first appeared in mechanics and was later extended to include electrostatics and electrodynamics. Joule performed experiments in 1840–1845 that showed how heat could also be included in the conservation of energy. The first law leads to the definitions of the internal energy U and enthalpy H. One of the important applications of the first law in chemistry is to the interpretation of the heat effects of chemical reactions. Furthermore, if heat capacities of reactants and products are known, the heat of reaction may be calculated at other temperatures after it has been measured at one.

1.1 SYSTEM, SURROUNDINGS, STATE OF A SYSTEM, AND STATE VARIABLES

A thermodynamic system is a part of the physical universe that is under consideration. It is separated from its surroundings by a boundary. If the boundary prevents any interaction with the surroundings, the system is an isolated system. If matter can pass across the boundary, we have an open system. If it cannot, we have a closed system. Heat may enter or leave a closed system, and work may be done by or on a closed system.

A system may be taken through a series of changes in which work and heat pass through the boundary so that there is a change in the surroundings as well as the system. If the boundary does not permit the flow of heat, any process occurring in the system is said to be adiabatic, and the boundary is called an adiabatic wall.

When a system is at equilibrium under a given set of conditions, it is said to be in a definite state. The state of a system may be identified from the fact that when it is in a definite state, each of its properties has a definite value. It is found that, for a fixed amount of a one-component fluid (gas or liquid), the state is completely defined by any two of the three variables—pressure, volume, and temperature. Such variables are referred to as state variables.

Thermodynamics is concerned with transformations from an initial state to a final state. In such a transformation heat may pass through the boundary of the system, and work may be done on the system or on the surroundings.

1.2 EXTENSIVE AND INTENSIVE THERMODYNAMIC QUANTITIES

If the size of a thermodynamic system is doubled without any other change, certain thermodynamic quantities that are used to describe the system also double. Examples are the volume of the system and its energy. Such thermodynamic quantities are referred to as *extensive*. Other thermodynamic properties, such as the temperature and pressure, are not affected by changing the size of the system, so they are referred to as *intensive* properties.

If an extensive thermodynamic property of a system is divided by a measure of the amount of substance (being chemists we will usually use the mole), an *intensive* property is obtained. For example, if a system consists of a pure substance, doubling the size doubles the volume, but the molar volume remains constant. Strictly speaking, we should use a different symbol for the volume and the molar volume, because these quantities have different dimensions. The volume may be expressed in cubic meters or liters, but the molar volume is expressed in cubic meters per mole or liters per mole. Unfortunately, using different symbols for extensive thermodynamic properties and their molar counterparts greatly increases the number of symbols and equations. Actually most chemical calculations are carried out with molar quantities. Therefore we will use a single set of symbols for thermodynamic quantities and let them represent the intensive, or molar, quantities. Thus the perfect gas law will be written $PV = RT$, where V represents volume *per mole*. Since the mole is a unit in the SI system, it is important to write it in expressing physical quantities. On occasions where V is used to represent the extensive property, this will be noted.

1.3 THE ZEROTH LAW OF THERMODYNAMICS

If two closed systems are brought together so that they are in thermal contact, changes may take place in the properties of both. Eventually a state is reached in which there is no further change, and this is the state of thermal equilibrium. Thus we can readily determine whether two systems are at the same temperature by bringing them into contact and seeing whether observable changes take place in the properties of either system. If no change occurs, they are at the same temperature.

Now let us consider three systems A, B, and C. It is an experimental fact that if system A is in thermal equilibrium with system C, and system B is also in thermal equilibrium with system C, then A and B are in thermal equilibrium with each other. It is not obvious that this should necessarily be true, and so this empirical fact is referred to as the *zeroth law of thermodynamics.*

This law puts the concept of temperature on a firm basis in the following way: if two systems are in thermal equilibrium they have the same temperature; if they are not in thermal equilibrium they have different temperatures. Now, how can a temperature scale be set up?

To establish a temperature scale we start with system B in a state defined by volume V_B and pressure P_B. The values of V_A and P_A of a system of fluid A that is in equilibrium with B are determined experimentally. There are many combinations of P_A and V_A for which there is equilibrium, and these pairs of values may be plotted on a graph of P_A versus V_A such as that in Fig. 1.1. According to the zeroth law of thermodynamics, this curve at constant temperature (an isotherm) is independent of the nature of system B, since the same result would be obtained by using instead of B any other system in equilibrium with it. If the thermal state of B is changed and the experiment repeated, another isotherm for fluid A will be obtained. Each isotherm obtained in this way may be assigned a temperature θ, and in this way a temperature scale may be set up. All systems having the same temperature θ will remain unchanged in properties when they are brought into thermal contact with each other through a wall that allows the two systems to have different pressures and different chemical compositions.

Many different temperature scales may be defined, but the simplest and most useful is that based on the behavior of perfect gases, as obtained by the extrapolation of the behavior of real gases to zero pressure. We will find later that this scale is identical with one based on the second law of thermodynamics, and that is independent of the properties of any particular substance (see Section 2.4). In Chapter 13 the perfect gas temperature scale will be identified with that which arises in statistical mechanics.

The pairs of the variables P and V that correspond to the same temperature may be determined (e.g., the curves in Fig. 1.1) and represented by means of a function

$$f(P, V) = \theta \tag{1.1}$$

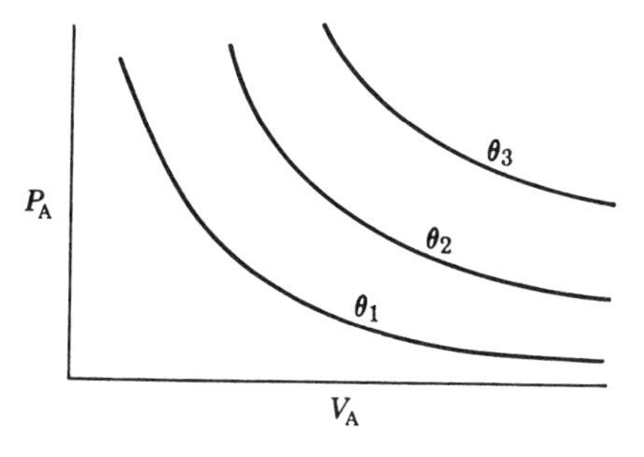

Fig. 1.1 Isotherms for fluid A. This plot, which is for a hypothetical fluid A, might look quite different for some other fluid.

where θ is the temperature. Such an equation is called the *equation of state* for the fluid. Real fluids all have different equations of state. According to this equation, there exists a function of the state of a fluid, called the temperature, that has the same value for fluids that are in thermal equilibrium with each other. In order to use the behavior of gases at low pressures to define θ, we start with Boyle's law (1662). As the pressure of a fixed amount of a gas is lowered, it follows the equation

$$PV = k \qquad \text{(at constant temperature)} \tag{1.2}$$

more and more closely. It is found that if the molar volume is used, the PV products of gases extrapolated to zero pressure all follow the same function of temperature.

$$\lim_{P \to 0} (PV)_\theta = f(\theta) \tag{1.3}$$

It is convenient to take this function as RT, where R is the gas constant and T is the perfect gas temperature.*

$$\lim_{P \to 0} (PV)_T = RT \tag{1.4}$$

All that is needed to complete the definition of the temperature scale is to specify T at some standard condition so that the gas constant R may be calculated. The size of the unit of perfect gas temperature, the Kelvin K in SI units,† is set by assigning the origin at absolute zero and 273.16 K (exactly) to the triple point of water (the temperature and pressure at which ice, liquid, and vapor are in equilibrium with each other in the absence of air).

The ice point (the temperature at which ice and water are in equilibrium in the presence of air at one atmosphere) sets the zero on the Celsius scale. The ice point is now defined as the temperature 0.0100 °C below that of the triple point of water, that is, 273.1500 K. Thus the Celsius temperature (t) is defined in terms of the thermodynamic temperature (T) by $t/°\text{C} = T/\text{K} - 273.15$.

1.4 THE GAS CONSTANT *R*

Very careful experiments with molecular oxygen show that as the pressure is reduced indefinitely the PV product for oxygen (31.9988 g mol^{-1}) approaches 22.413 83 L atm mol^{-1} at 0 °C (273.1500 K). We may calculate the gas constant using equation 1.4.

$$R = \frac{\lim_{P \to 0} (PV)_T}{T} = \frac{22.413\,83 \text{ L atm mol}^{-1}}{(273.1500 \text{ K})}$$

$$= 0.082\,0569 \text{ L atm K}^{-1} \text{ mol}^{-1} \tag{1.5a}$$

---

* It is not obvious that this temperature corresponds with our ideas of hotter and colder. However, the connection can be made; A. B. Pippard, *The Elements of Classical Thermodynamics*, Cambridge University Press, Cambridge, England, 1960.

† The unit is called the Kelvin because this temperature scale was introduced by Lord Kelvin in 1848. The K unit does not have a degree sign, but the degree sign is retained in the unit for the Celsius scale (°C).

A standard atmosphere is equal to the pressure required to support 76 cm of mercury at 0 °C at a point on the earth where the acceleration of gravity g is $9.806\,65\ \mathrm{m\,s^{-2}}$. The density of mercury at 0 °C is $13.5951\ \mathrm{g\,cm^{-3}}$, or $13.5951 \times 10^3\ \mathrm{kg\,m^{-3}}$.

Although the atmosphere is not an SI unit, we will sometimes find it convenient to use the atmosphere for measuring gas pressures.

The unit of pressure in the SI system is the pascal Pa, which is the pressure produced by a force of 1 N on an area of a square meter. Thus the pressure of a standard atmosphere in pascals may be calculated as follows.

$$\begin{aligned} P &= (0.76\ \mathrm{m})(13.5951 \times 10^3\ \mathrm{kg\,m^{-3}})(9.806\,65\ \mathrm{m\,s^{-2}}) \\ &= 101{,}325\ \mathrm{N\,m^{-2}} = 101{,}325\ \mathrm{Pa} \end{aligned}$$

Since pressure is force per unit area, the product of pressure and volume has the dimensions of force times distance, which is work or energy. Thus the gas constant may be expressed in the SI unit of energy, the joule J, by expressing the pressure in pascals.

$$\begin{aligned} R &= \frac{PV}{T} = \frac{(101{,}325\ \mathrm{N\,m^{-2}})(22.413\,83 \times 10^{-3}\ \mathrm{m^3\,mol^{-1}})}{(273.1500\ \mathrm{K})} \\ &= 8.314\,41\ \mathrm{J\,K^{-1}\,mol^{-1}} \end{aligned} \tag{1.5b}$$ *

1.5 WORK

In thermodynamics heat and work are algebraic quantities that can be positive or negative. Force is a vector quantity; that is, it has direction as well as magnitude. We will use boldface type for vectors.

Force is defined by

$$\boldsymbol{F} = m\boldsymbol{a} \tag{1.6}$$

where $\boldsymbol{F}$ is the force that will give a mass m an acceleration $\boldsymbol{a}$.

Work (w) is a scalar quantity defined by

$$w = \boldsymbol{F} \cdot \boldsymbol{l} \tag{1.7}$$

where $\boldsymbol{F}$ is the vector force, $\boldsymbol{l}$ is the vector length of path, and the dot indicates a scalar product (i.e., the product is taken of the magnitude of one vector by the projection of the second vector along the direction of the first). If the force vector of magnitude F and the vector length of magnitude l are separated by the angle θ, the work is given by $Fl\cos\theta$. In the SI system, the unit of work is the joule J; 1 J = 1 N m.

* The calorie is defined in terms of the joule. Since 1 cal = 4.184 J, the gas constant R is $8.314\,41\ \mathrm{J\,K^{-1}\,mol^{-1}}/4.184\ \mathrm{J\,cal^{-1}} = 1.987\,19\ \mathrm{cal\,K^{-1}\,mol^{-1}}$. The calorie is not a part of the SI system of units.

Work may also be expressed as the product of two factors: an intensity factor and a capacity factor. Examples are given in Table 1.1. The differential quantity of work done by a force F operating over a distance dl is $F\,dl$. Since pressure P is force per unit area, the force on a piston is PA, where A is the surface area perpendicular to the direction of the motion of the piston. Thus the differential quantity of work done by an expanding gas that causes the piston to move distance dl is $PA\,dl$. But $A\,dl = dV$, the increase in gas volume, and so the differential quantity of work is $P\,dV$.

Table 1.1 Intensity and Capacity Factors for Various Types of Work

Type of Work	Intensity Factor	Capacity Factor
Mechanical (J)	Force (N)	Change in distance (m)
Volume expansion (J)	Pressure (N m^{-2})	Change in volume (m^3)
Surface increase (J)	Surface tension (N m^{-1})	Change in area (m^2)
Electrical (J)	Potential difference (V)	Quantity of electricity (C = A s)
Gravitational (J)	Gravitational potential (height × acceleration) ($m^2\ s^{-2}$)	Mass (kg)

Work is often conveniently measured by the lifting of weights. The work required to lift a mass m in the earth's gravitational field, which has an acceleration g, is mgh, where h is the height through which the weight is lifted.

The work w required to lift a kilogram 0.1 m is

$$w = mgh = (1\text{ kg})(9.807\text{ m s}^{-2})(0.1\text{ m}) = 0.9807\text{ J} \tag{1.8}$$

Work is an algebraic quantity, and so it is important to adopt a sign convention. We will take a positive value of w to indicate that *work is done by an external force on the system*. A negative value of w indicates that a system does work on its surroundings, for example, when a gas expands against a piston.

1.6 JOULE'S EXPERIMENTS

Joule showed that under adiabatic conditions a given amount of work would heat the water in a calorimeter a certain number of degrees, independent of whether the work was used to turn a paddle wheel or was dissipated by an electrical current flowing through a resistance or by the friction of rubbing two objects together. Since a given change in state of the water in the calorimeter can be accomplished in different ways involving the same amount of work, or by different sequences of steps, the change in state is independent of the path and is dependent only on the total amount of work. This makes it possible to express the change in state of a system in an adiabatic process in terms of the work required, without stating the type of work or the sequence of steps used. The property of the system whose change is calculated in this way is called the *internal energy* U. Since the internal energy U of a system may be increased by doing work on it, we may calculate the increase in internal energy from the work w done on a system to change it from one state to another in an adiabatic process.

$$\Delta U = w \qquad \text{(in an adiabatic process)} \tag{1.9}$$

The symbol Δ indicates the value of the quantity in the final state minus the value of the quantity in the initial state; $\Delta U = U_2 - U_1$ where U_1 is the internal energy in the initial state and the U_2 is the internal energy in the final state. *If the system does work on its surroundings, w is negative and, furthermore, ΔU is negative* (i.e., *the internal energy of the system decreases*) *if the process is adiabatic.*

1.7 HEAT

A given change in state of a system can be accomplished in ways other than by the performance of work under adiabatic conditions. A change equivalent to that in the Joule experiment described in the preceding section may be obtained by immersing a hot object in the water. When this is done, heat flows from the so-called heat reservoir (the hot object) to the water. We should not say, however, that the water now has more "heat" any more than we would say it has more "work" after it has been heated with moving paddle wheels. In other words, "heat" and "work" are forms of energy *in transit*. After the experiment, the temperature of the water is higher, and it has a greater internal energy U.

Since the same change in state (as determined by measuring properties such as temperature, pressure, and volume) may be produced by doing work on the system or by allowing heat to flow in, the amount of heat q may be expressed in mechanical units. In experiments like that of Joule's, it is found that the expenditure of 1 J of work produces the same change of state as the transfer of 1 J of heat.

Heat is an algebraic quantity, and so it is important to adopt a sign convention. We will take a positive value of q to indicate that heat is absorbed by the system from its surroundings. A negative value of q means that the system gives up heat to its surroundings. The change in internal energy U produced by the transfer of heat q to a system when no work is done is given by

$$\Delta U = q \qquad \text{(no work done)} \tag{1.10}$$

In words, *the heat absorbed by a closed system in a process in which no work is done is equal to the increase in internal energy of the system.* Or, put another way, if no work is done, the heat evolved is equal to the decrease in the internal energy of the system.

1.8 THE FIRST LAW OF THERMODYNAMICS

Now let us consider a change from state 1 to state 2 in which a closed system does work or has work done on it and gains or loses heat by being brought in contact with a heat reservoir. Together, the system and the reservoir form a larger adiabatic system to which equation 1.9 can be applied to obtain

$$(U_2 - U_1) + (U_2' - U_1') = w \tag{1.11}$$

where $U_2 - U_1$ is the change in internal energy of the system and $U_2' - U_1'$ is the change in internal energy of the heat reservoir. No work is done on or by the reservoir. The heat q gained by the system is equal to the negative of the heat gained by the heat reservoir so, by equation 1.10, $(U_2' - U_1') = -q$. Thus equation 1.11 becomes

$$U_2 - U_1 = \Delta U = q + w \tag{1.12}$$

Equation 1.12 expresses the *first law of thermodynamics* for closed systems. The first law is not just equation 1.12, but includes the statement that the thermodynamic function U defined by this equation is a function of the state of the system only. It should be noted that the first law provides a means for determining changes in internal energy but not the absolute value of internal energy.

If there is no change of internal energy (as in the isothermal expansion of a perfect gas, Section 1.15), the work done must be the negative of the heat absorbed.

If ΔU is negative we may say that the system loses energy and that this energy is dissipated in heat that is evolved and work that is done by the system. The first law has nothing to say about how much heat is evolved and how much work is done except that equation 1.12 is obeyed. In other words, the entire decrease in internal energy could show up as work ($q = 0$). Another possibility is that even more than this amount of work would be done and heat would be absorbed ($q > 0$), so that equation 1.12 is obeyed. Although the first law has nothing to say about the relative amounts of heat and work, the second law does.

The first law is frequently stated in the form that energy may be transferred in one form or another, but it cannot be created or destroyed. Thus the total energy of an isolated system is constant.*

---

* It is sometimes incorrectly stated that mass can be converted into energy according to Einstein's relation $E = mc^2$, where c is the speed of light. The mass m in this equation is the relativistic mass that is related to the rest mass m_o by

$$m = m_o(1 - v^2/c^2)^{-1/2}$$

where v is the speed of the object. The correct interpretation of $E = mc^2$ is that energy E and mass m are related through the necessarily positive proportionality constant c^2. A given mass m is therefore equivalent to a certain energy. Since, according to the first law of thermodynamics, energy is conserved, Einstein's relation requires that mass be included in this conservation principle. R. P. Bauman, *J. Chem. Ed.*, **43**, 366 (1966), gives a clear illustration of the meaning of relativistic mass and its relation to energy. If a ball of mass m_o is kicked, its energy and mass are increased at the expense of the energy and mass of the kicker. As the ball bounces to rest, it gives up energy and mass to the earth. In nuclear fission the rest mass of the fragments is less than the rest mass of the original atom, but the mass of the surroundings is increased through collision with the fission fragments.

1.9 EXACT AND INEXACT DIFFERENTIALS

The internal energy U is a state function, like V, because it depends only on the state of the system. The integral of the differential of a state function along any arbitrary path is simply the difference between values of the function at two limits. For example, if a system goes from state a to state b, we can write

$$\int_a^b dU = U_b - U_a \tag{1.13}$$

Since the integral is path independent, the differential of a state function is called an *exact differential.*

The quantities q and w are not state functions. The integrals of their differentials in going from state a to state b *depend on the path chosen.* Therefore their differentials are called *inexact differentials.* We will use $đ$ instead of d to indicate inexact differentials. In going from state a to state b the work w done is represented by

$$\int_a^b đw = w \tag{1.14}$$

Note that the result of the integration is not written $w_b - w_a$, because the amount of work done depends on the particular path that is followed between state a and state b. For example, when a gas is allowed to expand, the amount of work obtained may vary from zero (if the gas is allowed to expand into a vacuum) to a maximum value that is obtained if the expansion is carried out reversibly, as described in Section 1.15.

If an infinitesimal quantity of heat $đq$ is absorbed by a system, and an infinitesimal amount of work $đw$ is done on the system, the infinitesimal change in the internal energy is given by

$$dU = đq + đw \tag{1.15}$$

It is interesting to note that the sum of two inexact differentials can be an exact differential. To illustrate this point further, we consider the following.

The differential $dz = y\,dx$ is not an exact differential

$$\int_a^b y\,dx = \text{area I} \tag{1.16}$$

because this area depends on the path between a and b, as may be seen from Fig. 1.2.

The differential $dz = y\,dx + x\,dy$ is an exact differential. Since $dz = d(xy)$,

$$\int_a^b dz = \int_a^b d(xy) = x_b y_b - x_a y_a \tag{1.17}$$

The reason $dz = y\,dx + x\,dy$ is an exact differential may be seen from Fig. 1.2. The integral of dz from state a to state b may be written

$$\int_a^b dz = \int_a^b y\,dx + \int_a^b x\,dy = \text{area I} + \text{area II} \tag{1.18}$$

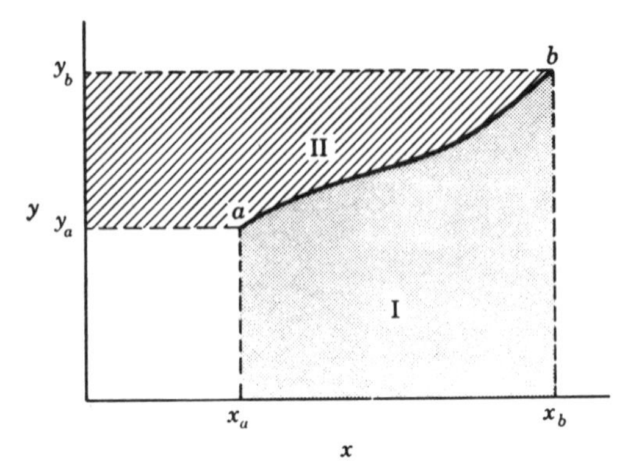

Fig. 1.2 Path of a system in going from state a to state b.

The sum of these areas is independent of the shape of the curve (path) between a and b. There is a simple test to see whether a differential is exact (see Section 2.13).

A cyclic process is a process in which a system is carried through a series of steps that eventually bring the system back to its *initial* conditions. The change in internal energy for a cyclic process is zero since dU is exact and, therefore, from equation 1.13,

$$\int_a^a dU = \oint dU = 0 \tag{1.19}$$

where the circle indicates integration around a cycle. The cyclic integrals of q and w are not in general equal to zero, and their values depend on the path followed.

1.10 ENTHALPY

Constant-pressure processes are more common in chemistry than constant-volume processes because most operations are carried out in open vessels. If only pressure-volume work is done and the pressure is constant, the work done on the system w equals $P\Delta V$, so that equation 1.12 may be written

$$\Delta U = q - P\Delta V \tag{1.20}$$

If the initial state is designated by 1 and the final state by 2, then

$$U_2 - U_1 = q - P(V_2 - V_1) \tag{1.21}$$

so that the heat absorbed is given by

$$q = (U_2 + PV_2) - (U_1 + PV_1) \tag{1.22}$$

Since the heat absorbed is given by the difference of two quantities that are functions of the state of the system, it is convenient to introduce a new state function, the *enthalpy* H, which is defined by

$$H = U + PV \tag{1.23}$$

Since the enthalpy is defined in terms of other thermodynamic quantities, the only gain is in terms of convenience. Equation 1.22 may be written $q = H_2 - H_1 = \Delta H$. In words, *the heat absorbed in a process at constant pressure is equal to the change in enthalpy if the only work done is pressure-volume work.*

When pressure-volume work is the only kind of work (electrical and other kinds being excluded), it is easy to visualize ΔU and ΔH; in a constant-volume calorimeter the evolution of heat is a measure of the decrease in internal energy U, and in a constant-pressure calorimeter the evolution of heat is a measure of the decrease in enthalpy H.

1.11 HEAT CAPACITY

For a chemically inert system of fixed mass the internal energy U may be taken to be a function of any two of T, V, and P. It is most convenient to take it as a function of T and V. Since U is a state function, the differential dU is given by

$$dU = \left(\frac{\partial U}{\partial T}\right)_V dT + \left(\frac{\partial U}{\partial V}\right)_T dV \tag{1.24}$$

The first term is the change in internal energy due to the temperature change alone, and the second term is the change in internal energy due to the volume change alone. Since only pressure-volume work is involved, the heat absorbed is given by

$$đq = dU + P\,dV = \left(\frac{\partial U}{\partial T}\right)_V dT + \left[P + \left(\frac{\partial U}{\partial V}\right)_T\right] dV \tag{1.25}$$

where the second form has been obtained by introducing equation 1.24. Thus the amount of heat absorbed depends on the volume change as well as the temperature change.

If the volume of the system is held constant, then

$$đq_V = \left(\frac{\partial U}{\partial T}\right)_V dT \tag{1.26}$$

The ratio $đq_V/dT$ may be measured experimentally at constant volume and is known as C_V, the heat capacity of the system at constant volume.

$$C_V = \frac{đq_V}{dT} = \left(\frac{\partial U}{\partial T}\right)_V \tag{1.27}$$

This equation may be applied to a system of any size but, in this book, we will always apply it to a mole. Thus C_V is a molar thermodynamic quantity and has the units calories per Kelvin per mole.

The change in internal energy of one mole of substance heated from T_1 to T_2 at constant volume is

$$\Delta U = \int_{T_1}^{T_2} C_V\,dT \tag{1.28}$$

For a chemically inert system of fixed mass, it is most convenient to take the enthalpy H as a function of temperature and pressure. Since H is a state function, the differential dH is given by

$$dH = \left(\frac{\partial H}{\partial T}\right)_P dT + \left(\frac{\partial H}{\partial P}\right)_T dP \tag{1.29}$$

If the pressure of the system is held constant, then

$$đq_P = \left(\frac{\partial H}{\partial T}\right)_P dT \tag{1.30}$$

The ratio $đq_P/dT$ is the heat capacity at constant pressure C_P. Thus

$$C_P = \frac{đq_P}{dT} = \left(\frac{\partial H}{\partial T}\right)_P \tag{1.31}$$

The heat capacity at constant pressure C_P is always larger than the heat capacity at constant volume C_V, because pressure-volume work is done when a substance is heated at constant pressure. The equation for the difference $C_P - C_V$ may be derived by dividing equation 1.25 by dT at constant pressure.

$$\frac{đq_P}{dT} = \left(\frac{\partial U}{\partial T}\right)_V + \left[P + \left(\frac{\partial U}{\partial V}\right)_T\right]\left(\frac{\partial V}{\partial T}\right)_P \tag{1.32}$$

$$C_P = C_V + \left[P + \left(\frac{\partial U}{\partial V}\right)_T\right]\left(\frac{\partial V}{\partial T}\right)_P \tag{1.33}$$

where equation 1.33 has been obtained by inserting equations 1.27 and 1.31.

The quantity $(\partial U/\partial V)_T$ may be measured in principle in an experiment devised by Joule. Imagine two gas bottles connected with a valve and immersed in a stirred liquid in a thermally isolated container. The two bottles constitute the system under consideration. The first bottle is filled with a gas under pressure, and the second is evacuated. When the valve is opened, gas rushes from the first bottle into the second. Joule found that there was no discernible change in the temperature of the stirred liquid as a result of this expansion, and so $đq = 0$. No work is done in this expansion, and so $đw = 0$ and $dU = đq + đw = 0$. Since the temperature is constant, equation 1.24 becomes

$$dU = \left(\frac{\partial U}{\partial V}\right)_T dV = 0 \tag{1.34}$$

Since $dV \neq 0$,

$$\left(\frac{\partial U}{\partial V}\right)_T = 0 \tag{1.35}$$

In words, the internal energy of the gas is independent of volume at constant temperature. Joule's experiment is not very sensitive because of the large heat capacity of the stirred liquid and the small heat capacity of the gas. The quantity $(\partial U/\partial V)_T$, although usually not large, is different from zero for real gases.

Equation 1.35 applies to a perfect gas, and at constant temperature the internal energy of a perfect gas is also independent of pressure.

$$\left(\frac{\partial U}{\partial P}\right)_T = 0 \tag{1.36}$$

The molecular interpretation of this relation is that there is no interaction between the molecules of a perfect gas, and so the energy does not change with the distance between molecules. Equations 1.35 and 1.36 for a perfect gas are not really separate from the perfect gas law, since they may be derived from the second law of thermodynamics by application of the perfect gas law (see Section 2.15).

Since for a perfect gas $(\partial U/\partial V)_T = 0$ and $(\partial V/\partial T)_P = R/P$, equation 1.33 becomes

$$C_P - C_V = R \tag{1.37}$$

for a mole of gas. This relationship may be visualized as follows. When a mole of perfect gas is heated at constant pressure, the work done in pushing back a piston is $P\,\Delta V = R\,\Delta T$. For a 1 K change in temperature the amount of work done is $R(1\text{ K})$, and this is just the extra energy required to heat a mole of perfect gas at constant pressure over that required to heat it 1 K at constant volume.

The general expression for $C_P - C_V$ for gases is given in the next chapter because its derivation uses the second law of thermodynamics (Problem 2.23).

1.12 HEAT CAPACITIES OF GASES

Values of C_P at 25 °C for about 200 substances are given in Table A.1 in the Appendix, and values from 298 to 3000 K are given for a smaller number of substances in Table A.2. The dependence of C_P on temperature is shown for a number of gases in Fig. 1.3. In general, the more complex the molecule, the greater its molar heat capacity, and the greater the increase with rising temperature.

Kinetic theory (Chapter 14) shows that the kinetic energy of a mole of perfect monatomic gas is $\frac{3}{2}RT$. This energy is independent of the pressure or the molar mass and so this is the internal energy relative to that at absolute zero.

$$U - U_0 = \tfrac{3}{2}RT \tag{1.38}$$

The enthalpy of a perfect gas is larger than the internal energy by PV (or RT), so that

$$H - H_0 = \tfrac{3}{2}RT + RT = \tfrac{5}{2}RT \tag{1.39}$$

Thus the heat capacities of a perfect monatomic gas are expected to be

$$C_V = \left(\frac{\partial U}{\partial T}\right)_V = \tfrac{3}{2}R = 12.472\text{ J K}^{-1}\text{ mol}^{-1} \tag{1.40}$$

$$C_P = \left(\frac{\partial H}{\partial T}\right)_P = \tfrac{5}{2}R = 20.786\text{ J K}^{-1}\text{ mol}^{-1} \tag{1.41}$$

at all temperatures.

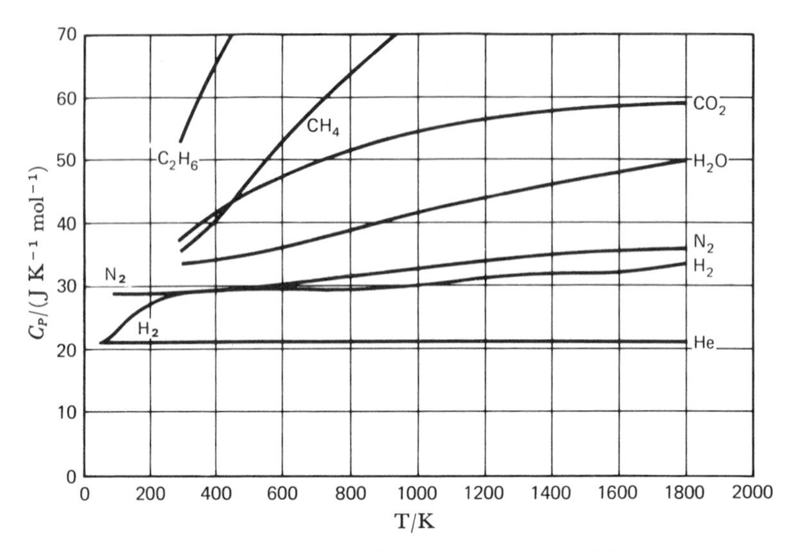

Fig. 1.3 Influence of temperature on the molar heat capacities of gases at constant pressure.

Tables A.1 and A.2 show that values of C_P for atomic gases are constant at 20.786 J K^{-1} mol^{-1} independent of temperature, except for cases where electrons in the atom can be excited to low-lying levels [see especially O(g)]. Polyatomic gases can absorb energy in rotational and vibrational motion, which we need to discuss now.

Equation 1.38 may be given the following interpretation: since three independent coordinates are needed to describe translational motion (e.g., the three Cartesian components of the velocity v), we can say that a molecule has three degrees of translational freedom and each contributes $\frac{1}{2}RT$ to the translational energy. This is an example of the principle of equipartition. According to this principle, energy is absorbed into each of the degrees of freedom of a molecule.

A molecule containing N nuclei has $3N$ degrees of freedom because $3N$ coordinates are required to locate the nuclei in space. Three coordinates are required to locate the center of mass of the molecule, and so there are $3N$-3 internal degrees of freedom. A diatomic molecule or a linear polyatomic molecule has two rotational degrees of freedom because two angles are required to orient the axis of the molecule with respect to a coordinate system. Thus, for a diatomic or linear molecule, there are $3N$-5 vibrational degrees of freedom. For a diatomic molecule there is a single vibration, and for a linear triatomic molecule there are four (see Section 11.10). For a nonlinear polyatomic molecule there are three rotational degrees of freedom because three angles are required to orient the molecule with respect to a coordinate system. For such molecules there are $3N$-6 vibrational degrees of freedom.

Each degree of translational or rotational freedom contributes $\frac{1}{2}R$ to C_V. Thus, for a rigid diatomic molecule, we would expect

$$C_V = \tfrac{5}{2}R = 20.786 \text{ J K}^{-1}\text{ mol}^{-1}$$
$$C_P = \tfrac{7}{2}R = 29.100 \text{ J K}^{-1}\text{ mol}^{-1}$$

These predictions of elementary kinetic theory are borne out quite well by the experimental values for N_2 and H_2 in Table A.1 at 25 °C. However, as shown in Fig. 1.3, the heat capacity of H_2 decreases below 300 K and approaches the value expected for a monatomic gas at its boiling point. Furthermore, the heat capacities of both H_2 and N_2 begin to increase above about 400 K.

According to classical kinetic theory, each vibrational degree of freedom contributes R to C_V. The reason for this larger contribution by vibrational degrees of freedom is that a vibration has both kinetic and potential energy associated with it, and each contributes $\frac{1}{2}R$. Thus, according to classical kinetic theory, the molar heat capacities of a vibrating diatomic molecule would be expected to be

$$C_V = \tfrac{7}{2}R = 29.100 \text{ J K}^{-1}\text{ mol}^{-1}$$

$$C_P = \tfrac{9}{2}R = 37.415 \text{ J K}^{-1}\text{ mol}^{-1}$$

Figure 1.3 shows that the heat capacities of H_2 and N_2 are not this large, even though their heat capacities above 400 K are larger than expected for a rigid diatomic molecule.

The values of C_P of polyatomic molecules increase with temperature, but even at 3000 K are smaller than the values expected from classical mechanics. Thus we conclude that the principle of equipartition applies to translational degrees of freedom at all temperatures and to rotational degrees of freedom at sufficiently high temperatures, but that the vibrational degrees of freedom are not fully utilized even at very high temperatures. This is an example of the failure of classical mechanics when applied to molecules. We will defer the discussion of the resolution of this problem by quantum mechanics to Chapters 8 and 13.

1.13 HEAT CAPACITIES OF SOLIDS

At room temperature and above, the heat capacities of solid elements heavier than potassium are about 26.8 J K^{-1} mol^{-1}. This relationship, which was first pointed out by Dulong and Petit in 1819, played an important role in the early determination of atomic masses because it could be used to select the multiple of the known equivalent weight (obtained from quantitative analysis) that is required to give the atomic mass.

However, solid elements at the beginning of the periodic table have lower heat capacities, and the heat capacities of all solids approach zero as the temperature is reduced toward absolute zero. It was found that if the heat capacities of solid elements are plotted versus T/Θ, where Θ was an empirical constant for each element, a common temperature dependence is found. That plot is shown in Fig. 1.4.

The high temperature limit of the heat capacity of an atomic solid can be given a simple interpretation in terms of classical mechanics. Each atom is considered to be a harmonic oscillator with three degrees of freedom (corresponding with the x, y, and z directions). Since a harmonic oscillator has both kinetic energy and potential energy connected with it, each degree of freedom contributes RT to the

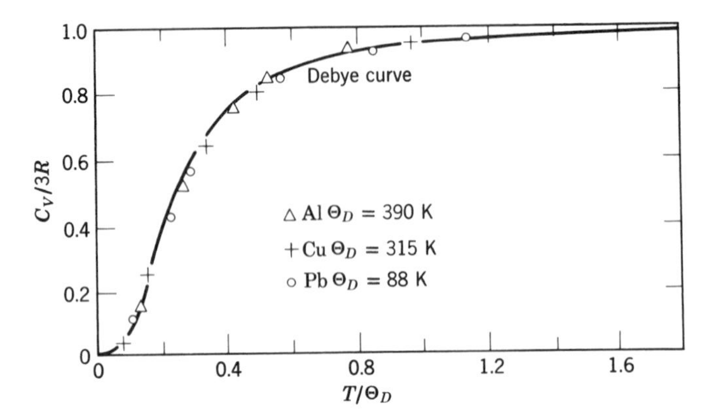

Fig. 1.4 Heat capacity data for several solid elements plotted versus T/Θ_D, where Θ_D is the Debye temperature (Section 13.16). (Data from: *A Compendium of the Properties of Materials at Low Temperature, Part II, Properties of Solids*, Wadd Technical Report 60-56, Part II, 1960. Reproduced with permission from *Statistical Physics* by F. Mandl. © Copyright 1971 John Wiley & Sons Ltd.)

energy and, therefore, R to the heat capacity. Since the atoms in the crystal each have three degrees of freedom, we expect $C_V = 3R$.

However, classical mechanics did not provide an explanation of the decrease of C_V to zero as the temperature approaches absolute zero. This effect was only explained later in terms of quantum mechanics.

Statistical mechanical calculations of the heat capacities of solids are discussed in Section 13.16. It is found that the heat capacity at constant volume is usually proportional to T^3 below about 15 K. This relation, which is confirmed by the Debye theory (equation 13.101), is useful for extrapolating heat capacity data to the neighborhood of absolute zero, where it is difficult to obtain experimental values.

1.14 REVERSIBLE PROCESSES

A *reversible* process is a process that may be reversed at any moment by changing an independent variable by an infinitesimal amount. Thus, in the reversible expansion of a gas, the expansion can be stopped at any point by increasing the pressure exerted by the piston by an infinitesimal amount. A reversible process is often spoken of as one that consists of a series of successive equilibria. Such processes are idealizations that may be closely approached, but not actually reached, in the laboratory. To carry out a finite process reversibly would require an infinite time. Reversible processes are of great conceptual importance because they yield the maximum amount of work that may be obtained from a given net change. This amount of work is just sufficient to return the system to its original conditions. When a process is carried out *irreversibly*, less work is obtained than would be required to return the system to its initial state.

Imagine a gas enclosed in a cylinder fitted with a frictionless and weightless piston and maintained at a constant temperature. The external pressure on the piston is decreased by an infinitesimal amount, dP, and the gas expands by an amount dV. In this expansion the pressure of the gas in the cylinder decreases until it becomes equal to the external pressure, and then the piston ceases to move. A second infinitesimal decrease in pressure produces a second expansion dV; as the pressure is decreased in successive amounts, the volume undergoes a series of increases. During each little expansion, the pressure throughout the gas is constant (within an infinitesimal amount). In each little expansion the work done on the system is the external pressure multiplied by $-dV$, and the total work obtainable in expanding the gas reversibly from the *initial* volume V_1 to the *final* volume V_2 is equal to the integral of the pressure times the differential of the volume.

$$w_{\text{rev}} = -\int_{V_1}^{V_2} P\,dV \tag{1.42}$$

Since the gas is at its equilibrium pressure (within an infinitesimal amount) at each stage in the expansion, we may substitute an expression for the dependence of pressure on volume obtained from equilibrium measurements. If the gas was allowed to expand rapidly, the pressure and temperature would not be uniform throughout the volume of the gas, and so such a substitution could not be made. *Only if the expansion is carried out reversibly at constant temperature can sufficient energy be obtained to reverse the process, compressing the gas to its original conditions.*

1.15 REVERSIBLE ISOTHERMAL EXPANSION OF A PERFECT GAS

The maximum work that can be obtained from the *isothermal* expansion of a perfect gas may readily be calculated. If the expansion is carried out reversibly at constant temperature, the pressure is always given by $P = RT/V$. Substituting in equation 1.42, we obtain

$$w_{\text{rev}} = -\int_{V_1}^{V_2} \frac{RT}{V}\,dV = -RT \ln \frac{V_2}{V_1} \tag{1.43)*}$$

since the temperature is constant.

In integration the lower limit always refers to the initial state and the upper limit to the final state. If the gas is compressed, the final volume is smaller and w_{rev} is positive. The positive value means that work is done on the gas. The expansion of a perfect gas by a factor of ten yields $w_{\text{rev}} = -RT \ln 10 = -5229$ J mol^{-1} at 273.15 K.

Since for a perfect gas at constant temperature $P_1V_1 = P_2V_2$, the work for 1 mol of perfect gas is also given by

$$w = RT \ln \frac{P_2}{P_1} \tag{1.44}$$

* ln represents natural logarithms and log represents base 10 logarithms; $\ln x = 2.303 \log x$.

Since the internal energy and enthalpy do not change in an isothermal expansion of a perfect gas, $w = -q$, and so

$$q = -RT \ln \frac{P_2}{P_1} = -RT \ln \frac{V_1}{V_2} \tag{1.45}$$

1.16 REVERSIBLE ADIABATIC EXPANSION OF A PERFECT GAS

An *adiabatic* process is one in which there is no loss or gain of heat, that is, one in which the system under investigation is thermally isolated from its environment so that $q = 0$. In an adiabatic expansion work is done at the expense of the internal energy of the gas and the temperature drops. Thus, when a gas expands adiabatically to a larger volume and a lower pressure, the volume is smaller than it would be after an isothermal expansion to the same pressure. Plots of pressure versus volume for adiabatic and isothermal expansions are shown in Fig. 1.5.

The work done by the isothermal reversible expansion of the gas, represented by the area under *AB*, is larger than the work done by the reversible adiabatic expansion, represented by the area under *AC*. The energy for doing the additional work in the isothermal expansion is provided by heat absorbed from the constant-temperature reservoir. The energy for doing work in the adiabatic expansion comes only from the cooling of the gas itself.

Now let us consider the reversible adiabatic expansion of 1 mol of perfect gas. Since for an adiabatic process $đq = 0$, then, by the first law,

$$dU = đw = -P\,dV \tag{1.46}$$

We have seen earlier (Section 1.12) that the internal energy of a perfect gas depends only on the temperature and that $dU = C_V\,dT$. Substituting this for dU in equation 1.46 and eliminating P by use of the perfect gas law yields

$$C_V\,dT = -P\,dV$$
$$C_V \frac{dT}{T} = -R \frac{dV}{V} \tag{1.47}$$

Assuming that C_V is independent of temperature,

$$C_V \int_{T_1}^{T_2} \frac{dT}{T} = -R \int_{V_1}^{V_2} \frac{dV}{V}$$
$$C_V \ln \frac{T_2}{T_1} = R \ln \frac{V_1}{V_2} \tag{1.48}$$

This equation is a good approximation only if the temperature range is small enough so that C_V does not change very much.

Another form of the relation for a reversible adiabatic expansion of a perfect gas

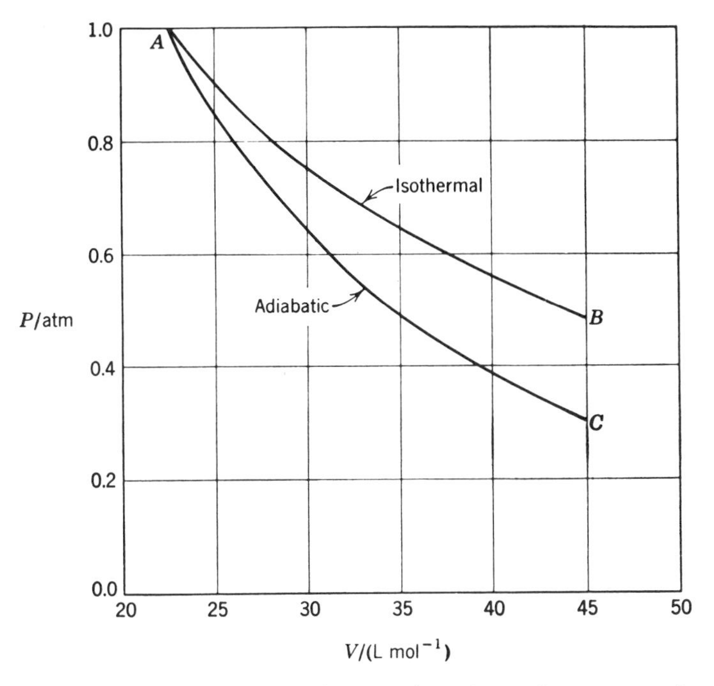

Fig. 1.5 Isothermal and adiabatic expansion of a perfect monatomic gas.

may be obtained by differentiating $PV = RT$ to obtain $P\,dV + V\,dP = R\,dT$ and substituting this into equation 1.47.

$$(C_V + R)\frac{dV}{V} + C_V\frac{dP}{P} = 0 \tag{1.49}$$

Since $C_P - C_V = R$,

$$\gamma\frac{dV}{V} + \frac{dP}{P} = 0 \tag{1.50}$$

where $\gamma = C_P/C_V$. If the temperature range is sufficiently small so that γ is essentially constant, this equation may be integrated to obtain

$$\frac{P_1}{P_2} = \left(\frac{V_2}{V_1}\right)^{\gamma} \tag{1.51a}$$

If we substitute the perfect gas law we obtain

$$\frac{T_1}{T_2} = \left(\frac{V_2}{V_1}\right)^{\gamma - 1} \tag{1.51b}$$

so that

$$\frac{T_1}{T_2} = \left(\frac{P_1}{P_2}\right)^{(\gamma - 1)/\gamma} \tag{1.51c}$$

The reversible work for an adiabatic expansion may be calculated using $dU = C_V\,dT$. Since $q = 0$, the first law yields

$$\Delta U = w = \int_{T_1}^{T_2} C_V\,dT \tag{1.52}$$

Example 1.1 Figure 1.5 shows that when a perfect monatomic gas is allowed to expand adiabatically from 22.4 L at 1 atm and 0 °C to a volume of 44.8 L, the pressure drops to 0.32 atm. Confirm this pressure and calculate the temperature at C. How much work is done in the adiabatic expansion?

$$\gamma = \frac{(\frac{5}{2}R)}{(\frac{3}{2}R)} = \frac{5}{3}$$

$$P_2 = P_1\left(\frac{V_1}{V_2}\right)^{\gamma} = (1\text{ atm})\left(\frac{22.4\text{ L}}{44.8\text{ L}}\right)^{5/3} = 0.315\text{ atm}$$

$$T_2 = T_1\left(\frac{V_1}{V_2}\right)^{\gamma-1} = (273.15\text{ K})\left(\frac{22.4\text{ L}}{44.8\text{ L}}\right)^{2/3} = 172.07\text{ K or } -101.08\text{ °C}$$

$$w = \int_{T_1}^{T_2} C_V\,dT = \tfrac{3}{2}R(172.07\text{ K} - 273.15\text{ K}) = -1261\text{ J mol}^{-1}$$

1.17 EQUATIONS OF STATE FOR GASES

An equation of state gives the relation between pressure, volume, and temperature for a fixed quantity of a substance. The perfect gas law is the simplest example. Much more complicated equations are required to represent pressure-volume-temperature data for real gases. One of the ways of presenting experimental data is to plot the *compressibility factor* PV/RT versus pressure at constant temperature.

The compressibility factors for H_2 and O_2 at 0 °C are plotted versus pressure in Fig. 1.6. The behavior of a perfect gas is represented by the horizontal dashed line. In the limit of zero pressure the compressibility factor of every gas is unity because it behaves perfectly.

A compressibility factor of less than unity indicates that the gas is more compressible than a perfect gas. All gases show a minimum in the plot of compressibility factor versus pressure if the temperature is low enough. Hydrogen and helium, which have very low boiling points, exhibit this minimum only at temperatures much below 0 °C.

The effect of temperature on the change of compressibility factor of N_2 with pressure is shown in Fig. 1.7. At 0 °C and low pressures, N_2 is more compressible than a perfect gas, but at high pressures it is less compressible.

The dependence of the compressibility factor Z on the concentration of the gas $(1/V)$ may be represented by a power series that is referred to as the virial equation.

$$Z = \frac{PV}{RT} = 1 + \frac{B}{V} + \frac{C}{V^2} + \frac{D}{V^3} + \cdots \tag{1.53}$$

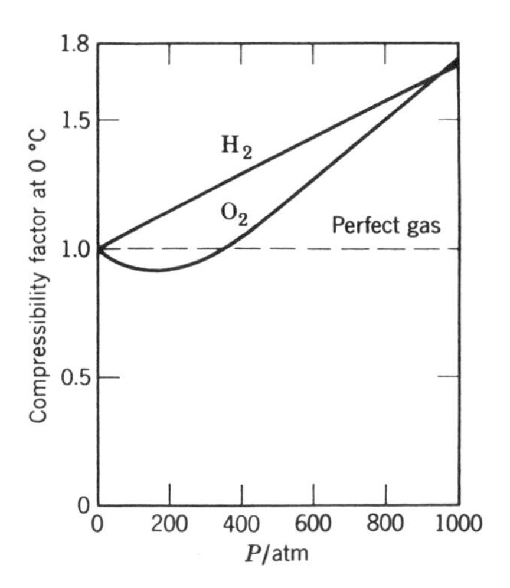

Fig. 1.6 Influence of high pressure on the compressibility factor, PV/RT, for two gases.

For a particular real gas the second, third, fourth, . . . virial coefficients B, C, D, . . . are functions only of the temperature. Statistical mechanics shows that the second virial coefficient B depends on the forces between pairs of molecules, the third C to triplets, and so on. The word virial is derived from the Latin word for force. The virial expansion is not convergent at high gas concentrations.

The second virial coefficient is shown as a function of temperature for several gases in Fig. 1.8.

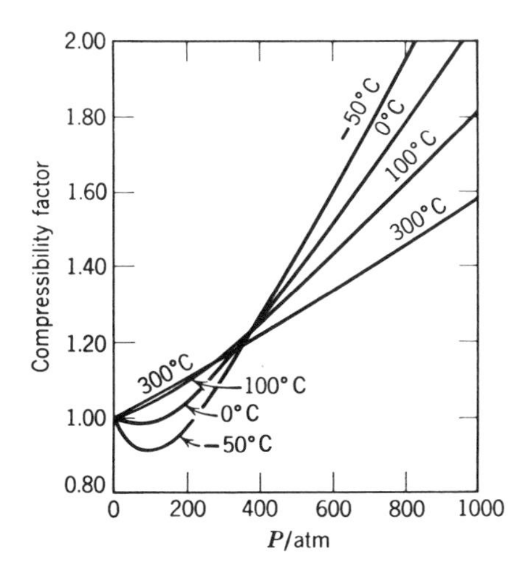

Fig. 1.7 Influence of pressure on the compressibility factor, PV/RT, for nitrogen at different temperatures.

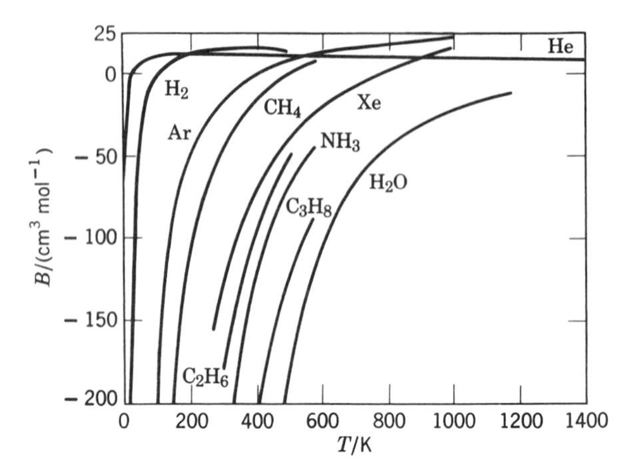

Fig. 1.8 Second virial coefficient B. (From K. E. Bett, J. S. Rowlinson, and G. Saville, *Thermodynamics for Chemical Engineers*, p. 141, The MIT Press, Cambridge, Mass., 1975. © University of London (Athlone Press) 1975.)

More complicated equations of state have been developed. The Beattie-Bridgeman equation, which is probably the most precise, involves five adjustable parameters. Instead of discussing the more complicated alternative equations, we will discuss a very simple one, the van der Waals' equation, and the idea of a reduced equation of state that can be derived from it.

1.18 THE VAN DER WAALS EQUATION

In 1879 van der Waals pointed out that since gas molecules themselves occupy a significant fraction of the volume at higher gas densities, this effect might be taken into account by subtracting an excluded volume b from the actual molar volume V in the perfect gas law.

$$P(V - b) = RT \tag{1.54}$$

This equation can represent compressibility factors greater than unity, but cannot yield compressibility factors less than unity. In addition, van der Waals recognized that gas molecules attract each other and that real gases are therefore more compressible than perfect gases. The forces that lead to condensation are still referred to as van der Waals' forces, and their origin is discussed in Section 10.13. Van der Waals provided for intermolecular attraction by adding to the observed pressure P in the equation of state a term a/V^2, where a is a constant whose value depends on the gas.

Van der Waals' full equation is

$$\left(P + \frac{a}{V^2}\right)(V - b) = RT \tag{1.55}$$

for 1 mol of gas.

When the molar volume V is large, b becomes negligible in comparison with V and a/V^2 become negligible with respect to P; and van der Waals' equation reduces to the perfect gas law, $PV = RT$.

Van der Waals' constants for a few gases are listed in Table 1.2. They can be calculated from experimental measurements of P, V, and T or from the critical constants, as shown later in equation 1.60.

Table 1.2 Van der Waals' Constants

Gas	a/L^2 atm mol^{-2}	b/L mol^{-1}	Gas	a/L^2 atm mol^{-2}	b/L mol^{-1}
H_2	0.244 4	0.026 61	CH_4	2.253	0.042 78
He	0.034 12	0.023 70	C_2H_6	5.489	0.063 80
N_2	1.390	0.039 13	C_3H_8	8.664	0.084 45
O_2	1.360	0.031 83	$C_4H_{10}(n)$	14.47	0.122 6
Cl_2	6.493	0.056 22	C_4H_{10}(iso)	12.87	0.114 2
NO	1.340	0.027 89	$C_5H_{12}(n)$	19.01	0.146 0
NO_2	5.284	0.044 24	CO	1.485	0.039 85
H_2O	5.464	0.030 49	CO_2	3.592	0.042 67

Isotherms for isopentane are shown in Fig. 1.9. At the higher temperatures the pressure-volume curve is nearly that for a perfect gas. As the temperature is reduced, the isotherms have more complicated shapes, and at 461 K there is an inflection point at P_c and V_c. Below this temperature, called the critical temperature T_c, gas and liquid can exist in equilibrium at temperatures, pressures, and molar volumes within the light gray region. Above the critical temperature, the gas cannot be liquified at any pressure.

At the right of the diagram outside the two-phase region gas alone is present. At the left of the diagram outside the two-phase region liquid alone is present; and, since the liquid is much less compressible than the gas, the isotherms are much steeper than for the gas. The liquid region is marked with the darker gray. Along the horizontal lines gas and liquid exist together. A gas in equilibrium with the corresponding liquid is generally referred to as a vapor. The region in which vapor and liquid are coexistent is marked with light gray. The vapor pressures of isopentane at various temperatures are given by the horizontal lines in Fig. 1.9.

If liquid is converted to gas by a series of steps that take it through the two-phase region, a boundary or meniscus will be formed between the gas and liquid phases. However, the liquid may be converted to gas smoothly and continuously without the appearance of a meniscus. If a sample of liquid at point X is heated at constant volume to point Y, there is complete continuity between liquid and gaseous states.

Multiplying out the terms in van der Waals' equation 1.55 and rearranging in descending powers of V, we have

$$V^3 - V^2\left(b + \frac{RT}{P}\right) + V\frac{a}{P} - \frac{ab}{P} = 0 \tag{1.56}$$

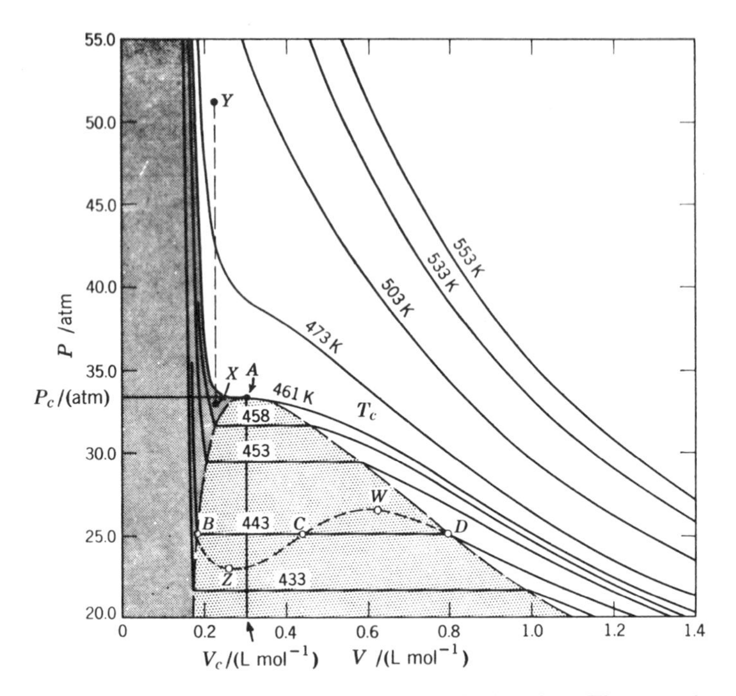

Fig. 1.9 Isotherms for isopentane showing the critical region. The two-phase region is light gray, and the liquid region is darker gray. The gas region is white.

At temperatures below the critical temperature this cubic equation has three real solutions, each value of P giving three values of V. This equation is shown graphically by the dashed line $DWCZB$ on the 443 K isothermal in Fig. 1.9, where the three values of V are the intersections B, C, and D on the horizontal line corresponding to a fixed value of the pressure. This dashed calculated line then appears to give a continuous transition from the gaseous phase to the liquid phase but, in reality, the transition is abrupt and discontinuous, both liquid and vapor existing along the straight horizontal lines. The theoretical dashed line $DWCZB$ does not correspond to normal physical conditions; for example, the slope of the curve at C is positive, a fact that would lead to the unnatural condition than an increase in pressure produces an increase in volume. It is possible, however, to have pressures of gas in an unstable condition represented by the beginning of the dashed line DW, before the supercooled vapor has a chance to liquify and bring the pressure down to that of the horizontal line. It is also possible to have liquid under metastable conditions along BZ. At the critical temperature (461 K for isopentane) and pressure there is only one real root, the critical volume V_c.

The values of van der Waals' constants may be calculated from the critical constants for a gas. As may be seen in Fig. 1.9, there is a horizontal inflection point in the P versus V curve at the critical point so that $(\partial P/\partial V)_{T_c} = 0$ and

$(\partial^2 P/\partial V^2)_{T_c} = 0$. At the critical temperature van der Waals' equation may be written

$$P = \frac{RT_c}{V - b} - \frac{a}{V^2} \tag{1.57}$$

Differentiating with respect to molar volume,

$$\left(\frac{\partial P}{\partial V}\right)_{T_c} = \frac{-RT_c}{(V - b)^2} + \frac{2a}{V^3} \tag{1.58}$$

$$\left(\frac{\partial^2 P}{\partial V^2}\right)_{T_c} = \frac{2RT_c}{(V - b)^2} - \frac{6a}{V^4} \tag{1.59}$$

At the critical point these derivatives are both equal to zero, and V is replaced by V_c and P by P_c.

Equations 1.57, 1.58, and 1.59 may be solved together to obtain

$$a = 3P_cV_c^2 \qquad b = \frac{V_c}{3} \qquad R = \frac{8P_cV_c}{3T_c} \tag{1.60}$$

Thus the van der Waals' constants may be calculated for a gas if the critical constants are known. The van der Waals' equation does not give a very accurate representation of the P–V–T relationship for real gases, but it does provide an introduction to the useful idea of a reduced equation of state. We can define a set of reduced variables, the reduced pressure P_r, the reduced molar volume V_r, and the reduced temperature T_r by

$$P_r \equiv \frac{P}{P_c} \qquad V_r \equiv \frac{V}{V_c} \qquad T_r \equiv \frac{T}{T_c} \tag{1.61}$$

By use of equation 1.60 the van der Waals' equation may be written in terms of the critical constants, and then equations 1.61 may be used to obtain

$$\left(P_r + \frac{3}{V_r^2}\right)(V_r - \tfrac{1}{3}) = \tfrac{8}{3}T_r \tag{1.62}$$

This is referred to as a *reduced equation of state*. If the van der Waals' equation were exact, all gases would follow this same equation of state, because all constants connected with the individual nature of the gas have disappeared. This equation does not represent experimental data any better than the original van der Waals' equation, but it leads to the idea that if different substances are compared at equal fractions of their critical temperatures, pressures, and volumes, they will appear to behave similarly. This principle of corresponding states is not exact, but it is the single most important basis for the development of correlation and estimation methods.*

---

* R. C. Reid, J. M. Prausnitz, and T. K. Sherwood, *The Properties of Gases and Liquids*, McGraw-Hill Book Co., New York, 1977.

1.19 THERMOCHEMISTRY

Thermochemistry deals with the heat absorbed or evolved in a chemical reaction, in a phase change, or in the dilution of a solution. Exothermic reactions evolve heat and have negative values of ΔH or ΔU, and endothermic reactions absorb heat and have positive values of ΔH or ΔU. The heat evolved or absorbed by a chemical reaction is measured with a calorimeter. In the most common type, the reaction is allowed to take place in a reaction chamber surrounded by a weighed quantity of water in an insulated vessel, and the rise in temperature is measured with a sensitive thermometer. The product of the rise in temperature and the total heat capacity of the water and calorimeter is equal to the heat evolved. The heat capacity of the surrounding water is obtained by weighing the water and multiplying by its known specific heat; the heat capacity of the calorimeter is determined by carrying out a reaction of known heat evolution in the calorimeter or by introducing a known quantity of heat with an electric heater.

There are two types of calorimetric experiments: constant volume and constant pressure. In a constant-volume calorimeter no work is done and so the heat absorbed is equal to the increase in internal energy ΔU. In a constant-pressure calorimeter work is done so that the heat absorbed is equal to the increase in enthalpy ΔH.

Heats of combustion are useful in calculating other thermochemical data. Also, they have practical as well as theoretical importance. The purchaser of coal is interested in its heat of combustion per ton. The dietician must know, among other factors, the number of calories obtainable from the combustion of various foods. In nutrition the term "calorie" refers to kilocalorie.

Lavoisier and Laplace recognized in 1780 that the heat absorbed in decomposing a compound must be equal to the heat evolved in its formation under the same conditions. Thus, if the reverse of a chemical reaction is written, the sign of ΔH is changed. Hess pointed out in 1840 that the overall heat of a chemical reaction at constant pressure is the same, regardless of the intermediate steps involved. These principles are both corollaries of the first law of thermodynamics and are a consequence of the fact that the enthalpy is a state function. This makes it possible to calculate the enthalpy changes for reactions that cannot be studied directly. For example, it is not practical to measure the heat evolved when carbon burns to carbon monoxide in a limited amount of oxygen, because the product will be an uncertain mixture of carbon monoxide and carbon dioxide. However, carbon may be burned completely to carbon dioxide in an excess of oxygen and the heat of reaction measured. Thus, for graphite at 25 °C,

$$C(s) + O_2(g) = CO_2(g) \qquad \Delta H^\circ = -393.509 \text{ kJ mol}^{-1} \tag{1.63}$$

The heat evolved when carbon monoxide burns to carbon dioxide can be readily measured also.

$$CO(g) + \tfrac{1}{2}O_2(g) = CO_2(g) \qquad \Delta H^\circ = -282.984 \text{ kJ mol}^{-1}* \tag{1.64}$$

---

* The mole refers to the equation *as balanced*. Thus this is the ΔH° per mole of CO reacting or per half mole of O_2 reacting.

Writing these equations in such a way as to obtain the desired reaction, adding, and canceling, we have:

$$C(s) + O_2(g) = CO_2(g) \qquad \Delta H^\circ = -393.509 \text{ kJ mol}^{-1}$$
$$CO_2(g) = CO(g) + \tfrac{1}{2}O_2(g) \qquad \Delta H^\circ = 282.984 \text{ kJ mol}^{-1}$$
$$C(s) + \tfrac{1}{2}O_2(g) = CO(g) \qquad \Delta H^\circ = -110.525 \text{ kJ mol}^{-1}$$

In this way an accurate value can be obtained for the heat of combustion of graphite to CO.

These data may be represented in the form of an enthalpy level diagram, as shown in Fig. 1.10. In addition, this diagram shows the enthalpy changes that would be involved in vaporizing graphite to atoms and dissociating oxygen into atoms.

1.20 ENTHALPY OF FORMATION

A chemical reaction with two reactants A_1 and A_2 and two products A_3 and A_4 may be represented as follows.

$$\nu_1 A_1 + \nu_2 A_2 = \nu_3 A_3 + \nu_4 A_4 \tag{1.65}$$

where the ν's are stoichiometric coefficients. We will adopt the convention that the stoichiometric coefficients are positive for products and negative for reactants so that a general chemical reaction may be written as follows:

$$0 = \sum \nu_i A_i \tag{1.66}$$

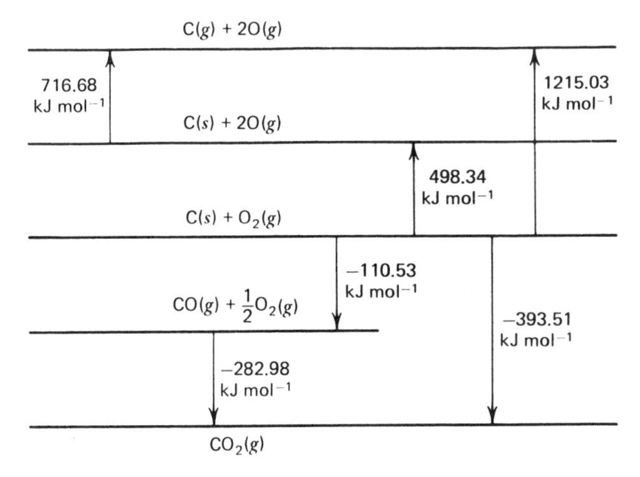

Fig. 1.10 Enthalpy level diagram for the system $C(s) + O_2(g)$. The differences in level are standard enthalpy changes at 25 °C.

This symbolism is convenient because it allows the reactants and products to be numbered.

The enthalpy change for a reaction is equal to the sum of the enthalpies of the products minus the sum of the enthalpies of the reactants at the temperature concerned.

$$\Delta H = \sum \nu_i H_i \tag{1.67}$$

Absolute enthalpies of substances are not known but, since equation 1.67 deals with differences between enthalpies, enthalpies relative to an arbitrary standard state may be used, provided the same standard state is used for reactants and products. It is convenient to calculate the enthalpy of a substance with respect to the elements from which it was formed. These enthalpies are referred to as enthalpies of formation ΔH_f°, where the superscript degree sign reminds us that the substance under consideration and the elements from which it was formed are in their standard states. For a gas the standard state is the hypothetical perfect gas at 1 atm pressure, at each temperature. For a solid or liquid the standard state is the pure substance under a pressure of 1 atm, at each temperature. If there is more than 1 solid form of an element, one must be selected as a reference. The reference form is the most stable form of the element at 25 °C, 1 atm pressure. Thus the reference form of hydrogen is $H_2(g)$ instead of $H(g)$, the reference form of carbon is graphite, and the reference form of sulfur is rhombic sulfur. The standard enthalpy of formation of an element in its reference form is taken to be zero at *all* temperatures. The standard state for a solute in aqueous solution is usually taken as the hypothetical ideal solution of unit molality.

Standard enthalpy changes for reactions may be calculated using enthalpies of formation as follows.

$$\Delta H^\circ = \sum \nu_i \, \Delta H_{f,i}^\circ \tag{1.68}$$

Enthalpies of formation ΔH_f° at 25 °C for some 200 substances are given in Table A.1 of the Appendix. These values are from a much larger table, *Selected Values of Chemical Thermodynamic Properties 270*, published by the National Bureau of Standards. Enthalpies of formation of a smaller number of substances from 0 to 3000 K are given in Table A.2. By use of this table it is possible to calculate enthalpy changes at high temperatures for many reactions.

You may calculate ΔH° for any reaction in this way, but the reaction will not necessarily occur in the direction written. The question as to whether or not the reaction may occur is answered by calculation based on the second law of thermodynamics.

Example 1.2 The reaction of heated coal (approximated here by graphite) with superheated steam absorbs heat. This heat is usually provided by burning some of the coal. Calculate ΔH° (500 K) for both reactions.

(a) $C(\text{graphite}) + H_2O(g) = CO(g) + H_2(g)$

(b) $C(\text{graphite}) + O_2(g) = CO_2(g)$

How many moles of carbon are required to produce 1 mol of hydrogen?

Using Table A.2 for reaction (a),

$$\Delta H^\circ \text{ (500 K)} = -110.022 - (-243.612) = 133.599 \text{ kJ mol}^{-1}$$

Using Table A.2 for reaction (b),

$$\Delta H^\circ \text{ (500 K)} = -393.677 \text{ kJ mol}^{-1}$$

Thus 133.599/393.677 = 0.339 mol of graphite has to be burned to provide heat for the first reaction. Thus, 1.339 mol of graphite is required to produce a mole of hydrogen at a constant temperature of 500 K.

Example 1.3 Calculate ΔH_0°, ΔH° (298 K), and ΔU° (298 K) for

$$H_2(g) = 2H(g)$$

Using Table A.2,

$$\Delta H_0^\circ = 2(216.037 \text{ kJ mol}^{-1}) = 432.073 \text{ kJ mol}^{-1}$$

This value is often referred to as the H—H bond energy. In Chapter 10 we will see how this value can be calculated theoretically; there this dissociation energy is represented by D°.

Using Table A.1,

$$\Delta H^\circ \text{ (298 K)} = 2(217.965 \text{ kJ mol}^{-1}) = 435.931 \text{ kJ mol}^{-1}$$

$$\begin{aligned} \Delta U^\circ \text{ (298 K)} &= \Delta H^\circ \text{ (298 K)} - \Delta nRT \\ &= 435.931 \text{ kJ mol}^{-1} - (8.314 \times 10^{-3} \text{ kJ K}^{-1} \text{ mol}^{-1})(298 \text{ K}) \\ &= 433.458 \text{ kJ mol}^{-1} \end{aligned}$$

In this calculation Δn is the change in the number of moles of gas in the reaction. Perfect gas behavior is assumed.

1.21 ENTHALPY CHANGES OF SOLUTION REACTIONS

When a solute is dissolved in a solvent, heat may be absorbed or evolved; in general, the heat of solution depends on the concentration of the final solution. The *integral heat of solution* is the enthalpy change for the solution of 1 mol of solute in n mol of solvent. The solution process may be represented by a chemical equation such as

$$HCl(g) + 5H_2O(l) = HCl \text{ in } 5H_2O \qquad \Delta H^\circ \text{ (298 K)} = -64.06 \text{ kJ mol}^{-1} \tag{1.69}$$

where "HCl in $5H_2O$" represents a solution of 1 mol of HCl in 5 mol of H_2O. The integral heats of solution of HCl, NaOH, and NaCl are plotted versus the number of moles of water per mole of solute in Fig. 1.11. As the amount of water is increased, the integral heats of solution approach asymptotic values.

The symbol *aq* is used to represent an aqueous solution that is so dilute that additional dilution produces no thermal effect. As an illustration,

$$HCl(g) = HCl(aq) \qquad \Delta H^\circ \text{ (298 K)} = -75.15 \text{ kJ mol}^{-1} \qquad 1.70)$$

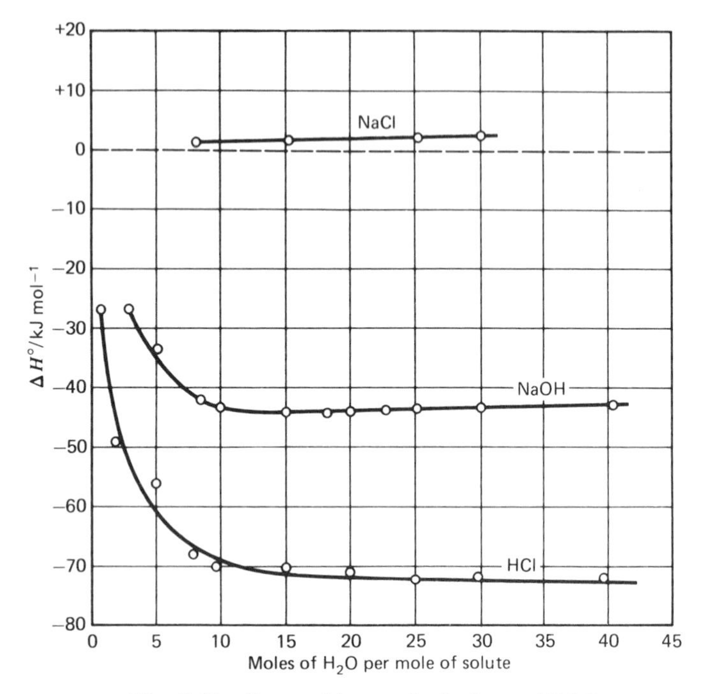

Fig. 1.11 Integral heats of solution at 25 °C.

When a solute is dissolved in a solvent that is chemically quite similar to it and there are no complications of ionization or solvation, the heat of solution may be nearly equal to the heat of fusion of the solute. It might be expected that heat would always be absorbed in overcoming the attraction between the molecules or ions of the solid solute when the solute is dissolved. Another process that commonly occurs, however, is a strong interaction with the solvent, referred to as solvation, which evolves heat. In the case of water the solvation is called hydration.

The importance of this attraction of the solvent for the solute in the process of solution is illustrated by the dissolving of sodium chloride in water. In the crystal lattice of sodium chloride, positive sodium ions and negative chloride ions attract each other strongly. The energy required to separate them is so great that nonpolar solvents like benzene and carbon tetrachloride do not dissolve sodium chloride; but a solvent like water, which has a high dielectric constant and a large dipole moment, has a strong attraction for the sodium and chloride ions and solvates them with a large decrease in the energy of the system. When the energy required to separate the ions from the crystal is about the same as the solvation energy, as it is for dissolving NaCl in water, ΔH for the net process is close to zero. When NaCl is dissolved in water at 25 °C, there is only a small cooling effect; q is positive. When Na_2SO_4 is dissolved in water at 25 °C, there is an evolution of heat because the energy of hydration of the ions is greater than the energy required to separate the ions from the crystal.

The integral heat of dilution between two molalities m_1 and m_2 is the heat accompanying the dilution of an amount of solution of concentration m_1 containing 1 mol of solute with pure solvent to make a solution of concentration m_2.

Integral heats of solution, heats of dilution, and heats of reaction in solution may be calculated from tabulated values of heats of formation in solution. The enthalpy of formation of water is neglected in calculations if there is the same number of moles of water on both sides of the balanced chemical equation. Also, the enthalpy of formation of pure water is used for water in aqueous solution. This is done arbitrarily, much as the enthalpies of formation of the elements are taken equal to zero.

Example 1.4 Calculate $\Delta H°$ (298 K) for the reaction

$$\text{HCl in } 100\text{H}_2\text{O} + \text{NaOH in } 100\text{H}_2\text{O} = \text{NaCl in } 200\text{H}_2\text{O} + \text{H}_2\text{O}(l)$$

$$\Delta H° \text{ (298 K)} = -406.752 - 285.830 + 166.159 + 469.060$$
$$= -57.363 \text{ kJ mol}^{-1}$$

For dilute solutions it is found that the heat of reaction of strong bases, like NaOH and KOH, with strong acids, like HCl and HNO_3, is independent of the nature of the acid or base. This constancy of the heat of neutralization is a result of the complete ionization of strong acids and bases and the salts formed by neutralization. Thus, when a dilute solution of a strong acid is added to a dilute solution of a strong base, the only chemical reaction is

$$\text{OH}^-(aq) + \text{H}^+(aq) = \text{H}_2\text{O}(l) \qquad \Delta H° \text{ (298 K)} = -55.835 \text{ kJ mol}^{-1}$$

When a dilute solution of a weak acid or base is neutralized, the heat of neutralization is somewhat less because of the absorption of heat in the dissociation of the weak acid or base.

Since for strong electrolytes in dilute solution the thermal properties of the ions are essentially independent of the accompanying ions, it is convenient to use relative enthalpies of formation of individual ions. The sum of the enthalpies of formation of H^+ and OH^- ions may be calculated from

$$\text{H}_2\text{O}(l) = \text{H}^+(aq) + \text{OH}^-(aq) \qquad \Delta H° = 55.835 \text{ kJ mol}^{-1}$$
$$\text{H}_2(g) + \tfrac{1}{2}\text{O}_2(g) = \text{H}_2\text{O}(l) \qquad \Delta H° = -285.830 \text{ kJ mol}^{-1}$$
$$\text{H}_2(g) + \tfrac{1}{2}\text{O}_2(g) = \text{H}^+(aq) + \text{OH}^-(aq) \qquad \Delta H° = -229.994 \text{ kJ mol}^{-1}$$

The separate enthalpies of formation of H^+ and OH^- cannot be calculated and so, in order to construct a table of enthalpies of formation of individual ions, it is necessary to adopt an arbitrary convention. Enthalpies of formation of aqueous ions in Table A.1 are based on the convention that $\Delta H_f° = 0$ for $H^+(aq)$. In other words, by convention,

$$\tfrac{1}{2}\text{H}_2(g) = \text{H}^+(aq) + e \qquad \Delta H_f° = 0$$

where e is the electron. Therefore, the enthalpy of formation of OH^- is given by

$$\tfrac{1}{2}\text{H}_2(g) + \tfrac{1}{2}\text{O}_2(g) + e = \text{OH}^-(aq) \qquad \Delta H_f° = -229.994 \text{ kJ mol}^{-1}$$

On the basis of these values for the enthalpies of formation of H^+ and OH^-, the enthalpies of formation of other ions of strong electrolytes may be calculated.

From the enthalpy of formation of HCl(aq) it is possible to calculate the enthalpy of formation of $Cl^-(aq)$.

$$\tfrac{1}{2}H_2(g) + \tfrac{1}{2}Cl_2(g) = H^+(aq) + Cl^-(aq) \qquad \Delta H^\circ = -167.159 \text{ kJ mol}^{-1}$$

$$\tfrac{1}{2}Cl_2(g) + e = Cl^-(aq) \qquad \Delta H_f^\circ = -167.159 \text{ kJ mol}^{-1}$$

1.22 DEPENDENCE OF ENTHALPY CHANGE OF REACTION ON TEMPERATURE

If the enthalpy change for a reaction is known at one temperature, it can be calculated at another temperature if the heat capacities of reactants and products are known over the intervening temperature range.

For a general chemical reaction given in equation 1.65 the enthalpy change is given by equation 1.67. The rate of change of ΔH with temperature is obtained by differentiating equation 1.67 with respect to temperature at constant pressure.

$$\left[\frac{d(\Delta H)}{dT}\right]_P = \sum \nu_i \left(\frac{dH_i}{dT}\right)_P \tag{1.71}$$

Remembering that $(dH/dT)_P = C_P$, we see that

$$\left[\frac{d(\Delta H)}{dT}\right]_P = \sum \nu_i C_{P,i} = \Delta C_P \tag{1.72}$$

This equation may be stated in words as follows: the change in enthalpy of reaction at constant pressure per degree rise in temperature is equal to the change in heat capacity at constant pressure of the system as the result of the reaction.

Equation 1.72 may be integrated between two temperatures T_1 and T_2 to obtain the relation between the enthalpy changes at these two temperatures.

$$\int_{\Delta H_1}^{\Delta H_2} d(\Delta H) = \Delta H_2 - \Delta H_1 = \int_{T_1}^{T_2} \Delta C_P \, dT \tag{1.73}$$

By use of this equation it is possible to calculate ΔH for a reaction at another temperature if it is known at one temperature and if the values of C_P for the reactants and products are known in the intervening temperature range.

Equation 1.73 is applicable only if there are no changes in phase in going from T_1 to T_2; additional terms must be introduced for the enthalpy changes accompanying phase transformations such as melting or vaporization.

Example 1.5 What is the enthalpy change for the vaporization of water at 0 °C? This value may be estimated from Table A.1 by assuming that the heat capacities of $H_2O(l)$ and $H_2O(g)$ are independent of temperature from 0 to 25 °C.

$$H_2O(l) = H_2O(g)$$

$$\Delta H^\circ \text{ (298 K)} = -241.818 - (-285.830) = 44.011 \text{ kJ mol}^{-1}$$

$$\Delta H_2^\circ = \Delta H_1^\circ + \Delta C_P(T_2 - T_1)$$

$$\Delta H^\circ \text{ (273 K)} = \Delta H^\circ \text{ (298 K)} + [C_P(H_2O, g) - C_P(H_2O, l)](273 - 298)$$

$$= 44{,}011 - (33.577 - 75.291)(-25) = 42{,}970 \text{ kJ mol}^{-1}$$

References

K. E. Bett, J. S. Rowlinson, and G. Saville, *Thermodynamics for Chemical Engineers*, The MIT Press, Cambridge, Mass., 1975.

K. G. Denbigh, *The Principles of Chemical Equilibrium*, Cambridge University Press, Cambridge, 1971.

R. E. Dickerson, *Molecular Thermodynamics*, W. A. Benjamin Inc., New York, 1969.

G. Kirkwood and I. Oppenheim, *Chemical Thermodynamics*, McGraw-Hill Book Co., New York, 1961.

I. M. Klotz and R. M. Rosenberg, *Chemical Thermodynamics*, W. A. Benjamin Inc., New York, 1972.

G. N. Lewis, M. Randall, revised by K. S. Pitzer, and L. Brewer, *Thermodynamics*, McGraw-Hill Book Co., New York, 1961.

P. A. Rock, *Chemical Thermodynamics*, The Macmillan Co., London, 1969.

D. R. Stull, E. F. Westrum, and G. C. Sinke, *The Chemical Thermodynamics of Organic Compounds*, Wiley-Interscience, New York, 1969.

R. E. Wood, *Introduction to Chemical Thermodynamics*, Appleton-Century-Crofts, New York, 1970.

Problems

1.1 How much work is done when a person weighing 75 kg (165 lb) climbs the Washington monument, 555 ft high? How many kilojoules must be supplied to do this muscular work, assuming that 25% of the energy produced by the oxidation of food in the body can be converted into muscular mechanical work? *Ans.* 497.6 kJ.

1.2 A mole of liquid water is vaporized at 100 °C and 1 atm. The heat of vaporization is 40.69 kJ mol^{-1}. What are the values of (*a*) w_{rev}, (*b*) q, (*c*) ΔU, and (*d*) ΔH?
Ans. (*a*) -3.10, (*b*) 40.69, (*c*) 37.59, (*d*) 40.69 kJ mol^{-1}.

1.3 Considering H_2O to be a rigid nonlinear molecule, what value of C_P for the gas would be expected classically? If vibration is taken into account, what value is expected? Compare these values of C_P with the actual values at 298 and 3000 K in Table A.1.
Ans. 20.782, 45.724 J K^{-1} mol^{-1}.

1.4 The equation for the molar heat capacity of *n*-butane is

$$C_P = 19.41 + 0.233T$$

where C_P is given in J K^{-1} mol^{-1}. Calculate the heat necessary to raise the temperature of 1 mole from 25 to 300 °C at constant pressure. *Ans.* 33.31 kJ mol^{-1}.

1.5 Calculate H° (2000 K) $-$ H° (0 K) for $H(g)$ and compare it with the value in Table A.2. *Ans.* 41.572 kJ mol^{-1}.

1.6 One mole of nitrogen at 25 °C and 1 atm is expanded reversibly and isothermally to a pressure of 0.132 atm. (*a*) What is the value of w? (*b*) What is the value of w if the temperature is 100 °C? *Ans.* −5027, (*b*) −6292 J mol^{-1}.

1.7 A mole of argon is allowed to expand adiabatically from a pressure of 10 atm and 298.15 K to 1 atm. What is the final temperature and how much work can be done? *Ans.* 118.7 K. 2238 J mol^{-1}.

1.8 A tank contains 20 L of compressed nitrogen at 10 atm and 25 °C. Calculate the maximum work (in joules) that can be obtained when the gas is allowed to expand to 1 atm pressure (*a*) isothermally, and (*b*) adiabatically. *Ans.* (*a*) 46.7, (*b*) 24.46 kJ mol^{-1}.

1.9 Calculate the temperature increase and final pressure of helium if a mole is compressed adiabatically and reversibly from 44.8 L at 0 °C to 22.4 L. *Ans.* 160.4 K. 1.587 atm.

1.10 Calculate the second virial coefficient for hydrogen at 0 °C from the fact that the molar volumes at 50, 100, 200, and 300 atm are 0.4634, 0.2386, 0.1271, and 0.090 04 L mol^{-1}, respectively. *Ans.* 0.0138 L mol^{-1}.

1.11 The second virial coefficient B of methyl isobutyl ketone is −1580 cm^3 mol^{-1} at 120 °C. Compare its compressibility factor at this temperature with that of a perfect gas at 1 atm. *Ans.* 0.951.

1.12 Show that for a gas of spherical molecules b in the van der Waals equation is four times the molecular volume times Avogadro's constant.

1.13 In an adiabatic calorimeter, oxidation of 0.4362 g of naphthalene caused a temperature rise of 1.707 °C. The heat capacity of the calorimeter and water was 10,290 J K^{-1}. If corrections for oxidation of the wire and residual nitrogen are neglected, what is the enthalpy of combustion of naphthalene per mole? *Ans.* −5163 kJ mol^{-1}.

1.14 The following reactions might be used to power rockets.

$$(1)\ H_2(g) + \tfrac{1}{2}O_2(g) = H_2O(g)$$
$$(2)\ CH_3OH(l) + 1\tfrac{1}{2}O_2(g) = CO_2(g) + 2H_2O(g)$$
$$(3)\ H_2(g) + F_2(g) = 2HF(g)$$

(*a*) Calculate the enthalpy changes at 25 °C for each of these reactions per kilogram of reactants. (*b*) Since the thrust is greater when the molar mass of the exhaust gas is lower, divide the heat per kilogram by the molar mass of the product (or the average molar mass in the case of reaction 2) and arrange the above reactions in order of effectiveness on the basis of thrust. *Ans.* (*a*) −13.4, −7.98, −13.6 MJ kg^{-1}. (*b*) (1) > (3) > (2).

1.15 Calculate the enthalpy of formation of $PCl_5(s)$, given the heats of the following reactions at 25 °C.

$$2P(s) + 3Cl_2(g) = 2PCl_3(l) \qquad \Delta H° = -635.13\ \text{kJ mol}^{-1}$$
$$PCl_3(l) + Cl_2(g) = PCl_5(s) \qquad \Delta H° = -137.28\ \text{kJ mol}^{-1}$$

Ans. −454.85 kJ mol^{-1}.

1.16 Calculate the heat of hydration of $Na_2SO_4(s)$ from the integral heats of solution of $Na_2SO_4(s)$ and $Na_2SO_4 \cdot 10H_2O(s)$ in infinite amounts of H_2O, which are −2.34 kJ mol^{-1} and 78.87 kJ mol^{-1}, respectively. Enthalpies of hydration cannot be measured directly because of the slowness of the phase transition. *Ans.* −81.21 kJ mol^{-1}.

1.17 Calculate the integral heat of solution of one mole of $HCl(g)$ in $200H_2O(l)$.

$$HCl(g) + 200H_2O(l) = HCl \text{ in } 200H_2O$$

Ans. −73.965 kJ mol^{-1}.

1.18 Calculate the enthalpies of reaction at 25 °C for the following reactions in dilute aqueous solutions.

(*a*) $HCl(aq) + NaBr(aq) = HBr(aq) + NaCl(aq)$

(*b*) $CaCl_2(aq) + Na_2CO_3(aq) = CaCO_3(s) + 2NaCl(aq)$

Ans. (*a*) 0, (*b*) 13.05 kJ mol^{-1}.

1.19 Calculate $\Delta H°$ for the dissociation

$$O_2(g) = 2O(g)$$

at 0, 298, and 3000 K. In Section 10.1 the enthalpy change for dissociation at 0 K will be found to be equal to the spectroscopic dissociation energy $D°$.

Ans. 493.570, 498.340, 513.444 kJ mol^{-1}.

1.20 Compare the enthalpies of combustion of $CH_4(g)$ to $CO(g)$ and $H_2O(g)$ at 298 and 2000 K.

$$CH_4(g) + 2O_2(g) = CO_2(g) + 2H_2O(g)$$

Ans. −802.303, −807.513 kJ mol^{-1}.

1.21 What is the heat evolved in freezing water at −10 °C given that:

$H_2O(l) = H_2O(s)$ $\qquad \Delta H°\ (273\ K) = -6004\ J\ mol^{-1}$

$C_P(H_2O, l) = 75.3\ J\ K^{-1}\ mol^{-1}$ $\qquad C_P(H_2O, s) = 36.8\ J\ K^{-1}\ mol^{-1}$

Ans. −5619 J mol^{-1}

1.22 Calculate how high a 70-kg man could climb if he could convert the heat of combustion of 1 oz of chocolate (628 kJ) completely into work of vertical displacement.

1.23 The surface tension of water is 71.97×10^{-3} N m^{-1} or 71.97×10^{-3} J m^{-2} at 25 °C. Calculate the surface energy in joules of 1 mol of water dispersed as a mist containing droplets 1 μm (10^{-4} cm) in radius. The density of water may be taken as 1.00 g cm^{-3}.

1.24 A mole of ammonia gas is condensed at its standard boiling point of −33.4 °C by the application of a pressure infinitesimally greater than 1 atm. To evaporate ammonia at its boiling point requires the absorption of 23.30 kJ mol^{-1}. Calculate (*a*) w_{rev}, (*b*) q, (*c*) ΔH, and (*d*) ΔU.

1.25 For $CO(g)$, $CO_2(g)$ (*note:* linear molecule), $NH_3(g)$, and $CH_4(g)$, compare the values of C_P expected from classical kinetic theory with the values at 3000 K in Table A.2.

1.26 From the following data calculate the value of $(H°_{298} - H°_0)$ for $Al_2O_3(s)$, assuming the validity of the Debye T^3 law for the heat capacity below 10 K.

T/K	$C°_P$/J K^{-1} mol^{-1}	T	$C°_P$	T	$C°_P$	T	$C°_P$
10	0.009	90	9.69	180	43.79	270	72.37
20	0.076	100	12.84	190	47.53	280	74.84
30	0.263	110	16.32	200	51.14	290	77.19
40	0.691	120	20.06	210	54.60	298.16	79.01
50	1.492	130	23.96	220	57.92	273.16	73.16
60	2.779	140	27.96	230	61.10		
70	4.582	150	31.98	240	64.13		
80	6.895	160	35.99	250	67.01		
		170	39.94	260	69.76		

1.27 One mole of methane (considered to be a perfect gas) initially at 25 °C and 1 atm pressure is heated at constant pressure until the volume has doubled. The variation of the molar heat capacity with absolute temperature is given by

$$C_P = 22.34 + 48.1 \times 10^{-3} T$$

where C_P is in J K^{-1} mol^{-1}. Calculate (*a*) ΔH, and (*b*) ΔU.

1.28 One mole of perfect gas at 25 °C and 100 atm is allowed to expand reversibly and isothermally to 5 atm. Calculate (*a*) the work done on the gas in liter-atmospheres, (*b*) the heat absorbed in joules, (*c*) ΔU, and (*d*) ΔH.

1.29 One mole of hydrogen at 25 °C and 1 atm is compressed adiabatically and reversibly into a volume of 5 L. Assuming perfect gas behavior, calculate (*a*) the final temperature, (*b*) the final pressure, and (*c*) the work done on the gas.

1.30 One mole of argon at 25 °C and 1 atm pressure is allowed to expand reversibly to a volume of 50 L (*a*) isothermally and (*b*) adiabatically. Calculate the final pressure in each case, assuming perfect gas behavior.

1.31 (*a*) Calculate the maximum work of isothermal expansion of 10 g of helium from a volume of 10 L to a volume of 50 L at 25 °C. (*b*) Calculate the maximum work of adiabatic expansion, starting with the same conditions and allowing the gas to expand to 50 L.

1.32 The coefficient of thermal expansivity α is defined by

$$\alpha = \frac{1}{V}\left(\frac{\partial V}{\partial T}\right)_P$$

and the isothermal compressibility κ is defined by

$$\kappa = -\frac{1}{V}\left(\frac{\partial V}{\partial P}\right)_T$$

Calculate these quantities for a perfect gas.

1.33 Show that to a first approximation the equation of state of a gas that dimerizes to a small extent is given by

$$\frac{PV}{RT} = 1 - \frac{K_c}{V}$$

where K_c is the equilibrium constant for the formation of dimer.

$$2A = A_2 \qquad K_c = \frac{(A_2)}{(A)^2}$$

1.34 Assuming that CH_4 is a spherical molecule, calculate the molecular diameter from the van der Waals constant b. The molecular diameter obtained from viscosity measurements is 0.414 nm.

1.35 Using van der Waals equation, calculate the pressure exerted by 1 mol of carbon dioxide at 0 °C in a volume of (*a*) 1.00 L, (*b*) 0.05 L. (*c*) Repeat the calculations at 100 °C and 0.05 L.

1.36 Show that for a van der Waals gas the second B and third C virial coefficients are given by

$$B = b - \frac{a}{RT} \qquad C = b^2$$

1.37 The combustion of oxalic acid in a bomb calorimeter yields 2816 J g^{-1} at 25 °C. Calculate (*a*) $\Delta U°$ and (*b*) $\Delta H°$ for the combustion of 1 mol of oxalic acid (M = 90.0 g mol^{-1}).

1.38 A solution of hemoglobin containing 5 g of the protein ($M = 64{,}000$ g mol^{-1}) in 100 cm^3 of solution is completely oxygenated. Each mole of hemoglobin binds 4 mol of oxygen. The temperature of the solution rises 0.031 °C. What is the enthalpy of reaction per mole of oxygen bound? The heat capacity of the solution may be assumed to be 4.18 J K^{-1} cm^{-3}.

1.39 The enthalpy change for the combustion of toluene to $H_2O(l)$ and $CO_2(g)$ is -3910.0 kJ mol^{-1} at 25 °C. Calculate the enthalpy of formation of toluene.

1.40 Construct an energy level diagram like Fig. 1.10 from the following data at 298 K.

$$CH_4(g) + 2O_2(g) = CO_2(g) + 2H_2O(l) \qquad \Delta H° = -890.4 \text{ kJ mol}^{-1}$$
$$CH_4(g) = C(s) + 2H_2(g) \qquad \Delta H° = 74.9 \text{ kJ mol}^{-1}$$
$$C(s) + O_2(g) = CO_2(g) \qquad \Delta H° = -393.51 \text{ kJ mol}^{-1}$$
$$H_2(g) + \tfrac{1}{2}O_2(g) = H_2O(l) \qquad \Delta H° = -285.85 \text{ kJ mol}^{-1}$$

1.41 For 25 °C the following results are given.

$$Be_3N_2(s) + 3Cl_2(g) = 3BeCl_2(s) + N_2(g) \qquad \Delta H° = -897.1 \pm 0.4 \text{ kJ mol}^{-1}$$
$$Be(s) + Cl_2(g) = BeCl_2(s) \qquad \Delta H° = -494.1 \pm 2.5 \text{ kJ mol}^{-1}$$
$$3Be(s) + 2NH_3(g) = Be_3N_2(s) + 3H_2(g) \qquad \Delta H° = -495.4 \pm 1.7 \text{ kJ mol}^{-1}$$

Compare the two values for the standard enthalpy of formation of Be_3N_2, which can be obtained from the foregoing data.

1.42 The following heats of solution are obtained at 18 °C when a large excess of water is used:

$$CaCl_2(s) = CaCl_2(aq) \qquad \Delta H° = -75.3 \text{ kJ mol}^{-1}$$
$$CaCl_2 \cdot 6H_2O(s) = CaCl_2(aq) \qquad \Delta H° = 18.8 \text{ kJ mol}^{-1}$$

Calculate the heat of hydration of $CaCl_2$ to give $CaCl_2 \cdot 6H_2O$ by (*a*) $H_2O(l)$, and (*b*) $H_2O(g)$. The heat of vaporization of water at this temperature is 2452 J g^{-1}.

1.43 Calculate $\Delta H°$ (298 K) per gram of fuel (exclude oxygen) for:

(*a*) $H_2(g) + \tfrac{1}{2}O_2(g) = H_2O(g)$

(*b*) $CH_4(g) + 2O_2(g) = CO_2(g) + 2H_2O(g)$

(*c*) $CH_3OH(l) + \tfrac{3}{2}O_2(g) = CO_2(g) + 2H_2O(g)$

(*d*) $C_6H_{14}(g) + 9\tfrac{1}{2}O_2(g) = 6CO_2(g) + 7H_2O(g)$

1.44 What are the enthalpy changes for the following reactions at 298 K?

(*a*) $H_2(g) + F_2(g) = 2HF(g)$

(*b*) $H_2(g) + Cl_2(g) = 2HCl(g)$

(*c*) $H_2(g) + I_2(g) = 2HI(g)$

1.45 Hydrogen may be produced from methane at 1000 K by use of the reaction

$$CH_4(g) + H_2O(g) = CO(g) + 3H_2(g)$$

How much heat is absorbed or evolved by this reaction?

1.46 According to Table A.2, how much heat is required to raise the temperature of a mole of oxygen from 298 to 3000 K at constant pressure?

1.47 Calculate the heat of vaporization of water at 25 °C. The specific heat of water may be taken as 4.18 J K^{-1} g^{-1}. The heat capacity of water vapor at constant pressure in this temperature range is 33.5 J K^{-1} mol^{-1}, and the heat of vaporization of water is 100 °C is 2258 J g^{-1}.

CHAPTER 2

SECOND AND THIRD LAWS OF THERMODYNAMICS

The first law of thermodynamics states that when one form of energy is converted into another the total energy is conserved. It does not indicate any other restriction on this process. However, we know that many processes have a natural direction, and it is with the question of direction that the second law is concerned. For example, a gas expands into a vacuum but, although it would not violate the first law, the reverse never occurs. For a bar at uniform temperature to become hot at one end and cold at the other would not be a violation of the first law, yet we know this never occurs spontaneously. The second law establishes a criterion for predicting whether a process can occur spontaneously, and so it is of great importance to chemistry.

The quantity that tells us whether a chemical reaction or a physical change can occur spontaneously in an isolated system is the entropy S. The entropy is a function of the state of the system, as is the internal energy U.

The introduction of S completes the necessary set of thermodynamic quantities, but it is useful to define certain further quantities, especially the Gibbs energy G and the chemical potential μ, in terms of other thermodynamic quantities. The quantities G and μ are especially useful in discussing phase equilibria (Chapter 3) and chemical equilibria (Chapters 4, 5, and 6). This chapter closes with the third law of thermodynamics, which allows us to obtain the absolute value of the entropy of a substance.

2.1 SPONTANEOUS AND NONSPONTANEOUS CHANGES

We are familiar with the fact that many changes occur spontaneously, that is, when systems are simply left to themselves. For example, heat flows from a hotter body to a colder body, and chemical reactions proceed to equilibrium. It is a matter of experience that spontaneous changes do not reverse themselves; heat never flows spontaneously from a cold body to a hot body, and a chemical reaction that has reached equilibrium never spontaneously produces the original reactants. These latter changes are referred to as nonspontaneous changes or unnatural changes.

For any spontaneous change, it is possible to devise, in principle at least, a means for getting useful work. Thus a heat engine may be used to do work in the process of transferring heat from a hot reservoir to a cold reservoir, and a chemical reaction may be harnessed in a battery. Since work can be obtained from a spontaneous change, it is evident that in the occurrence of a spontaneous change, the system loses capacity to do work. Nonspontaneous changes can be made to occur only by supplying work from outside the system. For example, heat can be transferred from a cold object to a hot object by doing work on a "working fluid" in a refrigerator, and some chemical reactions can be reversed by electrical work (i.e., electrolysis). Since the work required can be supplied only by some other spontaneous change, it is apparent that a spontaneous change may be reversed only by harnessing, in some way, work from another spontaneous change.

The second law of thermodynamics provides the means for determining whether a given chemical reaction is spontaneous under a given set of conditions. Historically it was through the consideration of heat engines that the second law of thermodynamics was formulated in the 1860s. Therefore we begin our consideration of the natural direction of change with heat engines.

2.2 CARNOT HEAT ENGINE

A heat engine is a device that exchanges only heat and work with its surroundings and operates in a cycle. The arrangements for a Carnot heat engine are shown in Fig. 2.1. The arrows indicate that in one cycle this engine receives heat $|q_1|$ from the high-temperature reservoir, rejects heat $|q_2|$ to the low-temperature reservoir, and does work $|w|$ on its surroundings. The absolute value signs are used because the signs of these algebraic quantities are set by conventions that require in this case that q_1 is positive, q_2 is negative, and w is negative. In the operation of the Carnot heat engine a working fluid, which we will refer to as a gas, is taken through a sequence of four steps that return it to its initial state. The engine itself consists of an idealized cylinder with a piston that can slide without friction and can do work on the surroundings or have the surroundings do work on the system.

The cycle for the Carnot heat engine consists of the following four steps, which are represented in Fig. 2.2.

1. Reversible isothermal expansion of the gas from state 1 to state 2. During this expansion step, the piston does work $|w_{12}|$ on the surroundings, and heat q_2 is absorbed by the gas from the high-temperature reservoir. Note that according to the convention that w is work done on the system (i.e., the gas), w_{12} is negative. In order to assure that the heat transfer is reversible, the temperature of the gas is only infinitesimally lower than the temperature T_1 of the high-temperature heat reservoir.
2. Reversible adiabatic expansion of the gas from state 2 to state 3. For this step we can imagine that the piston and cylinder are thermally insulated so that no heat is gained or lost. The expansion continues until the temperature of the gas has dropped to T_2. During this expansion step, the piston does work $|w_{23}|$ on the surroundings.
3. Reversible isothermal compression of the gas from state 3 to state 4. During this compression step the surroundings do work w_{34} on the gas and heat $|q_2|$ flows

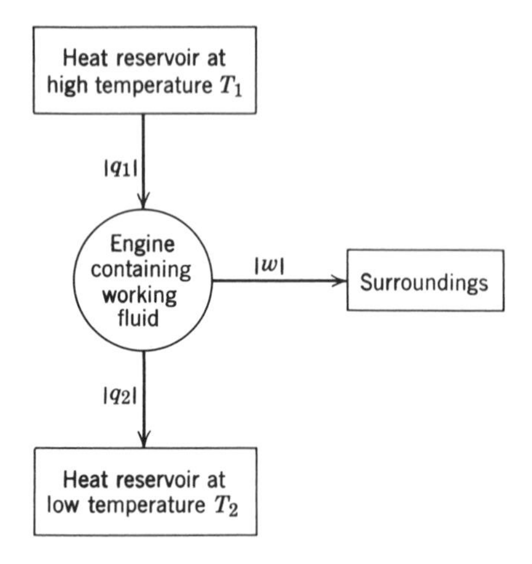

Fig. 2.1 Carnot heat engine.

out of the gas to the low-temperature heat reservoir. Note that according to the convention that q is heat absorbed by the gas, q_2 is negative. The gas is maintained at a temperature only infinitesimally higher than the temperature T_2 of the low-temperature heat reservoir so that the heat transfer is reversible.

4. Reversible adiabatic compression of the gas from state 4 to state 1. This step completes the cycle by bringing the gas back to its initial state at temperature T_1. During this compression step, the surroundings do work w_{41} on the gas, but no heat is gained or lost.

Since the gas is returned to its initial state at the end of the process, $\Delta U = 0$ for the cycle, and the first law of thermodynamics yields

$$\begin{aligned}\Delta U = 0 &= q_1 + q_2 + w_{12} + w_{23} + w_{34} + w_{41} \\ &= q_1 + q_2 + w \end{aligned} \tag{2.1}$$

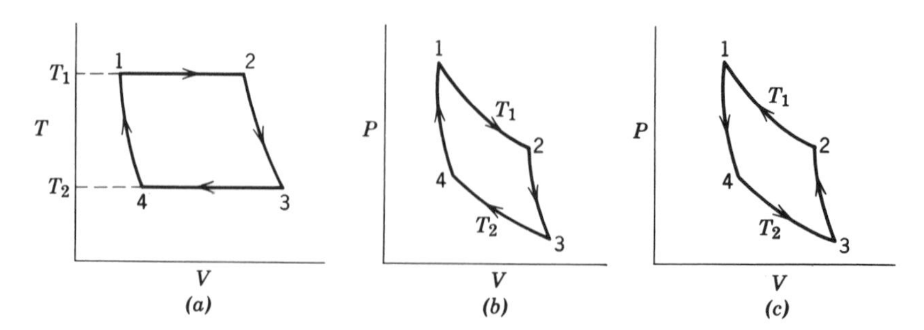

Fig. 2.2 The Carnot cycle. (*a*) Plot of T versus V for a Carnot engine. (*b*) Plot of P versus V for a Carnot engine. (*c*) Plot of P versus V for a Carnot refrigerator or Carnot heat pump.

where w is the net work done on the gas. The engine absorbs heat q_1 from the high-temperature reservoir and does work on the surroundings so that w is negative, but the first law does not tell us the relative amounts of work and of heat rejected at T_2.

In practice the low-temperature reservoir of a heat engine is the atmosphere so that the economic cost is mainly that of supplying q_1. Thus the efficiency η of a heat engine is defined as the ratio of the work done on the surroundings to the heat input at the higher temperature.

$$\eta \equiv -\frac{w}{q_1} = \frac{q_1 + q_2}{q_1} \tag{2.2}$$

where the negative sign is required by the fact that w is negative and the efficiency is, of course, positive. The second form of this equation is obtained by introducing equation 2.1. Since this is as far as we can go with the first law, we will have to look to the second law for any limitations on the conversion of heat to work by the cyclic process of a Carnot engine.

The work done by the surroundings on the gas in the four steps of the Carnot engine cycle is given by:

$$w_{12} = -\int_1^2 P\,dV \tag{2.3a}$$

$$w_{23} = -\int_2^3 P\,dV \tag{2.3b}$$

$$w_{34} = -\int_3^4 P\,dV \tag{2.3c}$$

$$w_{41} = -\int_4^1 P\,dV \tag{2.3d}$$

Thus the *magnitude* of the work for a step is given by area under the curve for the step in the P–V diagram. The net work done in a cycle is the sum of the four work terms, and the *magnitude* of the work for one complete cycle is equal to the area enclosed by the cycle in the P–V diagram.

2.3 CARNOT REFRIGERATOR AND CARNOT HEAT PUMP

Refrigeration may be provided by running the Carnot cycle backward, as shown in Fig. 2.2*c*. The first step is a reversible adiabatic expansion from state 1 to state 4. The second step is a reversible isothermal expansion that extracts heat q_2 from the heat reservoir at the low temperature. The third step is a reversible adiabatic compression from state 3 to state 2. The last step is a reversible isothermal compression in which heat $|q_1|$ is rejected to heat reservoir at the higher temperature T_1. In a household refrigerator the high-temperature reservoir is simply the room air. In contrast with the Carnot engine, net work has to be done on the gas. The work done by the surroundings on the gas in the four steps of the Carnot refrigerator cycle is given by:

$$w_{14} = -\int_1^4 P\,dV \tag{2.4a}$$

$$w_{43} = -\int_4^3 P\,dV \tag{2.4b}$$

$$w_{32} = -\int_3^2 P\,dV \tag{2.4c}$$

$$w_{21} = -\int_2^1 P\,dV \tag{2.4d}$$

The amount of work required to complete one cycle is equal to the area enclosed by the cycle in the P–V diagram. The first law relationship for the Carnot refrigerator is given by

$$\begin{aligned}\Delta U = 0 &= q_1 + q_2 + w_{14} + w_{43} + w_{32} + w_{21} \\ &= q_1 + q_2 + w \end{aligned} \tag{2.5}$$

where w is the net work done on the gas.

The coefficient of performance β of a refrigerator is defined as the ratio of the heat extracted from the low-temperature reservoir to the work done on the gas.

$$\beta = \frac{q_2}{w_{14} + w_{43} + w_{32} + w_{21}} = -\frac{q_2}{q_1 + q_2} \tag{2.6}$$

The Carnot heat pump uses the same cycle as the Carnot refrigerator, but the objective is to supply as much heat as possible to the reservoir at the higher temperature for a given amount of work performed on the gas on the piston. The low-temperature reservoir is simply the ground or a body of water that is cooler than the room to be heated. The coefficient of performance β' of a heat pump is the ratio of the heat supplied at the higher temperature to the work done on the gas.

$$\beta' = -\frac{q_1}{w_{14} + w_{43} + w_{32} + w_{21}} = \frac{q_1}{q_1 + q_2} \tag{2.7}$$

Both β and β' are positive quantities that may be greater than unity.

2.4 THE SECOND LAW OF THERMODYNAMICS

It is simply a generalization of experience to say that in transferring heat from a hot reservoir to a cold reservoir it is possible to obtain external work on the surroundings by use of a cyclic process and that the transfer of heat in the reverse direction requires work to be done on the system. From this viewpoint Clausius*

* R. Clausius, *Phil. Mag.*, **12**, 86 (1856).

stated the second law as follows: "Heat can never pass from a colder to a warmer body without some other change, connected therewith, occuring at the same time." We will see that there are other ways of stating the second law, but these different statements can all be shown to be equivalent. We cannot prove this statement of the second law; it is simply accepted as a postulate and is used to derive conclusions. These conclusions are always found to be correct when the second law is applied to systems containing many particles.* As we will see later, the second law is not derivable directly from mechanics, but is a statistical law that relies on the presence of large numbers of particles for its validity.

Clausius' statement of the second law may be used to prove the Carnot principle that no heat engine can be more efficient than a reversible engine operating between two given heat reservoirs and that all reversible heat engines operating between the same two heat reservoirs have the same efficiency, independent of the working fluid. We will not derive this principle from Clausius' statement of the second law, but want to emphasize how it leads to the definition of a thermodynamic temperature scale. Since all reversible engines operating between T_1 and T_2 have the same efficiency, $\eta = f(T_1, T_2)$, where f is a universal function independent of the working fluid. Since $\eta = 1 + q_2/q_1$, the ratio q_2/q_1 must also be a function of the temperatures of the two reservoirs, and so $q_2/q_1 = g(T_1, T_2)$. A number of different temperature scales that satisfy this relation can be devised but, by taking the temperature as directly proportional to the magnitude of the heat exchanged with a heat reservoir of a reversible Carnot engine, the thermodynamic temperature scale becomes proportional to the perfect gas temperature scale, as we will see in the next section.

$$\frac{|q_2|}{|q_1|} = \frac{T_2}{T_1} \tag{2.8}$$

This definition of the thermodynamic temperature leads to the following expression for the efficiency of a Carnot engine.

$$\eta = \frac{|w|}{|q_1|} = \frac{|q_1| - |q_2|}{|q_1|} = \frac{T_1 - T_2}{T_1} \tag{2.9}$$

The expression for the coefficient of performance β of a refrigerator becomes

$$\beta = \frac{|q_2|}{|w|} = \frac{|q_2|}{|q_1| - |q_2|} = \frac{T_2}{T_1 - T_2} \tag{2.10}$$

and the expression for the coefficient of performance β' of a heat pump becomes

$$\beta' = \frac{|q_1|}{|w|} = \frac{|q_1|}{|q_1| - |q_2|} = \frac{T_1}{T_1 - T_2} \tag{2.11}$$

---

* In Brownian motion, where only a few particles are observed with an ultramicroscope, it is sometimes found that the tiny particles move from the region of low concentration to the region of higher concentration in violation of what would be expected from the second law for a system containing many particles.

Example 2.1 (*a*) A steam engine in a nuclear power plant operates between 800 and 330 K. What is the maximum work that can be obtained from 1 kWh of heat? (*b*) A household refrigerator operates between 373 and 273 K. How many joules of heat can in principle be removed per kilowatt-hour of work? (*c*) A heat pump used to heat a building operates between an inside temperature of 295 K and an outside temperature of 273 K. How much work must in principle be expended for every kilowatt-hour of heat?

$$(a)\quad |w| = |q_1|\,\frac{T_1 - T_2}{T_1} = \frac{(1\text{ kWh})(470\text{ K})}{800\text{ K}} = 0.588\text{ kWh}$$

$$\begin{aligned}(b)\quad |q_2| &= |w_1|\,\frac{T_2}{T_1 - T_2} = \frac{(1\text{ kWh})(273\text{ K})}{100\text{ K}} \\ &= (2.73\text{ kWh})(10^3\text{ J s}^{-1}\text{ kW}^{-1})(60\text{ m h}^{-1})(60\text{ s m}^{-1}) \\ &= 9.828 \times 10^6\text{ J}\end{aligned}$$

$$(c)\quad |w| = |q_1|\,\frac{T_1 - T_2}{T_1} = \frac{(1\text{ kWh})(22\text{ K})}{295\text{ K}} = 0.0746\text{ kWh}$$

2.5 RELATION BETWEEN THE THERMODYNAMIC TEMPERATURE SCALE AND THE PERFECT GAS TEMPERATURE SCALE

Equation 2.9 for the efficiency of a Carnot engine can be obtained for the special case that the working fluid is a perfect gas by using equations derived in the preceding chapter. The reversible work for the isothermal steps are given by equation 1.43. The reversible work for the adiabatic steps is given by equation 1.52. Thus, the net work for one cycle of a Carnot heat engine with 1 mol of perfect gas is

$$\begin{aligned}w &= -RT_1 \ln\frac{V_2}{V_1} + \int_{T_1}^{T_2} C_V\,dT - RT_2 \ln\frac{V_4}{V_3} + \int_{T_2}^{T_1} C_V\,dT \\ &= -RT_1 \ln\frac{V_2}{V_1} + RT_2 \ln\frac{V_3}{V_4} \qquad (2.12)\end{aligned}$$

This equation may be simplified by use of the equations for the two adiabatic steps. From equation 1.51*b*,

$$T_1 V_2^{\gamma-1} = T_2 V_3^{\gamma-1} \qquad (2.13a)$$

$$T_1 V_1^{\gamma-1} = T_2 V_4^{\gamma-1} \qquad (2.13b)$$

where $\gamma = C_P/C_V$. Dividing equation 2.13*a* by equation 2.13*b*, we obtain

$$\frac{V_2}{V_1} = \frac{V_3}{V_4} \qquad (2.14)$$

Substituting this relation into equation 2.12, we obtain

$$w = -R(T_1 - T_2) \ln\frac{V_2}{V_1} \qquad (2.15)$$

Since the first step of the cycle is an isothermal expansion, equation 1.45 yields

$$q_1 = RT_1 \ln \frac{V_2}{V_1} \tag{2.16}$$

Substituting equations 2.15 and 2.16 into equation 2.2 for the efficiency η of a heat engine yields equation 2.9, which was obtained for any working fluid. The fact that we have obtained the same relationship as that derived from the second law shows that the temperature scales defined by a perfect gas and the second law are proportional to each other. They are made identical by taking the same defined point (273.16 K) for the triple point of water. As explained in Section 1.3, this makes the melting point of water at 1 atm pressure 273.15 K. The Kelvin unit is named in honor of Lord Kelvin, who was the first to define the thermodynamic temperature from the properties of reversible heat engines.

2.6 ENTROPY

Equating the two expressions (equations 2.2 and 2.9) for the efficiency of a Carnot heat engine, we obtain

$$\frac{q_1}{T_1} + \frac{q_2}{T_2} = 0 \tag{2.17}$$

This suggests the use of the quantity q/T as a state function, since the sum of the changes of a state function for a cycle must total zero. Clausius did just this in 1865 when he defined the entropy S by the relationship

$$dS \equiv \frac{đq_{\text{rev}}}{T} \tag{2.18}$$

where $đq_{\text{rev}}$ is the heat absorbed by a system in an infinitesimal *reversible* process at temperature T. The $đ$ reminds us that $đq$ is not an exact differential. The important point to realize here is that dS is an exact differential, and so the entropy S is a function of the state of the system. Because of equation 2.18, we may say that $1/T$ is the integrating factor for heat, since $đq_{\text{rev}}/T$ is an exact differential. The entropy change for a given change in state is *independent of path*. The entropy change from state 1 to state 2 is given by the integral of equation 2.18 along a reversible path.

$$\Delta S = S_2 - S_1 = \int_1^2 \frac{đq_{\text{rev}}}{T} \tag{2.19}$$

Any reversible path between state 1 and state 2 will give the same value of the integral. The entropy is an extensive property, like the volume V and the internal energy U, so that it depends on the mass of the system considered. Since S is an extensive property, we can apply the above relations to a system of any size. However, when we calculate entropy changes for chemical systems, we will use S to represent the intensive property with units of joules per Kelvin per mole.

Since S is a function of the state of a system, its integral for a reversible cyclic process (like the Carnot cycle) is zero.

$$\Delta S = \oint \frac{đq_{\text{rev}}}{T} = 0 \tag{2.20}$$

Equation 2.18 can be written $đq_{\text{rev}} = T\,dS$, so that

$$q_{\text{rev}} = \int T\,dS \tag{2.21}$$

Thus the heat input during a reversible process is equal to the area in a plot of T versus S between the initial and final states. The Carnot cycle for a heat engine is shown in a T–S diagram in Fig. 2.3. In the reversible isothermal expansion from state 1 to state 2 the entropy increases, and the heat absorbed is proportional to the area under the line from state 1 to state 2. In the reversible adiabatic expansion from state 2 to state 3 the entropy is constant. In the reversible isothermal expansion from state 3 to state 4 the entropy decreases, and the heat absorbed is proportional to the area under the line from state 3 to state 4. In the reversible adiabatic expansion from state 4 to state 1 the entropy is against constant. Thus the area within the rectangle is proportional to the net heat absorption in the cycle.

2.7 ENTROPY CHANGES IN IRREVERSIBLE PROCESSES

In an infinitesimal reversible adiabatic process $dS = 0$ and for a reversible adiabatic change from state 1 to state 2, $\Delta S = 0$. What happens to the entropy in an irreversible adiabatic process? To answer this question, we will consider two hypothetical irreversible adiabatic processes, one in which the entropy increases and one in which the entropy decreases. These two irreversible processes from state 1 to state 2 are shown by the dashed lines in Fig. 2.4*a* and 2.4*b*. The dashed lines are used to indicate that the intermediate stages of an irreversible process are unknown, and the path is arbitrarily drawn as a straight line. The paths shown

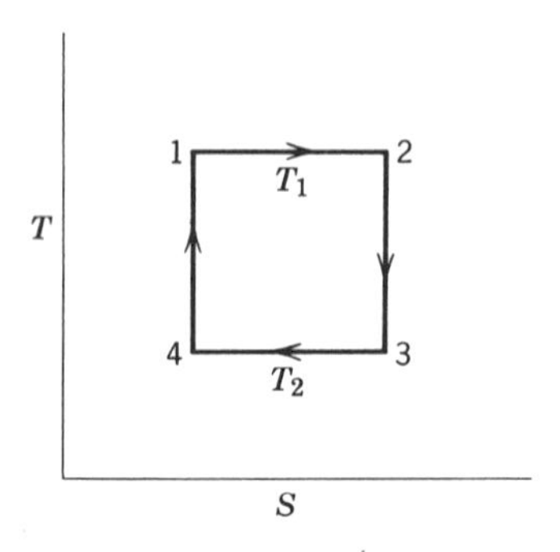

Fig. 2.3 Carnot cycle for a heat engine.

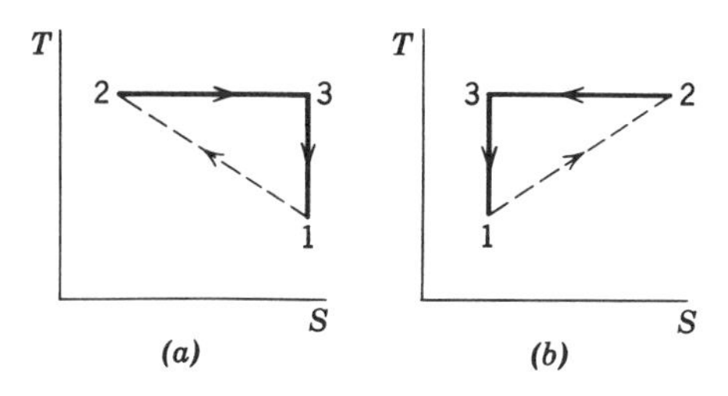

Fig. 2.4 Two possible irreversible adiabatic processes from state 1 to state 2 with reversible return paths. (From K. E. Bett, J. S. Rowlinson, and G. Saville, *Thermodynamics for Chemical Engineers*, The MIT Press, Cambridge, Mass., 1975.)

by the solid lines are reversible return paths. For both diagrams $\Delta U = 0$ for the cycle, and so the first law gives

$$w_{12} + w_{23} + w_{31} + q_{23} = 0 \tag{2.22}$$

since the irreversible processes and the 3-1 step in both cases are adiabatic. If q_{23} is positive, as it is in Fig. 2.4*a*, heat would be absorbed by the system in a cyclic process and completely converted to work. This is a violation of the second law and therefore the process represented in Fig. 2.4*a* cannot occur, according to the second law.

If q_{23} is negative heat would be evolved by the system in a cyclic process in which net work was performed on the system. Thus Fig. 2.4*b* simply describes the conversion of work into heat, which is permitted by the second law. Since the entropy increases in the irreversible adiabatic change from state 1 to state 2, we conclude that for an adiabatic process,

$$\Delta S_{\text{adiabatic}} \geqslant 0 \tag{2.23}$$

where the equal sign applies if the process is carried out reversibly and the inequality applies if the process is carried out irreversibly. The adiabatic change from state 1 to state 2 described in Fig. 2.4*b* is a natural change in an isolated system, and the fact that the entropy increases has given us a criterion for an irreversible (natural) process. Or, from a different point of view, we can conclude that when an irreversible change occurs in an isolated system, the entropy increases. When all possibilities for increasing the entropy in spontaneous change have been exhausted,

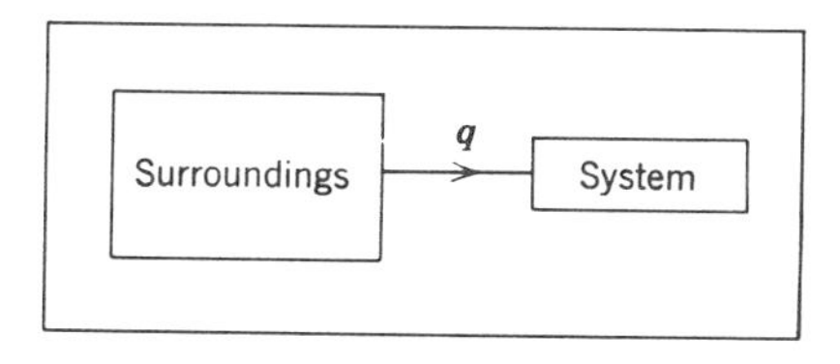

Fig. 2.5 An adiabatic system containing both a heat reservoir (referred to here as the surroundings) and the system of interest.

the entropy will have a maximum value. For any infinitesimal change at equilibrium in an isolated system, $dS = 0$.

In order to derive a related inequality for an irreversible isothermal process, we consider the "enlarged system" shown in Fig. 2.5, which includes both the system and its surroundings (specifically a heat reservoir at the same temperature as the system). The "enlarged system" may be treated as an adiabatic system, and so

$$\Delta S_{\text{system}} + \Delta S_{\text{surroundings}} \geqslant 0 \tag{2.24}$$

where the equal sign applies to a reversible process and the inequality applies to an irreversible process. If the transfer of heat from the heat reservoir to the system is reversible, as it will be, since they are at the same temperature within an infinitesimal temperature difference, the change in entropy of the surroundings is readily calculated.

$$\Delta S_{\text{surroundings}} = \frac{-q_{\text{system}}}{T_{\text{system}}} \tag{2.25}$$

Substituting this into equation 2.24 yields

$$\Delta S_{\text{system}} \geqslant \frac{q_{\text{system}}}{T_{\text{system}}} \tag{2.26}$$

where the equality applies to a reversible isothermal process and the inequality applies to an irreversible isothermal process. The importance of this inequality for chemistry is that if we can show that $\Delta S > q/T$ for an isothermal process, then that process can occur naturally (i.e., by itself). Such processes are generally referred to as spontaneous processes. It is important to remember that thermodynamics has nothing to say about the speed of a spontaneous process.

2.8 CALCULATION OF ENTROPY CHANGES

We will now consider some simple processes for which entropy changes are readily calculated. Two types of processes that may be carried out reversibly are phase transitions (e.g., the evaporation of a liquid into saturated vapor) and heating a substance.

The transfer of heat from one body to another at an infinitesimally lower temperature is a reversible change, since the direction of heat flow can be reversed by an infinitesimal change in the temperature of one of the bodies. The fusion of a solid at its melting point and the evaporation of a liquid at a constant partial pressure of the substance equal to its vapor pressure are examples of isothermal transformations that can be reversed by an infinitesimal change in temperature. The entropy change is easily calculated. Since T is constant, performing the integration of equation 2.18 yields

$$S_2 - S_1 = \Delta S = \frac{q_{\text{rev}}}{T} \tag{2.27}$$

where q_{rev} represents the heat absorbed in the reversible change. Since the pressure is constant, the reversible heat is equal to the change in enthalpy ΔH, so that

$$S_2 - S_1 = \Delta S = \frac{\Delta H}{T} \tag{2.28}$$

This equation may also be used to calculate the entropy change of sublimation or the entropy change for a transition between two forms of a solid. Since the heat gained by the system is equal to that lost by the surroundings, the entropy change for the surroundings is the negative of the entropy change for the system; for both the system and surroundings taken together, ΔS is zero if the transfer of heat is carried out reversibly, as required by equation 2.24.

Example 2.2 *n*-Hexane boils at 68.7 °C, and the heat of vaporization at constant pressure is 28,850 J mol^{-1} at this temperature. If liquid is vaporized into the saturated vapor at this temperature, the process is reversible and the entropy change per mole is given by

$$\Delta S = \frac{\Delta H}{T} = \frac{28{,}850 \text{ J mol}^{-1}}{341.8 \text{ K}} = 84.41 \text{ J K}^{-1} \text{ mol}^{-1}$$

(See Trouton's rule in Section 3.7.)

The molar entropy of a vapor is always greater than that of the liquid with which it is in equilibrium, and the molar entropy of the liquid is always greater than that of the solid at the melting point. According to the disorder concept of entropy to be discussed later in Section 2.24, in which the entropy is a measure of the disorder of the system, the molecules of the gas are more disordered than those of the liquid, and the molecules of the liquid are more disordered than those of the solid.

The increase in entropy of a system due to an increase in temperature can be calculated, since the temperature change can be carried out in a reversible manner. If the heating is carried out at constant pressure, the heat absorbed in each infinitesimal step is equal to the heat capacity C_P multiplied by the differential increase in temperature dT, and so

$$dS = \frac{C_P\, dT}{T} \tag{2.29}$$

Integrating equation 2.29 between the limits T_1 and T_2 gives

$$\int_{S_1}^{S_2} dS = \int_{T_1}^{T_2} \frac{C_P\, dT}{T} \tag{2.30}$$

If C_P is independent of temperature,

$$S_2 - S_1 = C_P \ln \frac{T_2}{T_1} \tag{2.31}$$

The fact that the entropy is always larger at the higher temperature agrees with the notion of increased disorder of the motion of the molecules at the higher temperature. If the heating is carried out at constant volume, C_V is used instead of C_P.

If the heat capacities change with temperature, an empirical equation representing C_P as a function of temperature may be inserted into equation 2.30 before integration, or the integration may be carried out numerically. The change in entropy may be obtained by plotting C/T versus T or C versus $\ln T$ and determining the area under the curve from temperature T_1 to temperature T_2.

$$S_2 - S_1 = \int_{T_1}^{T_2} C \, d \ln T \tag{2.32}$$

Example 2.3 How much does the standard entropy of helium increase when it is heated from 298 to 1000 K at constant pressure?

$$S = \int_{298}^{1000} \frac{C_P}{T} \, dT$$

Since C_P is constant for helium, integration yields

$$S = C_P \ln \tfrac{1000}{298} = \tfrac{5}{2} R \ln \tfrac{1000}{298} = 25.165 \text{ J K}^{-1} \text{ mol}^{-1}$$

2.9 ENTROPY CHANGE FOR AN IRREVERSIBLE PROCESS

The entropy change for an irreversible process may be calculated by considering a path by which the process can be carried out in a series of *reversible* steps. This is illustrated for the freezing of water below its freezing point.

The freezing of a mole of supercooled water at −10 °C is an irreversible change, but it can be carried out reversibly by means of the following three steps for which the entropy changes are indicated.

$$H_2O(l) \text{ at } -10\ °C \rightarrow H_2O(l) \text{ at } 0\ °C \qquad \Delta S = \int_{263}^{273} C_{liq} \frac{dT}{T}$$

$$H_2O(l) \text{ at } 0\ °C \rightarrow H_2O(s) \text{ at } 0\ °C \qquad \Delta S = \frac{\Delta H}{T}$$

$$H_2O(s) \text{ at } 0\ °C \rightarrow H_2O(s) \text{ at } -10\ °C \qquad \Delta S = \int_{273}^{263} C_{ice} \frac{dT}{T}$$

For the crystallization of liquid water at 0 °C, $\Delta H = -6004 \text{ J mol}^{-1}$. The heat capacity of water may be taken to be $75.3 \text{ J K}^{-1} \text{ mol}^{-1}$, and that of ice may be taken to be $36.8 \text{ J K}^{-1} \text{ mol}^{-1}$ over this range. Then the total entropy change of the water when 1 mole of liquid water at −10 °C changes to ice at −10 °C is simply the sum of the foregoing entropy changes:

$$\begin{aligned}\Delta S &= (75.3 \text{ J K}^{-1} \text{ mol}^{-1}) \ln \tfrac{273}{263} + \frac{(-6004 \text{ J mol}^{-1})}{273 \text{ K}} \\ &\quad + (36.8 \text{ J K}^{-1} \text{ mol}^{-1}) \ln \tfrac{263}{273} \\ &= -20.54 \text{ J K}^{-1} \text{ mol}^{-1}\end{aligned}$$

The decrease in entropy corresponds to the increase in structural order when water freezes.

The statement that the entropy of an isolated system increases in a spontaneous process may be illustrated by considering supercooled water at −10 °C in contact with a large heat reservoir at this temperature. The entropy change for the isolated system upon freezing includes the entropy change of the reservoir as well as the entropy change of the water. If the heat reservoir is large, the heat evolved by the water upon freezing is absorbed by the reservoir with only an infinitesimal change in temperature. Since the heat of fusion of water at −10 °C is 5619 J mol^{-1}, the entropy change of the reservoir in this reversible process is

$$\Delta S = \frac{(5619\ \text{J mol}^{-1})}{263\ \text{K}} = 21.37\ \text{J K}^{-1}\ \text{mol}^{-1}$$

The transfer of heat to a reservoir at the same temperature is a reversible process. The entropy change of the water is −20.54 J K^{-1} mol^{-1}, and the total entropy change of the system water plus reservoir is

$$\Delta S = 21.37 - 20.54 = 0.83\ \text{J K}^{-1}\ \text{mol}^{-1}$$

Thus the total entropy of the isolated system, including water and reservoir, increases, as required by inequality 2.24.

2.10 CRITERIA OF CHEMICAL EQUILIBRIUM

Early attempts by Berthelot to discover a thermodynamic criterion for spontaneous chemical reactions led him in 1879 to the false conclusion that reactions that evolve heat are spontaneous. The discovery of spontaneous reactions that absorb heat proved that this idea was wrong. According to the second law, a process will be spontaneous if its occurrence in an isolated system will lead to an increase in entropy of the system. In this section we will see how this statement leads to even more useful criteria of spontaneous chemical reactions.

It is of considerable practical importance to know whether a system is in equilibrium or just in a metastable state. By equilibrium we mean that a system is in such a state that it can undergo no spontaneous change under the given conditions. Thus, at equilibrium, any infinitesimal change that might take place in the system must be *reversible*, since any *irreversible* change would result in a displacement of the original equilibrium.

Let us consider a system in contact with a reservoir at temperature T in which an infinitesimal *irreversible* process occurs and the only work done is pressure-volume work. The quantity of heat $đq$ is exchanged with the reservoir and, since the process is irreversible, the entropy change dS for the system is greater than $đq/T$.

$$dS > \frac{đq}{T} \tag{2.33}$$

Since $T\,dS$ is greater than $đq$, $đq - T\,dS$ is negative.

$$đq - T\,dS < 0 \tag{2.34}$$

Because the only work done is pressure-volume work, $đq = dU + P\,dV$. Substituting this into equation 2.34, we have

$$dU + P\,dV - T\,dS < 0 \tag{2.35}$$

This inequality is always applicable if a spontaneous change occurs and the only work involved is pressure-volume work. If the volume and entropy of the system are held constant, then

$$(dU)_{V,S} < 0 \tag{2.36}$$

Thus, for any irreversible process in a system of constant volume that does not change its entropy, the internal energy decreases. This is the familiar condition for a conservative mechanical system that the stable state is the one of lowest energy.

The volume and internal energy of a system may be kept constant by isolating the system. For an isolated system, equation 2.33 becomes

$$dS > 0 \tag{2.37}$$

and so the entropy must increase in such an irreversible process. It must be remembered that inequalities 2.35, 2.36, and 2.37 apply only to systems where the only work done is of the pressure-volume type.

If the system is not isolated, there are entropy changes in the surroundings that must also be considered. If the volume is constant during the infinitesimal irreversible process, inequality 2.35 becomes

$$(dU - T\,dS)_V < 0 \tag{2.38}$$

which may also be written

$$d(U - TS)_{T,V} < 0 \tag{2.39}$$

The quantity $U - TS$ is referred to as the Helmholtz energy and is represented by A.

$$A \equiv U - TS \tag{2.40}$$

Thus, from equation 2.39,

$$(dA)_{T,V} < 0 \tag{2.41}$$

for a spontaneous process. Thus, in an irreversible process at constant T and V, the Helmholtz energy A *decreases*.

Physical processes or chemical reactions are usually carried out in the laboratory at constant pressure and temperature. When P and T are constant, inequality 2.35 may be written

$$d(U + PV - TS)_{T,P} < 0 \tag{2.42}$$

The quantity $U + PV - TS$ is referred to as the Gibbs energy* and is represented by the symbol G.

$$G \equiv U + PV - TS = H - TS \tag{2.43}$$

from equation 2.42

$$(dG)_{T,P} < 0 \tag{2.44}$$

Thus, in an irreversible process at constant T and P in which only pressure-volume work is done, the Gibbs energy decreases.

If the processes just discussed were reversible, the inequalities would all be replaced with equal signs because of equation 2.18. The conditions for irreversibility and reversibility for processes involving only pressure-volume work are summarized in Table 2.1. Each line of this table represents a mathematical way of stating the second law of thermodynamics. Since the Gibbs energy decreases in an irreversible process at constant T and P, it becomes a minimum at the final equilibrium state, where $dG = 0$ for any infinitesimal change. We can *imagine* a process occurring at equilibrium; for example, we may imagine the evaporation of an infinitesimal amount of water from the liquid into a vapor phase that is saturated with water vapor at constant temperature and pressure. For such a process, $dG = 0$.

Table 2.1 Criteria for Irreversibility and Reversibility for Processes Involving No Work or Only Pressure-Volume Work

For Irreversible Processes	For Reversible Processes
$(dS)_{V,U} > 0$	$(dS)_{V,U} = 0$
$(dU)_{V,S} < 0$	$(dU)_{V,S} = 0$
$(dA)_{T,V} < 0$	$(dA)_{T,V} = 0$
$(dG)_{T,P} < 0$	$(dG)_{T,P} = 0$
$(dH)_{P,S} < 0$	$(dH)_{P,S} = 0$

These same relations may be applied to finite changes as well as infinitesimal changes, replacing the d's by Δ's. It must be remembered, however, that spontaneous changes always go to the minimum (as in the case of the Gibbs energy at constant T and P) or to the maximum (as in the case of the entropy of an isolated system) and not to some other condition, even though the change to some other condition satisfies the required inequality expressed using a Δ.

Although these criteria show whether a certain change is spontaneous, it does not necessarily follow that the change will take place with an appreciable speed. Thus, a mixture of 1 mol of carbon and 1 mol of oxygen at 1 atm pressure and 25 °C has a Gibbs energy greater than that of 1 mol of carbon dioxide at 1 atm and 25 °C,

* This is sometimes referred to as the free energy or the Gibbs free energy. The Gibbs energy G is named in honor of Professor J. Willard Gibbs of Yale University, whose many important generalizations in thermodynamics have given him a position as one of the great geniuses of science.

and so it is possible for the carbon and oxygen to combine to form carbon dioxide at this constant temperature and pressure. Although carbon may exist for a very long time in contact with oxygen, the reaction is theoretically possible. The reverse of a thermodynamically spontaneous change is, of course, a nonspontaneous change. Thus the decomposition of carbon dioxide to carbon and oxygen at room temperature, which involves an increase in Gibbs energy, is nonspontaneous. It can occur only with the aid of an outside agency.

This discussion has been restricted to systems that do not have the capability of doing work other than pressure-volume work. If the system contains an electrochemical cell, then electrical work could be done and the criteria for equilibrium are altered.

The infinitesimal Gibbs energy change is, in general,

$$dG = dU + P\,dV + V\,dP - T\,dS - S\,dT \tag{2.45}$$

At constant pressure and temperature,

$$dG = dU + P\,dV - T\,dS \tag{2.46}$$

Substituting $dU = đq + đw$, we have

$$dG = đq + đw + P\,dV - T\,dS \tag{2.47}$$

If a change is carried out by a reversible process and heat $đq$ is transferred from a reservoir at the same temperature T as the system, then $đq = T\,dS$ and $đw = đw_{rev}$. For this case equation 2.47 can be written

$$-dG = -đw_{rev} - P\,dV \tag{2.48}$$

Thus, for a reversible process at constant temperature and pressure, the decrease in Gibbs energy is equal to the maximum work that can be done by the system in excess of the pressure-volume work.

If a change is carried out by an irreversible process and heat $đq$ is transferred from a reservoir at the same temperature T as the system, then $T\,dS > đq$. Substitution into equation 2.47 yields

$$-dG > -đw_{irrev} - P\,dV \tag{2.49}$$

For a finite change

$$-\Delta G > -w_{irrev} - P\,\Delta V \tag{2.50}$$

Thus, for an irreversible process at constant temperature and pressure, the decrease in Gibbs energy is greater than the maximum work done by the system in excess of the pressure-volume work.

Since $\Delta G = \Delta A + P\,\Delta V$ at constant pressure, equation 2.48 shows that

$$-\Delta A = -w_{rev} \tag{2.51}$$

for a process at constant temperature and pressure. Thus the decrease in Helmholtz energy is equal to the maximum amount of work that can be done by the system in an isothermal process.

2.11 THE GIBBS ENERGY AS A CRITERION OF EQUILIBRIUM AT CONSTANT TEMPERATURE AND PRESSURE

At constant temperature, equation 2.43 may be applied to a physical process or chemical reaction to obtain

$$\Delta G = \Delta H - T\Delta S \tag{2.52}$$

Whether or not a process is spontaneous at constant temperature and pressure depends on two terms, ΔH and $T\Delta S$. A change is favored if ΔH is negative or ΔS is positive; these changes correspond with a decrease in energy and an increase in disorder, respectively. If ΔH is large in magnitude and $T\Delta S$ is small in magnitude, the sign of ΔG will be determined solely by the sign of ΔH. If $T\Delta S$ is large in magnitude and ΔH is small in magnitude, the sign of ΔG will be determined solely by the sign of ΔS. This always occurs at very high temperatures. In other cases both terms may be important. In any case the spontaneous process leads to the minimum possible value of $H - TS$ for the system at constant temperature and pressure.

To visualize the roles of H and TS in determining the equilibrium position, consider the vaporization of a solid in a closed space. The tendency of the system to achieve a low enthalpy would by itself lead to the complete condensation of the vapor phase onto the solid, this being the phase of lower enthalpy. The tendency of the system to achieve a high entropy would by itself lead to the complete vaporization of the solid to the gaseous state, this being the phase of higher entropy. The dependence of H and TS on the fraction of the substance in the vapor phase may be calculated and is illustrated in a general way in Fig. 2.6.* The enthalpy of the

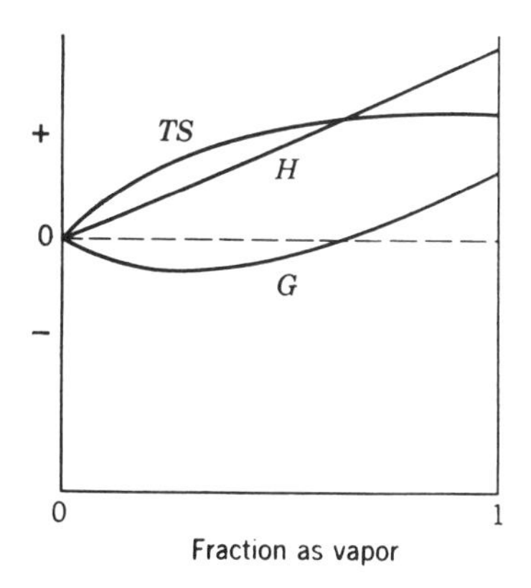

Fig. 2.6 Thermodynamic quantities for a crystal-vapor system at constant temperature and pressure.

---

* K. G. Denbigh, *The Principles of Chemical Equilibrium*, Cambridge University Press, Cambridge, England, 1971, p. 84.

system increases linearly with the fraction in the vapor phase, but the entropy increases more rapidly at first. Thus $H - TS$ will show a minimum value with some fraction of the substance in the vapor phase, and this is the position of the equilibrium. If the concentration in the vapor phase is lower than the equilibrium value, vaporization will occur spontaneously; if the concentration in the vapor phase is higher, condensation will occur spontaneously. Either of these processes leads to a decrease in the Gibbs energy.

2.12 FUNDAMENTAL EQUATIONS FOR CLOSED SYSTEMS

For a closed system of constant composition we have, in addition to the mechanical properties P and V, the fundamental properties T, U, and S, introduced by the zeroth, first, and second laws, respectively. In addition, there are three properties, H, A, and G, defined in terms of the others. For a reversible process in a closed system of constant composition that can only perform pressure-volume work, the first (equation 1.12) and second (equation 2.18) laws may be combined to yield

$$dU = T\,dS - P\,dV \tag{2.53}$$

This equation is sometimes referred to as the *fundamental equation of thermodynamics*. It may be expressed in three other ways by use of

$$H = U + PV \tag{2.54}$$

$$A = U - TS \tag{2.55}$$

$$G = U + PV - TS \tag{2.56}$$

The differentials of these equations are

$$dH = dU + P\,dV + V\,dP \tag{2.57}$$

$$dA = dU - T\,dS - S\,dT \tag{2.58}$$

$$dG = dU + P\,dV + V\,dP - T\,dS - S\,dT \tag{2.59}$$

Introducing equation 2.53 into each of these yields

$$dH = T\,dS + V\,dP \tag{2.60}$$

$$dA = -S\,dT - P\,dV \tag{2.61}$$

$$dG = -S\,dT + V\,dP \tag{2.62}$$

Equation 2.53 plus these three equations relates the eight thermodynamic properties with each other for a system at equilibrium.

Since equation 2.53 relates the change in internal energy dU to the change in entropy dS and in volume dV, S and V are called the *natural variables* for U. The natural variables for H are S and P, the natural variables for A are T and V, and the natural variables for G are T and P. As we will see later, the Gibbs energy G

is an especially useful property for chemistry, since many processes are carried out at constant temperature and pressure, its natural variables.

Since the internal energy U is a state function and may be expressed as a function of S and V, its differential is exact and is given by

$$dU = \left(\frac{\partial U}{\partial S}\right)_V dS + \left(\frac{\partial U}{\partial V}\right)_S dV \tag{2.63}$$

Comparing equations 2.53 and 2.63, we see that

$$\left(\frac{\partial U}{\partial S}\right)_V = T \tag{2.64}$$

and

$$\left(\frac{\partial U}{\partial V}\right)_S = -P \tag{2.65}$$

These equations illustrate the fact that the derivative of an extensive property with respect to another extensive property is an intensive property.

Since dH, dA, and dG are exact differentials (see next section), these equations are convenient sources of the following derivatives.

$$\left(\frac{\partial H}{\partial S}\right)_P = T \tag{2.66}$$

$$\left(\frac{\partial H}{\partial P}\right)_S = V \tag{2.67}$$

$$\left(\frac{\partial A}{\partial T}\right)_V = -S \tag{2.68}$$

$$\left(\frac{\partial A}{\partial V}\right)_T = -P \tag{2.69}$$

$$\left(\frac{\partial G}{\partial T}\right)_P = -S \tag{2.70}$$

$$\left(\frac{\partial G}{\partial P}\right)_T = V \tag{2.71}$$

As an illustration of the usefulness of these derivatives, let us consider the last two. Since the entropy of a system is always positive, G decreases with increasing temperature at constant pressure. Since S is greater for a gas than for the corresponding solid, the temperature coefficient of the Gibbs energy for a gas is much more negative than for the corresponding solid. Since the volume of a system is always positive, G increases with increasing P at constant T. Since V is greater for a gas than for the corresponding solid, the pressure coefficient of G is much larger for a gas than for the corresponding solid.

2.13 EXACT AND INEXACT DIFFERENTIALS

The total differential dz of a quantity z may be determined by the differentials dx and dy in two other quantities x and y. In general,

$$dz = M(x, y)\, dx + N(x, y)\, dy \tag{2.72}$$

where M and N are functions of the variables x and y. If we move from a point (x_1, y_1) to another point (x_2, y_2), there will be a change in z of Δz. If Δz depends on the path taken between the points, then dz is said to be an inexact differential. If Δz does not depend on the path taken between the points, then dz is said to be an exact differential. Thermodynamic quantities like U, S, and G form exact differentials, since their values are dependent on the conditions and not the path by which the system got there.

Fortunately there is a simple test for exactness. This relation, which we are going to derive, is also very useful for obtaining relations between the derivatives of thermodynamic functions. If z has a definite value at each point in the x–y plane, then it must be a function of x and y. If $z = f(x, y)$, then

$$dz = \left(\frac{\partial z}{\partial x}\right)_y dx + \left(\frac{\partial z}{\partial y}\right)_x dy \tag{2.73}$$

Comparing equations 2.72 and 2.73,

$$M(x, y) = \left(\frac{\partial z}{\partial x}\right)_y \tag{2.74}$$

$$N(x, y) = \left(\frac{\partial z}{\partial y}\right)_x \tag{2.75}$$

Since the mixed second derivatives are equal,

$$\left[\frac{\partial}{\partial y}\left(\frac{\partial z}{\partial x}\right)_y\right]_x = \left[\frac{\partial}{\partial x}\left(\frac{\partial z}{\partial y}\right)_x\right]_y \tag{2.76}$$

then

$$\left(\frac{\partial M}{\partial y}\right)_x = \left(\frac{\partial N}{\partial x}\right)_y \tag{2.77}$$

This equation must be satisfied if dz is an exact differential. It is Euler's criterion for exactness.

2.14 MAXWELL RELATIONS

Application of equation 2.77 to equations 2.53 and 2.60 to 2.62 yields the Maxwell relations.

$$\left(\frac{\partial T}{\partial V}\right)_S = -\left(\frac{\partial P}{\partial S}\right)_V \tag{2.78}$$

$$\left(\frac{\partial T}{\partial P}\right)_S = \left(\frac{\partial V}{\partial S}\right)_P \tag{2.79}$$

$$\left(\frac{\partial S}{\partial V}\right)_T = \left(\frac{\partial P}{\partial T}\right)_V \tag{2.80}$$

$$-\left(\frac{\partial S}{\partial P}\right)_T = \left(\frac{\partial V}{\partial T}\right)_P \tag{2.81}$$

As an example of the usefulness of these equations, equation 2.80 is applied in the next section.

2.15 THERMODYNAMIC EQUATION OF STATE

The fundamental equation (equation 2.53) for a closed system of constant composition expressed in terms of the internal energy U may be written in the form

$$\left(\frac{\partial U}{\partial V}\right)_T = T\left(\frac{\partial S}{\partial V}\right)_T - P \tag{2.82}$$

Introducing equation 2.80, we have

$$\left(\frac{\partial U}{\partial V}\right)_T = T\left(\frac{\partial P}{\partial T}\right)_V - P \tag{2.83}$$

which is referred to as the *thermodynamic equation of state*. This equation is perfectly general and applies to the thermodynamic properties of any closed system of constant composition. For a perfect gas, $P = RT/V$, so that

$$\left(\frac{\partial P}{\partial T}\right)_V = \frac{R}{V} \tag{2.84}$$

Substituting this relation in equation 2.83 yields

$$\left(\frac{\partial U}{\partial V}\right)_T = 0 \tag{2.85}$$

Thus the internal energy of a perfect gas is independent of the volume. It is important to realize that equation 2.85 is not an independent criterion of a perfect gas, but is a consequence of the perfect gas law and the second law of thermodynamics. The internal energy of a perfect gas is also independent of the pressure at constant temperature, and the enthalpy H is independent of the volume and pressure at constant temperature. In molecular terms there are negligible interactions between molecules of a perfect gas.

Alternatively, we can use equations 2.83 and 2.85 to derive the perfect gas law and thereby show that the temperature introduced in the definition of entropy is the same as that in the perfect gas law.

2.16 INFLUENCE OF TEMPERATURE ON GIBBS ENERGY

The change in Gibbs energy with temperature is related to the enthalpy change in a simple way that was first derived independently by Gibbs and by Helmholtz. Substitution of equation 2.70 into equation 2.43 yields

$$G = H + T\left(\frac{\partial G}{\partial T}\right)_P \tag{2.86}$$

This equation involves both the Gibbs energy and the temperature derivative of the Gibbs energy, and it is more convenient to transform it so that only a temperature derivative appears. This may be accomplished by first differentiating G/T with respect to temperature at constant pressure.

$$\left[\frac{\partial(G/T)}{\partial T}\right]_P = -\frac{G}{T^2} + \frac{1}{T}\left(\frac{\partial G}{\partial T}\right)_P \tag{2.87}$$

Eliminating G from the right-hand side by use of equation 2.86, we have

$$\left[\frac{\partial(G/T)}{\partial T}\right]_P = \frac{-H}{T^2} \tag{2.88}$$

Since $\partial(1/T)/\partial T = -T^{-2}$,

$$\left[\frac{\partial(G/T)}{\partial(1/T)}\right]_P = \left[\frac{\partial(G/T)}{\partial T}\right]_P \frac{\partial T}{\partial(1/T)} = H \tag{2.89}$$

This equation may also be written in terms of ΔG and ΔH to obtain

$$\left[\frac{\partial(\Delta G/T)}{\partial(1/T)}\right]_P = \Delta H \tag{2.90}$$

Thus ΔH for a reaction may be obtained from a plot of $\Delta G/T$ versus $1/T$ as well as from calorimetric measurements. Since ΔH changes with temperature, the slope is taken at a specific temperature. This equation, which is referred to as the Gibbs-Helmholtz equation is important for the calculation of ΔG at another temperature if it is known at one temperature and ΔH is known (see Section 4.7).

2.17 THERMODYNAMICS OF A PERFECT GAS

In Sections 1.15 and 1.16 we applied the first law to perfect gases, and we are now in the position to derive further relations for perfect gases using the second law. First, we will derive the relation for the pressure dependence of the Gibbs energy of a perfect gas, and then we will use that relation to obtain the pressure dependence of the other thermodynamic quantities.

To derive the expression for the Gibbs energy of a perfect gas, we start with equation 2.71 which, at constant temperature, can be written

$$dG = V\,dP \tag{2.91}$$

at a constant temperature. For 1 mol of a perfect gas,

$$dG = RT\frac{dP}{P} \tag{2.92}$$

Integration from a standard pressure P° to any pressure P at a constant temperature yields

$$\int_{G^\circ}^{G} dG = RT\int_{P^\circ}^{P} d\ln P \tag{2.93}$$

$$G = G^\circ + RT\ln\frac{P}{P^\circ} \tag{2.94}$$

The standard Gibbs energy G° has different values at different temperatures. The logarithmic dependence of the molar Gibbs energy of a perfect gas on the pressure of the gas is illustrated in Fig. 2.7.

Alternatively, equation 2.93 may be integrated between any two pressures to obtain

$$\Delta G = G_2 - G_1 = RT\ln\frac{P_2}{P_1} \tag{2.95}$$

All the other thermodynamic quantities may be expressed in terms of G and its derivatives with respect to temperature and pressure. These relations, which can be readily derived, are

$$S = -\left(\frac{\partial G}{\partial T}\right)_P \tag{2.96a}$$

$$A = G - P\left(\frac{\partial G}{\partial P}\right)_T \tag{2.96b}$$

$$U = G - T\left(\frac{\partial G}{\partial T}\right)_P - P\left(\frac{\partial G}{\partial P}\right)_T \tag{2.96c}$$

$$H = G - T\left(\frac{\partial G}{\partial T}\right)_P \tag{2.96d}$$

$$V = \left(\frac{\partial G}{\partial P}\right)_T \tag{2.96e}$$

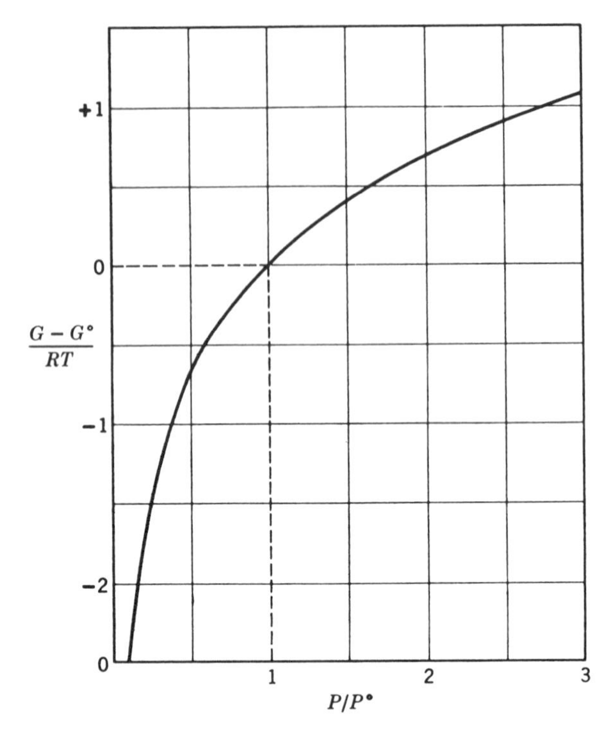

Fig. 2.7 Dependence of the molar Gibbs energy of a perfect gas on the pressure of the gas.

Substituting equation 2.94 for 1 mol of perfect gas into these relations yields

$$S = S^\circ - R \ln \frac{P}{P^\circ} \tag{2.97}*$$

$$A = A^\circ + RT \ln \frac{P}{P^\circ} \tag{2.98}$$

$$U = U^\circ = H^\circ - RT \tag{2.99}$$

$$H = H^\circ = G^\circ + TS^\circ \tag{2.100}$$

$$V = \frac{RT}{P} \tag{2.101}$$

where $S^\circ = -(\partial G^\circ/\partial T)_P$ and $A^\circ = G^\circ + TS^\circ - RT$. As discussed in Section 2.15, the internal energy U and enthalpy H of a perfect gas are independent of pressure and volume.

* The statistical interpretation of this relationship is that as the pressure increases, the number of positions in space available to each molecule decreases and consequently the entropy decreases.

Example 2.4 According to Table A.1, the standard entropy of $O_2(g)$ at 298 K is 205.029 J K^{-1} mol^{-1}. What is the entropy at $\frac{10}{1}$ atm at 298 K?

$$S = S^\circ - R \ln \frac{P}{P^\circ} = 205.029 - (8.314) \ln 0.1 = 224.175 \text{ J K}^{-1} \text{ mol}^{-1}$$

Example 2.5 One mole of a perfect gas at 27 °C expands isothermally and reversibly from 10 atm to 1 atm against a pressure that is gradually reduced. Calculate q and w and each of the thermodynamic quantities ΔU, ΔH, ΔG, ΔA, and ΔS.

Since the process is carried out isothermally and reversibly,

$$w_{\max} = -RT \ln \frac{V_2}{V_1} = -RT \ln \frac{P_1}{P_2} = -(8.314 \text{ J K}^{-1} \text{ mol}^{-1})(300.1 \text{ K}) \ln \tfrac{10}{1}$$

$$= -5746 \text{ J mol}^{-1}$$

$$\Delta A = w_{\max} = -5746 \text{ J mol}^{-1}$$

Since the internal energy of a perfect gas is not affected by a change in volume,

$$\Delta U = 0$$

$$q = \Delta U - w = 0 + 5746 = 5746 \text{ J mol}^{-1}$$

$$\Delta H = \Delta U + \Delta(PV) = 0 + 0 = 0$$

since PV is constant for a perfect gas at constant temperature.

$$\Delta G = \int_{10}^{1} V\,dP = RT \ln \tfrac{1}{10} = (8.314 \text{ J K}^{-1} \text{ mol}^{-1})(300.1 \text{ K})(-2.303)$$

$$= -5746 \text{ J mol}^{-1}$$

$$\Delta S = \frac{q_{\text{rev}}}{T} = \frac{5746 \text{ J mol}^{-1}}{300.1 \text{ K}} = 19.15 \text{ J K}^{-1} \text{ mol}^{-1}$$

Also,

$$\Delta S = \frac{\Delta H - \Delta G}{T} = \frac{0 - (-5746 \text{ J mol}^{-1})}{300.1 \text{ K}} = 19.15 \text{ J K}^{-1} \text{ mol}^{-1}$$

Example 2.6 One mole of a perfect gas expands isothermally at 27 °C into an evacuated vessel so that the pressure drops from 10 to 1 atm; that is, it expands from a vessel of 2.462 L into a connecting vessel such that the total volume is 24.62 L. Calculate the change in thermodynamic quantities.

This process is isothermal, but it is not reversible.

$w = 0$ because the system as a whole is closed and no external work can be done.

$\Delta U = 0$ because the gas is a perfect gas.

$$q = \Delta U - w = 0 + 0 = 0$$

ΔU, ΔH, ΔG, ΔA, and ΔS are the same as in example 2.5 because the initial and final states are the same.

2.18 FUGACITY

In order to treat the thermodynamics of real gases it is convenient to keep the form of equation 2.94, which applies only to a perfect gas. Therefore the fugacity f is defined so that

$$G = G^\circ + RT \ln \left(\frac{f}{f^\circ} \right) \tag{2.102}$$

where f° is the fugacity of the gas in its standard state. Thus the fugacity is a measure of the molar Gibbs energy of a real gas. The fugacity has the same unit as pressure and approaches the pressure as the pressure approaches zero. In short,

$$\lim_{P \to 0} \left(\frac{f}{P} \right) = 1 \tag{2.103}$$

Figure 2.8 gives a plot of fugacity versus pressure for a hypothetical gas that shows how the fugacity approaches the pressure as the pressure approaches zero. The standard state of a real gas is a hypothetical state of unit fugacity and unit pressure obtained from the initial slope of the plot of fugacity versus pressure. If the fugacity of a real gas is known, its thermodynamic properties may be calculated using the forms of equations we have derived above.

The fugacity of a gas may be calculated from P–V–T data. For an isothermal change with a perfect gas,

$$dG = V\,dP = \frac{RT\,dP}{P} \tag{2.104}$$

For a real gas the differential of the Gibbs energy for an isothermal process is given by

$$dG = V\,dP = \frac{RT\,df}{f} \tag{2.105}$$

Subtracting $RT\,dP/P$ from both sides of this equation yields

$$RT\,d \ln \left(\frac{f}{P} \right) = \left(V - \frac{RT}{P} \right) dP \tag{2.106}$$

Integrating at constant temperature from zero pressure to the pressure P' at which we want to obtain the fugacity yields

$$RT \ln \left(\frac{f}{P} \right)_{P=P'} - RT \ln \left(\frac{f}{P} \right)_{P=0} = \int_0^{P'} \left(V - \frac{RT}{P} \right) dP \tag{2.107}$$

Since f/P approaches unity as the pressure approaches zero,

$$RT \ln \left(\frac{f}{P'} \right) = \int_0^{P'} \left(V - \frac{RT}{P} \right) dP \tag{2.108}$$

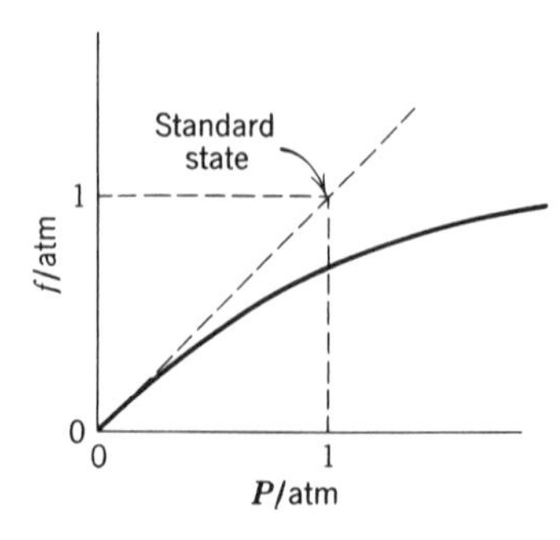

Fig. 2.8 Plot of fugacity versus pressure for a hypothetical gas. The standard state is an idealized state of unit fugacity and unit pressure.

Thus the fugacity is readily calculated from data on the molar volume V of a gas as a function of pressure. If this information is not available, estimates may be made using the principle of corresponding states (see Section 1.18).

2.19 FUNDAMENTAL EQUATIONS FOR OPEN SYSTEMS

If substances are added to a system or taken away from it or if a chemical reaction occurs in a system, the thermodynamic properties of the system change. Such a system is referred to as an open system. In this section we will consider homogeneous open systems because they are simpler than open systems with more than one phase. If a homogeneous system contains k different substances its Gibbs energy may be considered to be a function of $T, P, n_1, n_2, \ldots, n_k$, where n_i is the amount of substance i in the system. The total differential of G is

$$dG = \left(\frac{\partial G}{\partial T}\right)_{P,n_i} dT + \left(\frac{\partial G}{\partial P}\right)_{T,n_i} dP + \sum_{i=1}^{k} \left(\frac{\partial G}{\partial n_i}\right)_{T,P,n_j} dn_i \tag{2.109}$$

where the subscript n_i means that the amounts of all substances are held constant and the subscript n_j means that the amounts of all substances are held constant except for the one being varied (i.e., $j \neq i$). The first two derivatives are given by equations 2.70 and 2.71, and so

$$dG = -S\,dT + V\,dP + \sum_{i=1}^{k} \mu_i\,dn_i \tag{2.110}$$

where

$$\mu_i = \left(\frac{\partial G}{\partial n_i}\right)_{T,P,n_j} \tag{2.111}$$

is referred to as the *chemical potential* of the ith component. The definition shows that the chemical potential of a component of a homogeneous mixture is equal to the ratio of the increase in Gibbs energy upon the addition of an infinitesimal amount of that substance to the amount. Since the chemical potential is the derivative of one extensive property by another, it is an *intensive* property; that is, like the temperature and pressure, the chemical potential is independent of the size of a system. The chemical potential receives its name from the fact that, like the electric potential and the gravitational potential, it is a driving force. It is the chemical potential that determines whether a substance will undergo a chemical reaction or diffuse from one part of a system to another.

By use of the definitions of U, H, and A, equation 2.111 may be used to obtain the other fundamental equations for open systems.

$$dU = T\,dS - P\,dV + \sum_{i=1}^{k} \mu_i\,dn_i \tag{2.112}$$

$$dH = T\,dS + V\,dP + \sum_{i=1}^{k} \mu_i\,dn_i \tag{2.113}$$

$$dA = -S\,dT - P\,dV + \sum_{i=1}^{k} \mu_i\,dn_i \tag{2.114}$$

Thus the chemical potential μ_i also is given by

$$\mu_i = \left(\frac{\partial U}{\partial n_i}\right)_{S,V,n_j} = \left(\frac{\partial H}{\partial n_i}\right)_{S,P,n_j} = \left(\frac{\partial A}{\partial n_i}\right)_{T,V,n_j} \tag{2.115}$$

Equation 2.110 may be expressed in integrated form as follows. Let us suppose that the system being considered is increased in size with the temperature, pressure, and the relative proportions of the components being held constant. Since the relative proportions of the components are unchanged,

$$G = \sum_{i=1}^{k} \mu_i n_i \tag{2.116}$$

This equation shows that the Gibbs energy of a system is the sum of the contributions of the various components. Since equilibrium is reached at constant temperature and pressure when G attains its *minimum* value, we will differentiate this equation (Section 4.1) to obtain a simple expression for chemical equilibrium in terms of the chemical potentials of the components.

Using the definitions of U, H, and A,

$$U = TS - PV + \sum_{i=1}^{k} n_i \mu_i \tag{2.117}$$

$$H = TS + \sum_{i=1}^{k} \mu_i n_i \tag{2.118}$$

$$A = -PV + \sum_{i=1}^{k} \mu_i n_i \tag{2.119}$$

A useful equation for dealing with solutions is obtained by differentiating equation 2.116 to obtain

$$dG = \sum_{i=1}^{k} \mu_i \, dn_i + \sum_{i=1}^{k} n_i \, d\mu_i \tag{2.120}$$

and substituting equation 2.110 to obtain

$$S\,dT - V\,dP + \sum_{i=1}^{k} n_i \, d\mu_i = 0 \tag{2.121}$$

This equation, which is referred to as the Gibbs-Duhem equation, shows that the possible variations of the intensive variables T, P, and $\mu_1, \ldots, \mu_k$ for a system are restricted. For a two-component system at constant temperature and pressure that contains 1 mol of material,

$$X_1 \, d\mu_1 + X_2 \, d\mu_2 = 0 \tag{2.122a}$$

$$X_1 \, d\mu_1 + (1 - X_1) \, d\mu_2 = 0 \tag{2.122b}$$

where X_1 is the mole fraction of component 1 and $(1 - X_1)$ is the mole fraction of component 2. Thus the change in the chemical potential of component 2 is not independent of the change in the chemical potential of component 1.

2.20 PARTIAL MOLAR QUANTITIES

The extensive thermodynamic properties of a homogeneous mixture V, U, H, S, A, and G can be considered to be functions of T, P, n_1, n_2, . . ., n_k. As an example, the total differential of the volume of a homogeneous binary solution may be written

$$dV = \left(\frac{\partial V}{\partial T}\right)_{P,n_1,n_2} dT + \left(\frac{\partial V}{\partial P}\right)_{T,n_1,n_2} dP + \left(\frac{\partial V}{\partial n_1}\right)_{T,P,n_2} dn_1 + \left(\frac{\partial V}{\partial n_2}\right)_{T,P,n_1} dn_2$$

$$= \left(\frac{\partial V}{\partial T}\right)_{P,n_1,n_2} dT + \left(\frac{\partial V}{\partial P}\right)_{T,n_1,n_2} dP + \bar{V}_1\, dn_1 + \bar{V}_2\, dn_2 \qquad (2.123)$$

The partial molar volume of a component is defined by

$$\bar{V}_i = \left(\frac{\partial V}{\partial n_i}\right)_{T,P,n_j} \qquad (2.124)$$

The subscript n_j means that the number of moles of each component is held constant except for component i. This definition may be stated in words by saying that $\bar{V}_i$ is the change in V per mole of i added, when an infinitesimal amount of this component is added to the solution at constant temperature and pressure. Alternatively, it may be said that $\bar{V}_i$ is the change in V when 1 mol of i is added to an infinite amount of the solution at constant temperature and pressure.

Equation 2.123 may be integrated (as we did to obtain equation 2.116) at constant composition, temperature, and pressure to obtain

$$V = \bar{V}_1 n_1 + \bar{V}_2 n_2 \qquad (2.125)$$

The volume V of the solution may be calculated for a given concentration, using this equation, if the partial molar volumes $\bar{V}_1$ and $\bar{V}_2$ at the given concentration are known.

The volume of an ideal solution is simply the sum of the volumes of the components. This is not true for many real solutions, however; for example, if 100 cm^3 of sulfuric acid is added to 100 cm^3 of water, the final volume is 182 cm^3, and there is an evolution of a considerable quantity of heat. The sulfuric acid and the water interact and the sulfuric acid ionizes, with the result that the volumes are not additive.

The partial molar volumes of the components of a binary solution may be calculated from measurements of the densities of solutions over a range of concentrations. The method of intercepts is one of the most graphic for visualizing partial molar quantities. For this method the volume of a mole of solution (i.e., a total of 1 mol of the two components) is plotted versus the mole fraction of one of the components, as shown in Fig. 2.9. The molar volume of the solution is given by

$$V = \bar{V}_1 X_1 + \bar{V}_2 X_2 = \bar{V}_1 + (\bar{V}_2 - \bar{V}_1) X_2 \qquad (2.126)$$

where the last form is obtained by introducing $X_1 = 1 - X_2$. If the mole fraction of component 2 is a, the slope of the plot of V versus X_2 is $\bar{V}_2(a) - \bar{V}_1(a)$, as shown in Fig. 2.9. The equation for the dashed tangent line at this mole fraction is

$$V = \bar{V}_1(a) + [\bar{V}_2(a) - \bar{V}_1(a)] X_2 \qquad (2.127)$$

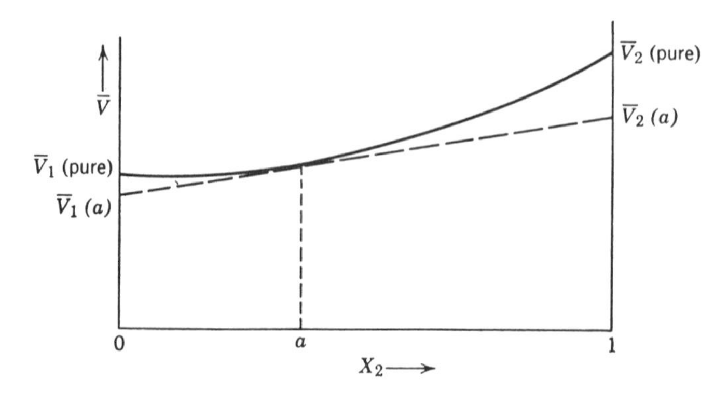

Fig. 2.9 Molar volume of a binary solution versus the mole fraction of one of the components. The intercepts of the sloped dashed line are the partial molar volumes of the two components at $X_2 = a$.

By putting $X_2 = 0$ in this equation, $V = \bar{V}_1(a)$ so that the intercept at $X_2 = 0$ is the partial molar volume of component 1 when $X_2 = a$. By putting $X_2 = 1$ in this equation, $V = \bar{V}_2(a)$ so that the intercept at $X_2 = 1$ is the partial molar volume of component 2 when $X_2 = a$.

Thus we may obtain the partial molar volumes of the two components at $X_2 = a$ by observing the intercepts of the tangent line on the two vertical coordinate lines of the graph. It can be seen that as the composition of the solution approaches pure component 1 or pure component 2, the partial molar volume of that component approaches the molar volume of the pure component.

The method for determining partial molar volumes illustrated in Fig. 2.9 is not very accurate because of the inherent difficulty in obtaining the slope of a curve; however, this method of visualizing the partial molar volumes will be found useful later. Other partial molar thermodynamic quantities may be calculated in the same way.

The various partial molar thermodynamic quantities for a component of a homogeneous mixture are related to each other in the same way as the extensive properties for a pure substance. For example, by differentiating $G = H - TS$ for a system with respect to n_i and keeping T, P, and n_j constant, we obtain

$$\bar{G}_i = \bar{H}_i - T\bar{S}_i \tag{2.128}$$

Thus all the equations written before can be rewritten in terms of intensive properties.

In order to obtain expressions for the temperature and pressure derivatives of the chemical potential of a component of a mixture, we start with equation 2.110. Taking mixed partial derivatives, as we did in obtaining the Maxwell relations (equations 2.78 to 2.81), we obtain

$$-\left(\frac{\partial S}{\partial n_i}\right)_{P,T,n_j} = \left(\frac{\partial \mu_i}{\partial T}\right)_{P,n_i,n_j} \tag{2.129}$$

$$\left(\frac{\partial V}{\partial n_i}\right)_{P,T,n_j} = \left(\frac{\partial \mu_i}{\partial P}\right)_{T,n_i,n_j} \tag{2.130}$$

Using the notation for partial molar quantities,

$$-\bar{S}_i = \left(\frac{\partial \mu_i}{\partial T}\right)_{P,n_i,n_j} \tag{2.131}$$

$$\bar{V}_i = \left(\frac{\partial \mu_i}{\partial P}\right)_{T,n_i,n_j} \tag{2.132}$$

In words the first equation says that when the temperature is changed at constant pressure and composition, the differential change in μ_i is proportional to the negative of the partial molar entropy of component i.

Equation 2.132 says that when the pressure is changed at constant temperature and composition, the differential change in μ_i is proportional to the partial molar volume of the component.

2.21 THERMODYNAMICS OF A MIXTURE OF PERFECT GASES

Thermodynamics alone cannot lead to the conclusion that a mixture of perfect gases will behave as a perfect gas. Nevertheless, it is found that at low pressures mixtures of perfect gases do behave as perfect gases. Such mixtures are referred to as *ideal* mixtures to indicate the additional nonthermodynamic assumption involved.

The chemical potential μ_i for component i of an ideal gas mixture is

$$\mu_i = \mu_i^\circ + RT \ln\left(\frac{P_i}{P^\circ}\right) \tag{2.133}$$

where P_i is the partial pressure, P° is the standard pressure, and μ_i° is the standard chemical potential of pure gas i at a pressure of P°. The partial pressure P_i of any species i in a gas mixture is defined by

$$P_i \equiv X_i P \tag{2.134}$$

where X_i is the mole fraction and P is the total pressure. The sum of the partial pressures of all the species in a gas mixture is equal to the total pressure because

$$P_1 + P_2 + P_3 + \cdots = (X_1 + X_2 + X_3 + \cdots)P = P \tag{2.135}$$

and $\sum_i X_i = 1$.

Thus the equation for the chemical potential of component i in an ideal gas mixture may be written

$$\mu_i = \mu_i^\circ + RT \ln\left(\frac{X_i P}{P^\circ}\right) \tag{2.136}$$

In order to consider changes of composition, but not of total pressure, it is convenient to replace the standard chemical potential by that of the pure component i

at the prevailing pressure P, which we represent by μ_i^*. From equation 2.136, with $X_i = 1$,

$$\mu_i^* = \mu_i^\circ + RT \ln \left(\frac{P}{P^\circ}\right) \tag{2.137}$$

Subtracting this from equation 2.133,

$$\mu_i = \mu_i^* + RT \ln X_i \tag{2.138}$$

This equation defines an ideal mixture. (We will use this equation to define ideal liquid and solid mixtures as well.) The concept of an ideal mixture is useful because it provides a standard with which real mixtures may be compared.

As in the case of a pure component (see Section 2.17), we will obtain the thermodynamic properties of an ideal gaseous mixture by finding the expression for the Gibbs energy and calculating the other thermodynamic properties from it. Combining equations 2.116 and 2.138,

$$G = \sum_i n_i\mu_i = \sum_i n_i\mu_i^* + RT \sum_i n_i \ln X_i \tag{2.139}$$

$$S = -\left(\frac{\partial G}{\partial T}\right)_{P,n_i} = \sum_i n_i\bar{S}_i^* - R \sum_i n_i \ln X_i \tag{2.140}$$

$$H = G + TS = \sum_i n_i\bar{H}_i^* \tag{2.141}$$

$$V = \left(\frac{\partial G}{\partial P}\right)_{T,n_i} = \sum_i n_i\left(\frac{\partial \mu_i^*}{\partial P}\right)_{T,n_i} = \sum_i n_i\bar{V}_i^* = \frac{RT}{P}\sum_i n_i \tag{2.142}$$

Thus the partial molar volume $\bar{V}_i^*$ of a perfect gas in an ideal gas mixture is equal to the volume that 1 mol of pure gas would occupy at a pressure equal to that of the mixture. All perfect gases in an ideal mixture have the same partial molar volume. The total volume of an ideal gas mixture follows the perfect gas law, with the number of moles equal to the total number of moles of gas present.

These equations may be readily applied to calculate the changes in thermodynamic properties on mixing. The mixing of two perfect gases, each initially at 1 atm pressure, is described in Fig. 2.10. Each thermodynamic property for gases before mixing and at the pressure of the final mixture is given by the constant terms in equations 2.139 to 2.142. Thus,

$$\Delta G_{\text{mix}} = RT \sum_i n_i \ln X_i \tag{2.143}$$

$$\Delta S_{\text{mix}} = -R \sum_i n_i \ln X_i \tag{2.144}$$

$$\Delta H_{\text{mix}} = 0 \tag{2.145}$$

$$\Delta V_{\text{mix}} = 0 \tag{2.146}$$

Since mole fractions are less than unity, the logarithmic terms are negative, and $\Delta G_{\text{mix}} < 0$. This corresponds with the fact that the mixing of gases is a spontaneous

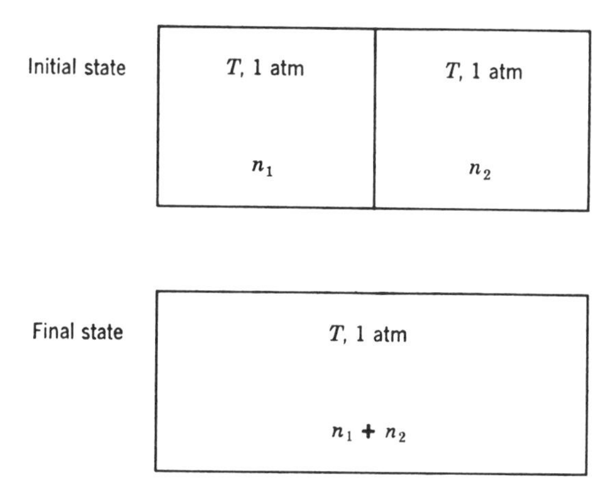

Fig. 2.10 Mixing of perfect gases. The partition between n_1 mol of gas 1 at T and 1 atm and n_2 mol of gas 2 at T and 1 atm is withdrawn so that the gases can mix.

process at constant temperature and pressure. In other words, if two gases at the same pressure and temperature are brought into contact, they will spontaneously diffuse into each other until the gas phase is macroscopically homogeneous. Since there is no interaction between molecules of perfect gases, there is no change in energy on mixing. The mixing occurs spontaneously only due to a change in entropy.

The Gibbs energy change for mixing two perfect gases is plotted versus the mole fraction of one of the gases in Fig. 2.11*a*. The greatest Gibbs energy change on mixing is obtained for $X_1 = X_2 = \frac{1}{2}$. The dependence of the entropy of mixing on the mole fraction of one of the components is shown in Fig. 2.11*b*.

When perfect gases are mixed at constant temperature and pressure, there is no heat effect. This corresponds with the fact that molecules of perfect gases do not attract or repel each other. Thus, from an energy standpoint, it makes no difference

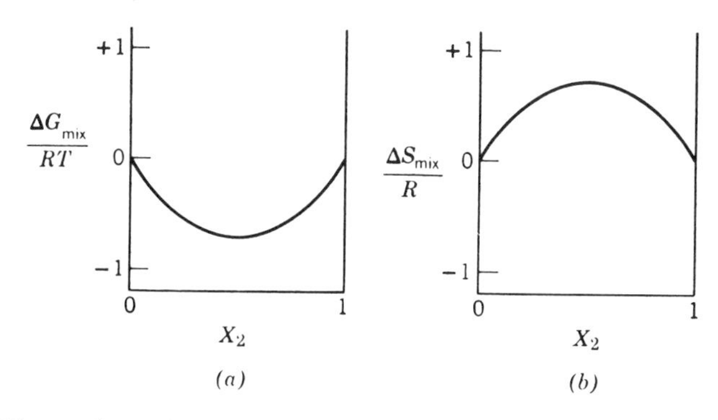

Fig. 2.11 Thermodynamic quantities for the mixing of two perfect gases to form 1 mol of mixture.

whether the gases are separated or mixed. The driving force for mixing arises exclusively from the change in entropy. From the viewpoint of statistical mechanics (Chapter 13), the mixed state is found at equilibrium because it is more probable, as discussed in Section 2.24 for diffusion in idealized crystals.

Example 2.7 Calculate the changes in the thermodynamic quantities G, S, H, and V for the mixing of $\frac{1}{2}$ mol of oxygen with $\frac{1}{2}$ mol of nitrogen at 25 °C and 1 atm, assuming that they are perfect gases.

$$\begin{aligned}\Delta G_{mix} &= RT(X_1 \ln X_1 + X_2 \ln X_2)\\ &= (8.314\ \mathrm{J\ K^{-1}\ mol^{-1}})(298.15\ \mathrm{K})(0.5 \ln 0.5 + 0.5 \ln 0.5)\\ &= -1718\ \mathrm{J\ mol^{-1}}\end{aligned}$$

$$\Delta S_{mix} = -\frac{\Delta G_{mix}}{T} = 5.763\ \mathrm{J\ K^{-1}\ mol^{-1}}$$

$$\Delta H_{mix} = 0$$

$$\Delta V_{mix} = 0$$

2.22 THIRD LAW OF THERMODYNAMICS

In 1902 Richards noted that as the temperature was reduced, $\Delta S°$ for some chemical reactions approached zero. In 1906 Nernst postulated that for isothermal reactions of pure crystalline solids, $\Delta S°$ approaches zero at absolute zero. In 1913 Planck went further in suggesting that $\Delta S°$ approaches zero at absolute zero because the entropy of a pure crystal of any substance is zero at absolute zero. We will see in Section 2.24 that this corresponds to a single quantum state ($\Omega = 1$) for a perfect crystal at absolute zero. We will use the following statement of the third law: *The entropy of a perfect crystal is zero at absolute zero.*

The third law is important because it makes possible the calculation of $\Delta G°$ for a chemical reaction purely from calorimetric measurements. In Chapter 1 we saw how $\Delta H°$ for a previously unstudied reaction could be obtained from calorimetric data on other reactions. In the next section we will see how entropies of reactants and products may be determined at any desired temperature by means of calorimetric measurements down to the neighborhood of absolute zero, assuming the entropy of the pure crystal is zero at absolute zero.

In Chapter 13 on statistical mechanics we will see that the entropies of relatively simple gases at any desired temperature may be calculated from molar masses and certain spectroscopic information. The theoretically calculated molar entropy of a gas at a certain temperature may be compared with the molar entropy obtained from calorimetric measurements, assuming that the entropy of the pure crystalline substance is zero at absolute zero. The fact that the entropies obtained in such different ways are in agreement within experimental error in all but a few exceptional cases provides powerful support for the third law.

When calorimetric entropies for H_2, CO, N_2O, and H_2O are compared with the values expected from measurements of chemical equilibria or statistical mechanical calculations, it is found that molar entropies of these crystals do not approach zero as the temperature approaches absolute zero. The case of H_2 is special in that H_2 has *o*- and *p*-nuclear spin states that do not come to equilibrium upon cooling. The molar entropy of crystals of carbon monoxide approaches 4.6 J K^{-1} mol^{-1} as the temperature approaches 0 K. This entropy of the crystal is apparently due to disorder in the arrangement of CO molecules. Thus, in solid CO, the molecules are arranged CO, OC, CO, CO, OC instead of CO, CO, CO, CO, CO. If the orientation were perfectly random, the crystal might be regarded as a mixed crystal with equal mole fractions of CO and OC. The entropy of the mixed crystal would then be the entropy change of mixing. Using equation 2.144, we see that

$$\Delta S_{\text{mix}} = -R(\tfrac{1}{2}\ln\tfrac{1}{2} + \tfrac{1}{2}\ln\tfrac{1}{2}) = 5.76\,\text{J K}^{-1}\,\text{mol}^{-1} \tag{2.147}$$

The molar entropy of NNO, a linear molecule, at 0 K is explained in the same way. In an ice crystal each HOH molecule is oriented so that its two H atoms are pointed toward two of the four O atoms that surround it tetrahedrally. The molar entropy (3.35 J K^{-1} mol^{-1}) approached by the ice crystal at 0 K is due to the random orientation of H_2O molecules in this lattice.

The zeros for the entropies of solid CO, N_2O, and H_2O for thermodynamic tabulations are not taken as the entropies of the actual crystals, but as the entropies of hypothetical perfect crystals.

There is a corollary to the third law that is very much in the spirit of the Clausius statement of the second law, since it states an impossibility. According to this corollary, it is impossible to reduce the temperature of a system to 0 K in a finite number of steps. This conclusion that absolute zero is unattainable may be derived from the third law.*

2.23 CALORIMETRIC DETERMINATION OF ENTROPIES

The entropy of a substance at any desired temperature relative to its entropy at absolute zero may be obtained by integrating dq_{rev}/T from absolute zero to the desired temperature. This requires heat capacity measurements over the whole temperature range and enthalpy of transition measurements for all transitions in this temperature range. Since measurements of C_P cannot be carried to 0 K, the Debye function (equation 13.101) is used to represent C_P below the temperature of the lowest measurements.

If data on the enthalpy of fusion at the melting point T_m and the enthalpy of vaporization at the boiling point T_b are available, the third law entropy at a temperature T above the boiling point T_b may be calculated from

$$\Delta S_T^\circ = \int_0^{T_m} \frac{C_P(s)}{T}\,dT + \frac{\Delta H_{\text{fus}}^\circ}{T_m} + \int_{T_m}^{T_b} \frac{C_P(l)}{T}\,dT + \frac{\Delta H_{\text{vap}}^\circ}{T_b} + \int_{T_b}^{T} \frac{C_P(g)}{T}\,dT \tag{2.148}$$

* E. A. Guggenheim, *Thermodynamics*, North-Holland Publishing Co., Amsterdam, 1967, p. 157.

If there are various solid forms with enthalpies of transition between the forms, the corresponding entropies of transition would have to be included in this sum.

As an illustration of the determination of the third law entropy of a substance, the measured heat capacities for SO_2 are shown as a function of T and of log T in Fig. 2.12*a** and 2.12*b*. Solid SO_2 melts at 197.64 K, and the heat of fusion is 7402 J mol^{-1}. Liquid SO_2 vaporizes at 263.08 K, and the heat of vaporization is 24,937 J mol^{-1}. The calculation of the entropy at 25 °C is summarized in Table 2.2.

Table 2.2 The Entropy of Sulfur Dioxide

T/K	Method of Calculation	$\Delta S°$/J K^{-1} mol^{-1}
0–15	Debye function (C_P = constant T^3)	1.26
15–197.64	Graphical, solid	84.18
197.64	Fusion, 7402/197.64	37.45
197.64–263.08	Graphical, liquid	24.94
263.08	Vaporization, 24,937/263.08	94.79
263.08–298.1	From C_P of gas	5.23
	$S°$ (298 K) =	247.85

Heat-capacity measurements down to these very low temperatures are made with special calorimeters in which the substance is heated electrically in a carefully insulated system and the input of electrical energy and the temperature are measured accurately.

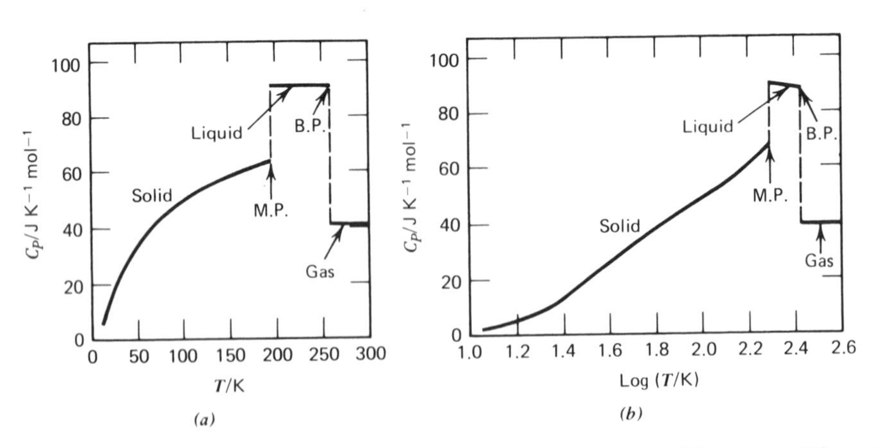

Fig. 2.12 Heat capacity of sulfur dioxide at a constant pressure of 1 atm at different temperatures.

* W. F. Giauque and C. C. Stephenson, *J. Am. Chem. Soc.*, **60**, 1389 (1938).

The attainment of very low temperatures in the laboratory involves successive application of different methods. Vaporization of liquid helium (normal b.p. 4.2 K) at low pressures produces temperatures down to about 0.3 K. Lower temperatures may be reached by use of adiabatic demagnetization. A paramagnetic (Section 12.1) salt such as gadolinium sulfate is cooled with liquid helium in the presence of a strong magnetic field. The salt is thermally isolated from its surroundings and the magnetic field is slowly removed. The salt undergoes a reversible adiabatic process in which the atomic spins become disordered. Since the energy must come from the crystal lattice, the salt is cooled. Temperatures of about 0.001 K may be reached in this way. Adiabatic demagnetization of nuclear spins can then be used to obtain temperatures of the order of a millionth of a degree Kelvin.

The entropies of substances determined in this way are often called conventional entropies because they are based on the convention that the entropy of a perfect crystal at absolute zero is zero. Conventional entropies exclude contributions from nuclear spins and from the mixing of isotopic species because these contributions to the entropy do not change in ordinary chemical reactions.

The entropies of a number of substances at 298.15 K are given in Table A.1 in the Appendix. Where the entropies have been determined calorimetrically, corrections have been made for imperfections of crystals encountered in the neighborhood of absolute zero in the few cases where this occurs. The standard states are, of course, the same as discussed in Section 1.20. The entropy of $H^+(aq)$ is arbitrarily assigned the value zero, and this makes it possible to calculate entropies of other aqueous ions.

The entropies of a smaller number of substances are given in Table A.2 for 0, 298, 500, 1000, 2000, and 3000 K. Except for graphite, the entropies at 0 K are for the hypothetical gas state.

The change in standard entropy for a chemical reaction at any temperature may be calculated from

$$\Delta S^\circ = \sum \nu_i S^\circ{}_i \tag{2.149}$$

2.24 THE STATISTICAL INTERPRETATION OF THE ENTROPY OF MIXING

As stated earlier, thermodynamics is not concerned with molecules or particular models of systems. To develop an intuitive feeling for thermodynamic functions, however, it is very helpful to think in terms of molecules. The calculation of thermodynamic properties from information about molecules is discussed in the chapter on statistical mechanics (Chapter 13). As an introduction to the ideas of statistical mechanics we consider here the mixing of two idealized crystals from a simple statistical point of view. In an idealized crystal we can count the possible arrangements of molecules, and we will arrive at the conclusion that the mixed state is more probable than the unmixed state. Boltzmann recognized that this is the reason the mixed state is observed at equilibrium. The following derivation is designed to clarify this point.

Since the entropy is an extensive property, the entropy of a system consisting of two parts having the same intensive variables is simply the sum of the entropies of the two parts of the system: $S = S_1 + S_2$. If the number of equally probable arrangements for one part of the system is Ω_1 and for the other is Ω_2, the number of equally probable arrangements for the whole system is $\Omega_1\Omega_2$, since any arrangement of the first system can be combined with any arrangement of the second to specify an arrangement for the whole system. If we assume that the entropy is given by a function $f(\Omega)$ of the number of equally probable arrangements, then the entropy of the system is given by

$$S = S_1 + S_2 \tag{2.150}$$

$$f(\Omega_1\Omega_2) = f(\Omega_1) + f(\Omega_2) \tag{2.151}$$

For this to be true there must be a logarithmic relation between S and Ω. Boltzmann postulated that

$$S = k \ln \Omega \tag{2.152}$$

where k is the Boltzmann constant, that is, the molecular gas constant R/N_A where N_A is the Avogadro constant. The general quantitative expression for Ω is given in Chapter 13 after a discussion of the possible quantum states of a system, but we can illustrate the use of equation 2.152 now by applying it to the mixing of two idealized crystals.

Imagine that two small crystals are brought into contact so that atoms can diffuse from one to the other. The crystals are assumed to be sufficiently alike so that this can occur without a change in energy; in other words, we assume that the lattice structures are the same and the various atom-atom interactions are identical so that the mixed crystals formed are of the ideal solid-solution type (Section 3.22). As illustrated in Fig. 2.13, there are at first four A atoms in crystal A and four B atoms in crystal B. Initially there is just this one possible arrangement for the atoms in this system.

$$\Omega_{4:0} = 1$$

where $\Omega_{4:0}$ represents the number of arrangements with four A atoms to the left of the dividing plane and no A atoms to the right.

We are interested in the number of different arrangements of atoms in the various possible mixed crystals. Suppose one atom of A diffuses to the right of the plane and one atom of B diffuses to the left of the plane. Since the atom A can be placed on any one of

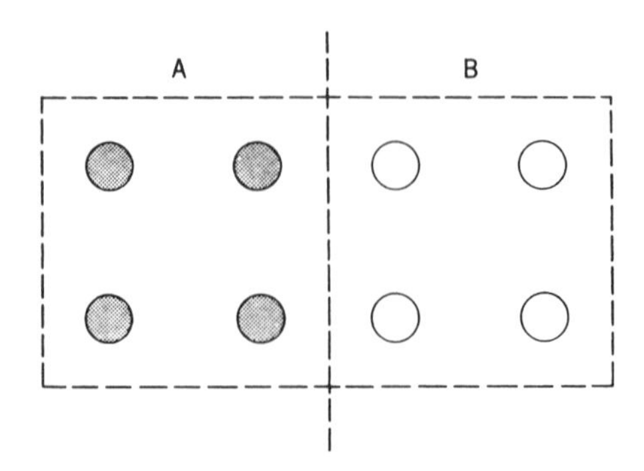

Fig. 2.13 Mixing of two idealized crystals. The black circles represent A atoms, and the open circles represent B atoms.

the four lattice points on the right and the atom B can be placed on any one of the four lattice points on the left, there are 16 distinguishable arrangements.

$$\Omega_{3:1} = 16$$

where $\Omega_{3:1}$ represents the number of arrangements with three A atoms to the left of the dividing plane and one to the right.

Suppose that two atoms of A diffuse to the right of the dividing plane and two atoms of B diffuse to the left. The first atom of A can occupy any one of the four sites and the second can occupy any one of the three remaining sites. This gives rise to 4×3 arrangements, but only $4 \times 3/2!$, where $2!$ is factorial 2, of these are distinguishable, since the two A atoms are assumed to be identical. Each of the $4 \times 3/2!$ arrangements of A atoms on the right side can be combined with any one of the $4 \times 3/2!$ arrangements of B atoms on the left side to give 36 different arrangements.

$$\Omega_{2:2} = \frac{4 \times 3}{2!} \cdot \frac{4 \times 3}{2!} = 36$$

Suppose that three atoms of A diffuse to the right of the dividing plane and three atoms of B diffuse to the left. Since the atom of A remaining on the left can be placed on any one of the four lattice points to the left and the atom of B can be placed on any one of the four lattice points to the right, there are again 16 different arrangements.

$$\Omega_{1:3} = 16$$

If all four atoms of A diffuse to the right there is just one arrangement.

$$\Omega_{0:4} = 1$$

Thus, after the two crystals have been in contact for awhile, the atoms will be found to be in one of the 70 possible different arrangements.

$$\begin{aligned} \Omega &= \Omega_{4:0} + \Omega_{3:1} + \Omega_{2:2} + \Omega_{1:3} + \Omega_{0:4} \\ &= 1 + 16 + 36 + 16 + 1 = 70 \end{aligned} \tag{2.153}$$

Since the energy is the same for each of these arrangements, there is no reason to believe that one is more probable than another. In fact, it is a basic postulate of statistical mechanics that all states of a system with the same total energy and volume are equally probable. Thus, at some later time, the probability of finding two A atoms to the right of the plane and two B atoms to the left is $\frac{36}{70}$, whereas the probability of finding the crystals in their initial states is $\frac{1}{70}$. The equally mixed state is more probable than any other because there are more different ways in which the atoms can be arranged in this state than any other. If we had considered two crystals with larger numbers of atoms, this effect would have been much more striking. This is why spontaneous processes occur at fixed energy; a system goes spontaneously to a state with a larger number of possible arrangements because it spends equal time in each of the arrangements available to it, and there are so many more of the mixed ones.

The change in entropy for the interdiffusion of these two idealized crystals is readily calculated with the Boltzmann relation. Initially $S_1 = k \ln 1$ and at equilibrium, $S_2 = k \ln 70$. Thus the change in entropy

$$\Delta S = S_2 - S_1 = k \ln \tfrac{70}{1}$$

is positive as required for a spontaneous change in an isolated system.

Now let us generalize on this simple example by considering N_1 molecules of component 1 and N_2 molecules of component 2. We wish to calculate the number of ways, Ω, in which the molecules may be distributed among the sites. There are $N_1 + N_2$ choices of sites for the first molecule, $N_1 + N_2 - 1$ for the second, $N_1 + N_2 - 2$ for the third, etc., and so the total number of possibilities is $(N_1 + N_2)(N_1 + N_2 - 1)(N_1 + N_2 - 2) \cdots = (N_1 + N_2)!$. However, since molecules of type 1 are not distinguishable from each other, we must correct the total number of possibilities for the number of ways in which molecules of type 1 may be interchanged with each other. Since these molecules occupy N_1 lattice sites, the first may be placed on any one of N_1, the second on any one of $N_1 - 1$, etc., so that there are

$$N_1(N_1 - 1)(N_1 - 2) \cdots = N_1!$$

possibilities.

We must divide the number of ways of arranging N_1 molecules of 1 and N_2 molecules of 2, that is, $(N_1 + N_2)!$ by $N_1!$ to correct for the indistinguishability of molecules of type 1 and by $N_2!$ to correct for the indistinguishability of molecules of type 2 to get the number Ω_{mixed} of different mixed states.

$$\Omega_{\text{mixed}} = \frac{(N_1 + N_2)!}{N_1!\, N_2!} \tag{2.154}$$

By substituting $N_1 = 4$ and $N_2 = 4$, we obtain 70, just as in equation 2.153.

The number of distinguishable arrangements of the molecules of the pure components before mixing is

$$\Omega_1 = \frac{N_1!}{N_1!} = 1 \qquad \Omega_2 = \frac{N_2!}{N_2!} = 1$$

so that the entropy of the system before mixing is $S = k \ln 1 + k \ln 1 = 0$. Thus the entropy change on mixing is given by

$$\Delta S_{\text{mix}} = k \ln \frac{(N_1 + N_2)!}{N_1!\, N_2!} \tag{2.155}$$

When N_1 and N_2 are very large numbers, as is usually the case for molecular systems, Stirling's approximation may be used to eliminate the factorials.

$$\ln N! = N \ln N - N \tag{2.156}$$

This leads to

$$\begin{aligned} \Delta S_{\text{mix}} &= k\{(N_1 + N_2) \ln (N_1 + N_2) - (N_1 + N_2) - [N_1 \ln N_1 - N_1] - [N_2 \ln N_2 - N_2]\} \\ &= -k\left[N_1 \ln \frac{N_1}{N_1 + N_2} + N_2 \ln \frac{N_2}{N_1 + N_2}\right] = -R[n_1 \ln X_1 + n_2 \ln X_2] \end{aligned} \tag{2.157}$$

where $n_1 = N_1/N_A$ and $n_2 = N_2/N_A$ are numbers of moles and X_1 and X_2 are mole fractions. This equation agrees with equation 2.144 derived for the mixing of perfect gases.

These same ideas about the number of possible microscopic arrangements of a system may also be used to understand the expansion of a gas. Suppose we have a perfect gas in a bulb that is connected with an evacuated bulb. When the stopcock is opened, the number of arrangements available to the system is vastly increased because of the increase in volume. The number of arrangements now available includes all those for the initial system (i.e., gas all in one bulb) as well as a much greater number of new arrangements. Over a long enough period of time all of the possible arrangements will actually occur. Therefore, there is a chance that at a later time all of the gas molecules will be back in the first bulb. For a macroscopic amount of gas, however, the number of arrangements is so large that the probability of observing this occurrence is negligibly small. Furthermore, we can never predict when a large spontaneous fluctuation will occur.

Why the gas expands is sufficiently important to bear repeating once more. When the stopcock is opened, the system can move over all of the possible arrangements. Each of these arrangements is equally probable. Since there are so many more arrangements that correspond to the mixed states, the chances are overwhelming that when we observe the system at a later time, the gas density will be found to be uniform in the two bulbs. The increase in entropy is determined by the increase in the number of arrangements.

Thermodynamics and equilibrium statistical mechanics do not deal with the rate of approach to equilibrium, but only with the equilibrium state. Some time is required even for a gas to expand into another container, and for some chemical reactions the rate of approach to equilibrium is very slow.

References

F. C. Andrews, *Thermodynamics: Principles and Applications*, Wiley-Interscience, New York, 1971.

K. E. Bett, J. S. Rowlinson, and G. Saville, *Thermodynamics for Chemical Engineers*, The MIT Press, Cambridge, Mass., 1975.

K. Denbigh, *The Principles of Chemical Equilibrium*, Cambridge University Press, Cambridge, 1971.

J. W. Gibbs, *The Collected Works of J. Willard Gibbs*, Yale University Press, New Haven, Conn., 1948.

N. A. Gokcen, *Thermodynamics*, Techscience Inc., Hawthorne, Calif., 1975.

E. A. Guggenheim, *Thermodynamics*, North-Holland Publishing Co., Amsterdam, 1967.

W. Kauzmann, *Thermal Properties of Matter Vol. II, Thermodynamics and Statistics: With Applications to Gases*, W. A. Benjamin, Inc., New York, 1967.

G. Kirkwood and I. Oppenheim, *Chemical Thermodynamics*, McGraw-Hill Book Co., New York, 1961.

I. M. Klotz and R. M. Rosenberg, *Chemical Thermodynamics*, W. A. Benjamin, Inc., New York, 1972.

G. N. Lewis and M. Randall, revised by K. S. Pitzer and L. Brewer, *Thermodynamics*, McGraw-Hill Book Co., New York, 1961.

P. A. Rock, *Chemical Thermodynamics*, The Macmillan Co., London, 1969.

D. R. Stull, E. F. Westrum, and G. C. Sinke, *The Chemical Thermodynamics of Organic Compounds*, Wiley-Interscience, New York, 1969.

F. T. Wall, *Chemical Thermodynamics*, W. H. Freeman & Co., San Francisco, 1965.

R. E. Wood, *Introduction to Chemical Thermodynamics*, Appleton-Century-Crofts, New York, 1970.

Problems

2.1 Theoretically, how high could a gallon of gasoline lift an automobile weighting 2800 lb against the force of gravity, if it is assumed that the cylinder temperature is 2200 K and the exit temperature 1200 K? (Density of gasoline = 0.80 g cm^{-3}; 1 lb = 453.6 g; 1 ft = 30.48 cm; 1 L = 0.2642 gal. Heat of combustion of gasoline = 46.9 kJ g^{-1}).

Ans. 17,000 ft.

2.2 (*a*) What is the maximum work that can be obtained from 100 J of heat supplied to a water boiler at 100 °C if the condenser is at 20 °C? (*b*) If the boiler temperature is raised to 150 °C by the use of superheated steam under pressure, how much more work can be obtained? *Ans.* (*a*) 214, (*b*) 93 J.

2.3 Calculate the increase in entropy of a mole of silver that is heated at constant pressure from 0 to 30 °C, if the value of C_P in this temperature range is considered to be constant at 25.48 J K^{-1} mol^{-1}. *Ans.* 2.657 J K^{-1} mol^{-1}.

2.4 Calculate the change in entropy of a mole of aluminum which is heated from 600 to 700 °C. The melting point of aluminum is 660 °C, the heat of fusion is 393 J g^{-1}, and the heat capacities of the solid and liquid may be taken as 31.8 and 34.3 J K^{-1} mol^{-1}, respectively. *Ans.* 14.92 J K^{-1} mol^{-1}.

2.5 A mole of steam is condensed at 100 °C, and the water is cooled to 0 °C and frozen to ice. What is the entropy change of the water? Consider that the average specific heat of liquid water is 4.2 J K^{-1} g^{-1}. The heat of vaporization at the boiling point and the heat of fusion at the freezing point are 2258.1 and 333.5 J g^{-1}, respectively.
Ans. $\Delta S = -154.4$ J K^{-1} mol^{-1}.

2.6 Two blocks of the same metal are of the same size but are at different temperatures, T_1 and T_2. These blocks of metal are brought together and allowed to come to the same temperature. Show that the entropy change is given by

$$\Delta S = C_P \ln \left[\frac{(T_1 + T_2)^2}{4T_1 T_2}\right]$$

if C_P is constant. How does this equation show that the change is spontaneous?
Ans. ΔS is positive.

2.7 Calculate the entropy changes for the following processes: (*a*) melting of 1 mol of aluminum at its melting point, 660 °C ($\Delta H_{fus} = 7.99$ kJ mol^{-1}); (*b*) evaporation of 1 mol of liquid oxygen at its boiling point, -182.97 °C ($\Delta H_{vap} = 6.820$ kJ mol^{-1}); (*c*) heating 1 mol of hydrogen sulfide from 50 to 100 °C at constant pressure ($C_P = 29.92 + 0.013\,89T$). *Ans.* (*a*) 8.58, (*b*) 75.61, (*c*) 5.000 J K^{-1} mol^{-1}.

2.8 Calculate the entropy change in joules for a hundredfold isothermal expansion of a mole of perfect gas. *Ans.* 38.3 J K^{-1} mol^{-1}.

2.9 In the reversible isothermal expansion of a perfect gas at 300 K from 1 to 10 L, where the gas has an initial pressure of 20 atm, calculate (*a*) ΔS for the gas, and (*b*) ΔS for all systems involved in the expansion. *Ans.* (*a*) 15.56, (*b*) 0 J K^{-1}.

2.10 Calculate $\Delta G°$ for

$$H_2O(g, 25\ °C) = H_2O(l, 25\ °C)$$

The vapor pressure of water at 25 °C is 3168 Pa. *Ans.* -8.59 kJ mol^{-1}.

2.11 Calculate the change in Gibbs energy for the process

$$H_2O(l, -10\ °C) = H_2O(s, -10\ °C)$$

The vapor pressure of water at -10 °C is 286.5 Pa, and the vapor pressure of ice at -10 °C is 260.0 Pa. The process may be carried out by the following reversible steps:

1. A mole of water is transferred at -10 °C from liquid to saturated vapor ($P = 286.5$ Pa). $\Delta G = 0$, since the two phases are in equilibrium.
2. The water vapor is allowed to expand from 286.5 to 260.0 Pa at -10 °C.
3. A mole of water is transferred at -10 °C from vapor at $P = 260.0$ Pa to ice at -10 °C.

Ans. -212.5 J mol^{-1}.

2.12 Equations 2.80 and 2.81 are important because they express the rate of change of entropy with respect to pressure at constant temperature and the rate of change of entropy with volume at constant temperature in terms of readily measured quantities. Express these derivatives in terms of α and κ, the coefficients of thermal expansion and compressibility (Problem 2.23), respectively. (It is necessary to use the cyclic rule

$$\left(\frac{\partial P}{\partial T}\right)_V\left(\frac{\partial T}{\partial V}\right)_P\left(\frac{\partial V}{\partial P}\right)_T = -1$$

to obtain one of the results.) *Ans.* $\left(\frac{\partial S}{\partial P}\right)_T = -V\alpha$. $\left(\frac{\partial S}{\partial V}\right)_T = \frac{\alpha}{\kappa}$

2.13 The standard entropy of $H_2(g)$ at 298 K is given in Table A.1. What is the entropy at 10 and 100 atm, assuming perfect gas behavior? *Ans.* 111.429, 92.285 J K^{-1} mol^{-1}.

2.14 At 298 K, $S° = 205.029$ J K^{-1} mol^{-1} for $O_2(g)$. What is the entropy of $O_2(g)$ at 100 atm, assuming that it is a perfect gas? *Ans.* 166.742 J K^{-1} mol^{-1}.

2.15 The change in Gibbs energy for the conversion of aragonite to calcite at 25 °C is -1046 J mol^{-1}. The density of aragonite is 2.93 g cm^{-3} at 25 °C and the density of calcite is 2.71 g cm^{-3}. At what pressure at 25 °C would these two forms of $CaCO_3$ be in equilibrium? *Ans.* 3780 atm.

2.16 A mole of helium is compressed isothermally and reversibly at 100 °C from a pressure of 2 to 10 atm. Calculate (*a*), q, (*b*) w, (*c*) ΔG, (*d*) ΔA, (*e*) ΔH, (*f*) ΔU, and (*g*) ΔS.
Ans. (*a*) -4993, (*b*) -4993, (*c*) 4993, (*d*) 4993, (*e*) 0, (*f*) 0 J mol^{-1}, (*g*) -13.38 J K^{-1} mol^{-1}.

2.17 From tables giving $\Delta G_f°$, $\Delta H_f°$, and C_P for $H_2O(l)$ and $H_2O(v)$ at 298 K, calculate (*a*) the vapor pressure of $H_2O(l)$ at 25 °C, and (*b*) the standard boiling point.
Ans. (*a*) 3.169×10^3 Pa, (*b*) 373.5 K.

2.18 Calculate the entropy changes for the gas plus reservoir in Examples 2.5 and 2.6.
Ans. $\Delta S = 0$, $\Delta S = 19.16$ J K^{-1} mol^{-1}.

2.19 One mole of a perfect gas is allowed to expand reversibly and isothermally (25 °C) from a pressure of 1 atm to a pressure of 0.1 atm. (*a*) What is the change in Gibbs energy? (*b*) What would be the change in Gibbs energy if the process occurred irreversibly?
Ans. (*a*) -5708, (*b*) -5708 J mol^{-1}.

2.20 (*a*) Calculate the work done against the atmosphere when 1 mol of toluene is vaporized at its boiling point, 111 °C. The heat of vaporization at this temperature is 361.9 J g^{-1}. For the vaporization of 1 mol, calculate (*b*) q, (*c*) ΔH, (*d*) ΔU, (*e*) ΔG, and (*f*) ΔS.
Ans. (*a*) 3193, (*b*) 33,342, (*c*) 33,342, (*d*) 30,149, (*e*) 0 J mol^{-1}, (*f*) 86.8 J K^{-1} mol^{-1}.

2.21 One mole of a perfect gas at 300 K has an initial pressure of 15 atm and is allowed to expand isothermally to a pressure of 1 atm. Calculate (*a*) the maximum work that can be obtained from the expansion, (*b*) ΔU, (*c*) ΔH, (*d*) ΔG, and (*e*) ΔS.
Ans. (*a*) -6.754, (*b*) 0, (*c*) 0, (*d*) -6.754 kJ mol^{-1}, (*e*) 22.51 J K^{-1} mol^{-1}.

2.22 One mole of ammonia (considered to be a perfect gas) initially at 25 °C and 1 atm pressure is heated at constant pressure until the volume has trebled. Calculate (*a*) q, (*b*) w, (*c*) ΔH, (*d*) ΔU, and (*e*) ΔS. Given: $C_P = 25.895 + 32.999 \times 10^{-3}T - 30.46 \times 10^{-7}T^2$.
Ans. (*a*) 26.4, (*b*) -4.96, (*c*) 26.4, (*d*) 21.4 kJ mol^{-1}, (*e*) 46.99 J K^{-1} mol^{-1}.

2.23 Show that, by use of the second law, the difference in heat capacities at constant pressure and constant volume

$$C_P - C_V = \left[P + \left(\frac{\partial U}{\partial V}\right)_T\right]\left(\frac{\partial V}{\partial T}\right)_P$$

may be written in terms of quantities

$$\alpha = \frac{1}{V}\left(\frac{\partial V}{\partial T}\right)_P \quad \text{and} \quad \kappa = -\frac{1}{V}\left(\frac{\partial V}{\partial P}\right)_T$$

that may be determined more easily from experiment, as

$$C_P - C_V = \frac{TV\alpha^2}{\beta}$$

2.24 Calculate the partial molar volume of zinc chloride in 1-molal $ZnCl_2$ solution using the following data.

% by weight of $ZnCl_2$	2	6	10	14	18	20
Density/g cm^{-3}	1.0167	1.0532	1.0891	1.1275	1.1665	1.1866

Ans. 29.3 $cm^3\ mol^{-1}$.

2.25 Calculate ΔG and ΔS for the formation of a quantity of air containing 1 mol of gas by mixing nitrogen and oxygen at 298.15 K. Air may be taken to be 80% nitrogen and 20% oxygen by volume. *Ans.* -1239 J mol^{-1}. 4.159 J K^{-1} mol^{-1}.

2.26 Calculate the molar entropy of liquid chlorine at its melting point, 172.12 K, from the following data obtained by W. F. Giauque and T. M. Powell.

T/K	15	20	25	30	35	40	50	60
C_P/J K^{-1} mol^{-1}	3.72	7.74	12.09	16.69	20.79	23.97	29.25	33.47

T/K	70	90	110	130	150	170	172.12
C_P/J K^{-1} mol^{-1}	36.32	40.63	43.81	47.24	51.04	55.10	M.P.

The heat of fusion is 6406 J mol^{-1}. Below 15 K it may be assumed that C_P is proportional to T^3. *Ans.* 107.75 J K^{-1} mol^{-1}.

2.27 Using molar entropies from Table A.1, calculate $\Delta S°$ for the following reactions at 25 °C.

(*a*) $H_2(g) + \frac{1}{2}O_2(g) = H_2O(l)$

(*b*) $H_2(g) + Cl_2(g) = 2HCl(g)$

(*c*) Methane$(g) + \frac{1}{2}O_2(g) =$ methanol(l)

Ans. (*a*) -163.17, (*b*) 20.067, (*c*) -161.9 J K^{-1} mol^{-1}.

2.28 What is $\Delta S°$ for $H_2(g) = 2H(g)$ at 298, 1000, and 3000 K?

Ans. 98.634, 113.403, 122.410 J K^{-1} mol^{-1}.

2.29 Calculate the theoretical maximum efficiency with which heat can be converted into work in the following hypothetical turbines: (*a*) steam at 100 °C with exit at 40 °C; (*b*) mercury vapor at 300 °C with exit at 140 °C; (*c*) steam at 400 °C and exit at 150 °C; (*d*) air at 800 °C and exit at 400 °C; (*e*) air at 1000 °C and exit at 400 °C, special alloys being used; and (*f*) helium at 1500 °C and exit at 400 °C, heat from atomic energy being used.

2.30 When a perfect gas is allowed to expand isothermally in a piston, $\Delta U = q + w = 0$. Thus the work done by the system on the surroundings is equal to the heat transferred from the reservoir to the gas, the efficiency of turning heat into work is 100%. Explain why this is not a violation of the second law.

2.31 Calculate the differences between the molar entropies of Hg(l) and Hg(s) at -50 °C. The melting point of mercury is -39 °C, and the heat of fusion is 2343 J mol^{-1}. The heat capacity per gram atom of Hg(l) may be taken as $29.7–0.0067\,T$, and that of Hg(s) as 26.8 J K^{-1} mol^{-1}.

2.32 Derive the expression for the entropy change of a van der Waals' gas that is allowed to expand from volume V_1 to V_2 at constant temperature.

2.33 Calculate the increase in entropy of nitrogen when it is heated from 25 to 1000 °C (*a*) at constant pressure, and (*b*) at constant volume. Given: $C_P = 26.9835 + 5.9622 \times 10^{-3}T - 3.377 \times 10^{-7}T^2$.

2.34 Compute the entropy difference between 1 mol of liquid water at 25 °C and 1 mol of water vapor at 100 °C and 1 atm. The average specific heat of liquid water may be taken as 4.2 J K^{-1} g^{-1}, and the heat of vaporization is 2259 J g^{-1}.

2.35 Show that the process

$$H_2O(l, -5\ °C) = H_2O(s, -5\ °C)$$

is a spontaneous process in an isolated system containing in addition to the water a thermostat at −5 °C. The heat of fusion of water is 333.5 J g^{-1} at 0 °C, and the specific heats for water and ice may be taken as 4.2 J K^{-1} g^{-1} and 2.1 J K^{-1} g^{-1}, respectively.

2.36 What is the expression for the entropy change and Gibbs energy change of mixing of three components to form an ideal solution?

2.37 Show that

$$\left(\frac{\partial U}{\partial S}\right)_V = \left(\frac{\partial H}{\partial S}\right)_P \qquad \left(\frac{\partial H}{\partial P}\right)_S = \left(\frac{\partial G}{\partial P}\right)_T$$

2.38 The vapor pressure of water and ice at −5 °C are 421.7 and 401.7 Pa, respectively. Calculate ΔG for the transformation of water to ice at −5 °C.

2.39 Assuming the density of water is independent of pressure in the range 1 to 50 atm, what is the change in Gibbs energy of a mole of water when the pressure is raised this amount?

2.40 Derive the relation for $C_P - C_V$ for a gas that follows van der Waals' equation.

2.41 If z is a function of x and y, it is readily shown that

$$\left(\frac{\partial x}{\partial y}\right)_z \left(\frac{\partial y}{\partial z}\right)_x \left(\frac{\partial z}{\partial x}\right)_y = -1$$

which is referred to as the cyclic rule. Use this rule to show that for a gas

$$\left(\frac{\partial P}{\partial T}\right)_V = \frac{\alpha}{\kappa}$$

where α is the coefficient of thermal expansivity and κ is the isothermal compressibility (Problem 2.23).

2.42 Show that the entropy of a van der Waals' gas can be expressed as

$$S(V, T) = nR \ln (V - nb) + f(T)$$

where $f(T)$ is a function of T only (in other words, f is independent of V). Why—physically—is the volume contribution to S less than that for a perfect gas?

2.43 At 298 K, $S° = 130.574$ J K^{-1} mol^{-1} for $H_2(g)$. What is the entropy of $H_2(g)$ at 10^{-2} atm?

2.44 A mole of perfect gas is compressed isothermally from 1 to 5 atm at 100 °C. (*a*) What is the Gibbs energy change? (*b*) What would have been the Gibbs energy change if the compression had been carried out at 0 °C?

2.45 At 50 °C the partial pressure of $H_2O(g)$ over a mixture of $CuSO_4 \cdot 3H_2O(s)$ and $CuSO_4 \cdot H_2O(s)$ is 4.0×10^3 Pa and over a mixture of $CuSO_4 \cdot 3H_2O(s)$ and $CuSO_4 \cdot 5H_2O(s)$ is 6.3×10^3 Pa. Calculate the change in Gibbs energy for the reaction

$$CuSO_4 \cdot 5H_2O(s) = CuSO_4 \cdot H_2O(s) + 4H_2O(g)$$

2.46 The heat of vaporization of liquid oxygen at 1 atm is 6820 J mol^{-1} at its boiling point, −183 °C. For the reversible evaporization of 1 mol of liquid oxygen, calculate (*a*) q, (*b*) ΔU, (*c*) ΔG, and (*d*) ΔS.

2.47 One mole of a perfect gas in 22.4 L is expanded isothermally and reversibly at 0 °C to a volume of 224 L and $\frac{1}{10}$ atm. Calculate (*a*) w, (*b*) q, (*c*) ΔH, (*d*) ΔG, and (*e*) ΔS for the gas.

One mole of a perfect gas in 22.4 L is allowed to expand irreversibly into an evacuated vessel such that the final total volume is 224 L. Calculate (*f*) w, (*g*) q, (*h*) ΔH, (*i*), ΔG, and (*j*) ΔS for the gas.

Calculate (*k*) ΔS for the system and its surroundings involved in the reversible isothermal expansion and calculate (*l*) ΔS for the system and its surroundings involved in the irreversible isothermal expansion.

2.48 One mole of steam is compressed reversibly to liquid water at the boiling point 100 °C. The heat of vaporization of water at 100 °C and 1 atm is 2258.1 J g^{-1}. Calculate w and q and each of the thermodynamic quantities ΔH, ΔU, ΔG, ΔA, and ΔS.

2.49 An equation for the pressure P as a function of height h in the atmosphere may be derived easily if it is assumed that the temperature T, acceleration g of gravity, and molar mass M of air are constant. The molar Gibbs energy of the gas is a function of pressure and height and must be the same throughout the atmosphere. Thus

$$dG = \left(\frac{\partial G}{\partial P}\right)_h dP + \left(\frac{\partial G}{\partial h}\right)_P dh = 0$$

Show that $(\partial G/\partial h)_P = Mg$, and then that

$$P = P_0 e^{-Mgh/RT}$$

2.50 Calculate $(\partial U/\partial V)_T$ for a van der Waals' gas.

2.51 The apparent specific volume v in milliliters per gram is equal to the volume of the solution minus the volume of pure solvent it contains divided by the weight of solute.

$$v = \frac{100 - (100\rho_s - g)/\rho_0}{g}$$

where g is the number of grams of solute in 100 mL of solution, ρ_s is the density of the solution, and ρ_0 is the density of the solvent. The density of a solution of serum albumin containing 1.54 g of protein per 100 cm^3 is 1.0004 g cm^{-3} at 25 °C ($\rho_0 = 0.977\,07$ g cm^{-3}). Calculate the apparent specific volume.

2.52 A solution of magnesium chloride, $MgCl_2$, in water containing 41.24 g L^{-1} has a density of 1.0311 g cm^{-3} at 20 °C. The density of water at this temperature is 0.998 23 g cm^{-3}. Calculate (*a*) the apparent specific volume (see the equation in Problem 2.51), and (*b*) the apparent molar volume of $MgCl_2$ in this solution.

2.53 When 1 mol of water was added to an infinitely large amount of an aqueous methanol solution having a mole fraction of methanol of 0.40, the volume of the solution increased 17.35 mL. When 1 mol of methanol was added to such a solution, the volume increased 39.01 mL. Calculate the volume of solution containing 0.40 mol of methanol and 0.60 mol of water.

2.54 Calculate ΔG and ΔS for mixing 2 mol of H_2 with 1 mol of O_2 at 25 °C under conditions where no chemical reaction occurs.

2.55 Calculations with the Debye formula show the molar entropy of silver iodide to be 6.3 J K^{-1} mol^{-1} at 15 K. From the following data for the molar heat capacity at constant pressure, calculate the molar entropy of silver iodide at 298.1 K.

T/K	C_P	T/K	C_P	T/K	C_P	T/K	C_P
21.00	15.98	64.44	39.16	145.67	49.87	258.79	54.60
30.53	21.88	88.58	41.84	170.86	50.84	273.23	55.48
42.70	29.66	105.79	46.82	198.89	52.26	287.42	56.40
52.15	34.31	126.53	48.53	228.34	53.39	301.37	57.07

2.56 Calculate the molar entropy of carbon disulfide at 25 °C from the following heat-capacity data and the heat of fusion, 4389 J mol^{-1}, at the melting point (161.11 K).

T/K	15.05	20.15	29.76	42.22	57.52	75.54	89.37
C_P/J K^{-1} mol^{-1}	6.90	12.01	20.75	29.16	35.56	40.04	43.14

T/K	99.00	108.93	119.91	131.54	156.83	161–298
C_P/J K^{-1} mol^{-1}	45.94	48.49	50.50	52.63	56.62	75.48

2.57 Using the following data, calculate the molar entropy of gaseous isobutene at 25 °C [S. S. Todd and G. S. Parks, *J. Am. Chem. Soc.*, **58**, 134 (1936)].

$$S_{90K} = 45.23 \text{ J K}^{-1} \text{ mol}^{-1}$$
$$\text{F.P.} = -140.7\ ^\circ\text{C}$$
$$\Delta H_{fus} = 105.52 \text{ J g}^{-1}$$
$$\text{B.P.} = -7.1\ ^\circ\text{C}$$
$$\Delta H_{vap} = 403.76 \text{ J g}^{-1}$$
$$C_P(\text{gas}) \text{ in range 226–298 K} = 83.7 \text{ J K}^{-1} \text{ mol}^{-1}$$

Specific Heat (J g^{-1} K^{-1})

T/K	93.3	105.5	118.9	139.2	166.1	179.8	210.2	253.1
	1.0452	1.1502	1.2786	1.9025	1.9334	1.9585	2.0334	2.1644

2.58 If C_P in the temperature range below 15 K is given by

$$C_P = C_{15}\left(\frac{T}{15}\right)^3$$

show that

$$S_{15} = \frac{C_{15}}{3}$$

2.59 Calculate the entropy changes for the following reactions at 25 °C.

(*a*) $H^+(aq) + OH^-(aq) = H_2O(l)$
(*b*) $Ag^+(aq) + Cl^-(aq) = AgCl(s)$
(*c*) $HS^-(aq) = H^+(aq) + S^{2-}(aq)$

2.60 What is ΔS° (298 K) for

$$H_2O(l) = H^+(aq) + OH^-(aq)$$

Why is this change negative and not positive?

2.61 Calculate ΔS° for the reaction

$$CH_4(g) + 2O_2(g) = CO_2(g) + 2H_2O(g)$$

at 298 and 1000 K.

CHAPTER 3

PHASE EQUILIBRIA

A *phase* is a part of a system, uniform throughout in chemical composition and physical properties, which is separated from other homogeneous parts of the system by boundary surfaces. The phase behavior exhibited by pure substances is quite varied and complicated, but this information may be organized and predictions may be made by use of thermodynamics. The understanding of phase behavior was greatly increased by the *phase rule* derived by Gibbs. The Clausius and Clausius-Clapeyron equations deal with the change of equilibrium pressure with temperature.

Two-component, liquid-liquid systems may show very complicated behavior. However, some of these systems follow the same equations as ideal mixtures of gases do. Therefore, the concept of ideal solutions provides a standard with which real solutions may be compared. Deviations from ideal behavior are represented by the use of activity coefficients.

Phase diagrams representing the separation of solid and liquid phases from two-component mixtures are also discussed. This includes freezing point lowering, ideal solubility, the formation of congruently and incongruently melting compounds, and the formation of solid solutions. The chapter ends with discussions of several phase diagrams for systems of three components.

3.1 CRITERIA OF EQUILIBRIUM IN TERMS OF INTENSIVE PROPERTIES

Several criteria of equilibrium among phases can be stated in terms of the intensive properties T, P, and μ. We are familiar with the fact that for two phases, such as ice and water, to be in equilibrium they must be at the same temperature and pressure. This can be proved by considering that an infinitesimal quantity of heat dq is reversibly transferred from a phase α to a phase β with which it is in equilibrium. The condition for equilibrium in an isolated system is that the entropy of the system not be changed by the transfer. Thus

$$dS = 0 \qquad \text{or} \qquad dS_\alpha + dS_\beta = 0 \tag{3.1}$$

where the phases are designated by subscripts. Since the process is reversible,

$$\frac{-dq}{T_\alpha} + \frac{dq}{T_\beta} = 0 \tag{3.2}$$

or

$$T_\alpha = T_\beta \tag{3.3}$$

The fact that the pressure must be the same in two phases at equilibrium may be proved by considering that phase α increases in volume by an infinitesimal volume dV and phase β decreases by the same amount. If the temperature and volume of the whole system are held constant, $dA = 0$, so that

$$dA_\alpha + dA_\beta = 0 \qquad \text{or} \qquad -P_\alpha\,dV + P_\beta\,dV = 0 \tag{3.4}$$

and so

$$P_\alpha = P_\beta \tag{3.5}$$

An additional restriction may be derived by considering the transfer of a small quantity of substance dn_i from phase α to phase β, which are in equilibrium. If the temperature and pressure of the whole system are kept constant then, since $dG = 0$,

$$dG_\alpha + dG_\beta = 0 \tag{3.6}$$

or, using equation 2.110 (Section 2.19),

$$-\mu_{i\alpha}\,dn_i + \mu_{i\beta}\,dn_i = 0 \tag{3.7}$$

or

$$\mu_{i\alpha} = \mu_{i\beta} \tag{3.8}$$

Thus the chemical potential of a component is the same in all phases at equilibrium.

If phases α and β are not in equilibrium and a small quantity dn_i of species i is transferred from phase α to phase β in the direction of approaching equilibrium, we have at constant temperature and pressure

$$dG_\alpha + dG_\beta < 0 \qquad \text{or} \qquad -\mu_{i\alpha}\,dn_i + \mu_{i\beta}\,dn_i < 0 \tag{3.9}$$

If dn_i is positive then

$$\mu_{i\alpha} > \mu_{i\beta} \tag{3.10}$$

Thus, a substance will tend to pass spontaneously from the phase where it has the higher chemical potential to the phase where it has the lower chemical potential. Also, a substance will diffuse spontaneously from a region where its concentration and chemical potential are higher into a more dilute solution where its chemical potential is lower. In this respect the chemical potential is like other kinds of potential, electrical, gravitational, etc., in that the spontaneous change is always in the direction from high to low potential.

3.2 PHASE RULE

In 1876, Gibbs* derived a simple relationship between the number of phases in equilibrium, the number of components, and the number of intensive independent variables that must be specified in order to describe the state of the system completely.

---

* J. W. Gibbs, *Trans. Conn. Acad. Arts Sci.*, 1876–1878; *The Collected Works of J. Willard Gibbs*, Vol. 1, Yale University Press, New Haven, reprinted 1948.

The *number of components* c in a system is the smallest number of substances in terms of which the compositions of each of the phases in the system may be described separately. The number of components may be smaller than the number of substances s that may be added to form the system, because there may be relationships between the concentrations of various substances at equilibrium that make it unnecessary to specify the concentrations of all s substances in describing the system. There are two types of relations between concentrations of components: chemical equilibrium expressions and initial conditions. For each independent chemical equilibrium expression, the number of independent concentrations is reduced by one. For example, if solid calcium oxide, solid calcium carbonate, and gaseous carbon dioxide are in equilibrium, the number of independent components is reduced by one by the equilibrium expression for

$$CaCO_3(s) = CaO(s) + CO_2(g) \tag{3.11}$$

If molecular hydrogen and oxygen are in equilibrium with water there are at most two independent components (H_2O and O_2, H_2O and H_2, or H_2 and O_2), because the concentration of the third is fixed by the equilibrium expression

$$K_P = \frac{P_{H_2}P_{O_2}^{1/2}}{P_{H_2O}} \tag{3.12}$$

If the initial conditions are specified, the number of components is reduced to one. For example, if the hydrogen and oxygen are formed only from water, there is an additional relationship, $P_{H_2} = 2P_{O_2}$.

If there are s substances, n independent equilibrium constant expressions, and m relations between concentrations due to initial conditions, the number of components c is given by

$$c = s - n - m \tag{3.13}$$

There are often several possible and equally satisfactory choices of components. The choices are arbitrary, but the *number* of components is an important characteristic of a system.

The *number of degrees of freedom or variance* v of a system is the smallest number of independent variables (pressure, temperature, and concentrations of the various phases) that must be specified to describe completely the state of the system. As we have seen before, to describe the state of a fixed amount of a pure gas it is necessary to specify only two variables, T and P, or P and V, or V and T, because the third variable can be calculated from the equation of state. Thus a pure gas has two degrees of freedom or a variance $v = 2$.

Consider a system in equilibrium that consists of p phases. When we refer to the number of phases, we mean the number of different kinds of phases; for example, a system containing liquid water and many pieces of ice, but no gas phase, has only two phases. If a phase contains c components, its composition may be specified by stating $(c - 1)$ concentrations—one less than the number of components because the concentration of one component can be obtained from $\sum X_i = 1$, where X_i represents the mole fraction of component i. Thus the total number of concentrations to be specified for the whole system is $(c - 1)$ for each of the p phases or $(c - 1)p$ concentrations. In general, there are two more variables that have to be considered, temperature and pressure, so that the total number of independent variables is $(c - 1)p + 2$. We do not have to talk about the temperatures and pressures of the different phases separately because they are in equilibrium and so they are all at the same pressure and temperature. If temperature or pressure were held constant, the number of independent variables would be $(c - 1)p + 1$. On the other hand, if the system were affected by both temperature and pressure and another independent variable, such as magnetic field strength, the number of variables would be $(c - 1)p + 3$.

Next we consider the number of relationships that must be satisfied at equilibrium. The chemical potential μ for each component is the same in each phase α, β, γ, etc., and so $\mu_{i,\alpha} = \mu_{i,\beta} = \mu_{i,\gamma} = \cdots$ for component i. There are p phases but only $(p - 1)$ equilibrium relationships of the type $\mu_{i\alpha} = \mu_{i\beta}$ for each component. For example, if there are two phases, there is only one equilibrium relationship for each component that gives its distribution between the two phases. Altogether there are c components, each one of which can be involved in an equilibrium between phases. Thus there is a total of $c(p - 1)$ equilibrium relations.

The number of degrees of freedom v is equal to the total number of variables minus the total number of equilibrium relations between these variables; that is, v is the additional number of variables that must be specified to define the system completely. Thus, for systems in which pressure, temperature, and concentration are the only variables,

$$v = [p(c - 1) + 2] - c(p - 1)$$

or

$$v = c - p + 2 \qquad (3.14)$$

This is the important phase rule of Gibbs.

It can be seen from this equation that the greater the number of components in a system, the greater the number of degrees of freedom or variance. On the other hand, the greater the number of phases, the smaller the number of variables such as temperature, pressure, and concentration that must be specified to describe the system completely.

For a given number of components the number of phases is a maximum when the variance is zero. For a one-component system, the maximum number of phases that can be present at equilibrium is $p = c - v + 2 = 3$. For a two-component system, the maximum number of phases is 4, etc.

3.3 *P–V–T* SURFACE FOR WATER

As an example of the pressure-volume-temperature behavior of a pure substance, the three-dimensional diagram for H_2O is given in Fig. 3.1. Each point on this surface represents an equilibrium state. Projections of this surface on the *P–T* plane and the *P–V* plane are shown. There are three two-phase regions on the surface:

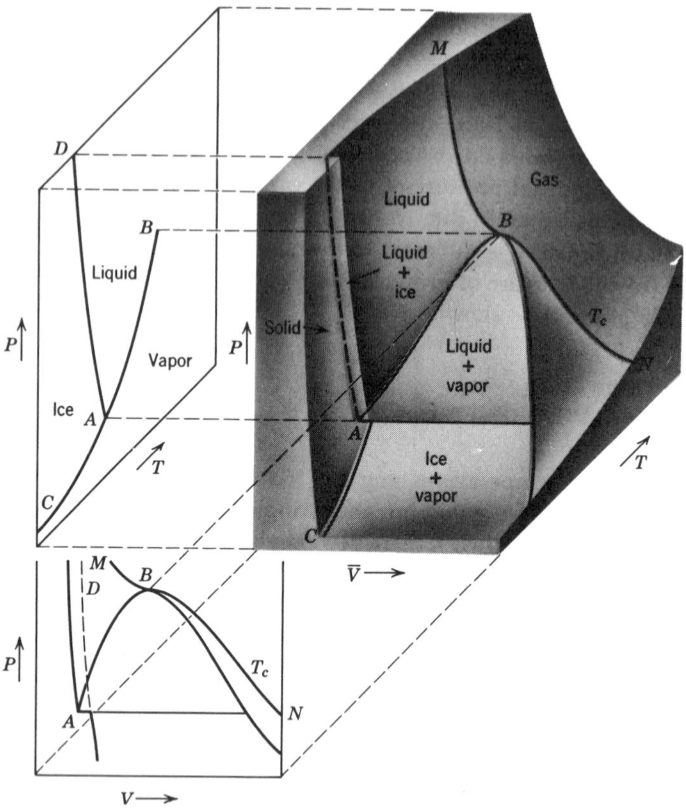

Fig. 3.1 Plot of the pressure-volume-temperature relation for a pure substance like water. The diagram is not drawn to scale.

liquid plus vapor, ice plus vapor, and liquid plus ice. These are ruled surfaces—that is, they may be thought of as being generated by a moving straight line, in this case one perpendicular to the P–T plane. These three surfaces intersect at the triple point A. Thus, at the triple point, vapor, liquid, and solid are in equilibrium.

The projection of the three-dimensional surface on the P–T plane is shown to the left of the main diagram in Fig. 3.1. The vapor pressure curve goes from the triple point A to the critical point B at 374 °C and 218 atm. The sublimation pressure curve goes from the triple point A to absolute zero. The melting curve rises from the triple point with $(\partial P/\partial T)$ negative for water because it expands on freezing. Most substances contract on freezing, and for them $(\partial P/\partial T)$ is positive. The value of $(\partial P/\partial T)$ is given by the Clapeyron equation, which we will discuss shortly.

Investigations conducted by Bridgman to determine the course of the melting curve AD have revealed the existence of seven different crystalline modifications of ice, all of which, with the exception of ordinary ice, are denser than water. The first of these new forms of ice makes its appearance at a pressure of 2047 atm and the last at a pressure of 21,680 atm. If one solid form may be changed into another solid form, as in Bridgman's experiments with ice, the transformation is called an enantiotropic change.

The triple point of water in the absence of air is at 0.0100 °C and 611 Pa. In the presence of air at 1 atm the three phases are in equilibrium at 0 °C, which is one of the defined points on the Celsius scale. The total pressure in this case is 1 atm, but the partial pressure of water vapor is only 611 Pa. The lowering of the triple point by air is due to two effects: (1) the solubility of air in liquid water at 1 atm pressure is sufficient to lower the freezing point 0.0024 °C (Section 3.19), and (2) the increase of pressure from 611 Pa to 1 atm lowers the freezing point 0.0075 °C, as will be shown shortly.

For a one-component system the phase rule, equation 3.14, yields $v = 3 - p$. Thus in the single-phase regions—labeled ice, liquid, and vapor in Fig. 3.1—the variance is two, and two independent variables have to be specified to describe the system. For example, if the system consists of water vapor, it is necessary to specify the temperature and pressure to define the system. Along the lines in the P–T diagram two phases are in equilibrium, and so $v = 1$, and it is necessary to specify only one variable to describe the system. For example, if the system consists of water vapor in equilibrium with liquid water, it is necessary to specify only the temperature *or* pressure, because for a given temperature there is only one equilibrium pressure and for a given pressure there is only one equilibrium temperature. If there are three phases in equilibrium, $v = 0$, and so for a one-component system three phases can coexist only at a particular temperature and pressure. Such a system is said to be *invariant.*

3.4 CRITICAL PHENOMENA

The P–V projection of the three-dimensional surface for water is also shown in Fig. 3.1. The corresponding diagram for isopentane is given in Fig. 1.9, which has been discussed in connection with the van der Waals equation. For a pure substance the critical state may be defined by either of the following two criteria: (1) the critical state is the state of temperature and pressure at which the gas and liquid phases become so nearly alike that they can no longer exist as separate phases; or (2) the critical temperature of a pure liquid is the highest temperature at which gas and liquid phases can exist as separate phases. The critical pressure is the pressure at the critical point, and the critical volume is the molar volume under these conditions.

Only the first of these definitions is applicable to mixtures.* More complicated behavior is encountered with mixtures, since the coexisting liquid and vapor phases are, in general, of different composition.

The critical temperature of a pure substance is usually obtained by observing the

---

* L. G. Roof, *J. Chem. Educ.*, **34**, 492 (1957).

temperature at which the meniscus between the gas and liquid phases under pressure disappears when the system is warmed and reappears when it is cooled. If an amount of liquid is sealed in a tube so that the average density of liquid and vapor corresponds with the critical density ρ_c, heating the tube through the critical point will show the phenomenon of critical opalescence. Critical opalescence arises from fluctuations that extend over distances comparable with the wavelength of visible light and that therefore cause light scattering (see Section 20.7). Such fluctuations arise because at the critical point the compressibility $[\kappa = -V^{-1}(\partial V/\partial P)_T]$ is infinite so that fluctuations in density are no longer restrained by the rise in pressure with height in the fluid. In the neighborhood of the critical point a fluid is so compressible that the acceleration of gravity sets up large differences in density between the top and bottom of a container. Gravity can produce density differences as large as 10% in a column of fluid only a few centimeters high. This makes it difficult to determine PV isotherms near the critical point.

Equations of state used to represent the P–V–T properties of pure substances assume that pressure is an analytic function of volume and temperature at and near the critical point. However, it is now known that this assumption is not justified, and data in the neighborhood of the critical point can only be represented by use of a special nonanalytic function known as a scaling equation. In the immediate neighborhood of the critical point, the changes in various thermodynamic properties may be represented as some power of $(T_c - T)$, where T_c is the critical temperature. Table 3.1 defines three of these functions and gives the values for CO_2 and for a van der Waals fluid.

Table 3.1 Critical Indices

Definition	Exponent for CO_2	Exponent for van der Waals Fluid
$\frac{1}{C_V} = \text{const}\ (T_c - T)^{\alpha}$	0.065	0
$(\rho^l - \rho^g)_T = \text{const}\ (T_c - T)^{\beta}$	0.347	0.5
$\left(\frac{\partial P}{\partial V}\right)_T = -\text{const}\ (T_c - T)^{\gamma}$	1.241	1

The experimental results can be accounted for by extension of a theory developed in 1925 by Ising to explain ferromagnetism. He assumed that the magnetic spins are localized on the lattice sites of a regular array and are capable of only two orientations in opposite directions. He assumed that parallel spins attract each other and antiparallel spins repel, but that there are interactions only between nearest neighbors. Thus the cooperative effect of ferromagnetism arises through chains of nearest-neighbor interactions. In the critical fluid case the spins pointing in one direction are replaced by molecules, and spins pointing in the opposite direction are replaced by "holes" or empty sites.

Critical phenomena are also shown by liquid solutions, biopolymers, liquid crystals, alloys, superconductors, and ferromagnetic metals.

3.5 EXISTENCE OF PHASES IN A ONE-COMPONENT SYSTEM

In order to understand the change from solid to liquid to gas phases when a solid is heated at constant pressure, we may consider a plot of molar Gibbs energy versus temperature at constant pressure for the various phases, as shown in Fig. 3.2*a*. As we have already seen (Section 3.1), the stable phase is that with the lowest value of the chemical potential. For a one-component system, the molar Gibbs energy is equal to the chemical potential. If two or three phases of a single component have the same molar Gibbs energy, they will coexist at equilibrium as at the melting point T_m, boiling point T_b, or triple point. Below the melting point T_m the solid has the lowest Gibbs energy and is therefore the stable phase. Above the boiling point T_b the gas phase has the lowest Gibbs energy, and it is the stable phase. Between T_m and T_b, the liquid is the stable phase. It may be seen from this figure that the phase transitions are sudden and there are no indications of a drastic change as the temperature is changed toward the transition point.*

The slopes of the lines giving the Gibbs energies of solid, liquid, and gas in Fig. 3.2*a* are given by (see equation 2.70)

$$\left(\frac{\partial G}{\partial T}\right)_P = -S \tag{3.15}$$

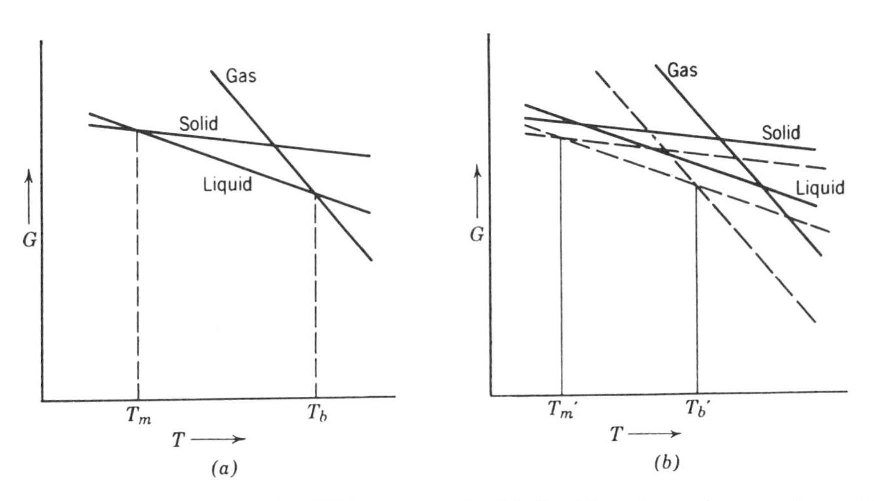

Fig. 3.2 Dependence of molar Gibbs energy of solid, liquid, and gas phases on temperature at constant pressure. The dashed lines in (*b*) are for a lower pressure. The plots should be slightly concave downward, since the entropy increases with increasing temperature, but they have been drawn as straight lines here for simplicity.

---

* There is another type of transition, called a lambda transition, in which the heat capacity changes rapidly as the transition temperature is approached. This also occurs at the critical point, but not elsewhere along the vapor-liquid equilibrium line.

Since the entropy is positive the slopes are negative, and since $S_g > S_l > S_s$, the slope is more negative for the gas than the liquid and more negative for the liquid than for the solid.

At a lower pressure the plots of G versus T are displaced, as shown in Fig. 3.2*b*. The effect of pressure on the Gibbs energy of a pure substance at constant temperature is given by (see equation 2.71)

$$\left(\frac{\partial G}{\partial P}\right)_T = V \tag{3.16}$$

Since the molar volume is always positive, the chemical potential μ decreases as the pressure is decreased at constant temperature. Since $V_g \gg V_l, V_s$, this effect is much larger for a gas than for a liquid or solid. As shown in Fig. 3.2*b*, reducing the pressure lowers the boiling point and normally the melting point. The effect on the boiling point is much larger because of the large difference in the molar volumes of gas and liquid. As a result the range of temperature over which the liquid is the stable phase has been reduced. It is evident that at a sufficiently low pressure the curve for the chemical potential of the gas will intercept the solid curve below the temperature where the solid and liquid have the same chemical potential. At this low pressure the solid will sublime instead of melt; that is, it passes directly into the vapor without going through the liquid state, as illustrated by dry ice.

At some particular pressure the solid, liquid, and vapor curves will intersect at a point; the temperature and pressure at which the three phases coexist is referred to as the triple point.

3.6 THE CLAPEYRON EQUATION

If two phases of a pure substance are in equilibrium with each other they have the same molar Gibbs energy at that temperature and pressure. When the temperature is changed at constant pressure, or the pressure is changed at constant temperature, one of the phases will disappear. However, if the temperature and pressure are both changed in such a way as to keep the two chemical potentials equal to each other, the two phases will continue to coexist. The necessary relation for dP/dT was derived by Clapeyron.

If two phases α and β of a pure substance are in equilibrium the chemical potentials, or molar Gibbs energies, are equal.

$$G_\alpha = G_\beta \tag{3.17}$$

If the pressure and temperature are changed so that equilibrium is maintained, it is necessary that

$$dG_\alpha = dG_\beta \tag{3.18}$$

Since G depends only on P and T, this equation may be written

$$\left(\frac{\partial G_\alpha}{\partial P}\right)_T dP + \left(\frac{\partial G_\alpha}{\partial T}\right)_P dT = \left(\frac{\partial G_\beta}{\partial P}\right)_T dP + \left(\frac{\partial G_\beta}{\partial T}\right)_P dT \tag{3.19}$$

Utilizing equations 2.70 and 2.71, this becomes

$$V_\alpha \, dP - S_\alpha \, dT = V_\beta \, dP - S_\beta \, dT \tag{3.20}$$

or

$$\frac{dP}{dT} = \frac{S_\beta - S_\alpha}{V_\beta - V_\alpha} = \frac{\Delta S}{\Delta V} = \frac{\Delta H}{T \Delta V} \tag{3.21}$$

This equation is referred to as the Clapeyron equation, and it may be applied to vaporization, sublimation, fusion, or the transition between two solid phases of a pure substance. The molar enthalpies of sublimation, fusion, and vaporization at a given temperature are related by

$$\Delta H_{\text{sub}} = \Delta H_{\text{fus}} + \Delta H_{\text{vap}} \tag{3.22}$$

since the heat required to vaporize a given amount of the solid is the same whether this process is carried out directly or by first melting the solid and then vaporizing the liquid.

In using equation 3.21 it is necessary to express the enthalpy change of the process in the same units as the product of pressure and volume change.

Example 3.1 What is the change in the boiling point of water at 100 °C per Pa change in atmospheric pressure?

The enthalpy of vaporization is 40.69 kJ mol^{-1}, the molar volume of liquid water is $0.019 \times 10^{-3}\ \text{m}^3\ \text{mol}^{-1}$, and the molar volume of steam is $30.199 \times 10^{-3}\ \text{m}^3\ \text{mol}^{-1}$, all at 100 °C and 1 atm.

$$\frac{dP}{dT} = \frac{\Delta H_{\text{vap}}}{T(V_v - V_l)} = \frac{(40{,}690\ \text{J mol}^{-1})}{(373.15\ \text{K})(30.180 \times 10^{-3}\ \text{m}^3\ \text{mol}^{-1})}$$

$$= 3613\ \text{Pa K}^{-1}$$

Thus $dT/dP = 2.768 \times 10^{-4}\ \text{K Pa}^{-1}$.

Example 3.2 Calculate the change in pressure required to change the freezing point of water 1 °C. At 0 °C the heat of the fusion of ice is 333.5 J g^{-1}, the density of water is 0.9998 g cm^{-3}, and the density of ice is 0.9168 g cm^{-3}. The reciprocals of the densities, 1.0002 and 1.0908, are the volumes in cubic centimeters of 1 g. The volume change upon freezing $(V_l - V_s)$ is therefore $-9.06 \times 10^{-8}\ \text{m}^3\ \text{g}^{-1}$. For small changes ΔH_{fus}, T, and $(V_l - V_s)$ are virtually constant, so that

$$\frac{\Delta P}{\Delta T} = \frac{\Delta H_{\text{fus}}}{T(V_l - V_s)} = \frac{333.5\ \text{J g}^{-1}}{(273.15\ \text{K})(-9.06 \times 10^{-8}\ \text{m}^3\ \text{g}^{-1})}$$

$$= -1.348 \times 10^7\ \text{Pa K}^{-1}$$

The change in the freezing point of water per atmosphere pressure is

$$\frac{\Delta T}{\Delta P} = \frac{101{,}325\ \text{Pa atm}^{-1}}{-1.348 \times 10^7\ \text{Pa K}^{-1}} = -0.0075\ \text{K atm}^{-1}$$

shows that an increase in pressure of 1 atm lowers the freezing point 0.0075 K. The negative sign indicates that an increase in pressure causes a decrease in temperature.

3.7 THE CLAUSIUS-CLAPEYRON EQUATION

For vaporization and sublimation Clausius showed how the Clapeyron equation may be simplified by assuming that the vapor obeys the perfect gas law and by neglecting the molar volume of the liquid V_l in comparison with the molar volume of the vapor V_v. Substituting RT/P for V_v, we have

$$\frac{dP}{dT} = \frac{\Delta H_{\text{vap}}}{TV_v} = \frac{P\,\Delta H_{\text{vap}}}{RT^2} \tag{3.23}$$

On rearrangement equation 3.23 becomes

$$\frac{dP}{P} = d\ln\left(\frac{P}{P^+}\right) = \frac{\Delta H_{\text{vap}}}{RT^2}\,dT \tag{3.24}$$

where P^+ is the unit of pressure used. Integrating on the assumption that ΔH_{vap} is independent of temperature and pressure yields

$$\int d\ln\left(\frac{P}{P^+}\right) = \frac{\Delta H_{\text{vap}}}{R}\int T^{-2}\,dT \tag{3.25}$$

$$\ln\left(\frac{P}{P^+}\right) = -\frac{\Delta H_{\text{vap}}}{RT} + C \tag{3.26}$$

where C is the integration constant. This suggests that a plot of $\ln(P/P^+)$ versus $1/T$ should be linear, and this is borne out by data on both vaporization and sublimation, as shown in Fig. 3.3. Over a wide temperature range there are significant deviations from linearity because the vapor does not follow the perfect gas law, and ΔH_{vap} varies with temperature.

Frequently, it is more convenient to use the equation obtained by integrating between limits, P_2 at T_2 and P_1 at T_1, as follows.

$$\int_{P_1/P^+}^{P_2/P^+} d\ln\left(\frac{P}{P^+}\right) = \frac{\Delta H_{\text{vap}}}{R}\int_{T_1}^{T_2} T^{-2}\,dT \tag{3.27}$$

$$\ln\frac{P_2}{P_1} = \frac{\Delta H_{\text{vap}}}{R}\left[-\frac{1}{T_2} - \left(-\frac{1}{T_1}\right)\right] \tag{3.28}$$

$$\ln\frac{P_2}{P_1} = \frac{\Delta H_{\text{vap}}(T_2 - T_1)}{RT_1T_2} \tag{3.29}$$

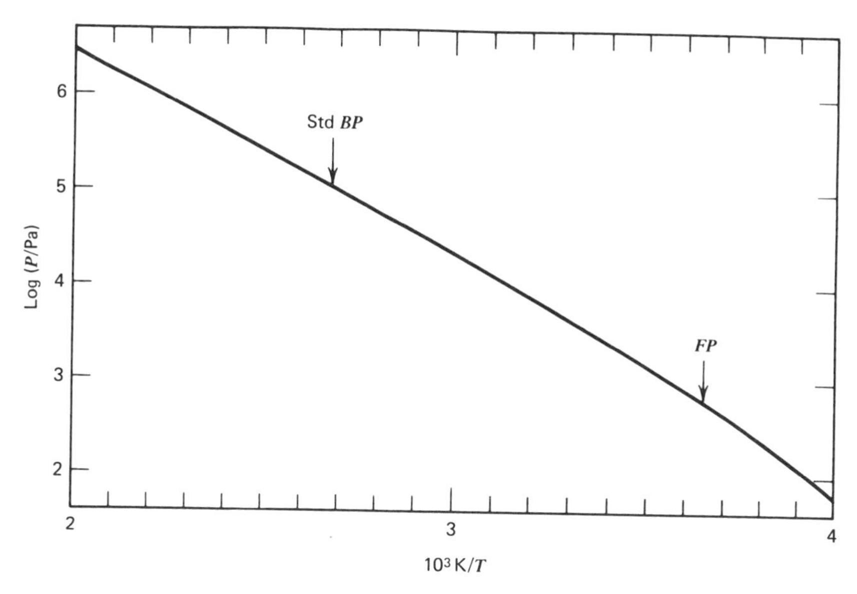

Fig. 3.3 Vapor pressure of water.

Using this equation, it is possible to calculate the heat of vaporization or the heat of sublimation from the vapor pressures at two different temperatures.

If the enthalpy of vaporization of a liquid is not known, an approximate value may be estimated using Trouton's rule that the molar entropy of vaporization at the standard boiling point (the boiling point at 1 atm pressure) is a constant, about 88 J K^{-1} mol^{-1}.

$$\Delta S_{\text{vap}} = \frac{\Delta H_{\text{vap}}}{T_b} \cong 88 \text{ J K}^{-1} \text{ mol}^{-1} \tag{3.30}$$

The relative constancy of the entropy of vaporization from liquid to liquid is readily understood in terms of the Boltzmann hypothesis relating entropy to disorder. The change from liquid to vapor leads to increased disorder. The entropy of vaporization is zero at the critical temperature because the liquid and gas are indistinguishable and the enthalpy of vaporization is zero. Most liquids behave alike not only at their critical temperatures but also at equal fractions of their critical temperatures. Hence, different liquids should have about the same entropy of vaporization at their boiling point, provided that there is no association or dissociation upon vaporization. For substances like water and alcohols, which form hydrogen bonds (Section 10.8), the entropy of vaporization is greater than 88 J K^{-1} mol^{-1}. Hydrogen and helium, which boil at only a little above absolute zero, might well be expected to show large departures from this rule. Acetic acid and carboxylic acids, in general, have abnormally low heats of vaporization, since the vapor consists of double molecules and still more energy would be required to break them up into single molecules comparable with those of other gases.

3.8 TWO COMPONENTS: LIQUID AND VAPOR

For a two-component system, $v = 4 - p$. Since there is at least one phase, the maximum number of degrees of freedom is 3, so that the system may be represented by a three-dimensional plot. As shown in Fig. 3.4, the plot for a two-component liquid-vapor system consists of a pair of surfaces (P, T, X) and (P, T, Y), where X is the mole fraction of component 1 in the liquid phase and Y is the mole fraction of component 1 in the vapor phase. Below the lower surface there is a single vapor phase but, as the pressure is raised, condensation starts when the surface is reached; this is the reason it is called the *dew point surface*. Above the upper surface there is a single liquid phase but, as the pressure is reduced, bubbles of vapor start to form when the surface is reached; this is the reason it is called the *bubble point surface*. These surfaces are joined at the vapor pressure lines of pure component 1 and pure component 2 on sides of the figure. These vapor pressure lines end at the respective critical points c_1 and c_2.

Considering the system represented in Fig. 3.4 under conditions such that there is a single phase, the system is trivariant; and the temperature, pressure, and mole fraction, X, of one of the components must be specified to describe the system completely. If two phases are present, as will be the case between surfaces in a P–T–X plot, only two variables need be specified to define the system. For example, if the pressure and temperature are specified, the compositions of the liquid and vapor phases are given by the phase diagram. The relative amounts of the two phases are not fixed by specifying only the temperature and pressure, but the phase rule is not concerned with the relative amounts of phases. Since two variables have to be specified, we say that the system is bivariant.

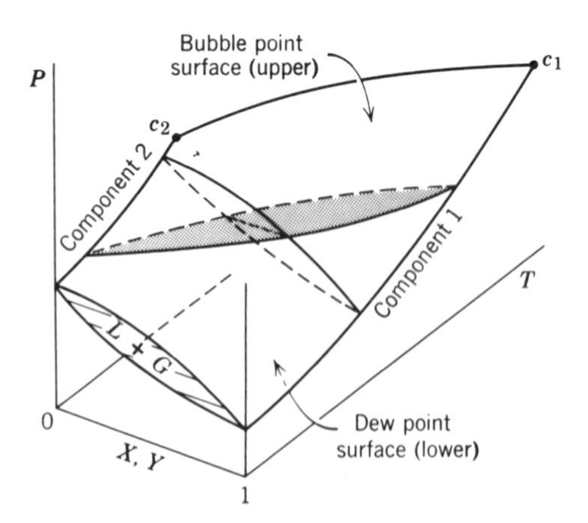

Fig. 3.4 Three-dimensional plot for a two-component liquid and vapor system. The plot consists of two surfaces (bubble point surface and dew point surface). (From K. E. Bett, J. S. Rowlinson, and G. Saville, *Thermodynamics for Chemical Engineers*, The MIT Press, Cambridge, Mass., 1975. © University of London (Athlone Press) 1975.)

Figure 3.5 shows cross sections of Fig. 3.4 at constant temperature, pressure, and composition. The lower line of each pair in Fig. 3.5*a* is a plot of pressure versus the mole fraction Y_1 of component 1 in the vapor phase that exists at pressures below the line; this line is the dew point line. The upper line of each pair, the bubble point line, is a plot of pressure versus X_1, the mole fraction of component 1 in the liquid phase that exists above the line. In Fig. 3.5*a* the horizontal tie lines in the regions in which liquid and vapor are both present remind us that vapor with composition Y_1 at the left end of the tie line is in equilibrium with liquid of composition X_1 at

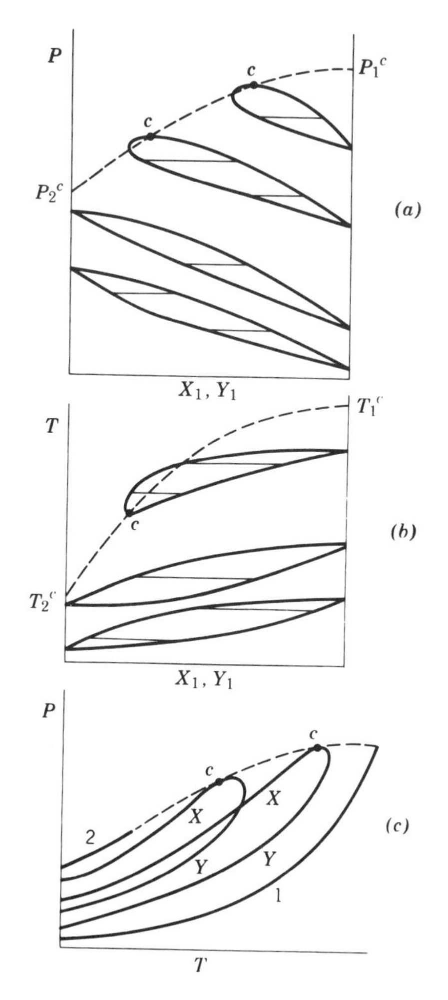

Fig. 3.5 Cross sections through the preceding figure in (*a*) the P–X, Y plane at four different temperatures, (*b*) the T–X, Y plane at three different pressures, and (*c*) the P–T plane at three different compositions. (From K. E. Bett, J. S. Rowlinson, and G. Saville, *Thermodynamics for Chemical Engineers*, The MIT Press, Cambridge, Mass., 1975. © University of London (Athlone Press) 1975.)

the right end of the tie line. Points c and c are critical points of binary mixtures at two different temperatures between the critical temperature of components 1 and 2.

Figure 3.5*b* shows cross sections at constant pressure. Such plots, which are referred to as boiling point diagrams, are useful in analyzing the distillation of binary mixtures, since distillations are carried out at constant pressure.

Figure 3.5*c* shows cross sections at two different compositions and the vapor pressure curves for pure components 1 and 2. Points c and c designate the critical points of these two mixtures. Let us consider passage through one of these tongues at constant pressure. At low temperatures a liquid phase exists. As the temperature is raised, a vapor phase appears when line X is crossed, grows in relative amount, and disappears when line Y is crossed. But more remarkable behavior is observed if the pressure is raised at a constant temperature slightly higher than the critical temperature c for that composition. As the pressure is raised, condensation starts when the Y curve is reached. Condensation continues as the pressure is raised a little more, but then the liquid begins to *evaporate as the pressure is raised further and liquid disappears as the pressure is raised to a point above the curve.* (Note that we are referring to the Y curve to the right of c.) This unusual process is referred to as *retrograde condensation.*

3.9 RAOULT'S LAW

We now consider cross sections of Fig. 3.4 at constant temperature, but for a particular system, benzene and toluene. As shown in Fig. 3.6*a*, the bubble point line is straight. This line gives the total vapor pressure of the mixture; that is, the sum of the partial pressures of benzene and toluene.

It is evident that the vapor pressure of benzene over solutions of benzene and toluene is directly proportional to the mole fraction of benzene in the solution, and the proportionality constant is the vapor pressure of pure benzene. A similar

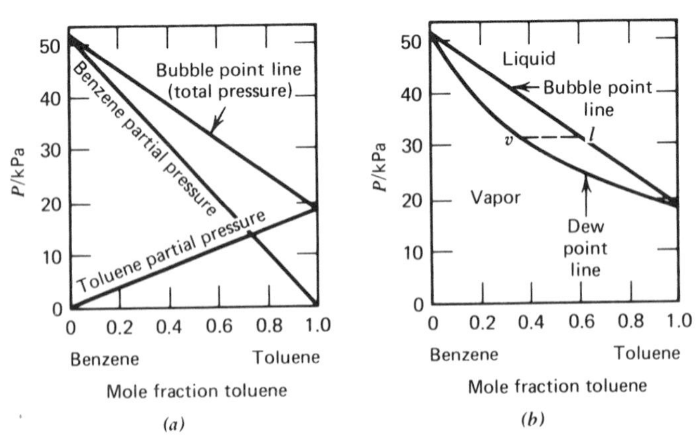

Fig. 3.6 Benzene-toluene at 60 °C. (*a*) Partial and total pressures. (*b*) Liquid and vapor compositions.

statement may be made about the vapor pressure of toluene. This generalization, discovered by Raoult in 1884, is referred to as Raoult's law. Raoult's law may be written

$$P_1 = X_1 P_1^\circ \tag{3.31}$$

$$P_2 = X_2 P_2^\circ \tag{3.32}$$

where P_1° and P_2° are the vapor pressures of pure component 1 and pure component 2, respectively, at the equilibrium temperature. Actually, air is present, and the total pressure is 1 atm. Raoult's law is obeyed by pairs of quite similar liquids A and B where the A–A, A–B, and B–B interactions are all about the same.

The mole fraction of a component in the vapor is equal to its pressure fraction in the vapor. Since for solutions obeying Raoult's law the partial pressures of the components in the vapor may be readily calculated using equations 3.31 and 3.32, the mole fraction of a component in the vapor may be calculated using

$$X_{1,\text{vap}} = \frac{P_1}{P_1 + P_2} = \frac{X_1 P_1^\circ}{X_1 P_1^\circ + X_2 P_2^\circ} = \frac{X_1 P_1^\circ}{X_1(P_1^\circ - P_2^\circ) + P_2^\circ} \tag{3.33}$$

Example 3.3 The vapor pressures of pure benzene and toluene at 60 °C are 51.3 and 18.5 kPa, respectively. Calculate the partial pressures of benzene and toluene, the total vapor pressure of the solution, and the mole fraction of toluene in the vapor above a solution with 0.60 mole fraction toluene.

$$P_{\text{benzene}} = (0.40)(51.3 \text{ kPa}) = 20.5 \text{ kPa}$$
$$P_{\text{toluene}} = (0.60)(18.5 \text{ kPa}) = 11.1 \text{ kPa}$$
$$P_{\text{total}} = 31.6 \text{ kPa}$$
$$X_{\text{toluene,vap}} = \frac{11.1}{31.6} = 0.351$$

The total vapor pressure of liquid of this composition is shown as point l in Fig. 3.6b, and the composition of equilibrium vapor is shown as point v. The dashed line connecting these points is referred to as a tie line because it ties together the compositions of phases that are in equilibrium with each other.

3.10 THERMODYNAMICS OF IDEAL LIQUID MIXTURES

In Section 2.21 an ideal gas mixture was defined as one for which the components have chemical potentials given by

$$\mu_i = \mu_i^* + RT \ln X_i \tag{3.34}$$

This was done because of the simplicity of this relationship and the fact that relations derived from it apply to real gases at low pressures. In this section we will use equation 3.34 to derive the corresponding equation for liquid mixtures. We will find that Raoult's law is not obeyed exactly by ideal liquid mixtures.

In order for liquid and vapor phases to be in equilibrium they must be at the same temperature and pressure, and the chemical potential of each component must be the same in the liquid and vapor phases. This last condition is represented by

$$\mu_i(\text{liq}, T, P, X_i) = \mu_i(\text{gas}, T, P, Y_i) \tag{3.35}$$

If the vapors of the components behave as perfect gases, the chemical potential of component i in the vapor phase may be written

$$\mu_i(\text{gas}, T, P, Y_i) = \mu_i^+(T) + RT \ln (P_i/P^+) \tag{3.36}$$

where P_i is the partial pressure and P^+ is an arbitrary reference pressure. A similar equation may be written for the chemical potential of the vapor of component i in contact with the pure liquid phase.

$$\mu_i^\circ(\text{gas}, T, P_i^\circ) = \mu_i^+(T) + RT \ln (P_i^\circ/P^+) \tag{3.37}$$

This is the chemical potential of the vapor of component i at a pressure equal to its vapor pressure, P_i°. The arbitrary reference chemical potential μ_i^+ may be eliminated between equations 3.36 and 3.37 to obtain

$$\mu_i(\text{gas}, T, P, Y_i) = \mu_i^\circ(\text{gas}, T, P_i^\circ) + RT \ln (P_i/P_i^\circ) \tag{3.38}$$

If component i follows Raoult's law, $P_i/P_i^\circ = X_i$, and equation 3.38 becomes

$$\mu_i(\text{gas}, T, P, Y_i) = \mu_i^\circ(\text{gas}, T, P_i^\circ) + RT \ln X_i \tag{3.39}$$

where X_i is the mole fraction of i in the liquid phase. Inserting equation 3.39 into equation 3.35, we obtain

$$\mu_i(\text{liq}, T, P, X_i) = \mu_i^\circ(\text{gas}, T, P_i^\circ) + RT \ln X_i \tag{3.40}$$

Since for the pure components $\mu_i^\circ(\text{gas}, T, P_i^\circ) = \mu_i^\circ(\text{liq}, T, P_i^\circ)$,

$$\mu_i(\text{liq}, T, P, X_i) = \mu_i^\circ(\text{liq}, T, P_i^\circ) + RT \ln X_i \tag{3.41}$$

The right-hand side of equation 3.41 refers to the partial pressure of pure liquid i, P_i°. Since the effect of pressure on the chemical potential of a liquid is small, it is a good approximation to write

$$\mu_i^\circ(\text{liq}, T, P_i^\circ) \approx \mu_i^\circ(\text{liq}, T, P) \tag{3.42}$$

where P is the total pressure. Then, if the approximation sign is taken as an equality,

$$\mu_i(\text{liq}, T, P, X_i) = \mu_i^\circ(\text{liq}, T, P) + RT \ln X_i \tag{3.43}$$

and

$$G = \sum_i n_i\mu_i = \sum_i n_i\mu_i^\circ + RT \sum_i n_i \ln X_i \tag{3.44}$$

Since the Gibbs energy of the separate pure liquid component is

$$G^\circ = \sum_i n_i\mu_i^\circ \tag{3.45}$$

the Gibbs energy of mixing is

$$\Delta G_{\text{mix}} = G - G^\circ = RT \sum_i n_i \ln X_i \tag{3.46}$$

Since

$$\Delta S = -\left(\frac{\partial \Delta G}{\partial T}\right)_{P,n_i} \tag{3.47}$$

$$\Delta S_{\text{mix}} = -R \sum_i n_i \ln X_i \tag{3.48}$$

$$\Delta H_{\text{mix}} = \Delta G_{\text{mix}} + T\Delta S_{\text{mix}} = 0 \tag{3.49}$$

$$\Delta V_{\text{mix}} = \left(\frac{\partial \Delta G}{\partial P}\right)_{T,n_i} = 0 \tag{3.50}$$

These are the same equations that were obtained for ideal gas mixtures. Liquid mixtures that follow these equations are called ideal mixtures.

3.11 BOILING POINT DIAGRAMS OF BINARY LIQUID MIXTURES

The preceding discussions have been concerned with vapor-pressure isotherms. We now consider plots of boiling temperature at a given pressure versus mole fraction.

A solution boils when the sum of the partial pressures of the components becomes equal to the applied pressure. The boiling points of benzene-toluene solutions are given by the lower line in Fig. 3.7.

The relationship between boiling point diagrams and vapor pressure diagrams is shown in Fig. 3.4, where the cross section at constant pressure is shown in gray. Figure 3.5*b* shows boiling point diagrams at three pressures, one of which is above the critical pressure of one of the components. Note that in boiling point diagrams the curve for the vapor lies above that for the liquid.

The boiling point diagram can be calculated for two liquids that form ideal solutions, provided the vapor pressures are known for the two pure liquids at temperatures between their boiling points. This is illustrated in example 3.3.

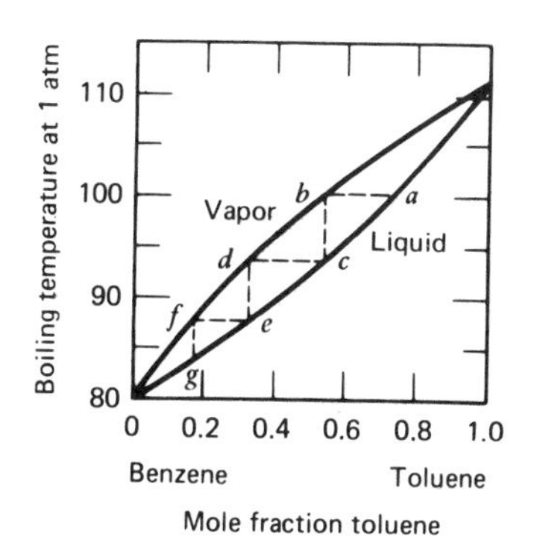

Fig. 3.7 Benzene-toluene boiling points; liquid and vapor compositions. The liquid boils at the temperature given by the lower curve.

Example 3.4 Calculate the composition of the benzene-toluene solution that will boil at 1 atm pressure at 90 °C, assuming that the solution is ideal. Also calculate the vapor composition. At 90 °C benzene has a vapor pressure of 136.3 kPa, and toluene has a vapor pressure of 54.1 kPa. The mole fraction of benzene in the liquid that will boil at 90 °C is obtained from

$$101.325 \text{ kPa} = (136.3 \text{ kPa})X_B + (54.1 \text{ kPa})(1 - X_B) \qquad X_B = 0.575$$

The mole fraction of benzene in the vapor is equal to its pressure fraction in the vapor, which is given by

$$X_{B,vap} = \frac{(136.3 \text{ kPa})X_B}{101.325 \text{ kPa}} = 0.773$$

Other points on the liquid and vapor curves in Fig. 3.7 may be calculated in the same way. For nonideal solutions the points have to be obtained experimentally.

3.12 FRACTIONAL DISTILLATION

When a binary solution is partially vaporized, the component that has the higher vapor pressure is concentrated in the vapor phase, thus producing a difference in composition between the liquid and the equilibrium vapor. This vapor may be condensed, and the vapor obtained by partially vaporizing this condensate is still further enriched in the more volatile component. In *fractional distillation* this process of successive vaporization and condensation is carried out in a fractionating column. Figure 3.7 shows that a solution of 0.75 mole fraction toluene and 0.25 mole fraction benzene boils at 100 °C under 1 atm pressure, as indicated by point *a*. The equilibrium vapor is richer in the more volatile compound, benzene, and has the composition *b*. This vapor may be condensed by lowering the temperature along the line *bc*. If a small fraction of this condensed liquid is vaporized, the first vapor formed will have the composition corresponding to *d*. This process of vaporization and condensation may be repeated many times, with the result that a vapor fraction rich in benzene is obtained.

Each vaporization and condensation represented by the line *abcde* corresponds to an idealized process in that only a small fraction of the vapor is condensed and only a small fraction of the condensate is revaporized. It is more practical to effect the separation by means of a distillation column, such as the bubble-cap column illustrated in Fig. 3.8.

Each layer of liquid on the plates of the column is equivalent to the boiling liquid in a distilling flask, and the liquid on the plate above it is equivalent to the condenser. The vapor passes upward through the bubble caps, where it is partially condensed in the liquid and mixed with it. Part of the resulting solution is vaporized in this process and is condensed in the next higher layer, while part of the liquid overflows and runs down the tube to the next lower plate. In this way there is a continuous flow of redistilled vapor coming out the top and a continuous flow of recondensed liquid returning to the boiler at the bottom. To make up for this loss of material from the distilling column, fresh solution is fed into the column, usually at the middle. The column is either well insulated or surrounded by a controlled

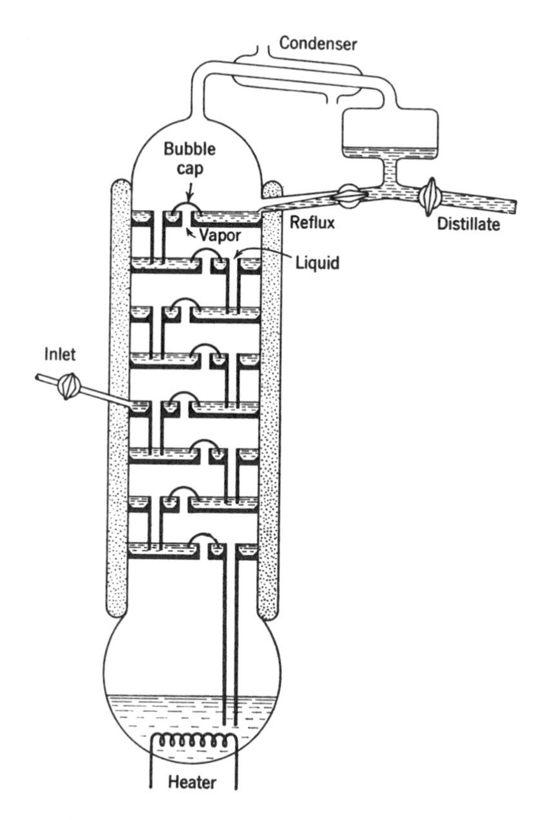

Fig. 3.8 Bubble-cap fractionating column.

heating jacket so that there will not be too much condensation on the walls. The whole system reaches a steady state in which the composition of the solution on each plate remains unchanged as long as the composition of the liquid in the distilling pot remains unchanged.

A distillation column may alternatively be packed with material that provides efficient contact between liquid and vapor and occupies only a small volume, so that there is free space to permit a large throughput of vapor. Helices of glass, spirals of screen, and different types of packing* are used with varying degrees of efficiency.

The efficiency of a column is expressed in terms of the equivalent number of theoretical plates. The number of *theoretical plates* in a column is equal to the number of successive infinitesimal vaporizations at equilibrium required to give the separation that is actually achieved. The number of theoretical plates depends somewhat on the reflux ratio, the ratio of the rate of return of liquid to the top of

---

* F. Daniels, J. W. Williams, P. Bender, R. A. Alberty, C. D. Cornwell, and J. E. Harriman, *Experimental Physical Chemistry*, McGraw-Hill Book Co., New York, 1970; T. P. Carney, *Laboratory Fractional Distillation*, The Macmillan Co., New York, 1949.

the column to the rate of distilling liquid off. The number of theoretical plates in a distillation column under actual operating conditions may be obtained by counting the number of equilibrium vaporizations required to achieve the separation actually obtained with the column.

Suppose that in distilling a solution of benzene and toluene with a certain distillation column it is found that distillate of composition g is obtained when the composition of the liquid in the boiler is given by a, in Fig. 3.7. Such a distillation is equivalent to three simple vaporizations and condensations, as indicated by steps *abc*, *cde*, and *efg*. Since the distilling pot itself corresponds to one theoretical plate, the column has two theoretical plates.

3.13 OSMOTIC PRESSURE

When a solution is separated from the solvent by a semipermeable membrane that is permeable to solvent but not to solute, the solvent flows through the membrane into the solution, where the chemical potential of the solvent is lower. This process is known as osmosis. This flow of solvent through the membrane can be prevented by applying a sufficiently high pressure to the solution. The osmotic pressure Π is the pressure difference across the membrane required to prevent spontaneous flow in either direction across the membrane.

The phenomenon of osmotic pressure was described by Abbé Nollet in 1748, and Pfeffer, a botanist, made the first direct measurements in 1877. Van't Hoff analyzed Pfeffer's data on the osmotic pressure of sugar solutions and found empirically that an equation quite analogous to the perfect gas law gave approximately the behavior of dilute solutions, namely, $\Pi V = RT$, where V is the volume of solution containing a mole of solute. The origin of the pressure is quite different from that for a gas, however, and the equation of the form of the perfect gas equation is applicable only in the limit of low concentrations.

At equilibrium the chemical potential $\mu_1^\circ(P, T)$ of pure solvent at pressure P is equal to the chemical potential of the solvent in the solution at pressure $P + \Pi$.

$$\mu_1^\circ(P, T) = \mu_1(P + \Pi, T, X_1) \tag{3.51}$$

The osmotic pressure Π that is applied to the solution exactly compensates for the lowering of the chemical potential of the solvent that is caused by the solute. For an ideal solution, equation 3.51 may be written

$$\mu_1^\circ(P, T) = \mu_1^\circ(P + \Pi, T) + RT \ln X_1 \tag{3.52}$$

where $\mu_1^\circ(P + \Pi, T)$ is the chemical potential of the pure solvent at temperature T and pressure $P + \Pi$. According to equation 2.132,

$$d\mu_1 = \bar{V}_1 \, dP \qquad \text{(constant } T \text{ and composition)} \tag{3.53}$$

where $\bar{V}_1$ is the partial molar volume of the solvent. Thus the effect on the chemical potential of the solvent of raising the pressure is given by

$$\mu_1^\circ(P + \Pi, T) = \mu_1^\circ(P^\circ, T) + \int_P^{P+\Pi} V_1^\circ \, dP \tag{3.54}$$

Assuming that $\bar{V}_1$ is constant and equal to V_1°, the molar volume of the solvent, we obtain

$$\mu_1^\circ(P + \Pi, T) = \mu_1^\circ(P^\circ, T) + V_1^\circ\Pi \tag{3.55}$$

Substituting equation 3.55 into equation 3.52 yields

$$V_1^\circ\Pi = -RT \ln X_1 = -RT \ln (1 - X_2) \tag{3.56}$$

This equation is, of course, only applicable to ideal solutions, since equation 3.52 has been used in the derivation.

At sufficiently high dilution the logarithmic term may be expanded according to

$$\ln (1 + x) = x - \tfrac{1}{2}x^2 + \tfrac{1}{3}x^3 - \cdots \qquad (-1 < x < 1) \tag{3.57}$$

When only the first term in the series is retained, equation 3.56 becomes

$$V_1^\circ\Pi = RTX_2 \tag{3.58}$$

Since the solution is dilute, $X_2 = n_2/(n_1 + n_2) \cong n_2/n_1$ and $V_1^\circ = V/n_1$, where V is the volume of the solution. Thus equation 3.58 may be written

$$\Pi V = n_2RT \tag{3.59}$$

or

$$\Pi = \frac{cRT}{M} \tag{3.60}$$

where c is the concentration of solute in grams per unit volume and M is the molar mass of the solute. This is the approximate equation that van't Hoff found empirically. It is evident from the approximations introduced why this equation cannot hold for concentrated solutions.

Since the osmotic pressure depends on the number of molecules, the molar mass determined in this way is the number average molar mass defined in equation 20.41 if the solute has a distribution of molar mass.

3.14 VAPOR PRESSURE OF NONIDEAL SOLUTIONS

Large deviations from Raoult's law are often found. Figure 3.9 shows four types of isothermal diagrams that may be encountered. Figure 3.9*a* shows negative deviations from Raoult's law. The dashed lines show the values of the two partial pressures and the total pressure calculated with Raoult's law. Note that the total vapor pressure curve actually has a minimum. At the composition corresponding with the minimum, the vapor and liquid phases have the same composition. Solutions of acetone and chloroform show this kind of isothermal diagram because of the formation of a weak hydrogen bond between the oxygen of the acetone and the hydrogen of the chloroform. A hydrogen bond is a bond between two molecules, or two parts of one molecule, that results from the sharing of a proton between

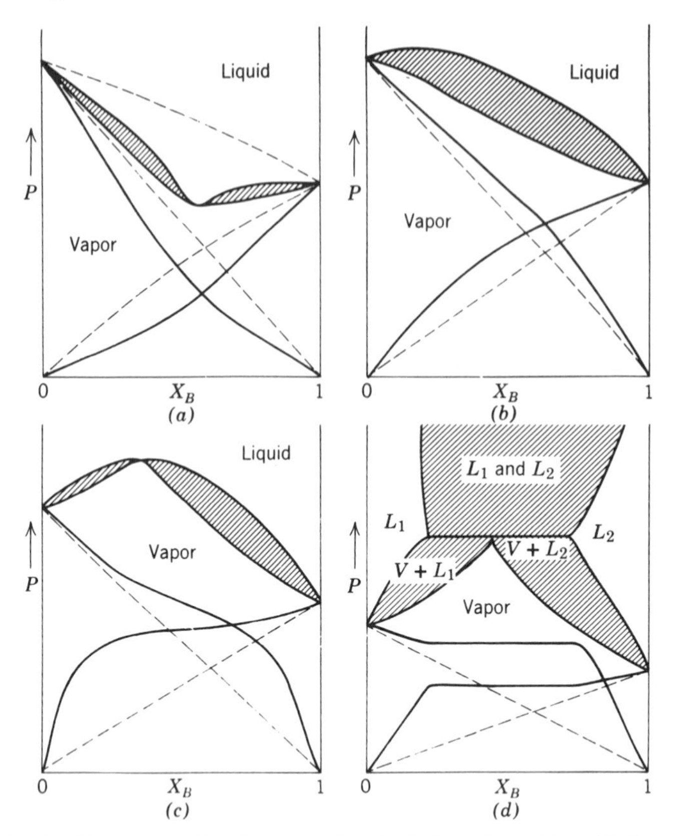

Fig. 3.9 (*a*) Liquid mixture showing negative deviations from Raoult's law. (*b*) Liquid mixture showing small positive deviations from Raoult's law. (*c*) Liquid mixture showing large positive deviations from Raoult's law. (*d*) Liquid mixture showing a range of immiscibility.

two atoms, one of which is usually fluorine, oxygen, or nitrogen (Section 10.8). Thus:

$$\begin{array}{ccccc} & Cl & & & CH_3 \\ & | & & & | \\ Cl— & C & —H\cdots\cdot & O= & C \\ & | & & & | \\ & Cl & & & CH_3 \end{array}$$

Thus the vapor pressures of both components are less than would be expected if there were no interaction and the mixture obeyed Raoult's law.

Figure 3.9*b* shows small positive deviations from Raoult's law, and Fig. 3.9*c* shows larger deviations that lead to a maximum in the total vapor pressure curve. If the positive deviations from Raoult's law are large enough, the molecules of the two types squeeze each other out, and immiscibility of the two liquids results. If the vapor pressure of a component above a rather dilute solution approaches that of the pure component, the conditions are favorable for phase separation. Phase separation occurs when the Gibbs energy of the two-phase system is lower than that of the homogeneous system. Figure 3.9*d* shows the type of phase diagram that results if the positive deviations from Raoult's law are so large that there is a range of immiscibility of the two liquids.

It is important to notice that in all of these four diagrams the vapor pressure of the component present at higher concentration approaches the values given by Raoult's law as its mole fraction approaches unity. Other types of deviations from Raoult's law are also found. A component may show positive deviations in dilute solutions and negative deviations in concentrated solutions or vice versa.*

Solutions that exhibit a maximum or a minimum in the vapor pressure curves exhibit a minimum or a maximum in the boiling point curves. When a boiling point curve has a maximum or a minimum, the solutions having the maximum or minimum boiling points are called azeotropes. These solutions distill without change in composition because the liquid and the vapor have the same composition. Many examples of azeotropic solutions are known.† At 1 atm, ethanol, which boils at 78.3 °C, and water form a minimum-boiling azeotrope that boils at 78.174 °C and contains 4.0% water by weight. Hydrochloric acid, which boils at −80 °C, and water form a maximum-boiling azeotrope at 108.584 °C that contains 20.222% HCl by weight.

3.15 HENRY'S LAW

In all the examples shown in Fig. 3.9 there is a region at low concentration where the partial pressure of the solute (the component at low concentration) is directly proportional to its mole fraction.

$$P_2 = X_2 K_2 \tag{3.61}$$

The subscript 2 indicates that the solute is being considered. This equation is referred to as Henry's law, and the constant K_2 is referred to as the Henry's law constant. In dilute solutions the environment of the minor component is constant, and its escaping tendency is proportional to its mole fraction. For nonideal solutions Henry's law holds for the solute in the same range where Raoult's law holds for the solvent. For ideal solutions $K_2 = P_2^\circ$, and Henry's law becomes identical with Raoult's law.

The value of the Henry's law constant K_2 is obtained by plotting the ratio P_2/X_2 versus X_2 and extrapolating to $X_2 = 0$. Such a plot is shown later in Fig. 3.11.

It is convenient to express the solubilities of gases in liquids by use of Henry's law constants. A few gas solubilities at 25 °C are summarized in this way in Table 3.2.

---

* M. L. McGlashan, *J. Chem. Ed.*, **40**, 516 (1963).

† L. H. Horsley and co-workers, *Azeotropic Data, Advances in Chemistry Series*, American Chemical Society, Washington, D.C., 1963.

Up to a pressure of 1 atm Henry's law holds within 1 to 3% for many slightly soluble gases.

Table 3.2 Henry's Law Constants ($K_2/10^9$ Pa) for Gases at 25 °C[a]

	Solvent	
Gas	Water	Benzene
H_2	7.12	0.367
N_2	8.68	0.239
O_2	4.40	
CO	5.79	0.163
CO_2	0.167	0.0114
CH_4	4.19	0.0569
C_2H_2	0.135	
C_2H_4	1.16	
C_2H_6	3.07	

[a] $K_2 = P_2/X_2$. The partial pressure of the gas is expressed in pascals.

Example 3.5 Using the Henry's law constant, calculate the solubility of carbon dioxide in water at 25 °C at a partial pressure of CO_2 over the solution of 1 atm. Assume that a liter of solution contains practically 1000 g of water.

$$K = \frac{P}{X} = 0.167 \times 10^9 \text{ Pa} = \frac{101{,}325 \text{ Pa}}{(CO_2)}\left((CO_2) + \frac{1000}{18.02}\right)$$

Since (CO_2) may be considered negligible in comparison with the number of moles of water, 1000/18.02,

$$(CO_2) = \frac{(101{,}325 \text{ Pa})(5.49 \text{ mol L}^{-1})}{0.167 \times 10^9 \text{ Pa}} = 3.37 \times 10^{-2} \text{ mol L}^{-1}$$

The solubility of a gas in liquids usually decreases with increasing temperature, since heat is generally evolved in the solution process. There are numerous exceptions, however, especially with the solvents liquid ammonia, molten silver, and many organic liquids. It is a common observation that a glass of cold water, when warmed to room temperature, shows the presence of many small air bubbles.

The solubility of an unreactive gas is due to intermolecular attractive forces between gas molecules and solvent molecules. There is a good correlation between the solubilities of gases in solvents at room temperature and their boiling points. Substances with low boiling points (He, H_2, N_2, Ne, etc.) have weak intermolecular attractions and are therefore not very soluble in liquids.

The solubility of gases in water is usually decreased by the addition of other solutes, particularly electrolytes. The extent of this "salting out" varies considerably with different salts, but with a given salt the relative decrease in solubility is nearly the same for different gases. The solubility of liquids and solids in water also shows this salting-out phenomenon.

3.16 ACTIVITY AND ACTIVITY COEFFICIENT

The concept of an ideal solution forms such a useful basis for comparison that it is advantageous in dealing with nonideal solutions to use equations of the same form as for ideal solutions. This is accomplished by introducing the activity. The activity a_B of a substance B in a liquid or solid mixture is defined by

$$\mu_B = \mu_B^* + RT \ln a_B \tag{3.62}$$

where μ_B^* is the chemical potential of the pure substance B at the same temperature and pressure. According to this definition, the activity is a dimensionless quantity and

$$\lim_{X_B \to 1} a_B = 1 \qquad \text{(constant } T \text{ and } P\text{)} \tag{3.63}$$

It is useful to consider that the activity of a substance is the product of the concentration of the substance on some concentration scale and an activity coefficient.* In this section we will use only the mole fraction scale and define the activity coefficient f_B of a substance B in a liquid or solid mixture by

$$\mu_B = \mu_B^* + RT \ln f_B X_B \tag{3.64}$$

where μ_B^* is the chemical potential of the pure substance B at the same pressure and temperature. Thus,

$$a_B = f_B X_B \tag{3.65}$$

The chemical potential μ_B^*, in the standard state where $a_B = 1$, is a function of temperature and pressure only, whereas f_B is a function of concentration as well as temperature and pressure. To complete the definition of the activity coefficient f_B, it is necessary to specify the conditions under which f_B becomes equal to unity. There are two ways of doing this because Raoult's law is approached as $X_B \to 1$ and Henry's law is approached as $X_B \to 0$.

---

* Activity coefficients may also be defined on the m scale (mol kg^{-1}) and c scale (mol L^{-1}). On the m scale the activity of a solute substance B is defined by

$$a_B = \frac{\gamma_B m_B}{m^\circ}$$

where m° is the standard value of the molality (1 mol/kg of solvent). According to this definition,

$$\lim_{m_B \to 0} \gamma_B = 1$$

On the c scale the activity of solute substance B is defined by

$$a_B = \frac{y_B c_B}{c^\circ}$$

where c° is the standard value of the molar concentration (1 mol L^{-1}). According to this definition,

$$\lim_{c_B \to 0} y_B = 1$$

Convention I. If the components of the solution are liquids, the activity coefficient of each component may be taken to approach unity as its mole fraction approaches unity.

$$\lim_{X_B \to 1} f_B = 1 \qquad \text{(at constant } P \text{ and } T\text{)} \tag{3.66}$$

Since the logarithmic term in equation 3.64 vanishes under these limiting conditions, μ_B^* is equal to the Gibbs energy of a mole of pure B at the temperature and pressure under consideration. If both components follow equation 3.34 over the whole range of concentration (as they do in an ideal solution), their activity coefficients will be equal to unity over the whole range of concentration if Convention I is used.

Convention II. It is convenient to use this convention if it is not possible to vary the mole fractions of both components up to unity. For example, one component may be a gas or a solid. For such solutions a different convention is applied for solvent and solute. The activity coefficient of the solvent is given by Convention I.

$$\lim_{X_{\text{solvent}} \to 1} f_{\text{solvent}} = 1 \tag{3.67}$$

The activity coefficient for the solute is taken to approach unity as its mole fraction approaches zero.

$$\lim_{X_{\text{solute}} \to 0} f_{\text{solute}} = 1 \tag{3.68}$$

If the activity coefficient of the solute is to approach unity at infinite dilution, μ_i^* for the solute in equation 3.34 must be the chemical potential of solute in a hypothetical standard state in which the solute at unit concentration has the properties that it would have at infinite dilution.

3.17 CALCULATION OF ACTIVITY COEFFICIENTS FOR BINARY LIQUID MIXTURES

When solution and vapor phases are at equilibrium, the chemical potentials for each component are equal in the two phases.

$$\mu_{i,\text{soln}} = \mu_{i,\text{vap}} \tag{3.69}$$

Assuming the vapor phase is ideal, so that equation 2.133 can be used, and using equation 3.64,

$$\mu_i^* + RT \ln f_i X_i = \mu_i^\circ + RT \ln P_i \tag{3.70}$$

By rearranging, we obtain

$$P_i = f_i X_i e^{(\mu_i^* - \mu_i^\circ)/RT} \tag{3.71}$$

Comparing equations 3.71 and 3.61, we see that the Henry law constant K_i is given by

$$K_i = f_i e^{(\mu_i^* - \mu_i^\circ)/RT} \tag{3.72}$$

or, since $\mu_i^* = \mu_i^\circ + RT \ln P_i^\circ$, where P_i° is the vapor pressure of pure component i

$$K_i = f_i P_i^\circ \tag{3.73}$$

and

$$P_i = f_i X_i P_i^\circ \tag{3.74}$$

which is Raoult's law if $f_i = 1$. It is evident from equation 3.74 that for solutions that show positive deviations from Raoult's law, f_i is greater than unity, and for solutions that show negative deviations from Raoult's law, f_i is less than unity.

The activity coefficients of ether and acetone in ether-acetone solutions may be calculated from the data of Fig. 3.10. If these substances formed ideal solutions, the partial pressure of acetone above a solution containing 0.5 mole fraction acetone would be given by point B, which is $0.5P_2^\circ$, acetone being referred to as component 2. The actual partial pressure is represented by point C. Solving equation 3.74 for the activity coefficient of acetone, we have

$$f_2 = \frac{P_2}{X_2 P_2^\circ} = \frac{22.4 \text{ kPa}}{18.9 \text{ kPa}} = 1.19 \tag{3.75}$$

Similarly, according to Convention I, the activity coefficient of ether (component 1) at 0.5 mole fraction is given by

$$f_1 = \frac{P_1}{X_1 P_1^\circ} = \frac{52.1 \text{ kPa}}{43.1 \text{ kPa}} = 1.21 \tag{3.76}$$

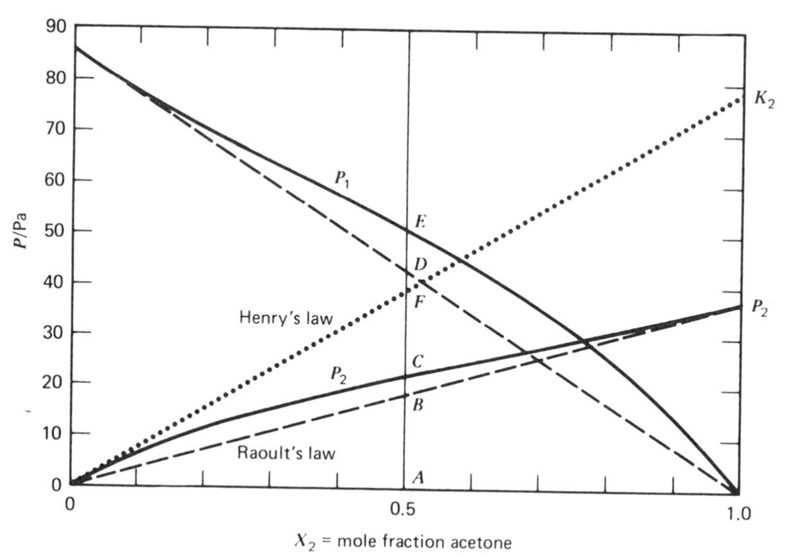

Fig. 3.10 Partial pressures of ether-acetone solutions at 30 °C.

The activity coefficients of both components, calculated in this way at other concentrations by use of Convention I, are summarized in Table 3.3. It will be noted that, as the mole fraction of either component approaches unity, its activity coefficient approaches unity, since the vapor pressure asymptotically approaches that given by Raoult's law.

Table 3.3 Activity Coefficients for Acetone-Ether Solutions at 30 °C

Mole Fraction Acetone	Convention I						Convention II[a]	
	Ether			Acetone			Acetone	
X_2	P_1/kPa	$X_1P_2^\circ$/kPa	f_1	P_2/kPa	$X_2P_2^\circ$/kPa	f_2	K_2X_2/kPa	f_2
0	86.1	86.1	1.00	0	0	...	0	(1.000)
0.2	71.3	68.9	1.04	12.0	7.5	1.60	15.7	0.77
0.4	58.7	51.7	1.14	19.7	15.1	1.31	31.4	0.63
0.5	52.1	43.1	1.21	22.4	18.9	1.19	39.2	0.57
0.6	44.3	34.4	1.28	25.3	22.7	1.12	47.1	0.54
0.8	26.9	17.3	1.56	31.3	30.1	1.04	62.7	0.50
1.0	0	0	...	37.7	37.7	1.00	78.4	(0.48)

[a] The activity coefficients for ether are the same as those calculated by Convention I.

The calculation of the activity coefficients of acetone on the basis of Convention II is accomplished by use of the line passing through F in Fig. 3.10. This line is tangent to the vapor-pressure curve for acetone in the limit as the mole fraction of acetone approaches zero, and its slope is equal to the value of the Henry's law constant for acetone in ether, obtained by extrapolating the apparent Henry's law constant defined by

$$K_2' = \frac{P_2}{X_2} \tag{3.77}$$

to infinite dilution of the acetone. The extrapolation of this ratio for the data of Fig. 3.10 is illustrated in Fig. 3.11, where values of P_2/X_2 are plotted versus X_2. It is found that the Henry's law constant at infinite dilution (K_2) has a value of 78.3 kPa at this temperature of 30 °C.

If acetone obeyed Henry's law with this value of the constant over the entire concentration range, its vapor pressure at 0.5 mole fraction would be given by point F in Fig. 3.10. The actual partial pressure is given by point C. The deviation from Henry's law may be expressed by the activity coefficient f_2 in $P_2 = f_2K_2X_2$. The activity coefficient of acetone at mole fraction 0.5 is given by

$$f_2 = \frac{P_2}{K_2X_2} = \frac{22.4 \text{ kPa}}{39.2 \text{ kPa}} = 0.57 \tag{3.78}$$

Thus, when Convention II is used, the extent to which the activity coefficient differs from unity is a measure of the deviation from Henry's law. The activity coefficients of acetone calculated in this way are also summarized in Table 3.3. The activity

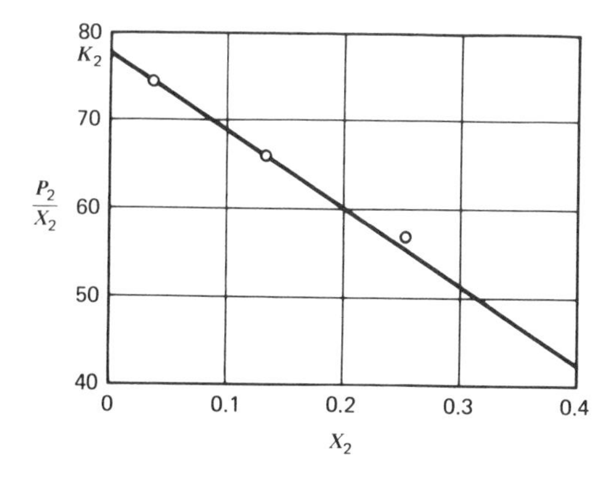

Fig. 3.11 Evaluation of the Henry's law constant, K_2, for acetone in ether-acetone solutions at 30 °C.

coefficients for the "solvent" ether remain the same as calculated before with Convention I.

Although the numerical values of the activity coefficients of acetone depend on which convention is employed, the same result is obtained in any thermodynamic calculation using these activity coefficients, independent of whether Convention I or II is chosen. These thermodynamic calculations involve two different concentrations, and the standard reference state cancels out. The magnitude of the activity coefficient also depends on the concentration scale used; for example, the molal scale would require a different standard state from that needed with the mole-fraction scale.

3.18 TWO-COMPONENT SYSTEMS CONSISTING OF SOLID AND LIQUID PHASES

The simplest type of binary system consisting of only solid and liquid phases is encountered where the components are completely miscible in the liquid state and completely immiscible in the solid state, so that only the pure solid phases separate out on cooling solutions. Such a phase diagram is illustrated in Fig. 3.12. The diagram is for a constant pressure sufficiently high that no vapor phase is present in this temperature range. Under these conditions the phase rule becomes $v = c - p + 1$, so that $v = 2$ when one phase is present, $v = 1$ when two phases are present, and $v = 0$ when three phases are present. Figure 3.12 shows how such a diagram may be determined by studying the rate of cooling of solutions of various compositions.

When a liquid consisting of one component is cooled, the plot of temperature versus time has a nearly constant slope. At the temperature at which the solid crystallizes out, however, the cooling curve becomes horizontal if the cooling is

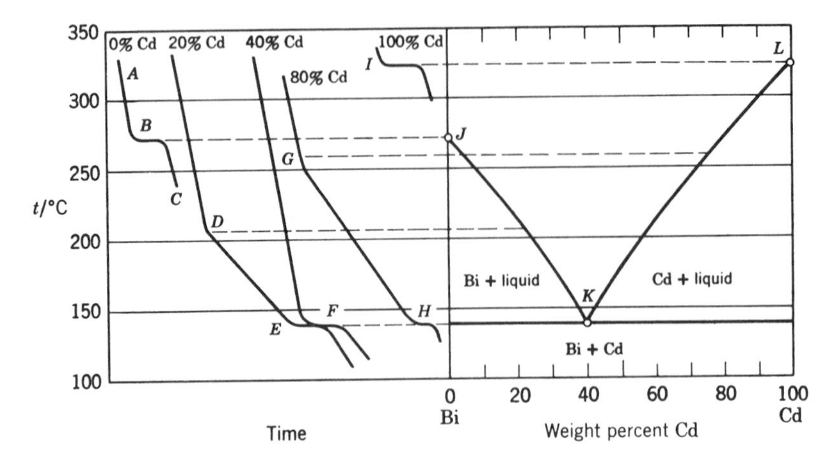

Fig. 3.12 Cooling curves and the temperature-concentration phase diagram for the system bismuth-cadmium.

slow enough. The halt in the cooling curve results from the heat evolved when the liquid solidifies. This is shown by the cooling curves for bismuth (labeled 0% Cd) and cadmium in Fig. 3.12 at 273 °C and 323 °C, respectively.

When a *solution* is cooled, there is a change in slope of the cooling curve at the temperature at which one of the components begins to crystallize out. The change in slope is due to the evolution of heat by the progressive crystallization of the solid as the solution is cooled and to the change in heat capacity. Such changes in slope are evident in the cooling curves for 20% cadmium and 80% cadmium. These curves also show horizontal sections, both at 140 °C. At this temperature both solid cadmium and solid bismuth come out together. The temperature at which this occurs is called a *eutectic temperature.* A solution of cadmium and bismuth containing 40% cadmium shows a single plateau F at 140 °C, and so this is the eutectic composition.

The temperatures at which new phases appear, as indicated by the cooling curves, are then transferred to the temperature-composition diagram, as shown at the right in Fig. 3.12. In the area above JKL there is one liquid phase, and $v = c - p + 1 = 2 - 1 + 1 = 2$. Along JK, bismuth freezes out, and along LK, cadmium freezes out. Thus along line JK and in the area under it and down to the eutectic temperature K there are two phases, solid bismuth and a solution having a composition that is determined by the temperature. Since $v = 2 - 2 + 1 = 1$, the system is univariant. When either the temperature of the composition of the liquid phase is specified, the other may be found from the diagram on the line JK. Also, along the line KL and in the area under it there are two phases, solid cadmium and solution, and accordingly $v = 1$.

At the eutectic point K there are three phases—solid bismuth, solid cadmium, and liquid solution containing 40% cadmium. Then $v = 2 - 3 + 1 = 0$, and so this is an invariant point. There is only one temperature and one composition of solution at which these three phases can exist together at equilibrium at a given constant pressure.

The area below the eutectic temperature K is a two-phase area in which solid bismuth and solid cadmium are present, and $v = 2 - 2 + 1 = 1$. Only the temperature need be specified to describe the system completely at a given constant pressure. The ratio of bismuth to cadmium may change, but there is only a mixture of pure solid bismuth and pure solid cadmium, and there is no need to specify any concentration. The eutectic is not a phase; it is a mixture of two solid phases and has a fine grain structure.

3.19 FREEZING POINT LOWERING AND IDEAL SOLUBILITY

It is evident from Fig. 3.12 that the addition of cadmium lowers the freezing point of bismuth along line JK, and that the addition of bismuth lowers the freezing point of cadmium along line LK. Alternatively we may consider that JK is the solubility curve for bismuth in cadmium, and LK is the solubility curve for cadmium in bismuth. If the solutions are ideal and if the phases that separate are pure solids, the equations for these lines may be readily derived. When there is equilibrium between solid and solution phases, the chemical potentials of component 1 (the one forming the solid) must be the same in both phases.

$$\mu_1(s) = \mu_1(\text{soln}) \tag{3.79}$$

$$= \mu_1^* + RT \ln X_1 \tag{3.80}$$

where X_1 is the mole fraction of 1 in solution. Rearranging we obtain

$$\frac{\mu_1(s)}{T} - \frac{\mu_1^*}{T} = R \ln X_1 \tag{3.81}$$

Then by differentiating equation 3.81 with respect to absolute temperature at constant pressure, we have

$$\left[\frac{\partial(\mu_1(s)/T)}{\partial T}\right]_P - \left[\frac{\partial(\mu_1^*/T)}{\partial T}\right]_P = R\frac{\partial \ln X_1}{\partial T} \tag{3.82}$$

Using equation 2.88, we see that

$$\frac{-H_s}{T^2} + \frac{H_1^*}{T^2} = R\frac{\partial \ln X_1}{\partial T} = \frac{\Delta H_{\text{fus},1}}{T^2} \tag{3.83}$$

since $H_1^* - H_s = \Delta H_{\text{fus},1}$. Integrating from mole fraction X_1 at temperature T to $X_1 = 1$ at the freezing point of the pure component $T_{0,1}$, assuming that $\Delta H_{\text{fus},1}$ is independent of T, we then have

$$\int_{X_1}^{X_1=1} d \ln X_1 = \int_{T}^{T_{1,0}} \frac{\Delta H_{\text{fus},1}}{RT^2}\, dT \tag{3.84}$$

$$-\ln X_1 = \frac{\Delta H_{\text{fus},1}(T_{0,1} - T)}{RTT_{0,1}} \tag{3.85}$$

or

$$T = \frac{T_{0,1}}{1 - \frac{RT_{0,1}}{\Delta H_{\text{fus},1}} \ln X_1} \tag{3.86}$$

This equation gives the temperature T at which pure solid 1 is in equilibrium with liquid solution of mole fraction X_1. The freezing point of the component that freezes out is $T_{0,1}$, and its heat of fusion is $\Delta H_{\text{fus},1}$. Since equation 3.86 contains no parameters for the solvent, we can see that the solubility in mole fraction units is the same in all solvents that form ideal solutions.

Exercise I Assuming that bismuth and cadmium form ideal solutions, plot lines JK and LK, using equation 3.86. The heat of fusion of cadmium is 6.07 kJ mol^{-1}, and the heat of fusion of bismuth is 10.5 kJ mol^{-1}.

Equation 3.85 may be written as follows for small freezing point depressions.

$$-\ln X_1 = \frac{\Delta H_{\text{fus},1}\, \Delta T_f}{RT_{0,1}^2} \tag{3.87}$$

This equation has been derived for ideal solutions but is applicable to nonideal solutions, provided that the mole fraction of the solvent is very close to unity. For dilute solutions, $-\ln X_1$ can be represented by the first several terms of a power series in X_2, the mole fraction of the solute, as shown in equation 3.57. For sufficiently low concentrations of solute, the second and higher terms of this series are negligible, and so equation 3.87 may be written

$$\Delta T_f = \frac{RT_{0,1}^2}{\Delta H_{\text{fus},1}} X_2 \tag{3.88}$$

In a discussion of the depression of the freezing point, the concentration of the solute is generally given in terms of molal concentration m (i.e., moles of solute per 1000 g of solvent) rather than of mole fraction. The relation between these concentrations is

$$X_2 = \frac{n_2}{n_1 + n_2} = \frac{m}{1000/M_1 + m} \xrightarrow{m \to 0} \frac{m}{1000/M_1} \tag{3.89}$$

where M_1 is the molar mass of the solvent. The last form is applicable to dilute solutions for which the number of moles of solute is negligible in comparison with the number of moles of solvent.

Substituting the last form of equation 3.89 into equation 3.88 and rearranging, we obtain

$$\Delta T_f = \frac{RT_{0,1}^2 M_1 m}{1000\, \Delta H_{\text{fus},1}} = K_f m \tag{3.90}$$

and

$$K_f = \frac{RT_{0,1}^2 M_1}{1000\,\Delta H_{\text{fus},1}} \tag{3.91}$$

$$\Delta T_f = K_f \frac{w_2}{M_2} \frac{1000}{w_1} \tag{3.92}$$

Example 3.6 Calculate the freezing point constant K_f for water. The enthalpy of fusion is 333.5 J g^{-1} at 273.1 K.

$$K_f = \frac{RT_{0,1}^2 M_1}{1000\,\Delta H_{\text{fus},1}} = \frac{(8.314\text{ J K}^{-1}\text{ mol}^{-1})(273.1\text{ K})^2(18.02\text{ g mol}^{-1})}{(1000\text{ g kg}^{-1})(1802\text{ g mol}^{-1})(333.5\text{ J g}^{-1})}$$

$$= 1.86\text{ K (mol kg}^{-1})^{-1}$$

According to this value of K_f, 1 mol of solute added to 1000 g of water will lower the freezing point 1.86 K, but the relation holds only for dilute solutions. Even a 1-molal solution is too concentrated, and the depression will be something less than 1.86 K.

Studies of freezing point depression are useful for determining the molar mass of solutes, but care must be taken not to supercool the solution. The foregoing relations apply only to ideal solutions. Information about activity coefficients may be obtained by studying the freezing points of more concentrated solutions.

3.20 CONGRUENTLY MELTING COMPOUND

The components of a binary system may react to form a solid compound that exists in equilibrium with liquid over a range of composition. If the formation of a compound leads to a maximum in the temperature-composition diagram, as illustrated by Fig. 3.13 for the zinc-magnesium system, we say there is a congruently melting compound. The composition that corresponds to the maximum temperature is the composition of the compound. On the mole percent scale such maxima may be achieved at 50%, 33%, 25%, etc., corresponding to integer ratios of the components of 1:1, 1:2, 1:3, etc. Figure 3.13 looks very much like two-phase diagrams of the type we have discussed placed side by side, but there is a difference. The liquidus curve has a horizontal tangent (zero slope) at the melting point of the congruently melting compound $MgZn_2$, while the slope is not zero at the melting points of the pure components.* This means that if a congruently melting compound *AB* exists in an *A–B* system, additions of very small amounts of *A* and *B* to the compound will not lower the melting or freezing point.

Example 3.7 Six-tenths mole of Mg and 0.40 mol of Zn are heated to 650 °C, represented by point *J* in Fig. 3.13. Describe what happens when this solution is cooled

* A. F. Berndt and D. J. Diestler, *J. Phys. Chem.*, **72**, 2263 (1968).

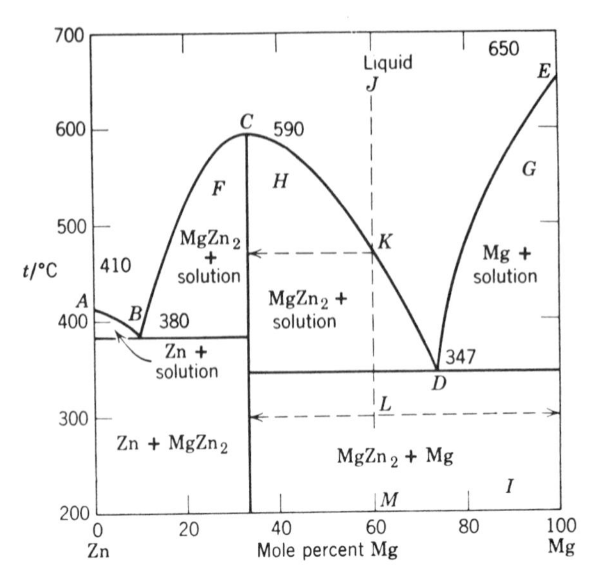

Fig. 3.13 Temperature-composition diagram, showing a maximum for the system zinc-magnesium.

down to 200 °C, as indicated by the vertical line. (The experiment would have to be done in an inert atmosphere to prevent oxidation by air.) At 470 °C point *K* is reached and solid $MgZn_2$ separates from solution. The freezing point is gradually lowered as the solution becomes richer in Mg. Finally, at 347 °C, when the liquid is 74 mole % in Mg and 26 mole % in Zn, the whole solution freezes, and solid $MgZn_2$ and solid Mg come out together.

From this temperature down there is no further change in the phases. At all temperatures below 347 °C there are pure solids Mg and $MgZn_2$.

3.21 INCONGRUENTLY MELTING COMPOUND

Instead of melting, a compound may decompose into another compound and a solution at a definite temperature. This melting point, called an incongruent melting point, is illustrated in Fig. 3.14, which shows part of the phase diagram for the sodium sulfate-water system.

When pure $Na_2SO_4 \cdot 10H_2O$ is heated, it undergoes a transition at 32.38 °C to give anhydrous Na_2SO_4 and solution of composition *C*. The line *BC* gives the solubility of $Na_2SO_4 \cdot 10H_2O$ in water, and the line *CD* gives the solubility of Na_2SO_4 in water. Figure 3.14 explains the discontinuity in the solubility curve *BCD* for sodium sulfate in water.

When the three phases Na_2SO_4, $Na_2SO_4 \cdot 10H_2O$, and saturated solution are in equilibrium with each other at constant pressure (point *C*), the system is invariant.

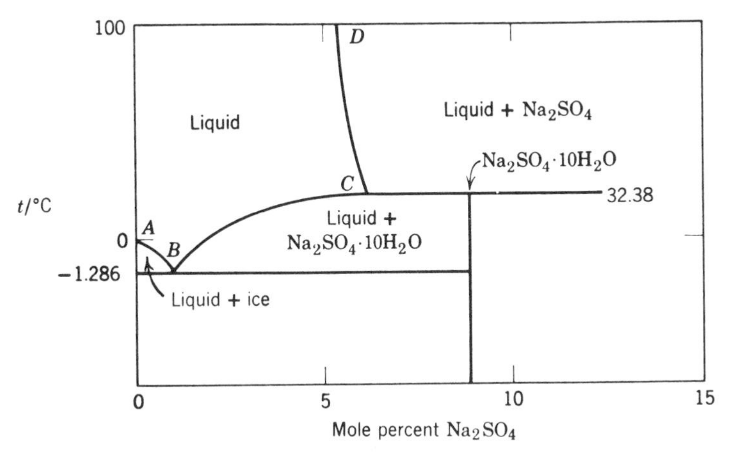

Fig. 3.14 Part of the phase diagram for Na_2SO_4—H_2O showing incongruent melting of $Na_2SO_4 \cdot 10H_2O$ to rhombic anhydrous Na_2SO_4.

3.22 SOLID SOLUTIONS

Often pure solid freezes out of a solution, but for some systems a solid solution freezes out. A continuous series of solid solutions may be formed, as illustrated in Fig. 3.15 for platinum and gold. The two lines in this diagram give the compositions of the liquid solutions (upper line) and solid solutions (lower line) that are in equilibrium with each other. When these diagrams are studied, it is convenient to remember that the liquid phase is richer in that component of mixture that has the lower melting point.

Above the upper line of Fig. 3.15 the two metals exist in liquid solutions; below the lower line the two metals exist in solid solutions. The upper curve is the freezing point curve for the liquid, and the lower one is the melting point curve for the solid. The space between the two curves represents mixtures of the two—one liquid

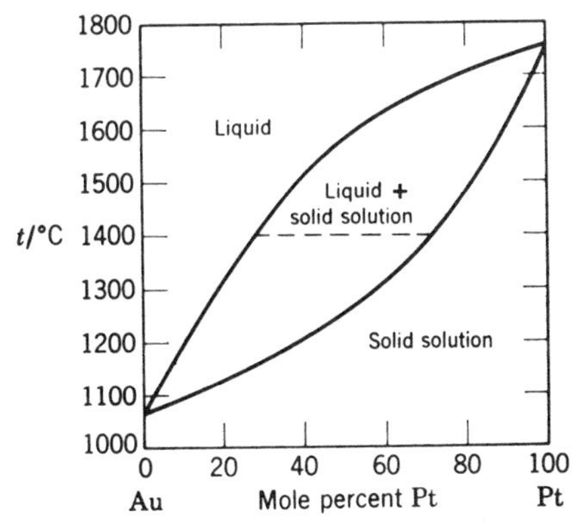

Fig. 3.15 Phase diagram for gold-platinum showing solid solutions.

solution and one solid solution in equilibrium. For example, a mixture containing 50 mole % gold and 50 mole % platinum, when brought to equilibrium at 1400 °C, will consist of two phases, a solid solution containing 70 mole % platinum and a liquid solution containing 28 mole % platinum. If the original mixture contained 60 mole % platinum, there would still be the same two liquid and solid solutions at 1400 °C of the same compositions, 70 and 28 mole %, but there would be a relatively greater amount of the solid solution that contains 70 mole % platinum.

The fractional crystallization of solid solutions is seriously complicated by the fact that the attainment of equilibrium is much slower in solid solutions than in liquid solutions. It takes a considerable length of time, particularly at low temperatures, for a change in concentration at the surface to affect the concentration at a point in the interior of the solid solution.

In view of the use of the freezing point as a criterion of purity, it is important to note that when solid solutions are formed the freezing point may be *raised* by the presence of the other component.

Figure 3.15 is analogous to the phase diagram for two miscible liquids and vapor, as shown in Fig. 3.7. Systems exhibiting solid solution behavior may show maxima or minima in their melting curves that have nothing to do with the formation of compounds.

Many properties of alloys, ceramics, and structural materials depend on the presence of solid solutions. The hardening and tempering of steel involve the existence of solid solutions of carbon in different iron-carbon compounds. The solid solution stable at the high temperatures is hard; to retain this hardness, the proper compositions and temperatures are obtained, as indicated by the phase diagrams, and the steel is quenched quickly in oil or water, so that it does not have time to form the solid solution, which is stable at lower temperatures. Reheating the steel to a somewhat lower temperature gives an opportunity for partial conversion to the softer solid solution, which is stable at the lower temperature. In this way the steel may be given different degrees of hardening.

Sometimes partial miscibility is encountered in the solid state just as it is in the liquid state. The silver-copper system is an example of partial miscibility of solids. As shown by Fig. 3.16, at 800 °C copper dissolves in solid silver to the extent of 6%

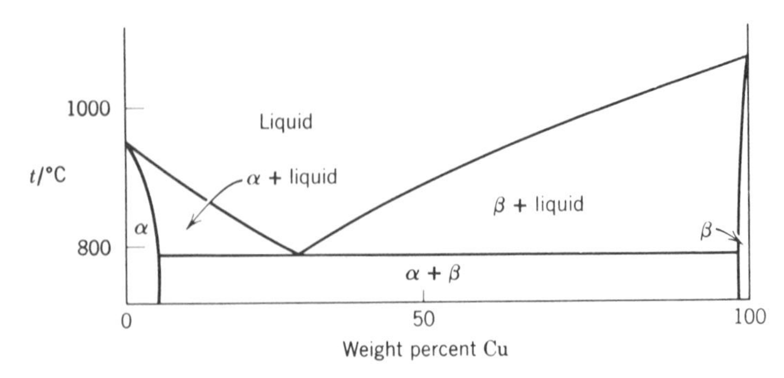

Fig. 3.16 Phase diagram for silver-copper showing partial miscibility of solid solutions.

by weight, and silver dissolves in copper to the extent of 2% by weight. At the eutectic point, pure copper and silver do not crystallize out, but saturated solid solutions do. The regions α and β represent continuously variable solid solutions, and at constant pressure the variance v is 2 in these regions because there is a single phase.

3.23 SYSTEMS OF THREE COMPONENTS

For systems of three components the phase rule yields $v = 5 - p$. If there is a single phase, then $v = 4$, and so a complete geometrical representation would require the use of four-dimensional space. If the pressure is constant, a three-dimensional representation may be used. If both the temperature and pressure are constant, then $v = 3 - p$, and so the system may be represented in two dimensions and

$p = 1$	$v = 2$	bivariant
$p = 2$	$v = 1$	univariant
$p = 3$	$v = 0$	invariant

A system of three components has two independent composition variables, say X_2 and X_3. Thus the composition of a three-component system can be plotted in Cartesian coordinates with X_2 on one axis and X_3 on the other, bounded by the line $X_2 + X_3 = 1$. Since this plot is not symmetrical with respect to the three components, it is more common to plot compositions on an equilateral triangle in which each apex represents a pure component.

In an equilateral triangle, the sum of the distances from any given point to the three sides, along the perpendiculars to the sides, is equal to the height of the triangle. The distance from each apex to the center of the opposite side of the equilateral triangle is divided into 100 parts, corresponding to percentage composition, and the composition corresponding to a given point is readily obtained by measuring the perpendicular distance to the three sides. For example, in Fig. 3.17 point O represents a mixture with a gross composition of 50% by weight acetic acid, 10% by weight vinyl acetate, and 40% by weight water.

Of the many possible kinds of ternary systems, we will consider only certain types formed by three liquids and by a liquid and two solids.

If two pairs of the liquids are completely miscible and one pair is partially miscible, a diagram of the type illustrated in Fig. 3.17 is obtained. This figure represents the system water–acetic acid–vinyl acetate at 25 °C and atmospheric pressure.*

When water is added to vinyl acetate along the line BC, the water dissolves at first, forming a homogeneous solution. However, as more water is added, saturation is reached at composition x, and there are two liquid phases, vinyl acetate saturated

* J. C. Smith, *J. Phys. Chem.*, **45**, 1301 (1941).

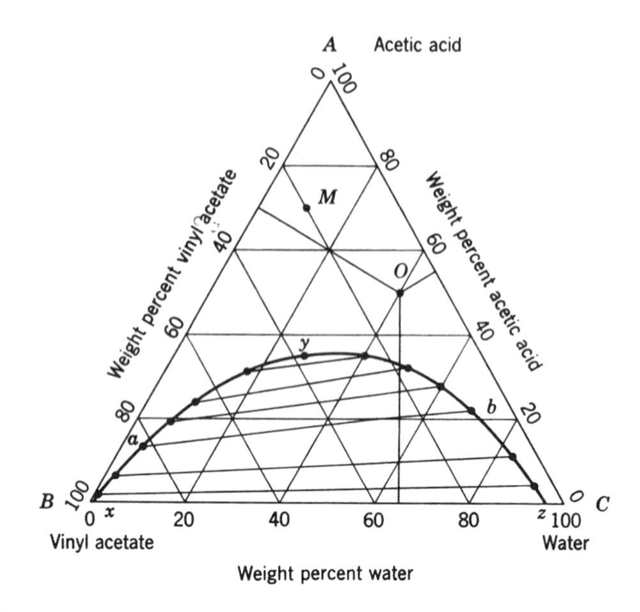

Fig. 3.17 Three-component phase diagram at 25 °C showing regions of miscibility and immiscibility.

with water and a little water saturated with vinyl acetate, having the composition z. As more water is added, the amount of the z phase increases and that of the x phase decreases, but the composition of each phase remains always the same. Finally, when the percent of water exceeds that given by z, there is only one liquid phase, an unsaturated solution of vinyl acetate in water. At all compositions between x and z there are two liquid phases with compositions x and z.

If acetic acid, which is miscible with vinyl acetate and with water in all proportions, is added, it is distributed between the two layers, forming two ternary solutions of vinyl acetate, water, and acetic acid that are in equilibrium with each other, provided that the gross composition of the mixture falls in the region below the xyz curve. For example, if the gross composition lies on the line ab, the two phases that are in equilibrium are represented by points a and b.

Other tie lines are shown for other gross compositions; usually tie lines are not parallel to each other or to a side of the triangle. The compositions of the two phases that are in equilibrium with each other, corresponding to the intersection of the tie line with the curves xy and zy, have to be determined experimentally. As more acid is added, the two phases become more alike, and the tie lines become shorter. Ultimately, when the compositions of the two solutions become identical, the tie line shrinks to the single point y. Point y is a *critical point*, since further addition of acetic acid will result in the formation of a single homogeneous phase. Any point under the curve represents a ternary mixture that will separate into two liquid phases; any point above the curve represents a single homogeneous liquid phase.

If there are two liquid phases, as in the area below the line *xyz*, the variance is 1 at fixed T and P, so that it is necessary to specify the percentage of one component in only one phase to describe the system completely. The percentage of the other components in this phase can be obtained from the intersection of this percentage with the line *xyz*, and the composition of the other phase can be obtained from the intersection of the other end of the tie line with the line *xyz*. For example, if one phase in the two-phase system in Fig. 3.17 contains 5% water, the composition of this phase is given by point *a* and the composition of the other phase by point *b*.

When a substance is added to a two-phase liquid mixture, it is generally distributed with different equilibrium concentrations in the two phases. The distribution of acetic acid between the water-rich and vinyl acetate-rich phases may be calculated from the data of Fig. 3.17. It is apparent from this figure that the ratio of the concentrations of acetic acid in the two phases, as given by the ends of the tie lines, changes with the amount of acetic acid added. However, if the amount of solute added is sufficiently small, it is often found that the distribution coefficient, which is defined as the ratio of the concentrations of the solute in the two phases, is relatively independent of concentration. In some cases the distribution coefficient depends markedly on concentration because the solute exists in dissociated or associated forms in one of the phases. For example, hydrochloric acid dissolves in water to give H^+ and Cl^- ions, but in benzene it is not dissociated into ions. Other solutes, as, for example, benzoic acid, associate in a nonpolar solvent like benzene to give double molecules as determined by boiling point or freezing point measurements, but they do not associate in a polar solvent like water or ether. The association is due to the formation of hydrogen bonds.

Substances having slightly different distribution coefficients may be separated by a column operation in which one liquid phase is held by a finely divided solid with a large surface area and the other liquid phase flows through the column. This process is referred to as *partition chromatography* and is closely related to chromatography experiments, depending on adsorption (Section 7.14). In column operation the equivalent of many theoretical plates, each equivalent to a batch extraction, is obtained. The components of a mixture emerge at the bottom of the column one by one at different times if their distribution coefficients are sufficiently different.

3.24 SYSTEMS INVOLVING TWO SOLIDS AND A LIQUID

Figure 3.18 shows an example of such a system at 25 °C and atmospheric pressure in which no compounds are formed.* Solutions along *RP* are saturated with $Pb(NO_3)_2$, and solutions along *PS* are saturated with $NaNO_3$. At the intersection of these two solubility curves (point *P*), the solution is saturated with respect to

* The determination of this particular phase diagram has been described in detail by E. L. Heric, *J. Chem. Educ.*, **35**, 510 (1958).

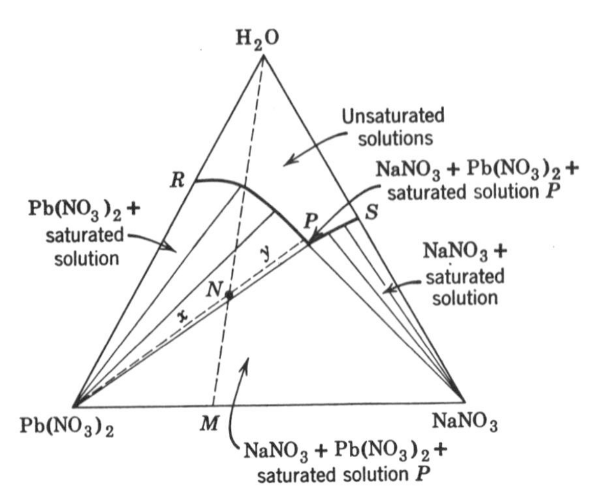

Fig. 3.18 Phase diagram for the system lead nitrate-sodium nitrate-water at 25 °C.

both $Pb(NO_3)_2$ and $NaNO_3$; there are three phases in equilibrium, and so this point is invariant if the temperature and pressure are constant. A few tie lines are shown in the two-phase regions.

A diagram such as Fig. 3.18 is useful in deciding how to obtain the maximum amount of pure substance from a mixture. For example, if water is added to mixture M, the gross composition moves along the dashed line toward the H_2O apex. If only a small amount of water is added, the phases $Pb(NO_3)_2$, $NaNO_3$, and solution of composition P will be present. If sufficient water is added to reach point N, the only solid phase at equilibrium will be $Pb(NO_3)_2$. If the solution is heated to get all the $NaNO_3$ into solution and then cooled to 25 °C, the solid phase will be pure $Pb(NO_3)_2$. The composition of the mother liquor would be only slightly different from P. The relative amounts of $Pb(NO_3)_2$ and mother liquor are given by y/x. Thus it is seen that if more water is added the recovery of $Pb(NO_3)_2$ will be reduced.

In a phase diagram like Fig. 3.18, two phase regions are indicated by the tie lines that show the compositions of phases that are in equilibrium. The blank regions may represent either one phase or three phases, and so it is important to understand how to distinguish between these two types of areas. One-phase regions, like the solutions region, have curved boundaries, whereas boundaries of the three-phase regions are straight, so that these regions have a triangular shape.

References

W. E. Addison, *The Allotropy of the Elements*, American Elsevier, New York, 1968.

A. Alper, *Phase Diagrams*, Volumes I, II, and III, Academic Press, New York, 1970.

K. E. Bett, J. S. Rowlinson, and G. Saville, *Thermodynamics for Chemical Engineers*, The MIT Press, Cambridge, Mass., 1975.

R. S. Bradley and D. C. Munro, *High Pressure Chemistry*, Pergamon Press, New York, 1965.

K. G. Denbigh, *The Principles of Chemical Equilibrium*, Cambridge University Press, Cambridge, 1971.
R. R. Dreisbach, *Physical Properties of Chemical Compounds*, American Chemical Society, Washington, D.C., 1955.
A. Findlay, A. N. Campbell, and N. O. Smith, *The Phase Rule and Its Applications*, Dover Publications, New York, 1951.
A. W. Francis, *Liquid-Liquid Equilibriums*, Interscience Publishers, New York, 1963.
R. M. Garrels and C. L. Christ, *Solutions, Minerals, and Equilibria*, Harper and Row, New York, 1965.
E. A. Guggenheim, *Thermodynamics*, Wiley, New York, 1967.
L. H. Horsley, *Azeotropic Data, Advances in Chemistry Series*, American Chemical Society, Washington, D.C., 1963.
A. Reisman, *Phase Equilibria*, Academic Press, New York, 1970.
A. G. Williamson, *An Introduction to Non-Electrolyte Solutions*, Wiley, New York, 1967.

Problems

3.1 What is the maximum number of phases that can be in equilibrium at constant temperature and pressure in one-, two-, and three-component systems?

Ans. $p = 3, 4, 5$.

3.2 The critical temperature of carbon tetrachloride is 283.1 °C. The densities in grams per cubic centimeter of the liquid ρ_l and vapor ρ_v at different temperatures are as follows:

t/°C	100	150	200	250	270	280
ρ_l	1.4343	1.3215	1.1888	0.9980	0.8666	0.7634
ρ_v	0.0103	0.0304	0.0742	0.1754	0.2710	0.3597

What is the critical molar volume of CCl_4? It is found that the mean of the densities of the liquid and vapor does not vary rapidly with temperature and can be represented by

$$\frac{\rho_l + \rho_v}{2} = AT + B$$

where A and B are constants. The extrapolated value of the average density at the critical temperature is the critical density. The molar volume V_c at the critical point is equal to the molar mass divided by the critical density. *Ans.* 276 $cm^3\ mol^{-1}$.

3.3 Liquid mercury has a density of 13.690 g cm^{-3}, and solid mercury has a density of 14.193 g cm^{-3}, both being measured at the melting point, −38.87 °C, under 1 atm pressure. The heat of fusion is 9.75 J g^{-1}. Calculate the melting points of mercury under a pressure of (*a*) 10 atm, and (*b*) 3540 atm. The observed melting point under 3540 atm is −19.9 °C. *Ans.* (*a*) −38.81 °C, (*b*) −15 °C.

3.4 The heats of vaporization and of fusion of water are 2490 J g^{-1} and 333.5 J g^{-1} at 0 °C. The vapor pressure of water at 0 °C is 611 Pa. Calculate the sublimation pressure of ice at −15 °C, assuming that the enthalpy changes are independent of temperature.

Ans. 166 Pa.

3.5 *n*-Propyl alcohol has the following vapor pressures.

t/°C	40	60	80	100
P/kPa	6.69	19.6	50.1	112.3

Plot these data so as to obtain a nearly straight line, and calculate (*a*) the heat of vaporization, and (*b*) the boiling point at 1 atm. *Ans.* (*a*) 44.8 kJ mol^{-1}, (*b*) 96.5 °C.

3.6 For uranium hexafluoride the vapor pressures (in Pa) for the solid and liquid are given by

$$\ln P_s = 29.411 - 5893.5/T \qquad \ln Pl = 22.254 - 3479.9/T$$

Calculate the temperature and pressure of the triple point. *Ans.* 64 °C, 152.2 kPa.

3.7 If $\Delta C_P = C_{P,\text{vap}} - C_{P,\text{liq}}$ is independent of temperature, then

$$\Delta H_{\text{vap}} = \Delta H_{0,\text{vap}} + T\Delta C_P$$

where $\Delta H_{0,\text{vap}}$ is the hypothetical enthalpy of vaporization at absolute zero. Since ΔC_P is negative, ΔH_{vap} decreases as the temperature increases. Show that if the vapor is a perfect gas, the vapor pressure is given as a function of temperature by

$$\ln P = \frac{-\Delta H_{0,\text{vap}}}{RT} + \frac{\Delta C_P}{R}\ln T + \text{constant}$$

3.8 The sublimation pressures of solid Cl_2 are 352 Pa at −112 °C and 35 Pa at −126.5 °C. The vapor pressures of liquid Cl_2 are 1590 Pa at −110 °C and 7830 Pa at −80 °C. Calculate (*a*) ΔH_{sub}, (*b*) ΔH_{vap}, (*c*) ΔH_{fus}, and (*d*) the triple point.

Ans. (*a*) 31.4, (*b*) 22.2, (*c*) 9.3 kJ mol^{-1}, (*d*) −103 °C.

3.9 The normal boiling point of *n*-hexane is 69.0 °C. Estimate (*a*) its molar heat of vaporization, and (*b*) its vapor pressure at 60 °C. *Ans.* (*a*) 30.1 kJ mol^{-1}, (*b*) 0.751 atm.

3.10 Ethanol and methanol form very nearly ideal solutions. The vapor pressure of ethanol is 5.93 kPa, and that of methanol if 11.83 kPa, at 20 °C. (*a*) Calculate the mole fraction of methanol and ethanol in a solution obtained by mixing 100 g of each. (*b*) Calculate the partial pressures and the total vapor pressure of the solution. (*c*) Calculate the mole fraction of methanol in the vapor.

Ans. (*a*) $X_{C_2H_5OH} = 0.410$; $X_{CH_3OH} = 0.590$.
(*b*) $P_{C_2H_5OH} = 18.2$, $P_{CH_3OH} = 52.3$, $P_{\text{total}} = 9.40$ kPa.
(*c*) 0.741.

3.11 The vapor pressure of a solution containing 13 g of a nonvolatile solute in 100 g of water at 28 °C is 3.6492 kPa. Calculate the molar mass of the solute, assuming that the solution is ideal. The vapor pressure of water at this temperature is 3.7417 kPa.

Ans. 92.4 g mol^{-1}.

3.12 Use the Gibbs-Duhem equation to show that if one component of a binary liquid solution follows Raoult's law, the other component will, too.

3.13 What are the entropy change and Gibbs energy change on mixing to produce a benzene-toluene solution with $\frac{1}{3}$ mole fraction benzene at 25 °C?

Ans. 5.293 K^{-1} mol^{-1}. −1577 J mol^{-1}.

3.14 At 100 °C benzene has a vapor pressure of 180.9 kPa, and toluene has a vapor pressure of 74.4 kPa. Assuming that these substances form ideal binary solutions with each other, calculate the composition of the solution that will boil at 1 atm at 100 °C and the vapor composition. *Ans.* $X_{\text{benzene,liq}} = 0.253$, $X_{\text{benzene,vap}} = 0.451$.

3.15 The following table gives mole % acetic acid in aqueous solutions and in the equilibrium vapor at the boiling point of the solution at 1 atm.

B.P., °C	118.1	113.8	107.5	104.4	102.1	100.0
Mole % of acetic acid						
In liquid	100	90.0	70.0	50.0	30.0	0
In vapor	100	83.3	57.5	37.4	18.5	0

Calculate the minimum number of theoretical plates for the column required to produce an initial distillate of 28 mole % acetic acid from a solution of 80 mole % acetic acid.

Ans. 3.

3.16 If two liquids (1 and 2) are completely immiscible, the mixture will boil when the sum of the two partial pressures exceeds the applied pressure: $P = P_1^\circ + P_2^\circ$. In the vapor phase the ratio of the mole fractions of the two components is equal to the ratio of their vapor pressures.

$$\frac{P_1^\circ}{P_2^\circ} = \frac{X_1}{X_2} = \frac{g_1 M_2}{g_2 M_1}$$

where g_1 and g_2 are the masses of components 1 and 2 in the vapor phase, and M_1 and M_2 are their molar masses. The boiling point of the immiscible liquid system naphthalene-water is 98 °C under a pressure of 97.7 kPa. The vapor pressure of water at 98 °C is 94.3 kPa. Calculate the weight percent of naphthalene in the distillate. *Ans.* 20.7%.

3.17 Calculate the osmotic pressure of a 1 mol L^{-1} sucrose solution in water from the fact that at 30 °C the vapor pressure of the solution is 4.1606 kPa. The vapor pressure of water at 30 °C is 4.2429 kPa. The density of pure water at this temperature (0.99564 g cm^{-3}) may be used to estimate $\bar{V}_1$ for a dilute solution. To do this problem, Raoult's law is introduced into equation 3.56. *Ans.* 26.9 atm.

3.18 For a solution of *n*-propanol and water, the following partial pressures in kPa are measured at 25 °C. Draw a complete pressure-composition diagram, including the total pressure. What is the composition of the vapor in equilibrium with a solution containing 0.5 mole fraction of *n*-propanol?

$X_{n\text{-propanol}}$	P_{H_2O}	$P_{n\text{-propanol}}$	$X_{n\text{-propanol}}$	P_{H_2O}	$P_{n\text{-propanol}}$
0	3.168	0.00	0.600	2.65	2.07
0.020	3.13	0.67	0.800	1.79	2.37
0.050	3.09	1.44	0.900	1.08	2.59
0.100	3.03	1.76	0.950	0.56	2.77
0.200	2.91	1.81	1.000	0.00	2.901
0.400	2.89	1.89			

Ans. $X_{n\text{-propanol,vap}} = 0.406$.

3.19 Using the Henry law constants in Table 3.1, calculate the percentage (by volume) of oxygen and nitrogen in air dissolved in water at 25 °C. The air in equilibrium with the water at 1 atm pressure may be considered to be 20% oxygen and 80% nitrogen by volume. *Ans.* 33% oxygen, 67% nitrogen.

3.20 The following data on ethanol-chloroform solutions at 35 °C were obtained by G. Scatchard and C. L. Raymond [*J. Am. Chem. Soc.*, **60**, 1278 (1938)]:

$X_{EtOH,liq}$	0	0.2	0.4	0.6	0.8	1.0
$X_{EtOH,vap}$	0.0000	0.1382	0.1864	0.2554	0.4246	1.0000
Total pressure, kPa	39.345	40.559	38.690	34.387	25.357	13.703

Calculate the activity coefficients of ethanol in these solutions according to Convention I.
Ans. 2.045, 1.316, 1.065, 0.982, 1.000.

3.21 Using the data in Problem 3.18, calculate the activity coefficients of water and *n*-propanol at 0.20, 0.40, 0.60, and 0.80 mole fraction *n*-propanol, using Convention II and considering *n*-propanol to be the solvent. *Ans.* $X_1 = 0.20, 0.40, 0.60, 0.80$;
$f_1 = 3.12, 1.63, 1.19, 1.02$;
$f_2 = 0.314, 0.417, 0.574, 0.773$.

3.22 Calculate the solubility of naphthalene at 25 °C in any solvent in which it forms an ideal solution. The melting point of naphthalene is 80 °C, and the heat of fusion is 19.29 kJ mol^{-1}. The actual measured solubility of naphthalene in benzene is $X_1 = 0.296$.
Ans. $X_1 = 0.298$.

3.23 If 68.4 g of sucrose (M = 342 g mol^{-1}) is dissolved in 1000 g of water: (*a*) What is the vapor pressure at 20 °C? (*b*) What is the freezing point? The vapor pressure of water at 20 °C is 2.3149 kPa. *Ans.* (*a*) 2.3066 kPa, (*b*) −0.372 °C.

3.24 The phase diagram for magnesium-copper at constant pressure shows that two compounds are formed: $MgCu_2$ which melts at 800 °C, and Mg_2Cu which melts at 580 °C. Copper melts at 1085 °C, and Mg at 648 °C. The three eutectics are at 9.4% by weight Mg (680 °C), 34% by weight Mg (560 °C), and 65% by weight Mg (380 °C). Construct the phase diagram. State the variance for each area and eutectic point.

Ans. In the liquid region v = 2, in the two-phase regions v = 1, and at the eutectic points v = 0.

3.25 For the ternary system benzene-isobutanol-water at 25 °C and 1 atm the following compositions have been obtained for the two phases in equilibrium.

Water-Rich Phase		Benzene-Rich Phase	
Isobutanol, wt. %	Water, wt. %	Isobutanol, wt. %	Benzene, wt. %
2.33	97.39	3.61	96.20
4.30	95.44	19.87	79.07
5.23	94.59	39.57	57.09
6.04	93.83	59.48	33.98
7.32	92.64	76.51	11.39

Plot these data on a triangular graph, indicating the tie lines. (*a*) Estimate the compositions of the phases that will be produced from a mixture of 20% isobutanol, 55% water, and 25% benzene. (*b*) What will be the composition of the principal phase when the first drop of the second phase separates when water is added to a solution of 80% isobutyl alcohol in benzene? *Ans.* (*a*) H_2O layer: 5.23% isobutanol, 94.5% H_2O.
Benzene layer: 39.57% isobutanol, 57.09% benzene.
(*b*) 10% H_2O; 72% isobutanol; 18% benzene.

3.26 The following data are available from the system nickel sulfate-sulfuric acid-water at 25 °C. Sketch the phase diagram on triangular coordinate paper, and draw appropriate tie lines.

Liquid Phase		
$NiSO_4$, wt. %	H_2SO_4, wt. %	Solid Phase
28.13	0	$NiSO_4 \cdot 7H_2O$
27.34	1.79	$NiSO_4 \cdot 7H_2O$
27.16	3.86	$NiSO_4 \cdot 7H_2O$
26.15	4.92	$NiSO_4 \cdot 6H_2O$
15.64	19.34	$NiSO_4 \cdot 6H_2O$
10.56	44.68	$NiSO_4 \cdot 6H_2O$
9.65	48.46	$NiSO_4 \cdot H_2O$
2.67	63.73	$NiSO_4 \cdot H_2O$
0.12	91.38	$NiSO_4 \cdot H_2O$
0.11	93.74	$NiSO_4$
0.08	96.80	$NiSO_4$

3.27 The densities of liquid and vapor methyl ether in grams per cubic centimeter at various temperatures are as follows:

$t/°C$	30	50	70	100	120
ρ_l	0.6455	0.6116	0.5735	0.4950	0.4040
ρ_v	0.0142	0.0241	0.0385	0.0810	0.1465

Calculate the critical density and temperature. (See Problem 3.2.)

3.28 The heat of vaporization of water at 0 °C is 44.85 kJ mol^{-1}, and the heat of sublimation of ice is 50.71 kJ mol^{-1}. Given the fact that the triple point is at 0.0099 °C and a pressure of 611 Pa, calculate the vapor pressure of water at 15 °C and of ice at −15 °C using the Clausius-Clapeyron equation. These three points are used to construct a plot of log P versus $1/T$. The fusion curve may be drawn in as a vertical line above the triple point, since the effect of pressures in this range is negligible. The regions in the diagram are then labeled.

3.29 What is the boiling point of water on a mountain where the barometer reading is 88 kPa? The heat of vaporization of water may be taken to be 40.67 kJ mol^{-1}.

3.30 At 0 °C ice absorbs 333.5 J g^{-1} in melting; water absorbs 2490 J g^{-1} in vaporizing. (*a*) What is the heat of sublimation of ice at this temperature? (*b*) At 0 °C the vapor pressure of both ice and water is 611 Pa. What is the rate of change of vapor pressure with temperature dP/dT for ice and liquid water at this temperature? (*c*) Estimate the vapor pressures of ice and of liquid water at −5 °C.

3.31 (*a*) Calculate the vapor pressure of water at 200 °C using the Clausius-Clapeyron equation and assuming that ΔH_{vap} is 40.656 kJ mol^{-1} independent of temperature. (*b*) Calculate the vapor pressure, allowing for the fact that $\Delta C_P = -41.8$ J K^{-1} mol^{-1}. The directly measured value is 1554.7 kPa. (See Problem 3.7)

3.32 Ice has the unusual property of a melting point that is lowered by increasing pressure. Thus, one can skate on ice provided that the pressure exerted by one's skates is great enough to liquefy the ice under them. Would a 75-kg skater whose skates contact the ground with an area of 0.1 cm^2 be able to skate at −3 °C?

3.33 Estimate the vapor pressure of ice at the temperature of solid carbon dioxide (−78 °C at 1 atm pressure of CO_2), assuming that the heat of sublimation is constant. The heat of sublimation of ice is 2.83 kJ g^{-1}, and the vapor pressure of ice is 611 Pa at 0 °C.

3.34 The sublimation pressure of solid CO_2 is 133 Pa at −134.3 °C and 2660 Pa at −114.4 °C. Calculate the heat of sublimation.

3.35 The vapor pressure of solid benzene, C_6H_6, is 299 Pa at −30 °C and 327 Pa at 0° C, and the vapor pressure of liquid C_6H_6 is 6170 Pa at 10 °C and 15,800 Pa at 30 °C. From these data, calculate (*a*) the triple point of C_6H_6, and (*b*) the heat of fusion of C_6H_6.

3.36 The vapor pressure of toluene is 8.00 kPa at 40.3 °C and 2.67 kPa at 18.4 °C. Calculate (*a*) the heat of vaporization, and (*b*) the vapor pressure at 25 °C.

3.37 Show that the effect of external hydrostatic pressure P on the vapor pressure p of a liquid of partial molar volume $\bar{V}_l$ is given by

$$RT\frac{d \ln p}{dP} = \bar{V}_l$$

provided the vapor behaves as a perfect gas. *Hint:* When the pressure is changed at constant temperature, $dG_l = dG_{vap}$, so that $V_{gas}\,dp = \bar{V}_l\,dP$.

3.38 At 25 °C the vapor pressures of chloroform and carbon tatrachloride are 26.54 and 15.27 kPa, respectively. If the liquids form an ideal solution, (*a*) what is the composition of the vapor in equilibrium with a solution containing 1 mol of each; (*b*) what is the total vapor pressure of the mixture?

3.39 Ethylene dibromide and propylene dibromide form very nearly ideal solutions. Plot the partial vapor pressure of ethylene dibromide ($P^\circ = 22.9$ kPa), the partial vapor pressure of propylene dibromide ($P^\circ = 16.9$ kPa), and the total vapor pressure of the solution versus the mole fraction of ethylene dibromide at 80 °C. (*a*) What will be the composition of the vapor in equilibrium with a solution containing 0.75 mole fraction of ethylene dibromide? (*b*) What will be the composition of the liquid phase in equilibrium with ethylene dibromide-propylene dibromide vapor containing 0.50 mole fraction of each?

3.40 The vapor pressure of water at 25 °C is 3.1672 kPa. Calculate the vapor pressure of solutions containing (*a*) 6.01 g of urea, NH_2CONH_2, (*b*) 9.4 g of phenol, C_6H_5OH, and (*c*) 6.01 g of urea + 9.4 g cf phenol per 1000 g of water, assuming no chemical action between the two substances. (*d*) Calculate (*c*), assuming that a stable compound is formed containing 1 mol of the urea to 1 mol of phenol.

3.41 Benzene and toluene form very nearly ideal solutions. At 80 °C, the vapor pressures of benzene and toluene are as follows: benzene, $P^\circ = 100.4$ kPa; toluene, $P^\circ = 38.7$ kPa. (*a*) For a solution containing 0.5 mole fraction of benzene and 0.5 mole fraction of toluene, what is the composition of the vapor and the total vapor pressure at 80 °C? (*b*) What is the composition of the liquid phase in equilibrium at 80 °C with benzene-toluene vapor having 0.75 mole fraction benzene?

3.42 At 25 °C the vapor pressure of carbon tetrachloride, CCl_4, is 19.1 kPa and that of chloroform, $CHCl_3$, is 26.5 kPa. These liquids form very nearly ideal solutions with each other. If 1 mol of CCl_4 and 3 mol of $CHCl_3$ are mixed, what will be the mole fraction o CCl_4 in the vapor phase and the total vapor pressure of the solution?

3.43 At 1 atm pressure propane boils at −42.1 °C and *n*-butane boils at −0.5 °C; the following vapor-pressure data are available.

t/°C	−31.2	−16.3
P/kPa (propane)	160.0	298.6
P/kPa (*n*-butane)	26.7	53.3

Assuming that these substances form ideal binary solutions with each other, (*a*) calculate the mole fractions of propane at which the solution will boil at 1 atm pressure at −31.2 and −16.3 °C. (*b*) Calculate the mole fractions of propane in the equilibrium vapor at these temperatures. (*c*) Plot the temperature-mole fraction diagram at 1 atm, using these data.

3.44 The following table gives the mole percent of *n*-propanol ($M = 60.1$ g mol^{-1}) in aqueous solutions and in the vapor at the boiling point of the solution at 1 atm pressure.

B.P., °C	100.0	92.0	89.3	88.1	87.8	88.3	90.5	97.3
Mole % of *n*-propanol								
In liquid	0	2.0	6.0	20.0	43.2	60.0	80.0	100.0
In vapor	0	21.6	35.1	39.2	43.2	49.2	64.1	100.0

With the aid of a graph of these data, calculate the mole fraction of *n*-propanol in the first drop of distillate when the following solutions are distilled with a simple distilling flask that gives one theoretical plate: (*a*) 87 g of *n*-propanol and 211 g of water; (*b*) 50 g of *n*-propanol and 5.02 g of water.

3.45 Plot the following boiling point data for benzene-ethanol solutions and estimate the azeotropic composition.

B.P., °C	78	75	70	70	75	80
Mole fraction of benzene						
In liquid	0	0.04	0.21	0.86	0.96	1.00
In vapor	0	0.18	0.42	0.66	0.83	1.00

State the range of mole fractions of benzene for which pure benzene could be obtained by fractional distillation at 1 atm.

3.46 The vapor pressure of the immiscible liquid system diethylaniline-water is 1 atm at 99.4 °C. The vapor pressure of water at that temperature is 99.2 kPa. How many grams of steam are necessary to distill 100 g of diethylaniline? (See Problem 3.16.)

3.47 From the data given in the following table construct a complete temperature-composition diagram for the system ethanol-ethyl acetate for 1 atm pressure. A solution containing 0.8 mole fraction of ethanol, EtOH, is distilled completely at 1 atm. (*a*) What is the composition of the first vapor to come off? (*b*) That of the last drop of liquid to evaporate? (*c*) What would be the values of these quantities if the distillation were carried out in a cylinder provided with a piston so that none of the vapor could escape?

$X_{EtOH,liq}$	$X_{EtOH,vap}$	B.P., °C	$X_{EtOH,liq}$	$X_{EtOH,vap}$	B.P., °C
0	0	77.15	0.563	0.507	72.0
0.025	0.070	76.7	0.710	0.600	72.8
0.100	0.164	75.0	0.833	0.735	74.2
0.240	0.295	72.6	0.942	0.880	76.4
0.360	0.398	71.8	0.982	0.965	77.7
0.462	0.462	71.6	1.000	1.000	78.3

3.48 The Henry law constants for oxygen and nitrogen in water at 0 °C are 2.55×10^9 and 5.44×10^9 torr, respectively. Calculate the lowering of the freezing point of water by dissolved air with 80% N_2 and 20% O_2 by volume at 1 atm pressure.

3.49 Using the data in Problem 3.18, calculate the activity coefficients of water and *n*-propanol at 0.20, 0.40, 0.60, and 0.80 mole fraction of *n*-propanol, using Convention II and considering water to be the solvent.

3.50 Using the data of the following table, which gives vapor pressures in kPa at 35.2 °C, calculate the activity coefficients of acetone and chloroform at 35.2 °C, following Convention I, for mole fractions of chloroform of 0.20, 0.40, 0.60, and 0.80.

X_{CHCl_3}	0	0.2	0.4	0.6	0.8	1.00
P_{CHCl_3}	0	4.5	10.9	19.7	30.0	39.1
$P_{acetone}$	45.9	36.0	24.4	13.6	5.6	0

3.51 By use of the data of the following table, which gives pressures in kPa at 35.2 °C for carbon disulfide-acetone solutions, calculate the activity coefficients (following Convention I) for acetone and carbon disulfide at 35.2 °C for a solution containing 0.6 mole fraction of carbon disulfide.

X_{CS2}	0	0.2	0.4	0.6	0.8	1.0
P_{CS2}	0	37.3	50.4	56.7	61.3	68.3
$P_{acetone}$	45.9	38.7	34.0	30.7	25.3	0

3.52 Calculate the solubility of anthracene ($M = 178.2 \text{ g mol}^{-1}$) in toluene ($M = 92.1 \text{ g mol}^{-1}$) at 100 °C. The heat fusion of anthracene is 28.9 kJ mol^{-1}, and the melting point of anthracene is 217 °C. The actual solubility is 0.0592 on the mole-fraction scale. How do you explain the discrepancy?

3.53 Calculate the solubility of cadmium in bismuth at 250 °C. The melting point of cadmium is 323 °C and the heat of fusion at the melting point is 6.07 kJ mol^{-1}.

3.54 Calculate the temperature at which pure cadmium is in equilibrium with a Bi-Cd solution containing 0.846 mole fraction Cd. Given: $\Delta H_{\text{fus}} = 6.07$ kJ mol^{-1}.

3.55 The following cooling curves have been found for the system antimony–cadmium.

Cd, wt. %	0	20	37.5	47.5	50	58	70	93	100
First break in curve, °C	—	550	461	—	419	—	400	—	—
Continuing constant temperature, °C	630	410	410	410	410	439	295	295	321

Construct a phase diagram, assuming that no breaks other than these actually occur in any cooling curve. Label the diagram completely and give the formula of any compound formed. State the variance in each area and at each eutectic point.

3.56 The melting points of magnesium and nickel are 651 and 1450 °C, respectively. As Ni is added to Mg, the freezing point is lowered until a eutectic point is reached at 510 °C and 28% by weight Ni. The other phase that separates contains 54.7% by weight Ni. As the weight percent Ni is increased past 28%, the temperature of first-phase separation rises to a maximum of 1180 °C. Above 770 °C and 38% Ni, the solid phase that separates out contains 83% Ni. The phase containing 83% Ni melts sharply at 1180 °C. There is a eutectic point at 88% Ni and 1080 °C. Draw the phase diagram and indicate the phases present in each region.

3.57 The following data are obtained by cooling solutions of magnesium and nickel.

Ni, wt. %	0	10	28	38	60	83	88	100
Inflection in cooling curve, °C	—	608	—	770	1050	—	—	—
Plateau in cooling curve, °C	651	510	510	510	770	1180	1080	1450

It is found that in addition cooling solutions containing between 28 and 38% Ni deposit Mg_2Ni, whereas solutions containing between 38 and 82% Ni deposit $MgNi_2$. Plot the phase diagram.

3.58 The following are the compositions of the phases in equilibrium with each other in the system methylcyclohexane–aniline–*n*-heptane at 1 atm and 25 °C. Draw a triangular diagram for the system, including tie lines, and compute the exact composition of the first infinitesimal increment of the new liquid phase that forms when a sufficient quantity of pure aniline is added to a 40% solution of methylcyclohexane in *n*-heptane to give separation into two phases.

Hydrocarbon Layer		Aniline Layer	
Methylcyclohexane, wt. %	*n*-Heptane, wt. %	Methylcyclohexane, wt. %	*n*-Heptane, wt. %
0.0	92.0	0.0	6.2
9.2	83.0	0.8	6.0
18.6	73.4	2.7	5.3
33.8	57.6	4.6	4.5
46.0	45.0	7.4	3.6
59.7	30.7	9.2	2.8
73.6	16.0	13.1	1.4
83.3	4.4	15.6	0.6
88.1	0.0	16.9	0.0

3.59 The following data are available for the system Na_2SO_4-$Al_2(SO_4)_3$-H_2O at 42 °C. Draw the phase diagram on triangular coordinate paper, and draw appropriate tie lines.

Liquid Phase		
Na_2SO_4, wt. %	$Al_2(SO_4)_3$, wt. %	Solid Phase
33.20	0	Na_2SO_4
32.00	1.52	Na_2SO_4
31.79	1.87	Na_2SO_4
28.75	1.71	$Na_2SO_4 \cdot Al_2(SO_4)_3 \cdot 14H_2O$
24.47	2.84	$Na_2SO_4 \cdot Al_2(SO_4)_3 \cdot 14H_2O$
16.81	5.63	$Na_2SO_4 \cdot Al_2(SO_4)_3 \cdot 14H_2O$
10.93	10.49	$Na_2SO_4 \cdot Al_2(SO_2)_3 \cdot 14H_2O$
4.72	17.11	$Na_2SO_4 \cdot Al_2(SO_4)_3 \cdot 14H_2O$
1.75	18.59	$Al_2(SO_4)_3$
0	16.45	$Al_2(SO_4)_3$

3.60 At 25 °C the solubility of KNO_3 in pure H_2O is 46.2% by weight, the solubility of $NaNO_3$ in pure H_2O is 52.2% by weight, and $NaNO_3$, KNO_3, and saturated solution are in equilibrium when the composition of the solution is H_2O, 31.3%, KNO_3, 28.9%, and $NaNO_3$, 39.8%. No crystalline hydrates or double salts are formed. Sketch this system on a triangular diagram, labeling the areas in which you would expect to find (*a*) only solution; (*b*) a mixture of solution and solid KNO_3; (*c*) a mixture of solution and solid $NaNO_3$; and (*d*) a mixture of solid KNO_3, solid $NaNO_3$, and solution.

CHAPTER 4
CHEMICAL EQUILIBRIUM

In the preceding chapters we applied thermodynamics mainly to physical processes. We now apply these ideas to chemical reactions, starting with reactions of perfect gases. The thermodynamic treatment of chemical equilibrium illustrates the usefulness of the concept of chemical potential.

The idea of the reversibility of chemical reactions was first stated clearly in 1799 by C. Berthollet, while he was acting as scientific adviser to Napoleon in Egypt. He noted the deposits of sodium carbonate in certain salt lakes and concluded that they were produced by the high concentration of sodium chloride and dissolved calcium carbonate, the reverse of the laboratory experiment in which sodium carbonate reacts with calcium chloride to precipitate calcium carbonate. In 1863 the influence that the concentrations of ethyl alcohol and acetic acid have on the concentration of ethyl acetate was reported by M. Berthelot and Saint-Gilles.

In 1864 Guldberg and Waage showed experimentally that in chemical reactions a definite equilibrium is reached that can be approached from either direction. They were apparently the first to realize that there is a mathematical relation between the concentrations of reactants and products at equilibrium. In 1877 van't Hoff suggested that in the equilibrium expression for the hydrolysis of ethyl acetate, the concentration of each reactant should appear to the first power, each reactant having a coefficient of unity in the balanced chemical equation.

4.1 DERIVATION OF THE GENERAL EQUILIBRIUM EXPRESSION

We showed earlier (in Section 1.20) that a generalized chemical reaction may be represented by

$$0 = \sum_i \nu_i \mathrm{A}_i \tag{4.1}$$

where the stoichiometric coefficients ν_i are pure numbers that are positive for products and negative for reactants. At any time during the reaction the Gibbs energy of the reaction mixture is given by equation 2.116, which is

$$\begin{aligned} G &= n_1\mu_1 + n_2\mu_2 + \cdots \\ &= \sum_i n_i\mu_i \end{aligned} \tag{4.2}$$

where n_i represents the number of moles of substance i at a particular extent of reaction. When a chemical reaction occurs, the changes in the number of moles of the various reactants and products are related through the stoichiometric coefficients of the balanced chemical equation 4.1. If a reaction system initially contains n_{io} moles of i, the number n_i of moles of i at a later time may be written

$$n_i = n_{io} + \nu_i \xi \tag{4.3}$$

where ξ(*xi*) is the extent of reaction, expressed in moles. The extent of reaction is defined so that it is initially zero and is always positive.

We may express the Gibbs energy of a reaction system in terms of the extent of reaction ξ by substituting equation 4.3 into equation 4.2.

$$\begin{aligned} G &= n_{1o}\mu_1 + n_{2o}\mu_2 + \cdots + (\nu_1\mu_1 + \nu_2\mu_2 + \cdots)\xi \\ &= \sum_i n_{io}\mu_i + \xi \sum_i \nu_i\mu_i \end{aligned} \tag{4.4}$$

The derivative of the Gibbs energy with respect to the extent of reaction at constant temperature and pressure is

$$\left(\frac{dG}{d\xi}\right)_{T,P} = \sum_i \nu_i\mu_i \tag{4.5}$$

At equilibrium at constant temperature and pressure the Gibbs energy of the system has its minimum value and so this derivative is equal to zero, and

$$\sum_i \nu_i\mu_i = 0 \tag{4.6}$$

where the chemical potentials in this equation are the values at equilibrium. This general condition applies to all chemical reactions, whether they involve gases, liquids, solids, or solutions.

As shown in Section 2.18, the chemical potential of a substance may be expressed in the most general way in terms of its fugacity f_i, which may be regarded as a kind of effective pressure.

$$\mu_i = \mu_i^\circ + RT \ln \frac{f_i}{f_i^\circ} \tag{4.7}$$

where μ_i° is the chemical potential in the standard state and f_i° is the fugacity in the standard state. G. N. Lewis called the quantity f_i/f_i° the *activity* a_i.

$$\mu_i = \mu_i^\circ + RT \ln a_i \tag{4.8}$$

Substituting this relationship into equation 4.6 yields

$$\sum_i \nu_i\mu_i^\circ + RT \sum_i \nu_i \ln a_i = 0 \tag{4.9}$$

$$\sum_i \nu_i\mu_i^\circ = -RT \ln \prod_i a_i^{\nu_i} \tag{4.10}$$

where $\prod_i$ indicates a product is to be taken.

Since $\sum_i \nu_i \mu_i^\circ$ is a function only of temperature, $\prod_i a_i^{\nu_i}$ must be a function only of temperature and is called the equilibrium constant K. Thus

$$\Delta G^\circ = -RT \ln K \tag{4.11}$$

where $\Delta G^\circ = \sum_i \nu_i \mu_i^\circ$ is the change in Gibbs energy when unmixed reactants in their standard states are converted into unmixed products in their standard states and the equilibrium constant is defined by

$$K = \prod_i a_i^{\nu_i} \tag{4.12}$$

where the activities are the values at equilibrium. Note that the equilibrium constant defined in this way is a pure number, since activities are pure numbers. We usually do not have the activity data that would permit us to use this form.

If the reactants and products are not in their standard states, the change in Gibbs energy for the reaction in which the unmixed reactants at temperature T are converted to the unmixed products at the same temperature is

$$\Delta G = \sum_i \nu_i \mu_i \tag{4.13}$$

Substituting equation 4.8, we obtain

$$\Delta G = \sum_i \nu_i \mu_i^\circ + RT \ln \prod_i a_i^{\nu_i} = \Delta G^\circ + RT \ln \prod_i a_i^{\nu_i} \tag{4.14}$$

This equation is useful for calculating whether or not a reaction can occur spontaneously when the reactants are provided at certain activities and the products are withdrawn at certain other activities. For perfect gases the activities of reactants and products in this expression may be replaced by partial pressures and for ideal solutions by mole fractions.

4.2 CHEMICAL EQUILIBRIUM OF PERFECT GASES

Substituting the expression for the chemical potential of a perfect gas (equation 2.133) into the general expression for chemical equilibrium (equation 4.6) yields

$$\sum_i \nu_i \mu_i^\circ + RT \sum_i \nu_i \ln \left(\frac{P_i}{P^\circ}\right) = 0 \tag{4.15a}$$

$$\sum_i \nu_i \mu_i^\circ = -RT \ln \prod_i \left(\frac{P_i}{P^\circ}\right)^{\nu_i} \tag{4.15b}$$

$$\Delta G^\circ = -RT \ln K_P \tag{4.16}$$

so that

$$K_P = \prod_i \left(\frac{P_i}{P^\circ}\right)^{\nu_i} \tag{4.17}$$

The equilibrium constant K_P defined in this way is a pure number.* In equation 4.16, ΔG° is the change in standard Gibbs energy when the indicated numbers of moles in the *unmixed* reactants in their standard states at temperature T and 1 atm pressure react to form *unmixed* products in their standard states at the same temperature and pressure. The Gibbs energy change for the reaction from mixed reactants to mixed products or the Gibbs energy change from the unmixed reactants to the equilibrium mixture is not equal to ΔG°.

The equilibrium constant may also be expressed in terms of molar concentrations. For perfect gas mixtures,

$$P_i = \frac{n_i RT}{V} = c_i RT \tag{4.18}$$

Substituting this into equation 4.17 yields

$$K_P = \prod_i \left(\frac{c_i RT}{P^\circ}\right)^{\nu_i} \tag{4.19}$$

In order to define a dimensionless equilibrium constant in terms of concentration, we introduce the standard concentration c°, which is usually expressed in moles per liter. Introducing this standard concentration into each term of equation 4.19,

$$K_P = \prod_i \left[\left(\frac{c_i}{c^\circ}\right)\left(\frac{c^\circ RT}{P^\circ}\right)\right]^{\nu_i} = \left(\frac{c^\circ RT}{P^\circ}\right)^{\Sigma_i \nu_i} \prod_i \left(\frac{c_i}{c^\circ}\right)^{\nu_i} \tag{4.20}$$

$$= \left(\frac{c^\circ RT}{P^\circ}\right)^{\Sigma_i \nu_i} K_c \tag{4.21}$$

where the equilibrium constant expressed in terms of concentration

$$K_c = \prod_i \left(\frac{c_i}{c^\circ}\right)^{\nu_i} \tag{4.22}$$

is a function only of temperature for a perfect gas mixture. If $c^\circ = 1$ mol L^{-1} = 10^3 mol m^{-3} and $P^\circ = 1$ atm $= 101{,}325$ N m^{-2}, $c^\circ RT/P^\circ = 24.46$ at 298.15 K.

We may also calculate a standard Gibbs energy change for a reaction from K_c.

$$\Delta G^\circ = -RT \ln K_c \tag{4.23}$$

This value of ΔG° has a different interpretation from that calculated using equation 4.16. When K_c is used, ΔG° is the change in Gibbs energy when the indicated numbers of moles of *unmixed* reactants at temperature T and at concentrations of 1 mol/L react to form *unmixed* products at concentrations of 1 mol/L and the same temperature.

---

* Later, in Section 13.14, when we calculate K from statistical mechanics, we will find that we obtain a pure number. Nevertheless, the numerical value obtained depends on the standard pressure P° used in the calculation of K. The standard pressure used in chemical thermodynamics is 1 atm $= 101{,}325$ N m^{-2}.

Example 4.1 For the ammonia synthesis reaction

$$N_2(g) + 3H_2(g) = 2NH_3(g)$$

equation 4.17 yields

$$K_P = \frac{(P_{NH_3}/P^\circ)^2}{(P_{N_2}/P^\circ)(P_{H_2}/P^\circ)^3}$$

The numerical value of K_P depends on the choice of P°. At 400 °C, $K_P = 1.64 \times 10^{-4}$ when the standard pressure is 1 atm. What is the value of K_c

$$K_c = \frac{(c_{NH_3}/c^\circ)^2}{(c_{N_2}/c^\circ)(c_{H_2}/c^\circ)^3}$$

for a standard concentration c° of 1 mol L^{-1} and what is the interpretation of ΔG° calculated from this value of K_c? Using equation 4.21,

$$\begin{aligned} K_c &= 1.64 \times 10^{-4}\left(\frac{P^\circ}{c^\circ RT}\right)^{-2} \\ &= 1.64 \times 10^{-4}\left[\frac{101{,}325 \text{ Pa}}{(10^3 \text{ mol m}^{-3})(8.314 \text{ J K}^{-1} \text{ mol}^{-1})(673 \text{ K})}\right]^{-2} \\ &= 0.500 \\ \Delta G^\circ &= -RT \ln K_c = -(8.314 \text{ J K}^{-1} \text{ mol}^{-1})(673 \text{ K}) \ln 0.500 \\ &= 3880 \text{ J mol}^{-1} \end{aligned}$$

This is the change in Gibbs energy when 1 mol of N_2 in the perfect gas state at the concentration of 1 mol/L and 3 mol of H_2 in the perfect gas state at 1 mol/L react to form 2 mol of NH_3 in the perfect gas state at 1 mol/L.

Sometimes it is convenient to define K_P by $\prod_i P_i^{\nu_i}$ and K_c by $\prod_i c_i^{\nu_i}$. The equilibrium constants obtained in this way have dimensions if $\Delta\nu_i \neq 0$.

Assuming that the partial pressure of a gas in a mixture is equal to its mole fraction X_i times the total pressure P of reactants and products, K_P for a gas reaction is given by

$$K_P = \prod_i \left(\frac{P_i}{P^\circ}\right)^{\nu_i} = \prod_i \left(\frac{X_i P}{P^\circ}\right)^{\nu_i} = \left(\frac{P}{P^\circ}\right)^{\sum_i \nu_i} \prod_i X_i^{\nu_i} \tag{4.24}$$

Thus the mole fractions of individual reactants are independent of total pressure if $\sum_i \nu_i = 0$. It is frequently convenient to use equation 4.24 in calculating the equilibrium composition of a reaction mixture.

Example 4.2 A gas stream containing 2 mol of hydrogen per mole of carbon monoxide is passed over a catalyst for the reaction

$$CO(g) + 2H_2(g) = CH_3OH(g)$$

If equilibrium is reached at 500 K, where $K_P = 6.25 \times 10^{-3}$, what is the equilibrium mole fraction of $CH_3OH(g)$ at a total pressure of 100 atm?

The following relations exist between the mole fractions.

$$X_{CO} + X_{H_2} + X_{CH_3OH} = 1 \qquad X_{H_2} = 2X_{CO} \qquad 3X_{CO} + X_{CH_3OH} = 1$$

The equilibrium constant may be expressed in terms of X_{CO} as follows.

$$K_P = \left(\frac{P}{P^\circ}\right)^{-2} \frac{X_{CH_3OH}}{X_{CO}X_{H_2}^2} = \left(\frac{P}{P^\circ}\right)^{-2} \frac{1 - 3X_{CO}}{X_{CO}(2X_{CO})^2} = \frac{100^{-2}(1 - 3X_{CO})}{4X_{CO}^3}$$

$$= 6.25 \times 10^{-3}$$

$$X_{CO} = 0.134$$

$$X_{CH_3OH} = 0.599$$

4.3 THERMODYNAMICS OF A SIMPLE GAS REACTION

In order to illustrate why gas reactions do not go to completion, let us consider a simple isomerization of perfect gas A to perfect gas B.

$$A(g) = B(g) \tag{4.25}$$

According to equation 4.2, the Gibbs energy of the reaction mixture at any extent of reaction is

$$G = n_A\mu_A + n_B\mu_B \tag{4.26}$$

where n_A is the number of moles of A and n_B is the number of moles of B. If the reaction is started with 1 mol of A, the numbers of moles of A and B at a later time are given in terms of the extent of reaction ξ by

$$n_A = 1 - \xi \tag{4.27}$$

$$n_B = \xi \tag{4.28}$$

Thus

$$G = (1 - \xi)\mu_A + \xi\mu_B \tag{4.29}$$

Using equation 2.133, the chemical potentials of A and B in the ideal gaseous mixture are given by

$$\begin{aligned}\mu_A &= \mu_A^\circ + RT\ln(P_A/P^\circ)\\ &= \mu_A^\circ + RT\ln X_A + RT\ln(P/P^\circ)\\ &= \mu_A^\circ + RT\ln(1 - \xi) + RT\ln(P/P^\circ)\end{aligned} \tag{4.30}$$

$$\begin{aligned}\mu_B &= \mu_B^\circ + RT\ln X_B + RT\ln(P/P^\circ)\\ &= \mu_B^\circ + RT\ln\xi + RT\ln(P/P^\circ)\end{aligned} \tag{4.31}$$

where P is the total pressure at equilibrium. Substituting these equations into equation 4.29,

$$G = (1 - \xi)\mu_A^\circ + \xi\mu_B^\circ + RT\ln(P/P^\circ) + RT[(1 - \xi)\ln(1 - \xi) + \xi\ln\xi] \tag{4.32}$$

$$= \mu_A^\circ - (\mu_A^\circ - \mu_B^\circ)\xi + RT\ln(P/P^\circ) + \Delta G_{mix}^\circ \tag{4.33}$$

where ΔG_{mix}° is the Gibbs energy of mixing $(1 - \xi)$ moles of A with ξ moles of B. Figure 4.1 gives a plot of G versus ξ. The first two terms in equation 4.32 give the

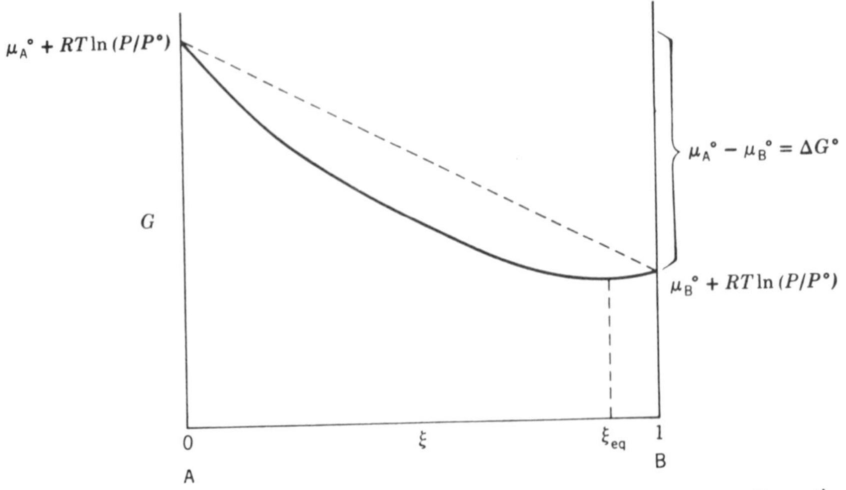

Fig. 4.1 Gibbs energy of the reaction system A(*g*) = B(*g*) versus extent of reaction ξ at constant temperature and pressure.

linear function represented by the dashed line. It is the mixing term that causes the minimum in the plot of G versus extent of reaction ξ. At constant temperature and pressure, the criterion of equilibrium is that the Gibbs energy is a minimum (see Section 2.11). Thus, starting with A, the Gibbs energy can decrease along the curve until ξ_{eq} moles of B have been formed. Starting with B the Gibbs energy can decrease until $(1 - \xi_{eq})$ moles of A have formed. Even though B has the lower value of the molar Gibbs energy ($\mu_A^\circ > \mu_B^\circ$), the system can achieve a lower Gibbs energy by having some A present at equilibrium with the resulting Gibbs energy of mixing. Generalizing from this example we can say that no chemical reaction of gases really goes to completion; nevertheless it may be very difficult to detect reactants at equilibrium if the products have a very much lower Gibbs energy.

4.4 DETERMINATION OF EQUILIBRIUM CONSTANTS

If the initial concentrations of the reactants are known and only a single reaction occurs, it is necessary to determine the concentration of only one reactant or product at equilibrium to be able to calculate the concentrations or pressures of the others by means of the balanced chemical equation. Chemical methods can be used for such analyses only when the reaction may be stopped at equilibrium, as by a very sudden chilling to a temperature where the rate of further chemical change is negligible, or by destruction of a catalyst. Otherwise, the concentrations will shift during the chemical analysis.

Measurements of physical properties, such as density, pressure, light absorption, refractive index, and electrical conductivity, are especially useful for the determination of the concentrations of reactants at equilibrium since, for these methods, it is unnecessary to "stop" the reaction.

It is essential to know that equilibrium has been reached before the analysis of the mixture can be used for calculating the equilibrium constant. The following criteria for the attainment of equilibrium at constant temperature are useful.

1. The same equilibrium constant should be obtained when the equilibrium is approached from either side.
2. The same equilibrium constant should be obtained when the concentrations of reacting materials are varied over a wide range.

The determination of the density of a partially dissociated gas provides one of the simplest methods for measuring the extent to which the gas is dissociated. When a gas dissociates, more molecules are produced, and at constant temperature and pressure the volume increases. The density at constant pressure then decreases, and the difference between the density of the undissociated gas and that of the partially dissociated gas is directly related to the degree of dissociation.

Let α represent the fraction dissociated, so that $1 - \alpha$ denotes the fraction remaining undissociated. If we start with 1 mol of gas the number of moles of gaseous products if the reaction goes to completion is $1 + \sum \nu_i$. Therefore the number of moles of gas present at equilibrium is

$$(1 - \alpha) + \left(1 + \sum \nu_i\right)\alpha = 1 + \alpha \sum \nu_i \tag{4.34}$$

Since the density of a perfect gas at constant pressure and temperature is *inversely* proportional to the number of moles for a given weight, the ratio of the density ρ_1 of the undissociated gas to the density ρ_2 of the partially dissociated gas is given by the expression

$$\frac{\rho_1}{\rho_2} = 1 + \alpha \sum \nu_i \tag{4.35}$$

$$\alpha = \frac{\rho_1 - \rho_2}{\rho_2 \sum \nu_i} \tag{4.36}$$

It is always advantageous to check an equation with some simple calculations. With reference to equation 4.36, if there is no dissociation, then $\alpha = 0$ and $\rho_1 = \rho_2$; if dissociation is complete, then $\alpha = 1$, $\rho_2 \sum \nu_i = \rho_1 - \rho_2$ and $\rho_1 = (1 + \sum \nu_i)\rho_2$.

Molar masses may be substituted in equation 4.36 for the densities of gases to which they are proportional at constant temperature and pressure, giving

$$\alpha = \frac{M_1 - M_2}{M_2 \sum \nu_i} \tag{4.37}$$

where M_1 is the molar mass of the undissociated gas, and M_2 is the average molar mass of the gases when the gas is partially dissociated.

The dissociation of nitrogen tetroxide is represented by the equation

$$N_2O_4(g) = 2NO_2(g)$$

and

$$K_P = \frac{(P_{NO_2}/P^\circ)^2}{P_{N_2O_4}/P^\circ} \tag{4.38}$$

If α represents the degree of dissociation, $(1 - \alpha)$ is proportional to the number of moles of undissociated N_2O_4; 2α is proportional to the number of moles of NO_2; and $(1 - \alpha) + 2\alpha$ or $1 + \alpha$ is proportional to the total number of moles.

If the total pressure of N_2O_4 plus NO_2 is P, the partial pressures are:

$$P_{N_2O_4} = \frac{1 - \alpha}{1 + \alpha} P \qquad \text{and} \qquad P_{NO_2} = \frac{2\alpha}{1 + \alpha} P$$

Then,

$$K_P = \frac{\left(\dfrac{2\alpha}{1 + \alpha}\dfrac{P}{P^\circ}\right)^2}{\dfrac{1 - \alpha}{1 + \alpha}\dfrac{P}{P^\circ}} = \frac{4\alpha^2(P/P^\circ)}{1 - \alpha^2} \tag{4.39}$$

In this reaction there is an increase in volume at constant pressure; each mole of gas that dissociates produces 2 mol of gas. According to the principle of Le Châtelier (Section 4.7) an increase of pressure will cause the reaction to shift toward N_2O_4, since the equilibrium shifts in the direction that tends to minimize the effect of the applied change.

Example 4.3 If 1.588 g of nitrogen tetroxide gives a total pressure of 1 atm when partially dissociated in a 500-cm³ glass vessel at 25 °C, what is the degree of dissociation α? What is the value of K_P? What is the degree of dissociation at a total pressure of 0.5 atm?

$$M_2 = \frac{RT}{P}\frac{g}{V} = \frac{(8.314 \text{ J K}^{-1} \text{ mol}^{-1})(298.1 \text{ K})(1.588 \times 10^{-3} \text{ kg})}{(101{,}325 \text{ Pa})(0.5 \times 10^{-3} \text{ m}^3)}$$

$$= 77.70 \text{ g mol}^{-1}$$

$$\alpha = \frac{92.01 - 77.70}{77.69} = 0.1842$$

$$K_P = \frac{4\alpha^2(P/P^\circ)}{1 - \alpha^2} = \frac{(4)(0.1842)^2(1)}{1 - (0.1842)^2} = 0.141$$

The degree of dissociation at 0.5 atm is obtained as follows:

$$K_P = 0.141 = \frac{4\alpha^2(0.5)}{1 - \alpha^2} \qquad 0.141(1 - \alpha^2) = 2\alpha^2 \qquad \alpha = 0.257$$

If the product molecules are different, the calculation of K_P is altered; this is illustrated for the reaction

$$PCl_5(g) = PCl_3(g) + Cl_2(g)$$

When 1 mol of PCl_5 dissociates, there will be at equilibrium $1 - \alpha$ mol of PCl_5, α mol of PCl_3, and α mol of Cl_2, where α is the degree of dissociation. The total number of moles is

$$(1 - \alpha) + \alpha + \alpha = 1 + \alpha$$

If P is the total pressure due to PCl_5, PCl_3, and Cl_2, the partial pressures are

$$P_{PCl_5} = \frac{1 - \alpha}{1 + \alpha} P \qquad P_{PCl_3} = \frac{\alpha}{1 + \alpha} P \qquad P_{Cl_2} = \frac{\alpha}{1 + \alpha} P$$

and

$$K_P = \frac{(P_{PCl_3}/P^\circ)(P_{Cl_2}/P^\circ)}{P_{PCl_5}/P^\circ} = \frac{\left(\frac{\alpha}{1 + \alpha}\frac{P}{P^\circ}\right)\left(\frac{\alpha}{1 + \alpha}\frac{P}{P^\circ}\right)}{[(1 - \alpha)/(1 + \alpha)](P/P^\circ)} = \frac{\alpha^2(P/P^\circ)}{1 - \alpha^2} \tag{4.40}$$

It should be noted that this equation differs from equation 4.39 by the lack of the factor 4. As in equation 4.39, the pressure P is the total pressure of gases involved in the equilibrium and does not include the partial pressures of inert gases.

This reaction is an interesting one to discuss in more detail. Qualitatively, it can be seen that increasing the total pressure of the three reactants will decrease the degree of dissociation, because the undissociated PCl_5 occupies the smaller volume. If chlorine is added, P_{Cl_2} increases, and since K_P remains constant, P_{PCl_3} must diminish and P_{PCl_5} must increase. The degree of dissociation is decreased also by the addition of PCl_3. In general, the dissociation of any substance is repressed by the addition of its dissociation products. The addition of an inert gas at constant volume has no effect on the dissociation, because the partial pressures of the gases involved in the reaction are not affected by the presence of another gas provided that the gases behave ideally.

4.5 EQUILIBRIA BETWEEN GASES AND SOLIDS

The expression for chemical equilibrium in equation 4.6 is perfectly general. If pure solids (or pure immiscible liquids) are involved, their chemical potentials are constants independent of the extent of reaction, as long as they are not used up before equilibrium is reached. For example, for the reaction

$$CaCO_3(s) = CaO(s) + CO_2(g) \tag{4.41}$$

at equilibrium, equation 4.6 becomes

$$\mu^\circ_{CaCO_3(s)} = \mu^\circ_{CaO(s)} + \mu^\circ_{CO_2(g)} + RT \ln\left(\frac{P_{CO_2}}{P^\circ}\right) \tag{4.42}$$

$$\Delta G^\circ = \mu^\circ_{CaO(s)} + \mu^\circ_{CO_2(g)} - \mu^\circ_{CaCO_3(s)} = -RT \ln\left(\frac{P_{CO_2}}{P^\circ}\right) = -RT \ln K_P \tag{4.43}$$

so that

$$K_P = \frac{P_{CO_2}}{P^\circ} \tag{4.44}$$

There is chemical equilibrium only if all substances in the chemical equation are present in the phases shown. The equilibrium constant of such a reaction is independent of the amount of pure solid (or liquid) phase, provided only that it is present at equilibrium. Table 4.1 gives the equilibrium pressures of CO_2 above $CaCO_3$ + CaO at various temperatures.

Table 4.1 Equilibrium Constant for Reaction 4.41

$t/°C$	500	600	700	800
K_P	9.3×10^{-5}	2.42×10^{-3}	2.92×10^{-2}	0.220
$t/°C$	897	1000	1100	1200
K_P	1.000	3.871	11.50	28.68

If the partial pressure of CO_2 over $CaCO_3$ is maintained lower than K_P at a constant temperature, all the $CaCO_3$ is converted into CaO and CO_2. On the other hand, if the partial pressure of CO_2 is maintained higher than K_P, all the CaO is converted into $CaCO_3$. In this respect equilibria involving pure solids (or liquids) are different from other chemical equilibria, which would simply go to a new equilibrium position and not to completion if the partial pressure of one of the reactants or products is maintained constant.

If solid or liquid solutions are formed (e.g., if CaO and $CaCO_3$ were somewhat mutually soluble), the position of the equilibrium would depend on the concentration or, more specifically, the activities of the components in the equilibrium solid solution.

4.6 GIBBS ENERGY OF FORMATION

Instead of tabulating equilibrium constants for a large number of reactions, it is more convenient to tabulate the standard Gibbs energy of formation ΔG_f° for pure substances (see the definition of the standard enthalpy of formation in Section 1.20). The standard Gibbs energy of formation of a substance is the standard Gibbs energy change for the reaction in which the substance in its standard state is formed from elements in their standard states. Thus the standard Gibbs energies of formation of elements in their reference forms are zero at all temperatures. The standard Gibbs energy change for a reaction may be calculated from the standard Gibbs energy of formation by use of the equation

$$\Delta G^\circ = \sum_i \nu_i \, \Delta G_{f,i}^\circ \tag{4.45}$$

The form of the equilibrium constant expression that corresponds to this value of ΔG° depends on the states of the reactants and products used in the table.

The standard Gibbs energies of formation at 298 K are given for a large number of substances in Table A.1, and the standard Gibbs energies of formation of a smaller number of substances at 0, 298, 500, 1000, 2000, and 3000 K are given in Table A.2. The convention that ΔH_f°, ΔG_f°, S°, and C_P° for $H^+(aq, m = 1)$ are zero causes the equation $\Delta G_f^\circ = \Delta H_f^\circ - T\Delta S_f^\circ$ to fail for individual ionic species. However, the equation $\Delta G^\circ = \Delta H^\circ - T\Delta S^\circ$ still applies to a balanced chemical equation, even though entropies of formation of individual ions are used in the calculation.

There are three ways that ΔG° for a reaction may be obtained: (1) ΔG° may be calculated from a measured equilibrium constant using equation 4.11, (2) ΔG° may be calculated from

$$\Delta G^\circ = \Delta H^\circ - T\Delta S^\circ \tag{4.46}$$

using ΔH° obtained calorimetrically and ΔS° obtained from third law entropies, and (3) for gas reactions ΔG° may be calculated using statistical mechanics (Chapter 13) and certain information about molecules obtained from spectroscopic data. Methods 2 and 3 make it possible to calculate equilibrium constants of reactions that have never been studied in the laboratory. In method 2 the necessary data are obtained solely from thermal measurements, including heat capacity measurements down to the neighborhood of absolute zero. The calculation of equilibrium constants using statistical mechanics is even more remarkable in that only properties of individual molecules are used.

Example 4.4 Calculate the degree of dissociation of $H_2(g)$ at 3000 K and 1 atm.

$$H_2(g) = 2H(g)$$

Using Table A.2,

$$\Delta G^\circ = 2(46{,}170\ \text{J mol}^{-1}) = 92{,}340\ \text{J mol}^{-1}$$
$$= -(8.314\ \text{J K}^{-1}\ \text{mol}^{-1})(3000\ \text{K})\ln K_P$$
$$K_P = 2.47 \times 10^{-2} = \frac{4\alpha^2(P/P^\circ)}{1-\alpha^2} \qquad \alpha = 0.0783$$

A value of 0.072 was obtained experimentally by Langmuir.

Example 4.5 Calculate the equilibrium pressure for the conversion of graphite to diamond at 25 °C. The densities of graphite and diamond may be taken to be 2.25 and 3.51 g cm^3, respectively, independent of pressure, in calculating the change of ΔG with pressure.

$$\text{C(graphite)} = \text{C(diamond)}$$

Using Table A.1,

$$\Delta G^\circ = 2900\ \text{J mol}^{-1}$$

$$\Delta V = 12\left(\frac{1}{3.51} - \frac{1}{2.25}\right) \times 10^{-6}\ \text{m}^3\ \text{mol}^{-1} = -1.91 \times 10^{-6}\ \text{m}^3\ \text{mol}^{-1}$$

$$\int_1^2 d\,\Delta G = \int_1^P \Delta V\,dP = \Delta G_2 - \Delta G_1 = \Delta V(P_2 - P_1)$$

Since $(\partial \Delta G/\partial P)_T = \Delta V$,

$$P_2 = \frac{\Delta G_2 - \Delta G_1}{\Delta V} + P_1$$
$$= \frac{0 - 2900\ \text{J mol}^{-1}}{-1.91 \times 10^{-6}\ \text{m}^3\ \text{mol}^{-1}} + 101{,}325\ \text{Pa}$$
$$= 1.52 \times 10^9\ \text{Pa} \quad \text{or} \quad 15.0 \times 10^3\ \text{atm}$$

Example 4.6 Calculate the equilibrium constant for the reaction

$$2C(\text{graphite}) + 2H_2O(g) = CH_4(g) + CO_2(g)$$

at 1000 K using data from Table A.2.

$$\Delta G^\circ = [19.351 - 395.924 - 2(-192.631)] \text{ kJ mol}^{-1} = 8.689 \text{ kJ mol}^{-1}$$

$$= -RT \ln K_P = -(8.314 \text{ J K}^{-1} \text{ mol}^{-1})(1000 \text{ K}) \ln K_P$$

$$K_P = \frac{(P_{CO_2}/P^\circ)(P_{CH_4}/P^\circ)}{(P_{H_2O}/P^\circ)^2} = 0.352$$

Example 4.7 The first step of the Fischer-Tropsch process is to produce a mixture of CO and H_2 from coal and high-temperature steam. This reaction is endothermic (see Example 1.2), and so usually some coal is burned to maintain the high temperature required for the equilibrium constant to be greater than unity. The mixture of CO and H_2 is used for further synthetic reactions. Calculate the equilibrium constant at 1000 K for

$$C(\text{graphite}) + H_2O(g) = CO(g) + H_2(g)$$

Using data from Table A.2,

$$\Delta G^\circ(1000 \text{ K}) = [-200.242 - (-192.631)] \text{ kJ mol}^{-1} = -7.611 \text{ kJ mol}^{-1}$$

$$= -(8.314 \text{ J K}^{-1} \text{ mol}^{-1})(1000 \text{ K}) \ln K_P$$

$$K_P = \frac{(P_{H_2}/P^\circ)(P_{CO}/P^\circ)}{(P_{H_2O}/P^\circ)} = 2.50$$

where the standard pressure P° is 1 atm. Coal contains some hydrogen, oxygen, and other elements; it is chemically very complicated and reacts differently from graphite.

The reactions of the preceding two examples do not occur alone when graphite is in contact with water vapor at high temperatures. These reactions are of interest in the gasification and liquefaction of coal. In order to calculate the equilibrium composition of the gases in contact with carbon, it is necessary to take into account all the molecular species that may be present. Under conditions where the partial pressures of higher hydrocarbons are negligible, the carbon-hydrogen-oxygen system may be represented by

$$C(\text{graphite}) + H_2O(g) = CO(g) + H_2(g)$$

$$CO(g) + H_2O(g) = CO_2(g) + H_2(g)$$

$$C(\text{graphite}) + 2H_2(g) = CH_4(g)$$

By use of a computer it is possible to obtain equilibrium compositions over a range of temperature, pressure, and H/O ratio.* Figure 4.2 shows the equilibrium mole

* H. C. Hottel and J. B. Howard, *New Energy Technology*, The MIT Press, Cambridge, Mass., 1971, p. 103; R. E. Baron, J. H. Porter, and O. H. Hammond, Jr., *Chemical Equilibria in Carbon-Hydrogen-Oxygen Systems*, The MIT Press, Cambridge, Mass., 1976.

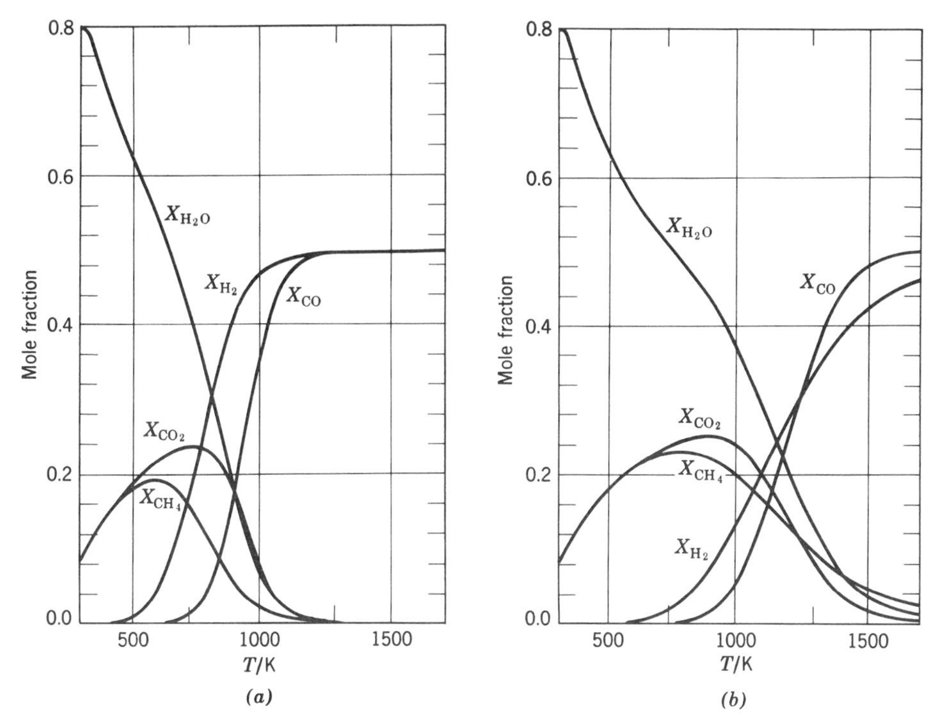

Fig. 4.2 Mole fraction of various gases in equilibrium with graphite at (*a*) 1 atm, and (*b*) 100 atm total pressure and H/O = 2. (Reprinted from *Chemical Equilibria in Carbon-Hydrogen-Oxygen Systems* by R. E. Baron, J. H. Porter, and O. H. Hammond by permission of The MIT Press, Cambridge, Mass. Copyright © 1976 by the Massachusetts Institute of Technology.)

fractions of the various gases in contact with graphite at 1 atm and 100 atm for H/O = 2, which corresponds with starting with graphite and H_2O.

4.7 EFFECT OF TEMPERATURE ON CHEMICAL EQUILIBRIUM

The effect of temperature on chemical equilibrium is determined by ΔH°, as shown by the Gibbs-Helmholtz equation 2.90. If $\Delta G^\circ = -RT \ln K$ is substituted into this equation, we obtain

$$\frac{\partial \ln K}{\partial(1/T)} = -\frac{\Delta H^\circ}{R} \tag{4.47}$$

Assuming that ΔH° is independent of temperature, the indefinite integral of this equation is

$$\ln K = \frac{-\Delta H^\circ}{RT} + C \tag{4.48}$$

where C is an integration constant. Comparison of this equation with equation 4.46 shows that it may be written

$$\ln K = -\frac{\Delta H^\circ}{RT} + \frac{\Delta S^\circ}{R} \tag{4.49}$$

According to this equation a plot of ln K versus $1/T$ has a slope of $-\Delta H^\circ/R$ and an intercept at $1/T = 0$ of $\Delta S^\circ/R$. If a straight line is not obtained, ΔH° and ΔS° depend on the temperature. The value of ΔH° at any temperature may be calculated from the slope of the plot at that temperature.

Data are given in Table 4.2 for the equilibrium constants at different temperatures for the reaction $N_2(g) + O_2(g) = 2NO(g)$.

Table 4.2 Equilibrium Constants for the Reaction $N_2(g) + O_2(g) = 2NO(g)$

T/K	1900	2000	2100	2200	2300	2400	2500	2600
$K_P/10^{-4}$	2.31	4.08	6.86	11.0	16.9	25.1	36.0	50.3

In Fig. 4.3 the values of ln K_P are plotted against $1/T$, and it is evident that a straight line is produced. The standard enthalpy change for the reaction ΔH° in the range 1900–2600 K can be calculated from the slope of the line as follows.

$$\begin{aligned}\Delta H^\circ &= -\text{slope} \times R = -(-2.19 \times 10^4\ \text{K})(8.314\ \text{J K}^{-1}\ \text{mol}^{-1}) \\ &= 182\ \text{kJ mol}^{-1}\end{aligned}$$

The intercept of Fig. 4.3 at $1/T = 0$ can be calculated from the experimental value of K_P at some temperature and the slope. The intercept may be used to calculate the standard entropy change ΔS° according to equation 4.49.

$$\frac{\Delta S^\circ}{R} = 12.36$$

$$\Delta S^\circ = (12.36)(8.314\ \text{J K}^{-1}\ \text{mol}^{-1}) = 102.8\ \text{J K}^{-1}\ \text{mol}^{-1}$$

If ΔH° is independent of temperature, the integral of equation 4.47 from T_1 to T_2 yields

$$\ln \frac{K_2}{K_1} = \frac{\Delta H^\circ (T_2 - T_1)}{RT_1T_2} \tag{4.50}$$

Example 4.8 Calculate the enthalpy change for the reaction $N_2(g) + O_2(g) = 2NO(g)$ from the equilibrium constants given in Table 4.2 for 2000 K and 2500 K.

$$\ln \frac{K_{2500\ \text{K}}}{K_{2000\ \text{K}}} = \ln \frac{3.60 \times 10^{-3}}{4.08 \times 10^{-4}} = \frac{\Delta H^\circ (2500\ \text{K} - 2000\ \text{K})}{(8.314\ \text{J K}^{-1}\ \text{mol}^{-1})(2500\ \text{K})(2000\ \text{K})}$$

and

$$\Delta H^\circ = 181\ \text{kJ mol}^{-1}$$

According to Le Châtelier's principle, when an equilibrium system is perturbed, the equilibrium will always be displaced in such a way as to oppose the applied change. When the temperature of an equilibrium system is raised, this change cannot be prevented by the system, but what happens is that the equilibrium shifts

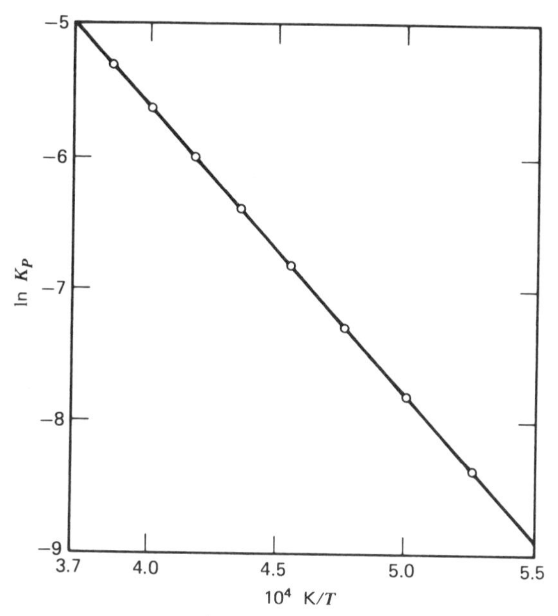

Fig. 4.3 Ln K_P plotted against reciprocal absolute temperature for the reaction $N_2(g)$ + $O_2(g)$ = $2NO(g)$. The standard enthalpy change for the reaction is calculated from the slope of the straight line.

in such a way that more heat is required to heat the reaction mixture to the higher temperature than would have been required if the mixture were inert. In other words, when the temperature is raised, the equilibrium shifts in the direction that causes an absorption of heat. This conclusion should be verified by thinking about the signs of the quantities in equation 4.49.

To state Le Châtelier's principle in a completely unambiguous way it is necessary to introduce more advanced thermodynamic concepts. The correct formulations and the difficulties with simpler formulations have been discussed by de Heer.*

4.8 GIBBS ENERGY FUNCTION

In calculating equilibrium constants at high temperatures it is most convenient to use the Gibbs energy function defined by $(G_T^\circ - H_{298}^\circ)/T$.† This quantity is used because it is readily calculated for simple gas molecules from spectroscopic data and because it varies more slowly with temperature than ΔG_f°. The Gibbs energy function may also be calculated from heat capacity measurements down to

* J. de Heer, *J. Chem. Educ.*, **34**, 375 (1957); **35**, 133 (1958).

† Note that the previous edition of this book used a Gibbs energy function $(G^\circ - H_0^\circ)/T$ with a different reference temperature.

absolute zero. The JANAF Thermochemical Tables give values every 100 K up to 6000 K. A few entries from the JANAF Thermochemical Tables are given in Table A.2.

The Gibbs energy function for a substance is related to other thermodynamic quantities by

$$S_T^\circ = -\frac{(G_T^\circ - H_T^\circ)}{T} = \frac{(H_T^\circ - H_{\text{ref}}^\circ)}{T} - \frac{(G_T^\circ - H_{\text{ref}}^\circ)}{T} \tag{4.51}$$

Thus

$$\frac{(G_T^\circ - H_{\text{ref}}^\circ)}{T} = \frac{(H_T^\circ - H_{\text{ref}}^\circ)}{T} - S_T^\circ \tag{4.52}$$

The reference temperature is arbitrary, but is taken as 298.15 K in the JANAF Thermochemical Tables.

In order to calculate the equilibrium constant for a reaction using the Gibbs energy function it is also necessary to know ΔH_{298}°. The difference between the Gibbs energy function of the reactants and products may be written

$$\Delta\left(\frac{G_T^\circ - H_{298}^\circ}{T}\right) = \frac{\Delta G_T^\circ}{T} - \frac{\Delta H_{298}^\circ}{T} \tag{4.53}$$

$$\frac{\Delta G_T^\circ}{T} = \frac{\Delta H_{298}^\circ}{T} + \Delta\left(\frac{G_T^\circ - H_{298}^\circ}{T}\right) \tag{4.54}$$

Since $\Delta G_T^\circ = -RT \ln K$,

$$\ln K = \frac{-1}{R}\left[\frac{\Delta H_{298}^\circ}{T} + \Delta\left(\frac{G_T^\circ - H_{298}^\circ}{T}\right)\right] \tag{4.55}$$

The Gibbs energy function may be calculated from heat capacity measurements down to the neighborhood of absolute zero.

$$\frac{(G_T^\circ - H_{298}^\circ)}{T} = -S_T^\circ + \frac{(H_T^\circ - H_{298}^\circ)}{T} \tag{4.56}$$

The entropy at temperature T is calculated from

$$S_T^\circ = \int_0^T \frac{C_P^\circ}{T}\, dT + \sum \frac{\Delta H_{\text{tr}}^\circ}{T_{\text{tr}}} \tag{4.57}$$

and the enthalpy term is calculated from

$$H_T^\circ - H_{298}^\circ = \int_{298}^T C_P^\circ\, dT + \Delta H_{\text{tr}}^\circ \tag{4.58}$$

where $\Delta H_{\text{tr}}^\circ$ is the enthalpy change for a phase transition.

Example 4.9 Methane may be produced from CO and H_2 provided the temperature is not too high. Use the Gibbs energy function to calculate the equilibrium constant at 650 K for the reaction

$$CO(g) + 3H_2(g) = CH_4(g) + H_2O(g)$$

$$\ln K_P = \frac{-1}{R}\left[\frac{\Delta H_{298}^\circ}{T} + \Delta\left(\frac{G_T^\circ - H_{298}^\circ}{T}\right)\right]$$

The values of the Gibbs energy function at 650 K may be obtained by linear interpolation from Table A.2.

$$\ln K_P = \frac{-1}{8.314}\left\{\frac{-74{,}873 - 241{,}789 + 110{,}529}{1000} + [-196.138 - 196.661 + 204.422 + 3(137.336)]\right\}$$

$$= -2.100$$

$$K_P = \frac{(P_{H_2O}/P^\circ)(P_{CH_4}/P^\circ)}{(P_{CO}/P^\circ)(P_{H_2}/P^\circ)^3} = 0.122$$

where $P^\circ = 1$ atm. Since this reaction is exothermic, the reverse may be used to produce hydrogen from methane at higher temperatures.

4.9 THEORETICAL CALCULATION OF EQUILIBRIUM CONSTANTS

To suggest at this stage that the equilibrium constant for a gas reaction may be calculated from spectroscopic data may be a bit mysterious. The following derivation is intended to make this a little less mysterious. As we will see later in connection with quantum mechanics, a molecule may exist in any of a series of energy levels. It is the distribution of energy levels that determines the equilibrium state of the system. This may be most simply illustrated for an isomerization reaction of perfect gases: $A(g) = B(g)$. The energy level patterns of A and B are represented schematically in Fig. 4.4. For real molecules the energy level pattern would be much more complicated. For A and B the energies in the lowest state (i.e., ϵ_{A0} and ϵ_{B0}) are taken as zero for the measurement of energies of higher states. Molecules A and B in their lowest energy states have different energies, and this

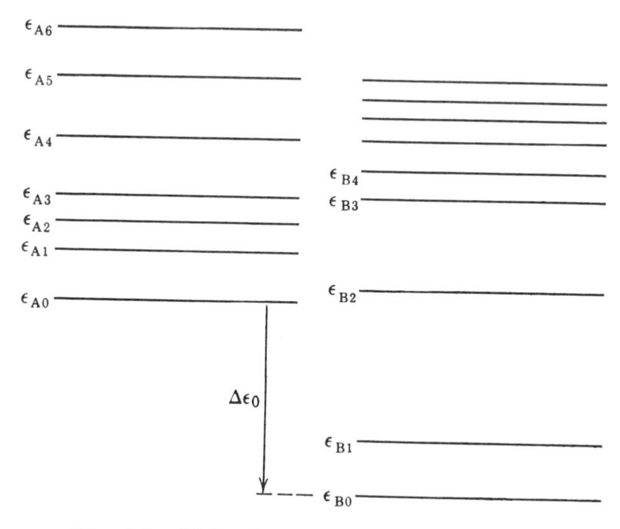

Fig. 4.4 Molecular energy levels of A and B.

difference is represented by $\Delta\epsilon_0$. The energy change at absolute zero is $\Delta\epsilon_0$, and the $A(g) = B(g)$ reaction is exothermic at absolute zero.

Molecules of A in the ith energy state are in equilibrium with those in the lowest energy state. Since the individual states of an A molecule differ in energy, but not in volume or entropy, $\Delta G^\circ_{0i} = U_{Ai} - U_{A0} = N_A(\epsilon_{Ai} - \epsilon_{A0}) = N_A\epsilon_{Ai}$, where Avogadro's constant N_A is introduced to convert molecular energies to energy per mole. Since $\Delta G^\circ_{0i} = -RT \ln K_{0i}$, we can write $N_{Ai}/N_{A0} = K_{0i} = e^{-\Delta G^\circ_{0i}/RT} = e^{-\varepsilon_{Ai}/kT}$, where k is the Boltzmann constant. If all energy states are in equilibrium, we can write $N_{Ai} = N_{A0}e^{-\varepsilon_{Ai}/kT}$ for each state. The total number N_A of molecules in A levels is given by

$$\begin{aligned} N_A &= N_{A0} + N_{A1} + N_{A2} + \cdots \\ &= N_{A0} + N_{A0}e^{-\varepsilon_{A1}/kT} + N_{A0}e^{-\varepsilon_{A2}/kT} + \cdots \\ &= N_{A0}(1 + e^{-\varepsilon_{A1}/kT} + e^{-\varepsilon_{A2}/kT} + \cdots) \\ &= N_{A0} \sum_{i=0}^{\infty} e^{-\varepsilon_{Ai}/kT} \end{aligned} \tag{4.59}$$

Similarly, for B,

$$N_B = N_{B0} \sum_{i=0}^{\infty} e^{-\varepsilon_{Bi}/kT} \tag{4.60}$$

Summations of the type given in equations 4.59 and 4.60 are referred to as molecular partition functions.

Since there is no change in volume in this gaseous isomerization reaction, the equilibrium constant K_P is simply equal to the ratio of the number of B molecules to A molecules at equilibrium: $K_P = N_B/N_A$. Substituting equations 4.59 and 4.60, we have

$$K_P = \frac{N_{B0} \sum_{i=0}^{\infty} e^{-\varepsilon_{Bi}/kT}}{N_{A0} \sum_{i=0}^{\infty} e^{-\varepsilon_{Ai}/kT}} \tag{4.61}$$

If we assume that B molecules in their lowest energy state are in equilibrium with A molecules in their lowest energy state, then $\Delta G_{00} = N_A\,\Delta\epsilon_0 = -RT \ln K_{00}$ and

$$K_{00} = \frac{N_{B0}}{N_{A0}} = e^{-\Delta\varepsilon_0/kT} \tag{4.62}$$

Since $\Delta\epsilon_0$ is negative in Fig. 4.4, the ground level for B will be more highly populated than for A. Substituting equation 4.62 into equation 4.61, we have

$$K = e^{-\Delta\varepsilon_0/kT} \frac{\sum_{i=0}^{\infty} e^{-\varepsilon_{Bi}/kT}}{\sum_{i=0}^{\infty} e^{-\varepsilon_{Ai}/kT}} \tag{4.63}$$

The equilibrium constant for a reaction may be calculated if the energy levels of the reactants and products are known and $\Delta\epsilon_0$ is known. Exact calculations for simple molecules will be made in Chapter 13.

We have already discussed how the enthalpy change and entropy change determine the equilibrium constant for a reaction. The first factor $e^{-\Delta\varepsilon_0/kT}$ is an enthalpy factor that favors the product B if $\Delta\epsilon_0$ is negative. This factor becomes less important as the temperature is raised. The second factor, the ratio of molecular partition functions, favors the product B if it has *more* low-lying energy levels than A. This ratio may be considered to be an entropy or probability factor.

4.10 CHEMICAL EQUILIBRIUM IN LIQUID SOLUTION

Different expressions are obtained from equation 4.6 for chemical equilibrium in solution, depending on the way the chemical potential is related to composition. On the mole fraction scale the chemical potential of a component of a liquid solution may be expressed by

$$\mu_i = \mu_i^* + RT \ln f_i X_i \tag{4.64}$$

(given earlier in equation 3.64), where f_i is the activity coefficient of component i, X_i is its mole fraction, and μ_i^* is the chemical potential in the standard state. Substituting equation 4.64 into equation 4.6 yields

$$K = \prod_i (f_i X_i)^{\nu_i} \tag{4.65}$$

and $\Delta G^\circ = \sum_i \nu_i \mu_i^*$. Equation 4.65 is the equilibrium expression of choice for solutions whenever it is not obvious which component is solvent and which is solute. Table A.1 gives standard Gibbs energies of formation for pure liquids. For ideal solutions (Section 3.10) the activity coefficients f_i are equal to unity and the equilibrium constant expression becomes

$$K = \prod_i X_i^{\nu_i} \tag{4.66}$$

Example 4.10 One mole of acetic acid is mixed with 1 mol of ethanol at 25 °C, and after equilibrium is reached a titration with standard alkali solution shows that 0.667 mol of acetic acid has reacted to form ethyl acetate according to

$$CH_3CO_2H(l) + C_2H_5OH(l) = CH_3CO_2C_2H_5(l) + H_2O(l)$$

Calculate the apparent equilibrium constant in terms of mole fraction.

$$K = \frac{X_{CH_3CO_2C_2H_5} X_{H_2O}}{X_{CH_3CO_2H} X_{C_2H_5OH}} = \frac{(0.667/2)(0.667/2)}{[(1.000 - 0.667)/2][(1.000 - 0.667)/2]} = 4.00$$

The total number of moles appears in the denominator of each term and therefore cancels. When 0.500 mol of ethanol is added to 1.000 mol of acetic acid at 25 °C, how much ester x will be formed at equilibrium?

$$K = 4.00 = \frac{x^2}{(1.000 - x)(0.500 - x)} \qquad x = 0.422 \quad \text{or} \quad 1.577 \text{ mol}$$

Two solutions of a quadratic equation are possible, but in such problems one solution is incompatible with the physical-chemical facts. In this example it is impossible to produce more moles of ester than the original number of moles of ethanol, and so the value 1.577 is impossible. Actually 0.422 mol of ester and 0.422 mol of water are formed, and (0.500 − 0.422) or 0.078 mol of ethanol and (1.000 − 0.422) or 0.578 mol of acetic acid remain unreacted. An experimental value of 0.414 mol of ester was obtained in the laboratory. Additional determinations are as follows.

Moles ethanol added to 1 mol of acetic acid	0.080	0.280	2.240	8.000
Moles of ethyl acetate calculated from K	0.078	0.232	0.864	0.945
Moles of ethyl acetate found experimentally	0.078	0.226	0.836	0.966

In example 4.10 the number of moles of reactants and of products is the same, and the calculation is simplified. In reactions where $\sum \nu_i$ is not zero the total number of moles will not cancel out.

The chemical potential of a component of a liquid solution may also be expressed on the molal concentration scale.

$$\mu_i = \mu_i^\circ + RT \ln \left(\frac{\gamma_i m_i}{m^\circ} \right) \tag{4.67}$$

The molal concentration m_i of component i is the number of moles of i per kilogram of solvent. The activity coefficient on the molal scale is represented by γ_i. Since the activity coefficient is dimensionless, $\gamma_i m_i$ has to be divided by m°, the standard concentration of 1 mol kg^{-1}, to obtain a dimensionless measure of the activity. This expression for the chemical potential leads to

$$K_m = \prod_i \left(\frac{\gamma_i m_i}{m^\circ} \right)^{\nu_i} \tag{4.68}$$

In Table A.1 the standard state for a solute in aqueous solution is taken as the hypothetical ideal solution of unit molality.

The chemical potential of a component of a liquid solution may also be expressed on the molar concentration scale.

$$\mu_i = \mu_i^\circ + RT \ln \left(\frac{y_i c_i}{c^\circ} \right) \tag{4.69}$$

The molar concentration c_i of component i is the number of moles of i per liter of solution. The activity coefficient on the molar scale is represented by y_i, and the standard concentration, 1 mol L^{-1}, is represented by c°. This expression for the chemical potential leads to

$$K_c = \prod_i \left(\frac{y_i c_i}{c^\circ} \right)^{\nu_i} \tag{4.70}$$

This form of the equilibrium constant is frequently used for calculations in dilute aqueous solutions, as in Chapter 6.

Example 4.11 What are the values of ΔG°, ΔH°, ΔS°, and K_m at 298.15 K for

$$H_2O(l) = H^+(aq) + OH^-(aq)$$

Using Table A.1,

$$\Delta G^\circ = (-157.293 + 237.178) \text{ kJ mol}^{-1} = 79.885 \text{ kJ mol}^{-1}$$

$$\Delta H^\circ = (-229.994 + 285.830) \text{ kJ mol}^{-1} = 55.835 \text{ kJ mol}^{-1}$$

$$\Delta S^\circ = (-10.753 - 69.915) \text{ J K}^{-1} \text{ mol}^{-1} = -80.668 \text{ J K}^{-1} \text{ mol}^{-1}$$

$$K_m = e^{-\Delta G^\circ/RT} = 1.008 \times 10^{-14} = (m_{H^+}/m^\circ)(m_{OH^-}/m^\circ)$$

where concentrations are expressed in moles per kilogram of solvent.

References

K. E. Bett, J. S. Rowlinson, and G. Saville, *Thermodynamics for Chemical Engineers*, The MIT Press, Cambridge, Mass., 1975.

K. G. Denbigh, *The Principles of Chemical Equilibrium*, Cambridge University Press, Cambridge, 1971.

R. E. Dickerson, *Molecular Thermodynamics*, W. A. Benjamin Co., New York, 1969.

I. M. Klotz and R. M. Rosenberg, *Chemical Thermodynamics*, W. A. Benjamin Co., New York, 1972.

G. N. Lewis, M. Randall, revised by K. S. Pitzer, and L. Brewer, *Thermodynamics*, McGraw-Hill Book Co., New York, 1961.

A. B. Pippard, *Elements of Classical Thermodynamics*, Cambridge University Press, Cambridge, 1960.

P. A. Rock, *Chemical Thermodynamics*, The Macmillan Co., London, 1969.

D. R. Stull and H. Prophet, *JANAF Thermochemical Tables*, 2nd ed., NSRDS-NBS37, U.S. Government Printing Office, Washington, D.C., 1971.

J. Szekely, J. W. Evans, and H. Y. Sohn, *Gas-Solid Reactions*, Academic Press, New York, 1976.

F. T. Wall, *Chemical Thermodynamics*, W. H. Freeman and Co., San Francisco, 1965.

Problems

4.1 For the reaction

$$N_2(g) + 3H_2(g) = 2NH_3(g)$$

$K_P = 1.64 \times 10^{-4}$ at 400 °C. Calculate (*a*) $\Delta G°$ and (*b*) ΔG when the pressures of N_2 and H_2 are maintained at 10 and 30 atm, respectively, and NH_3 is removed at a partial pressure of 3 atm. (*c*) Is the reaction spontaneous under the latter conditions?

Ans. (*a*) 48.77, (*b*) −8.91 kJ mol^{-1}. (*c*) Yes.

4.2 A 1:3 mixture of nitrogen and hydrogen was passed over a catalyst at 450 °C. It was found that 2.04% by volume of ammonia was formed when the total pressure was maintained at 10 atm [A. T. Larson and R. L. Dodge, *J. Am. Chem. Soc.*, **45**, 2918 (1923)]. Calculate the value of K_P for $\frac{3}{2}H_2(g) + \frac{1}{2}N_2(g) = NH_3(g)$ at this temperature.

Ans. $K_P = 6.44 \times 10^{-3}$.

4.3 Water vapor is passed over coal (assumed to be pure graphite in this problem) at 1000 K. Assuming that the only reaction occurring is the water gas reaction,

$$C\,(\text{graphite}) + H_2O(g) = CO(g) + H_2(g) \qquad K_P = 2.49$$

calculate the equilibrium pressures of H_2O, CO, and H_2 at a total pressure of 1 atm. (Actually the water gas shift reaction

$$CO(g) + H_2O(g) = CO_2(g) + H_2(g)$$

occurs in addition, but it is considerably more complicated to take this subsequent reaction into account.)

Ans. 0.08, 0.46, 0.46 atm.

4.4 How many moles of phosphorus pentachloride must be added to a liter vessel at 250 °C to obtain a concentration of 0.1 mol of chlorine per liter?

Given:

$$K_P = \frac{(P_{PCl_3}/P°)(P_{Cl_2}/P°)}{(P_{PCl_5}/P°)} = 1.78$$

Ans. 0.341 mol.

4.5 Calculate the total pressure that must be applied to a mixture of three parts of hydrogen and one part nitrogen to give a mixture containing 10% ammonia at 400 °C. At 400 °C, $K_P = 1.64 \times 10^{-4}$ for the reaction $N_2(g) + 3H_2(g) = 2NH_3(g)$.

Ans. 29.7 atm.

4.6 The value of K_P for $N_2O_4(g) = 2NO_2(g)$ is 0.141 at 25 °C. What is the value of K_c?

Ans. 5.76×10^{-3}.

4.7 At 1273 K and at a total pressure of 30 atm the equilibrium in the reaction

$$CO_2(g) + C(s) = 2CO(g)$$

is such that 17 mole % of the gas is CO_2. (*a*) What percentage would be CO_2 if the total pressure were 20 atm? (*b*) What would be the effect on the equilibrium of adding N_2 to the reaction mixture in a closed vessel until the partial pressure of N_2 is 10 atm? (*c*) At what pressure of the reactants will 25% of the gas be CO_2?

Ans. (*a*) 12.6%, (*b*) no effect, (*c*) 54 atm.

4.8 At 2000 °C water is 2% dissociated into oxygen and hydrogen at a total pressure of 1 atm. (*a*) Calculate K_P for

$$H_2O(g) = H_2(g) + \tfrac{1}{2}O_2(g)$$

(*b*) Will the degree of dissociation increase or decrease if the pressure is reduced? (*c*) Will the degree of dissociation increase or decrease if argon gas is added, holding the total pressure equal to 1 atm? (*d*) Will the degree of dissociation change if the pressure is raised by addition of argon at constant volume to the closed system containing partially dissociated water vapor? (*e*) Will the degree of dissociation increase or decrease if oxygen gas is added while holding the total pressure constant at 1 atm?

Ans. (*a*) 2.03×10^{-3}, (*b*) increase, (*c*) increase, (*d*) no change, (*e*) decrease.

4.9 In the synthesis of methanol by

$$CO(g) + 2H_2(g) = CH_3OH(g)$$

at 500 K, calculate the total pressure required for a 90% conversion to methanol if CO and H_2 are initially in a 1:2 ratio. Given: $K_P = 6.25 \times 10^{-3}$. *Ans.* 228 atm.

4.10 An evacuated tube containing 5.96×10^{-3} mol L^{-1} of solid iodine is heated to 973 K. The experimentally determined pressure is 0.490 atm [M. L. Perlman and G. K. Rollefson, *J. Chem. Phys.*, **9**, 362 (1941)]. Assuming perfect-gas behavior, calculate K_P for $I_2(g) = 2I(g)$. *Ans.* 1.74×10^{-3}.

4.11 At 55 °C and 1 atm the average molar mass of partially dissociated N_2O_4 is 61.2 g mol^{-1}. Calculate (*a*) α and (*b*) K_P for the reaction $N_2O_4(g) = 2NO_2(g)$. (*c*) Calculate α at 55 °C if the total pressure is reduced to 0.1 atm.

Ans. (*a*) 0.503, (*b*) 1.36, (*c*) 0.879.

4.12 For the reaction $N_2O_4(g) = 2NO_2(g)$, K_P at 25 °C is 0.141. What pressure would be expected if 1 g of liquid N_2O_4 were allowed to evaporate into a liter vessel at this temperature? Assume that N_2O_4 and NO_2 are perfect gases. *Ans.* 0.347 atm.

4.13 Under what total pressure at equilibrium must PCl_5 be placed at 250 °C to obtain a 30% conversion into PCl_3 and Cl_2? For the reaction $PCl_5(g) = PCl_3(g) + Cl_2(g)$, $K_P = 1.78$ at 250 °C. *Ans.* 18.0 atm.

4.14 The equilibrium constant for the reaction $SO_2(g) + \tfrac{1}{2}O_2(g) = SO_3(g)$ at 727 °C is given by

$$K_P = \frac{P_{SO_3}}{P_{SO_2}P_{O_2}^{1/2}} = 1.85$$

What is the ratio P_{SO_3}/P_{SO_2} (*a*) when the partial pressure of oxygen at equilibrium is 0.3 atm, (*b*) when the partial pressure of oxygen at equilibrium is 0.6 atm? (*c*) What is the effect on the equilibrium if the total pressure of the mixture of gases is increased by forcing in nitrogen at constant volume?

Ans. (*a*) 1.01, (*b*) 1.43, (*c*) no effect if the gases behave ideally.

4.15 A liter reaction vessel containing 0.233 mol of N_2 and 0.341 mol of PCl_5 is heated to 250 °C. The total pressure at equilibrium is 28.95 atm. Assuming that all the gases are ideal, calculate K_P for the only reaction that occurs.

$$PCl_5(g) = PCl_3(g) + Cl_2(g)$$

Ans. 1.80.

4.16 At 250 °C PCl_5 is 80% dissociated at a pressure of 1 atm, and so $K_P = 1.78$. What is the percentage dissociation at equilibrium after sufficient nitrogen has been added at constant pressure to produce a nitrogen partial pressure of 0.9 atm? The total pressure is maintained at 1 atm.

Ans. 97.3%.

4.17 Calculate (*a*) K_P and (*b*) $\Delta G°$ for the following reaction at 20 °C.

$$CuSO_4 \cdot 4NH_3(s) = CuSO_4 \cdot 2NH_3(s) + 2NH_3(g)$$

The equilibrium pressure of NH_3 is 62 torr. *Ans.* (*a*) 6.66×10^{-3}, (*b*) 12.2 kJ mol^{-1}.

4.18 Air (20% O_2) will not oxidize silver at 200 °C and 1 atm pressure. Make a statement about the equilibrium constant for the reaction

$$2Ag(s) + \tfrac{1}{2}O_2(g) = Ag_2O(s)$$

Ans. $K < 2.24$.

4.19 The dissociation of ammonium carbamate takes place according to the reaction

$$(NH_2)CO(ONH_4)(s) = 2NH_3(g) + CO_2(g)$$

When an excess of ammonium carbamate is placed in a previously evacuated vessel, the partial pressure generated by NH_3 is twice the partial pressure of the CO_2, and the partial pressure of $(NH_2)CO(ONH_4)$ is negligible in comparison. Show that

$$K_P = (P_{NH_3})^2 P_{CO_2} = \tfrac{4}{27}P^3$$

where P is the total pressure.

4.20 From the $\Delta G_f°$ of $Br_2(g)$ at 25 °C, calculate the vapor pressure of $Br_2(l)$. The pure liquid at 1 atm and 25 °C is taken as the standard state. *Ans.* 28.6 kPa.

4.21 Calculate the acid dissociation constant for acetic acid in water at 25 °C using values of the Gibbs energy of formation from Table A.1. *Ans.* 1.75×10^{-5}.

4.22 Calculate the equilibrium constants for the following reactions at 25 °C.

(*a*) $2CH_4(g) = C_2H_6(g) + H_2(g)$
(*b*) $CH_4(g) + H_2O(g) = CH_3OH(g) + H_2(g)$
(*c*) $C_2H_6(g) = C_2H_4(g) + H_2(g)$
(*d*) $C_2H_4(g) + H_2O(g) = C_2H_5OH(g)$

Ans. (*a*) 9.54×10^{-13}, (*b*) 2.77×10^{-21}, (*c*) 2.02×10^{-18}, (*d*) 26.3.

4.23 In order to produce more hydrogen from "synthesis gas" ($CO + H_2$) the water gas shift reaction is used.

$$CO(g) + H_2O(g) = CO_2(g) + H_2(g)$$

Calculate K_P at 1000 K and the equilibrium extent of reaction starting with an equimolar mixture of CO and H_2O.

Ans. 1.44, 0.546.

4.24 Calculate the equilibrium constant K_P for the production of H_2 from CH_4 at 1000 K using the reaction

$$CH_4(g) + H_2O(g) = CO(g) + 3H_2(g)$$

Ans. 25.6.

4.25 Calculate the equilibrium constants for

$$C(\text{graphite}) + 2H_2(g) = CH_4(g)$$

at 500 and 1000 K. *Ans.* 2.69×10^3, 9.75×10^{-2}.

4.26 What are the percentage dissociations of $H_2(g)$, $O_2(g)$, and $I_2(g)$ at 2000 K and a total pressure of 1 atm? *Ans.* 0.081, 0.033, 94.7%.

4.27 The following reaction takes place in the presence of aluminum chloride.

$$\text{Cyclohexane}(l) = \text{Methylcyclopentane}(l)$$

At 25 °C $K_c = 0.143$, and at 45 °C $K_c = 0.193$. From these data calculate: (*a*) $\Delta G°$ at 25 °C, (*b*) $\Delta H°$, and (*c*) $\Delta S°$. *Ans.* (*a*) 4820, (*b*) 11,800 J mol^{-1}, (*c*) 23.41 J K^{-1} mol^{-1}.

4.28 The following data apply to the reaction $Br_2(g) = 2Br(g)$.

T/K	1123	1173	1223	1273
$K_P/10^{-3}$	0.403	1.40	3.28	7.1

Determine by graphical means the enthalpy change when 1 mol of Br_2 dissociates completely at 1200 K. *Ans.* 199.2 kJ mol^{-1}.

4.29 Mercuric oxide dissociates according to the reaction $2HgO(s) = 2Hg(g) + O_2(g)$. At 420 °C the dissociation pressure is 5.16×10^4 Pa, and at 450 °C it is 10.8×10^4 Pa. Calculate (*a*) the equilibrium constants, and (*b*) the enthalpy of dissociation per mole of HgO. *Ans.* (*a*) 0.0196, 0.1794 atm^3, (*b*) 154 kJ mol^{-1}.

4.30 The vapor pressure of water above mixtures of $CuCl_2 \cdot H_2O(s)$ and $CuCl_2 \cdot 2H_2O(s)$ is given as a function of temperature in the following table.

t/°C	17.9	39.8	60.0	80.0
P/atm	0.0049	0.0247	0.120	0.322

(*a*) Calculate $\Delta H°$ for the reaction

$$CuCl_2 \cdot 2H_2O(s) = CuCl_2 \cdot H_2O(s) + H_2O(g)$$

(*b*) Calculate $\Delta G°$ for the reaction at 60 °C. (*c*) Calculate $\Delta S°$ for the reaction at 60 °C. *Ans.* (*a*) 57.3 kJ mol^{-1}, (*b*) 5900 J mol^{-1}, (*c*) 154.4 J K^{-1} mol^{-1}.

4.31 (*a*) Calculate the equilibrium constants of the following reactants at 298 K.

$$C(\text{graphite}) + 2H_2(g) = CH_4(g) \qquad 2C(\text{graphite}) + 3H_2(g) = C_2H_6(g)$$

(*b*) Assuming that $\Delta H°$ is independent of temperature, at what temperature will these reactions have equilibrium constants of unity.

Ans. (*a*) 7.79×10^8, 5.77×10^5. (*b*) 927, 487 K.

4.32 Using values for the ion product for water at 25 °C from Table 18.3, calculate $\Delta H°$ for

$$H_2O(l) = H^+(aq) + OH^-(aq)$$

and compare it with the value calculated from Table A.1. *Ans.* 56.5, 55.8 kJ mol^{-1}.

4.33 The following reaction is nonspontaneous at room temperature and endothermic.

$$3C(\text{graphite}) + 2H_2O(g) = CH_4(g) + 2CO(g)$$

As the temperature is raised the equilibrium constant will become equal to unity at some point. Estimate this temperature using data from Table A.2. *Ans.* 1023 K.

4.34 Calculate the degree of dissociation of $H_2O(g)$ into $H_2(g)$ and $O_2(g)$ at 2000 K and 1 atm. (Since the degree of dissociation is small, the calculation may be simplified by assuming that $P_{H_2O} = 1$ atm.) *Ans.* 0.0055.

4.35 The measured density of an equilibrium mixture of N_2O_4 and NO_2 at 15 °C and 1 atm is 3.62 g L^{-1}, and the density at 75 °C and 1 atm is 1.84 g L^{-1}. What is the enthalpy change of the reaction $N_2O_4(g) = 2NO_2(g)$? *Ans.* 75 kJ mol^{-1}.

4.36 The equilibrium constant for the association of benzoic acid to a dimer in dilute benzene solutions is as follows at 43.9 °C.

$$2C_6H_5COOH = (C_6H_5COOH)_2 \qquad K_c = 2.7 \times 10^2$$

Molar concentrations are used in expressing the equilibrium constant. Calculate $\Delta G°$, and state its meaning. *Ans.* -14.8 kJ mol^{-1}. This is the decrease in Gibbs energy when 2 mol of monomer at unit activity on the molar scale is converted to 1 mol of dimer at unit activity on the molar scale.

4.37 In determining equilibrium constants for reactions with large or small constants, the analytical method usually puts a limit on the magnitude of the constant that can be experimentally determined. For a reaction of the type A = B it is found that there is less than 1 part per 1000 of B at equilibrium at 25 °C. Calculate the *minimum* value for $\Delta G°_{298}$ for this reaction.

4.38 For the reaction

$$C(s) + 2H_2(g) = CH_4(g)$$

at 1000 °C, $K_P = 0.263$. Calculate the total pressure at equilibrium when 0.100 mol of CH_4 is placed in a volume of 2 L at 1000 °C.

4.39 Calculate the percent methane at equilibrium at 1000 K if hydrogen is added to carbon and the total pressure is maintained at 50 atm. Given:

$$\text{C (graphite)} + 2H_2(g) = CH_4(g) \qquad K_P = \frac{P_{CH_4}}{P_{H_2}^2} = 9.85 \times 10^{-2}$$

4.40 At 500 °C, $K_P = 5.5$ for the reaction

$$CO(g) + H_2O(g) = CO_2(g) + H_2(g)$$

If a mixture of 1 mol of CO and 5 mol of H_2O is passed over a catalyst at this temperature, what will be the equilibrium mole fraction of H_2O?

4.41 A mixture of N_2 and H_2 is passed over a catalyst at 400 °C to obtain the equilibrium conversion to NH_3. The equilibrium constant at this temperature for $N_2(g) + 3H_2(g) = 2NH_3(g)$ is $K_P = 1.64 \times 10^{-4}$. (*a*) Calculate the total pressure required to produce 5.26 mole % NH_3 if an equimolar mixture of N_2 and H_2 is passed over the catalyst. (*b*) If a mixture of 2 mol of N_2 per mol of H_2 is passed over the catalyst, what total pressure will be required to produce 5.26 mole % NH_3 at equilibrium? The fact that a higher pressure is required when a larger proportion of N_2 is used may appear to be in violation of Le Châtelier's principle, but it is not in conflict with the correct statement of this principle [J. de Heer, *J. Chem. Educ.*, **34**, 375 (1957)].

4.42 For the reaction

$$CH_4(g) + 2H_2S(g) = CS_2(g) + 4H_2(g)$$

$K_P = 2.05 \times 10^9$ at 25 °C. Calculate (*a*) K_P and (*b*) K_c at this temperature for

$$2H_2(g) + \tfrac{1}{2}CS_2(g) = H_2S(g) + \tfrac{1}{2}CH_4(g)$$

4.43 For the reaction

$$CO(g) + 2H_2(g) = CH_3OH(g)$$

$K_P = 6.25 \times 10^{-3}$ at 500 K. What is the value of K_c?

4.44 For the formation of nitric oxide

$$N_2(g) + O_2(g) = 2NO(g)$$

K_P at 2126.9 °C is 2.5×10^{-3}. (*a*) In an equilibrium mixture containing 0.1 atm partial pressure of N_2 and 0.1 atm partial pressure of O_2, what is the partial pressure of NO?

(*b*) In an equilibrium mixture of N_2, O_2, NO, CO_2, and other inert gases at 2126.9 °C and 1 atm total pressure, 80% by volume of the gas is N_2 and 16% is O_2. What is the percent by volume of NO? (*c*) What is the total partial pressure of the inert gases?

4.45 An equimolar mixture of CO(*g*), $H_2(g)$, and $H_2O(g)$ at 1000 K is compressed. At what total pressure will solid carbon start to precipitate out if there is chemical equilibrium? (See Problem 4.3.)

4.46 Since the synthesis of ammonia

$$N_2(g) + 3H_2(g) = 2NH_3(g)$$

is exothermic, lower temperatures favor higher equilibrium concentrations of ammonia. However, the reaction rate is so slow below about 400 °C that the process is uneconomical with known catalysts. The following table gives the percentage of ammonia at equilibrium at 400 °C for a H_2/N_2 ratio of 3.

P/atm	1	10	100	1000
%NH_3	0.44	3.85	25.12	79.82

What are the values of K_P at each pressure? Why do these values of K_P vary with the pressure?

4.47 The dissociation of N_2O_4 is represented by $N_2O_4(g) = 2NO_2(g)$. If the density of the equilibrium gas mixture is 3.174 g L^{-1} at a total pressure of 1 atm at 24 °C, what minimum pressure would be required to keep the degree of dissociation of N_2O_4 below 0.1 at this temperature?

4.48 When N_2O_4 is allowed to dissociate to form NO_2 at 25 °C at a total pressure of 1 atm, it is 18.5% dissociated at equilibrium, and so $K_P = 0.141$. (*a*) If N_2 is added to the system at constant volume, will the equilibrium shift? (*b*) If the system is allowed to expand as N_2 is added at a constant total pressure of 1 atm, what will be the equilibrium degree of dissociation when the N_2 partial pressure is 0.6 atm.

4.49 Calculate the equilibrium constant for the reaction

$$2NO_2(g) = N_2O_4(g)$$

at 25 °C using data in Table A.1.

4.50 At 55 °C $K_P = 1.36$ for $N_2O_4(g) = 2NO_2(g)$. (*a*) How many moles of N_2O_4 must be added to a 10-L vessel at 55 °C for the concentration of NO_2 to be 0.1 mol L^{-1}? (*b*) How many moles of N_2O_4 must be added to a previously evacuated 10-L vessel at 55 °C for the total pressure to be 2 atm?

4.51 At 250 °C, 1 L of partially dissociated phosphorus pentachloride gas, at 1 atm, weighs 2.690 g. Calculate the degree of dissociation α and the equilibrium constant K_P.

4.52 The reaction

$$2NOCl(g) = 2NO(g) + Cl_2(g)$$

comes to equilibrium at 1 atm total pressure and 227 °C when the partial pressure of the nitrosyl chloride, NOCl, is 0.64 atm. Only NOCl was present initially. (*a*) Calculate $\Delta G°$ for this reaction, (*b*) At what total pressure will the partial pressure of Cl_2 be 0.1 atm?

4.53 For the reaction

$$2HI(g) = H_2(g) + I_2(g)$$

at 698.6 K, $K_P = 1.83 \times 10^{-2}$. (*a*) How many grams of hydrogen iodide will be formed when 10 g of iodine and 0.2 g of hydrogen are heated to this temperature in a 3-L vessel? (*b*) What will be the partial pressures of H_2, I_2, and HI?

4.54 At 400 °C, $K_P = 78.1$ for the reaction

$$NH_3(g) = \tfrac{1}{2}N_2(g) + \tfrac{3}{2}H_2(g)$$

Show that the fraction α of NH_3 dissociated at a total pressure P is given by

$$\alpha = \frac{1}{\sqrt{1 + kP}}$$

and calculate the value of k in this equation.

4.55 Ten grams of calcium carbonate is placed in a container of 1-L capacity and heated to 800 °C. (*a*) How many grams of $CaCO_3$ remain undecomposed? (*b*) If the amount of $CaCO_3$ were 20 g, how much would remain undecomposed?

4.56 A gas mixture containing 97 mole % water and 3 mole % hydrogen is heated to 1000 K. Will the equilibrium mixture react with nickel at 1000 K to produce nickel oxide?

$$Ni(s) + \tfrac{1}{2}O_2(g) = NiO(s) \qquad \Delta G^\circ_{1000} = 148.1 \text{ kJ mol}^{-1}$$
$$H_2(g) + \tfrac{1}{2}O_2(g) = H_2O(g) \qquad \Delta G^\circ_{1000} = -192.631 \text{ kJ mol}^{-1}$$

4.57 The solubility of hydrogen in a molten iron alloy is found to be proportional to the square root of the partial pressure of hydrogen. How can this be explained?

4.58 Calculate the equilibrium constant for the dissociation of gaseous oxygen at 2000 K.

$$O_2(g) = 2O(g)$$

4.59 Calculate ΔH°, ΔG°, and K at 25 °C for the reaction

$$C(\text{graphite}) + H_2O(g) = \tfrac{1}{2}CH_4(g) + \tfrac{1}{2}CO_2(g)$$

4.60 Is magnetite (Fe_3O_4) or hematite (Fe_2O_3) the more stable ore thermodynamically at 25 °C in contact with air?

4.61 Using ΔG_f° data in Table A.2, calculate K_P at 700 K for

$$2HI(g) = H_2(g) + I_2(g)$$

The experimental value at 698.3 K is 1.84×10^{-2} [A. H. Taylor, Jr. and R. H. Crist, *J. Am. Chem. Soc.*, **63**, 1381 (1941)].

4.62 What is K_P for

$$I_2(g) = 2I(g)$$

at 1000 K and what is the degree of dissociation at 1 atm? At 0.1 atm?

4.63 Calculate the equilibrium constant for

$$2C(\text{graphite}) + 2H_2O(g) = CH_4(g) + CO_2(g)$$

at 500 K and the equilibrium extent of reaction.

4.64 The equilibrium constant K_P for $H_2(g) = 2H(g)$ is 1.52×10^{-7} at 1800 K and 3.10×10^{-6} at 2000 K. Calculate ΔH° for this temperature range.

4.65 The average molar mass M of equilibrium mixtures of NO_2 and N_2O_4 at 1 atm total pressure are given in the following table at three temperatures.

t/°C	25	45	65
M/g mol^{-1}	77.64	66.80	56.51

(*a*) Calculate the degree of dissociation of N_2O_4 and the equilibrium constant at each of these temperatures. (*b*) Plot log K_P against $1/T$ and calculate ΔH° for the dissociation of N_2O_4. (*c*) Calculate the equilibrium constant at 35 °C. (*d*) Calculate the degree of dissociation α for N_2O at 35 °C when the total pressure is 0.5 atm.

4.66 The partial pressure of oxygen in equilibrium with a mixture of silver oxide and silver is given by

$$\log P = 6.2853 - 2859/T$$

where P is expressed in atmospheres. Calculate the temperature at which silver oxide should begin to decompose when it is heated in air at 1 atm pressure.

4.67 Calculate $\Delta H°$ for the reaction

$$CaCO_3(s) = CaO(s) + CO_2(g)$$

in the range 1000 to 1200 °C from the data of Table 4.1.

4.68 For the reaction

$$Fe_2O_3(s) + 3CO(g) = 2Fe(s) + 3CO_2(g)$$

the following values of K_P are known.

t/°C	100	250	1000
K_P	1100	100	0.0721

At 1120 °C for the reaction $2CO_2(g) = 2CO(g) + O_2(g)$, $K_P = 1.4 \times 10^{-12}$ atm. What equilibrium partial pressure of O_2 would have to be supplied to a vessel at 1120 °C containing solid Fe_2O_3 just to prevent the formation of Fe?

4.69 (*a*) Calculate the equilibrium constants of the following reactions at 298 K.

$$3C\,(\text{graphite}) + 2H_2O(g) = CH_4(g) + 2CO(g)$$

$$5C\,(\text{graphite}) + 3H_2O(g) = C_2H_6(g) + 3CO(g)$$

(*b*) Assuming that $\Delta H°$ is independent of temperature, at what temperatures will these reactions have equilibrium constants of unity?

4.70 Calculate the partial pressure of CO_2 over $CaCO_3(s)$—$CaO(s)$ at 500 °C using Table A.1.

4.71 Given the equilibrium constant for the formation of ethyl acetate in example 4.10 and the standard Gibbs energies of formation of ethanol, acetic acid, and water in Table A.1, calculate the standard Gibbs energy of formation of ethyl acetate.

4.72 Amylene, C_5H_{10}, and acetic acid react to give the ester according to the reaction

$$C_5H_{10} + CH_3COOH = CH_3COOC_5H_{11}$$

What is the value of K_c if 0.006 45 mol of amylene and 0.001 mol of acetic acid dissolved in 845 cm^3 of a certain inert solvent react to give 0.000 784 mol of ester?

4.73 Using the Gibbs energy function, calculate the equilibrium constant at 1000 K for the reaction

$$CO(g) + H_2O(g) = CO_2(g) + H_2(g)$$

CHAPTER 5
ELECTROCHEMICAL CELLS

The measurement of the electromotive force of an electrochemical cell over a range of temperature makes it possible to obtain the thermodynamic quantities for the reaction that occurs in the cell. The activity coefficients of electrolytes may also be calculated from these measurements; this is illustrated in this chapter by the determination of the activity coefficient of hydrochloric acid.

Electrochemical cells are of practical interest in that they offer the means to convert the Gibbs energy change of a chemical reaction into work without the second-law losses of heat engines.

An understanding of the conversion of chemical energy into electrical energy is important for work with batteries, fuel cells, electroplating, corrosion, electrorefining (e.g., the production of aluminum), and electroanalytical techniques. This chapter discusses the thermodynamics of such processes. The kinetics of electrode reactions are discussed briefly in Chapter 16.

5.1 COULOMB'S LAW, ELECTRIC POTENTIAL, AND UNITS

Of the four kinds of interactions recognized in physics, strong nuclear interactions, weak interactions, electromagnetic interactions, and gravitation, only electromagnetic interactions are of importance in chemistry. The most elementary of these is the attraction or repulsion between two charges, Q_1 and Q_2, which is directed along their line of centers. According to Coulomb's law, the force F is given by

$$F = \frac{Q_1 Q_2}{4\pi\epsilon_0 \kappa r^2} \tag{5.1}$$

where r is the distance between the charges, κ is the dielectric constant, and the permittivity of free space ϵ_0 equals $8.854\,187\,82 \times 10^{-12}\ \mathrm{C^2\ N^{-1}\ m^{-2}}$. For doing calculations it is convenient to know that $1/4\pi\epsilon_0 = 8.987\,551\,78 \times 10^9\ \mathrm{C^{-2}\ N\ m^2}$.

In the SI system the ampere A is a defined quantity (see Appendix) and so the unit of electric charge is the coulomb C, which is an ampere second.

The electric field around a charged particle may be described by a scalar quantity called the electric potential. The difference between the electric potential at two points is equal to the work per unit charge required to move a charge from one point to another. Thus the unit of potential difference is joules per coulomb; this unit is referred to as a volt. $1\ \mathrm{V} = 1\ \mathrm{J\ C^{-1}}$. The choice of zero potential is arbitrary, but it is customary to define the potential as zero when the particles are at infinite distance. Thus the electric potential at a point is the work required to bring a unit positive charge from infinity to the point in question. The electric potential is given by

$$\varphi = -\int_{\infty}^{r} \frac{Q_2\, dr}{4\pi\epsilon_0 \kappa r^2} = \frac{Q_2}{4\pi\epsilon_0 \kappa r} \tag{5.2}$$

Table 5.1 summarizes the symbols and definitions of the units that we will be using.

Table 5.1 Electric Units in the SI System

Quantity	Unit	Symbol	Definition
Electric current I	ampere	A	See Appendix
Electric charge Q	coulomb	C	A s
Electric potential difference φ	volt	V	$\mathrm{J\ A^{-1}\ s^{-1} = J\ C^{-1}}$
Electric resistance R	ohm	Ω	$\mathrm{V\ A^{-1}}$

An electrochemical cell maintains an essentially constant potential difference across its terminals. This potential difference is sometimes called an electromotive force, but that is somewhat of a misnomer because it is not a force. The electromotive force of a cell is the energy supplied by the cell divided by the electric charge transported through the cell.

In order to measure the maximum value of the electromotive force, it is necessary to carry out the measurement reversibly using a potentiometer. For a reversible cell the direction of the current through the cell may be reversed by an infinitesimal change in the applied potential difference. It is not possible to measure the difference in electric potential between two different chemical substances because there are local interactions of the test charge with the chemically different surrounding media. Thus it is impossible to measure the difference in potential between an electrode and the solution with which it is in contact. The difference in potential between two electrodes is measured, and one of the electrodes is frequently referred to as a reference electrode (e.g., the hydrogen electrode or calomel electrode).

5.2 ELECTROCHEMICAL CELLS

Figure 5.1*a* shows a simple cell in which gaseous hydrogen is bubbled over a piece of platinized platinum immersed in the hydrochloric acid solution to form a hydrogen electrode. The other electrode consists of a silver wire coated with a deposit of silver chloride. When the two electrodes are connected through a resistor, as illustrated in the figure, a current flows, and we refer to the cell as a galvanic cell.

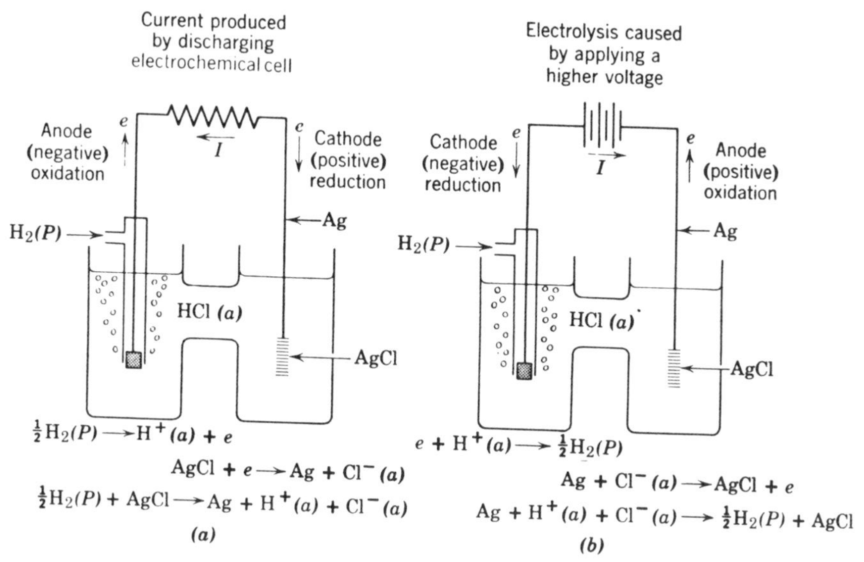

Fig. 5.1 Operation of the $H_2(P) \mid HCl(a) \mid AgCl \mid Ag$ cell as (*a*) a galvanic cell, or (*b*) an electrolysis cell. Activity is represented by *a*.

The hydrogen molecules give up electrons to the platinum to form hydrogen ions, and silver ions from the AgCl react with electrons (coming through the wire) to produce metallic silver. The difference in electrical potential between the electrodes is due to the fact that H_2 has a greater tendency to give up electrons in the presence of H^+ than Ag does in the presence of Cl^-.

The electrode at which the *oxidation* half-reaction occurs is the *anode*, and the electrode at which the *reduction* half-reaction occurs is the *cathode*. Thus, when a galvanic cell discharges spontaneously, electrons flow through the external circuit from anode to cathode, as illustrated in Fig. 5.1.*

The cell reaction may be reversed by applying a voltage higher than the reversible electromotive force of the cell to produce an electrolysis cell. The chemical processes at the electrodes are then reversed: the hydrogen electrode becomes the cathode, and the Ag-AgCl electrode the anode, as shown in Fig. 5.1*b*. Whether a cell is operating as a galvanic cell or as an electrolysis cell, the "anode" is the electrode at which oxidation occurs and electrons are given up to that electrode; the "cathode" is the electrode at which reduction occurs and electrons are taken up by a reactant in solution.

If in the experiment described in Fig. 5.1*a* the Galvanic cell was operated long enough to deliver 1 faraday (F) of electricity, the following processes would take place.

* Conventionally the "current" I is designated as a flow of positive current from the positive electrode through the wire to the negative electrode. In discussing electrochemical reactions the flow of negative electrons, which is in the opposite direction, is emphasized.

1. One faraday, 96,485 coulombs or ampere-seconds, which is Avogadro's number of electrons,* would travel through the external circuit to the Ag-AgCl electrode.
2. One-half mole of hydrogen gas would be consumed with the production of 1 mol of hydrogen ion.
3. One mole of AgCl would be converted into 1 mol of silver plus 1 mol of chloride ion.

As long as the circuit is closed, these reactions go on producing hydrochloric acid and metallic silver at the expense of H_2 and AgCl.

The electromotive force of a reversible cell may be measured under conditions such that by an infinitesimal change of the applied voltage the cell can be changed from a discharging battery to an electrolysis cell. The reversible electromotive force may be measured by use of a potentiometer. To obtain this reversible electromotive force, all steps in the cell reaction must be at equilibrium. Not all cells are reversible. In some, irreversible processes occur so that it is not possible to reverse the chemical reaction by an infinitesimal change in the applied voltage.

A junction between two electrolyte solutions contributes a potential to a cell. If, for example, a concentrated solution of hydrochloric acid forms a junction with a dilute solution, both hydrogen ions and chloride ions diffuse from the concentrated solution into the dilute solution. The hydrogen ion moves faster, and thus the dilute solution becomes positively charged because of an excess of hydrogen ions. The more concentrated solution is left with an excess of chloride ions and thus acquires a negative charge. The actual separation of charge is very small, but the potential difference produced is appreciable.

In general, it may be stated that the difference of potential resulting from the junction of two solutions is caused by the difference in the rates of diffusion of the two ions, *the more dilute solution acquiring a charge corresponding to that of the faster-moving ion.*

The difference in potential resulting from the junction of two liquids must be eliminated or corrected to obtain electrode potentials. The most convenient way of minimizing the liquid junction potential is by means of a salt bridge of potassium chloride connecting the two solutions of different electrolytes. Potassium chloride is often used because the mobilities (Section 18.3) of the two ions are about the same. When potassium chloride is not feasible as, for example, in a cell containing silver nitrate, a salt bridge of ammonium nitrate is substituted.

In the schematic representation of a cell a single vertical bar | represents a phase boundary or a junction between solutions, and double vertical bars || represent a salt bridge in which the liquid junction potential is assumed to be minimized.

5.3 THERMODYNAMICS OF ELECTROCHEMICAL CELLS

The Gibbs energy changes for reactions occurring in electrochemical cells may be readily calculated from the *reversible* electromotive force. If a cell is exactly balanced

* $F = N_A e = (6.022\ 045 \times 10^{23}\ \text{mol}^{-1})(1.602\ 189\ 2 \times 10^{-19}\ \text{C}) = 9.648\ 456 \times 10^4\ \text{C mol}^{-1}$.

against an external electromotive force so that no charging or discharging of the cell is taking place, and we imagine that an infinitesimal quantity of electricity is allowed to pass through the cell, the reversible electrical work at constant temperature and pressure, or Gibbs energy change, is equal to the product of the voltage and the quantity of electricity. The quantity of electrical charge corresponding to the molar quantities indicated in the balanced chemical equation is zF, where z is the charge number for the cell reaction and F is the Faraday constant (96,485 C mol^{-1}). The charge number z is a positive number equal to the number of electrons transferred in the cell reaction as written. When the chemical reaction occurs as written, the quantity of electric charge that flows is zF. If this quantity of electrical charge is transported through a potential difference of E volts, the amount of work required is given by zFE. Since this electrical change does not involve pressure-volume work and is carried out isothermally, the change in Gibbs energy is given by

$$\Delta G = -zFE \tag{5.3}$$

where E is the potential difference, which by convention is taken as positive. Since ΔG is negative for a spontaneous cell reaction and E for a spontaneously discharging cell is taken as positive, the negative sign must be used in equation 5.3. The electromotive force of a cell does not depend on the stoichiometric coefficients in the balanced chemical reaction, but the change in Gibbs energy ΔG does depend on z, which in turn depends on how the stoichiometric equation is written.

If the Faraday constant is expressed in coulombs per mole, the electrical work calculated from equation 5.3 is expressed in joules per mole. Since J = VC, the Faraday constant may also be written as 96,485 J mol^{-1} V^{-1}.

The entropy change for the cell reaction may be calculated from the temperature coefficient of the electromotive force, since equation 2.70 may be written

$$\left(\frac{\partial\, \Delta G}{\partial T}\right)_P = -\Delta S \tag{5.4}$$

Introducing equation 5.3, we have

$$zF\left(\frac{\partial E}{\partial T}\right)_P = \Delta S \tag{5.5}$$

The enthalpy change for the cell reaction may be calculated by substituting equations 5.3 and 5.5 into

$$\Delta H = \Delta G + T\Delta S = -zFE + zFT\left(\frac{\partial E}{\partial T}\right)_P \tag{5.6}$$

Thus, from measurements of the reversible electromotive force of a cell at a series of temperatures, it is possible to calculate ΔG, ΔS, and ΔH for the cell reaction. The great accuracy of electrical measurements often makes the determination of thermodynamic quantities by this method more exact than the direct determination of equilibrium constants or the calorimetric determination of enthalpies of reaction.

Example 5.1 The electromotive force of the cell

$$Cd \mid CdCl_2 \cdot 2\tfrac{1}{2}H_2O, \text{ sat. solution} \mid AgCl \mid Ag$$

at 25 °C is 0.675 33 V, and the temperature coefficient is -6.5×10^{-4} V K^{-1}. Calculate the values of ΔG, ΔS, and ΔH at 25 °C for the reaction

$$Cd(s) + 2AgCl(s) = 2Ag(s) + CdCl_2 \cdot 2\tfrac{1}{2}H_2O(s)$$

Two electrons are transferred, so that:

$$\begin{aligned}
\Delta G &= -(2)(96{,}485 \text{ C mol}^{-1})(0.675\,33 \text{ V}) \\
&= -130.318 \text{ kJ mol}^{-1} \\
\Delta S &= (2)(96{,}485 \text{ C mol}^{-1})(-6.5 \times 10^{-4} \text{ V K}^{-1}) \\
&= -125 \text{ J K}^{-1} \text{ mol}^{-1} \\
\Delta H &= -(2)(96{,}485)(0.675\,33) - (2)(96{,}485)(2.982)(6.5 \times 10^{-4}) \\
&= -167.580 \text{ kJ mol}^{-1}.
\end{aligned}$$

Direct calorimetric measurements give $\Delta H = -165.394$ kJ mol^{-1}.

5.4 FUNDAMENTAL EQUATION FOR AN ELECTROCHEMICAL CELL

As shown in Section 1.20, a chemical reaction may be represented by $\sum \nu_i A_i = 0$. Equation 4.14 may be rewritten by introducing $\Delta G = -zFE$ and $\Delta G^\circ = -zFE^\circ$ to obtain

$$E = E^\circ - \frac{RT}{zF} \ln \prod_i a_i^{\nu_i} \tag{5.7}$$

where the a's represent the activities of reactants and products under a given set of conditions. The standard electromotive force for the cell, E°, is the electromotive force for the cell in which the activities of the reactants and products of the cell reaction are each equal to unity. At 25 °C,

$$E = E^\circ - \frac{(8.314 \text{ J K}^{-1} \text{ mol}^{-1})(298.15 \text{ K})}{z(96{,}485 \text{ C mol}^{-1})} \ln \prod_i a_i^{\nu_i}$$

$$= E^\circ - \frac{(0.025\,69 \text{ V})}{z} \ln \prod_i a_i^{\nu_i} \tag{5.8}$$

This equation is often referred to as the Nernst equation. When the activities are all unity, the logarithmic term vanishes and $E = E^\circ$.

If the activities of the reactants and products correspond with those of an equilibrium mixture, $E = 0$, and equation 5.7 becomes

$$E^\circ = \frac{RT}{zF} \ln K \qquad \text{or} \qquad K = e^{zFE^\circ/RT} \tag{5.9}$$

where K is the equilibrium constant for the cell reaction.

Since special equations are involved in expressing the activities of electrolytes, we must now consider these expressions.

5.5 ACTIVITY OF ELECTROLYTES

When a solute does not dissociate, the activity is equal to the product of the concentration and an activity coefficient. Using Convention II (Section 3.16), the activity coefficient approaches unity at infinite dilution. When the solute is an electrolyte that is considered to be completely dissociated in solution, the expression for the activity becomes more complicated.

The chemical potential of a strong, completely dissociated, electrolyte MX is equal to the sum of the chemical potentials of the ions M^+ and X^-.

$$\mu_{MX} = \mu_{M^+} + \mu_{X^-} \tag{5.10}$$

$$\mu^\circ_{MX} + RT \ln a_{MX} = \mu^\circ_{M^+} + RT \ln a_{M^+} + \mu^\circ_{X^-} + RT \ln a_{X^-} \tag{5.11}$$

where μ°_{MX} is the chemical potential of MX at unit activity, $\mu^\circ_{M^+}$ is the chemical potential of the cation at unit activity, and $\mu^\circ_{X^-}$ is the chemical potential of the anion at unit activity. Since $\mu^\circ_{MX} = \mu^\circ_{M^+} + \mu^\circ_{X^-}$, equation 5.11 shows that

$$a_{MX} = (a_{M^+})(a_{X^-}) \tag{5.12}$$

The activities of the cation and anion may be expressed as products of the molal concentrations m and the activity coefficients γ_+ and γ_- of the cation and anion.

$$a_{M^+} = m\gamma_+ \qquad \text{and} \qquad a_{X^-} = m\gamma_- \tag{5.13}*$$

Then

$$a_{MX} = (m\gamma_+)(m\gamma_-) = m^2\gamma_\pm{}^2 \tag{5.14}$$

where $\gamma_\pm$ represents the mean ionic activity coefficient for the 1–1 electrolyte (i.e., one having a univalent cation and a univalent anion). It follows from equation 5.14 that

$$\gamma_\pm = (\gamma_+\gamma_-)^{1/2} \tag{5.15}$$

The mean ionic activity coefficient is important, since it can be determined experimentally, whereas the individual ion activity coefficients γ_+ and γ_- cannot. The mean ionic activity coefficient $\gamma_\pm$ approaches unity as the concentration of MX approaches zero.

The expression for the mean ionic activity coefficient becomes more complicated for molecules with polyvalent ions. If the strong electrolyte is $M_{\nu_+}X_{\nu_-}$, where ν_+ is the number of cations and ν_- is the number of anions,

$$\mu_{M_{\nu_+}X_{\nu_-}} = \nu_+\mu_M + \nu_-\mu_X \tag{5.16}$$

which leads to

$$a_{M_{\nu_+}X_{\nu_-}} = (a_M)^{\nu_+}(a_X)^{\nu_-} \tag{5.17}$$

---

* Strictly speaking, these equations should be written $a = \gamma(m/m^\circ)$, where m° is the standard value of the molality of 1 mol kg^{-1} but, to keep the equations simpler, we will not include m° in the equations of this chapter.

The activity coefficients of the cation and anion are equal to the ratios of their activities to their concentrations.

$$\gamma_+ = a_M/m_M = a_M/\nu_+ m \qquad \gamma_- = a_X/m_X = a_X/\nu_- m \tag{5.18}$$

Then, using these relations to eliminate a_M and a_X from equation 5.17,

$$a_{M_{\nu_+}X_{\nu_-}} = (\nu_+ m\gamma_+)^{\nu_+}(\nu_- m\gamma_-)^{\nu_-} = (m_\pm\gamma_\pm)^{\nu_++\nu_-} \tag{5.19}$$

By rearranging terms in this equation, it is seen that the mean ionic molality $m_\pm$ and mean ionic activity coefficient $\gamma_\pm$ are given by

$$m_\pm = m(\nu_+^{\nu_+}\nu_-^{\nu_-})^{1/(\nu_++\nu_-)} \tag{5.20}$$

$$\gamma_\pm = (\gamma_+^{\nu_+}\gamma_-^{\nu_-})^{1/(\nu_++\nu_-)} \tag{5.21}$$

The mean ionic molality of a 1–1 electrolyte like NaCl is equal to m, of a 2–1 electrolyte like $CaCl_2$ is equal to $4^{1/3}m$, of a 2–2 electrolyte like $CuSO_4$ is equal to m, and of a 3–1 electrolyte like $LaCl_3$ is equal to $27^{1/4}m$, as may be deduced from equation 5.21. The numbers 1, 2, and 3 refer to the number of charges on the cation and anion.

In equations where the activities of electrolytes occur, these activities may be replaced by expressions involving the molality m and the mean ionic activity coefficient $\gamma_\pm$.

Example 5.2 Write the expressions for the activities of NaCl, $CaCl_2$, $CuSO_4$, and $LaCl_3$ in terms of their molalities and mean ionic activity coefficients.

$a_{NaCl} = m^2\gamma_\pm{}^2 \qquad a_{CaCl_2} = 4m^3\gamma_\pm{}^3 \qquad a_{CuSO_4} = m^2\gamma_\pm{}^2 \qquad a_{LaCl_4} = 27m^4\gamma_\pm{}^4$

5.6 IONIC STRENGTH

Electrolytes containing ions with multiple charges have larger effects on the activity coefficients of ions than electrolytes containing only singly charged ions. In order to express electrolyte concentrations in a way that takes this into account, G. N. Lewis* introduced the ionic strength I defined by

$$I = \tfrac{1}{2}\sum_i m_i z_i^2 = \tfrac{1}{2}(m_1 z_1^2 + m_2 z_2^2 + \cdots) \tag{5.22}$$

where the summation is continued over all the different ionic species in the solution and m is the molal concentration. The greater effectiveness of ions of higher charge in reducing the activity coefficient is provided for by multiplying their concentrations by the square of their charges. According to equation 5.22, the ionic strength of a 1–1 electrolyte is equal to its molality. The ionic strength for a 1–2 electrolyte is $3m$ and for a 2–2 electrolyte is $4m$.

---

* G. N. Lewis and M. Randall, revised by K. S. Pitzer and L. Brewer, *Thermodynamics*, McGraw-Hill Book Co., New York, 1961, p. 335.

5.7 DEBYE-HÜCKEL THEORY*

The activity coefficient of an electrolyte depends markedly on the concentration. In dilute solutions the interaction between ions is a simple Coulombic attraction or repulsion, and this interaction extends considerably further through the solution than other intermolecular forces. At infinite dilution the distribution of ions in an electrolytic solution can be considered to be completely random because the ions are too far apart to exert any attraction on each other, and the activity coefficient of the electrolyte is unity. At higher concentrations, where the ions are closer together, however, the Coulomb attractive and repulsive forces become important. Because of this interaction of ions the concentration of positive ions is slightly higher in the neighborhood of a negative ion, and the concentration of negative ions is slightly higher in the neighborhood of a positive ion, than in the bulk solution. Because of the attractive forces between an ion and its surrounding ionic atmosphere, the activity coefficient of the electrolyte is reduced. This effect is greater for ions of high charge and is greater in solvents of lower dielectric constant where the electrostatic interactions are stronger.

Debye and Hückel were able to show that in dilute solutions the activity coefficient γ_i of an ion species i with z_i electronic charges is given by

$$\log \gamma_i = -Az_i^2 I^{1/2} \tag{5.23}$$

where I is the ionic strength and

$$A = \frac{1}{2.303}\left(\frac{2\pi N_A m_{\text{solv}}}{V}\right)^{1/2}\left(\frac{e^2}{4\pi\epsilon_0 \kappa kT}\right)^{3/2} \tag{5.24}$$

where m_{solv} is the mass of solvent in volume V.

Example 5.3 Calculate the value of the coefficient A in the Debye-Hückel equation for aqueous solutions at 298.15 K. The dielectric constant κ of water at this temperature is 78.54.

$$A = \frac{1}{2.303}\left[\frac{2\pi(6.022 \times 10^{23}\ \text{mol}^{-1})(997\ \text{kg})}{1.000\ \text{m}^3}\right]^{1/2}$$
$$\times \left[\frac{(1.602 \times 10^{-19}\ \text{C})^2(0.8988 \times 10^{10}\ \text{N m}^2\ \text{C}^{-2})}{(78.54)(1.3807 \times 10^{-23}\ \text{J K}^{-1})(298.15\ \text{K})}\right]^{3/2}$$
$$= 0.509\ \text{kg}^{1/2}\ \text{mol}^{-1/2}$$

Since the ionic strength I has the units mol kg^{-1}, the units of $I^{1/2}$ and A cancel, as they must, to give a logarithm.

Equation 5.23 gives the activity coefficient of a single ion, but the quantity that is accessible to experimental determination is the mean ionic activity coefficient, which for the electrolyte $M_{\nu_+}X_{\nu_-}$ is by equation 5.21

$$\gamma_\pm = (\gamma_+^{\nu_+}\gamma_-^{\nu_-})^{1/(\nu_+ + \nu_-)} \tag{5.25}$$

* P. Debye and E. Hückel, *Physik Z.*, **24**, 185, 305 (1923).

Taking the logarithm of equation 5.25, we have

$$\log \gamma_{\pm} = \frac{1}{(\nu_+ + \nu_-)} (\nu_+ \log \gamma_+ + \nu_- \log \gamma_-) \tag{5.26}$$

Substituting equation 5.23 for each activity coefficient, we have

$$\log \gamma_{\pm} = -0.509\sqrt{I}\left(\frac{\nu_+ z_+^2 + \nu_- z_-^2}{\nu_+ + \nu_-}\right) \tag{5.27}$$

Introducing $\nu_+ z_+ = \nu_- z_-$, we see that

$$\log \gamma_{\pm} = -0.509 z_+ z_- \sqrt{I} \tag{5.28}$$

The Debye-Hückel theory has been of great value in interpreting the properties of electrolyte solutions. It is a limiting law at low concentrations in the same sense that the perfect gas law is a limiting law at low pressures. At high values of the ionic strength the activity coefficient of an electrolyte usually increases with increasing ionic strength. Equation 5.28 is in excellent agreement with experiment up to an ionic strength of about 0.01, but large deviations are encountered even at this ionic strength if the product of the charge of the highest charged ion of the salt and the charge of the oppositely charged ion of the electrolyte medium is greater than about 4.

Applications of the Debye-Hückel theory are illustrated in the next section and in Section 6.2.

5.8 DETERMINATION OF THE ACTIVITY COEFFICIENT OF HYDROCHLORIC ACID

A cell without liquid junction contains a single electrolyte solution; thus the two electrodes must be chosen so that one is reversible with respect to a cation of the electrolyte and the other with respect to an anion of the electrolyte. For example, if the electrolyte is hydrochloric acid, one electrode would be the hydrogen electrode and the other a chlorine or silver chloride electrode. In the latter case the cell may be represented by

$$\mathrm{Pt} \mid \mathrm{H_2}(g) \mid \mathrm{HCl}(m) \mid \mathrm{AgCl} \mid \mathrm{Ag}$$

The cell reaction is

$$\tfrac{1}{2}\mathrm{H_2}(g) + \mathrm{AgCl}(s) = \mathrm{HCl}(m) + \mathrm{Ag}(s) \tag{5.29}$$

According to equation 5.7, the electromotive force for this cell is given by

$$E = E^\circ - \frac{RT}{zF} \ln \frac{a_{\mathrm{HCl}}}{P_{\mathrm{H_2}}^{1/2}} \tag{5.30}$$

assuming that H_2 may be treated as a perfect gas. If the pressure of hydrogen is 1 atm and equation 5.14 is introduced,

$$E = E^\circ - \frac{2.303RT}{zF} \log m^2\gamma_\pm{}^2 \tag{5.31}$$

The mean ionic activity coefficient of hydrochloric acid is represented by $\gamma_\pm$, and m is the molality. The electromotive force of the cell when the activity of hydrochloric acid is unity and the pressure of hydrogen is 1 atm is represented by E°, called the standard electromotive force of the cell.

As it stands, equation 5.31 contains two unknown quantities, E° and $\gamma_\pm$. These may be obtained by determining the electromotive force of this cell over a range of hydrochloric acid concentrations, including dilute solutions. Rearranging equation 5.31 and substituting numerical values for 25 °C gives

$$E + 0.1183 \log m = E^\circ - 0.1183 \log \gamma_\pm \tag{5.32}$$

The exponents in equation 5.31 have been placed in front of the logarithmic term, giving (2)(0.059 16) = 0.1183. Since at infinite dilution $m = 0$, $\gamma_\pm = 1$, and $\log \gamma_\pm = 0$, it can be seen that, when $E + 0.1183 \log m$ is plotted against m, the extrapolation of $E + 0.1183 \log m$ to $m = 0$ will give E°.

To make a satisfactory extrapolation, use is made of the Debye-Hückel theory to furnish a function that will give nearly a straight line. The following expression is used for the mean ionic activity coefficient of a 1–1 electrolyte in dilute aqueous solutions at 25 °C.

$$\log \gamma_\pm = -0.509\sqrt{m} + bm$$

where b is an empirical constant.

Substituting into equation 5.32 and rearranging terms, we have

$$E + 0.1183 \log m - 0.0602m^{1/2} = E' = E^\circ - (0.1183b)m \tag{5.33}$$

According to this equation, the left-hand side, which we will designate as E', will give a straight line when it is plotted against m, and the intercept at $m = 0$ is E°.

In Fig. 5.2, E' is plotted against m. The extrapolated value is 0.2224 V. This is the electromotive force that the cell would deliver with the hydrochloric acid at unit activity, and it is the standard electrode potential of the silver-silver chloride electrode, since the other electrode is the standard hydrogen electrode for which the standard electrode potential is zero by definition. Similar cells have been used for determining the standard electrode potentials for other electrodes.

The value of E° having been determined, the activity coefficient of hydrochloric acid at any other concentration may be calculated from the electromotive force of the cell containing hydrochloric acid at that concentration.

Example 5.4 Calculate the mean ionic activity coefficient of 0.1 mol kg^{-1} hydrochloric acid at 25 °C from the fact that the electromotive force of the cell described in this section is 0.3524 V at 25 °C. Substituting into equation 5.32, we have

$$0.3524 = 0.2224 - 0.1183 \log \gamma_\pm - 0.1183 \log 0.1$$

$$\log \gamma_\pm = \frac{(-0.3524 + 0.2224 + 0.1183)}{0.1183} = -0.0989$$

$$\gamma_\pm = 0.796$$

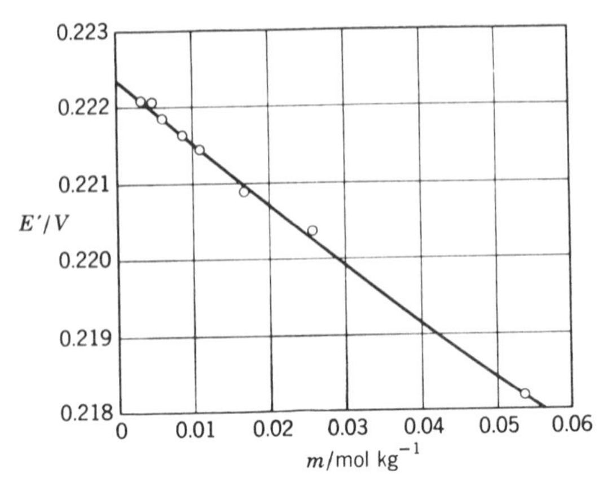

Fig. 5.2 Determination of the silver-silver chloride electrode potential by extrapolation of a function of the potential of the cell Pt | H_2(1 atm) | HCl(m) | AgCl | Ag to infinite dilution.

In this general manner the activity coefficients of the electrolytes plotted in Fig. 5.3 have been determined. It should be noted that at high concentrations of electrolytes, activity coefficients may be considerably greater than unity.

Activity coefficients may be calculated from a number of different types of

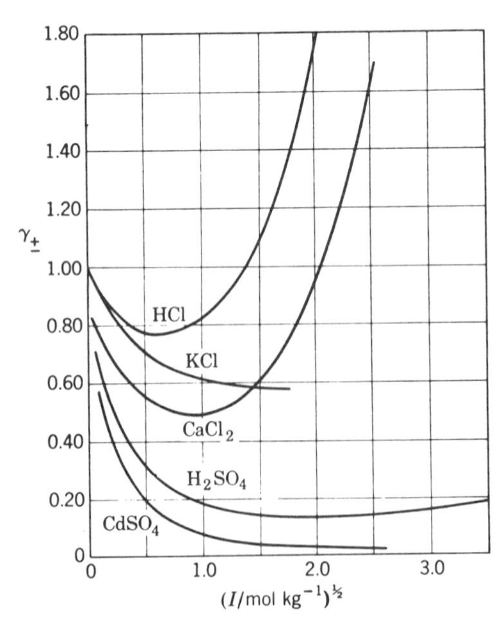

Fig. 5.3 Dependence of the mean ionic activity coefficient $\gamma_\pm$ on $I^{1/2}$ for electrolytes at 25 °C.

measurements, including vapor pressure, freezing point lowering, boiling point elevation, osmotic pressure, distribution coefficients, equilibrium constants, solubility, and electromotive force. All different methods for determining activity coefficients must lead to the same value for a given solution.

5.9 ELECTRODE POTENTIALS

The relative electrode potential of any electrode may be determined by combining the electrode with a standard hydrogen electrode, preferably without a liquid junction, and measuring the electromotive force of the cell as a function of concentration. If one electrode consists of a metal in contact with a solution containing its ions with charge $n+$, the cell may be represented by

$$\mathrm{Pt} \mid \mathrm{H_2}(g, 1\ \mathrm{atm}) \mid \mathrm{H^+} \parallel \mathrm{M}^{n+} \mid \mathrm{M} \tag{5.34}$$

The corresponding cell reaction is written according to conventions that are given shortly.

$$\tfrac{1}{2}\mathrm{H_2}(g) + \frac{1}{n}\mathrm{M}^{+n}(aq) = \mathrm{H^+}(aq) + \frac{1}{n}\mathrm{M}(c) \tag{5.35}$$

The term *standard electrode potential* is used to designate the potential that would be obtained with the constituents present at unit activity. *The standard electrode potential of an electrode is given a positive value if this electrode is more positive than the standard hydrogen electrode and a negative sign if it is more negative than the standard hydrogen electrode.* The electrode with the more positive electrode potential is attached to the positive terminal of the potentiometer. The positive electrode is the one at which reduction occurs, and the negative electrode is the one at which oxidation occurs if the cell is allowed to react spontaneously.

All electrodes may be arranged in a table according to their standard electrode potentials, which are determined as described in the preceding section by means of an extrapolation to find the electromotive force at unit activity. Table 5.2 gives standard electrode potential $E°$ of a number of electrodes at 25 °C. It is to be noted that these electrodes are written in the order ion-electrode and that *standard electrode potentials are standard reduction potentials*. The electrode reaction is written as a reduction, that is, the addition of electrons.

The magnitude of the standard electrode potential is a measure of the tendency of the half-reaction to occur in the direction of reduction. The lower a half-reaction is in the last column of Table 5.2, the greater is the tendency of the oxidized form to accept electrons and be reduced. The higher a half-reaction is in the table, the greater is the tendency of the reduced form to donate electrons and be oxidized. For example, the active metals sodium and potassium have very large negative standard electrode potentials and strong tendencies to donate electrons.

The reduced form of any element or ion at unit activity will reduce the oxidized form of any element or ion at unit activity, which has a less negative, that is, more positive, standard electrode potential. Table 5.2 of standard electrode potentials

Table 5.2 Standard Electrode Potentials at 25 °C[a]

Electrode	$E°$	Half Cell Reaction
$Li^+ \mid Li$	−3.045	$Li^+ + e = Li$
$K^+ \mid K$	−2.925	$K^+ + e = K$
$Na^+ \mid Na$	−2.714	$Na^+ + e = Na$
$Mg^{2+} \mid Mg$	−2.37	$\frac{1}{2}Mg^{2+} + e = \frac{1}{2}Mg$
$Th^{4+} \mid Th$	−1.90	$\frac{1}{4}Th^{4+} + e = \frac{1}{4}Th$
$Al^{3+} \mid Al$	−1.66	$\frac{1}{3}Al^{3+} + e = \frac{1}{3}Al$
$Zn^{2+} \mid Zn$	−0.763	$\frac{1}{2}Zn^{2+} + e = \frac{1}{2}Zn$
$Fe^{2+} \mid Fe$	−0.440	$\frac{1}{2}Fe^{2+} + e = \frac{1}{2}Fe$
$Cr^{3+}, Cr^{2+} \mid Pt$[b,c]	−0.41	$Cr^{3+} + e = Cr^{2+}$
$Cd^{2+} \mid Cd$	−0.403	$\frac{1}{2}Cd^{2+} + e = \frac{1}{2}Cd$
$Br^- \mid PbBr_2(s) \mid Pb$	−0.280	$\frac{1}{2}PbBr_2 + e = \frac{1}{2}Pb + Br^-$
$Ni^{2+} \mid Ni$	−0.250	$\frac{1}{2}Ni^{2+} + e = \frac{1}{2}Ni$
$I^- \mid AgI(s) \mid Ag$	−0.151	$AgI + e = Ag + I^-$
$Sn^{2+} \mid Sn$	−0.140	$\frac{1}{2}Sn^{2+} + e = \frac{1}{2}Sn$
$Pb^{2+} \mid Pb$	−0.126	$\frac{1}{2}Pb^{2+} + e = \frac{1}{2}Pb$
$D^+ \mid D_2 \mid Pt$	−0.0034	$D^+ + e = \frac{1}{2}D_2$
$H^+ \mid H_2 \mid Pt$	0.0000	$H^+ + e = \frac{1}{2}H_2$
$Ti^{4+}, Ti^{3+} \mid Pt$	0.04	$Ti^{4+} + e = Ti^{3+}$
$Br^- \mid AgBr(s) \mid Ag$	0.095	$AgBr + e = Ag + Br^-$
$Sn^{4+}, Sn^{2+} \mid Pt$	0.15	$\frac{1}{2}Sn^{4+} + e = \frac{1}{2}Sn^{2+}$
$Cu^{2+}, Cu^+ \mid Pt$	0.153	$Cu^{2+} + e = Cu^+$
$Cl^- \mid AgCl(s) \mid Ag$	0.2224	$AgCl + e = Ag + Cl^-$
$Cl^- \mid Hg_2Cl_2(s) \mid Hg$[d]	0.268	$\frac{1}{2}Hg_2Cl_2 + e = Hg + Cl^-$
$Cu^{2+} \mid Cu$	0.337	$\frac{1}{2}Cu^{2+} + e = \frac{1}{2}Cu$
$OH^- \mid O_2 \mid Pt$	0.401	$\frac{1}{4}O_2 + \frac{1}{2}H_2O + e = OH^-$
$H^+ \mid C_2H_4(g), C_2H_6(g) \mid Pt$	0.52	$H^+ + \frac{1}{2}C_2H_4(g) + e = \frac{1}{2}C_2H_6(g)$
$Cu^+ \mid Cu$	0.521	$Cu^+ + e = Cu$
$I^- \mid I_2(s) \mid Pt$	0.5355	$\frac{1}{2}I_2 + e = I^-$
$H^+ \mid$ quinhydrone$(s) \mid Pt$	0.6996	$\frac{1}{2}C_6H_4O_2 + H^+ + e = \frac{1}{2}C_6H_6O_2$
$Fe^{3+}, Fe^{2+} \mid Pt$	0.771	$Fe^{3+} + e = Fe^{2+}$
$Hg_2^{2+} \mid Hg$	0.789	$\frac{1}{2}Hg_2^{2+} + e = Hg$
$Ag^+ \mid Ag$	0.7991	$Ag^+ + e = Ag$
$Hg^{2+}, Hg_2^{2+} \mid Pt$	0.920	$Hg^{2+} + e = \frac{1}{2}Hg_2^{2+}$
$Br^- \mid Br_2(l) \mid Pt$	1.0652	$\frac{1}{2}Br_2(l) + e = Br^-$
$Cl^- \mid Cl_2(g) \mid Pt$	1.3595	$\frac{1}{2}Cl_2(g) + e = Cl^-$
$Pb^{2+} \mid PbO_2 \mid Pb$	1.455	$\frac{1}{2}PbO_2 + 2H^+ + e = \frac{1}{2}Pb^{2+} + H_2O$
$Au^{3+} \mid Au$	1.50	$\frac{1}{3}Au^{3+} + e = \frac{1}{3}Au$
$F^- \mid F_2(g) \mid Pt$	2.87	$\frac{1}{2}F_2(g) + e = F^-$
$HF(aq) \mid F_2(g) \mid Pt$	3.06	$H^+ + \frac{1}{2}F_2(g) + e = HF(aq)$

[a] All ions are at unit activity in water, and all gases are at 1 atm.

[b] The symbol Pt represents an inert electrode like platinum.

[c] The order of writing the ions in the electrolyte solution is immaterial.

[d] The electromotive force of the normal calomel electrode is 0.2802 V and of the calomel electrode containing saturated KCl is 0.2415 V.

may be used for calculations of the electromotive forces of electrochemical cells and the equilibrium constants of the corresponding chemical reactions. The following two rules are convenient for calculating the standard electromotive force of a cell, determining the cell reaction, and deciding whether the cell reaction is spontaneous as written. The standard electromotive force E° of a cell is the electromotive force when all constituents are at unit activity.

1. *The standard electromotive force of a cell is equal to the standard electrode potential of the right-hand electrode minus the standard electrode potential of the left-hand electrode.*

$$E^\circ = E^\circ_{\text{right}} - E^\circ_{\text{left}} \tag{5.36}$$

It is to be noted that the electrode at the right is automatically written in the order ion-electrode, as it is in Table 5.2, and that, according to the geometry of the cell, the electrode at the left must be written in the order electrode-ion.

For the cell

$$\text{Zn} \mid \text{Zn}^{2+} \parallel \text{Cu}^{2+} \mid \text{Cu}$$

$$E^\circ = 0.337 - (-0.763) = 1.100\text{ V} \qquad \text{at } 25\ ^\circ\text{C} \tag{5.37}$$

For the cell

$$\text{Cu} \mid \text{Cu}^{2+} \parallel \text{Zn}^{2+} \mid \text{Zn}$$

$$E^\circ = -0.763 - (0.337) = -1.100\text{ V} \qquad \text{at } 25\ ^\circ\text{C} \tag{5.38}$$

2. *The reaction taking place at the left electrode is written as an oxidation reaction, and the reaction taking place at the right electrode is written as a reduction reaction. The cell reaction is the sum of these two reactions.* For cell 5.37,

$$\text{Left electrode:} \quad \text{Zn} = \text{Zn}^{2+} + 2e \quad \text{(oxidation)} \tag{5.39}$$

$$\text{Right electrode:} \quad \text{Cu}^{2+} + 2e = \text{Cu} \quad \text{(reduction)} \tag{5.40}$$

$$\text{Cell reaction:} \quad \text{Zn} + \text{Cu}^{2+} = \text{Zn}^{2+} + \text{Cu} \tag{5.41}$$

These equations could equally well be multiplied through by $\frac{1}{2}$ so that the cell reaction would correspond to a one-electron change and would be written

$$\tfrac{1}{2}\text{Zn} + \tfrac{1}{2}\text{Cu}^{2+} = \tfrac{1}{2}\text{Zn}^{2+} + \tfrac{1}{2}\text{Cu} \tag{5.42}$$

For cell 5.38 the cell reaction would be written

$$\text{Cu} + \text{Zn}^{2+} = \text{Zn} + \text{Cu}^{2+} \tag{5.43}$$

or

$$\tfrac{1}{2}\text{Cu} + \tfrac{1}{2}\text{Zn}^{2+} = \tfrac{1}{2}\text{Zn} + \tfrac{1}{2}\text{Cu}^{2+} \tag{5.44}$$

Since $\Delta G^\circ = -zFE^\circ$, a cell reaction is spontaneous when E° is positive. Thus reaction 5.41 is spontaneous when Cu^{2+} and Zn^{2+} are at unit activity because $E^\circ = 1.100$ V and

$$\Delta G^\circ = -(2)(96{,}485\text{ C mol}^{-1})(1.100\text{ V}) = -212.267\text{ kJ mol}^{-1} \tag{5.45}$$

Reaction 5.43 is not spontaneous at unit activities of reactants and products because $E^\circ = -1.100$ V and $\Delta G^\circ = 212.267$ kJ mol^{-1}. Multiples of a cell reaction may be taken without changing the electromotive force E°, but ΔG° depends on how the chemical equation is written as indicated by the factor z in equation 5.3. For example, the Gibbs energy change for reaction 5.42 is

$$\Delta G^\circ = -(1)(96{,}485 \text{ C mol}^{-1})(1.100 \text{ V}) = -106.134 \text{ kJ mol}^{-1} \tag{5.46}$$

For a cell in which the reactants are not at unit activity, the criterion of spontaneous reaction is that $\Delta G = -zFE$ is negative.

Example 5.5 Calculate E° at 25 °C for the cell

$$\text{Cd} \mid \text{Cd}^{2+} \parallel \text{Cu}^{2+} \mid \text{Cu}$$

and determine the cell reaction and its equilibrium constant.

Reduction at right:	$Cu^{2+} + 2e = Cu(s)$	$E^\circ_{Cu^{2+} \mid Cu} = 0.337$ V
Oxidation at left:	$Cd(s) = Cd^{2+} + 2e$	$E^\circ_{Cd^{2+} \mid Cd} = -0.403$ V
	$Cd(s) + Cu^{2+} = Cd^{2+} + Cu(s)$	$E^\circ = 0.740$ V

Since E° is positive, the reaction is spontaneous as written when reactants and products are at unit activity, and Cd(s) will precipitate Cu(s) from a solution in which $a_{Cu^{2+}} = 1$ to produce Cd^{2+} at unit activity. The Gibbs energy change for this reaction is

$$\begin{aligned}\Delta G^\circ &= -zFE^\circ = -(2)(96{,}485 \text{ C mol}^{-1})(0.740 \text{ V}) \\ &= -142.8 \text{ kJ mol}^{-1}\end{aligned}$$

The equilibrium constant for the cell reaction is readily calculated from the Gibbs energy change using equation 5.9.

$$K = e^{-\Delta G^\circ/RT} = e^{-(-142{,}800)/(8.31441)(298.15)} = 1.04 \times 10^{25}$$

Alternatively, this problem may be solved by using data from Table A.1.

$$\Delta G^\circ = -77.58 - 64.98 = -142.6 \text{ kJ mol}^{-1}$$

$$E^\circ = \frac{-142{,}600 \text{ J mol}^{-1}}{-2(96{,}485 \text{ C mol}^{-1})} = 0.739 \text{ V}$$

$$K = e^{142{,}600/(8.31441)(298.15)} = 0.96 \times 10^{25}$$

Equation 5.8 can be used to calculate the electromotive force of a cell when the ions are not at unit activity. It is not possible to determine the activities of single ions but, for some purposes, it is possible to replace activities by concentrations. In other cases the activity coefficients for ions of the same charge type tend to cancel in equation 5.8. This is a good approximation when these ions are in the same solution or if they are in different solutions of the same ionic strength.

5.10 DETERMINATION OF pH

The concentrations of hydrogen ions in aqueous solutions range from about 1 mol L^{-1} in 1 mol L^{-1} HCl to about 10^{-14} mol L^{-1} in 1 mol L^{-1} NaOH. Because of this wide range of concentrations, Sorenson adopted an exponential notation in

1909. He defined pH as the negative exponent of 10 that gives the hydrogen-ion concentration. Thus

$$(H^+) = 10^{-pH}$$

or

$$pH = -\log (H^+) \qquad (5.47)$$

Electromotive-force methods are useful for studying hydrogen-ion concentrations; however, since the activity of a single ion cannot be determined, the modern definition of pH is expressed operationally in terms of the method of measuring it instead of by equation 5.47. The pH of a solution may be determined by use of a cell such as the following.

$$Pt \mid H_2(P) \mid H^+(a_{H^+}) \parallel Cl^- \mid Hg_2Cl_2 \mid Hg$$

The electromotive force of this cell may be considered to be made up of three contributions.

$$E = 0.2802 - 0.0591 \log \left(\frac{a_{H^+}}{P_{H_2}^{1/2}}\right) + E_{\text{liquid junction}} \qquad (5.48)$$

where the contribution by the normal calomel electrode is 0.2802 V at 25 °C. Although the activities of single ions cannot be determined, equation 5.48 is often used with the assumption that $E_{\text{liquid junction}} = 0$. If $P_{H_2} = 1$ atm,

$$E - 0.2802 = -0.0591 \log a_{H^+} \qquad (5.49)$$

Hydrogen-ion activities obtained in this way are of great practical use, even though they are based on an approximation.

Taking pH $= -\log a_{H^+}$, equation 5.49 becomes

$$E - 0.2802 = 0.0591 \text{ pH}$$

or

$$pH = \frac{E - 0.2802}{0.0591} \qquad (5.50)$$

A glass electrode consists of a reversible electrode, such as a calomel or Ag-AgCl electrode, in a solution of constant pH inside a thin membrane of a special glass. The thin glass bulb of this electrode is immersed in the solution to be studied along with a reference calomel electrode to form the cell indicated by

$$Ag \mid AgCl \mid Cl^-, H^+ \mid \text{glass membrane} \mid \text{solution} \parallel \text{calomel electrode}$$

It is found experimentally that the potential of such a glass electrode varies with the activity of hydrogen ions in the same way as the hydrogen electrode, that is, 0.0591 V per pH unit at 25 °C. An ordinary potentiometer cannot be used to measure the voltage of such a cell because of the high resistance of the glass membrane, and so an electronic voltmeter must be employed. Electronic devices have been developed that make it possible to measure pH values to ± 0.01 pH unit with easily portable apparatus. The pH meter, as it is often called, is calibrated by means of a buffer of known pH before it is used to measure the pH of an unknown solution. The theory and use of the glass electrode have been fully treated by Bates.*

The glass electrode has become the most useful electrode for determining the pH of a solution. It is not affected by oxidizing or reducing agents and is not easily poisoned. It is especially useful in biochemical investigations.

The pCO_2 electrode is a useful clinical tool. This electrode consists of a glass electrode in contact with a solution of fixed bicarbonate concentration that is separated from the sample solution by a polymer membrane permeable to CO_2. When CO_2 diffuses into the bicarbonate solution, it is hydrated to H_2CO_3 (Section 6.3) and then rapidly ionizes to $HCO_3^- + H^+$ with a consequent change in pH.

There is a growing number of ion-selective electrodes that use semipermeable membranes to obtain a cell potential that depends on a particular ion.

5.11 FUEL CELLS

Fuel cells offer the possibility of achieving high thermodynamic efficiency in the conversion of Gibbs energy into mechanical work. Internal combustion engines at best convert only the fraction $(T_2 - T_1)/T_2$ of the heat of combustion into mechanical work. In this relation, which comes from the second law of thermodynamics, T_2 is the temperature of the gas during expansion and T_1 is the temperature of the exhaust (see Section 2.4).

Fuel cells may be classified according to the temperature range in which they operate: low temperature (25 to 100 °C), medium temperature (100 to 500 °C), high temperature (500 to 1000 °C), and very high temperature (above 1000 °C). The advantage of using high temperatures is that catalysts for the various steps in the process are not so necessary. Polarization of a fuel cell reduces the current. Polarization is the result of slow reactions or processes such as diffusion in the cell.

Figure 5.4 indicates the construction of a hydrogen-oxygen fuel cell with a solid electrolyte, which is an ion exchange membrane. The membrane is impermeable to the reactant gases, but is permeable to hydrogen ions, which carry the current between the electrodes. In order to keep the resistance of the cell as low as possible, the membrane is kept as thin as possible. In order to facilitate the operation of the cell at 40 to 60 °C, the electrodes are covered with finely divided platinum that functions as a catalyst. Water is drained out of the cell during operation. Fuel cells of this general type have been used successfully in the space program and are quite efficient. Their disadvantages for large-scale commercial application are that hydrogen presents storage problems, and platinum is an expensive catalyst. Cheaper catalysts have been found for higher temperature operation of hydrogen-oxygen fuel cells.

Fuel cells that use hydrocarbons and air have been developed, but their power per unit weight is too low to make them practical in ordinary automobiles. Better catalysts are needed.

---

* R. G. Bates, *Determination of pH: Theory and Practice*, Wiley, New York, 1973.

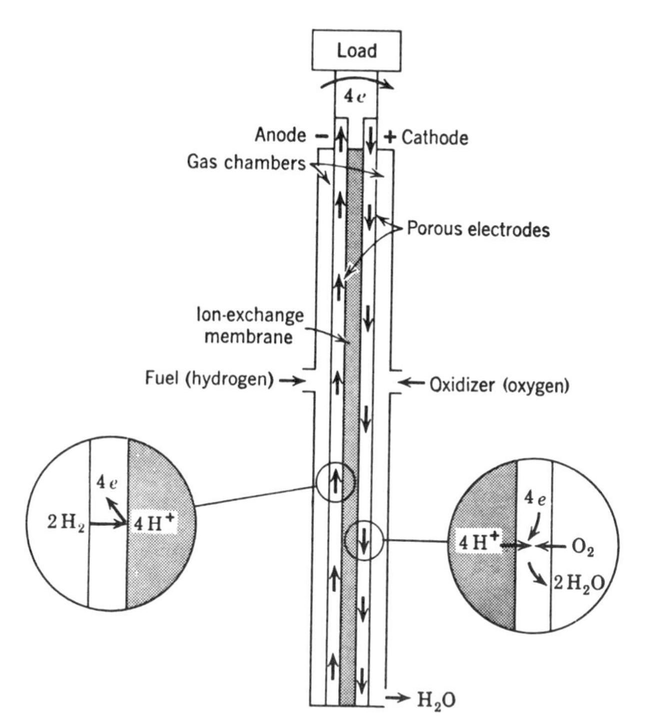

Fig. 5.4 Hydrogen-oxygen fuel cell with an ion exchange membrane. (From J. O'M. Bockris and S. Srinivasan, *Fuel Cells, Their Electrochemistry*. Copyright © 1969 by McGraw-Hill, Inc. Used with permission of McGraw-Hill Book Co.)

A hydrogen-oxygen fuel cell may have an acidic or alkaline electrolyte. The half-cell reactions are

$$
\begin{array}{ll}
H_2(g) = 2H^+ + 2e & E^\circ = 0 \\
\frac{1}{2}O_2(g) + 2H^+ + 2e = H_2O(l) & E^\circ = 1.229\ \text{V} \\
\hline
H_2(g) + \frac{1}{2}O_2(g) = H_2O(l) & E^\circ = 1.229\ \text{V}
\end{array}
$$

or

$$
\begin{array}{ll}
H_2(g) + 2OH^- = 2H_2O(l) + 2e & E^\circ = -0.828\ \text{V} \\
\frac{1}{2}O_2(g) + H_2O(l) + 2e = 2OH^- & E^\circ = 0.401\ \text{V} \\
\hline
H_2(g) + \frac{1}{2}O_2(g) = H_2O(l) & E^\circ = 1.229\ \text{V}
\end{array}
$$

Example 5.6 Calculate the standard electromotive force of the hydrogen-oxygen fuel cell from the standard Gibbs energy of formation of $H_2O(l)$ at 25 °C.

$$\Delta G^\circ = -zFE^\circ$$

$$-237{,}191\ \text{J mol}^{-1} = -2(96{,}485\ \text{C mol}^{-1})E^\circ$$

$$E^\circ = 1.229\ \text{V}$$

To maximize the power per unit mass of an electrochemical cell, the electronic and electrolytic resistances of the cell must be minimized. Since fused salts have lower electrolytic resistances than aqueous solutions, high-temperature electrochemical cells are of special interest for practical applications. High temperatures also allow the use of liquid metal electrodes, which make possible higher current densities that solid electrodes.

In this chapter we have been concerned with the thermodynamics of electrochemical cells. Practical applications involve kinetic considerations, which we will describe in Section 16.10. Photoelectrochemical cells are discussed in Section 17.11.

References

R. G. Bates, *Determination of pH: Theory and Practice*, Wiley, New York, 1973.

A. J. de Bethune and N. A. S. Loud, *Standard Aqueous Electrode Potentials and Temperature Coefficients at 25°*, C. A. Hampel, Skokie, Ill., 1964.

J. O'M. Bockris and D. M. Drazic, *Electro-Chemical Science*, Taylor and Francis Ltd., London, 1972.

J. O'M. Bockris and A. K. N. Reddy, *Modern Electrochemistry*, Vols. 1 and 2, Plenum Publ. Corp., New York, 1970.

J. O'M. Bockris and S. Srinivasan, *Fuel Cells: Their Electrochemistry*, McGraw-Hill Book Co., New York, 1969.

E. A. Guggenheim, *Thermodynamics*, Wiley, New York, 1967.

H. S. Harned and B. B. Owen, *The Physical Chemistry of Electrolytic Solutions*, Reinhold Publishing Corp., New York, 1958.

A. McDougall, *Fuel Cells*, Wiley, New York, 1976.

J. S. Newman, *Electrochemical Systems*, Prentice-Hall, Englewood Cliffs, N.J., 1973.

R. A. Robinson and R. H. Stokes, *Electrolyte Solutions*, Academic Press, New York, 1959.

Problems

5.1 How much work is required to bring two protons from an infinite distance of separation of 0.1 nm? Calculate the answer in joules using the protonic charge 1.602×10^{-19} C. What is the work in kJ mol^{-1} for 1 mol of proton pairs?

Ans. 23.07×10^{-19} J. 1389 kJ mol^{-1}.

5.2 A small dry battery of zinc and ammonium chloride weighing 85 g will operate continuously through a 4-Ω resistance for 450 min before its voltage falls below 0.75 V. The initial voltage is 1.60, and the effective voltage over the whole life of the battery is taken to be 1.00. Theoretically, how many kilometers above the earth could this battery be raised by the energy delivered under these conditions? *Ans.* 8.1 km.

5.3 The voltage of the cell

$$\mathrm{Pb} \mid \mathrm{PbSO_4} \mid \mathrm{Na_2SO_4 \cdot 10H_2O(sat)} \mid \mathrm{Hg_2SO_4} \mid \mathrm{Hg}$$

is 0.9647 at 25 °C. The temperature coefficients is 1.74×10^{-4} V K^{-1}. (*a*) What is the cell reaction? (*b*) What are the values of ΔG, ΔS, and ΔH?

Ans. (*a*) $\mathrm{Pb}(s) + \mathrm{Hg_2SO_4}(s) = \mathrm{PbSO_4}(s) + 2\mathrm{Hg}(l)$.

(*b*) -186.16 kJ mol^{-1}, 33.58 J K^{-1} mol^{-1}, -176.15 kJ mol^{-1}.

5.4 Derive an expression for the activity of a 1–2 electrolyte (like Na_2SO_4) in terms of the mean ionic activity coefficient and the molality. *Ans.* $a_{Na_2SO_4} = 4m^3\gamma_\pm{}^3$.

5.5 What is the ionic strength of each of the following solutions: (*a*) 0.1 mol kg^{-1} NaCl, (*b*) 0.1 mol kg^{-1} $Na_2C_2O_4$, (*c*) 0.1 mol kg^{-1} $CuSO_4$, (*d*) a solution containing 0.1 mol kg^{-1} Na_2HPO_4 and 0.1 mol kg^{-1} NaH_2PO_4? *Ans.* (*a*) 0.1, (*b*) 0.3, (*c*) 0.4, (*d*) 0.4.

5.6 For 0.002 mol L^{-1} $CaCl_2$ at 25 °C use the Debye-Hückel limiting law to calculate the activity coefficients of Ca^{++} and Cl^-. What is the mean ionic activity coefficient for the electrolyte? *Ans.* $\gamma_{Ca^{++}} = 0.696$. $\gamma_{Cl^-} = 0.913$. $\gamma_\pm = 0.834$.

5.7 The cell Pt | H_2(1 atm) | HBr(*m*) | AgBr | Ag has been studied by H. S. Harned, A. S. Keston, and J. G. Donelson [*J. Am. Chem. Soc.*, **58**, 989 (1936)]. The following table gives the electromotive forces obtained at 25 °C.

m	0.01	0.02	0.05	0.10
E	0.3127	0.2786	0.2340	0.2005

Calculate (*a*) $E°$ and (*b*) the activity coefficient for a 0.10 mol kg^{-1} solution of hydrogen bromide. *Ans.* (*a*) 0.0707 V. (*b*) 0.801.

5.8 Given the cell at 25 °C,

$$Pb \mid Pb^{2+}(a = 1) \parallel Ag^+(a = 1) \mid Ag$$

(*a*) calculate the voltage; (*b*) write the cell reaction; and (*c*) calculate the Gibbs energy change. (*d*) Which electrode is positive?

Ans. (*a*) 0.925 V, (*b*) $\frac{1}{2}Pb(s) + Ag^+(a = 1) = \frac{1}{2}Pb^{2+}(a = 1) + Ag(s)$, (*c*) -89.2 kJ mol^{-1}, (*d*) silver electrode.

5.9 (*a*) Calculate the voltage of the following cell at 25 °C.

$$Zn \mid Zn^{2+}(a = 0.0004) \parallel Cd^{2+}(a = 0.2) \mid Cd$$

(*b*) Write the cell reaction. (*c*) Calculate the value of the Gibbs energy change involved in the reaction.

Ans. (*a*) 0.440 V, (*b*) $Zn(s) + Cd^{2+}(a = 0.2) = Zn^{2+}(a = 0.0004) + Cd(s)$, (*c*) -84.9 kJ mol^{-1}.

5.10 (*a*) Diagram the cell for the reaction

$$H_2(g, 1 \text{ atm}) + I_2(s) = 2HI(a = 1)$$

(*b*) Calculate $E°$. (*c*) Calculate $\Delta G°$. (*d*) Calculate K. (*e*) What differences would there be if the reaction had been written

$$\tfrac{1}{2}H_2(g, 1 \text{ atm}) + \tfrac{1}{2}I_2(s) = HI(a = 1)$$

Ans. (*a*) Pt | H_2(*g*, 1 atm) | HI(a = 1) | I_2(*s*) | Pt,
(*b*) 0.5355 V, (*c*) -103.34 kJ mol^{-1}, (*d*) 1.27×10^{18},
(*e*) *a* and *b* are the same; $\Delta G° = -51.67$ kJ mol^{-1}, $K = 1.13 \times 10^9$.

5.11 (*a*) Calculate the equilibrium constant at 25 °C for the reaction

$$Fe^{2+} + Ag^+ = Ag + Fe^{3+}$$

(*b*) Calculate the concentration of silver ion at equilibrium (assuming that concentrations may be substituted for activities) for an experiment in which an excess of finely divided metallic silver is added to an 0.05 mol kg^{-1} solution of ferric nitrate.

Ans. (*a*) 3.0, (*b*) 0.044 mol kg^{-1}.

5.12 Devise an electromotive force cell for which the cell reaction is

$$AgBr(s) = Ag^+ + Br^-$$

Calculate the equilibrium constant (usually called the solubility product) for this reaction at 25 °C. *Ans.* $10^{-11.90}$.

5.13 What are the equilibrium constants for the following reactions at 25 °C?

(*a*) $H^+(aq) + Li(c) = Li^+(aq) + \frac{1}{2}H_2(g)$

(*b*) $2H^+(aq) + Pb(c) = Pb^{2+}(aq) + H_2(g)$

(*c*) $3H^+(aq) + Au(c) = Au^{3+}(aq) + \frac{3}{2}H_2(g)$

Ans. (*a*) 3.0×10^{51}, (*b*) 1.8×10^4, (*c*) 8.6×10^{-77}.

5.14 For the reaction

$$\tfrac{1}{2}Cu(s) + \tfrac{1}{2}Cl_2(g) = \tfrac{1}{2}Cu^{2+} + Cl^-$$

at 25 °C, calculate (*a*) the standard Gibbs energy change, (*b*) the equilibrium constant, and (*c*) the standard electromotive force of the cell in which this reaction occurs.

Ans. (*a*) -98.7 kJ mol^{-1}. (*b*) 1.80×10^{17}. (*c*) 1.000 V.

5.15 What is the theoretical decomposition voltage of NaCl in aqueous solution if the reactants and products are in their standard states?

$$NaCl(aq) + H_2O(l) = NaOH(aq) + \tfrac{1}{2}Cl_2(g) + \tfrac{1}{2}H_2(g)$$

Ans. 2.188 V.

5.16 (*a*) Calculate the equilibrium constant at 25 °C for the reaction

$$Sn^{4+} + 2Ti^{3+} = 2Ti^{4+} + Sn^{2+}$$

(*b*) When 0.01 mol of Sn^{2+} ion is added to 1.0 mol of Ti^{4+} ion in 1000 g of water, what will be the concentration of Ti^{3+} ions (if it is assumed for the calculation that the activities are equal to the concentrations)? *Ans.* (*a*) 5.23×10^3. (*b*) 1.16×10^{-2} mol kg^{-1}.

5.17 (*a*) Diagram the cell that corresponds to the reaction $Ag^+ + Cl^- = AgCl$. Calculate at 25 °C (*b*) $E°$, (*c*) $\Delta G°$, and (*d*) K.

Ans. (*a*) Ag | AgCl | Cl^- ‖ Ag^+ | Ag. (*b*) 0.5767 V. (*c*) -55.64 kJ mol^{-1}. (*d*) 5.60×10^9.

5.18 Show that for the concentration cell

$$X \mid X^-(a_1) \parallel X^-(a_2) \mid X$$

where X^- is a negative ion, the equation for the electromotive force of the cell is

$$E = -\frac{RT}{F}\ln\frac{a_2}{a_1}$$

5.19 Given the cell at 25 °C,

$$Pt \mid Cl_2 \mid Cl^-(a = 0.1) \parallel Cl^-(a = 0.001) \mid Cl_2 \mid Pt$$

(*a*) Write the cell reaction. (*b*) Which electrode is negative? (*c*) What is the voltage of the cell? (*d*) Is the reaction spontaneous?

Ans. (*a*) $Cl^-(a = 0.1) = Cl^-(a = 0.001)$. (*b*) Left. (*c*) 0.118 32 V. (*d*) Yes.

5.20 When a hydrogen electrode and a normal calomel electrode are immersed in a solution at 25 °C a potential of 0.664 V is obtained. Calculate (*a*) the pH, and (*b*) the hydrogen-ion activity. *Ans.* (*a*) 6.48, (*b*) 3.31×10^{-7}.

5.21 Ammonia may be used as the anodic reactant in a fuel cell. The reactions occurring at the electrodes are

$$NH_3(g) + 3OH^-(aq) = \tfrac{1}{2}N_2(g) + 3H_2O(l) + 3e$$

$$O_2(g) + 2H_2O(l) + 4e = 4OH^-(aq)$$

What is the electromotive force of this fuel cell at 25 °C? *Ans.* 1.17 V.

5.22 Calculate the electromotive force of a methane-O_2 fuel cell at 25 °C.

Ans. 1.0597 V.

5.23 Calculate the electromotive force of

$$Li(l) \mid LiCl(l) \mid Cl_2(g)$$

at 900 K for $P_{Cl_2} = 1$ atm. This high-temperature battery is attractive because of its high electromotive force and low atomic masses. Lithium chloride melts at 883 K and lithium at 453.69 K. [The ΔG_f° for LiCl(l) at 900 K in JANAF Thermochemical Tables is -335.140 kJ mol^{-1}.]
Ans. 3.474 V.

5.24 How much work in kJ mol^{-1} can in principle be obtained when an electron is brought to 0.5 nm from a proton?

5.25 Calculate the energy in kJ mol^{-1} required to separate a positive and negative charge from 0.3 nm to infinity in (a) a vacuum, (b) a solvent of dielectric constant 10, and (c) water at 25 °C, which has a dielectric constant of approximately 80.

5.26 (a) Write the reaction that occurs when the cell

$$Zn \mid ZnCl_2(0.555 \text{ mol kg}^{-1}) \mid AgCl \mid Ag$$

delivers current and calculate (b) ΔG, (c) ΔS, and (d) ΔH at 25 °C for this reaction. At 25 °C $E = 1.015$ V and $(\partial E/\partial T)_P = -4.02 \times 10^{-4}$ V K^{-1}.

5.27 The solubility of Ag_2CrO_4 in water is 8.00×10^{-5} mol kg^{-1} at 25 °C, and its solubility in 0.04 mol kg^{-1} $NaNO_3$ is 8.84×10^{-5} mol kg^{-1}. What is the mean ionic activity coefficient of Ag_2CrO_4 in 0.04 mol kg^{-1} $NaNO_3$?

5.28 Give the expressions for the mean ionic activity coefficients of LiCl, $AlCl_3$, and $MgSO_4$ in terms of the single ion activities and the molality.

5.29 A solution of NaCl has an ionic strength of 0.24 mol kg^{-1}. (a) What is its concentration? (b) What concentration of Na_2SO_4 would have the same ionic strength? (c) What concentration of $MgSO_4$?

5.30 Design cells without liquid junction that could be used to determine the activity coefficients of aqueous solutions of (a) NaOH and (b) H_2SO_4. Give the equations relating voltage to the mean ionic activity coefficient.

5.31 Derive the expression for the electromotive force of the cell

$$Pt \mid H_2(g, 1 \text{ atm}) \mid HK_2PO_4(m_1), Na_2HPO_4(m_2) \parallel NaX(m_3) \mid AgX(s) \mid Ag$$

Substitute the equilibrium expression for the second dissociation of phosphoric acid and describe how the thermodynamic dissociation constant K for that dissociation could be obtained from electromotive force measurements at constant temperature.

5.32 What are ΔG° and E° for $Mg(c) + Cl_2(g) = MgCl_2(m = 1)$ at 25 °C?

5.33 The most electropositive metal is lithium, and the most electronegative element is fluorine. Calculate the reversible electromotive force of the Li-F_2 cell at 25 °C.

5.34 Given the cell at 25 °C,

$$Cd \mid Cd^{2+}(a = 1) \parallel I^-(a = 1) \mid I_2(s) \mid Pt$$

(a) Write the cell reaction. (b) Calculate E. (c) Calculate ΔG°. (d) Which electrode is positive?

5.35 (a) Design cell in which the reaction cell

$$\tfrac{1}{2}Cl_2(g, 1 \text{ atm}) + Br^-(a = 1) = Cl^-(a = 1) + \tfrac{1}{2}Br_2(l)$$

Calculate (b) the voltage at 25 °C, and (c) the Gibbs energy change of the reaction at 25 °C.

5.36 The value of E° for the electrode $Pt \mid O_2(g) \mid OH^-$ cannot be measured directly because the electrode is not reversible. Calculate E° and ΔG° at 25 °C for the electrode reaction

$$\tfrac{1}{4}O_2(g) + \tfrac{1}{2}H_2O + e = OH^-(a = 1)$$

from the known Gibbs energy changes for the reactions

$$H_2(g) + \tfrac{1}{2}O_2(g) = H_2O(l)$$
$$H_2O(l) = H^+(a = 1) + OH^-(a = 1)$$
$$\tfrac{1}{2}H_2(g) = H^+(a = 1) + e$$

5.37 (*a*) Calculate the voltage at 25 °C of the cell

$$\text{Pt} \mid Ti^{3+}(a = 0.3),\ Ti^{4+}(a = 0.5) \parallel Ce^{3+}(a = 0.7),\ Ce^{4+}(a = 0.002) \mid \text{Pt}$$

(*b*) Write the cell reaction. (*c*) Calculate ΔG for the cell reaction as written. (*d*) Calculate $E°$. (*e*) Calculate $\Delta G°$. (*f*) Calculate K.

5.38 Calculate the electromotive force of the following cell at 25 °C,

$$\text{Li} \mid \text{LiCl}(0.1 \text{ mol kg}^{-1}) \mid Cl_2(1 \text{ atm}) \mid \text{Pt}$$

assuming that the mean ionic activity coefficient is 0.77.

5.39 For the cell in the preceding problem calculate watt-hours per kilogram of reactants, assuming constant electromotive force.

5.40 Calculate the equilibrium constant at 25 °C for the reaction

$$2H^+ + D_2(g) = H_2(g) + 2D^+$$

from the electrode potential for $D^+ \mid D_2 \mid$ Pt, which is -3.4 mV at 25 °C.

5.41 At 25 °C the potential of the cell

$$\text{Ag} \mid \text{AgI} \mid \text{KI}(1 \text{ mol kg}^{-1}) \parallel AgNO_3(0.001 \text{ mol kg}^{-1}) \mid \text{Ag}$$

is 0.72 V. The mean ionic activity coefficient of 1 mol kg^{-1} KI may be taken as 0.65, and of 0.001 mol kg^{-1} $AgNO_3$ as 0.98. (*a*) What is the solubility of AgI? (*b*) What is the solubility of AgI in pure water?

5.42 What is the electromotive force of the following cell at 25 °C?

$$\text{Cu} \mid Cu^{2+}(a_1 = 0.01) \parallel Cu^{2+}(a_2 = 0.10) \mid \text{Cu}$$

5.43 A thallium amalgam of 4.93% Tl in mercury and another amalgam of 10.02% Tl are placed in separate legs of a glass cell and covered with a solution of thallous sulfate to form a concentration cell. The voltage of the cell is 0.029 480 V at 20 °C and 0.029 971 V at 30 °C. (*a*) Which is the negative electrode? (*b*) What is the heat of dilution per mole of Tl when Hg is added at 30 °C to change the concentration from 10.02 to 4.93%? (*c*) What is the voltage of the cell at 40 °C?

5.44 A hydrogen electrode and a normal calomel electrode give a voltage of 0.435 when placed in a certain solution at 25 °C. (*a*) What is the pH of the solution? (*b*) What is the value of a_{H^+}?

5.45 The oxidation of methanol may be utilized to produce an electric current in a fuel cell. Calculate the reversible electromotive force at 25 °C.

5.46 Calculate the standard electromotive force of an ethane-oxygen fuel cell at 25 °C.

5.47 Calculate the voltage at 25 °C for the fuel cell based on the reaction

$$H_2(1 \text{ atm}) + \tfrac{1}{2}O_2(1 \text{ atm}) = H_2O(l)$$

using (*a*) the Gibbs energy of formation of water, and (*b*) the oxygen electrode potential calculated in Problem 5.36.

5.48 A water electrolysis cell operated at 25 °C consumes 25 kWh/lb of hydrogen produced. Calculate the cell efficiency using $\Delta G°$ for the decomposition of water.

CHAPTER 6

EQUILIBRIA OF BIOCHEMICAL REACTIONS

Biochemical reactions often involve a coupling of different types of reactions. For example, the hydrolysis of adenosine triphosphate (ATP) is coupled with acid dissociations and magnesium and calcium ion bindings. The thermodynamics of biochemical reactions are handled in a different way from the thermodynamics of reactions in the gas phase or in solutions of nonelectrolytes. In biochemistry it is convenient to express equilibrium constants in terms of total concentrations of reactants and products, including various ionized and complexed species. In this case equilibrium constants and other thermodynamic quantities are functions of pH and the concentrations of metal ions at constant temperature and pressure.

Proteins have remarkable capacities to bind other substances in aqueous solutions. Their binding capacity may be markedly altered by changes in pH and by other substances present in the solution. In contrast with the binding properties of small molecules, the affinity of hemoglobin for oxygen is increased as more oxygen is bound. This is a consequence of changes in the internal structure of hemoglobin in the binding process.

6.1 THERMODYNAMIC FUNCTIONS FOR ACID DISSOCIATIONS

The equilibrium constant K for the dissociation of a weak acid

$$HA = H^+ + A^- \tag{6.1}$$

may be expressed in terms of activities a or in terms of concentrations.

$$K = \frac{a_{H^+} a_{A^-}}{a_{HA}} = \frac{(H^+)(A^-)y_{\pm}^2}{(HA)y_{HA}} \tag{6.2}$$

where $y_{\pm}$ is the mean ionic activity coefficient for the dissociated acid $H^+ + A^-$ on the moles per liter concentration scale. For many practical purposes, such as preparing buffers, interpreting titration curves, and calculating biochemical equilibria, it is more convenient to use the apparent acid dissociation constant K_{app}, which is expressed in terms of concentrations. By rearranging equation 6.2, we obtain

$$K_{app} = K\frac{y_{HA}}{y_{\pm}^2} = \frac{(H^+)(A^-)}{(HA)} \tag{6.3}$$

The apparent acid dissociation constant K_{app} for a given weak acid depends on the salt concentration. As the electrolyte and acid concentrations approach zero, K_{app} approaches K. In this chapter we will use a K without the subscript to represent both the limiting value of the dissociation constant and the value at a particular ionic strength. In using the dissociation constants of weak acids it is often convenient to write equation 6.3 as

$$\mathrm{pH} = \mathrm{p}K + \log \frac{(\mathrm{A}^-)}{(\mathrm{HA})} \tag{6.4}$$

where $\mathrm{pH} = -\log (\mathrm{H}^+)$ and $\mathrm{p}K = -\log K$.

The proton does not exist free in solution, but is in combination with the solvent. In water the proton may form a hydronium ion, H_3O^+, or other complex species. A hydronium ion H_3O^+ may be hydrated by three water molecules so that $H_9O_4{}^+$ is formed (Section 18.4). Mass spectroscopic studies show that this is a stable species in the gas phase. Since the state of the proton in aqueous solution is not exactly known, the symbol H^+ will be used to represent the hydrated hydrogen ion.

Table 6.1 gives the pK values and other thermodynamic functions at 25 °C for equilibrium constants expressed in terms of activities. Equilibrium constants for these acids at zero ionic strength have been determined by use of extrapolations, as described in the next section. From the temperature dependence of K it is possible to calculate ΔH°, ΔS°, and ΔC_P° for the dissociation.

The fact that ΔS° for dissociation of most weak acids is negative is surprising at first. Since the dissociation products have more freedom to move around than the undissociated acid, it might be expected that the entropy of the products would be greater than the entropy of the undissociated acid and that ΔS° would have a positive sign. However, the experimental results show that there is a decrease in entropy and, correspondingly, an increase in "order" in the dissociation. This results from a participation of water molecules in the reaction, which is not indicated by the balanced equation in 6.1. Water molecules, being dipoles, tend to be oriented in the neighborhood of ions. Thus the dissociation of an uncharged acid results in the orientation of a number of water molecules about the dissociation products, and the consequent decrease in S° overshadows the increase in S° resulting from the formation of two particles from one. The effects that lead to a negative value of ΔS° tend to oppose dissociation. For a number of weak acids in Table 6.1 the value of ΔH° is very small, and so the standard entropy change large determines the value of the pK according to $2.3RT\mathrm{p}K = \Delta H^\circ - T\Delta S^\circ$.

There is almost no entropy change in the dissociation of ammonium ion

$$\mathrm{NH_4}^+ = \mathrm{H}^+ + \mathrm{NH_3}$$

because there is no change in the number of ions in the reaction, so that the dissociation causes little change in water structure. Water molecules are not so organized around the methylammonium ion as around $NH_4{}^+$ and, therefore, there is a decrease in entropy upon dissociation. The entropy decrease upon dissociation is even greater for the dimethylammonium and trimethylammonium ions, suggesting decreasing water organization as methyl substitution increases.

Table 6.1 Thermodynamic functions for acid dissociations at 25 °C[a]

	pK	ΔG°	ΔH°	ΔS°	ΔC_P°
Water (K_w)	13.997	79.868	56.563	−78.2	−197
Acetic acid	4.756	27.137	−0.385	−92.5	−155
Chloroacetic acid	2.861	16.322	−4.845	−71.1	−167
Butyric acid	4.82	27.506	−2.900	−102.1	0
Succinic acid, pK_1	4.207	24.016	3.188	−69.9	−134
Succinic acid, pK_2	5.636	32.188	−0.452	−109.2	−218
Carbonic acid, pK_1	6.352	36.259	9.372	−90.4	−377
Carbonic acid, pK_2	10.329	58.961	15.075	−147.3	−272
Phosphoric acid, pK_1	2.148	12.259	−7.648	−66.9	−155
Phosphoric acid, pK_2	7.198	41.099	4.130	−123.8	−226
Glycerol-2-phosphoric acid, pK_1	1.335	7.615	12.103	−66.1	−326
Glycerol-2-phosphoric acid, pK_2	6.650	37.945	−1.724	−133.1	−226
Ammonium ion	9.245	52.777	52.216	−1.7	0
Methylammonium ion	10.615	60.601	54.760	−19.7	33
Dimethylammonium ion	10.765	49.618	49.618	−39.7	96
Trimethylammonium ion	9.791	55.890	36.882	−63.6	184
Tris(hydroxymethyl)aminomethane	8.076	46.099	45.606	−1.3	0
Glycine, pK_1	2.350	13.410	4.837	−28.9	−134
Glycine, pK_2	9.780	55.815	44.141	−39.3	−50
Glycylglycine, pK_1	3.148	17.322	3.607	−54.0	−167
Glycylglycine, pK_2	8.252	47.112	44.350	−8.4	−42

[a] ΔG° and ΔH° are in kJ mol^{-1}. ΔS° and ΔC_P° are in J K^{-1} mol^{-1}. These values apply at zero ionic strength and are obtained by extrapolation of experimental data at higher ionic strengths. From J. Edsall and J. Wyman, *Biophysical Chemistry*, Academic Press, New York, 1958.

6.2 EFFECT OF IONIC STRENGTH ON DISSOCIATION CONSTANTS

At low ionic strength values (several hundredths for monovalent weak acids and several thousandths for highly charged weak acids), the dependence of pK on ionic strength may be calculated using the Debye-Hückel theory (Section 5.7) and simple extensions of that theory. According to the Güntelberg equation,

$$\log y_i = \frac{-z_i^2 A I^{1/2}}{1 + I^{1/2}} \tag{6.5}$$

where y_i is the activity coefficient of ion i with charge z_i (in units of proton charge), I is ionic strength on the molar scale, and A is the Debye-Hückel constant (0.5091 at 25 °C). This equation differs from the Debye-Hückel theory by the $1 + I^{1/2}$ in the denominator. Although this is a useful improvement, coefficients of $I^{1/2}$ other than unity give better results for particular weak acid-electrolyte combinations.

For the dissociation of an acid with n negative charges

$$HA^{-n} = H^+ + A^{-(n+1)} \tag{6.6}$$

the thermodynamic dissociation constant $K_{I=0}$ is given by

$$K_{I=0} = \frac{a_{H^+}(A^{-(n+1)})y_{-(n+1)}}{(HA^{-n})y_{-n}} = K_I \frac{y_{-(n+1)}}{y_{-n}} \tag{6.7}$$

Substituting equation 6.5 and rearranging yields

$$pK_I = pK_{I=0} - \frac{(2n+1)AI^{1/2}}{(1+I^{1/2})} \tag{6.8}$$

Example 6.1 Estimate pK_1 and pK_2 for H_3PO_4 at 25 °C and 0.01 ionic strength.

$$pK_I = pK_{I=0} - \frac{(2n+1)(0.5091)(0.01)^{1/2}}{(1+0.01^{1/2})} = pK_{I=0} - (2n+1)0.046$$

$$pK_1 = 2.148 - 0.046 = 2.102$$
$$pK_2 = 7.198 - 3(0.046) = 7.060$$

6.3 CARBON DIOXIDE, CARBONIC ACID, AND BICARBONATE ION THERMODYNAMICS

The equilibria discussed in this section are of interest because of the importance of CO_2 in the regulation of the pH of blood. The kinetics of these reactions are discussed in Section 16.6.

Only a small fraction of the CO_2 dissolved in water is hydrated to carbonic acid.

$$CO_2 + H_2O = H_2CO_3 \tag{6.9}$$

The hydration constant K_h is written

$$K_h = \frac{(H_2CO_3)}{(CO_2)} \tag{6.10}$$

At 25 °C, $K_h = 0.002\ 58$. The standard enthalpy change and the standard entropy change are both unfavorable for the hydration reaction, as shown by the values in Table 6.2; nevertheless, CO_2 has to be hydrated in order for the dissolved CO_2 produced by metabolism in our tissues to be converted to HCO_3^- and be transported to our lungs. The entropy change for reaction 6.9 is negative, because joining two molecules to form one reduces the translational and rotational degrees of freedom of the system. ΔC_P is negative because $H_2O + CO_2$ can absorb more heat per degree increase in temperature than H_2CO_3.

Carbonic acid is a somewhat stronger acid than acetic acid.

$$H_2CO_3 = H^+ + HCO_3^- \tag{6.11}$$

$$K_{H_2CO_3} = \frac{(H^+)(HCO_3^-)}{(H_2CO_3)} \tag{6.12}$$

Table 6.2 Thermodynamic Quantities at 25 °C and Zero Ionic Strength[a,b]

Equilibrium Constant	pK	ΔG°	ΔH°	ΔS°	ΔC_P°
$K_h = \frac{(H_2CO_3)}{(CO_2)}$	2.59	14.770	4.730	−33	−264
$K_{H_2CO_3} = \frac{(H^+)(HCO_3^-)}{(H_2CO_3)}$	3.77	21.630	4.230	−59	−117
$K_1 = \frac{(H^+)(HCO_3^-)}{(CO_2) + (H_2CO_3)}$	6.352	36.260	9.370	−90.4	−377
$K_2 = \frac{(H^+)(CO_3^{2+})}{(HCO_3^-)}$	10.329	58.960	15.080	−147.3	−272

[a] From J. T. Edsall, CO_2: *Chemical, Biochemical, and Physiological Aspects*, NASA SP–188, 1969.
[b] ΔG° and ΔH° in kJ mol^{-1}; ΔS° and ΔC_P° in J K^{-1} mol^{-1}.

At 25 °C, the acid dissociation constant $K_{H_2CO_3} = 1.72 \times 10^{-4}$ (pK = 3.77). However, this is *not* the acid dissociation constant you will find if you look in tables, and it is *not* the one you would use in the laboratory to do calculations on bicarbonate buffers. The reason is that in the laboratory you are not ordinarily able to distinguish between dissolved CO_2 and H_2CO_3 and simply lump these two species together by counting all dissolved CO_2 as H_2CO_3.

The first acid dissociation constant K_1 for carbonic acid is defined as

$$K_1 = \frac{(H^+)(HCO_3^-)}{(H_2CO_3) + (CO_2)} = \frac{(H^+)(HCO_3^-)}{(H_2CO_3)[1 + (CO_2)/(H_2CO_3)]} \tag{6.13}$$

Thus

$$K_1 = \frac{K_{H_2CO_3}}{1 + 1/K_h} \tag{6.14}$$

Calculating K_1 from the values of K_h and $K_{H_2CO_3}$ at 25 °C that are given above, we obtain $K_1 = 4.45 \times 10^{-7}$ (pK = 6.352). This is the pK you would find in titrating dissolved CO_2.

The first dissociation of carbonic acid is of special interest because it is the primary buffer system in blood. The Henry law constant (Section 3.15) for the distribution of CO_2 between the gas and aqueous phases at 38 °C is 3.3×10^{-5} mol L^{-1} torr^{-1}. Thus, since the normal partial pressure of carbon dioxide in the alveoli of the lungs is 5.3 kPa, the normal concentration of CO_2 in our blood is 0.0013 mol L^{-1}. The normal concentration of bicarbonate ion in our blood is 0.026 mol L^{-1}, and so equation 6.13 may be written

$$\mathrm{pH} = \mathrm{p}K_1 + \log \frac{(HCO_3^-)}{(CO_2) + (H_2CO_3)} \qquad 7.4 = 6.1 + \log \frac{0.026}{0.0013} \tag{6.15}$$

since 6.1 is the value of pK_1 at 0.15 mol L^{-1} ionic strength and $(H_2CO_3) \ll (CO_2)$. If we hyperventilate (breathe hard), reaction 6.9 shifts to the left; this causes reaction 6.11 to shift to the left and, as a result, the pH rises. If the concentration of HCO_3^- in our blood is too high, the blood pH is too high and the body increases its rate of excretion of HCO_3^- in the urine. In this way the pH of blood is maintained in the relatively narrow range of 7.0 to 7.6.

6.4 RELATION BETWEEN MICROSCOPIC DISSOCIATION CONSTANTS AND MACROSCOPIC DISSOCIATION CONSTANTS

The acid dissociations of amino acids offer interesting examples of the relations between what are called microscopic dissociation constants and macroscopic dissociation constants. Microscopic dissociation constants are defined in terms of particular ionic species and are not ordinarily accessible experimentally. In the dissociation of glycine

$$\begin{array}{ccccc} & K_{11} \; {\scriptstyle -H^+ / +H^+} & {}^+H_3NCH_2CO_2^- & {\scriptstyle -H^+ / +H^+} \; K_{22} & \\ {}^+H_3NCH_2CO_2H & & \Big\Updownarrow K_z & & H_2NCH_2CO_2^- \\ & K_{12} \; {\scriptstyle +H^+ / -H^+} & H_2NCH_2CO_2H & {\scriptstyle -H^+ / +H^+} \; K_{21} & \end{array} \tag{6.16}$$

the microscopic constants are K_{11}, K_{12}, K_{21}, K_{22}, and K_z.

When glycine is titrated, we do not have information on the relative amounts of the dipolar ion, ${}^+H_3NCH_2CO_2^-$, and the uncharged molecule, $H_2NCH_2CO_2H$, and so we define the first acid dissociation constant K_1, a macroscopic dissociation constant, by

$$K_1 = \frac{(H^+)[({}^+H_3NCH_2CO_2^-) + (H_2NCH_2CO_2H)]}{({}^+H_3NCH_2CO_2H)}$$

$$= K_{11} + K_{12} = 10^{-2.35} \tag{6.17}$$

where K_{11} and K_{12} are the microscopic dissociation constants for the two possible dissociations of a first proton. second acid-dissociation constant is given by

$$K_2 = \frac{(H^+)(H_2NCH_2CO_2^-)}{[({}^+H_3NCH_2CO_2^-) + (H_2NCH_2CO_2H)]}$$

$$= \frac{1}{1/K_{22} + 1/K_{21}} = 10^{-9.78} \tag{6.18}$$

where K_{21} and K_{22} are the microscopic dissociation constants for the two possible dissociations of a second proton. This same formulation would apply to any dibasic acid. In addition, the principle of detailed balancing (Section 15.11) yields a further relationship between the values of the four microscopic constants.

$$K_{11}K_{22} = K_{12}K_{21} \tag{6.19}$$

Since K_1 and K_2 are known (values are given for 25 °C in Table 6.1), we need a value of only one of the four microscopic constants in order to calculate values of the other three. Assuming that K_{12} has the same value as the dissociation constant of the methyl ester of glycine,

$$^{+}H_3NCH_2CO_2CH_3 = H^+ + H_2NCH_2CO_2CH_3 \qquad K_{12} = 10^{-7.70} \tag{6.20}$$

Thus $K_{11} = 10^{-2.35}$ from equation 6.17 and

$$\frac{K_{21}}{K_{22}} = \frac{10^{-2.35}}{10^{-7.70}} \tag{6.21}$$

from equation 6.19. Substituting this ratio in equation 6.18 yields $K_{21} = 10^{-4.43}$, so that $K_{22} = 10^{-9.78}$.

Thus the ratio K_z of dipolar ion to neutral molecule is

$$K_z = \frac{K_{11}}{K_{12}} = \frac{10^{-2.35}}{10^{-7.70}} = 10^{5.35} \tag{6.22}$$

The dissociation of the most acidic form of glycine accordingly goes almost exclusively by the top path; this is not true for all amino acids.

An amino acid is said to be *isoelectric* at the pH at which there are equal concentrations of the positively and negatively charged forms. Setting

$$(^{+}H_3NCH_2CO_2H) = (H_2NCH_2CO_2{}^{-})$$

and substituting from equations 6.17 and 6.18 yields

$$(H^+)^2_{\text{isoelectric}} = K_1K_2$$

which may be written

$$\text{pH}_I = \tfrac{1}{2}(\text{p}K_1 + \text{p}K_2) \tag{6.23}$$

where pH_I is the pH of the isoelectric point. For glycine the isoelectric point is

$$\text{p}I = \tfrac{1}{2}(2.35 + 9.78) = 6.06$$

Equation 6.19 may be written in the form

$$\text{p}K_{21} - \text{p}K_{11} = \text{p}K_{22} - \text{p}K_{12} \qquad 4.43 - 2.35 = 9.78 - 7.70 \tag{6.24}$$

This equation says that the strengthening of the carboxyl dissociation by the positive charge on the amino group is equal to the weakening of the dissociation of the $NH_3{}^+$ group by the negative charge on the carboxyl group.

Example 6.2 For a dibasic acid $HO_2C(CH_2)_nCO_2H$, where n is so large that the two carboxyl dissociations are independent, what is the relation between K_1 and K_2?

Using equations 6.17 and 6.18,

$$K_1 = K_{11} + K_{12} = 2K \qquad K_2 = \frac{1}{(1/K_{22} + 1/K_{21})} = \frac{K}{2}$$

since $K_{11} = K_{12} = K_{22} = K_{21}$. Thus $K_1 = 4K_2$. The 4 in this equation is referred to as a statistical factor, and we will learn more about statistical factors in Section 6.8.

6.5 THERMODYNAMICS OF BIOCHEMICAL REACTIONS

The thermodynamic quantities for biochemical reactions are handled in a different way from other reactions. The fact that the reactants and products exist in various degrees of ionization and complexation is usually ignored, and a single symbol is used to represent the sum of the concentrations of these various species. For example, the hydrolysis of adenosine triphosphate (Fig. 6.1) to adenosine diphosphate and inorganic phosphate is usually represented by

$$\text{ATP} + \text{H}_2\text{O} = \text{ADP} + \text{P} \tag{6.25}$$

and the equilibrium constant and standard Gibbs energy of hydrolysis are represented by

$$K' = \frac{(\text{ADP})(\text{P})}{(\text{ATP})} \tag{6.26}$$

$$\Delta G^{\circ\prime} = -RT \ln K' \tag{6.27}$$

The equilibrium constant is designated by a prime because it is the value obtained in the laboratory without knowing the relative concentrations of the various ionized and complexed forms of the reactants and products. The concentration of water is omitted because it is a constant in the dilute aqueous solutions that are used.

In reaction 6.25 and in the equilibrium expression, the symbol ATP refers to the sum of all of the various ionized and complexed forms that exist in the equilibrium solution, and the same comment applies to ADP and P. As a consequence, the

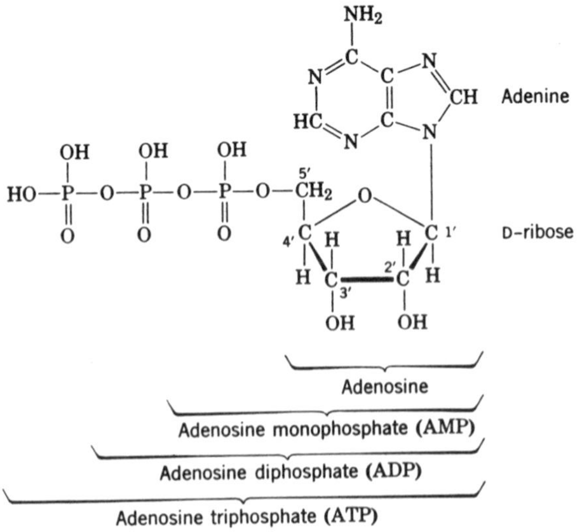

Adenine

D-ribose

Adenosine

Adenosine monophosphate (AMP)

Adenosine diphosphate (ADP)

Adenosine triphosphate (ATP)

Fig. 6.1 Structure of adenosine triphosphate, ATP.

value of K' at a given temperature and electrolyte concentration is a function of pH and concentration of metal ions that are complexed by ATP, ADP, or P. Since K' is a function of pH and metal ion concentration, the values of the various thermodynamic quantities are also functions of pH and metal ion concentration. This type of equilibrium constant will be related to the more familiar types of equilibrium constants after we have considered the acid dissociations and metal ion complexation reactions of ATP and related compounds.

The most important thermodynamic quantity for living things is the Gibbs energy since, at constant temperature and pressure, this thermodynamic quantity determines whether or not a reaction occurs spontaneously. The standard Gibbs energies of hydrolysis of a number of phosphate esters are given in Table 6.3. These are the Gibbs energy changes when 1 mol of the ester at a concentration of 1 mol/L is hydrolyzed to give the indicated products, each at one molar concentration, at 25 °C, pH 7 and pMg 4, where pMg $= -\log(\mathrm{Mg}^{2+})$. The Gibbs energy changes for phosphate transfer reactions may be obtained by adding and subtracting these reactions. Since the equilibrium constant may be calculated for any pair of reactions, such a table summarizes equilibrium data for a very large number of phosphate transfer reactions. Similar tables may be constructed for the transfer of pyrophosphate (pyrophosphoric acid is $H_4P_2O_7$) or any other group.

There is an analogy between the transfer of phosphate in these reactions and the transfer of electrons in reactions studied in electrochemistry or protons in acid-base equilibria.

Example 6.3 What is the equilibrium constant K' at pH 7 and pMg 4 and 25 °C for

$$\text{Creatine phosphate} + \text{ADP} = \text{Creatine} + \text{ATP}$$

Since

$$\text{Creatine phosphate} + H_2O = \text{Creatine} + \text{P} \qquad \Delta G^{\circ\prime} = -43.5\ \text{kJ mol}^{-1}$$

$$\text{ADP} + \text{P} = \text{ATP} + H_2O \qquad \Delta G^{\circ\prime} = 39.8\ \text{kJ mol}^{-1}$$

For the given reaction, $\Delta G^{\circ\prime} = -43.5 + 38.9 = -3.7\ \text{kJ mol}^{-1}$. Using $\Delta G^{\circ\prime} = -RT \ln K'$,

$$K' = \frac{(\text{Cr})(\text{ATP})}{(\text{CrP})(\text{ADP})} = 4.6$$

As shown in this example, a reaction with a more negative Gibbs energy change may be used to drive a phosphorylation reaction with a less negative Gibbs energy change. Such reactions are said to be coupled, and the coupling is provided by the enzyme (Section 16.8) that catalyzes the phosphate transfer reaction. Although we can calculate $\Delta G^{\circ\prime}$ by adding two reactions in Example 6.3, this does not mean that the coupled reaction occurs through these steps. In order for there to be coupling the mechanism has to be of the following type.

$$\text{Creatine phosphate} + \text{E} = \text{EP} + \text{Creatine}$$

$$\text{EP} + \text{ADP} = \text{ATP} + \text{E}$$

$$\overline{\text{Creatine phosphate} + \text{ADP} = \text{Creatine} + \text{ATP} \qquad \Delta G^{\circ\prime} = -3.7\ \text{kJ mol}^{-1}}$$

Table 6.3 Standard Gibbs Energies $\Delta G^{\circ\prime}$ of Hydrolysis at 25 °C, pH 7, pMg 4 at 0.2 mol L^{-1} Ionic Strength (P represents inorganic phosphate)

	kJ mol^{-1}
Phosphoenolpyruvate + H_2O = Enol pyruvate + P	−61.9
Creatine phosphate + H_2O = Creatine + P	−43.5
Acetyl phosphate + H_2O = Acetate + P	−43.1
CoA-S-phosphate + H_2O = CoA-SH + P	−37.7
ATP + H_2O = ADP + P	−39.7
ADP + H_2O = AMP + P	−36.8
Pyrophosphate (PP) + H_2O = 2P	−34.3
Arginine phosphate + H_2O = Arginine + P	−29.3
Glucose-6-phosphate + H_2O = Glucose + P	−12.6
Fructose-1-phosphate + H_2O = Fructose + P	−12.6
AMP + H_2O = Adenosine + P	−12.6
Glycerol-1-phosphate + H_2O = Glycerol + P	−9.2

where E is enzyme and EP is phosphorylated enzyme. The large change in Gibbs energy for the hydrolysis of creatine phosphate is not lost, but is conserved in the enzyme phosphate compound and then in ATP. This is an absolutely essential process for living things because this is the way a thermodynamically spontaneous reaction can drive a nonspontaneous reaction that synthesizes a needed compound.

At pH values well above the neutral range ATP exists in aqueous solutions as a −4 ion. As the pH is reduced to pH 7, ATP picks up a proton and becomes a −3 ion. As the pH is reduced to pH 4, ATP picks up another proton and becomes a −2 ion. Since ATP^{4-} is so highly charged, it has a strong tendency to bind cations from the medium. Significant concentrations of complexes with Na^+ and K^+ are formed at 0.1 mol L^{-1} concentrations of the ions. Therefore, the acid dissociation constants of ATP are most conveniently determined in media having bulky cations such as $(n\text{-propyl})_4N^+$ that are bound much less strongly. The p*K*s for ATP, ADP,

Table 6.4 Thermodynamic Quantities for Acid Dissociation at 25 °C and 0.2 mol L^{-1} Ionic Strength

		p*K*	ΔG°/kJ mol^{-1}	ΔH°/kJ mol^{-1}	ΔS°/J K^{-1} mol^{-1}
$HATP^{3-} = H^+ + ATP^{4-}$	K_{1ATP}	6.95	39.66	−7.03	−156.5
$H_2ATP^{2-} = H^+ + HATP^{3-}$	K_{2ATP}	4.06	23.22	0.00	−77.8
$HADP^{2-} = H^+ + ADP^{3-}$	K_{1ADP}	6.88	39.29	−5.73	−151.0
$H_2ADP^{1-} = H^+ + HADP^{2-}$	K_{2ADP}	3.93	22.43	4.18	−61.1
$HAMP^{1-} = H^+ + AMP^{2-}$	K_{1AMP}	6.45	36.82	−3.56	−135.6
$H_2AMP = H^+ + HAMP^-$	K_{2AMP}	3.75	21.34	4.18	−57.7
$H_2PO_4^{1-} = H^+ + HPO_4^{2-}$	K_{2P}	6.78	38.70	3.35	−118.8
$HP_2O_7^{3-} = H^+ + P_2O_7^{2-}$	K_{1PP}	8.95	51.09	1.67	−165.7
$H_2P_2O_7^{2-} = H^+ + HP_2O_7^{3-}$	K_{2PP}	6.12	34.94	0.46	−115.5

Table 6.5 Thermodynamic Quantities for the Dissociation of Complexes with Mg^{2+} at 25 °C and 0.2 mol L^{-1} Ionic Strength

		pK	$\Delta G°$/kJ mol^{-1}	$\Delta H°$/kJ mol^{-1}	$\Delta S°$/J K^{-1} mol^{-1}
$MgATP^{2-} = Mg^{2+} + ATP^{4-}$	K_{MgATP}	4.00	22.84	−13.81	−123.0
$MgHATP^{1-} = Mg^{2+} + HATP^{3-}$	K_{MgHATP}	1.49	8.54	−7.95	−55.2
$MgADP^{1-} = Mg^{2+} + ADP^{3-}$	K_{MgADP}	3.01	17.20	−15.06	−108.4
$MgHADP^{0} = Mg^{2+} + HADP^{2-}$	K_{MgHADP}	1.45	8.28	−8.37	−55.6
$MgAMP^{0} = Mg^{2+} + AMP^{2-}$	K_{MgAMP}	1.69	9.67	−12.13	−73.2
$MgHPO_4{}^{0} = Mg^{2+} + HPO_7^{4-}$	K_{MgP}	1.88	10.71	−12.13	−76.6
$MgP_2O_7^{2-} = Mg^{2} + P_2O_4^{2-}$	K_{MgPP}	5.41	30.92	−14.64	−152.7
$MgHP_2O_7^{1-} = Mg^{2+} + HP_2O_7^{3-}$	K_{MgHPP}	3.06	17.45	−14.64	−107.5

adenosine-5-monophosphate AMP, and inorganic phosphate determined in 0.2 mol L^{-1} ionic strength $(n\text{-propyl})_4NCl$ at 25 °C are summarized in Table 6.4. Contrary to the usual convention, the first proton to go on the most basic ion is referred to as number 1. This is done because the values of the dissociation constants for the two strongest dissociations of ATP are not known and are not needed for the calculations in which we are interested.

The values of $\Delta H°$ were obtained by determining pKs at two or more temperatures. Since the enthalpy changes are small in all cases, it is the entropy changes that determine the differences in strength of these acids. The magnitude of the standard entropy change correlates with the charge and charge distribution of the base formed. The larger the charge of the base formed, the more negative $\Delta S°$ is as a result of the greater hydration of the more highly charged ions.

The various phosphate ions bind Mg^{2+} and Ca^{2+}, and the dissociation constants of the complexes with Mg^{2+} expressed as pK values are given in Table 6.5 along with the other thermodynamic quantities. A simple physical interpretation of the dissociation constants is that they are equal to the molar concentration of free Mg^{2+} ion at which half of the ligand (that is the anion in this case) is complexed. The dissociation constants of the magnesium complexes were determined by acid titrations in the presence of various concentrations of magnesium salts.

Heat is evolved in the dissociation of $MgATP^{2-}$ ($\Delta H° = -13.8$ kJ mol^{-1}), so that the energy term *favors* dissociation, but the entropy term is negative and *does not favor* dissociation ($\Delta S° = -123$ J K^{-1} mol^{-1}). The entropy change strongly favors *association*. Why? When association occurs, H_2O of hydration is liberated, and there is an increase in randomness.

6.6 THERMODYNAMICS OF THE HYDROLYSIS OF ATP

Equation 6.26 for K' may be written in terms of the concentrations of the various ionic species that are important in the range of pH or pMg of interest. In order to simplify the equations that will be derived, we will not consider all of the species in Tables 6.4 and 6.5, but only those that have to be considered when pH > 5.5 and pMg > 1.5, so that only one acid dissociation and one magnesium ion complex need be considered for each reactant and product. In this case equation 6.26 can be written

$$K' = \frac{[(\mathrm{ADP}^{3-}) + (\mathrm{MgADP}^{1-}) + (\mathrm{HADP}^{2-})] \times [(\mathrm{HPO}_4^{2-}) + (\mathrm{MgHPO}_4{}^0) + (\mathrm{H_2PO}_4^{1-})]}{[(\mathrm{ATP}^{4-}) + (\mathrm{MgATP}^{2-}) + (\mathrm{HATP}^{3-})]} \tag{6.28}$$

If (ADP^{3-}) is taken out of the first term in the numerator, (HPO_4^{2-}) is taken out of the second, and (ATP^{4-}) is taken out of the denominator, the equilibrium constants for the various acid and magnesium complex dissociations may conveniently be introduced to obtain

$$K' = \frac{(\mathrm{ADP}^{3-})(\mathrm{HPO}_4^{2-})}{(\mathrm{ATP}^{4-})} \frac{\left[1 + \frac{(\mathrm{Mg}^{2+})}{K_{\mathrm{MgADP}}} + \frac{(\mathrm{H}^+)}{K_{1\mathrm{ADP}}}\right]\left[1 + \frac{(\mathrm{Mg}^{2+})}{K_{\mathrm{MgP}}} + \frac{(\mathrm{H}^+)}{K_{2\mathrm{P}}}\right]}{[1 + (\mathrm{Mg}^{2+})/K_{\mathrm{MgATP}} + (\mathrm{H}^+)/K_{1\mathrm{ATP}}]} \tag{6.29}$$

The hydrolysis of ATP may be expressed in terms of particular ionized species by

$$\mathrm{ATP}^{4-} + \mathrm{H_2O} = \mathrm{ADP}^{3-} + \mathrm{HPO}_4^{2-} + \mathrm{H}^+ \tag{6.30}$$

for which the equilibrium constant expression is

$$K = \frac{(\mathrm{ADP}^{3-})(\mathrm{HPO}_4^{2-})(\mathrm{H}^+)}{(\mathrm{ATP}^{4-})} \tag{6.31}$$

For reaction 6.30, the thermodynamic quantities are $\Delta G^\circ = 1.17\ \mathrm{kJ\ mol^{-1}}$, $\Delta H^\circ = -19.66\ \mathrm{kJ\ mol^{-1}}$, and $\Delta S^\circ = -69.87\ \mathrm{J\ K^{-1}\ mol^{-1}}$. Substituting equation 6.31 into equation 6.29 yields

$$K' = \frac{K[1 + (\mathrm{Mg}^{2+})/K_{\mathrm{MgADP}} + (\mathrm{H}^+)/K_{1\mathrm{ADP}}][1 + (\mathrm{Mg}^{2+})/K_{\mathrm{MgP}} + (\mathrm{H}^+)/K_{2\mathrm{P}}]}{(\mathrm{H}^+)[1 + (\mathrm{Mg}^{2+})/K_{\mathrm{MgATP}} + (\mathrm{H}^+)/K_{1\mathrm{ATP}}]} \tag{6.32}$$

Since K is independent of (H^+) and (Mg^{2+}), K' is given as a function of these variables by equation 6.32. A function of two variables may be represented by a surface and, in this case, it is convenient to represent log K' as a function of pH and pMg. The contour lines on this surface are shown in Fig. 6.2. In calculating this surface* all of the ionic equilibria in Tables 6.4 and 6.5 have been taken into account. In this range of pH and pMg the value of K' varies over 3 powers of 10.

At high pH in the absence of Mg^{2+}, the value of K' increases by a factor of 10 for each increase of pH of one unit. This results from the fact that under these conditions a mole of H^+ is liberated for each mole of ATP^{4-} hydrolyzed, as shown in equation 6.30, and this leads to a (H^+) term in the denominator of equation 6.32.

The expression of K' in terms of equilibrium constants and concentrations of H^+

---

* R. A. Alberty, *J. Biol. Chem.*, **244**, 3290 (1969).

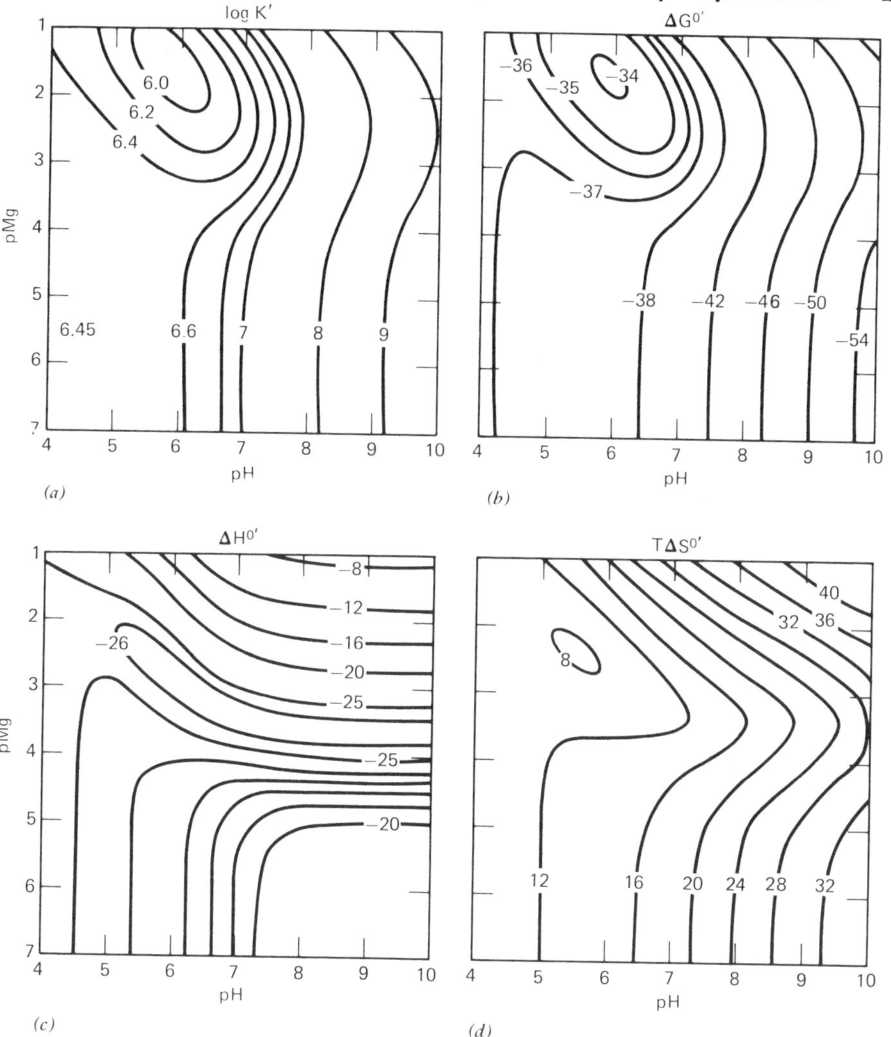

Fig. 6.2 Contour diagrams for thermodynamic quantities for ATP + H_2O = ADP + P at 25 °C as a function of pH and pMg. The values of (*a*) log K', (*b*) $\Delta G^{\circ\prime}$, (*c*) $\Delta H^{\circ\prime}$, and (*d*) $T\Delta S^{\circ\prime}$ are in kJ mol^{-1}. The principle electrolyte present is 0.2 ionic strength tetra-*n*-propyl ammonium chloride.

and Mg^{2+} is the key to the calculation of the standard Gibbs energy change $\Delta G^{\circ\prime}$, the standard enthalpy change $\Delta H^{\circ\prime}$, and the standard entropy change $\Delta S^{\circ\prime}$.

$$\Delta G^{\circ\prime} = -RT \ln K' \tag{6.33}$$

$$\Delta H^{\circ\prime} = \left[\frac{\partial(\Delta G^{\circ\prime}/T)}{\partial(1/T)}\right]_{\text{pH,pMg}} = RT^2\left[\frac{\partial \ln K'}{\partial T}\right]_{\text{pH,pMg}} \tag{6.34}$$

$$\Delta S^{\circ\prime} = -\left[\frac{\partial \Delta G^{\circ\prime}}{\partial T}\right]_{\text{pH,pMg}} = R\left[\frac{\partial T \ln K'}{\partial T}\right]_{\text{pH,pMg}} \tag{6.35}$$

To obtain $\Delta G^{\circ\prime}$, we simply apply equation 6.33 to equation 6.32. To obtain $\Delta H^{\circ\prime}$, this expression for $\Delta G^{\circ\prime}$ is divided by T and the derivative is taken with respect to $1/T$, as indicated in equation 6.34. This latter step is most easily taken by differentiating with respect to T and multiplying this derivative by $dT/d(1/T) = -T^2$. The derivatives of the dissociation constants with respect to temperature yield enthalpies of dissociation to obtain

$$\Delta H^{\circ\prime} = \Delta H^{\circ} - \frac{[(\mathrm{Mg}^{2+})/K_{\mathrm{MgADP}}]\,\Delta H^{\circ}_{\mathrm{MgADP}} + [(\mathrm{H}^+)/K_{1\mathrm{ADP}}]\,\Delta H^{\circ}_{1\,\mathrm{ADP}}}{1 + (\mathrm{Mg}^{2+})/K_{\mathrm{MgADP}} + (\mathrm{H}^+)/K_{1\,\mathrm{ADP}}}$$

$$- \frac{[(\mathrm{Mg}^{2+})/K_{\mathrm{MgP}}]\,\Delta H^{\circ}_{\mathrm{MgP}} + [(\mathrm{H}^+)/K_{2\mathrm{P}}]\,\Delta H^{\circ}_{2\mathrm{P}}}{1 + (\mathrm{Mg}^{2+})/K_{\mathrm{MgP}} + (\mathrm{H}^+)/K_{2\mathrm{P}}}$$

$$+ \frac{[(\mathrm{Mg}^{2+})/K_{\mathrm{MgATP}}]\,\Delta H^{\circ}_{\mathrm{MgATP}} + [(\mathrm{H}^+)/K_{1\mathrm{ATP}}]\,\Delta H^{\circ}_{1\mathrm{ATP}}}{1 + (\mathrm{Mg}^{2+})/K_{\mathrm{MgATP}} + (\mathrm{H}^+)/K_{1\mathrm{ATP}}} \qquad (6.36)$$

where $\Delta H^{\circ} = -19.66$ kJ mol^{-1} is the standard enthalpy change for reaction 6.30. and the other ΔH° values are those of the acid and complex dissociations in Tables 6.4 and 6.5. The enthalpy of hydrolysis of ATP at constant pressure varies with pH and pMg because the enthalpies of the various dissociation reactions are involved to different extents at different pH and pMg values. The dependence of the enthalpy of hydrolysis of ATP at constant pressure on pH and pMg is shown in Fig. 6.2, When this reaction is carried out in a buffer solution in a calorimeter, an additional heat effect has to be taken into account; the H^+ produced reacts with the buffer to give an enthalpy change of $-n_{\mathrm{H}}\,\Delta H^{\circ}_{\mathrm{B}}$, where $\Delta H^{\circ}_{\mathrm{B}}$ is the enthalpy of dissociation of the buffer acid and n_{H} is the number of moles of acid produced. The Mg^{2+} produced may also react with a component of the solution to produce a change in enthalpy.

The expression for $\Delta S^{\circ\prime}$ may be obtained by substituting equations 6.33 and 6.36 into

$$\Delta G^{\circ\prime} = \Delta H^{\circ\prime} - T\Delta S^{\circ\prime} \qquad (6.37)$$

The roles these various thermodynamic quantities play in determining K' changes with pH and pMg.*

At pH 7.4 and pMg 4.0, $\Delta G^{\circ\prime} = -41.0$ kJ mol^{-1}, $\Delta H^{\circ\prime} = -24.7$ kJ mol^{-1}, and $T\Delta S^{\circ\prime} = 20.5$ kJ mol^{-1}, so that the change in enthalpy and the change in entropy both make significant contributions to the change in Gibbs energy. As the pH increases above this point, the change in Gibbs energy becomes more negative because of the H^+ produced in reaction 6.30.

The biochemical equilibria we have been discussing involve small molecules, but many biochemical equilibria involve macromolecules such as proteins and nucleic acids. As an example we will consider the binding of small molecules by a protein.

---

* R. A. Alberty, *J. Chem. Ed.*, **46**, 713 (1969).

6.7 BINDING TO MULTIPLE SITES

A protein molecule P may bind a number n of molecules of a ligand A. The successive equilibria may be represented by

$$PA = P + A \qquad K_1 = \frac{(P)(A)}{(PA)} \tag{6.38}$$

$$PA_2 = PA + A \qquad K_2 = \frac{(PA)(A)}{(PA_2)} \tag{6.39}$$

$$\cdots$$

$$PA_n = PA_{n-1} + A \qquad K_n = \frac{(PA_{n-1})(A)}{(PA_n)} \tag{6.40}$$

The fraction Y of the binding sites, which is occupied, may be expressed in terms of the successive dissociation constants and the free concentration of A by

$$Y = \frac{(PA) + 2(PA_2) + \cdots + n(PA_n)}{n[(P) + (PA) + (PA_2) + \cdots + (PA_n)]}$$

$$= \frac{(A)/K_1 + 2(A)^2/K_1K_2 + \cdots + n(A)^n/K_1K_2\cdots K_n}{n[1 + (A)/K_1 + (A)^2/K_1K_2 + \cdots + (A)^n/K_1K_2\cdots K_n]} \tag{6.41}$$

This equation also applies to the binding of a number of molecules (or ions) of a ligand to another small molecule or metal ion. For example, it represents the titration curve of a polybasic acid. However, a different convention is usually used for numbering the acid dissociation constants for a polybasic acid; specifically, K_n in equation 6.40 is referred to as K_1 and K_1 in equation 6.38 is referred to as K_n when applied to a polybasic acid. We will apply equation 6.41 to the binding of oxygen by hemoglobin.

Myoglobin is an oxygen storage protein that is found in muscle tissue in many species. Its molar mass is 16,000 g mol^{-1}, and each molecule contains one heme group, one atom of iron, and binds one molecule of oxygen when it is saturated. Hemoglobin is the oxygen transport protein in many species. It has a molar mass of 64,000 g mol^{-1}, and each molecule contains four heme groups, four atoms of iron, and binds four molecules of oxygen when it is saturated. Hemoglobin may be reversibly dissociated into four molecules of molar mass 16,000 g mol^{-1}, each of which contains one heme and one atom of iron. These smaller protein molecules are of two types, represented by α and β, and hemoglobin has the composition $\alpha_2\beta_2$. Many enzymes have a similar subunit structure, and they and hemoglobin have remarkable binding properties, which are a consequence of this subunit structure.

The oxygen binding properties of myoglobin and hemoglobin are distinctively different, as shown by Fig. 6.3. Hemoglobin's S-shaped (sigmoid) binding curve is a great advantage for its physiological function, because the amount of oxygen bound changes rapidly between the partial pressure of oxygen in the lungs (about 13.3 kPa) and in the tissues (about 6.0 kPa).

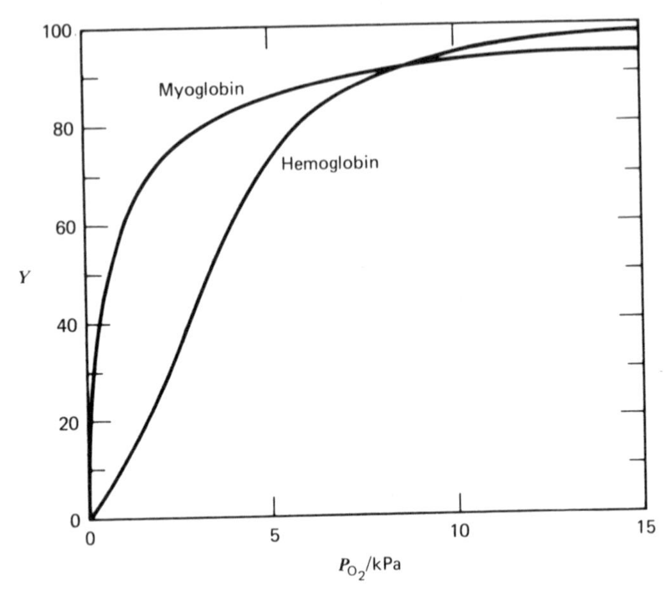

Fig. 6.3 Fractional saturation Y of myoglobin and hemoglobin with oxygen at pH 7.4 and 38 °C.

The shape of the binding curve for myoglobin is exactly what is expected for the simple equation

$$MbO_2 = Mb + O_2 \tag{6.42}$$

where Mb represents a molecule of myoglobin. The dissociation constant is defined by

$$K = \frac{(Mb)P_{O_2}}{(MbO_2)} \tag{6.43}$$

where P_{O_2} is partial pressure oxygen in the gas phase. Conservation of myoglobin requires that

$$(Mb)_o = (Mb) + (MbO_2) \tag{6.44}$$

where $(Mb)_o$ is the total molar concentration of myoglobin. Combining equations 6.43 and 6.44 yields

$$Y = \frac{(MbO_2)}{(Mb)_o} = \frac{P_{O_2}/K}{1 + P_{O_2}/K} \tag{6.45}$$

where Y is the fractional saturation. This is the equation of the myoglobin line in Fig. 6.3. At low P_{O_2} the fractional saturation is proportional to P_{O_2}, and at high P_{O_2}, it approaches unity.

The binding of oxygen by hemoglobin may be represented by equation 6.41. Since $n = 4$,

$$Y = \frac{P_{O_2}/K_1 + 2P_{O_2}^2/K_1K_2 + 3P_{O_2}^3/K_1K_2K_3 + 4P_{O_2}^4/K_1K_2K_3K_4}{4[1 + P_{O_2}/K_1 + P_{O_2}^2/K_1K_2 + P_{O_2}^3/K_1K_2K_3 + P_{O_2}^4/K_1K_2K_3K_4]} \tag{6.46}$$

where P_{O_2} has been used instead of the concentration (A) of the ligand.

Equation 6.46 can represent the sigmoid binding curve for hemoglobin but, remarkably, we find that the successive dissociation constants *decrease* instead of increase. This is contrary to the usual observation that the dissociation constants for successive ligand molecules increase.

When binding at the middle of the titration curve increases more rapidly with ligand concentration than can be accounted for by equation 6.45, it is said to be *cooperative*. Cooperativity arises when there is a mechanism for increasing the affinity of other sites once one or more sites are occupied by ligand. The origin of this effect in hemoglobin was quite mysterious until the structures of oxygenated and deoxygenated hemoglobin were determined by X-ray diffraction. Myoglobin was the first protein for which the detailed molecular structure was obtained by X-ray diffraction (Kendrew, 1959). When the structure of hemoglobin was obtained, it was found that each of its four subunits have three-dimensional configurations much like that of myoglobin. When oxygen is bound, groups near the heme shift slightly, and these structural changes affect the configurations of the four subunits so that the binding properties of the other heme groups are enhanced.*

Since it is difficult to determine the values of the four dissociation constants in equation 6.46, the empirical Hill equation

$$Y = \frac{1}{1 + K_h/P_{O_2}^h} \tag{6.47}$$

is frequently used to characterize binding. The Hill equation may be rearranged as follows:

$$\frac{Y}{1 - Y} = \frac{P_{O_2}^h}{K_h} \tag{6.48}$$

$$\log \frac{Y}{1 - Y} = -\log K_h + h \log P_{O_2} \tag{6.49}$$

The Hill coefficient h, which may be obtained from a plot of $\log [Y/(1 - Y)]$ versus $\log P$, is not necessarily an integer and, for normal human hemoglobin at pH 7, a value of about 2.8 is obtained. For a variety of binding systems relatively linear Hill plots are obtained for values of Y in the range 0.1 to 0.9, but deviations usually occur at the extremes unless $h = 1$. At the extremes the plot usually approaches a slope of unity.

Hemoglobin has another remarkable property that makes it even more effective in the oxygen transport system. When hemoglobin is oxygenated at pH 7.4, it dissociates 0.6 mol of H^+ for each oxygen molecule bound. A corollary of this so-called Bohr effect is that the affinity of hemoglobin for oxygen depends on pH. In the neighborhood of pH 7.4, the partial pressure of oxygen required to half saturate hemoglobin decreases as the pH is increased. The explanation of the Bohr effect is that the binding of oxygen affects the acid dissociation constants of certain acid groups in hemoglobin. This effect is of considerable physiological importance

---

* See R. E. Dickerson and I. Geis, *The Structure and Action of Proteins*, Harper and Row, New York, 1969.

because in the lungs H^+ liberated by hemoglobin upon oxygenation, reacts with HCO_3^- to make H_2CO_3. The carbonic acid then dehydrates and CO_2 diffuses into the air space of the lungs. If hemoglobin did not dissociate H^+, the blood would become alkaline in the lungs as CO_2 was exhaled. In the capillaries this process is reversed; hemoglobin absorbs H^+ as it loses oxygen, and this converts H_2CO_3, produced metabolically, to HCO_3^-.

6.8 BINDING TO MULTIPLE IDENTICAL AND INDEPENDENT SITES*

You might think that if the n binding sites are identical and independent, the various dissociation constants would be equal. But that is not true, as we saw in Example 6.2. If the sites are identical and independent, the relationships between the successive equilibrium constants may be derived as follows by considering the different forms of the various complexes. There are n forms of PA that we will represent by ${}_1PA$, ${}_2PA$, . . . , ${}_nPA$, where the subscript at the left indicates the site at which A is bound. The equilibrium constants for the formation of these various specific complexes are all equal.

$$K = \frac{(P)(A)}{({}_1PA)} = \frac{(P)(A)}{({}_2PA)} = \cdots = \frac{(P)(A)}{({}_nPA)} \tag{6.50}$$

This equilibrium constant K, which characterizes an individual binding site when there is more than one, is referred to as an *intrinsic* dissociation constant. Equation 6.50 indicates that the concentrations of the n forms of PA are equal.

The first dissociation constant obtained experimentally is defined by

$$K_1 = \frac{(P)(A)}{[({}_1PA) + ({}_2PA) + \cdots + ({}_nPA)]} = \frac{(P)(A)}{n({}_1PA)} = \frac{K}{n} \tag{6.51}$$

Thus the first dissociation constant obtained experimentally (by a method that does not distinguish between ${}_1PA$, ${}_2PA$, etc.) is one nth of the intrinsic dissociation constant K.

There are $n(n-1)/2$ forms of the PA_2 complex. The first A molecule may be placed on any one of the n sites and the second may be placed on any one of the remaining $(n-1)$ sites. The number of combinations $n(n-1)$ has to be divided by 2 to eliminate duplication. The various forms of PA_2 may be represented by ${}_{1,2}PA_2$, ${}_{1,3}PA_2$, etc., where the subscripts at the left indicate the two sites at which the two molecules of A are bound. These various complexes are equally probable, and so we may write the second dissociation constant as

$$K_2 = \frac{(PA)(A)}{(PA_2)} = \frac{[({}_1PA) + ({}_2PA) + \cdots + ({}_nPA)](A)}{[({}_{1,2}PA_2) + ({}_{1,3}PA_2) + \cdots]}$$

$$= \frac{n({}_1PA)(A)}{[n(n-1)/2]({}_{1,2}PA_2)} = \frac{2K}{n-1} \tag{6.52}$$

* C. Tanford, *Physical Chemistry of Macromolecules*, Wiley, New York, 1961, p. 532.

Table 6.6 Dissociation Constants for the Binding of Oxygen by Hemoglobin at pH 7 and 25 °C

	Experimental Value (kPa)	Relative Values	Relative Values from Equation 6.53	Enhancement E	$\Delta G°$ kJ mol^{-1}
K_1	2.73	1	1	1	0
K_2	3.12	1.14	8/3	2.3	−2.09
K_3	0.603	0.221	6	27	−8.20
K_4	0.417	0.152	16	105	−11.55

Continuing this process leads to a general relation between the dissociation constants K_i and the intrinsic dissociation constant K for the case that binding sites are identical and independent.

$$K_i = \frac{i}{n - i + 1} K \tag{6.53}$$

When this relationship is substituted into equation 6.41 for $n = 4$, we obtain

$$Y = \frac{(\mathrm{A})/K + 3(\mathrm{A})^2/K^2 + 3(\mathrm{A})^3/K^3 + (\mathrm{A})^4/K^4}{1 + 4(\mathrm{A})/K + 6(\mathrm{A})^2/K^2 + 4(\mathrm{A})^3/K^3 + (\mathrm{A})^4/K^4}$$

$$= \frac{[1 + (\mathrm{A})/K]^3(\mathrm{A})/K}{[1 + (\mathrm{A})/K]^4} = \frac{1}{1 + K/(\mathrm{A})} \tag{6.54}$$

Thus the fractional saturation Y follows equation 6.45 for a single binding equilibrium if the binding sites are identical and independent. By simply using titrations it is not possible to distinguish between a multisite protein with identical and independent sites and a single-site protein with a dissociation constant equal to the intrinsic constant for the multisite protein. These two types of proteins may, however, be distinguished by determining the molar mass and measuring the amount of ligand bound per unit mass of protein.

Equation 6.53 is useful in analyzing the binding of oxygen by hemoglobin. Since there is overlap of the successive oxygen binding steps, it is difficult to obtain values of K_1, K_2, K_3, and K_4, but values obtained at pH 7 by Szabo and Karplus* are given in Table 6.6. The ratios of these dissociation constants to K_1 may be compared with the ratios to be expected if the four binding sites were identical and independent. If the sites were identical and independent, equation 6.53 indicates that we would expect $K_1 = K/4$, $K_2 = 2K/3$, $K_3 = 3K/2$, and $K_4 = 4K$, where K is the intrinsic dissociation constant. The next to the last column of the table gives the ratio of the actual relative values of the dissociation constants to the relative dissociation constants expected for identical and independent sites. These enhancements E are expressed in the last column by the Gibbs energy calculated from $\Delta G° = -RT \ln E$ to give some indication of the magnitude of the enhancement of the binding due to the cooperative effect in the hemoglobin molecule.

* A. Szabo and M. Karplus, *J. Mole. Biol.*, **72**, 163 (1972).

References

R. G. Bates, *Determination of pH*, Wiley, New York, 1973.
R. P. Bell, *Acids and Bases*, Methuen and Co., London, 1969.
C. R. Cantor and P. R. Schimmel, *Biophysical Chemistry*, W. H. Freeman & Co., San Francisco, 1979.
R. E. Dickerson and I. Geis, *The Structures and Action of Proteins*, Harper and Row, New York, 1969.
J. Edsall and J. Wyman, *Biophysical Chemistry*, Academic Press, New York, 1958.
H. S. Harned and B. B. Owen, *The Physical Chemistry of Electrolytic Solutions*, Reinhold Publ. Corp., New York, 1958.
I. Klotz, *Energy Changes in Biochemical Reactions*, Academic Press, New York, 1967.
A. L. Lehninger, *Bioenergetics*, W. A. Benjamin, Inc., New York, 1965.

Problems

6.1 Calculate the pH of (*a*) a 0.1 mol L^{-1} solution of *n*-butyric acid, and (*b*) a solution containing 0.05 mol L^{-1} butyric acid and 0.05 mol L^{-1} sodium butyrate. Using these data, sketch the titration curve for 0.1 mol L^{-1} butyric acid that is titrated with a strong base so concentrated that the volume of the solution may be considered to remain constant ($K = 1.48 \times 10^{-5}$ at 25 °C). *Ans.* (*a*) 2.91, (*b*) 4.82.

6.2 A buffer contains 0.01 mol of lactic acid (pK = 3.60) and 0.05 mol of sodium lactate per liter. (*a*) Calculate the pH of this buffer. (*b*) Five milliliters of 0.5 mol L^{-1} hydrochloric acid is added to 1 L of the buffer. Calculate the change in pH. (*c*) Calculate the pH change to be expected if this quantity of acid is added to 1 L of a solution of a strong acid of the same initial pH. *Ans.* (*a*) 4.30, (*b*) 0.12, (*c*) 1.7.

6.3 Calculate the number of moles per liter of Na_2HPO_4 and NaH_2PO_4 that should be used to prepare a 0.10 mol L^{-1} ionic strength (equation 5.22) buffer of pH 7.30. At this ionic strength the second pK of phosphoric acid may be taken as 6.84.
Ans. (Na_2HPO_4) = 0.0299 mol L^{-1}, (NaH_2PO_4) = 0.0104 mol L^{-1}.

6.4 In order to determine the ionization constant of the weak monobasic acid dimethyl arsinic acid, a solution was titrated with a solution of sodium hydroxide, using a pH meter. After 17.3 cm^3 of NaOH had been added, the pH was 6.23. It was found that 27.6 cm^3 was required to neutralize the acid solution completely. Calculate the pK value. *Ans.* 6.00.

6.5 According to Table A.1, what are the values of ΔG°, ΔH°, and ΔS° at 298 K for

$$H_2O(l) = H^+(aq) + OH^-(aq)$$

Show that the same value of ΔS° is obtained from ΔG° and ΔH° as by using $\Delta S^\circ = \sum \nu_i S_i^\circ$.
Ans. 79.885, 55.836 kJ mol^{-1}. -80.67 J K^{-1} mol^{-1}.

6.6 Given the following thermodynamic quantities, calculate the pH of the midpoint of titration of the first acid group of carbonic acid at 25 °C.

	ΔH° kJ mol^{-1}	ΔS° J K^{-1} mol^{-1}
$CO_2 + H_2O = H_2CO_3$	4.73	−33
$H_2CO_3 = H^+ + HCO_3^-$	4.23	−59

Ans. 6.38.

6.7 Estimate pK_1 and pK_2 for H_3PO_4 at 25 °C and 0.1 mol L^{-1} ionic strength. The values at zero ionic strength are

$$pK_1 = 2.148 \qquad pK_2 = 7.198$$

Ans. 2.026, 6.831.

6.8 Will 0.01 mol L^{-1} creatine phosphate react with 0.01 mol L^{-1} adenosine diphosphate to produce 0.04 mol L^{-1} creatine and 0.02 mol L^{-1} adenosine triphosphate at 25 °C, pH 7, pMg 4? What concentration of ATP can be formed if the other reactants are maintained at the indicated concentrations? *Ans.* No. 1.1×10^{-2} mol L^{-1}.

6.9 In a series of biochemical reactions the product in one reaction is a reactant in the next. This has the effect that spontaneous reactions drive nonspontaneous reactions. For example, reaction 2 follows reaction 1.

1. *L*-malate = fumarate + H_2O $\qquad \Delta G^{\circ\prime} = 2.9$ kJ mol^{-1}
2. fumarate + ammonia = asparate $\qquad \Delta G^{\circ\prime} = -15.6$ kJ mol^{-1}

The $\Delta G^{\circ\prime}$ values are for pH 7 and 37 °C, and the state of ionization of the reactants is ignored. In reaction 1, the activity of H_2O is to be taken as 1. If the ammonia concentration is 10^{-2} mol L^{-1}, calculate (asparate)/(*L*-malate) at equilibrium. *Ans.* 1.3.

6.10 Biochemistry textbooks give $\Delta G^{\circ\prime} = -20.1$ kJ mol^{-1} for the hydrolysis of ethyl acetate at pH 7 and 25 °C. Experiments in acid solution show that

$$\frac{(CH_3CH_2OH)(CH_3CO_2H)}{(CH_3CO_2CH_2CH_3)} = 14$$

where concentrations are in moles per liter. What is the value of $\Delta G^{\circ\prime}$ obtained from this equilibrium quotient? The pK of acetic acid = 4.60 at 25 °C. *Ans.* -20.3 kJ mol^{-1}.

6.11 The cleavage of fructose 1,6-diphosphate (FDP) to dihydroxyacetone phosphate (DHP) and glyceraldehyde 3-phosphate (GAP) is one of a series of reactions most organisms use to obtain energy. At 37 °C and pH 7, $\Delta G^{\circ\prime}$ for the reaction FDP = DHP + GAP is 23.97 kJ mol^{-1}. What is $\Delta G'$ in an erythrocyte in which (FDP) = 3×10^{-6} mol L^{-1}, (DHP) = 138×10^{-6} mol L^{-1}, and (GAP) = 18.5×10^{-6} mol L^{-1}?

Ans. 5.77 kJ mol^{-1}.

6.12 From the data of Table 6.3 calculate $\Delta G^{\circ\prime}$ for

$$ATP + H_2O = AMP + PP$$

Ans. -42.3 kJ mol^{-1}.

6.13 0.01 mol L^{-1} NaH_2PO_4 dissolved in 0.2 mol L^{-1} $(CH_3CH_2)_4NCl$ is titrated with $(CH_3CH_2)_4NOH$ at 25 °C. The midpoint of the titration curve is at pH 6.80. An identical solution is made 0.05 mol L^{-1} in $MgCl_2$ and the titration is repeated. This time the midpoint is pH 6.37. What is the dissociation constant of $MgHPO_4$?

Ans. 2.96×10^{-2}.

6.14 At pH 7 and pMg 4 what value of pCa is required to put half the ATP in the form $CaATP^{-2}$? At 0.2 mol L^{-1} ionic strength and 25 °C the following constants are known.

$$HATP^{3-} = H^+ + ATP^{4-} \qquad pK = 6.95$$

$$MgATP^{2-} = Mg^{2+} + ATP^{4-} \qquad pK = 4.00$$

$$CaATP^{2-} = Ca^{2+} + ATP^{4-} \qquad pK = 3.60$$

Ans. 3.14.

6.15 Given $\Delta G^\circ = 49.4$ kJ mol^{-1} for

$$ATP^{4-} + H_2O = AMP^{2-} + P_2O_4^{7-} + 2H^+$$

calculate $\Delta G^{\circ\prime}$ at pH 7 and 25 °C and 0.2 mol L^{-1} ionic strength. See Table 6.4.

Ans. -41.0 kJ mol^{-1}.

6.16 The hydrolysis of adenosine triphosphate ATP to adenosine diphosphate ADP and inorganic phosphate at pH 8 and 25 °C

$$ATP^{-4} + H_2O = ADP^{-3} + HPO_4^{-2} + H^+$$

has a standard enthalpy change of -13 kJ mol^{-1}. The standard enthalpy changes of acid dissociation of $HATP^{-3}$, $HADP^{-2}$, and $H_2PO_4^{-1}$ are -8, 0, and $+8$ kJ mol^{-1}, respectively. Calculate the standard enthalpy change for the reaction

$$HATP^{-3} + H_2O = HADP^{-2} + H_2PO_4^{-}$$

Ans. -29 kJ mol^{-1}.

6.17 Derive the equations for the number r_A of molecules of A bound per molecule of P and the number r_B of molecules of B bound per molecule of P for the reactions

$$PA \overset{K_A}{=} P + A \qquad PB \overset{K_B}{=} P + B$$

where the Ks represent dissociation constants. Show that

$$\frac{\partial r_A}{\partial \ln (B)} = \frac{\partial r_B}{\partial \ln (A)}$$

This relation expresses the linkage between the binding of A and B.

Ans. $r_A = \dfrac{(A)/K_A}{1 + (A)/K_A + (B)/K_B} \qquad r_B = \dfrac{(B)/K_B}{1 + (A)/K_A + (B)/K_B}$.

6.18 If n molecules of a ligand L combine with a molecule of protein to form PL_n without intermediate steps, derive the relation between the fractional saturation Y and the concentration of L.

Ans. $Y = \dfrac{(L)^n/K}{1 + (L)^n/K}$.

6.19 The percent saturation of a sample of human hemoglobin was measured at a series of oxygen partial pressures at 20 °C, pH 7.1, 0.3 mol L^{-1} phosphate buffer, and 3×10^{-4} mol L^{-1} heme.

P_{O_2}/Pa	Percent Saturation
393	4.8
787	20
1183	45
2510	78
2990	90

Calculate the values of h and K in the Hill equation. *Ans.* 2.4, 4×10^7.

6.20 Assuming that a protein molecule has n_α independent sites with intrinsic dissociation constant K_α and n_β independent sites with intrinsic dissociation constant K_β, what is the expression for the number r of moles of A bound per mole of protein?

Ans. $r = \dfrac{n_\alpha}{1 + K_\alpha/(A)} + \dfrac{n_\beta}{1 + K_\beta/(A)}$.

6.21 The ionization constants at 25 °C for acetic acid, lactic acid, and bromoacetic acid are 1.8×10^{-5}, 1.4×10^{-4}, and 1.4×10^{-3}, respectively. Calculate the degree of dissociation α for a 0.01 mol L^{-1} solution of each of these acids (*a*) by the approximate method (assuming $1 - \alpha = 1$), and (*b*) without approximation.

6.22 Sketch the titration curve of aspartic acid indicating the approximate pH values obtained when monosodium aspartate and disodium aspartate are dissolved in water.

6.23 A buffer contains 0.04 mol of Na_2HPO_4 per liter and 0.02 mol of NaH_2PO_4 per liter. (*a*) Calculate the pH, using pK = 6.84, which is the pK corresponding to the ionic strength of the solution. (*b*) One centimeter of 1 mol L^{-1} HCl is added to 1 L of the buffer. Calculate the change in pH. (*c*) Calculate the pH change to be expected if this quantity of HCl is added to 1 L of pure water that has a pH of 7.

6.24 Given the following thermodynamic quantities, calculate the dissociation constant for $NH_4OH = NH_4^+ + OH^-$ at 25 °C.

	ΔH° kJ mol^{-1}	ΔS° J K^{-1} mol^{-1}
$NH_4^+ = NH_3 + H^+$	52.22	−1.7
$H_2O = H^+ + OH^-$	56.57	−78.2

6.25 Show that the slope of the titration curve of a monobasic weak acid is given by

$$\frac{d\alpha}{d\text{pH}} = \frac{2.303K(\text{H}^+)}{[K + (\text{H}^+)]^2}$$

where α is the degree of neutralization.

6.26 For the acid dissociation of acetic acid ΔH° is approximately zero at room temperature in H_2O. For the acidic form of aniline, which is approximately as strong an acid as acetic acid, ΔH° is approximately 21 kJ mol^{-1}. Calculate ΔS° for each of the following reactions.

$$CH_3CO_2H = H^+ + CH_3CO_2^- \qquad \text{p}K = 4.75$$

$$C_6H_5NH_3^+ = H^+ + C_6H_5NH_2 \qquad \text{p}K = 4.63$$

How do you interpret these entropy changes? What compensates for the increase in entropy expected from the increase in number of molecules in the reaction?

6.27 Using the Debye-Hückel theory, estimate the apparent pK for acetic acid at 0.01 mol L^{-1} ionic strength. At 25 °C the thermodynamic pK value is 4.76. It is assumed that the activity coefficient for undissociated acetic acid is unity at this value of the ionic strength.

6.28 Calculate the normal pH of human blood from the fact that the normal concentration of HCO_3^- is 0.026 mol L^{-1} and the normal partial pressure of CO_2 in the alveoli of the lungs is 5.3 kPa. The pK_1 value for H_2CO_3 at 38 °C and 0.15 mol L^{-1} ionic strength is 6.1. The Henry law constant K for the solubility of CO_2 in water is 2.48×10^{-4} mol L^{-1} kPa^{-1} where $(CO_2) = KP_{CO_2}$.

6.29 In the living cell two reactions may be coupled together by having a common intermediate. This is true for the following two reactions that are enzyme catalyzed,

creatine + inorg. phosphate = creatine phosphate $\qquad \Delta G^{\circ\prime} = 46$ kJ mol^{-1}

ATP + H_2O = ADP + inorg. phosphate $\qquad \Delta G^{\circ\prime} = -33$ kJ mol^{-1}

where ATP and ADP are adenosinetriphosphate and adenosinediphosphate, respectively. The phosphate ionizations are ignored and the $\Delta G^{\circ\prime}$ values are for pH 7.5 and 25 °C. If in a steady state in a living cell (ATP) = 10^{-3} mol L^{-1} and (ADP) = 10^{-4} mol L^{-1}, calculate the maximum value of the ratio (creatine phosphate)/(creatine).

6.30 Will cyclic AMP at 0.01 mol L^{-1} react with ADP and P at 0.01 mol L^{-1} at pH 7 and 25 °C to form ATP and AMP, each at 0.01 mol L^{-1}?

$$\text{CAMP} + H_2O = \text{AMP} \qquad \Delta G^{\circ\prime} = -41.8 \text{ kJ mol}^{-1}$$
$$\text{ADP} + \text{P} = \text{ATP} + H_2O \qquad \Delta G^{\circ\prime} = 39.8 \text{ kJ mol}^{-1}$$

6.31 How many grams of ATP have to be hydrolyzed to ADP to lift 100 lb 100 ft if the available Gibbs energy can be converted into mechanical work with 100% efficiency? It is assumed that (ATP) = (ADP) = (P) = 0.01 mol L^{-1} and that $\Delta G^{\circ\prime}$ is -39.8 kJ mol^{-1} at 25 °C.

6.32 What is the maximum concentration of ATP that can be formed enzymatically from acetyl phosphate and ADP each at 0.01 mol L^{-1} and pH 7 and pMg 4 at 25 °C, assuming that the ambient concentration of acetate is also 0.01 mol L^{-1}? Given

$$\text{acetyl P} + H_2O = \text{acetate} + \text{P} \qquad \Delta G^{\circ\prime} = -43.1 \text{ kJ mol}^{-1}$$
$$\text{ADP} + \text{P} = \text{ATP} + H_2O \qquad \Delta G^{\circ\prime} = 39.8 \text{ kJ mol}^{-1}$$

6.33 The pK for the dissociation of $CaATP^{2-}$ at 25 °C in 0.2 mol L^{-1} (n-propyl)$_4$NCl is 3.60. The pK for

$$HATP^{3-} = H^+ + ATP^{4-}$$

is 6.95. Calculate the apparent pK of this ATP ionization when ATP is titrated in a solution containing 0.1 mol L^{-1} $CaCl_2$. Assume that the Ca^{2+} concentration is much larger than the total ATP concentration.

6.34 If $\Delta G^{\circ\prime}$ for the hydrolysis of acetyle phosphate ($CH_3CO_3PO_3H_2$) is -43.1 kJ mol^{-1} at 25 °C and pH 7, what is the value at pH 4? It may be assumed that the pK of acetyl phosphate in the neighborhood of pH 7 is identical with pK_2 of orthophosphate.

6.35 At sufficiently high pH (pH > 8) and sufficiently high pMg (pMg > 5), the value of $\Delta G^{\circ\prime}$ for

$$\text{pyrophosphate} + H_2O = 2 \text{ phosphate}$$

can be calculated with only the following information. At 25 °C and 0.2 ionic strength,

$$P_2O_7^{4-} + H_2O = 2HPO_4^{2-} \qquad \Delta G^{\circ} = -44.56 \text{ kJ mol}^{-1}$$
$$HP_2O_7^{3-} = H^+ + P_2O_7^{4-} \qquad \Delta G^{\circ} = 51.09 \text{ kJ mol}^{-1}$$
$$MgP_2O_7^{2-} = Mg^{2+} + P_2O_7^{4-} \qquad \Delta G^{\circ} = 30.92 \text{ kJ mol}^{-1}$$

What is the value of the equilibrium constant

$$\frac{(\text{phosphate})^2}{(\text{pyrophosphate})}$$

at pH 9 and pMg 6?

6.36 At sufficiently high pH (pH > 7.5) and sufficiently high pMg (pMg > 3.5), the effect of (Mg^{2+}) on the heat of hydrolysis of ATP and on the maximum non-PV work at 25 °C may be calculated with only the following information.

	ΔH° kJ mol^{-1}	ΔS° J K^{-1} mol^{-1}
$ATP^{4-} + H_2O = ADP^{3-} + HPO_4^{2-} + H^+$	-19.7	-69.9
$MgATP^{2-} = Mg^{2+} + ATP^{4-}$	-13.8	-123.0

Calculate $\Delta H^{\circ\prime}$ and $\Delta G^{\circ\prime}$ at pH 8, pMg 4, and pH 8, pMg 8.

6.37 Given the following system of reactions,

$$\begin{array}{ccc} \mathrm{AH} + \mathrm{B} & \overset{K_1}{\rightleftharpoons} & \mathrm{CH} \\ K_{\mathrm{AH}} \upharpoonleft\downharpoonright & & \upharpoonleft\downharpoonright K_{\mathrm{CH}} \\ \mathrm{A} + \mathrm{B} & \underset{K_2}{\rightleftharpoons} & \mathrm{C} \end{array}$$

where the *K*s represent dissociation constants. (*a*) Calculate the dependence on hydrogen ion concentration of the apparent equilibrium constant

$$K' = \frac{[(\mathrm{C}) + (\mathrm{CH})]}{[(\mathrm{A}(+ (\mathrm{AH})](\mathrm{B})}$$

(*b*) What is the relationship between the four dissociation constants?

6.38 The partial pressure of oxygen required to half saturate hemoglobin at pH 7.4 is 3.7 kPa. If the partial pressure of oxygen in the alveolar spaces of the lungs is 13.3 kPa, and the partial pressure in the capillaries is 5.3 kPa, what percetnage of the total oxygen carrying capacity of hemoglobin is being used if h in the Hill equation is 2.7?

6.39 A protein binds five ligands (A) at identical and independent sites. What are the relative values of the five successive dissociation constants? What are the relative concentrations of the five complexes with respect to the molar concentrations of the unbound protein when $K = (\mathrm{A})$? (Note that the relative concentrations are given by the terms in $\{1 + [(\mathrm{A})/K]^n\}$.)

6.40 Suppose that a particular hemoprotein has two oxygen binding sites that are independent. The fractional saturation Y may be expressed in terms of microscopic dissociation constants k_1 and k_2 by

$$Y = \frac{1}{2}\left[\frac{1}{1 + k_1/P_{\mathrm{O}_2}} + \frac{1}{1 + k_2/P_{\mathrm{O}_2}}\right]$$

What is the expression for Y in terms of the usual thermodynamic dissociation constants K_1 and K_2? How are K_1 and K_2 related to k_1 and k_2?

CHAPTER 7

SURFACE THERMODYNAMICS

In our earlier consideration of thermodynamic properties we ignored surface effects. But molecules or atoms in a surface are in a different environment from molecules or atoms in the bulk phase, and if material is finely divided, surface effects may be quite significant. A cubic meter of material divided into 10^{-6} m cubes has a surface area of 6 km^2.

The unsymmetrical force field at a surface gives rise to a surface tension parallel with the surface, a tendency of asymmetric molecules to be oriented in a surface, and a capacity to bind other molecules at the surface either physically or chemically.

There are many practical applications of surface thermodynamics in understanding the lowering of surface tension by solutes, adsorption by solids, chromatography, colloids, and surface catalysis.

7.1 SURFACE TENSION

The molecules at the surface of a liquid are attracted into the body of the liquid because the attraction of the underlying molecules is greater than the attraction by the vapor molecules on the other side of the surface. This inward attraction causes the surface to contract if it can and gives rise to a force in the plane of the surface. Surface tension is responsible for the formation of spherical droplets, the rise of water in a capillary, and the movement of a liquid through a porous solid. Solids also have surface tensions, but it is harder to measure them. Crystals tend to form with faces with the lowest surface tensions.

The surface tension of a liquid, γ, is the force per unit length on the surface that opposes the expansion of the surface area. This definition is illustrated by the idealized experiment in Fig. 7.1, where the movable bar is pulled with force F to expand a liquid film that is stretched like a soap-bubble film on a wire frame. The surface tension is

$$\gamma = \frac{F}{2l} \tag{7.1}$$

where l is the length of the bar, and the factor 2 is introduced because there are two liquid surfaces, one at the front and one at the back.*

* The SI unit of surface tension is N m^{-1} or, since 1 J = 1 N m, J m^{-2}. In the older literature surface tensions are usually expressed in dyne cm^{-1}. A dyne is 10^{-5} N. The surface tension of water at 25 °C is (71.97 dyne cm^{-1})(10^{-5} N dyne^{-1})(10^2 cm m^{-1}) = 71.97 mN m^{-1} or 71.97 mJ m^{-2}.

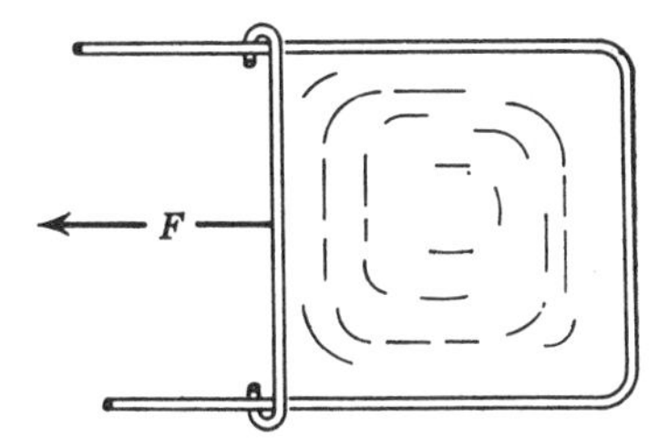

Fig. 7.1 Idealized experiment for the determination of the surface tension of a liquid.

The surface tension of a liquid may be measured by a variety of methods.* Since the equilibrium shape of liquid surfaces is determined by a balance of surface tension and gravitational forces, analysis of drop or bubble shape may be used to determine surface tension. The rise of liquid in a capillary or the pull on a thin vertical plate partially immersed in the liquid may be determined and used to calculate the surface tension quite accurately. Less accurate values of the surface tension may be obtained from measurements on moving liquid surfaces. These methods include studies of liquid jets, ripples, drop weight, and the force required to rupture a surface.

The surface tension of a liquid decreases as the temperature rises and becomes very small a few degrees below the critical temperature. It is zero at the critical temperature.

The surface tensions of liquid metals and molten salts are large in comparison with those of organic liquids. For example, the surface tension of mercury at 0 °C is 480.3 mN m^{-1}, and that of silver at 800 °C is 800 mN m^{-1}.

The interface between two mutually saturated immiscible liquids contracts because of the interfacial tension. In principle, this interfacial tension can be measured by all the methods used to measure surface tension but, since interfacial tensions are even more sensitive to impurities than surface tensions, such measurements are more difficult to make.

Work is required to increase the area of a surface. The work required to increase the area of the film illustrated in Fig. 7.1 is $w = Fd = 2\gamma ld = 2\gamma\mathscr{A}$, where d is the displacement of the moving bar, $\mathscr{A}$ is the area, and the factor 2 again comes from the fact that the liquid film has two surfaces. Thus, in general, the work done to increase the area of a surface is $\gamma\mathscr{A}$. Since in SI work is expressed in joules and area in square meters, it is evident that the surface tension may also be expressed in joules per square meter.

7.2 PRESSURE DROP ACROSS A CURVED SURFACE

The pressure drop across a curved surface is responsible for the rise, or depression, of a liquid surface in a capillary. This effect also leads to the dependence of the

* A. W. Adamson, *Physical Chemistry of Surfaces*, 3rd ed., Interscience-Wiley, New York, 1977.

vapor pressure of a liquid on the curvature of its surface. A liquid exerts a lower vapor pressure in a small bubble in the bulk phase than at a plane surface, and a small droplet of liquid exerts a higher vapor pressure.

Consider a spherical vapor bubble of radius r in a liquid at pressure P. The surface tension γ of the liquid tends to cause the bubble to contract, and at equilibrium the pressure inside the bubble is $P + \Delta P$. In a hypothetical contraction of the bubble to radius r–dr the work done against the excess pressure ΔP must be equal to the decrease in surface energy. Since the area of the liquid-vapor interface is $4\pi r^2$, the change in surface area would be $d\mathscr{A} = 8\pi r\,dr$ and the decrease in surface energy would be $\gamma\,d\mathscr{A} = 8\pi\gamma r\,dr$. Since the volume change is $dV = 4\pi r^2\,dr$, the PV work done in the contraction is $\Delta P\,dV = \Delta P 4\pi r^2\,dr$. Thus

$$\Delta P 4\pi r^2\,dr = 8\pi\gamma r\,dr \tag{7.2}$$

$$\Delta P = \frac{2\gamma}{r} \tag{7.3}$$

Since the radius of curvature in this case is positive, the pressure is greater on the concave side of the interface.

Example 7.1 What pressure in atmospheres is required to prevent water from rising in a 10^{-4} cm diameter capillary at 25 °C at 1 atm?

$$\Delta P = \frac{2\gamma}{r} = \frac{2(71.97 \times 10^{-3}\ \mathrm{N\ m^{-1}})}{(0.5 \times 10^{-6}\ \mathrm{m})(101{,}325\ \mathrm{N\ m^{-2}\ atm^{-1}})} = 2.84\ \mathrm{atm}$$

Equation 7.3 may be used to derive the relation between the surface tension and the rise, or depression, of a liquid in a capillary that is illustrated in Fig. 7.2. If the contact angle θ that the liquid makes with the wall is less than 90°, the liquid will rise; if the contact angle is greater than 90°, the surface of the liquid will be depressed. If the capillary is of sufficiently small diameter, the meniscus will be a section of a sphere, and the radius of curvature of the liquid surface in the capillary is given by

$$r = \frac{r_t}{\cos\theta} \tag{7.4}$$

where r_t is the radius of the capillary tube. The difference between the pressure in the liquid at the curved surface and the pressure at the flat surface of the liquid is given by equation 7.3.

$$\Delta P = \frac{2\gamma\cos\theta}{r_t} \tag{7.5}$$

Therefore, the atmospheric pressure pushes the liquid up the tube until the pressure difference between the liquid at the curved surface and the liquid at the plane surface are balanced by the hydrostatic pressure due to the liquid column of height h.

$$\frac{2\gamma\cos\theta}{r_t} = hg(\rho_l - \rho_v) \tag{7.6}$$

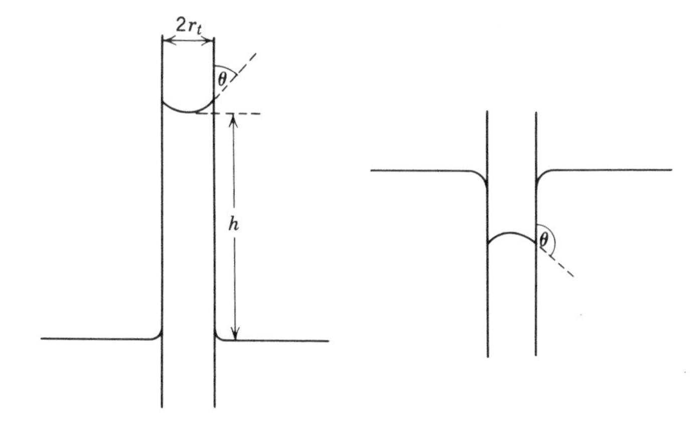

Fig. 7.2 Rise or depression of a liquid in a capillary.

where ρ_l is the density of the liquid, ρ_v is the density of the vapor, and g is the acceleration of gravity. Since the density of the vapor is usually much less than the density of the liquid, ρ_v may be neglected to obtain

$$\gamma = \frac{hg\rho r_t}{2 \cos \theta} \tag{7.7}$$

where ρ_l is the density of the liquid. The mass of liquid supported is slightly greater than that calculated from measuring height to the bottom of the meniscus, and it can be shown that for small capillaries $r_t/3$ should be added to the height measured to the bottom of the meniscus in the capillary to obtain h. For mercury in glass, θ is approximately 140°, and so mercury is depressed in the capillary.

7.3 THE KELVIN EQUATION

The vapor pressure of a small droplet exceeds that of a plane surface of the liquid, and the vapor pressure of a concave surface of a liquid (as in a bubble within a liquid) is less than that of a plane surface. The magnitude of this effect for droplets and bubbles calculated on the assumption that surface tension is independent of radius of curvature is shown in Fig. 7.3.

To derive the effect of curvature of a surface on vapor pressure, we need the relationship between the differential change in area of the interface and the differential change in Gibbs energy of the system. To obtain this relationship, we start with the first law in the form of $dU = dq + dw$. For a system at equilibrium $dq = T\,dS$ and

$$dw = -P\,dV + \gamma\,d\mathscr{A} \tag{7.8}$$

so that the combined first and second laws for an open one-component system is

$$dU = T\,dS - P\,dV + \gamma\,d\mathscr{A} + \mu\,dn \tag{7.9}$$

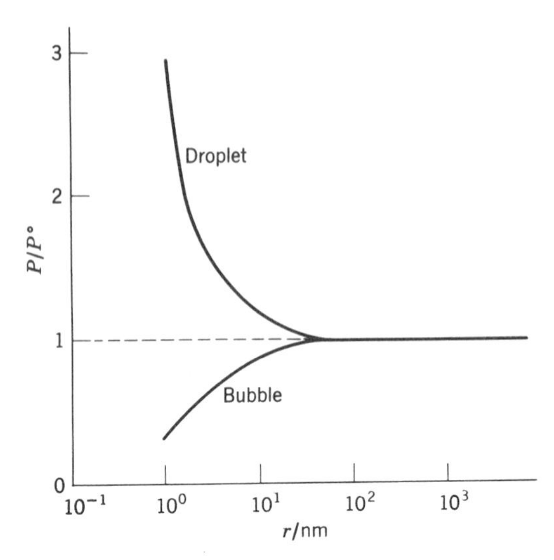

Fig. 7.3 Effect of curvature of a surface on the vapor pressure of water at 25 °C. (From R. Aveyard and D. A. Haydon, *An Introduction to the Principles of Surface Chemistry*, Cambridge University Press, Cambridge, England, 1973.)

where μ is the chemical potential and dn is the change in the number of moles in the system. The Gibbs energy G of the system is given by

$$G = U + PV - TS \tag{7.10}$$

so that

$$\begin{aligned} dG &= dU + P\,dV + V\,dP - T\,dS - S\,dT \\ &= -S\,dT + V\,dP + \gamma\,d\mathscr{A} + \mu\,dn \end{aligned} \tag{7.11}$$

where the second form has been obtained by substituting equation 7.9. The last term in this equation gives the change in G when dn moles of a substance are added to the droplet without changing the area of the surface. This is the case for a planar surface, and so we will use μ_{planar} to represent $(\partial G/\partial n)_{T,P,\mathscr{A}}$. When material is added to a spherical droplet, there is an increase in area of the surface. For a sphere $dV = 4\pi r^2\,dr$ and $d\mathscr{A} = 8\pi r\,dr$, so that $d\mathscr{A} = 2dV/r$. For an addition of dn there is an increase in area of

$$d\mathscr{A} = \frac{2\,dV}{r} = \frac{2V_m\,dn}{r} \tag{7.12}$$

where V_m is the molar volume of the liquid. Combining this equation with the preceding one yields

$$dG = -S\,dT + V\,dP + \left(\frac{2V_m\gamma}{r} + \mu_{\text{planar}}\right) dn \tag{7.13}$$

Thus the chemical potential of the liquid in the droplet is given by

$$\mu = \frac{2V_m\gamma}{r} + \mu_{\text{planar}} \tag{7.14}$$

If the vapor behaves like a perfect gas,

$$\mu = \mu_{\text{planar}} + RT \ln \frac{P}{P^\circ} \tag{7.15}$$

where P° is the vapor pressure for a planar surface. Combining this equation with the preceding one yields the Kelvin equation.

$$\ln\left(\frac{P}{P^\circ}\right) = \frac{2V_m\gamma}{rRT} \tag{7.16}$$

This equation gives the vapor pressure P of a droplet of radius r. We can think of small droplets as having higher vapor pressures than the bulk liquid because molecules are not drawn into the interior by so many near neighbors.

In order for a vapor to condense in the absence of foreign surfaces it is necessary for small clusters of molecules to form and to grow and finally coalesce to form the bulk phase. This does not happen if the pressure of the vapor is only slightly higher than the equilibrium vapor pressure, because the very small droplets that are formed first have a higher vapor pressure. However, when the pressure has been increased sufficiently over the equilibrium value, general condensation of droplets occurs.*

Example 7.2 Water vapor is rapidly cooled to 25 °C to find the degree of supersaturation required to nucleate water droplets spontaneously. It is found that the vapor pressure of water must be four times its equilibrium vapor pressure. (*a*) Calculate the radius of a water droplet formed at this degree of supersaturation. (*b*) How many water molecules are there in the droplet?

(*a*) $$r = \frac{2V_m\gamma}{RT\ln(P/P^\circ)} = \frac{2(18\times10^{-6}\ \text{m}^3\ \text{mol}^{-1})(0.071\,97\ \text{N m}^{-1})}{(8.314\ \text{J K}^{-1}\ \text{mol}^{-1})(298\ \text{K})\ln 4} = 0.75\ \text{nm}$$

(*b*) $$N = \frac{\frac{4}{3}\pi r^3\rho}{(M/N_A)} = \frac{\frac{4}{3}\pi(0.75\times10^{-9}\ \text{m})^3(1\times10^3\ \text{kg m}^{-3})}{(18\times10^{-3}\ \text{kg mol}^{-1})/(6.022\times10^{23}\ \text{mol}^{-1})} = 59$$

The derivation of equation 7.16 is based on the assumption that the surface tension is independent of the curvature of the surface. However, when the radius of curvature becomes comparable to molecular dimensions, this will no longer be a good assumption.

If the sign of one side of this equation is changed, it yields the vapor pressure of the concave surface of a liquid in a capillary or a small bubble. We can think of the

* Similar phenomena are involved in the freezing of liquids. Water without impurities or foreign surfaces may be cooled to −40 °C before nucleation begins spontaneously.

surface of the liquid in a capillary that it wets or in a small bubble as having a lower vapor pressure than the bulk liquid because molecules in the surface are drawn into the interior by more near neighbors than in a flat surface. The Kelvin equation helps us to understand why liquids have a tendency to superheat at their boiling points. If a small bubble starts to form at the boiling point, equation 7.16 (with a sign change) is not satisfied, and the bubble will be squeezed out of existence by the force of surface tension. At a temperature above the boiling point, the vapor pressure will be enough higher that a bubble of a certain radius will be thermodynamically stable.

7.4 CONTACT ANGLE AND ADHESION

Figure 7.4 shows a liquid in equilibrium with a solid surface and gas. For some liquid-solid pairs the contact angle θ is zero, and the liquid spreads until it covers the whole solid surface. For other liquid-solid pairs the contact angle θ is greater than 90°, and the liquid stands up in little droplets and does not wet the solid surface.

At equilibrium the three interfacial tensions must balance along the line of contact. Thus

$$\gamma_{lg} \cos \theta + \gamma_{sl} = \gamma_{sg} \tag{7.17}$$

or

$$\cos \theta = \frac{\gamma_{sg} - \gamma_{sl}}{\gamma_{lg}} \tag{7.18}$$

This equation does not hold if $\gamma_{sg} > \gamma_{sl} + \gamma_{lg}$, because in this case the system has the lowest Gibbs energy if the solid is completely wet. It also does not hold if $\gamma_{sl} > \gamma_{sg} + \gamma_{lg}$ so that the solid is not wet at all.

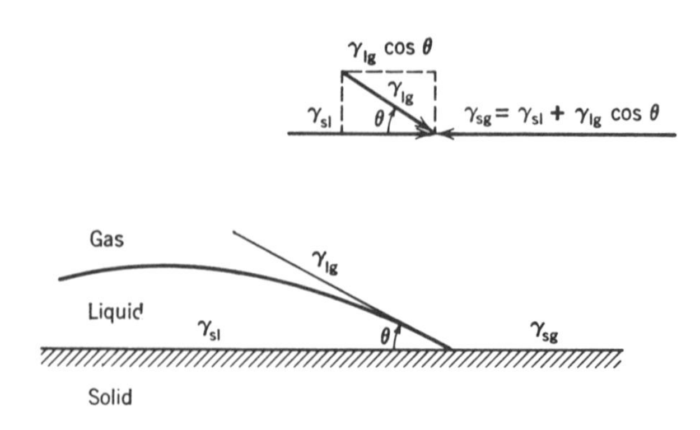

Fig. 7.4 A liquid in equilibrium with solid and vapor.

7.5 SURFACE CONCENTRATION

In a system containing two or more components, a substance may be concentrated or depleted in the neighborhood of a surface. If a substance is concentrated in the neighborhood of a surface, it is said to be positively adsorbed. If the concentration of the substance is lower in the interfacial region that in bulk, it is said to be negatively adsorbed.

If a substance is adsorbed, the amount of the substance in the system differs from what would be calculated for a hypothetical system with homogeneous bulk phases and an infinitely sharp dividing surface. Such a surface is referred to as a Gibbs surface because Gibbs was the first to make the analysis we are going to introduce here. Figure 7.5 shows the Gibbs surface for a two-component system. The thickness of the region of changing concentration is a few molecular diameters, except in the neighborhood of the critical point, where it is thicker. We can imagine this system as a solvent (component 1) in equilibrium with its vapor with a slightly volatile solute (component 2), which has a tendency to concentrate in the surface. The surface excess amount n_i^σ of a component may be positive, negative, or zero. A negative adsorption, of course, means that the surface layer is depleted in the component.

The surface excess amount n_i^σ of a component i is defined by

$$n_i^\sigma = n_i - V^\alpha c_i^\alpha - V^\beta c_i^\beta \tag{7.19}$$

where n_i is the total amount of component i in the system, c_i^α and c_i^β are the concentrations in the two bulk phases α and β, and V^α and V^β are the volumes of

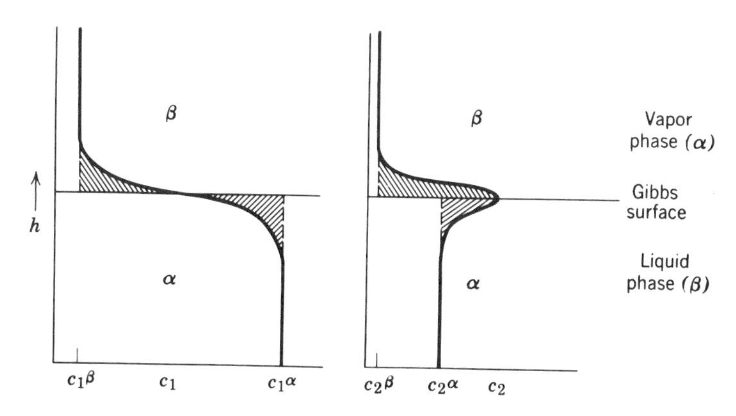

Fig. 7.5 (a) Concentration of component 1, and (b) concentration of component 2 as a function of distance perpendicular to a flat surface. The surface excess amount per unit area is zero for component 1, which is arbitrarily taken as the solvent, and positive for component 2.

the two phases defined by the Gibbs surface. In terms of the molar concentrations c_i of a component through the surface region,

$$n_i^\sigma = \int_{\substack{\text{phase } \alpha \\ \text{up to the} \\ \text{Gibbs surface}}} (c_i - c_i^\alpha)\, dV + \int_{\substack{\text{phase } \beta \\ \text{up to the} \\ \text{Gibbs surface}}} (c_i - c_i^\beta)\, dV \tag{7.20}$$

The surface excess amount is, of course, dependent on the location of the Gibbs surface. For a fluid-fluid interface it is not possible to obtain unambiguously the position of the Gibbs surface in terms of experimental quantities. The usual convention is to locate the Gibbs surface so that the surface excess amount of the "solvent" is zero.

In Fig. 7.5*a* the Gibbs surface is located so that $n_1^\sigma = 0$; in other words, the Gibbs surface is located so that the two integrals in equation 7.20 are equal but opposite in sign. In Fig. 7.5*b* the integrals add so that the surface excess amount per unit area is given by the sum of the two shaded portions of the diagram.

If the area $\mathscr{A}$ of the surface is known, the surface excess concentration Γ_i of a component is given by $\Gamma_i = n_i^\sigma/\mathscr{A}$. If the Gibbs surface is chosen such that $n_1^\alpha = 0$, the surface concentration of the solute is then represented by $\Gamma_2^{(1)}$ to indicate that this is its adsorption relative to the solvent (component 1). In terms of experimental quantities,

$$\Gamma_2^{(1)} = \frac{1}{\mathscr{A}}(n_2 - V^\alpha c_2^\alpha - V^\beta c_2^\beta) \tag{7.21}$$

where the volumes of the bulk phases α and β are defined by the Gibbs surface.

$$V^\alpha = \frac{n_1 - Vc_1^\beta}{c_1^\alpha - c_1^\beta} \tag{7.22}$$

$$V^\beta = \frac{Vc_1^\alpha - n_1}{c_1^\alpha - c_1^\beta} \tag{7.23}$$

The total volume V of the system is given by

$$V = V^\alpha + V^\beta \tag{7.24}$$

Exercise: Write the corresponding equation for $\Gamma_1^{(1)}$ for the same location of the Gibbs surface and show that $\Gamma_1^{(1)} = 0$.

7.6 SURFACES OF MULTICOMPONENT SYSTEMS

Equation 7.9 may be extended to multicomponent systems by writing it as

$$dU = T\,dS - P\,dV + \gamma\,d\mathscr{A} + \sum_i \mu_i\,dn_i \tag{7.25}$$

Instead of talking about the internal energy U of a multicomponent system involving surfaces, it is useful to talk about the surface internal energy U^σ, which is defined by

$$U = U^\alpha + U^\beta + U^\sigma \tag{7.26}$$

where U^α and U^β are the internal energies of the bulk phases α and β, so that

$$dU = dU^\alpha + dU^\beta + dU^\sigma \tag{7.27}$$

where

$$dU^\sigma = T\,dS^\sigma + \gamma\,d\mathscr{A} + \sum_i \mu_i\,dn_i^\sigma \tag{7.28}$$

where the $P\,dV$ term has been dropped because the surface phase has zero volume with the Gibbs definition. No superscript has been included on the intensive properties T and μ_i since, at equilibrium, these quantities are the same throughout the system.

The Helmholtz energy for the surface is given by

$$A^\sigma = U^\sigma - TS^\sigma \tag{7.29}$$

so that

$$dA^\sigma = dU^\sigma - T\,dS^\sigma - S^\sigma\,dT \tag{7.30}$$

Substitution of equation 7.28 yields

$$dA^\sigma = -S^\sigma\,dT + \gamma\,d\mathscr{A} + \sum_i \mu_i\,dn_i^\sigma \tag{7.31}$$

The surface tension γ is related to the derivative of the surface excess Helmholtz energy by $\gamma = (\partial A^\sigma/\partial\mathscr{A})_{T,n_i^\sigma}$.

7.7 GIBBS ADSORPTION EQUATION

The Gibbs adsorption equation is a thermodynamic expression that relates the surface excess concentration of a component to both the surface tension and the bulk activity of the component. Thus the Gibbs adsorption equation may be used to calculate the surface excess concentration from measurements of the surface tension of solutions of known concentrations. Alternatively, if the surface concentration can be readily measured, as in adsorption on a solid, the Gibbs adsorption equation may be used to calculate the effect on the surface tension.

In order to derive the Gibbs adsorption equation we start with the equation for the surface excess Helmholtz energy for a two-component system.

$$A^\sigma = \gamma\mathscr{A} + n_2^\sigma\mu_2 \tag{7.32}$$

The term involving the chemical potential of the solvent has been omitted, since the Gibbs surface is located so that $n_1^\sigma = 0$. The differential of this equation is

$$dA^\sigma = \gamma\,d\mathscr{A} + \mathscr{A}\,d\gamma + \mu_2\,dn_2^\sigma + n_2^\sigma\,d\mu_2 \tag{7.33}$$

Equation 7.31 for a two-component system is

$$dA^\sigma = -S^\sigma\, dT + \gamma\, d\mathscr{A} + \mu_2\, dn_2^\sigma \tag{7.34}$$

Subtracting this equation from equation 7.33 yields

$$0 = \mathscr{A}\, d\gamma + n_2^\sigma\, d\mu_2 + S^\sigma\, dT \tag{7.35}$$

This is the Gibbs adsorption equation, but it is usually put into a simpler form by restricting it to constant temperature and eliminating $d\mu_2$ in favor of the activity by inserting

$$d\mu_2 = RT\, d \ln a_2 \tag{7.36}$$

where a_2 is the activity of the adsorbate in the bulk phase. Thus

$$-\mathscr{A}\, d\gamma = n_2^\sigma RT\, d \ln a_2 \tag{7.37}$$

Dividing by $\mathscr{A}$ yields

$$\Gamma_2^{(1)} = \frac{n_2^\sigma}{\mathscr{A}} = -\frac{1}{RT}\left(\frac{d\gamma}{d \ln a_2}\right)_T = -\frac{a_2}{RT}\left(\frac{d\gamma}{da_2}\right)_T \tag{7.38}$$

Thus the surface excess concentration may be calculated from measurements of the surface tension γ and knowledge of the activity a_2 of the solute in the bulk phase. For a solute that lowers the surface tension the surface excess concentration $\Gamma_2^{(1)}$ is positive; and for a solute that raises the surface tension the surface concentration $\Gamma_2^{(1)}$ is negative. Soaps have a higher concentration in the neighborhood of the surface than in the bulk aqueous phase, and inorganic salts have lower concentrations in the neighborhood of the surface than in the bulk aqueous phase.

For a one-component system equation 7.35 is $\mathscr{A}\, d\gamma = -S^\sigma\, dT$. If the entropy per unit area is represented by s^σ,

$$\frac{\partial \gamma}{\partial T} = -s^\sigma \tag{7.39}$$

Since the surface tension decreases with increasing temperature, the surface excess entropy s^σ is positive; in other words, the entropy of a unit area of liquid in the surface region is larger than that of the same amount of bulk liquid.

7.8 SURFACE TENSION OF SOLUTIONS

The addition of a solute to a solvent may lower the surface tension considerably; but if the solute causes an increase in surface tension, the effect is small because the solute is forced out of the surface layer, as indicated by the Gibbs equation 7.38. Solutes are classified as "capillary active" or "capillary inactive" on the basis of their effect on the surface tension. For the aqueous solution-air interface, inorganic electrolytes, salts of organic acids, bases of low molar mass, and certain non-

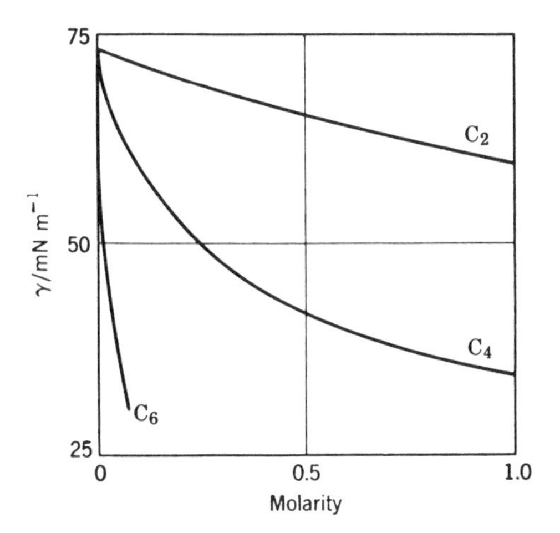

Fig. 7.6 Influence of concentration on the surface tension of aqueous solutions of fatty acids, including acetic (C_2), butyric (C_4), and hexanoic (C_6).

volatile nonelectrolytes such as sugar and glycerin are *capillary inactive*. *Capillary-active* solutes are organic acids, alcohols, esters, ethers, amines, ketones, etc. The effect of capillary-active substances on the surface tension of water may be very great, as illustrated in Fig. 7.6. Soaps and detergents are especially effective in lowering the surface tension or interfacial tension. They form surface films on dirt particles in washing. Since the addition of fatty acids, for example, lowers the surface tension (surface free energy), fatty acids tend to be concentrated spontaneously in the surface layer.

7.9 SURFACE FILMS

If the second component in a system is very insoluble and concentrated in the surface, the lowering of the surface tension can be measured directly by a method devised by Langmuir. In this method the force on a floating barrier between film-covered surface and clean-water surface is measured. A horizontal tray coated with a lacquer, which is not wet by water, is filled with clean water to a level slightly higher than the edges. Movable strips laid across the tray are used to sweep impurities off the surface. A small amount of film-forming substance is added to the surface as, for example, a solution of stearic acid in a volatile solvent that evaporates after spreading rapidly over the surface. The film is then compressed between a light floating barrier and one of the movable strips, and the pressure of the film against the floating barrier is measured by use of a torsion balance.

The force on the movable barrier may be looked at in another way. On the clean-water side the force per unit length pulling the barrier in the direction to reduce the area of the clear water surface is γ_0. On the side with the film the force per unit length pulling the barrier in the direction to reduce the area of the film-covered surface is γ. The measured surface pressure π on the floating barrier is simply the difference between those surface tensions; $\pi = \gamma_0 - \gamma$.

Some substances that are insoluble in water will spread on a water surface. A monomolecular film is usually produced and surplus material remains as a lens of liquid or a crystal in equilibrium with the monolayer on the surface.

The significance of surface pressure may be described by analogy with bulk phenomena, but there is no exact correspondence. We can, however, describe an insoluble monolayer by an equation of state that interrelates the surface pressure, area, and temperature. Care has to be taken because surface pressures may not be equilibrium values and spread films may not be homogeneous.

Insoluble monolayers that can withstand high surface pressures and are very incompressible are described as *condensed* monolayers. At the other extreme, monolayers that occupy large areas per molecule and are very compressible are described as *gaseous* monolayers.

Condensed monolayers are formed by aliphatic acids, alcohols, amines, and amides of sufficiently long chain length. For these monolayers the limiting area per molecule does not depend on chain length.

Stearic acid forms a condensed monolayer on water. No significant pressure is detected in a monolayer of stearic acid until the film has been confined to a certain area; then the surface pressure rises rapidly and further decreasing the area causes the film to crumple. Thus stearic acid films behave like two-dimensional solids. The movement of the film when it is pushed or blown may be made visible by dusting the surface with some inert powder.

The reason that stearic acid forms a surface film may be explained as follows. The carboxyl group, being a polar group, has a strong affinity for water, whereas the hydrocarbon chain does not; as a result, stearic acid is very insoluble in water. At the surface the carboxyl "heads" can be dissolved in the water phase, whereas the hydrocarbon "tails" stick up out of the surface. The correctness of this interpretation is confirmed by calculating the cross-sectional area of the stearic acid molecule and its length, assuming that the film is a monolayer.

Example 7.3 It is found that 0.106 mg of stearic acid covers 500 cm² of water surface at the point where the surface pressure just begins to rise sharply. Given the molar mass (284 g mol⁻¹) and density (0.85 g cm⁻³) of stearic acid, estimate the cross-sectional area a per stearic acid molecule and the thickness t of the film.

$$500\ \text{cm}^2 = \frac{(0.106 \times 10^{-3}\ \text{g})}{(284\ \text{g mol}^{-1})}(6.02 \times 10^{23}\ \text{mol}^{-1})a$$

$$a = 22 \times 10^{-16}\ \text{cm}^2$$

$$(500\ \text{cm}^2)t = \frac{0.106 \times 10^{-3}\ \text{g}}{0.85\ \text{g cm}^{-3}}$$

$$t = 25 \times 10^{-8}\ \text{cm or } 2.5\ \text{nm}$$

7.10 SURFACE EQUATION OF STATE

There are two ways of representing data on a surface film. First, the behavior of a surface film may be represented by a *surface equation of state*. This is the equation that represents the surface pressure π of a substance as a function of the surface concentration Γ. Since the surface equation of state is the two-dimensional analog of the three-dimensional equation of state, it may be written

$$\pi = \gamma_0 - \gamma = RTf(\Gamma) \tag{7.40}$$

where T is the thermodynamic temperature, R is the perfect gas constant, and $f(\Gamma)$ is a function of the surface excess concentration.

Second, a surface film may be represented by an *adsorption isotherm*. This is an equation that relates the pressure or concentration of an adsorbing species in the bulk phase to its surface concentration Γ at constant temperature. An adsorption isotherm has the general form

$$P = Kf'(\Gamma) \tag{7.41}$$

where P is the pressure or concentration in the bulk phase of the substance being adsorbed and K is a proportionality constant. One type of equation may be converted into the other by use of the Gibbs adsorption equation. For films on liquid surfaces it is easier to determine the surface equation of state directly; for surface films on solids it is easier to determine the adsorption isotherm directly.

The relationship between the surface equation of state and the adsorption isotherm may be illustrated for a perfect nonlocalized monolayer. Such a monolayer behaves like a perfect two-dimensional gas and has the equation of state

$$\pi\mathscr{A} = \left(\frac{N}{N_A}\right)RT \tag{7.42}$$

where N is the number of molecules contained in a film of area $\mathscr{A}$ and N_A is the Avogadro constant. This behavior is generally shown by low molecular weight fatty acids. For other substances the film pressure π varies with area per mole in the surface in more complicated ways reminiscent of nonideal gases. Equation 7.42 may be derived from statistical mechanics on the assumption that the surface monolayer consists of N indistinguishable mutually noninteracting molecules.

At constant temperature equation 7.35 may be written

$$-d\gamma = \Gamma_2^{(1)}\, d\mu_2 \tag{7.43}$$

for a two-component system. Since $\pi = \gamma_0 - \gamma$,

$$d\pi = \Gamma_2^{(1)}\, d\mu_2 \tag{7.44}$$

Assuming that the solute is in equilibrium with a perfect gas phase for which

$$\mu_2 = \mu_2^\circ + RT\ln\left(\frac{P_2}{P^\circ}\right) \tag{7.45}$$

and

$$d\mu_2 = RT\,d\left[\ln\left(\frac{P_2}{P^\circ}\right)\right] \tag{7.46}$$

equation 7.44 becomes

$$d\pi = \Gamma_2^{(1)}RT\,d\left[\ln\left(\frac{P_2}{P^\circ}\right)\right] \tag{7.47}$$

Since

$$\Gamma_2^{(1)} = \frac{N}{\mathscr{A} N_A} \tag{7.48}$$

then

$$d\pi = \frac{NRT}{\mathscr{A} N_A} d\left[\ln\left(\frac{P_2}{P^\circ}\right)\right] \tag{7.49}$$

The corresponding differential obtained from the perfect two-dimensional gas law, equation 7.42, is

$$d\pi = \frac{RT}{\mathscr{A} N_A} dN \tag{7.50}$$

Equating these two expressions for $d\pi$ and integrating yields

$$\frac{P_2}{P^\circ} = KN \tag{7.51}$$

where K is a constant of integration. Equation 7.51 is analogous to Henry's law. Thus, for a perfect nonlocalized monolayer, the amount adsorbed is directly proportional to the pressure.

7.11 ADSORPTION BY SOLIDS

Adsorption occurs on the surface of a solid because of the attractive forces of the atoms or molecules in the surface of the solid. The potential energy of a surface plus a molecule decreases as the molecule approaches the surface, and this may be represented by a curve like that shown in Fig. 14.8, which represents the potential energy of two atoms as a function of distance. The adsorbed molecules may be considered to form a two-dimensional phase. In the two-dimensional phase, the molecule may still retain two degrees of translational freedom; in that case, the adsorbed film is said to be mobile. If the molecule loses all its translational freedom, the film is immobile.

It is convenient to distinguish between *physical adsorption* and *chemisorption*. The forces causing physical adsorption are the same type as those that cause the condensation of a gas to form a liquid and are generally referred to as van der Waals forces. The heat evolved in the physical-adsorption process is of the order of magnitude of the heat evolved in the process of condensing the gas, and the amount adsorbed may correspond to several monolayers. Physical adsorption is readily reversed by lowering the pressure of the gas or the concentration of the solute, and the extent of physical adsorption is smaller at higher temperatures.

Chemisorption involves the formation of chemical bonds. It is therefore more specific than physical adsorption, but there are borderline cases in which it is not possible to make a sharp distinction between these two kinds of adsorption. In chemisorption the bonding may be so tight that the original species may not be recovered. For example, when oxygen is adsorbed on graphite, heating the surface in a vacuum yields carbon monoxide. The rate of chemisorption may be fast or slow, depending on the activation energy (Section 15.12). For example, physical adsorption of a gas may occur at a low temperature and the gas may become

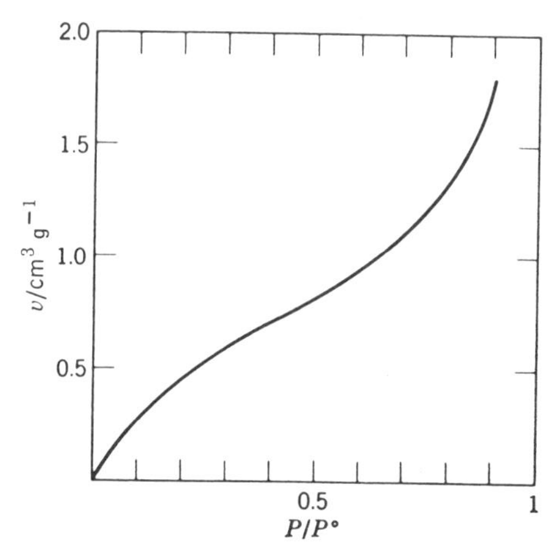

Fig. 7.7 Adsorption isotherm of nitrogen on finely divided potassium chloride at 89.9 K. The ordinate gives the volume (at 0 °C and 1 atm) adsorbed on 1 g of solid absorbent, and the abscissa gives the pressure of nitrogen as a fraction of the saturated vapor pressure. [From A. G. Keenan and J. M. Holmes, *J. Phys. Colloid Chem.*, **53**, 1309 (1949).]

chemisorbed when the temperature is raised. It is difficult to obtain a perfectly clean solid surface. Chemisorbed gases cannot be removed by simply exposing a solid surface to a vacuum.

The extent of adsorption depends greatly on the specific nature of the solid and of the molecules being adsorbed and is a function of pressure (or concentration) and temperature. If the gas being physically adsorbed is below its critical point, it is customary to plot the amount adsorbed per gram of adsorbent versus $P/P°$, where $P°$ is the vapor pressure of the bulk liquid adsorbate at the temperature of the experiment. Such an adsorption isotherm is shown in Fig. 7.7. The amount of gas adsorbed was determined by measuring the volume v of gas taken up by the adsorbent at various pressures and at a constant temperature of 89.9 K. As $P/P°$ approaches unity, the amount adsorbed increases rapidly because at $P/P° = 1$, bulk condensation can occur. If the temperature of the experiment is above the critical temperature of the gas, $P/P°$ cannot be calculated. When the adsorption of various gases is compared above their critical temperature, it is generally found that the amount adsorbed is smaller the lower the critical temperature.

To discuss the theory of adsorption we first consider a very simple model treated by Langmuir.

7.12 LANGMUIR THEORY OF ADSORPTION

Langmuir* considered the surface of a solid to be made up of elementary spaces, each of which could adsorb one gas molecule. He assumed that all the elementary

* I. Langmuir, *J. Am. Chem. Soc.*, **38**, 2267 (1916); **40**, 1361 (1918).

spaces are identical in their affinity for a gas molecule and that the presence of a gas molecule on one space does not affect the properties of neighboring spaces.

If θ is the fraction of the surface occupied by gas molecules, the rate of evaporation from the surface is $r\theta$, where r is the rate of evaporation from the completely covered surface at a certain temperature. The rate of adsorption of molecules on the surface is proportional to the fraction of the area that is not covered $(1 - \theta)$ and to the pressure of the gas. Thus the rate of condensation is expressed by $k(1 - \theta)P$, where k is a constant at a given temperature and includes a factor to allow for the fact that not every gas molecule that strikes an unoccupied space will stick.

At equilibrium the rate of evaporation of the adsorbed gas is equal to the rate of condensation.

$$r\theta = k(1 - \theta)P \tag{7.52}$$

$$\theta = \frac{kP}{r + kP} = \frac{1}{1 + r/kP} \tag{7.53}$$

or

$$P = \left(\frac{r}{k}\right)\frac{\theta}{1 - \theta} \tag{7.54}$$

Since the volume v of gas adsorbed is proportional to θ, equation 7.53 may be written as

$$v = \frac{v_m}{1 + k'/P} \tag{7.55}$$

where v_m is the volume of gas adsorbed when the entire surface is covered and $k' = r/k$. Thus v is directly proportional to P at very low pressures where $k'/P \gg 1$. As the pressure is increased, the volume adsorbed increases and approaches the value v_m asymptotically.

It is convenient to determine the constants k' and v_m by plotting $1/v$ versus $1/P$, since

$$\frac{1}{v} = \frac{1}{v_m} + \frac{k'}{v_m P} \tag{7.56}$$

Data such as that of Fig. 7.7 do not show asymptotic saturation and do not give a linear plot except at low pressures. Thus adsorption on solids is more complicated than the Langmuir theory indicates.

The derivation of the Langmuir adsorption isotherm involves five implicit assumptions: (1) the adsorbed gas behaves ideally in the vapor phase, (2) the adsorbed gas is confined to a monomolecular layer, (3) the surface is homogeneous, that is, the affinity of each binding site for gas molecules is the same, (4) there is no lateral interaction between adsorbate molecules, and (5) the adsorbed gas molecules are localized, that is, they do not move around on the surface. Although the first assumption is good at low pressure, the second one nearly always breaks down as the pressure of the adsorbed gas is increased. As the gas pressure approaches the saturation vapor pressure, the vapor condenses without limit on all surfaces if the contact angle θ is zero, as shown by Fig. 7.7. The third assumption is poor because real surfaces are heterogeneous; the affinity for gas molecules is different on different crystalline faces, and different kinds of binding sites are introduced by edges, cracks, and crystal imperfections. Heterogeneity leads to a decrease in binding energy as the surface coverage increases. The incorrectness of the fourth assumption was first shown experimentally when it was found that in certain cases the heat of adsorption may increase with the surface concentration of adsorbed molecules. This effect, which is the opposite of that expected to result from surface heterogeneity, is caused by lateral attractions of adsorbed molecules. The fifth assumption is incorrect because there are several kinds of evidence that surface films may be mobile.

In physical adsorption molecules of vapor may be adsorbed to the depth of many monolayers. An adsorption isotherm equation has been derived by Brunauer, Emmett, and Teller (BET)* to provide for multilayer adsorption. It is assumed that the surface possesses uniform, localized sites and that adsorption at one site does not affect adsorption at neighboring sites, just as in the Langmuir theory. It is further assumed that molecules can be adsorbed in second, third, . . ., and nth layers with the surface area available for the nth layer equal to the coverage of the $(n - 1)$th layer. The energy of adsorption in the first layer, E_1, is assumed to be constant, and the energy of adsorption in succeeding layers is assumed to be E_L, the energy of liquefaction of the gas.

By use of these assumptions it is possible to derive an equation that yields the volume of gas that is adsorbed when the entire adsorbent surface is covered with a complete unimolecular layer.

The surface area occupied by a single molecule of adsorbate on the surface may be estimated from the density of the liquefied adsorbate. For example, the area occupied by a nitrogen molecule at −195 °C is estimated to be 16.2×10^{-20} m^2 on the assumption that the molecules are spherical and that they are close packed in the liquid. Thus, from the measured value of v_m, the surface area of the adsorbent may be calculated. This method is widely used in the determination of the surface area of solid catalysts and adsorbents. The area values determined in this way seem, in general, to be perfectly satisfactory in spite of the rough approximations of the theory.

Example 7.4 The volume of nitrogen gas v_m (measured at 1 atm and 0 °C) required to cover a sample of silica gel with a unimolecular layer is 129 cm^3 g^{-1} of gel. Calculate the surface area per gram of the gel if each nitrogen molecule occupies 16.2×10^{-20} m^2.

$$\frac{(0.129 \text{ L g}^{-1})(6.02 \times 10^{23} \text{ mol}^{-1})(16.2 \times 10^{-20} \text{ m}^2)}{(22.4 \text{ L mol}^{-1})} = 560 \text{ m}^2 \text{ g}^{-1}$$

* S. Brunauer, P. H. Emmett, and E. Teller, *J. Am. Chem. Soc.*, **60**, 309 (1938).

7.13 ADSORPTION ON POROUS SOLIDS

When adsorbents are porous, the adsorbate may actually condense in the pores. This process is called capillary condensation and, when it happens, hysteresis is observed in the adsorption isotherm. As discussed in Section 7.3, a liquid condenses in a capillary at a pressure less than the vapor pressure of the adsorbate at the temperature of the adsorption experiment. Figure 7.8*a* illustrates the type of isotherm that would be obtained for a porous solid for which all the pores had the same radius r_t, where r_t is in the range where equation 7.16 is applicable. At pressure P', which is calculated from r_t, the capillaries fill up. In practice there is a distribution of pore size and, as a consequence, the second step in the isotherm is less steep.

Figure 7.8*b* illustrates the hysteresis that may occur in the isotherm of a porous adsorbent. As indicated by the arrows, less material is adsorbed at a given pressure when the pressure is rising than when the pressure is falling. This phenomenon, which is usually reproducible, is explained in terms of the "ink bottle" theory. When the pore entrance is narrower than the inside of the pore, the pore fills at the pressure that corresponds with the radius of the widest part of the pore. However, on desorption, the pore does not empty until the pressure corresponds with the radius of the neck of the pore.

7.14 CHROMATOGRAPHY*

In 1906 Tswett discovered that, if a solution of chlorophyll from leaves is poured on the top of a column of a suitable adsorbent and a solvent is used to wash the pigments down the column, the colored band is resolved into a series of bands moving at different rates. Each component of the original mixture is represented by a band and may be obtained in a pure form by collecting the various solutions as they come from the column or by cutting the column into segments and eluting the pigment from the adsorbent in each segment. This method is widely used for separation of substances difficult to separate by other methods. It has proved of great value in the detection and separation of biological materials. When used for separating radioactive materials, it is possible to detect and separate extremely small quantities of such materials.

Chromatography is a versatile and powerful tool for achieving separations. The separation process may be based on adsorption, distribution between gas and liquid or between two liquid phases, or differences in the rate of diffusion into a gel, depending on size; this group of processes is referred to as sorption. A mixture is passed through a column that contains a finely divided stationary phase. Molecules in the mixture go through a series of random steps alternating between sorption

* H. G. Cassidy, *Fundamentals of Chromatography*, Wiley-Interscience, New York, 1957; J. C. Giddings, *Dynamics of Chromatography, Part I, Principles and Theory*, Marcel Dekker, Inc., New York, 1965; L. Fischer, *An Introduction to Gel Chromatography*, North-Holland, London, 1969.

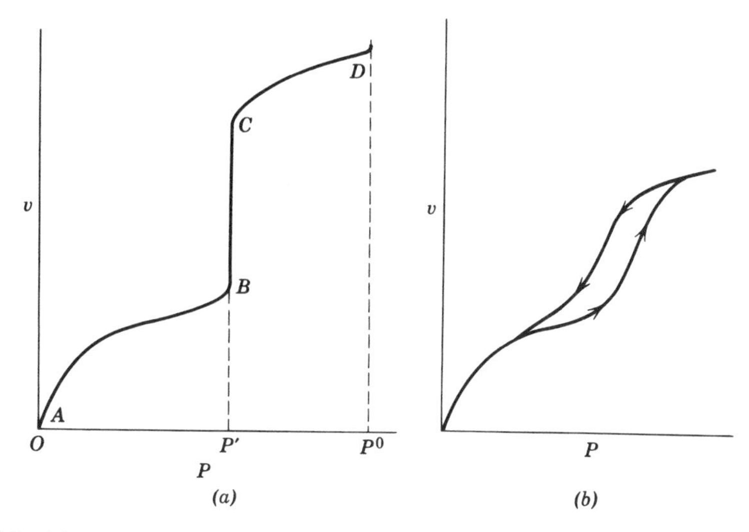

Fig. 7.8 (*a*) Adsorption isotherm of a porous solid with uniform cylindrical pores. (*b*) Hysteresis in the adsorption isotherm of a porous solid.

and desorption. The more time a molecule spends in the sorbed state, the more slowly it moves through the column. Thus the position of an emergent peak in a chromatogram is primarily controlled by its equilibrium sorption. The width of a peak depends on kinetic and mass transfer (diffusion, flow, turbulence) processes. The slower the flow in a chromatographic column, the nearer each step of the separation is to equilibrium, but the more each peak is broadened by diffusion. The flow rate may be increased if equilibration can be speeded. Equilibration can be speeded by using a more finely divided sorbent, but then higher pressures are required to get suitable flow rates.*

7.15 COLLOIDS

In 1861 in the course of his investigations on diffusion in solutions, Thomas Graham drew a distinction between "colloids" such as proteins, gums, and polysaccharides, and "crystalloids," which are substances of lower molar mass. He noted that colloids did not diffuse through membranes whereas crystalloids did, and he gave the name *dialysis* to this process of separation by diffusion. The essential difference between colloids and crystalloids is their relative particle sizes. It is generally considered that colloids range in diameter from about 1 to 1000 nm. Simple molecules and crystalloids are less than 1 nm in diameter. Particles larger than 1000 nm can be seen with a microscope, and they are regarded as beyond the colloidal size range.

---

*J. C. Giddings, *J. Chem. Ed.*, **44**, 704 (1967).

It is convenient to distinguish between colloids consisting of (1) small particles having the same structure as the bulk solid or liquid, (2) aggregates formed from smaller molecules, and (3) molecules so large in size that their dimensions are in the colloidal size range. A dispersion of a finely divided solid, such as gold, or liquid, such as benzene, is an example of the first type. Soaps and detergents are examples of the second type. They consist of organic molecules, with both hydrophobic ("water-hating") and hydrophilic ("water-loving") parts, which aggregate to form micelles. In these micelles, which may contain as many as 100 molecules, the hydrophobic parts are together on the inside and the hydrophilic parts are on the outside. Proteins and high polymers are examples of the third type. These substances consist of molecules held together by covalent bonds; the important characteristic they have in common with colloidal particles is their size. Proteins and other biological macromolecules like deoxynucleic acid (DNA) are of importance for understanding the chemical processes involved in living organisms. Synthetic high polymers are of increasing importance in industry.

Lyophobic colloids are those having little attraction for the solvent medium; examples are colloidal gold and inorganic precipitates. Lyophilic colloids are those having a strong attraction for the solvent medium; examples are proteins in water and polystyrene in benzene.

The most important factors in stabilizing colloids are electric charge, adsorption of large molecules, and a strong attraction for the solvent.* Either the existence of a net electric charge that causes the particles to repel each other or a coating of large molecules, which prevents the particles from adhering to each other, may be sufficient to keep the dispersed particles in the colloidal state. In lyophobic colloids the stability of the colloid depends chiefly on the fact that the charged particles repel each other.

References

N. K. Adam, *The Physics and Chemistry of Surfaces*, Dover Publications, New York, 1968.

A. W. Adamson, *Physical Chemistry of Surfaces*, 3rd ed., Wiley-Interscience, New York 1976.

R. Aveyard and D. A. Haydon, *An Introduction to the Principles of Surface Chemistry*, Cambridge University Press, Cambridge, England, 1973.

J. J. Bikerman, *Surface Chemistry*, Academic Press, New York, 1958.

J. H. de Boer, *The Dynamical Character of Adsorption*, 2nd ed., Clarendon Press, Cambridge, England, 1968.

S. Gregg and K. Sing, *Adsorption, Surface Area and Porosity*, Academic Press, New York 1967.

N. B. Hannay, *Solid-State Chemistry*, Prentice-Hall, Inc., Englewood Cliffs, N.J., 1967, Chapter 10.

S. Ross and J. P. Oliver, *On Physical Adsorption*, Interscience Publishers, New York, 1964.

---

* E. J. W. Verwey and J. Th. G. Overbeek, *Theory of the Stability of Lyophobic Colloids*, Elsevier Publishing Co., New York, 1948.

G. A. Somorjai, *Principles of Surface Chemistry*, Prentice-Hall, Inc., Englewood Cliffs, N.J., 1972.
E. J. W. Verwey and J. Th. Overbeek, *Theory of the Stability of Lyophobic Colloids*, Elsevier Publ. Co., New York, 1948.

Problems

7.1 Estimate the surface tension of a metal given that the heat of sublimation is 4×10^5 J mol^{-1} and that there are 10^{15} atoms cm^{-2} in the surface. It may be assumed that the structure is close packed so that each atom has 12 nearest neighbors. With this information only an approximate answer can be calculated. *Ans.* 3 J m^{-2}.

7.2 The surface tension of toluene at 20 °C is 0.0284 N m^{-1}, and its density at this temperature is 0.866 g cm^{-3}. What is the radius of the largest capillary that will permit the liquid to rise 2 cm? *Ans.* 0.0335 cm.

7.3 Calculate the vapor pressure of a water droplet at 25 °C that has a radius of 2 nm. The vapor pressure of a flat surface of water is 3167 Pa at 25 °C. *Ans.* 5350 Pa.

7.4 The surface tensions of 0.05 and 0.127 mol kg^{-1} solutions of phenol in water are 67.7 and 60.1 mN m^{-1}, respectively, at 20 °C. What is the surface concentration $\Gamma_2^{(1)}$ in the range 0 to 0.05 mol kg^{-1} and 0.05 to 0.127 mol kg^{-1}, assuming that the phenol concentrations can be treated as activities? The surface tension of water at 20 °C is 72.7 mN m^{-1}. *Ans.* 1.03×10^{-6}, 3.6×10^{-6} mol m^{-2}.

7.5 (*a*) The surface tension of water against air at 1 atm is given in the following table for various temperatures.

t/°C	20	22	25	28	30
γ/N m^{-1}	0.072 75	0.072 44	0.071 97	0.071 50	0.071 18

Calculate the surface enthalpy at 25 °C in joules per square centimeter. (*b*) If a finely divided solid whose surface is covered with a very thin layer of water is dropped into a container of water at the same temperature, heat will be evolved. Calculate the heat evolution for 10 g of a powder having a surface area of 200 m^2 g^{-1} [W. D. Harkins and G. Jura, *J. Am. Chem. Soc.*, **66**, 1362 (1944)].

Ans. (*a*) 1.161×10^{-5} J cm^{-2}, (*b*) 233 J.

7.6 A solution of palmitic acid (M = 256 g mol^{-1}) in benzene contains 4.24 g of acid per liter. When this solution is dropped on a water surface, the benzene evaporates and the palmitic acid forms a monomolecular film of the solid type. If we wish to cover an area of 500 cm^2 with a monolayer, what volume of solution should be used? The area occupied by one palmitic acid molecule may be taken to be 21×10^{-20} m^2.

Ans. 0.0239 cm^3.

7.7 A protein with a molar mass of 60,000 g mol^{-1} forms a perfect gaseous film on water. What area of film per milligram of protein will produce a pressure of 0.005 N m^{-1} at 25 °C? *Ans.* 82.7 cm^2.

7.8 The acid $CH_3(CH_2)_{13}CO_2H$ forms a nearly perfect gaseous monolayer on water at 25 °C. Calculate the weight of acid per 100 cm^2 required to produce a film pressure of 10^{-3} N m^{-1}. *Ans.* 9.86×10^{-7} g.

7.9 A monomolecular film obeys the following equation of state

$$\pi(\sigma - \beta) = kT$$

where β may be interpreted as the effective area occupied by a single molecule. This equation may also be written

$$\pi = \frac{kT}{\beta} \frac{\theta}{1 - \theta}$$

where θ is the fraction of the surface occupied. Derive the relation between pressure in the vapor phase and θ.

Ans. $P = K\left(\frac{\theta}{1 - \theta}\right)e^{1/(1-\theta)}$.

7.10 What volume of oxygen, measured at 25 °C and 1 atm, is required to form an oxide film on 1 m² of a metal with atoms in a square array 0.1 nm apart? *Ans.* 2 mL.

7.11 Two gases, A and B, compete for the binding sites on the surface of an adsorbent. Show that the fraction of the surface covered by A molecules is

$$\theta_A = \frac{b_A P_A}{1 + b_A P_A + b_B P_B}$$

where b_A and b_B are constants.

7.12 The following table gives the number of milliliters (v) of nitrogen (reduced to 0 °C and 1 atm) adsorbed per gram of active carbon at 0 °C at a series of pressures.

P/Pa	524	1731	3058	4534	7497
v/cm³ g⁻¹	0.987	3.04	5.08	7.04	10.31

Plot the data according to the Langmuir isotherm, and determine the constants.

Ans. $k' = 2.1 \times 10^4$ Pa, $v_m = 35$ cm³ g⁻¹.

7.13 Calculate the surface area of a catalyst that adsorbs 103 cm³ of nitrogen (calculated at 1 atm and 0 °C) per gram in order to form a monolayer. The adsorption is measured at −195°C, and the effective area occupied by a nitrogen molecule on the surface is 16.2×10^{-20} m² at this temperature. *Ans.* 449 m².

7.14 The pressures of nitrogen required to cause the adsorption of 1.0 cm³ g⁻¹ (25 °C, 1 atm) of gas on P-33 graphitized carbon black are 24 Pa at 77.5 K and 290 Pa at 90.1 K. Calculate the enthalpy of adsorption at this fraction of surface coverage using the Clausius-Clapeyron equation. *Ans.* 11.6 kJ mol⁻¹.

7.15 Adsorption of nitrogen, at its boiling point −195.8 °C, is often used in estimating surface areas of solids. At this temperature nitrogen has a density of 0.808 g cm⁻³ and a surface tension of 8.85 mN m⁻¹. If the isotherm is of the type shown in Fig. 7.8*b* and hysteresis is encountered at about $P/P^\circ = 0.5$, what does this imply about the radii of pores in the solid? *Ans.* About 0.7 nm.

7.16 Acetone has a density of 0.790 g cm⁻³ at 20 °C and rises to a height of 2.56 cm in a capillary tube having a radius of 0.0235 cm. What is the surface tension of the acetone at this temperature?

7.17 Mercury does not wet a glass surface. Calculate the capillary depression if the diameter of the capillary is (*a*) 0.1 mm, and (*b*) 2 mm. The density of mercury is 13.5 g cm⁻³. The surface tension of mercury at 25 °C is 0.520 N m⁻¹.

7.18 When *n*-butanol vapor is cooled at 0 °C, it is found that the degree of supersaturation has to be about 4 in order for droplets to nucleate spontaneously. The surface tension of *n*-butanol at 0 °C is 0.0261 N m⁻¹. (*a*) What are the radii of droplets formed at this degree of supersaturation. (*b*) How many molecules are there in a droplet?

7.19 Suppose that you want to measure the reduction of the vapor pressure of water caused by the curvature of the surface produced in a small capillary. In order to obtain a 10% reduction of the vapor pressure at 25 °C, how small a capillary should be used? (The surface tension of water at 25 °C is 0.071 97 N m^{-1}.)

7.20 Go through the steps to derive equation 7.21.

7.21 (*a*) If a liquid surface is expanded, will there be a heating effect or a cooling effect? (*b*) Given the surface-tension data for water in Problem 7.5, calculate the surface enthalpy.

7.22 The surface tension of liquid nitrogen at 75 K is 0.009 71 N m^{-1}, and the temperature coefficient of the surface tension is -23×10^{-4} N m^{-1} K^{-1}. What is the surface enthalpy?

7.23 One hundred grams of oleic acid, $C_{17}H_{33}COOH$, which forms a condensed film, is poured on the surface of a clean lake, where the spreading film can be seen if the water is rippled by a gentle wind or marked with rain drops. The cross-sectional area of the molecule is about 22×10^{-20} m^2. What will be the maximum diameter in meters of a circular film produced in this way?

7.24 A certain substance forms a surface film that obeys the perfect two-dimensional gas law. Calculate the excess surface concentration required to cause a surface tension lowering of 0.01 N m^{-1} at 25 °C.

7.25 Derive the equation for the surface pressure for an adsorbate that follows the Langmuir adsorption isotherm.

7.26 The adsorption of nitrogen on mica is as follows.

P	0.28	0.61	1.73
x/m	12.0	19.0	28.2

(P is given in N cm^{-2}, and x/m in cubic millimeters of gas at 20 °C and 1 atm adsorbed on 24.3 g of mica, having a surface of 5750 cm^2.) (*a*) Determine the constants of the Langmuir equation. (*b*) Calculate the value of x/m at $P = 2.38$. (The experimental value is 30.8.)

7.27 The diameter of the hydrogen molecule is about 0.27 nm. If an adsorbent has a surface of 850 m^2 cm^{-3} and 95% of the surface is active, how much H_2 (measured at standard conditions) could be adsorbed by 100 mL of the adsorbent? It may be assumed that the adsorbed molecules just touch in a plane and are arranged so that four adjacent spheres have their centers at the corners of a square.

7.28 One gram of a certain activated charcoal has a surface area of 1000 m^2. If complete surface coverage is assumed, as a limiting case, how much ammonia, at 25 °C and 1 atm, could be adsorbed on the surface of 45 g of activated charcoal? The diameter of the NH_3 molecule is 3×10^{-10} m, and it is assumed that the molecules just touch each other in a plane so that four adjacent spheres have their centers at the corners of a square.

7.29 The following table gives data on the adsorption of benzene by graphitized carbon black (P-33) at three surface coverages and three temperatures.

v(cm^3 g^{-1} at 25 °C, 1 atm)	0.2	0.4	0.6
t/°C		Pressure (Pa)	
0	13	27	40
35.0	80	170	290
49.45	170	390	710

Calculate the heat of adsorption q at each degree of coverage of the surface using the Clausius-Clapeyron equation and extrapolate this heat to zero coverage. (S. Ross and J. P. Oliver, *On Physical Adsorption*, Wiley-Interscience, New York, 1964, p. 238.)

7.30 The adsorption of ammonia on charcoal is studied at 30 and 80 °C. It is found that the pressure required to cause the adsorption of a certain amount of NH_3 per gram of charcoal is 1.41×10^4 Pa at 30 °C and 7.46×10^4 Pa at 80 °C. Calculate the enthalpy of adsorption using the Clausius-Clapeyron equation.

7.31 Suppose the isotherm shown in Fig. 7.8*b* is obtained for the adsorption of water on a porous adsorbent at 25 °C. What capillary radius in the pores of the solid is implied if the break occurs at $P/P° = 0.5$, where $P°$ is the vapor pressure of water? The surface tension of water at 25 °C is 71.87 mN m^{-1}.

PART TWO

QUANTUM CHEMISTRY

In this part of the book we shift to the microscopic approach to the understanding of chemical and physical properties. The development of quantum mechanics in this century has led to the possibility of calculating energy levels and other properties of atoms and molecules. We will consider the electronic orbitals of the hydrogen atom in detail and describe how these calculations may be extended to atoms with two or more electrons. The calculations show why properties such as ionization potential, electron affinity, and atom size vary in a complicated periodic way. In fact, quantum mechanics provides the explanation of the structure of the periodic table.

Before considering the application of quantum mechanics to molecules, we consider the symmetry of molecules in their equilibrium configurations. The symmetry of a molecule may greatly simplify quantum mechanical calculations of its energy levels and geometry. Symmetry also determines whether a molecule can be optically active or have a dipole moment.

The application of quantum mechanics to molecules has made it possible to understand the nature of the chemical bond. The bonding in H_2^+ and H_2 are considered in some detail in Chapter 10. The application of quantum mechanics to molecules (other than H_2^+) involves approximation, but energy levels, bond lengths, and angles may be calculated quite accurately for many small molecules. Approximate calculations for larger molecules are also useful. Applications of molecular quantum mechanics are increasing rapidly because of the increasing power of electronic computers.

Quantum mechanics provides the basis for understanding spectroscopy. Spectroscopy is useful for identifying molecules and determining their concentrations, but spectroscopy is especially important in physical chemistry because it yields information about individual molecules. Microwave and far infrared spectra provide information on internuclear distances and bond angles. Infrared and Raman spectroscopy provide information about vibrational frequencies. Visible and ultraviolet spectroscopy provide information on dissociation energies and electronic excited states.

Nuclear magnetic resonance spectroscopy and electron spin resonance spectroscopy have become so important in the practice of chemistry that they are discussed in a chapter separate from the rest of spectroscopy. Nuclear magnetic resonance is made possible by nuclei that have spin, and electron spin resonance is made possible by unpaired electrons in molecules, free atoms, or radicals.

Statistical mechanics is the science that connects the properties of individual molecules with the thermodynamic properties of matter in bulk. For perfect gases it is possible to calculate thermodynamic properties from information obtained from spectroscopy. We will illustrate that by calculating equilibrium constants for reactions of diatomic molecules. This is of tremendous importance, because it means that equilibrium compositions of gas mixtures may often be calculated more easily than they can be measured. The calculation of thermodynamic properties for dense gases, liquids, and solids is more difficult because of molecular interactions.

CHAPTER 8
QUANTUM THEORY

In the late nineteenth century it became apparent that classical mechanics was unable to account for many experimental facts concerning the behavior of systems of atomic size. We have already referred to the heat capacities of gases in Chapter 1. Another serious problem was that classical theories gave the wrong spectral distribution for radiation from a heated cavity. In 1900, in his derivation of the correct equation for the intensity of radiation of different frequencies from a cavity, Planck assumed that electromagnetic radiation is quantized. Planck's idea of quantization was used by Einstein in 1905 to interpret the photoelectric effect successfully. In 1913 Bohr developed his theory of the hydrogen atom by assuming that the angular momentum of the orbital electron is quantized. In 1924 de Broglie used the idea of quantization of energy to obtain an expression for the wavelength of an electron that has been accelerated by an electric field. In 1926 Schrödinger and Heisenberg independently developed quantum mechanics. Quantum mechanics has been of tremendous importance for the understanding of chemistry.

In this chapter we will consider the application of quantum mechanics to simple systems, including the hydrogen atom, which can be treated exactly, and to many electron atoms for which analytic solutions cannot be obtained. Applications of quantum mechanics to molecules are considered in Chapter 10, after an analysis of symmetry of molecules in Chapter 9.

8.1 CAVITY RADIATION

Hot gases produce line spectra, but hot solids produce continuous radiation. The electromagnetic radiation from different solids at the same temperature show rather different spectral distributions. It is found, however, that if the radiation inside an isothermal hollow body is viewed through a small hole in its wall, the intensity of radiation and distribution of wavelengths are independent of the material and of the size and shape of the cavity. Such radiation is often called black-body radiation, but we will refer to it as cavity radiation.

The first successful theory for the spectral distribution of cavity radiation was presented by Planck to the Berlin Physical Society on December 14, 1900. His derivation involved the bold assumption that radiation is emitted and absorbed by oscillating dipoles in the solid that can only have energies that are integer multiples of $h\nu$, where h is a constant (later termed Planck's constant) and ν is the frequency of radiation emitted and absorbed. Thus, the oscillators cannot radiate or absorb any amount of energy, but only quanta of size $h\nu$. The energy of a quantum of electromagnetic radiation may also be written hc/λ, where c is the velocity of light in a vacuum and λ is the wavelength.

The spectral concentration w_λ of the radiant energy density in a cavity is given by

$$w_\lambda = \frac{8\pi hc}{\lambda^5(e^{hc/\lambda kT} - 1)} \tag{8.1}$$

The spectral concentration of the radiant energy density is defined so that $w_\lambda\, d\lambda$ is the energy per unit volume in the wavelength range $d\lambda$. Thus w_λ has the SI units J m^{-4}. In order to calculate the rate with which energy is emitted per unit area from a hole in the cavity, it is necessary to multiply w_λ by $c/4$, where c is the velocity of light.* The quantity $cw_\lambda/4$ is represented by M_λ, which is referred to as the spectral concentration of the radiant exitance of wavelength λ. The spectral concentration of the radiant exitance is defined so that $M_\lambda\, d\lambda$ is the emission of energy per unit area per unit time in the range of wavelengths λ to $\lambda + d\lambda$. Thus M_λ has the SI units J m^{-3} s^{-1}. The relation derived by Planck is

$$M_\lambda = \frac{2\pi hc^2\lambda^{-5}}{\exp(hc/\lambda kT) - 1} = \frac{(3.7415 \times 10^{-16}\ \mathrm{J\ s^{-1}\ m^2})\lambda^{-5}}{\exp[(1.438\,79 \times 10^{-2}\ \mathrm{m\ K})/\lambda T] - 1} \tag{8.2}$$

where h is Planck's constant, c is the velocity of light ($2.997\,924\,58 \times 10^8$ m s^{-1}), and k is the Boltzmann constant ($1.380\,662 \times 10^{-23}$ J K^{-1}). From spectral data on cavity radiation he was able to calculate the value of what is now known as Planck's constant. The best value is

$$h = 6.626\,176 \times 10^{-34}\ \mathrm{J\ s}$$

The spectral concentration of the radiant exitance M_λ is plotted versus wavelength in Fig. 8.1 for $T = 1000$, 1500, and 2000 K. The visible region of the spectrum extends from 0.4×10^{-6} m (violet) to 0.8×10^{-6} m (red).

The radiant exitance M is the total energy emitted per unit time per unit area of opening from the cavity. The radiant exitance is given by

$$M = \int_0^\infty M_\lambda\, d\lambda \tag{8.3}$$

Thus the radiant exitance is proportional to the areas under the curves in Fig. 8.1. By carrying out the integration it may be shown that M is proportional to the fourth power of the absolute temperature.

$$M = \sigma T^4 \tag{8.4}$$

The Stefan-Boltzmann constant σ is given by

$$\sigma = \frac{2\pi^5 k^4}{15h^3c^2} = 5.6697 \times 10^{-8}\ \mathrm{J\ s^{-1}\ m^{-2}\ K^{-4}} \tag{8.5}$$

---

* In equation 14.48 we see that the flow of particles through unit area is equal to $\frac{1}{4}$ their average velocity multiplied by the number of particles per unit volume.

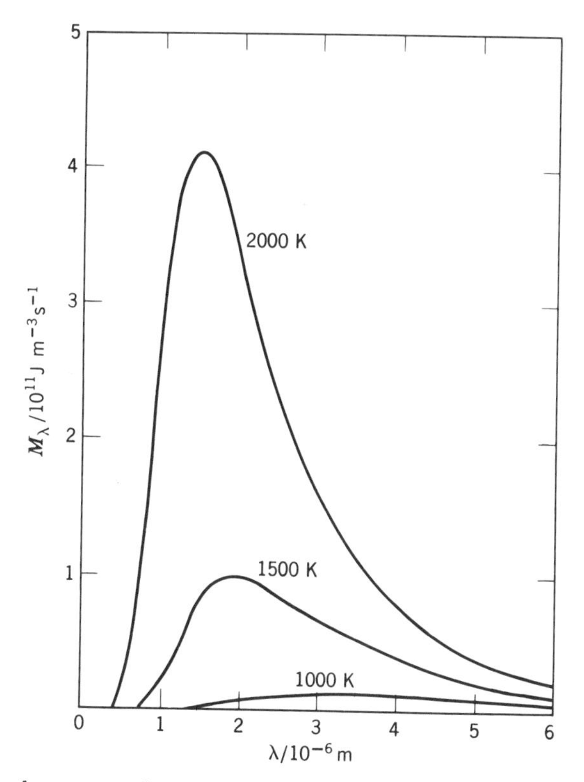

Fig. 8.1 Spectral concentration of the radiant exitance M_λ from a cavity at three temperatures. The ordinate is the radiant power per unit area in a small range of wavelengths, divided by the width of the range of wavelengths. The area under a curve is the radiant exitance M in J m^{-2} s^{-1}.

In 1905 the new idea of quantization received further support from Einstein, who used it to explain the photoelectric effect. Einstein considered light to be made up of particles (photons), with each particle having an energy of $E = h\nu$. When light is absorbed by a metal, the total energy of a photon $h\nu$ is given to a single electron within the metal. If this quantity of energy is sufficiently large, the electron may penetrate the potential barrier at the surface of the metal and still retain some energy as kinetic energy. The kinetic energy retained by the electron depends on the energy and, therefore, the frequency of the photon that ejected it. The number of electrons ejected depends on the number of incident photons, and therefore the intensity of light.

8.2 LINE SPECTRA

Whereas cavity radiation is continuous, other types of spectra contain lines. The nature of line spectra could not be explained by classical theories. Attempts to find regularity in spectra had shown that the frequencies of spectral lines could be calculated by taking differences between quantities called "terms." The various lines in a spectrum are accounted for by taking differences between a relatively small number of term values.

The simplest spectrum is that of hydrogen atoms, and Fig. 8.2 illustrates a small region of it. In 1885 Balmer discovered that the wavelengths λ of the lines in the visible region of the emission spectrum of hydrogen atoms could be expressed by a simple relation that may be written as

$$\frac{1}{\lambda} = \tilde{\nu} = R\left(\frac{1}{2^2} - \frac{1}{n_2^{\,2}}\right) \tag{8.6}$$

where n_2 is an integer greater than 2 and R is the Rydberg constant, 109 677.58 cm^{-1}. The value of R may be determined very accurately because of the high precision with which the wavelengths of spectral lines can be measured.

The reciprocal of the wavelength is represented by $\tilde{\nu}$ and referred to as the *wave number*. Wave numbers are usually expressed in cm^{-1}. In spectroscopy it is more convenient to use wave numbers than wavelengths because wave numbers are proportional to energy, and spectroscopy involves transitions between different energy levels.

It will be noticed in equation 8.6 that n_2 cannot be less than 2, because then $\tilde{\nu}$ would be a meaningless, negative number, and it cannot be 2, because then $\tilde{\nu}$ becomes zero. As n_2 becomes larger than 2, the corresponding value of $\tilde{\nu}$ becomes larger. When n_2 is already large, however, further increases cause $\tilde{\nu}$ to increase only very slightly and, as n_2 approaches infinity, $\tilde{\nu}$ approaches $\frac{1}{4}R$ as a limit.

The success of the Balmer formula led to further exploration, and other series of lines were discovered in the atomic hydrogen spectrum that could be represented by the equation

$$\tilde{\nu} = R\left(\frac{1}{n_1^{\,2}} - \frac{1}{n_2^{\,2}}\right) \tag{8.7}$$

where n_1 is also an integer. The series for which $n_1 = 1$ (Lyman series) is in the ultraviolet; the series for which $n_1 = 3$ (Paschen series), 4 (Brackett series), or 5 (Pfund series) are in the infrared region. It is important to note that every line in

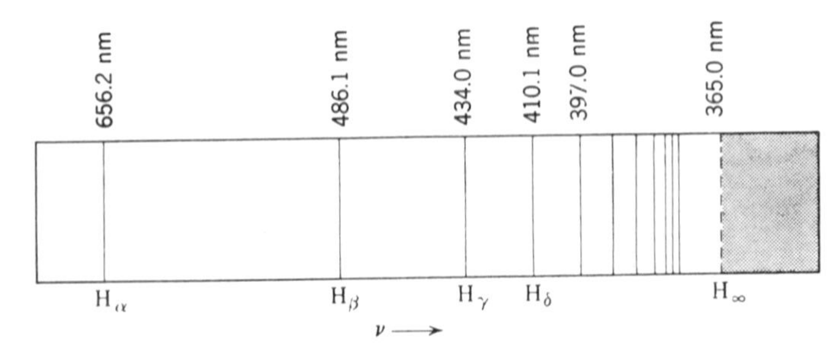

Fig. 8.2 Balmer series of lines in the emission spectrum of atomic hydrogen. The wavelengths are given in nanometers (1 nm = 10^{-9} m). (For the energy level diagram, see Fig. 8.3.)

the spectrum can be represented as a difference of two terms, R/n_1^2 and R/n_2^2. The spectra of other atoms are more complicated but, in general, it is found possible to represent the lines of the spectrum as differences between term values. This concept may be readily understood by the application of the principle of conservation of energy which requires that

$$h\nu = E_2 - E_1 \tag{8.8}$$

where E_2 is the energy of the atom or molecule before emission of a photon $h\nu$ and E_1 is the energy after emission. This equation is basic to all types of spectroscopy.

A successful theory of the spectrum of the hydrogen atom was developed in 1913 by Bohr. Bohr made a complete break with classical mechanics by assuming that in the hydrogen atom the angular momentum of the orbital electron can only have values that are integral multiples of a quantum of angular momentum of magnitude $\hbar$, referred to as *h*-bar, which is $h/2\pi$. Bohr assumed that an electron moves in a circular orbit around the positively charged nucleus. We now know that orbital electrons do not behave in this way but, nevertheless, Bohr was able to derive a correct expression for the energy levels of hydrogenlike atoms (i.e., atoms with one electron). He was also able to calculate the sizes of hydrogenlike atoms; he calculated the radius of the inner orbit of the hydrogen atom to be 0.0529 nm.

The electronic energy levels in the hydrogen atom, as calculated from the Bohr theory, are summarized in Fig. 8.3. The Lyman series of lines is produced by

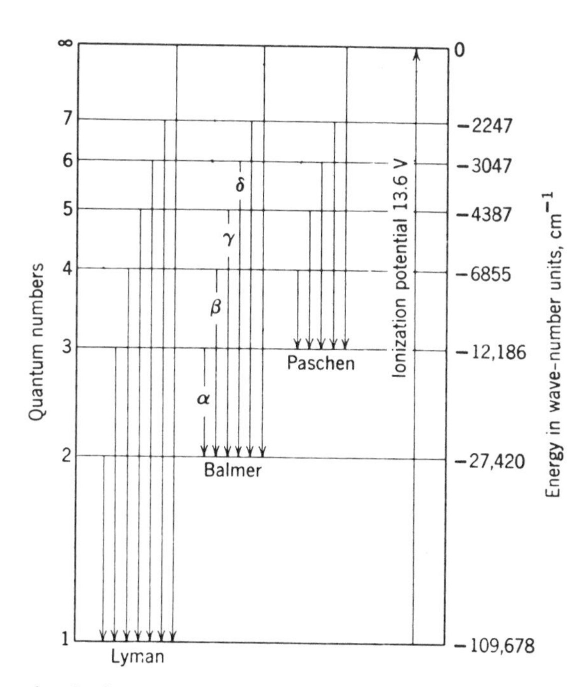

Fig. 8.3 Energy levels for the hydrogen atom as calculated from the Bohr theory. Ionization potentials are discussed in Section 8.21.

electrons jumping from orbits with quantum numbers 2, 3, 4, . . . into the lowest permitted orbit ($n_1 = 1$). The Balmer series of lines is produced by electrons falling from larger orbits into the second orbit ($n_1 = 2$), etc. The energies of the various orbits may be expressed in several ways. The energies in wave numbers given at the right in Fig. 8.3 are the wave numbers for radiation produced when an electron falls from an infinite distance into a given orbit with no initial kinetic energy. The wave number $\tilde{\nu}$ of any line in the spectrum may be obtained by subtracting the values at the right for the two energy levels involved. Thus the second line in the Balmer series is due to an electron falling from the fourth orbit into the second, and its wave number if $27\,420 - 6855 = 20\,565\ \text{cm}^{-1}$.

8.3 THE DE BROGLIE RELATION

In 1923 de Broglie suggested that the motion of electrons might have a wave aspect. He arrived at the relationship between wavelength λ and linear momentum by reasoning in analogy with photons. According to Einstein's special theory of relativity, the energy of a particle is given by

$$E = mc^2 \tag{8.9}$$

where m is the mass of the particle and c is the speed of light. Using $E = h\nu$, we get $mc^2 = h\nu = hc/\lambda$, so that for a photon

$$\lambda = \frac{h}{mc} = \frac{h}{p} \tag{8.10}$$

where p is the momentum. De Broglie suggested that for an electron

$$\lambda = \frac{h}{mv} = \frac{h}{p} \tag{8.11}$$

where v is the velocity of the electron. The wavelengths of particles calculated in this way are called de Broglie wavelengths. The wavelike nature of matter suggested by de Broglie was verified in 1928 by Davisson and Germer, who obtained a diffraction pattern from electrons impinging on the face of a nickel crystal.

8.4 HEISENBERG UNCERTAINTY PRINCIPLE

In 1927 Heisenberg* showed that there is a fundamental limit to the accuracy with which certain physical measurements may be made simultaneously. This limitation applies to many combinations of dynamical variables (coordinates, velocities,

* W. Heisenberg, *Z. Physik*, **43**, 172 (1927).

angular momenta, energy, time) that have the physical dimensions of *action*, that is mass·length^2·time^{-1} or energy·time, and can be expressed by the relations

$$\Delta q\,\Delta p \geqslant \hbar/2 \tag{8.12}$$

$$\Delta E\,\Delta t \geqslant \hbar/2 \tag{8.13}$$

where $\hbar = h/2\pi$, Δq is the root-mean-square uncertainty in position, Δp is the root-mean-square uncertainty in momentum, Δt is the root-mean-square uncertainty in time, and ΔE is the root-mean-square uncertainty in energy. Equation 8.13 indicates that if the energy of a system has a precise value, the lifetime in that state is infinite, and we can say that the system is in a *stationary state*. On the other hand, if the characteristic lifetime of a state is Δt, the uncertainty in the energy is $\geqslant \hbar/2\,\Delta t$.

Because of the small value of h, this uncertainty is not detectable for macroscopic objects, but for electrons, atoms, and molecules Heisenberg's relations are significant. The Heisenberg uncertainty relation indicates that it is meaningless to ask about the exact position and exact velocity of an electron in an atom.

It is important to realize that the uncertainties in equations 8.12 and 8.13 are not "experimental errors" dependent, say, on the quality of one's laboratory equipment, but are inherent in any measurement process. For example, suppose the position of a moving electron is measured by shining light on it, that is, by studying the scattering of photons. It would be necessary to use short wavelengths to locate the electron precisely, but this would make the momenta h/λ of the photons large, so that those that struck the electron would knock it significantly off course, and its original velocity would become very uncertain. Thus increasing certainty in the position of the electron implies decreasing certainty in its momentum, in agreement with equation 8.12.

8.5 SCHRÖDINGER EQUATION

Quantum mechanics was developed independently in 1926 by W. Heisenberg and Erwin Schrödinger. Heisenberg's approach is referred to as matrix mechanics and Schrödinger's approach is referred to as wave mechanics. Although the two methods appear different, they can be shown to be mathematically equivalent. We will consider only the Schrödinger formulation that uses ideas about wave motion.

The time-independent Schrödinger equation for a single particle of mass m moving in one dimension is

$$\frac{\hbar^2}{2m}\frac{d^2\psi(x)}{dx^2} + [E - V(x)]\psi(x) = 0 \tag{8.14}$$

where $\psi(x)$ is the wave function that is a function of distance, $V(x)$ is the potential energy function of the particle, and E is the total energy of the particle. This remarkable equation is not derived, but is regarded as a *postulate* of quantum mechanics, which is to be judged on the basis of results derived from it. The wave function $\psi(x)$ contains all of the information we can know about a system, in this case a single particle that can move in only one dimension. Born pointed out the way in which $\psi(x)$ gives information about the position of a particle. The probability of finding the particle between x and $x + dx$ is given by $|\psi(x)|^2\,dx$. The vertical bars denote the absolute value of $\psi(x)$. Since $\psi(x)$ may be a complex quantity (i.e., it may involve $i = \sqrt{-1}$), the square of the absolute value is given by the product of $\psi(x)$ with its complex conjugate $\psi^*(x)$, which is obtained by replacing i with $-i$ wherever it occurs. Since the probability of finding the particle between x and $x + dx$ is given by $\psi^*(x)\psi(x)\,dx$, the *probability density* is given by $\psi^*(x)\psi(x)$. Probability densities are discussed in connection with kinetic theory (Section 14.2).

The probability of finding the particle somewhere on the x-axis is given by

$$\int_{-\infty}^{\infty} \psi^*(x)\psi(x)\,dx = 1 \tag{8.15}$$

A wave function that satisfies this equation is said to be *normalized.* The wave function yields the probability of finding a particle in a certain range, but it does not tell us precisely where the particle is. We will see later that all the properties of any system can be calculated from its wave function.

The time-independent Schrödinger equation for a single particle of mass m moving in three dimensions is

$$-\frac{\hbar^2}{2m}\left[\frac{\partial^2\psi(x, y, z)}{\partial x^2} + \frac{\partial^2\psi(x, y, z)}{\partial y^2} + \frac{\partial^2\psi(x, y, z)}{\partial z^2}\right] + V(x, y, z)\psi(x, y, z) = E\psi(x, y, z) \tag{8.16}$$

where $\psi(x, y, z)$ is the wave function and $V(x, y, z)$ is the potential energy of the particle, both of which are functions of the three Cartesian coordinates. Equation 8.16 may be written in more compact form by use of the operator ∇^2 (referred to as del squared).

$$-\frac{\hbar^2}{2m}\nabla^2\psi(x, y, z) + V(x, y, z)\psi(x, y, z) = E\psi(x, y, z) \tag{8.17}$$

where

$$\nabla^2 \equiv \frac{\partial^2}{\partial x^2} + \frac{\partial^2}{\partial y^2} + \frac{\partial^2}{\partial z^2} \tag{8.18}$$

For $\psi^*\psi$ to be interpreted as a probability density, it must be single valued and have a finite integral over the space in which the particle is found. It is found that wave functions ψ with the desired properties exist only for certain values of the energy E. A particle in a box provides our first illustration of these ideas.

8.6 PARTICLE IN A ONE-DIMENSIONAL BOX

The simplest problem related to that of an electron in an atom or a molecule involves the calculation of the wave function for an electron constrained to move within a distance of length a in the x direction. The potential energy V is taken as zero within this region and infinite for other x values.

To determine the wave function for the particle between $x = 0$ and $x = a$ where $V = 0$, equation 8.14 may be written

$$\frac{\hbar^2}{2m}\frac{d^2\psi}{dx^2} + E\psi = 0 \tag{8.19}$$

The general solution of this equation is

$$\psi = A \sin \left(\frac{2mE}{\hbar^2}\right)^{1/2} x + A' \cos \left(\frac{2mE}{\hbar^2}\right)^{1/2} x \tag{8.20}$$

as may be shown by substitution into equation 8.19.

Since the potential outside the box is infinite, the probability of finding the particle outside must be zero.* To avoid a discontinuity at $x = 0$ and $x = a$, the wave function ψ must have a value of zero at these points. To satisfy the boundary condition at $x = 0$, the constant A' in equation 8.20 must be taken to be zero. The boundary condition at $x = a$ is satisfied† only if

$$\left(\frac{2mE}{\hbar^2}\right)^{1/2} a = n\pi \tag{8.21}$$

where n is an integer. According to this equation, the energy of the particle in the box is given by

$$E = \frac{h^2 n^2}{8ma^2} \tag{8.22}$$

A particle moving between two points on a line can have only the energies given by this equation for integer values of n, whereas a perfectly free particle can have any energy. Such discrete energy levels are characteristic of solutions of the Schrödinger equation for bound particles. No such discrete energy levels are expected on the basis of classical mechanics. It is apparent from equation 8.22 that the bigger the box or the heavier the particle, the more closely spaced are the energy levels.

A particle in a box cannot have zero energy because the lowest energy $h^2/8ma^2$ is given by equation 8.22 for $n = 1$. Although $n = 0$ satisfies the boundary conditions, the corresponding wave function is zero everywhere. Such "zero-point

---

* If the wave function did not have a value of zero outside of the box, equation 8.14 could not be satisfied except with infinite energy.

† There is a discontinuity in the slope of the plot of ψ versus x at $x = 0$ and $x = a$ resulting from the infinite jump in potential at these points. Infinite potential barriers do not exist in nature, and wave functions for real systems decrease to zero as distance is increased without discontinuities in the first derivative of the wave function.

energy" must be found whenever a particle is constrained to a finite region; if this were not so the uncertainty principle would be violated.* The next higher energy levels are at 4 times ($n = 2$) and 9 times ($n = 3$) this energy, as shown in Fig. 8.4*a*. The wave functions are superimposed on this plot, and we can see that the wavelength is equal to $2a/n$.

Figure 8.4*b* gives the probability densities $\psi^*\psi$ for a particle in a box. These are the probabilities per unit distance that the particle will be found at this point. The most probable position for a particle in the zero point level ($n = 1$) is in the center of the box.

The value of the constant A in equation 8.20 is calculated by *normalizing* the wave function. Equation 8.20 may be rewritten in terms of the quantum number by introducing equation 8.21.

$$\psi = A \sin \frac{\pi x n}{a} \tag{8.23}$$

The probability that the particle is somewhere between $x = 0$ and $x = a$ is, of

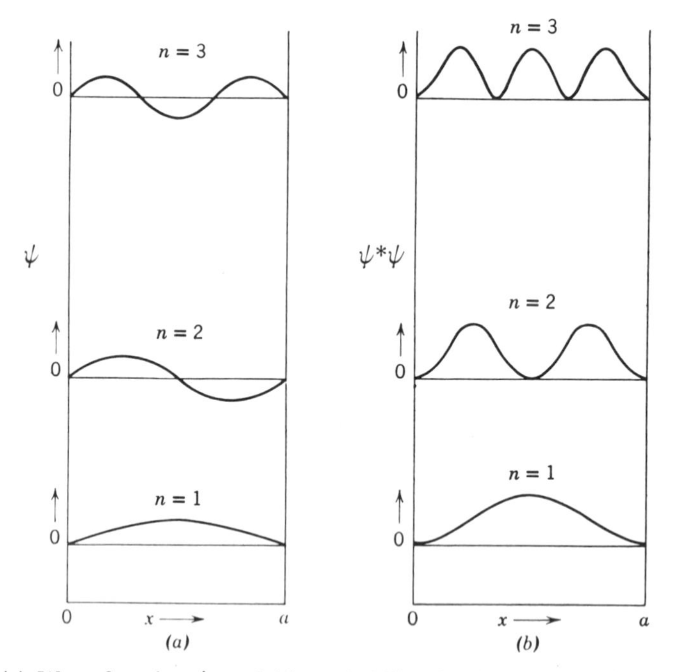

Fig. 8.4 (*a*) Wave function ψ, and (*b*) probability density function $\psi^*\psi$ for the lowest three energy levels for a particle in a box. The plots are placed at vertical heights that correspond to the energies of the levels. As the number of nodes goes up, the energy goes up.

* The zero point energy of the particle in a box is in agreement with the requirements of the Heisenberg uncertainty principle. Since $\Delta x \approx a$, then $\Delta p \approx \hbar/2a$, where $\approx$ means approximately equal to. Thus $E = p^2/2m \approx \hbar^2/8a^2m$, which is of the correct order of magnitude for $n = 1$.

course, unity, and this is expressed mathematically by integrating $\psi^*\psi$ over this distance.

$$1 = \int_0^a \psi^*\psi \, dx = A^2 \int_0^a \sin^2 \frac{\pi x n}{a} \, dx = \frac{A^2 a}{\pi} \int_0^\pi \sin^2 (n\alpha) \, d\alpha \tag{8.24}$$

where $\alpha = \pi x/a$. Since ψ is a real function, $\psi^*\psi$ is simply ψ^2. Since

$$\int_0^\pi \sin^2 (n\alpha) \, d\alpha = \frac{\pi}{2} \tag{8.25}$$

we find that $A = (2/a)^{1/2}$, so that the normalized wave function for a particle in a one-dimensional box is

$$\psi = \left(\frac{2}{a}\right)^{1/2} \sin \frac{\pi x n}{a} \tag{8.26}$$

The wave functions ψ_i have been normalized so that

$$\int_{-\infty}^{\infty} \psi_i^* \psi_j \, dx = 1 \qquad \text{if} \quad i = j \tag{8.27}$$

For particle-in-a-box wave functions

$$\int_{-\infty}^{\infty} \psi_i^* \psi_j \, dx = 0 \qquad \text{if} \quad i \neq j \tag{8.28}$$

Such wave functions are said to be orthogonal to each other. Relations 8.27 and 8.28 can be combined by writing

$$\int_{-\infty}^{\infty} \psi_i^* \psi_j \, dx = \delta_{ij} \tag{8.29}$$

where δ_{ij} is called the Kronecker delta, which is defined by

$$\delta_{ij} = \begin{cases} 0 & \text{for } i \neq j \\ 1 & \text{for } i = j \end{cases} \tag{8.30}$$

Wave functions that satisfy equation 8.29 are said to be orthonormal. Solutions of the Schrödinger equation corresponding to different energy eigenvalues for a system are always orthogonal.

8.7 PARTICLE IN A THREE-DIMENSIONAL BOX

The Schrödinger equation may readily be solved for a three-dimensional box with infinite potential everywhere outside. The Schrödinger equation in the form of equation 8.16 may be solved by writing the wave function as the product of three functions, each depending on just one coordinate.

$$\psi(x, y, z) = X(x)Y(y)Z(z) \tag{8.31}$$

By substituting this for ψ in equation 8.16 and then dividing by $X(x)Y(y)Z(z)$, we obtain

$$-\frac{\hbar^2}{2m}\left[\frac{1}{X(x)}\frac{d^2X(x)}{dx^2} + \frac{1}{Y(y)}\frac{d^2Y(y)}{dy^2} + \frac{1}{Z(z)}\frac{d^2Z(z)}{dz^2}\right] = E \tag{8.32}$$

since V is everywhere zero inside the box. If the energy is written as the sum of three contributions associated with the three coordinates,

$$E = E_x + E_y + E_z \tag{8.33}$$

and equation 8.32 can be separated into three equations because the functions $X(x)$, $Y(y)$, and $Z(z)$ are each functions of variables that can change independently of each other. For example, if y and z are held constant the second and third terms on the left-hand side of equation 8.32 will be zero. Since E is constant, the first term must be constant. We call this constant value E_x and find equation 8.34*a*. Similar arguments yield equations 8.34*b* and 8.34*c*. The important mathematical result here is that a partial differential equation has been converted into three ordinary differential equations that can be easily solved.

$$-\frac{\hbar^2}{2m}\left[\frac{1}{X(x)}\frac{d^2X(x)}{dx^2}\right] = E_x \tag{8.34a}$$

$$-\frac{\hbar^2}{2m}\left[\frac{1}{Y(y)}\frac{d^2Y(y)}{dy^2}\right] = E_y \tag{8.34b}$$

$$-\frac{\hbar^2}{2m}\left[\frac{1}{Z(z)}\frac{d^2Z(z)}{dz^2}\right] = E_z \tag{8.34c}$$

These equations are just like equation 8.19 and may be solved in the same way to obtain

$$X(x) = A_x \sin\frac{n_1\pi x}{a} = A_x \sin\left(\frac{2mE_x}{\hbar^2}\right)^{1/2} x \tag{8.35a}$$

$$Y(y) = A_y \sin\frac{n_2\pi y}{b} = A_y \sin\left(\frac{2mE_y}{\hbar^2}\right)^{1/2} y \tag{8.35b}$$

$$Z(z) = A_z \sin\frac{n_3\pi z}{c} = A_z \sin\left(\frac{2mE_z}{\hbar^2}\right)^{1/2} z \tag{8.35c}$$

where a, b, and c are the lengths of the sides in the x, y, and z directions, respectively, and n_1, n_2, and n_3 are quantum numbers. Thus there is a quantum number for each coordinate. The allowed energy levels are

$$E = \frac{h^2}{8m}\left(\frac{n_1^2}{a^2} + \frac{n_2^2}{b^2} + \frac{n_3^2}{c^2}\right) \tag{8.36}$$

where n_1, n_2, and n_3 may each have any integer value. Later in statistical mechanics (Section 13.7) we will use this equation for translational energy of a molecule in a container.

If the three-dimensional box has geometrical symmetry, a new feature arises. This is most easily illustrated for a cubical box for which equation 8.36 becomes

$$E = \frac{h^2}{8ma^2}(n_1^2 + n_2^2 + n_3^2) \tag{8.37}$$

For a one-dimensional box the state of the system could be specified by giving the value of the quantum number or the energy. For a cubical box this is no longer true, because a given energy may be achieved by different combinations of the three quantum numbers n_1, n_2, and n_3. The quantum numbers $n_1 = 2$, $n_2 = 1$, $n_3 = 1$; $n_1 = 1$, $n_2 = 2$, $n_3 = 1$;

and $n_1 = 1$, $n_2 = 1$, $n_3 = 2$ describe different states of the system (different wave functions), but these states have the same energy. Such an energy level is said to be *degenerate*, and the degeneracy is equal to the number of independent wave functions associated with a given energy level. An understanding of degenerate levels is important for later calculations.

8.8 OPERATORS

The one-particle, one-dimensional Schrödinger equation may be written in the form

$$\left[-\frac{\hbar^2}{2m}\frac{d^2}{dx^2} + V(x)\right]\psi(x) = E\psi(x) \tag{8.38}$$

We refer to the quantity in brackets as an operator that, when applied to the wave function $\psi(x)$, yields $E\psi(x)$, where E is the energy.

If the effect of operating on some function $f(x)$ with operator $\hat{A}$ is to obtain the same function multiplied by a constant k, then $f(x)$ is called an *eigenfunction* of $\hat{A}$ with *eigenvalue* k.

$$\hat{A}f(x) = kf(x) \tag{8.39}$$

Thus $\psi(x)$ in equation 8.38 is an eigenfunction, and E is an eigenvalue. A simple example of an eigenvalue problem is given in the following example.

Example 8.1 What are the eigenfunctions and eigenvalues of the operator d/dx?

$$\frac{d}{dx}f(x) = kf(x) \qquad \frac{df(x)}{f(x)} = k\,dx \qquad \ln f(x) = kx + c \qquad f(x) = e^c e^{kx} = c'e^{kx}$$

where c and c' are constants. For each different value of k there is an eigenfunction $c'e^{kx}$. Or, to put it another way, the eigenfunction $c'e^{kx}$ has the eigenvalue k.

There are many possible operators. The operator for multiplying by a constant c is represented by $\hat{c}$. The operator for multiplying by x is represented by $\hat{x}$. The operator for differentiation with respect to x is represented by $\hat{D}_x$.

Example 8.2 Satisfy yourself that

$$\hat{c}(x^2 + y^2) = cx^2 + cy^2 \qquad \hat{x}(x^2 + y^2) = x^3 + xy^2 \qquad \hat{D}_x(x^2 + y^2) = 2x$$

The operators occurring in quantum mechanics are all linear operators. Linear operators are such that

$$\hat{A}[f(x) + g(x)] = \hat{A}f(x) + \hat{A}g(x) \qquad \hat{A}[cf(x)] = c\hat{A}f(x) \tag{8.40a}$$

Taking the square root is an example of an operator that is not linear.

The sum of two operators that operate on a function of x is defined by

$$(\hat{A} + \hat{B})f(x) = \hat{A}f(x) + \hat{B}f(x) \tag{8.40b}$$

The product of two operators is defined by

$$\hat{A}\hat{B}f(x) = \hat{A}[\hat{B}f(x)] \tag{8.40c}$$

Operator $\hat{B}$ is applied to $f(x)$ and then operator $\hat{A}$ is applied to the resulting function. In general, $\hat{A}\hat{B}$ does not produce the same result as $\hat{B}\hat{A}$, in contrast with ordinary algebra. The commutator $[\hat{A}, \hat{B}]$ of the operators $\hat{A}$ and $\hat{B}$ is defined as

$$[\hat{A}, \hat{B}] = \hat{A}\hat{B} - \hat{B}\hat{A} \tag{8.41}$$

If $\hat{A}\hat{B} = \hat{B}\hat{A}$, the operators are said to commute.

Example 8.3 Show that if the eigenfunctions of two operators $\hat{A}$ and $\hat{B}$ are the same functions, $\hat{A}$ and $\hat{B}$ commute with each other. The eigenvalues of $\hat{A}$ and $\hat{B}$ are represented by a and b and the eigenfunctions are ψ_i, so that

$$\hat{A}\psi_i = a_i\psi_i \qquad \text{and} \qquad \hat{B}\psi_i = b_i\psi_i$$

The eigenfunctions of the operator $\hat{A}\hat{B}$ is obtained as follows.

$$\hat{A}\hat{B}\psi_i = \hat{A}(\hat{B}\psi_i) = \hat{A}b_i\psi_i = b_i\hat{A}\psi_i = b_ia_i\psi_i$$

The operator $\hat{B}\hat{A}$ has eigenvalue a_ib_i, as may be seen from

$$\hat{B}\hat{A}\psi_i = \hat{B}(\hat{A}\psi) = \hat{B}a_i\psi = a_i\hat{B}\psi = a_ib_i\psi$$

Since $a_ib_i = b_ia_i$, $\hat{A}$ and $\hat{B}$ commute with each other.

If operators A and B do commute, the observables they represent can in principle be measured simultaneously and precisely. If not, simultaneous measurement is impossible beyond a certain level of precision. The failure of x and p_x to commute leads, for example, to the Heisenberg uncertainty principle $\Delta x\, \Delta p_x \geqslant \hbar/2$. In general, the uncertainty relations for noncommuting operators are more complicated.

Operators are important in quantum mechanics because every observable quantity has a linear operator associated with it. To obtain a quantum mechanical operator, we start with the classical expression for the physical quantity of interest and form the operator by following a simple set of rules. We will begin our consideration with the energy operator.

We will consider only the quantum mechanics of "conservative" systems (i.e., those for which the potential energy depends only on the coordinates). For such systems the total energy H is the sum of the kinetic energy T and the potential energy V. The symbol H is used for the energy to indicate that the independent variables in the function are positions and momenta (not positions and velocities, for example). The use of momenta instead of velocities arises naturally in an alternative form of classical mechanics developed by Hamilton, which is fully equivalent to Newtonian mechanics but has advantages in adapting to curvilinear coordinate systems or other descriptions more complicated than Cartesian. The function

$$H = T + V \tag{8.43}$$

is known as the Hamiltonian function. The Hamiltonian function for a particle of mass m moving in the x direction is

$$H = \frac{p_x^2}{2m} + V(x) \tag{8.44}$$

where p_x is the momentum in the x direction.

The rules for converting a classical-mechanical function to the corresponding quantum mechanical operator are as follows.

1. Each Cartesian coordinate q is replaced by the operator multiplication by that coordinate.

$$\hat{q} = q \tag{8.45}$$

2. Each Cartesian component of linear momentum p_q is replaced by the operator.

$$\hat{p}_q = \frac{\hbar}{i}\frac{\partial}{\partial q} = -i\hbar\frac{\partial}{\partial q} \tag{8.46}$$

Applying these rules to the Hamiltonian function of equation 8.44 yields

$$\hat{H} = -\frac{\hbar^2}{2m}\frac{d^2}{dx^2} + V(x) \tag{8.47}$$

When this Hamiltonian operator is applied to one of its eigenfunctions ψ for a system, the eigenvalue obtained is the energy E, as shown by equation 8.14. Thus the time-independent Schrödinger equation may be written in the form

$$\hat{H}\psi = E\psi \tag{8.48}$$

If the particle can move in three dimensions, the classical Hamiltonian is

$$H = \frac{1}{2m}(p_x^2 + p_y^2 + p_z^2) + V(x, y, z) \tag{8.49}$$

Thus the Hamiltonian operator is given by

$$\hat{H} = \frac{-\hbar^2}{2m}\left(\frac{\partial^2}{\partial x^2} + \frac{\partial^2}{\partial y^2} + \frac{\partial^2}{\partial z^2}\right) + V(x, y, z) = -\frac{\hbar^2}{2m}\nabla^2 + V(x, y, z) \tag{8.50}$$

Classical expressions for other observables may be converted into quantum mechanical operators using rules 1 and 2, and the quantum mechanical operators may be used to calculate the possible eigenvalues. An example to be considered in Section 8.11 is angular momentum.

8.9 AVERAGE VALUES

To obtain the average value $\langle F \rangle$ of some observable F, we could measure F in a large number of separate systems each in the same stationary state ψ. The arithmetic mean of the observed values is taken for the average value. In quantum mechanics the term "expectation value" is often used as a synonym for average value.

If a system is in a state ψ the expectation value may be calculated from

$$\langle F \rangle = \int \psi^* \hat{F} \psi \, dx \, dy \, dz \tag{8.51}$$

where $\hat{F}$ is the linear operator corresponding to the observable F. The operator $\hat{F}$ is obtained from F by use of rules 1 and 2 of the preceding section.

If the application of an operator $\hat{F}$ to a function ψ yields a constant F_n, the constant is called an eigenvalue for the state described by the function ψ.

$$\hat{F}\psi = F_n \psi \tag{8.52}$$

When this occurs, the expectation value is equal to the eigenvalue.

$$\langle F \rangle = \int \psi^* \hat{F} \psi \, dx \, dy \, dz = \int \psi^* F_n \psi \, dx \, dy \, dz = F_n \int \psi^* \psi \, dx \, dy \, dz = F_n \tag{8.53}$$

The expectation value of x is

$$\langle x \rangle = \int_{-\infty}^{\infty} \psi^* x \psi \, dx \tag{8.54}$$

We will now discuss the solutions of the Schrödinger equation for three simple systems: (1) the simple harmonic oscillator, (2) the rigid rotor, and (3) the hydrogen atom. These illustrations show how the predictions of quantum mechanics differ from those of classical mechanics.

8.10 THE SIMPLE HARMONIC OSCILLATOR

The quantum mechanical treatment of a simple harmonic oscillator is required for the understanding of the vibration of molecules. To provide a background, the simple harmonic oscillator will be considered from a classical point of view before it is considered from a quantum mechanical point of view. In a simple harmonic oscillatory the force tending to restore the particle to its equilibrium position is directly proportional to the displacement from the equilibrium position. Thus

$$\text{Force} = -kx \tag{8.55}$$

where x is the distance measured from the equilibrium position and k is the force constant. The minus sign is required because the force is in the negative x direction if the displacement x from the equilibrium position is positive. When the particle is at the equilibrium distance, the force is zero.

A force of this type may be represented as the negative gradient of the potential energy V.

$$\text{Force} = -\frac{\partial V}{\partial x} = -kx \tag{8.56}$$

Integrating, we have

$$V = \tfrac{1}{2}kx^2 \tag{8.57}$$

if the zero of potential energy is taken at the origin, $x = 0$. Thus a plot of potential energy V versus displacement x is parabolic for a simple harmonic oscillator, as shown in Fig. 8.5*a*. A small object sliding without friction in a parabolic well would execute simple harmonic motion. As it slides through the minimum, the velocity is a maximum and the potential energy is a minimum. As it slides up the other side, kinetic energy is converted to potential energy. When the particle is at either one of its two highest points in the oscillation, the velocity is zero, and its energy is entirely potential energy.

The force in equation 8.56 may be equated to the mass of the particle times its acceleration. Thus

$$m\frac{d^2x}{dt^2} = -kx \tag{8.58}$$

The solution of this differential equation is

$$x = a \sin \left(\frac{k}{m}\right)^{1/2} t = a \sin 2\pi\nu_0 t \tag{8.59}$$

if $x = 0$ when $t = 0$, and

$$\nu_0 = \frac{1}{2\pi}\left(\frac{k}{m}\right)^{1/2} \tag{8.60}$$

is the fundamental vibration frequency. The frequency of vibration of a harmonic oscillator is independent of the amplitude of the vibration.

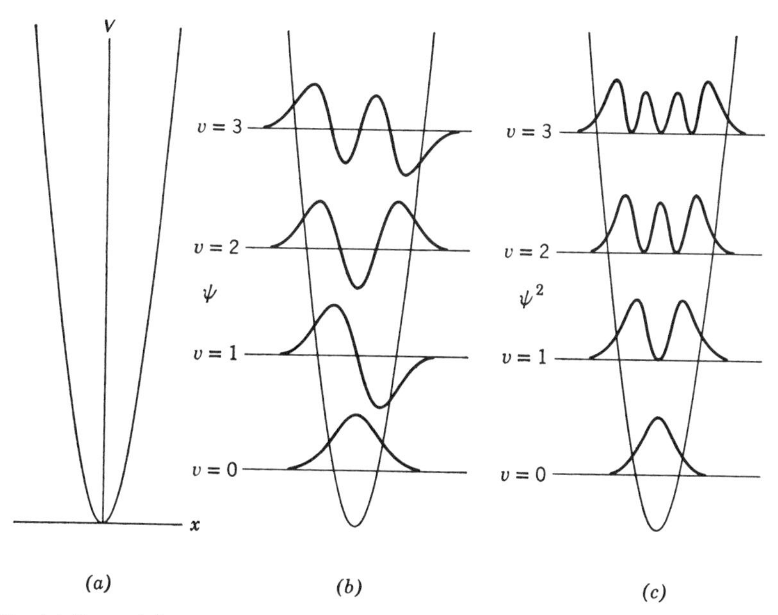

Fig. 8.5 (*a*) Potential energy curve for a classical harmonic oscillator. (*b*) Allowed energy levels and wave functions for a quantum mechanical harmonic oscillator. (*c*) Probability density functions for a quantum mechanical harmonic oscillator.

The energy (classical Hamiltonian) of a classical harmonic oscillator is given by the sum of the kinetic and potential energies.

$$H = \frac{p_x^2}{2m} + \tfrac{1}{2}kx^2 = \tfrac{1}{2}kA^2 \tag{8.61}$$

where A is the vibration amplitude.

In order to treat the harmonic oscillator from a quantum mechanical point of view, the Hamiltonian operator is obtained by substituting $(\hbar/i)(d/dx)$ for p_x in equation 8.61. Thus the Hamiltonian operator $\hat{H}$ for a simple harmonic oscillator is

$$\hat{H} = -\frac{\hbar^2}{2m}\frac{d^2}{dx^2} + \tfrac{1}{2}kx^2 \tag{8.62}$$

To obtain the allowed energy levels, we use this operator in equation 8.48 to obtain

$$\left(-\frac{\hbar^2}{2m}\frac{d^2}{dx^2} + \tfrac{1}{2}kx^2\right)\psi = E\psi \tag{8.63}$$

This equation is now solved to find the wave functions and corresponding eigenvalues E, which are the allowed energy levels.

The following eigenfunctions and eigenvalues are obtained from equation 8.63.

$$\begin{aligned}
\psi_0 &= \left(\frac{2a}{\pi}\right)^{1/4} e^{-ax^2} & E_0 &= \tfrac{1}{2}h\nu_0 \\
\psi_1 &= \left(\frac{2a}{\pi}\right)^{1/4} 2a^{1/2}xe^{-ax^2} & E_1 &= \tfrac{3}{2}h\nu_0 \\
\psi_2 &= \left(\frac{2a}{\pi}\right)^{1/4} (4ax^2 - 1)e^{-ax^2} & E_2 &= \tfrac{5}{2}h\nu_0 \\
&\vdots & &\vdots \\
\psi_v & & E_v &= (v + \tfrac{1}{2})h\nu_0
\end{aligned} \tag{8.64}$$

where $\nu_0 = (1/2\pi)\sqrt{k/m}$, the classical frequency for the harmonic oscillator, and $a = (\pi/h)\sqrt{km}$.

Exercise 1 Verify that ψ_0 and ψ_1 are solutions of the Schrödinger equation by insertion into equation 8.63. Also verify that these eigenfunctions are orthogonal.

The energy levels are equally spaced as shown in Fig. 8.5*b*, where the wave functions are also shown. Thus the quantum mechanical treatment of the harmonic oscillator yields rather different results from the classical treatment. According to classical mechanics, the oscillator may have any energy but, according to quantum mechanics, the possible energy levels are given by $E = (v + \tfrac{1}{2})h\nu_0$, where v is $0, 1, 2, \ldots$. According to classical mechanics, the oscillator may be at rest and have zero energy but, according to quantum mechanics, the lowest energy level permitted is $E = \tfrac{1}{2}h\nu_0$, called the zero point energy. This is an illustration of the Heisenberg uncertainty principle according to which $\Delta p_x \, \Delta x \geqslant \hbar/2$. If the particle were at rest at the origin, the uncertainties in p_x and x would each be zero, and the Heisenberg uncertainty principle would be violated (Section 8.4).

The wave functions given above have been normalized. The probability that the x coordinate of the harmonic oscillator is between x and $x + dx$ is given by $\psi^2\, dx$. If a large number of identically prepared systems are examined, the fraction having coordinates between x and $x + dx$ is equal to this probability. The *probability densities* ψ^2 are plotted in Fig. 8.5*c*, versus x for the first four energy levels. In the ground state ($v = 0$) the most probable internuclear distance occurs at the position of the minimum in the potential well. This is in distinct contrast with that for a classical simple harmonic oscillator, which would spend the longest times at the turning points. As the quantum number increases, however, the quantum mechanical probability density function approaches that for the classical harmonic oscillator. This is an example of Bohr's "correspondence principle" according to which the quantum mechanical result must approach the classical result in the limit of infinite quantum number.

8.11 ANGULAR MOMENTUM

Several examples of angular momentum are encountered in quantum mechanics: (1) a rotating molecule has angular momentum, (2) an orbital electron may have angular momentum, and (3) electrons and certain nuclei have intrinsic (spin) angular momentum. We will consider orbital angular momentum in this section and spin angular momentum in Section 8.14 and again in the chapter on magnetic resonance.

After reviewing angular momentum in classical mechanics, we will be able to consider angular momentum in quantum mechanics. The position $\boldsymbol{r}$ and linear momentum $\boldsymbol{p}$ of a classical particle of mass m are given by the vectors

$$\boldsymbol{r} = \boldsymbol{i}x + \boldsymbol{j}y + \boldsymbol{k}z \tag{8.65}$$

$$\boldsymbol{p} = m\frac{d\boldsymbol{r}}{dt} = \boldsymbol{i}m\frac{dx}{dt} + \boldsymbol{j}m\frac{dy}{dt} + \boldsymbol{k}m\frac{dz}{dt} = \boldsymbol{i}p_x + \boldsymbol{j}p_y + \boldsymbol{k}p_z \tag{8.66}$$

where $\boldsymbol{i}, \boldsymbol{j}$, and $\boldsymbol{k}$ are unit vectors in the x, y, and z directions, and p_x, p_y, and p_z are components of the linear momentum vector.

The angular momentum vector $\boldsymbol{L}$ is defined by

$$\boldsymbol{L} = \boldsymbol{r} \times \boldsymbol{p} = \begin{vmatrix} \boldsymbol{i} & \boldsymbol{j} & \boldsymbol{k} \\ x & y & z \\ p_x & p_y & p_z \end{vmatrix} \tag{8.67}$$

Thus the components of the classical angular momentum vector for a single particle are

$$L_x = yp_z - zp_y \tag{8.68a}$$

$$L_y = zp_x - xp_z \tag{8.68b}$$

$$L_z = xp_y - yp_x \tag{8.68c}$$

If no torque acts on the particle, its angular momentum is constant in classical mechanics. The quantum mechanical operators for the angular momentum are obtained by replacing the quantities in equations 8.68*a*, 8.68*b*, and 8.68*c* with their corresponding quantum mechanical operators; specifically $\hat{p}_x = (\hbar/i)(\partial/\partial x)$.

$$\hat{L}_x = -i\hbar\left(y\frac{\partial}{\partial z} - z\frac{\partial}{\partial y}\right) \tag{8.69a}$$

$$\hat{L}_y = -i\hbar\left(z\frac{\partial}{\partial x} - x\frac{\partial}{\partial z}\right) \tag{8.69b}$$

$$\hat{L}_z = -i\hbar\left(x\frac{\partial}{\partial y} - y\frac{\partial}{\partial x}\right) \tag{8.69c}$$

The operator for the square of the angular-momentum is given by

$$\hat{L}^2 = |\hat{\boldsymbol{L}}|^2 = \hat{L}\cdot\hat{L} = \hat{L}_x^2 + \hat{L}_y^2 + \hat{L}_z^2 \tag{8.70}$$

It can be shown* that the following pairs of operators do not commute, $\hat{L}_x, \hat{L}_y$; $\hat{L}_y, \hat{L}_z$; $\hat{L}_z, \hat{L}_x$, but that the following pairs do commute, $\hat{L}^2, \hat{L}_x$; $\hat{L}^2, \hat{L}_y$; $\hat{L}^2, \hat{L}_z$. You will remember (Section 8.8) that if operators do not commute, the state function cannot be simultaneously an eigenfunction of both operators and the quantities they represent cannot simultaneously have definite values. Therefore, only one component of the angular momentum can be specified in quantum mechanics. This is usually taken arbitrarily as the z component. Thus only one component of the angular momentum L_z and the square of its magnitude $L^2 = |\boldsymbol{L}|^2$, not its direction—except that it has a component L_z—can be specified. Since L_x and L_y cannot be specified, the vector $\boldsymbol{L}$ can be anywhere on the surface of a cone whose axis is the z axis, as shown in Fig. 8.6. The altitude of the cone is L_z and the slant height is L. This is all in distinct contrast with classical mechanics, where each of the three components of angular velocity has a definite value if angular momentum is conserved.

The eigenvalues and eigenfunctions of the operators $\hat{L}^2$ and $\hat{L}_z$ can be obtained, and it is found that for a single particle

$$\hat{L}^2\psi = l(l+1)\hbar^2\psi \qquad l = 0, 1, 2, \ldots \tag{8.71}$$

$$\hat{L}_z\psi = m\hbar\psi \qquad m = -l, -l+1, -l+2, \ldots, -1, 0, 1, l-2, l-1, l \tag{8.72}$$

where l is the angular momentum quantum number and m is the magnetic quantum number. Thus the allowed eigenvalues for the square of the magnitude of the angular momentum are given by $l(l+1)\hbar^2$, and the magnitude of the angular momentum is given by $[l(l+1)]^{1/2}\hbar$. The magnitude of the z component of the angular momentum is $|m|\hbar$. Note that the maximum value of L_z, $l\hbar$, still permits some ambiguity in L_x and L_y, since $L^2[=\hbar^2 l(l+1)]$ is greater than L_z^2.

Figure 8.6 shows the possible orientations of the angular momentum vector for $l = 1$ and $l = 2$. In the absence of an electric or magnetic field, there is a degeneracy

---

* I. N. Levine, *Quantum Chemistry*, Allyn and Bacon, Boston, 1974, p. 70.

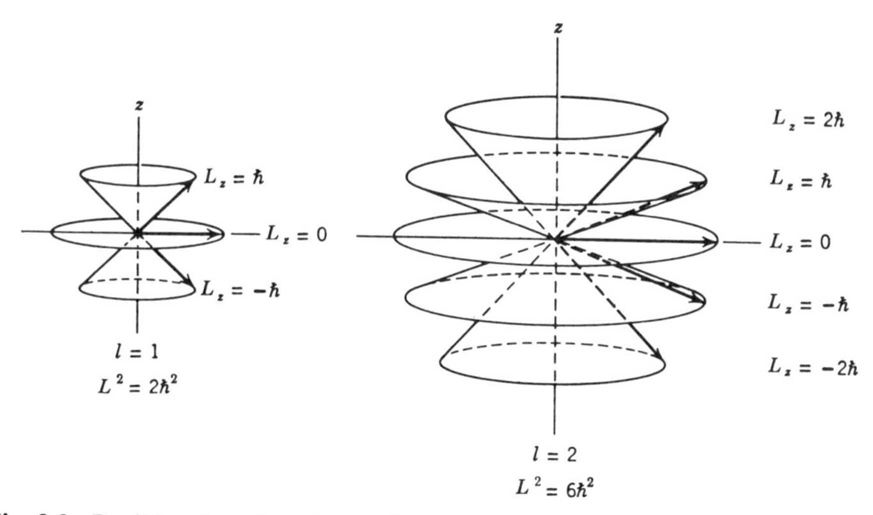

Fig. 8.6 Possible orientations for angular momentum vectors for $l = 1$ and $l = 2$.

of $2l + 1$ since, for any magnitude of the angular momentum vector, there are $2l + 1$ values of m.

8.12 THE TWO-PARTICLE RIGID ROTOR

The two-particle rigid rotor with masses m_1 and m_2 separated by distance $r_1 + r_2$ (see Fig. 8.7) can represent a diatomic molecule with a fixed distance between the two nuclei. The treatment of the translational and rotational motion of a two-particle rigid rotor can be reduced to two separate one-body problems. The translational motion of the system as a whole can be treated using the total mass of the two particles, and the rotational motion of the particles can be obtained by considering a hypothetical particle of reduced mass μ. The expression of the reduced mass μ is derived as follows. In Fig. 8.7 the center of mass is located at the origin of the Cartesian coordinate system, so that

$$m_1 r_1 = m_2 r_2 \tag{8.73}$$

The distance between the two particles is represented by r, so that

$$r_1 + r_2 = r \tag{8.74}$$

The moment of inertia I with respect to a line that passes through the center of mass is

$$I = m_1 r_1^2 + m_2 r_2^2 \tag{8.75}$$

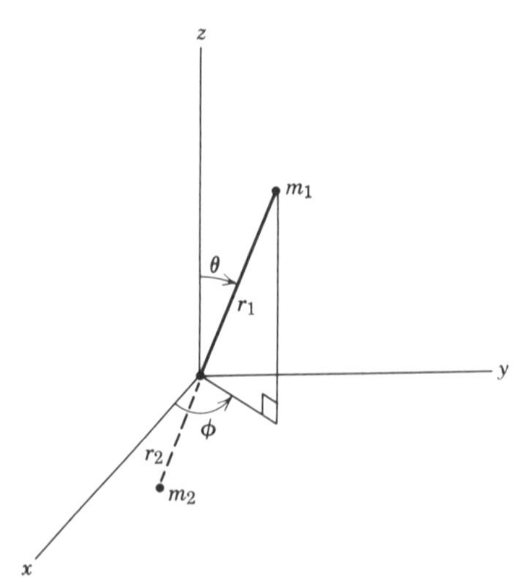

Fig. 8.7 Definition of coordinates for rigid rotor. The center of mass defined by $m_1r_1 = m_2r_2$ is located at the origin.

Eliminating r_1 and r_2 between equations 8.73, 8.74, and 8.75 yields

$$I = \frac{r^2}{\dfrac{1}{m_1} + \dfrac{1}{m_2}} = \mu r^2 \tag{8.76}$$

where μ is the reduced mass.

$$\mu = \frac{1}{\dfrac{1}{m_1} + \dfrac{1}{m_2}} = \frac{m_1 m_2}{m_1 + m_2} \tag{8.77}$$

The rigid rotor has only kinetic energy. The classical expression for the rotational kinetic energy T is given by

$$T = \frac{|\boldsymbol{L}|^2}{2\mu r^2} \tag{8.78}$$

where $\boldsymbol{L}$ is the angular momentum vector.

Now let us consider the two-particle rigid rotor from a quantum mechanical standpoint. To do that we convert the classical expression for the kinetic energy to the quantum mechanical Hamiltonian.

$$\hat{H} = \frac{\hat{L}^2}{2\mu r^2} = -\frac{\hbar^2}{2\mu} \nabla^2 \tag{8.79}$$

The Schrödinger equation yields

$$E = \frac{J(J+1)\hbar^2}{2\mu r^2} = \frac{J(J+1)\hbar^2}{2I} \qquad J = 0, 1, 2, \ldots \tag{8.80}$$

where J instead of l is used for the rotational quantum number. The energy depends only on the quantum number J, but the wave function depends on J and m (see equations 8.71 and 8.72). Since the values of m range from $-J$ to J, the rotational levels are $(2J + 1)$-fold degenerate. As shown in Fig. 8.6, the angular momentum vector can be oriented in $2J + 1$ different directions. Since r is constant for the rigid rotor, we can think of a particle of mass μ constrained to move on the surface of a sphere. The wave function is a function of θ and φ only. The eigenfunctions are the spherical harmonics $Y_1^m(\theta, \varphi)$, and the mathematical problem is identical with that for the angular part of the wave equation for the motion of an electron in the field of an atom. The degeneracy corresponds to the different possible orientations of the angular-momentum vector (see Section 8.11).

Example 8.4 What is the moment of inertia of a hydrogen atom if the distance between the proton and the electron is taken to be $5.291\,77 \times 10^{-11}$ m? As we will see in Example 8.6, this distance is 1 Bohr.

Using the rest masses of the electron and proton from the Appendix:

$$\mu = \frac{m_1 m_2}{m_1 + m_2} = \frac{(9.109\,534 \times 10^{-31}\ \text{kg})(1.672\,648\,5 \times 10^{-27}\ \text{kg})}{9.109\,534 \times 10^{-31}\ \text{kg} + 1.672\,648\,5 \times 10^{-27}\ \text{kg}}$$

$$= 9.104\,576 \times 10^{-31}\ \text{kg}$$

$$I = \mu r^2 = (9.104\,576 \times 10^{-31}\ \text{kg})(5.291\,77 \times 10^{-11}\ \text{m})^2$$

$$= 2.549\,539 \times 10^{-51}\ \text{kg m}^2$$

8.13 THE HYDROGEN ATOM

The simplest atomic systems are those consisting of a nucleus of mass M and charge Ze, where Z is the atomic number, and one electron with mass m_e and charge $-e$. The Schrödinger equation may be solved exactly for these hydrogenlike atoms, and these solutions are of tremendous importance for the treatment of atoms with two or more electrons and for molecules, for which closed mathematical solutions cannot be obtained. The complete treatment of the hydrogen atom is too complicated to give here, and so only the main features will be outlined.

The starting point for the quantum mechanical treatment is the classical Hamiltonian for the system consisting of an electron and nucleus of charge Ze. To simplify the expression for the Hamiltonian, we will assume that the electron moves around the stationary nucleus. Actually, the electron and nucleus move around their mutual center of mass. The nucleus of charge Ze is taken as the origin of the coordinate system, so the potential energy is given by Coulombs' law as $-Ze^2/4\pi\epsilon_0 r$, where r is the distance between nucleus and electron. The classical Hamiltonian is

$$H = \frac{1}{2m_e}(p_x^2 + p_y^2 + p_z^2) - \frac{Ze^2}{4\pi\epsilon_0 r} \tag{8.81}$$

where m_e is the mass of the electron. Converting this to the quantum mechanical Hamiltonian (Section 8.8) and applying this operator to ψ yields

$$\frac{\partial^2\psi}{\partial x^2} + \frac{\partial^2\psi}{\partial y^2} + \frac{\partial^2\psi}{\partial z^2} + \frac{2m_e}{\hbar^2}\left(E + \frac{Ze^2}{4\pi\epsilon_0 r}\right)\psi = 0 \tag{8.82}$$

In order to solve this equation it is convenient to transform the Cartesian coordinates to spherical coordinates r, θ, ϕ, which are defined in Fig. 8.8, since with $r = \sqrt{x^2 + y^2 + z^2}$ it is not possible to separate variables in Cartesian coordinates. This transformation yields

$$\frac{1}{r^2}\frac{\partial}{\partial r}\left(r^2\frac{\partial\psi}{\partial r}\right) + \frac{1}{r^2 \sin\theta}\frac{\partial}{\partial\theta}\left(\sin\theta\frac{\partial\psi}{\partial\theta}\right) + \frac{1}{r^2\sin^2\theta}\frac{\partial^2\psi}{\partial\phi^2} + \frac{8\pi^2 m_e}{h^2}\left(E + \frac{Ze^2}{4\pi\epsilon_0 r}\right)\psi = 0 \tag{8.83}$$

If the fact that the electron and the nucleus rotate about their common center of mass had been taken into account, the m_e in this equation would be replaced by the reduced mass μ (equation 8.77), which is given by $m_e M/(M + m_e)$, where m_e is the mass of the electron and M is the mass of the nucleus.

The ψ of equation 8.83 may be written as a product of three functions, one dependent on r, one dependent on θ, and one dependent on ϕ.

$$\psi(r, \theta, \phi) = R(r)\Theta(\theta)\Phi(\phi) \tag{8.84}$$

when this is substituted in equation 8.83, three ordinary differential equations result. Each of the resulting three differential equations may be solved, and each leads to the introduction of a quantum number that must be an integer. An

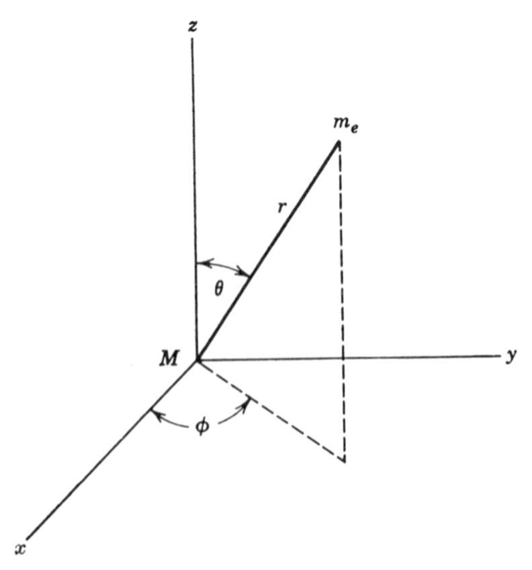

Fig. 8.8 Spherical coordinates used in describing the hydrogenlike atom.

analogous situation was encountered earlier where it was found that for a particle in a three-dimensional box, the wave function is the product of three functions, one for each coordinate. The radial equation [i.e., the one involving $R(r)$] yields the principal quantum number n. The principal quantum number n can have values of 1, 2, 3, ...

The equation for $\Theta(\theta)$ is a form of Legendre's equation and yields the angular momentum quantum number l.

$$l = 0, 1, 2, \ldots, (n - 1)$$

where n is the principal quantum number. Letter symbols are associated with values of l as follows.

$l =$	0	1	2	3
Symbol	s	p	d	f

The solution for the equation in $\Phi(\phi)$ is

$$\Phi(\phi) = \exp (im_l\phi) \tag{8.85}$$

Instead of working with this imaginary wave function, it is often more convenient to use real wave functions formed by taking linear combinations of complex wave functions. The justification of this procedure is that any linear combination of eigenfunctions of a degenerate energy level is also an eigenfunction of the Hamiltonian with the same eigenvalue. Thus

$$\Phi(\phi) = \sin m_l\phi \qquad \text{or} \qquad \cos m_l\phi \tag{8.86}$$

where m_l is the magnetic quantum number. For $\Phi(\phi)$ to be single valued and continuous, that is, $\Phi(\phi) = \Phi(\phi + 2\pi)$, the magnetic quantum number m_l can have only integer values $m_l = 0, \pm 1, \pm 2, \pm 3, \ldots, \pm l$. Since the operators $\hat{H}$, $\hat{L}^2$, and $\hat{L}_z$ all commute,

$$\hat{H}\psi = E_n\psi \qquad n = 1, 2, 3, \ldots \tag{8.87}$$

$$\hat{L}^2\psi = l(l + 1)\hbar^2\psi \qquad l = 0, 1, 2, \ldots, (n - 1) \tag{8.88}$$

$$\hat{L}_z\psi = m_l\hbar\psi \qquad m_l = 0, \pm 1, \pm 2, \ldots, \pm l \tag{8.89}$$

Although we cannot go into the steps of the calculation, equation 8.87 yields the following eigenvalues for the energy of the hydrogenlike atom.

$$E = -\frac{m_e e^4 Z^2}{2(4\pi\epsilon_0)^2\hbar^2 n^2} \qquad n = 1, 2, 3, \ldots \tag{8.90}$$

If $Z = 1$, this equation gives the energy of a hydrogen atom measured with respect to a proton and an electron separated by an infinite distance. Thus the energies of the various states of the hydrogen atom are inversely proportional to the square of the principal quantum number n. The energy is negative because the electron in a hydrogenlike atom has *less* energy than when it is free. Equation 8.90 gives the energy of a hydrogenlike atom with a nucleus of infinite mass. To calculate the energy for any other nuclear mass, the mass of the electron m_e is replaced by the reduced mass (Section 8.12) of the hydrogenlike atom.

Since the energy expression 8.90 does not involve the quantum numbers l and m, the levels are degenerate. The total degeneracy for any value of n will be the sum of all l, m values permitted for total n: $0 \leqslant l \leqslant (n-1)$, $-l \leqslant m \leqslant l$. This gives n^2 as the degeneracy of the nth energy level of the hydrogenlike atom.

Example 8.5 Calculate the energy of a hydrogen atom in its ground state ($n = 1$) using the mass of the electron instead of the reduced mass of the hydrogen atom.

$$E = -\frac{2\pi^2 m_e e^4}{(4\pi\epsilon_0)^2 h^2}$$

$$= -\frac{2\pi^2(9.109\,534 \times 10^{-31}\text{ kg})(1.602\,189\,2 \times 10^{-19}\text{ C})^4}{(4\pi 8.854\,187\,82 \times 10^{-12}\text{ C}^2\text{ N}^{-1}\text{ m}^{-2})^2(6.626\,176 \times 10^{-34}\text{ J s})^2}$$

$$= -2.179\,907 \times 10^{-18}\text{ J}$$

Since 1 eV is equal* to $1.602\,189\,2 \times 10^{-19}$ J,

$$E = \frac{-2.179\,907 \times 10^{-18}\text{ J}}{1.602\,189\,2 \times 10^{-19}\text{ J eV}^{-1}} = -13.605\,80\text{ eV}$$

The energy of a hydrogen atom in its ground state calculated with the reduced mass of a hydrogen atom is $-13.598\,396$ eV. This is in agreement with the experimental value of the ionization potential of a hydrogen atom in the ground state, which is known to a very high accuracy.

In the absence of magnetic or electric fields the energy of a hydrogenlike atom depends only on the principal quantum number n. Equation 8.90 applies to He^+ Li^{2+}, Be^{3+}, and all the other single-electron species, ignoring small spin-orbit couplings. As Z increases, the electrons become more tightly bound.

The energy levels for a one-electron atom with nuclear charge Z are given by equation 8.90, and hence the frequencies of the lines in the absorption or emission spectrum are given by

$$\tilde{\nu} = RZ^2\left(\frac{1}{n_1{}^2} - \frac{1}{n_2{}^2}\right) \tag{8.91}$$

where the Rydberg constant R is $1.096\,776 \times 10^7\text{ m}^{-1}$ for hydrogen and $1.097\,373\,177 \times 10^7\text{ m}^{-1}$ for a nucleus of infinite mass. For He^+ the frequencies or wave numbers of the lines are four times as great as those for hydrogen. The energy of ionization of the $1s$ orbital of hydrogen is 13.6 eV, of He^+ is $2^2 \cdot 13.6 = 54.4$ eV, and of Li^{+2} is $3^2 \cdot 13.6 = 122.4$ eV.

The orbital angular momentum is given by $\sqrt{l(l+1)}\hbar$. Thus s electrons ($l = 0$) have no orbital angular momentum, and p electrons ($l = 1$) have an angular momentum of $\sqrt{2}\hbar$. The angular momentum in a specified direction is given by

* The electron volt is the work done by moving an electronic charge through a potential difference of 1 V or, in other words, the energy acquired by an electron being accelerated by a potential difference of 1 V. The energy in joules may be calculated as follows.

$$\text{eV} = eE = (1.602\,189\,2 \times 10^{-19}\text{ C})(1\text{ V})$$
$$= 1.602\,189\,2 \times 10^{-19}\text{ J}$$

Table 8.1 Real Hydrogenlike Wave Functions[a]

n	l	m	Wave Function
1	0	0	$\psi_{1s} = \frac{1}{\sqrt{\pi}}\left(\frac{Z}{a_0}\right)^{3/2} e^{-\sigma}$
2	0	0	$\psi_{2s} = \frac{1}{4\sqrt{2\pi}}\left(\frac{Z}{a_0}\right)^{3/2} (2 - \sigma)e^{-\sigma/2}$
2	1	0	$\psi_{2p_z} = \frac{1}{4\sqrt{2\pi}}\left(\frac{Z}{a_0}\right)^{3/2} \sigma e^{-\sigma/2} \cos\theta$
2	1	± 1	$\psi_{2p_x} = \frac{1}{4\sqrt{2\pi}}\left(\frac{Z}{a_0}\right)^{3/2} \sigma e^{-\sigma/2} \sin\theta \cos\phi$
			$\psi_{2p_y} = \frac{1}{4\sqrt{2\pi}}\left(\frac{Z}{a_0}\right)^{3/2} \sigma e^{-\sigma/2} \sin\theta \sin\phi$
3	0	0	$\psi_{3s} = \frac{1}{81\sqrt{3\pi}}\left(\frac{Z}{a_0}\right)^{3/2} (27 - 18\sigma + 2\sigma^2)e^{-\sigma/3}$
3	1	0	$\psi_{3p_z} = \frac{\sqrt{2}}{81\sqrt{\pi}}\left(\frac{Z}{a_0}\right)^{3/2} (6 - \sigma)\sigma e^{-\sigma/3} \cos\theta$
3	1	± 1	$\psi_{3p_x} = \frac{\sqrt{2}}{81\sqrt{\pi}}\left(\frac{Z}{a_0}\right)^{3/2} (6 - \sigma)\sigma e^{-\sigma/3} \sin\theta \cos\phi$
			$\psi_{3p_y} = \frac{\sqrt{2}}{81\sqrt{\pi}}\left(\frac{Z}{a_0}\right)^{3/2} (6 - \sigma)\sigma e^{-\sigma/3} \sin\theta \sin\phi$
3	2	0	$\psi_{3d_{z^2}} = \frac{1}{81\sqrt{6\pi}}\left(\frac{Z}{a_0}\right)^{3/2} \sigma^2 e^{-\sigma/3}(3\cos^2\theta - 1)$
3	2	± 1	$\psi_{3d_{xz}} = \frac{\sqrt{2}}{81\sqrt{\pi}}\left(\frac{Z}{a_0}\right)^{3/2} \sigma^2 e^{-\sigma/3} \sin\theta \cos\theta \cos\phi$
			$\psi_{3d_{yz}} = \frac{\sqrt{2}}{81\sqrt{\pi}}\left(\frac{Z}{a_0}\right)^{3/2} \sigma^2 e^{-\sigma/3} \sin\theta \cos\theta \sin\phi$
3	2	± 2	$\psi_{3d_{x^2-y^2}} = \frac{1}{81\sqrt{2\pi}}\left(\frac{Z}{a_0}\right)^{3/2} \sigma^2 e^{-\sigma/3} \sin^2\theta \cos 2\phi$
			$\psi_{3d_{xy}} = \frac{1}{81\sqrt{2\pi}}\left(\frac{Z}{a_0}\right)^{3/2} \sigma^2 e^{-\sigma/3} \sin^2\theta \sin 2\phi$

[a] $\sigma = \frac{Z}{a_0} r$.

$m\hbar$. The angular momentum in a particular direction cannot be greater than the total angular momentum, and so $|m| \leqslant |l|$. Thus m may have any integer value between $+l$ and $-l$, including zero. Therefore $2l + 1$ values of m are possible. For $l = 3$, $m = -3, -2, -1, 0, 1, 2, 3$.

The wave functions for the hydrogenlike atoms through $n = 3$ are given in Table 8.1. It is convenient to write these equations in terms of the Bohr radius a_0, which, as we will see, is the most probable distance between the electron and proton in a hydrogen atom in the $1s$ state.

$$a_0 = \frac{\hbar^2(4\pi\epsilon_0)}{m_e e^2} \tag{8.92}$$

The term "orbital" is used for any one-electron wave function, such as those appearing in Table 8.1. We will later see that many-electron wave functions for more complex atoms or for molecules can be roughly approximated from orbitals of the individual electrons.

Example 8.6 Calculate the Bohr radius using the mass of the electron instead of the reduced mass for the hydrogen atom.

$$a_0 = \frac{h^2(4\pi\epsilon_0)}{4\pi^2 m_e e^2} = \frac{h^2\epsilon_0}{\pi m_e e^2}$$

$$= \frac{(6.626\,176 \times 10^{-34}\text{ J s})^2(8.854\,187\,82 \times 10^{-12}\text{ C}^2\text{ N}^{-1}\text{ m}^{-2})}{\pi(9.109\,534 \times 10^{-31}\text{ kg})(1.602\,189\,2 \times 10^{-19}\text{ C})^2}$$

$$= 5.291\,77 \times 10^{-11}\text{ m} = 0.052\,917\,7\text{ nm}$$

In considering atoms and molecules it is frequently convenient to use this distance as a unit length. In atomic units this distance is taken as 1 Bohr (1B).

To visualize the nature of these functions it is helpful to consider $R(r)$ and $\Theta(\theta)\Phi(\phi)$ separately.

The radial functions $R(r)$ are plotted in Fig. 8.9 for the wave functions in Table 8.1, with $Z = 1$. The radial function always contains the factor e^{-Zr/na_0}, where n is the principal quantum number. As Z is increased, the amplitude of the wave function falls off more rapidly with increasing r indicating that the electron is attracted more closely to the higher positively charged nuclei.

Figure 8.9*a* shows that the radial part $R(r)$ of the hydrogen atom wave functions for s orbitals makes the probability density for the electron greatest at the nucleus.

The radial wave functions have $n - l$ nodes ($R(r) = 0$), where n is the principal quantum number. The wave function changes sign at a node, but the square of the wave function does not. The existence of nodes is required so that the $1s$ and $2s$ and other orbitals will be orthogonal (Section 8.6); that is,

$$\int \psi_{1s}\psi_{2s}\,d\tau = 0 \tag{8.93}$$

where $d\tau$ represents the element of volume.

The probability of finding an electron in a volume $dx\,dy\,dz$ is $\psi^*\psi\,dx\,dy\,dz$. Thus $\psi^*\psi$ is the probability per unit volume. For spherically symmetric wave functions it is convenient to define a radical probability density as follows. The probability of finding an electron is a spherical shell between r and $r + dr$ is $\psi^*\psi(4\pi r^2\,dr)$, since $4\pi r^2\,dr$ is the volume of the spherical shell in which $\psi^*\psi$ is a constant. The term "radial probability density" is usually applied to the probability divided by 4π times the thickness of the shell; thus the radial probability density is $r^2\psi^*\psi$. The

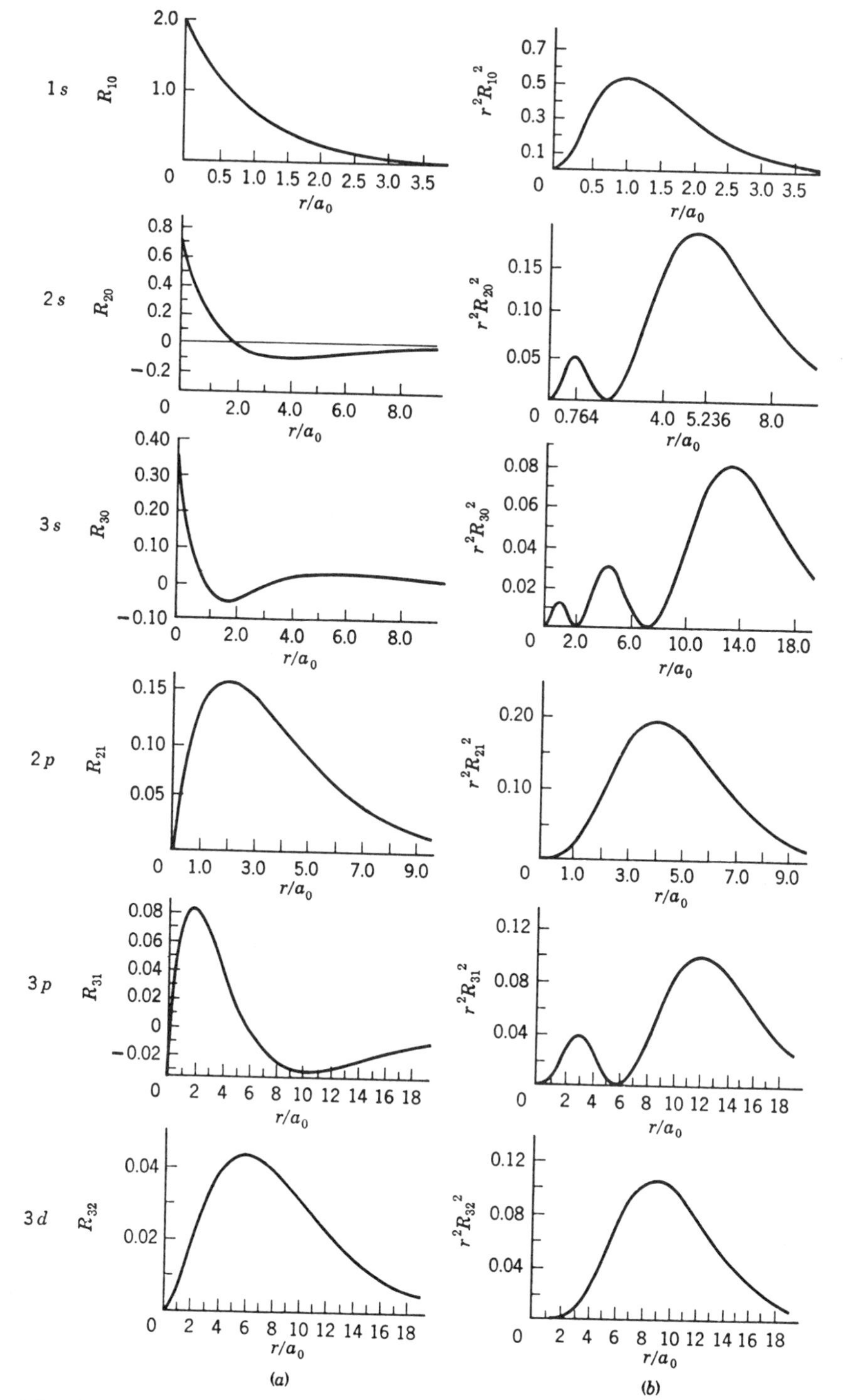

Fig. 8.9 (*a*) Electronic radial-wave functions $R(r)$ for the hydrogen atom. (*b*) Probability density for finding an electron at a distance between r and $r + dr$. (From M. Karplus and R. N. Porter, *Atoms and Molecules*. Copyright © 1970 by The Benjamin/Cummings Publishing Company, Inc., Menlo Park, Calif.)

radial distribution function for the $1s$ orbital has a maximum at a_0, as shown in Fig. 8.9*b*. This most probable distance for the electron agrees with the radius of the first Bohr orbit.

The total probability density $\psi^*\psi$ also depends on the angular variables θ and ϕ so that it can be written

$$\psi^*\psi = R^*R\Theta^*\Theta\Phi^*\Phi \tag{8.94}$$

where $\Theta^*\Theta\Phi^*\Phi$ represents the angular probability density. It is convenient to consider R^*R and $\Theta^*\Theta\Phi^*\Phi$ separately. Plots of r^2R^*R are shown in Fig. 8.9.

A three-dimensional wave function may be represented approximately by a contour surface of constant $\psi^*\psi$ within which most of the electron density is enclosed. Thus the surface encloses the region in which the electron is most likely to be found. Diagrams of this type are shown in Fig. 8.10.

The orientations of the p orbitals can be calculated by considering the magnitudes and signs of the trigonometric functions at several angles. In the absence of an electric or magnetic field electrons in p_x, p_y, and p_z, orbitals all have the same energy; that is, the energy depends only on the total quantum number n. In the presence of a magnetic field, however, electrons in the p orbital in the direction of the field have different energies, which is why m is called the magnetic quantum number. This is an example of the distinction between a quantum state and an energy level. For the hydrogen atom with quantum number $n = 2$ there are four states, all having the same energy in the absence of a magnetic or electric field. Such an energy level is said to be degenerate, and the degeneracy is the number of wave functions that have that particular energy associated with them.

The p orbitals do not have to point along the x, y, and z directions. Linear combinations of p_x, p_y, and p_z may be formed to point in any three mutually perpendicular directions. It will be seen in Chapter 10 that the directional character of certain chemical bonds results from the directed orientation of these and other orbitals.

There are five independent d orbitals. The $3d_{z^2}$ orbital has two large regions of electron density alone one axis, by convention the z axis, and a small donut-shaped orbital in the xy plane. The other four d orbitals have four equivalent lobes of electron density with two nodal planes separating them. Note that for lobes that are opposite each other the wave function has the same sign.

One of the deficiencies in the diagrams in Fig. 8.10 is that the nodal surfaces resulting from the radial functions $R(r)$ are not shown. Although the hydrogen atom has an infinite number of orbitals, most questions of chemical significance involve only the lowest energy orbitals.

The presentation of electron density as a function of r, θ, and ϕ would require four dimensions. One way to do this is to use the density of dots to represent the probability of finding an electron in a region of space and use stereo plots with a stereo viewer to see probability densities in three dimensions.*

---

* D. T. Cromer, *J. Chem. Ed.*, **45**, 626 (1968).

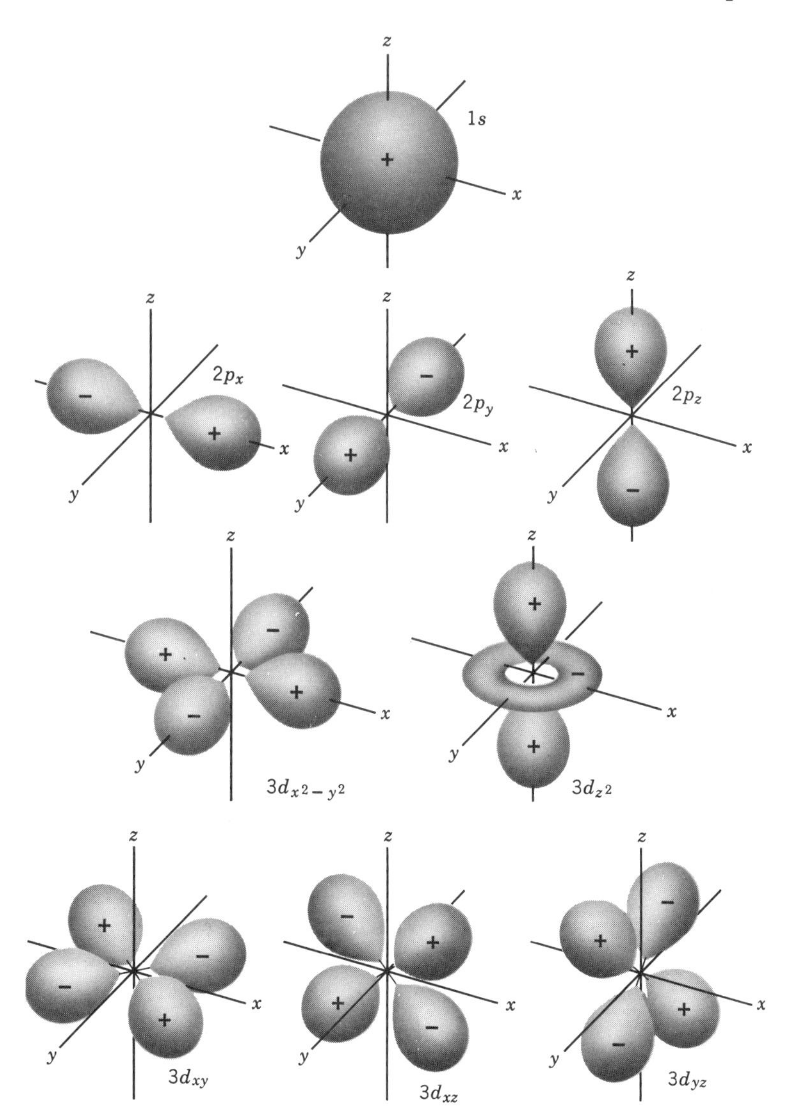

Fig. 8.10 Contour surfaces for constant $\psi^*\psi$ for one-electron atoms. The indicated signs are those of the wave functions. These signs are indicated because they will be of interest later when we discuss molecular orbitals. The probability density is, of course, always positive.

8.14 ELECTRON SPIN

The spectra of atoms show a fine structure that is not explained by the theory we have just discussed. For example, the sodium *D* line consists of two closely spaced lines. This indicates a doubling of the number of states available to the valence electron. This fine structure was interpreted in 1925 by Goudsmit and Uhlenbeck as being due to an intrinsic angular momentum of the electron that is independent of its orbital angular momentum. Later Dirac applied relativity theory to the quantum mechanical formulation of electron spin and demonstrated a theoretical basic for the intrinsic angular momentum of an electron. The term electron spin is used, but it is not really correct to think of the intrinsic angular momentum of the electron as being due to a spinning motion of the electron mass on its axis.

Since the spin angular momentum of an electron has no analog in classical mechanics, we cannot construct spin angular-momentum operators by first writing the classical Hamiltonian. As an analog to the angular-momentum operators $\hat{L}^2$, $\hat{L}_x$, $\hat{L}_y$, $\hat{L}_z$, there are spin angular-momentum operators $\hat{S}^2$, $\hat{S}_x$, $\hat{S}_y$, $\hat{S}_z$, and they obey the same commutation relations as the orbital angular-momentum operators. The eigenvalues of $\hat{S}^2$ and $\hat{S}_z$ are given by equations analogous to equations 8.71 and 8.72 for the operators $\hat{L}^2$ and $\hat{L}_z$,

$$\hat{S}^2\psi = s(s+1)\hbar^2 \qquad s = 0, \tfrac{1}{2}, 1, \tfrac{3}{2}, \ldots \tag{8.95}$$

$$\hat{S}_z\psi = m_s\hbar\psi \qquad m_s = -s, -s+1, \ldots, s-1, s \tag{8.96}$$

where s is the spin quantum number and m_s is the spin quantum number for the z component of the spin. The possibility of half-integer quantum numbers is characteristic of spin degrees of freedom, which have no classical analogue. For an electron the spin s is $\frac{1}{2}$ so that the magnitude of its total spin angular momentum is

$$[\tfrac{1}{2}(\tfrac{3}{2})\hbar^2]^{1/2} = \frac{\sqrt{3}}{2}\hbar$$

The component in the z direction is either $+\frac{1}{2}\hbar$ or $-\frac{1}{2}\hbar$, as illustrated in Fig. 8.11. We speak of these two eigenvalues as "spin up" and "spin down."

Since the spin eigenfunctions do not involve other kinds of coordinates, the two possible spin functions for an electron are represented by α and β. These spin functions have the following properties.

$$\hat{S}^2\alpha = \tfrac{1}{2}(\tfrac{1}{2}+1)\hbar^2\alpha = \tfrac{3}{4}\hbar^2\alpha \tag{8.97a}$$

$$\hat{S}^2\beta = \tfrac{1}{2}(\tfrac{1}{2}+1)\hbar^2\beta = \tfrac{3}{4}\hbar^2\beta \tag{8.97b}$$

$$\hat{S}_z\alpha = +\tfrac{1}{2}\hbar\alpha \tag{8.97c}$$

$$\hat{S}_z\beta = -\tfrac{1}{2}\hbar\beta \tag{8.97d}$$

The operator $\hat{S}^2$ and $\hat{S}_z$ commute with the Hamiltonian operator $\hat{H}$ and $\hat{L}^2$ and $\hat{L}_z$ so that the magnitude of the spin, the z component of the spin, the energy, the magnitude of the orbital angular momentum, and the z component of the orbital angular momentum can all have definite values.

Since there are two spin functions, there are twice as many wave functions for the hydrogen atom as we indicated earlier: $\psi\alpha$ and $\psi\beta$. Thus the wave function of the hydrogen atom depends on the four quantum numbers n, l, m, and m_s. For the hydrogen atom the only effect is to increase the degeneracy of the energy levels from n^2 to $2n^2$.

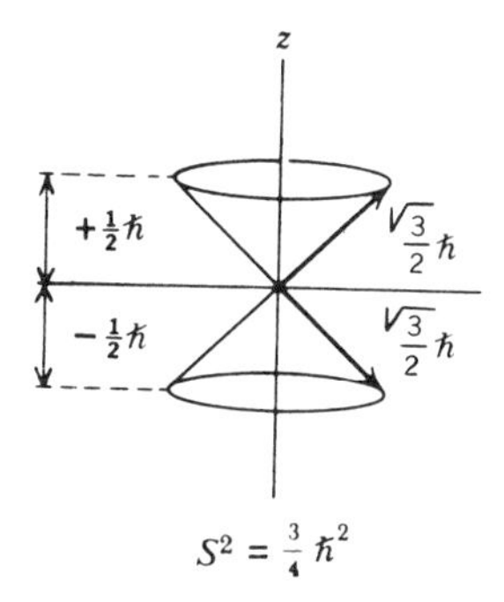

Fig. 8.11 Possible orientations of the spin angular momentum vectors for an electron for which $s = \frac{1}{2}$.

8.15 HELIUM ATOM

The helium atom has two electrons, and the coordinates used in writing the Hamiltonian are shown in Fig. 8.12. The two electrons repel each other with a potential energy $e^2/4\pi\epsilon_0 r_{12}$. The Hamiltonian operator is

$$\hat{H} = -\frac{\hbar^2}{2m_e}(\nabla_1{}^2 + \nabla_2{}^2) - \frac{1}{4\pi\epsilon_0}\left(\frac{2e^2}{r_1} + \frac{2e^2}{r_2} - \frac{e^2}{r_{12}}\right) \tag{8.98}$$

The first term is the kinetic energy operator for the two electrons, the next two terms represent the potential energy of each of the electrons in the field of the nucleus, and the last term represents the interelectronic potential energy.

In writing such equations it is convenient to introduce new units that are more appropriate for atomic dimensions and eliminate some constants. These units are referred to as atomic units (a.u.). The unit of mass is taken as the mass of the electron m_e. The unit of charge is taken as the charge of the electron e. The unit of

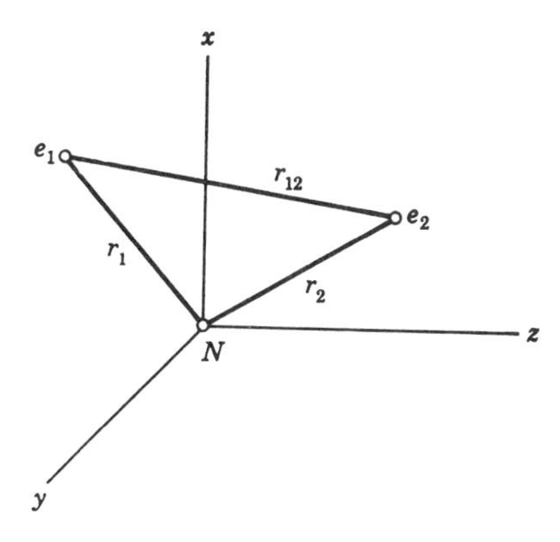

Fig. 8.12 Coordinates in the helium atom.

distance is taken as the Bohr radius a_0 of the hydrogen atom in its ground state (equation 8.92). The unit energy is the potential energy of two unit charges a unit distance apart. This unit of energy is referred to as the Hartree and is represented by E_h.

$$E_h = \frac{e^2}{4\pi\epsilon_0 a_0} \tag{8.99}$$

Example 8.7 Calculate the energy of a Hartree in electron volts.

$$E_h = \frac{e^2}{4\pi\epsilon_0 a_0} = \frac{(1.602\,189\,2 \times 10^{-19}\text{ C})^2}{4\pi(8.854\,187\,82 \times 10^{-12}\text{ C}^2\text{ N}^{-1}\text{ m}^{-2})(5.291\,770\,6 \times 10^{-11}\text{ m})}$$

$$= 4.359\,814\,4 \times 10^{-18}\text{ J} = \frac{4.359\,814\,4 \times 10^{-18}\text{ J}}{1.602\,189\,2 \times 10^{-19}\text{ J eV}^{-1}} = 27.211\,608\text{ eV}$$

Note that the Hartree is twice the energy of a hydrogen atom in its ground state, with the sign reversed (see Example 8.4).

In atomic units Planck's constant h has the value 2π. The Schrödinger equation becomes

$$\nabla^2\psi + 2(E - V)\psi = 0 \tag{8.100}$$

In writing the Hamiltonian in atomic units, $\hbar$, the electronic charge e, and the electron mass m_e are all taken as unity. In atomic units the Hamiltonian operator for the hydrogenlike atom is

$$\hat{H} = -\tfrac{1}{2}\nabla^2 - \frac{Z}{r} \tag{8.101}$$

The Hamiltonian operator for the helium atom is

$$\hat{H} = -\tfrac{1}{2}(\nabla_1{}^2 + \nabla_2{}^2) - \frac{2}{r_1} - \frac{2}{r_2} + \frac{1}{r_{12}} \tag{8.102}$$

The last term makes it impossible to separate the Schrödinger equation for the helium atom into two different equations. If the r_{12} term is omitted, the solution of the Schrödinger equation for the helium atom is of the form $\psi(1)\psi(2)$, where $\psi(1)$ is a one-electron function of the coordinates of electron 1 and $\psi(2)$ is a one-electron function of the coordinates of electron 2. These one-electron functions are taken, to a first approximation, to be orbitals of the hydrogen atom, characterized by the three quantum numbers n, l, and m_l. According to this approximation, the wave function of the ground state of the helium atom is

$$\psi = 1s(1)1s(2) = \frac{8}{\pi}\exp\left[-2(r_1 + r_2)\right] \tag{8.103}$$

where r_1 is the distance from the nucleus to electron 1 and r_2 is the distance from the nucleus to electron 2. The energy of the helium atom calculated using this wave function and neglecting the interelectronic repulsion is -4 Hartrees ($-4\ E_h$), in comparison with the experimental value of $-2.905\ E_h$. If the $1/r_{12}$ term in the Hamiltonian is not ignored, the ground state energy calculated with the approximate function 8.103 is $-2.75\ E_h$. In order to obtain a more accurate value for the ground state energy it is necessary to use the variation method.

8.16 THE VARIATION METHOD

The variation principle states that if a system has a Hamiltonian operator $\hat{H}$ and ψ is any normalized well-behaved function that satisfies the boundary conditions, then

$$\int \psi^* \hat{H} \psi \, d\tau \geqslant E_0 \tag{8.104}$$

where E_0 is the lowest energy eigenvalue of $\hat{H}$ and $d\tau$ represents the coordinates for the integration. Thus we can calculate an upper bound for the energy of the ground state.* If the wave function ψ has parameters that may be varied, the best wave function of this form will be obtained by varying the parameters so as to obtain the lowest energy. As more variable terms are introduced, a closer approximation may be obtained to the true energy, but at the price of increased cost of computation.

In the preceding section we saw that we can obtain the first approximation to the wave function for the helium atom by taking the product of two hydrogenlike wave functions.

A better value can be obtained from the variation method by introducing a single parameter into the "trial" wave function. Each electron tends to shield the other from the full nuclear charge. Therefore $Z'e$ is used to represent an effective nuclear charge, and the wave function is written

$$\psi = N \exp\left[-Z'(r_1 + r_2)\right] \tag{8.105}$$

where N is the normalization factor. Using this wave function with Hamiltonian 8.102 yields the following expression for the energy in terms of Z'.

$$E' = (Z')^2 - \tfrac{27}{8} Z' \tag{8.106a}$$

According to the variation theorem, the best value of Z' is the one that gives the lowest value of the energy. Differentiating with respect to Z',

$$\frac{dE'}{dZ'} = 2Z' - \tfrac{27}{8} = 0 \tag{8.106b}$$

so that $Z' = \frac{27}{16}$. Substituting this value of Z' into equation 8.106*a* yields $E' = -2.848\ E_h$, which is within 2% of the experimental value of $-2.905\ E_h$. Better agreement can be obtained by adding further terms to the trial wave function.

---

* For a proof of the variation principle, see I. N. Levine, *Quantum Chemistry*, Allyn and Bacon, Inc., Boston, 1974, p. 158.

As we will see in the next section, the wave function of a system containing identical particles must be antisymmetric with respect to the exchange of two electrons.

8.17 PAULI EXCLUSION PRINCIPLE

In classical mechanics identical particles can be distinguished from each other by specifying the path of each particle. For example, if n identical balls are rolling on a table, we can describe ball 1 as the one following a specified path, ball 2 as the one following another specified path, etc.

In quantum mechanics identical particles cannot be distinguished from each other because the uncertainty principle shows that we cannot experimentally follow exact paths. This has an important consequence, because the wave function of a system of interacting identical particles also must not distinguish the particles.

As an example let us consider the wave function for the ground state of the helium atom in which electrons 1 and 2 are both in the $1s$ orbital but have different spin functions. The wave function $1s(1)1s(2)\alpha(1)\beta(2)$ is not satisfactory because it implies that electrons 1 and 2 can be distinguished from each other (i.e., it is electron 1 whose spin is "up" and electron 2 whose spin is "down"). The following wave functions do not have this problem.

$$\psi_1 = 2^{-1/2}[1s(1)1s(2)\alpha(1)\beta(2) + 1s(2)1s(1)\alpha(2)\beta(1)] \qquad (8.107a)$$

$$\psi_2 = 2^{-1/2}[1s(1)1s(2)\alpha(1)\beta(2) - 1s(2)1s(1)\alpha(2)\beta(1)] \qquad (8.107b)$$

The $2^{-1/2}$ is required for normalization. The first wave function is said to be *symmetric* because the interchange of the two electrons does not change the function. The second wave function is said to be *antisymmetric* because the interchange of the two electrons changes the sign of the function.

$$1s(1)1s(2)\alpha(1)\beta(2) - 1s(2)1s(1)\alpha(2)\beta(1)$$
$$= -[1s(2)1s(1)\alpha(2)\beta(1) - 1s(1)1s(2)\alpha(1)\beta(2)] \qquad (8.108)$$

Wave functions 8.107*a* and 8.107*b* give different spin densities for the electrons in a helium atom in its ground state. They cannot both be correct. Experiment shows that only the antisymmetric wave function (8.107*b*) correctly describes the helium atom. The generalization of this experience may be stated as follows. *The wave function for any system of electrons must be antisymmetric with respect to the interchange of any two electrons.* This is referred to as the *Pauli principle*, or the *exclusion principle*.

In 1929 Slater developed a mathematical method for constructing approximate wave functions satisfying the antisymmetry requirement that can be written as determinants. In a *Slater determinant* the elements in a given column involve the same spin-orbital, while elements in the same row involve the same electron. The zeroth-order helium wave function (equation 8.107*b*) can be written in the form

$$\psi = \frac{1}{\sqrt{2}}\begin{vmatrix} 1s(1)\alpha(1) & 1s(1)\beta(1) \\ 1s(2)\alpha(2) & 1s(2)\beta(2) \end{vmatrix} \qquad (8.109)$$

Another way of stating the Pauli principle in a way that is useful in chemistry is to say that no two electrons in an atom may have the same four quantum numbers n, l, m_l, and m_s.

The requirement of the Pauli principle that no two electrons in an atom or molecule have all quantum numbers the same is provided for automatically by the Slater determinant. If two rows or two columns of a determinant are identical, the determinant vanishes. The other useful properties of determinants are the following: (1) the interchange of two rows or columns changes the sign of the determinant; and (2) the multiple of a row or column can be added to any other row or column without changing the value of the determinant.

Particles with half integral spin ($s = \frac{1}{2}, \frac{3}{2}, \ldots$) all require antisymmetric wave functions and are referred to as fermions because they must obey a kind of statistics called "Fermi-Dirac" statistics. Particles with integral spin ($s = 0, 1, 2, \ldots$) all require symmetric wave functions and are referred to as bosons because they follow a different statistical law called "Bose-Einstein" statistics.

8.18 FIRST EXCITED STATE OF HELIUM

The first excited state of helium provides the opportunity to learn more about spin functions of atoms. As a first approximation we may consider that one electron is in the $1s$ orbital and the other electron is in the $2s$ orbital. In order to provide for the indistinguishability of the two electrons, we can write two wave functions.

$$\psi_1 = 2^{-1/2}[1s(1)2s(2) + 1s(2)2s(1)] \tag{8.110}$$

$$\psi_2 = 2^{-1/2}[1s(2)2s(2) - 1s(2)2s(1)] \tag{8.111}$$

These spatial wave functions need to be combined with spin functions. When two electrons are in different orbitals, they may have the same spin or opposite spin; thus there are four possible spin functions.

$$\alpha(1)\alpha(2) \qquad \beta(1)\beta(2) \qquad \alpha(1)\beta(2) \qquad \alpha(2)\beta(1) \tag{8.112}$$

However, to provide for the indistinguishability of electrons the spin functions need to be written

$$\begin{array}{cc} \alpha(1)\alpha(2) & \beta(1)\beta(2) \\ 2^{-1/2}[\alpha(1)\beta(2) + \alpha(2)\beta(1)] & 2^{-1/2}[\alpha(1)\beta(2) - \alpha(2)\beta(1)] \end{array} \tag{8.113}$$

Each of these four spin functions may be used to multiply each of the two spatial functions (8.110 and 8.111), but only the following four wave functions are antisymmetric and useful in representing the first excited state of helium, according to the Pauli principle.

$$\psi_1 = (\tfrac{1}{2})[1s(1)2s(2) + 1s(2)2s(1)][\alpha(1)\beta(2) - \alpha(2)\beta(1)] \quad \text{singlet} \tag{8.114}$$

$$\left.\begin{aligned} \psi_2 &= (\tfrac{1}{2})[1s(1)2s(2) - 1s(2)2s(1)][\alpha(1)\beta(2) + \alpha(2)\beta(1)] \quad &(8.115)\\ \psi_3 &= (2^{-1/2})[1s(1)2s(2) - 1s(2)2s(1)]\alpha(1)\alpha(2) \quad &(8.116)\\ \psi_4 &= (2^{-1/2})[1s(1)2s(2) - 1s(2)2s(1)]\beta(1)\beta(2) \quad &(8.117) \end{aligned}\right\} \text{triplet}$$

The spatial part of ψ_1 is symmetric, but the spin part is antisymmetric so that the whole wave function is antisymmetric. The spatial parts of ψ_2, ψ_3, and ψ_4 are antisymmetric, but the spin parts are symmetric so that these wave functions are antisymmetric. Since, in the absence of a magnetic field, the energy of each state depends only on the spatial part of the wave function, ψ_2, ψ_3, and ψ_4 are degenerate and form a spin *triplet* (multiplicity of three). ψ_1 has a different energy and is not degenerate; it forms a spin *singlet* (multiplicity of one).

In the ground state of helium the electrons are paired, and the resultant electron spin is zero. However, when one of the electrons is excited to the $2s$ level, the electrons may be paired, as in the singlet state represented by ψ_1, or unpaired, as in the triplet state represented by ψ_2, ψ_3, and ψ_4. The three components of the triplet have spins in the z direction of 0, +1, and −1, respectively. In the presence of an external magnetic field the singlet level is not split into components, but the triplet level is split into three components.

8.19 ELECTRONIC STRUCTURE OF ATOMS

The exact calculation of wave functions for many electron atoms becomes difficult because of the many electron-electron repulsions that we have ignored for simplicity up to now. In 1927 Hartree suggested what is now known as the self-consistent field (SCF) method for coping with this problem in the calculation of wave functions for atoms. In this method it is assumed that each electron moves in a spherically symmetrical potential due to the nucleus and the averaged fields of all the other electrons except the one being considered. The calculation is started with some reasonable approximate wave functions. The average potential of a single electron owing to all the other electrons is calculated from the approximate wave function, and then the Schrödinger equation is solved from this one electron using this average potential due to the other electrons and the nucleus.

This effective one-electron problem will have a whole series of eigenvalues and eigenfunctions (orbitals). From these orbitals, a new wave function satisfying the Pauli principle is constructed. This wave function should be a better approximation than the original one. It is then used to calculate an effective field, and this process is continued until the calculated set of wave functions differs insignificantly from the previous set. This set of wave functions is then said to be self-consistent. A considerable amount of computing is required to calculate the wave functions for a many-electron atom. The self-consistent field theory treatment of a given atom yields a series of atomic orbitals, each characterized by four quantum numbers and a characteristic energy. In contrast with the case for hydrogenlike atoms, the orbital energies depend both on the principal quantum number n and the orbital quantum number l.

In 1930 Fock (and Slater) pointed out that instead of spatial orbitals it is necessary to use spin orbitals and to take an antisymmetric linear combination of products of spin orbitals. When this is done, the calculation is referred to as a Hartree-Fock SCF calculation. The Hartree-Fock orbitals are taken as linear combinations of a complete set of known functions, called *basis functions*.

The Hartree-Fock calculated energies usually agree with experimental values to about 1%. Their method provides for the interactions between electrons in an average way, but does not provide for their instantaneous interactions. Since electrons tend to stay away from each other, we may speak of electron correlation. The *correlation energy* is the difference between the exact energy and the Hartree-Fock energy. This energy, which is of the order of an electron volt or more, is large enough to be a serious problem in the calculation of differences in energies as, for example, in calculating enthalpy changes for reaction.

The effects of instantaneous electron interaction may be provided for by including excited configurations in a trial wave function, using the variation approach. This method is referred to as configuration interaction. By including more configurations in the wave function used in the variation method, a representation of the true wave function may be approached more closely.

Contour diagrams of the electron densities of the first 10 elements calculated by the Hartree-Fock method are shown in Fig. 8.13. These diagrams show that in spite of the large differences in numbers of orbital electrons, these atoms are roughly all of the same size in their ground states because the increasing nuclear charge results in the electrons being held more tightly.

8.20 THE PERIODIC TABLE AND THE AUFBAU PRINCIPLE

The ground state of an atom is that in which the electrons are in the lowest possible energy level consistent with the Pauli principle. Thus the electron configurations of successive elements in the periodic table are obtained by putting electrons in the levels shown in Fig. 8.14, starting with the lowest level. Each level may hold two electrons of opposite spin. When there are several equivalent orbitals of the same energy, the following rules, due to Hund, may be used to decide how the electrons are distributed between orbitals.

1. If the number of electrons is equal to or smaller than the number of equivalent orbitals, the electrons are assigned to different orbitals.
2. If two (or more) electrons singly occupy two (or more) equivalent orbitals, their spins will be parallel in the ground state.

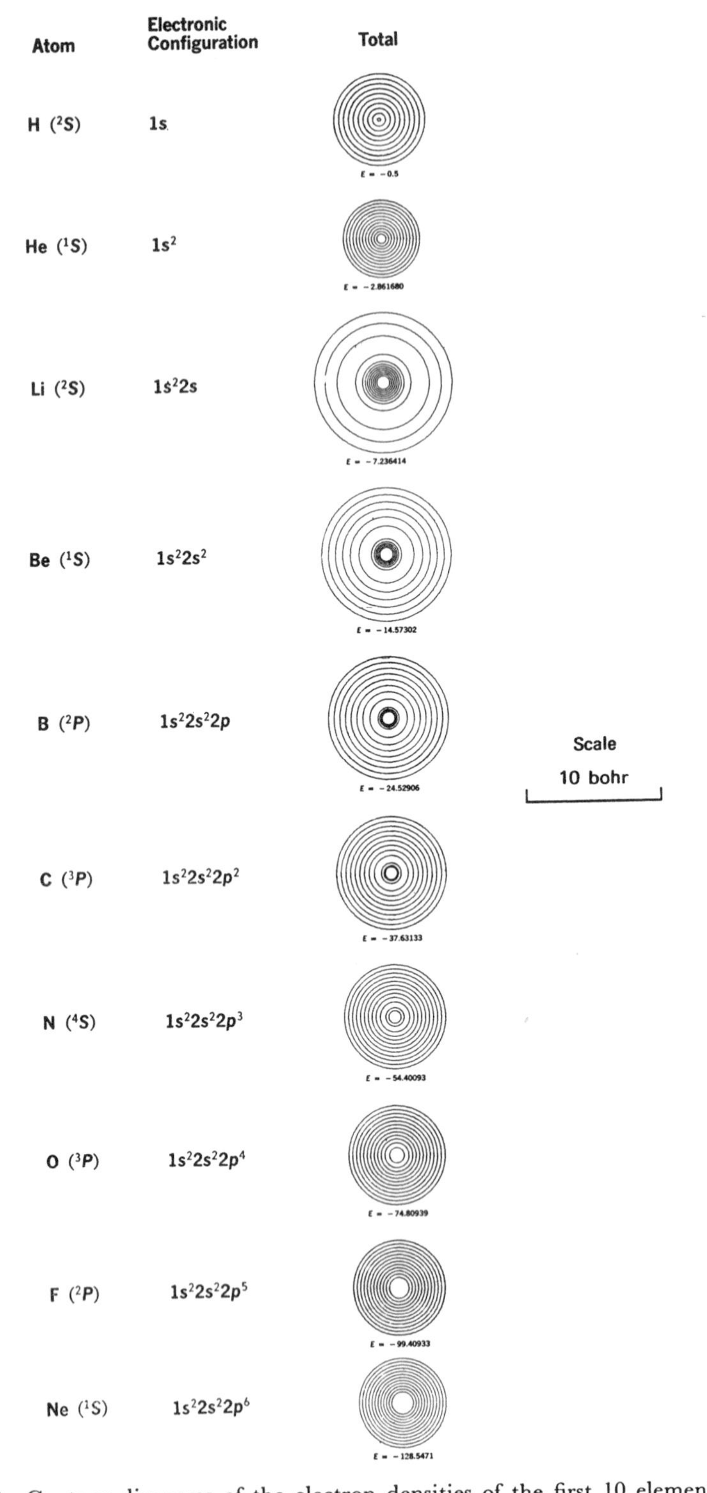

Fig. 8.13 Contour diagrams of the electron densities of the first 10 elements. (From *Atomic and Molecular Structure: 4 Wall Charts* by Arnold C. Wahl. Copyright © 1970 by McGraw-Hill, Inc. Used with permission of McGraw-Hill Book Co.) The outer contour line is at 4.9×10^{-4} electron/(bohr)^3 for each atom, and the inner contour line is at 1 electron/(bohr)^3, except for the hydrogen atom, where it represents 0.25 electron/(bohr)^3.

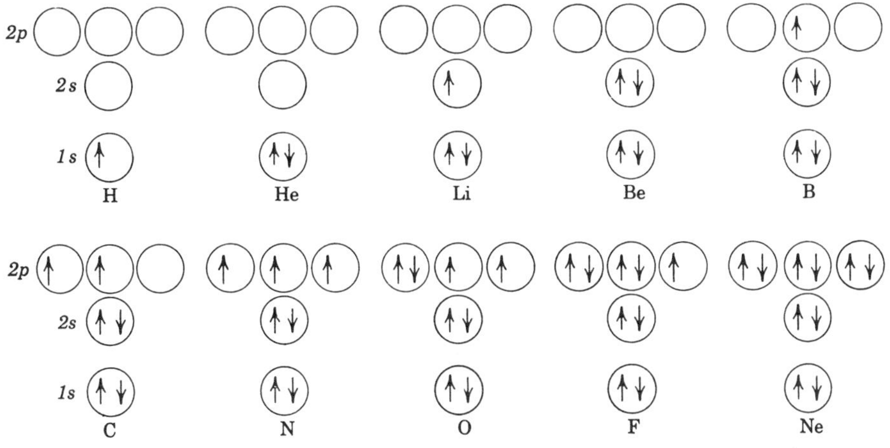

Fig. 8.14 Assignment of electrons to orbitals according to Hund's rules.

Assignment of electrons to orbitals in accord with Hund's rules tends to keep them as far apart as possible, on the average, and thus to minimize the contribution of interelectron repulsion to the energy.

Lithium has three electrons, two of which are in the $1s$ level and the third in the $2s$ level ($n = 2, l = 0$). Since the $2s$ electron is much further from the nucleus and is partially shielded from the $+3$ charge of the nucleus by the two inner electrons, the outer electron is easily removed to produce an ion with the electronic structure of helium. In going from lithium to neon, there are eight elements, ending with neon, which again has a stable structure with eight electrons with $n = 2$. The next element, sodium, has one $3s$ electron ($n = 3, l = 0$). This electron is shielded from the $+11$ nuclear charge by 10 inner electrons, so that it is loosely bound.

A description of the orbitals of an atom that are occupied is called the electron configuration. The electron configuration is given by using exponents to indicate the number of electrons in the $1s$, $2s$, $2p$, etc., orbitals. For example, hydrogen in the ground state is represented by $1s$, helium by $1s^2$, lithium by $1s^2\ 2s$, boron by $1s^2\ 2s^2\ 2p$, and sodium by $1s^2\ 2s^2\ 2p^6\ 3s$.

The form of the periodic table shown in Fig. 8.15* shows how the various orbitals are filled in forming the elements. In the hydrogen atom the energy of the electron depends only on its principal quantum number n but, in many electron atoms, the inner electrons screen the outer electrons from the full nuclear charge. As a result of the screening effect s orbitals have a lower energy than p orbitals, which have a lower energy than d orbitals, for a given value of n. The energies of the p orbitals are higher than those of the corresponding s orbitals because the p electrons spend most of their time further from the nucleus and are not attracted by the nucleus as strongly. This separation of levels, strictly speaking, differs from

---

* H. C. Longuet-Higgins, *J. Chem. Ed.*, **34**, 30 (1957).

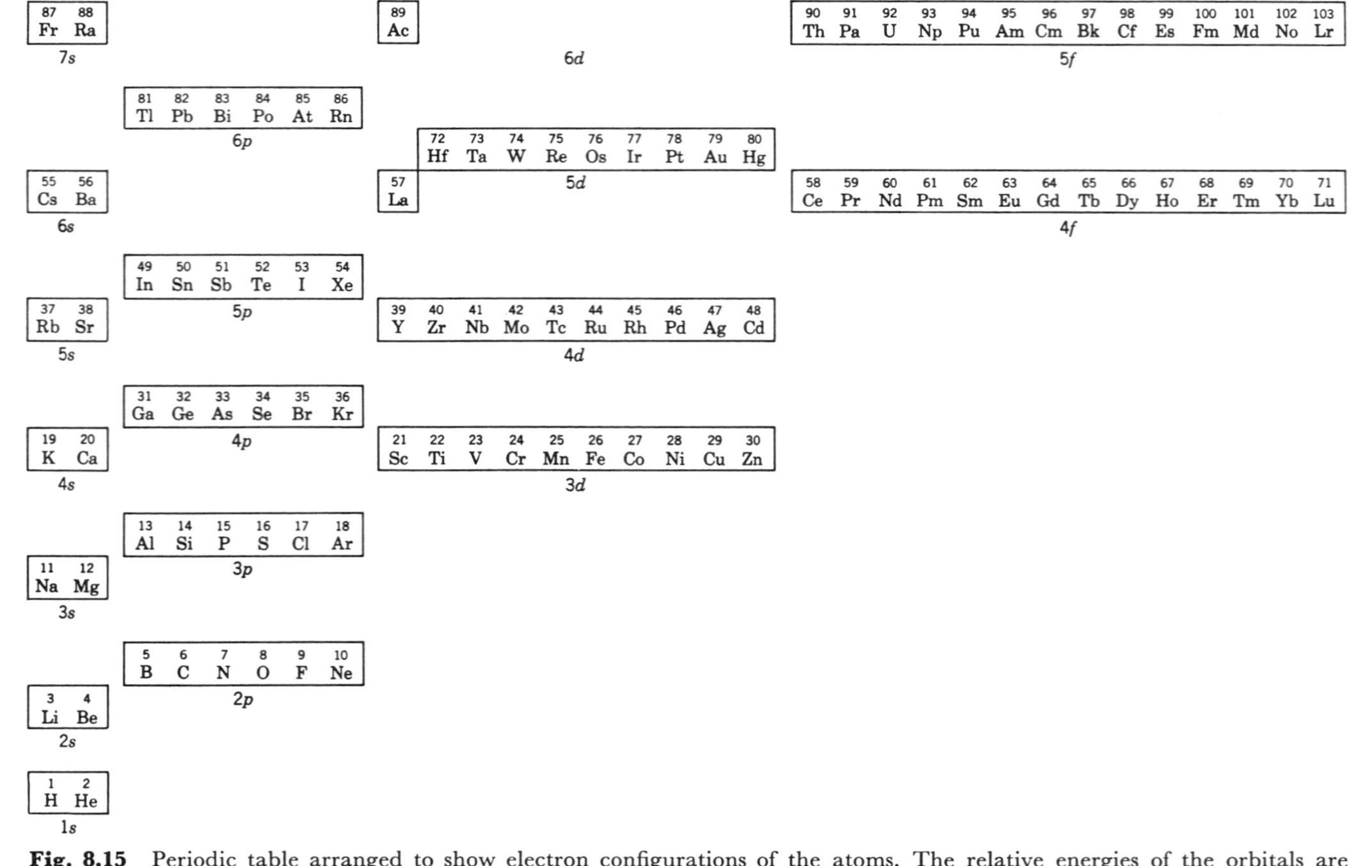

Fig. 8.15 Periodic table arranged to show electron configurations of the atoms. The relative energies of the orbitals are shown by their vertical positions. We can think of the electrons being fed successively into higher and higher energy orbitals.

atom to atom. However, the relative energies are shown approximately by the vertical heights in Fig. 8.15.

At lanthanum La(57) an electron goes into the 5*d* orbital, but in the next element, cerium Ce(58), an electron goes into the 4*f* orbital. This starts the lanthanide series (elements 57–71) in which 14 electrons are put into the 4*f* orbital. Because the chemical properties are largely determined by the outer valence electrons, all these elements are much alike. At actinium Ac(89) a similar thing happens; the 6*d* orbital is started but is not completed. Elements 89 to 103 form the actinide series in which the 5*f* orbitals is filled up.

8.21 IONIZATION POTENTIAL AND ELECTRON AFFINITY

The ionization potential is the voltage corresponding to the minimum energy required to remove an electron completely from a gaseous atom or molecule. The ionization potential may be determined by bombarding a gas with electrons that have been accelerated by a difference in electric potential between a grid and the hot filament that emits the electrons. If the accelerated electrons have insufficient kinetic energy to cause a shift from one energy level to another in the atoms or molecules they strike, the collisions are said to be elastic. As the potential is increased, the accelerated electrons gain sufficient energy to excite an orbital electron from one energy level to the next higher level. Light may be emitted when the electron returns to an empty lower level. This is known as the Auger effect. As the potential is further increased, new spectral lines appear. The potentials required to cause emission of light are called *resonance potentials*. The relation between the accelerating potential E and the frequency of the light emitted is

$$Ee = h\nu \tag{8.118}$$

where e is the charge on the electron.

If the accelerating potential exceeds the ionization potential, an electron may be driven from the atom or molecule, and this potential is called an *ionization potential*. The ionization potential of an atom or ion may be calculated from spectroscopic data, since this potential is given by the convergence limit of the atomic terms (Section 8.2). The singly charged positive ion produced may be ionized further by bombarding with electrons of still higher energy, that is, the second, third, . . ., ionization potentials correspond with the ejection of the second, third, . . ., electrons.

A plot of the first ionization potentials of gaseous atoms versus atomic number is given in Fig. 8.16. The ionization potentials change in a periodic way because of the progressive filling up of shells with electrons. The principal maxima in this plot are given by the inert gases, and the principal minima by the alkali metal atoms.

The alkali metal atoms are easily ionized, since they have a single electron in the outer orbital and the effective nuclear charge is low. The attraction of the nucleus for the outermost electron of the alkali metal atoms is quite effectively shielded by the electrons of the inner orbits. In the series lithium, sodium, potassium, rubidium,

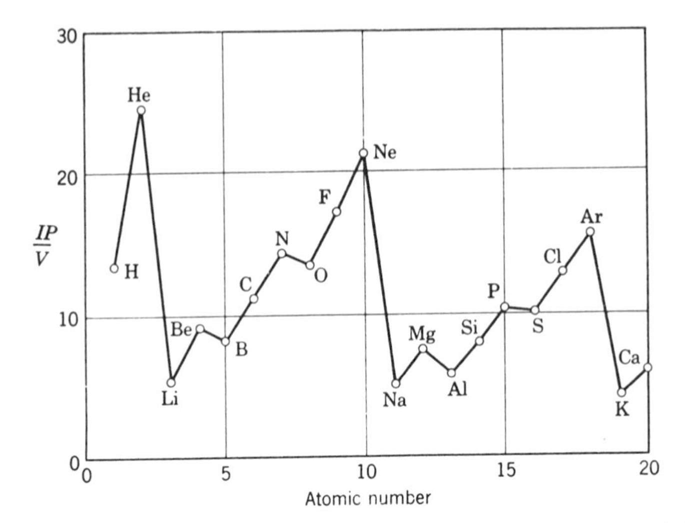

Fig. 8.16 First ionization potentials of gaseous atoms early in the periodic table.

and cesium the ionization potential decreases because of the increase in size of the outer orbit containing a single electron.

In contrast, the ionization potentials of the halogens are almost as great as those of the inert gases. The electrons in the outer orbits of the halogen atoms are shielded from the nuclear charge mainly by the electrons in inner orbits, since the electrons in the outer orbits are all approximately the same distance from the nucleus. A direct result of this incomplete shielding of the nuclear charge, as far as electrons in the outer orbit are concerned, is the fact that the halogen atoms readily take on an additional electron to form negative ions.

The electron affinity *EA* is defined as the energy released in the process

$$A + e \rightarrow A^- \tag{8.119}$$

If we consider the reverse of this process we can see that the electron affinity of A is the ionization potential of A^-. The electron affinity increases with increasing atomic number in a row of the periodic table; the electron affinity of Li is 0.6 eV, and the electron affinity of F is 3.45 eV. The affinities of chlorine, bromide, and iodine for an additional electron are 3.71, 3.49, and 3.19 eV. Oxygen atoms and sulfur atoms also have affinities for an additional electron (3.07 and 2.8 eV, respectively).

References

J. M. Anderson, *Introduction to Quantum Chemistry*, W. A. Benjamin, Inc., New York, 1969.
J. C. Davis, Jr., *Advanced Physical Chemistry*, The Ronald Press Co., New York, 1965.
C. F. Fischer, *The Hartree-Fock Method for Atoms*, Wiley, New York, 1977.
D. V. George, *Principles of Quantum Chemistry*, Pergamon Press, Inc., New York, 1972.
M. W. Hanna, *Quantum Mechanics in Chemistry*, W. A. Benjamin, Inc., New York, 1969.

G. Herzberg, *Atomic Spectra and Atomic Structure*, Prentice-Hall, Inc., Englewood Cliffs, N.J., 1937.
M. Karplus and R. N. Porter, *Atoms and Molecules*, W. A. Benjamin, Inc., New York, 1970.
I. N. Levine, *Quantum Chemistry*, Allyn and Bacon, Boston, 1974.
L. Pauling and E. B. Wilson, *Introduction to Quantum Mechanics*, McGraw-Hill Book Co., New York, 1935.

Problems

8.1 A hollow box with an opening of 1 cm^2 area is heated electrically. (*a*) What is the total energy emitted per second at 800 K? (*b*) How much energy is emitted per second if the temperature is 1600 K? (*c*) How long would it take the radiant energy emitted at this temperature, 1600 K, to melt 1000 g of ice?
Ans. (*a*) 2.33×10^4 J m^{-2} s^{-1}, (*b*) 3.73×10^5 J m^{-2} s^{-1}, (*c*) 8950 s.

8.2 Calculate the ratio of the intensities of light of 500-nm wavelength from cavities of 100 and 5000 K. *Ans.* 1.00×10^{-10}.

8.3 Calculate the wavelengths (in micrometers) of the first three lines of the Paschen series for atomic hydrogen. *Ans.* 1.8756, 1.2822, 1.0941 μm.

8.4 In the Balmer series for atomic hydrogen, what is the wavelength of the series limit? *Ans.* 364.7 nm.

8.5 Calculate the wavelength of light emitted when an electron falls from the $n = 100$ orbit to the $n = 99$ orbit of the hydrogen atom. Such species are known as high Rydberg atoms. They are detected in astronomy and are more and more studies in the laboratory. *Ans.* 0.0449 cm.

8.6 What potential difference is required to accelerate a singly charged gas ion in a vacuum so that it has (*a*) a kinetic energy equal to that of an average gas molecule at 25 °C, and (*b*) an energy equivalent to 83.7 kJ mol^{-1}? *Ans.* (*a*) 0.0385, (*b*) 0.867 V.

8.7 Calculate the velocity of an electron that has been accelerated by a potential difference of 1000 V. *Ans.* 1.87×10^7 m s^{-1}.

8.8 Electrons are accelerated by a 1000-V potential drop. (*a*) Calculate the de Broglie wavelength. (*b*) Calculate the wavelength of the X rays that could be produced when these electrons strike a solid. *Ans.* (*a*) 0.0387, (*b*) 1.24 nm.

8.9 Calculate the approximate quantum number corresponding with the translational energy of a 1-g bullet fired with a velocity of 300 m s^{-1} at a target 100 m away. What is the de Broglie wavelength? *Ans.* 9.06×10^{34}, 2.21×10^{-33} m.

8.10 (*a*) Calculate the energy levels for $n = 1$ and $n = 2$ for an electron in a potential well of width 0.5 nm with infinite barriers on either side. The energies should be expressed in kJ mol^{-1}. (*b*) If an electron makes a transition from $n = 2$ to $n = 1$, what will be the wavelength of the radiation emitted? *Ans.* (*a*) 145.1, 580.4 kJ mol^{-1}. (*b*) 274.7 nm.

8.11 Calculate the degeneracies of the first three levels for a particle in a cubical box. *Ans.* 1, 3, 3.

8.12 Show that the function $\psi = 8e^{5x}$ is an eigenfunction of the operator d/dx. What is the eigenvalue? *Ans.* 5.

8.13 Show that a simple harmonic oscillator in its ground state obeys the uncertainty principle by computing

$$\Delta x = \sqrt{\langle x^2 \rangle - \langle x \rangle^2} \quad \text{and} \quad \Delta p = \sqrt{\langle p^2 \rangle - \langle p \rangle^2}$$

8.14 How many kJ mol^{-1} are required to excite a hydrogen atom from $n = 1$ to $n = 2$? How many times larger than the translational energy of a hydrogen atom at room temperature is this? *Ans.* 983 kJ mol^{-1}, 266.

8.15 What are the wavelengths of the first line in the Balmer series for H (atomic mass 1.007 825) and D (atomic mass 2.014 10)? *Ans.* 656.4696, 656.2910 nm.

8.16 Calculate the Bohr radius with the reduced mass of the hydrogen atom. *Ans.* 0.052 946 5 nm.

8.17 Calculate the average distance between the electron and nucleus of a hydrogenlike atom in the 1*s* state. *Ans.* $\frac{3}{2}\frac{a_0}{Z}$.

8.18 Show that for a 1*s* orbital of a hydrogenlike atom the most probable distance from proton to electron is a_0/Z.

8.19 For the wave function

$$\psi = \begin{vmatrix} \psi_A(1) & \psi_A(2) \\ \psi_B(1) & \psi_B(2) \end{vmatrix}$$

show that (*a*) the interchange of two columns changes the sign of the wave function, (*b*) the interchange of two rows changes the sign of the wave function, and (*c*) the two electrons cannot have the same spin orbital. *Ans.* (*c*) $\psi = 0$.

8.20 What are the electron configurations for H^-, Li^+, O^{2-}, F^-, Na^+, and Mg^{2+}?
Ans. $1s^2$, $1s^2$, $1s^2\,2s^2\,2p^6$, $1s^2\,2s^2\,2p^6$, $1s^2\,2s^2\,2p^6$, $1s^2\,2s^2\,2p^6$.

8.21 Apply Hund's rules to obtain the electron configurations for Si, P, S, Cl, Ar.
Ans. Si $1s^2\,2s^2\,2p^6\,3s^2\,3p^2$ Cl $1s^2\,2s^2\,2p^6\,3s^2\,3p^5$
P $1s^2\,2s^2\,2p^6\,3s^2\,3p^3$ Ar $1s^2\,2s^2\,2p^6\,3s^2\,3p^6$
S $1s^2\,2s^2\,2p^6\,3s^2\,3p^4$

8.22 Calculate the ionization potential for H(*g*) from the energy given in Example 8.5. *Ans.* 13.598 40 V.

8.23 The first ionization potential for atomic lithium is 5.39 V ($Li = Li^+ + e$). The second ionization potential is 75.62 V ($Li^+ = Li^{+2} + e$). Calculate the wavelengths for the convergence limits indicated by these potentials. *Ans.* 230, 16.4 nm.

8.24 If 4.184×10^4 J is lost per minute by radiation from the door of an electrically heated furnace at 900 K, how many more watts of electricity must be applied to offset the losses due to radiation if the furnace is heated to 1100 K?

8.25 Plot the intensity of cavity radiation versus wavelength for a temperature of 10,000 K. What is the wavelength of maximum emission?

8.26 An electric heater of 10 cm^2 has a temperature of 800 K. How many joules of radiant heat are emitted per minute, if it is assumed that the heater is a cavity?

8.27 Calculate the frequency and the wavelength in nanometers for the line in the Paschen series of the hydrogen spectrum that is due to a transition from the sixth quantum level to the third.

8.28 There is a Brackett series in the hydrogen spectrum where $n_1 = 4$. Calculate the wavelengths, in nanometers, of the first two lines of this series.

8.29 Calculate the velocity of an electron that has been accelerated by a potential difference of 1.00 V.

8.30 Calculate the de Broglie wavelength for thermal neutrons at a temperature of 100° C.

8.31 Calculate the de Broglie wavelength of electrons that have been accelerated by 50,000 V.

8.32 Calculate the de Broglie wavelength of a hydrogen atom with a translationa energy corresponding to room temperature.

8.33 For a hydrogen atom in a one-dimensional box calculate the value of the quantum number of the energy level for which the energy is equal to $\frac{3}{2}kT$ at 25 °C (*a*) for a box 1 nm long, and (*b*) for a box 1 cm long.

8.34 Derive the expression for the energy of a particle in a one-dimensional box using the de Broglie formula.

8.35 Calculate the first three energy levels in kilojoules per mole for an electron in a potential well 0.5 nm in width with infinitely high potential outside.

8.36 For a particle in a cubical box calculate $E(8ma^2/h^2)$ for the first 10 states. What is the degeneracy for each energy level?

8.37 Show that the function $\psi = xe^{-ax^2}$ is an eigenfunction of the operator $d^2/dx^2 - 4a^2x^2$. What is the eigenvalue?

8.38 Show that the function $\psi = Ke^{r/k}$ is an eigenfunction of the operator d/dr. What is the eigenvalue?

8.39 Calculate $\langle r \rangle$ for a $2s$ electron in a hydrogen atom.

Given:
$$\int_0^\infty x^n e^{a-x}\, dx = \frac{n!}{a^{n+1}}$$

for $n > -1$, $a > 0$.

8.40 For a simple harmonic oscillator in its ground state, show that one obtains the classical results for $\langle x \rangle$ and $\langle p \rangle$.

8.41 Sketch the possible orientations for angular momentum vectors for $l = 3/2$.

8.42 Positronium consists of an electron and a positron (the mass of a positron is the same as that of an electron). (*a*) Calculate the wavelength of the radiation emitted when the electron falls from the orbital $n = 2$ to $n = 1$. (*b*) Calculate the ionization potential.

8.43 Calculate the value of the Rydberg constant R for hydrogen and compare it with the experimentally determined value.

8.44 For the H atom show that ψ_{1s} and ψ_{2s} are orthogonal. (See the integral in Problem 8.39.)

8.45 Show that the wave function for a $1s$ hydrogenlike orbital is normalized. (See the integral in Problem 8.39.)

8.46 Use the variation method to obtain an upper bound to the ground state energy of a particle in a one-dimensional box and compare the result with the true value given in equation 8.22. A trial variation function that satisfies the boundary conditions that $\psi = 0$ at $x = 0$ and $x = l$ is

$$\psi = x(l - x)$$

8.47 Write the Slater determinant for the ground state of the Li atom $1s^2\,2s$.

8.48 Show that the following wave function for the hydrogen atom is antisymmetric to the interchange of the two electrons

$$\begin{vmatrix} 1s\alpha(1) & 1s\beta(2) \\ 1s\alpha(2) & 1s\beta(2) \end{vmatrix}$$

8.49 In a hydrogen atom the $2s$ and $2p$ orbitals have the same energy. However, in a boron atom, the $2s$ electrons have a lower energy than the $2p$. Explain this in terms of the shape of the orbitals.

8.50 Considering only the first 18 elements of the periodic table, list those whose outer electrons are in spherically symmetrical orbits and those whose outer electrons are not.

8.51 Calculate the ionization potentials for He^+, Li^{2+}, Be^{3+}, B^{4+}, and C^{5+}.

8.52 Calculate the ionization potential for He^+ and the wavelengths of the first two lines of the Balmer series.

8.53 The first ionization potential of atomic hydrogen is 13.60 V. Calculate the wavelength of the light produced when a free electron without kinetic energy returns to the inner orbit.

8.54 Using Table A.1, calculate $\Delta H°$ at 25 °C and absolute zero for $H(g) = H^+(g) + e$. Express these enthalpy changes in electron volts and compare them with the values in this chapter.

CHAPTER 9
SYMMETRY

Ideas about symmetry are of great importance in connection with both theoretical and experimental studies of atomic and molecular structure. The basic principles of symmetry are applied in quantum mechanics, spectroscopy, and structural determinations by X-ray, neutron, and electron diffraction. Nature exhibits a great deal of symmetry, and this is especially evident when we examine molecules in their equilibrium configurations. By equilibrium configuration we refer to that with the atoms fixed in their mean positions. When symmetry is present, certain calculations are simplified if symmetry is taken into account. Aspects of symmetry also determine whether a molecule can be optically active or whether it may have a dipole moment. Single molecules, unlike crystalline solids (see Chapter 19), are not restricted in the kinds of symmetry that they may possess.

9.1 SYMMETRY OPERATIONS AND SYMMETRY ELEMENTS

A symmetry operation is the movement of an object such that the final appearance is indistinguishable from the initial appearance. Each symmetry operation is represented by an operator. The symmetry operators $\hat{\imath}$, $\hat{C}_n$, $\hat{\sigma}_h$, $\hat{\sigma}_v$, $\hat{S}_n$, and $\hat{E}$ are described briefly in Table 9.1.

A symmetry element is a point, line, or plane with respect to which a symmetry operation is carried out. The symbols for the symmetry elements are i, C_n, σ_h, σ_v, S_n, and E.

9.2 THE INVERSION OPERATION AND THE CENTER OF SYMMETRY

A molecule has a center of symmetry i if a straight line from any atom projected through the center of symmetry of the molecule encounters an equivalent atom equidistant from the center. The center of symmetry (or inversion center) is the symmetry element, and the operation is inversion through the center by which

Table 9.1 Symmetry Elements and Associated Operations

Symbol for Element	Element	Symbol for Operator	Operation
i	*Center of symmetry* (or inversion center).	$\hat{\imath}$	is projection through the center of symmetry to an equal distance on the other side from the center.
C_n	*Proper rotation axis.*	$\hat{C}_n$	is counterclockwise rotation about the C_n axis by $2\pi/n$ (or $360°/n$).
σ_h	*Horizontal symmetry plane* perpendicular to principal C_n axis (i.e., proper axis of highest symmetry).	$\hat{\sigma}_h$	is reflection across the plane of symmetry.
σ_v	*Vertical symmetry plane* containing the principal C_n axis.	$\hat{\sigma}_v$	is reflection across the plane of symmetry.
σ_d	*Diagonal symmetry plane* containing the principal C_n axis; the plane bisects the angle formed by two horizontal C_2 axes that are perpendicular to the principal axis C_n of highest symmetry.	$\hat{\sigma}_d$	is reflection across the plane of symmetry.
S_n	*Improper rotation axis* (also referred to as a *rotation-reflection axis or alternating axis*).	$\hat{S}_n$	is counterclockwise rotation about the S_n axis by $2\pi/n$ followed by reflection in a plane perpendicular to the axis (viz., the *combined operation* of a C_n rotation followed by reflection across a σ_h mirror plane).
E	*Identity element.*	$\hat{E}$	is the operator that leaves the system unchanged.

one-half of the molecule can be generated from the other half. The action of the inversion operation is to transform the coordinates (x, y, z) into their respective negatives $(-x, -y, -z)$. The inversion operation may be represented by

$$\hat{\imath}\cdot\begin{bmatrix} x \\ y \\ z \end{bmatrix} = \begin{bmatrix} -x \\ -y \\ -z \end{bmatrix} \tag{9.1}$$

If the inversion operator is applied twice we obtain the original configuration, as shown by

$$\hat{\imath}\cdot\hat{\imath}\cdot\begin{bmatrix} x \\ y \\ z \end{bmatrix} = \hat{\imath}\cdot\begin{bmatrix} -x \\ -y \\ -z \end{bmatrix} = \begin{bmatrix} x \\ y \\ z \end{bmatrix} \tag{9.2}$$

Thus the successive application of $\hat{\imath}$ an even number of times produces the identity operator $\hat{E}$. It is convenient to include $\hat{E}$ as an operator even though it does not change the configuration of a molecule.

Since $\hat{\imath}^k = \hat{\imath}$ when k is odd, and $\hat{\imath}^k = \hat{E}$ when k is even, the center of symmetry generates only *one distinct* operation. In molecules with a center of symmetry, the atoms may be thought of as occurring in centrosymmetric pairs with the exception of an unshifted atom if one lies at the center of symmetry. Centrosymmetric molecules include C_6H_6 (Fig. 9.1*a*), SF_6 (Fig. 9.1*b*), the *staggered* conformation of C_2H_6 (Fig. 9.3*b*), CO_2, and C_2H_4.

9.3 THE ROTATION OPERATION AND THE SYMMETRY AXIS

A symmetry axis is a line about which rotation through an angle $2\pi/n$ radians brings a structure into coincidence with itself. The rotation operator is represented by $\hat{C}_n$, where n is referred to as the order of the rotation, and the rotation is conventionally taken as positive in the counterclockwise direction. If the z axis is a twofold rotation axis, the action of the $\hat{C}_2$ rotation operator is to transform the coordinate (x, y, z) to $(-x, -y, z)$. The $\hat{C}_2$ rotation operation may be represented by

$$\hat{C}_2\cdot\begin{bmatrix} x \\ y \\ z \end{bmatrix} = \begin{bmatrix} -x \\ -y \\ z \end{bmatrix} \tag{9.3}$$

As shown in Fig. 9.2*a*, a water molecule has a C_2 axis passing through the oxygen and bisecting the angle between the O—H bonds. The ammonia molecule NH_3 (Fig. 9.2*b*) has a C_3 axis passing through the nitrogen. The benzene molecule (Fig. 9.1*a*) has a C_6 axis perpendicular to the plane of the ring and six C_2 axes lying in the plane of the ring. Any linear molecule, such as HCl, has a C_∞ axis (Fig. 9.2*c*), because the appearance of the molecule is not changed by a rotation of any angle (infinity in number) about the internuclear axis.

If the $\hat{C}_2$ operator is applied twice in succession, the identity operator is obtained.

$$\hat{C}_2\hat{C}_2 = \hat{C}_2{}^2 = \hat{E}$$

If the $\hat{C}_3$ operator is applied twice in succession, a 240° rotation is obtained; if it is applied three times, the identity operator is obtained.

$$\hat{C}_3\hat{C}_3 = \hat{C}_3{}^2 \qquad \hat{C}_3\hat{C}_3\hat{C}_3 = \hat{C}_3{}^3 = \hat{E}$$

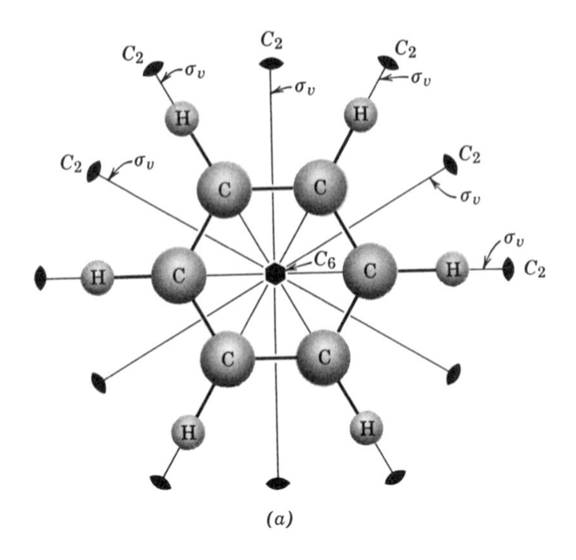

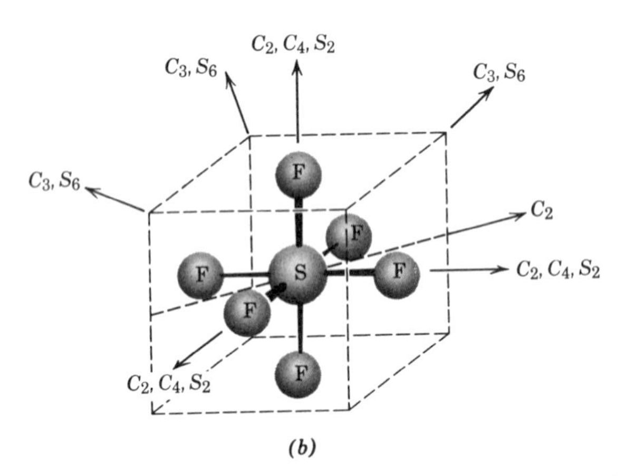

Fig. 9.1 Molecules with a center of symmetry. (*a*) C_6H_6 molecule with D_{6h} point-group symmetry. (*b*) SF_6 molecule with O_h point-group symmetry. This centrosymmetric molecule possesses three equivalent C_4 axes at right angles to one another. Three S_4 and three C_2 axes are coincident with the three C_4 axes. Four C_3 axes (and four coincident S_6 axes) are located along the four body-diagonals of the cube. A C_2 axis is labeled with ⬬, and a C_6 axis is labeled with ⬢.

The $\hat{C}_3{}^2$ operator is a new operator; thus the C_3 element of symmetry generates two operators, $\hat{C}_3$ and $\hat{C}_3{}^2$.

The operations generated by a fourfold axis are

$$\begin{matrix} \hat{C}_4{}^1 & \hat{C}_4{}^2 & \hat{C}_4{}^3 & \hat{C}_4{}^4 \\ & \hat{C}_2{}^1 & & \hat{E} \end{matrix}$$

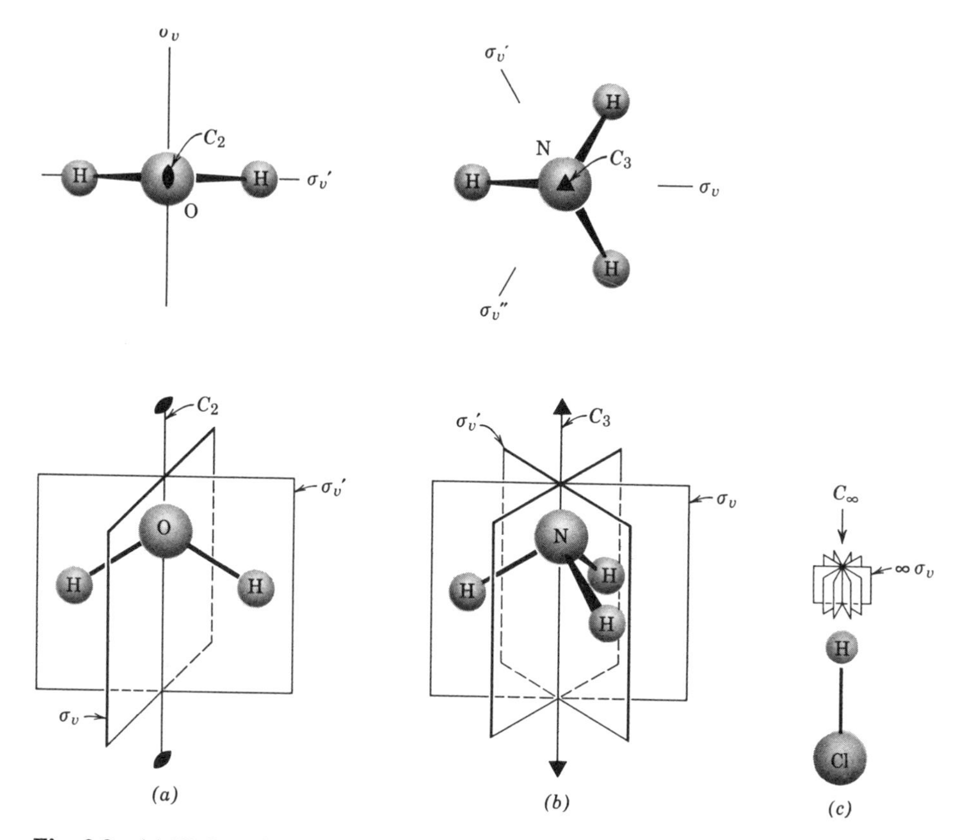

Fig. 9.2 (*a*) H_2O molecule with C_{2v} point-group symmetry. (*b*) NH_3 molecule with C_{3v} point-group symmetry. (*c*) HCl molecule with $C_{\infty v}$ point-group symmetry.

where equivalent operations are listed below. Since a C_2 axis is always coincident with a C_4 axis, only the operations $\hat{C}_4^{\,1}$ and $\hat{C}_4^{\,3}$ are *distinct* operations of the C_4 axis.

Example 9.1 How many distinct operators are implied by a C_6 axis?

Ans.	$\hat{C}_6^{\,1}$	$\hat{C}_6^{\,2}$	$\hat{C}_6^{\,3}$	$\hat{C}_6^{\,4}$	$\hat{C}_6^{\,5}$	$\hat{C}_6^{\,6}$
		$\hat{C}_3^{\,1}$	$\hat{C}_2^{\,1}$	$\hat{C}_3^{\,2}$		$\hat{E}$

Thus two operators ($\hat{C}_6^{\,1}$ and $\hat{C}_6^{\,5}$) are characteristic only of a C_6 axis and cannot be represented in any other way.

In discussing symmetry operations, it is convenient to orient the molecules in a right-hand Cartesian coordinate system. The thumb, index, and middle fingers of the right hand are pointed in three mutually perpendicular directions, and these are taken as the x, y, and z directions, respectively. The center of mass of the molecule under consideration is located at the origin of the Cartesian coordinate system, and its *principal axis* is aligned with the z axis. The *principal axis* is defined as the C_n axis with highest order n; if there are several rotational axes of the same highest symmetry (e.g., three twofold axes at right angles to one another), the z axis is taken along the one passing through the greatest number of atoms.

9.4 THE REFLECTION OPERATION AND THE SYMMETRY PLANE

If reflection of all of the nuclei through a plane in a molecule gives a configuration physically indistinguishable from the original one, the molecule is said to have a symmetry plane. The symmetry plane is represented by σ, and the reflection operator is represented by $\hat{\sigma}$. Since the operator $\hat{\sigma}$ gives a configuration equivalent to the original and since the application of the same $\hat{\sigma}$ twice to a molecule produces its original configuration, it follows that a symmetry plane generates only *one distinct* operation in that $\hat{\sigma}^k = \hat{\sigma}$ when k is odd, and $\hat{\sigma}^k = \hat{E}$ when k is even.

If the xz plane is a symmetry plane the reflection operation $\hat{\sigma}$ may be represented by

$$\hat{\sigma} \cdot \begin{bmatrix} x \\ y \\ z \end{bmatrix} = \begin{bmatrix} x \\ -y \\ z \end{bmatrix} \tag{9.4}$$

A symmetry plane (and corresponding operation of reflection) perpendicular to the direction of the principal C_n axis (i.e., the normal of the symmetry plane is coincident with the C_n axis of highest order n) is called a *horizontal* symmetry plane and denoted as σ_h. Molecules with a horizontal symmetry plane include C_6H_6 (Fig. 9.1a) (for which σ_h is perpendicular to the C_6 axis and contains all the atoms of this planar molecule) and the *eclipsed* conformation of ethane (Fig. 9.3a), which has a σ_h perpendicular to a C_3 principal axis.

Symmetry planes that contain the principal C_n axis are called *vertical* symmetry planes and *generally* are symbolized as σ_v. The particular symmetry planes that bisect the angles formed by pairs of horizontal C_2 axes are designated as σ_d. As is evident from Fig. 9.2a, a H_2O molecule has two *vertical* symmetry planes, σ_v and σ_v', which are perpendicular to each other. One of these symmetry planes (σ_v) comprises the plane of the molecule, and the other (σ_v') is perpendicular to it. The twofold axis lies in the intersection of the two symmetry planes. Ammonia has three σ_v containing the C_3 axis, and benzene has six σ_v containing the C_6 principal axis. The linear molecule HCl has an *infinite* number of *vertical* symmetry planes of type σ_v, all of which include the C_∞ rotational axis. Homonuclear diatomic molecules such as H_2 or Cl_2 have a σ_h in addition. Figure 9.3b shows the *staggered* conformation of

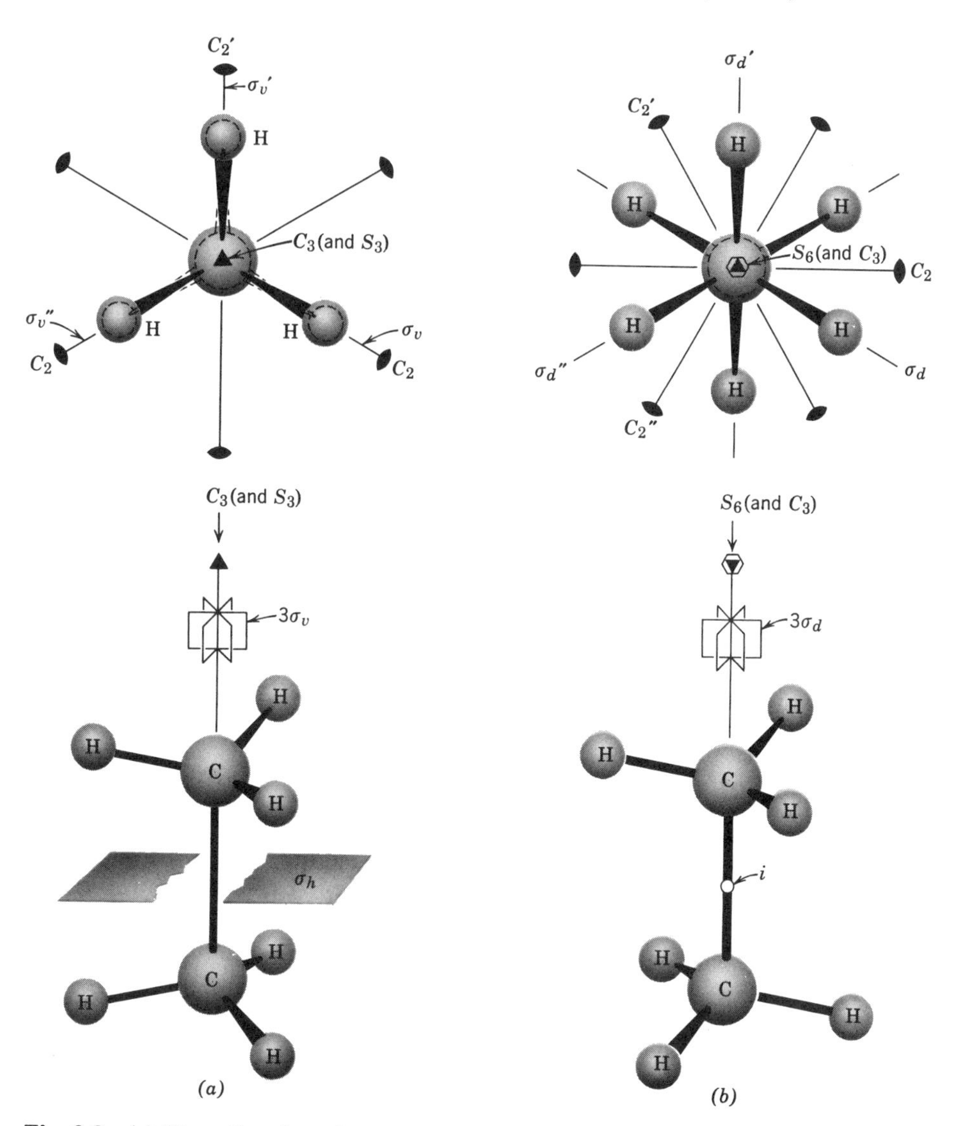

Fig. 9.3 (*a*) The eclipsed conformation of ethane, C_2H_6, with D_{3h} point symmetry. The view along the C_3 (and S_3) axis of the C—C bond is shown at the top. (*b*) The staggered conformation of ethane, C_2H_6, with D_{3d} point symmetry. The view along S_6 (and C_3) axis of the C—C bond is shown at the top.

ethane to possess three vertical σ_d which contain the principal C_3 axis and which bisect the three horizontal C_2 axes. In the *eclipsed* conformation of ethane (Fig. 9.3*a*), however, each of the three vertical symmetry planes contains one of the three horizontal C_2 axes (as well as the C_3 principal axis). Consequently, these symmetry planes are called σ_v.

9.5 THE OPERATION OF IMPROPER ROTATION AND THE IMPROPER AXIS

The operation of improper rotation consists of a rotation by $2\pi/n$ radians about an axis followed by reflection in a plane perpendicular to the axis. Thus the improper rotation operator $\hat{S}_n$ is the product of two operators.

$$\hat{S}_n = \hat{\sigma}\hat{C}_n \tag{9.5}$$

This means that the operators $\hat{C}_n$ and $\hat{\sigma}$ are applied successively. The improper axis is represented by S_n.

An S_1 element is equivalent to a *plane of symmetry* (σ) as the operation involving rotation of 360° followed by reflection across the symmetry plane perpendicular to the axis of rotation can be simply represented in terms of reflection across the mirror plane.

An S_2 element is equivalent to a *center of symmetry* (i), since the S_2 operation consisting of a counterclockwise rotation about an axis of $2\pi/2$ or 180° followed by reflection across a horizontal symmetry plane perpendicular to this axis yields the same configuration as an inversion through a center of symmetry located at the intersection of the axis of rotation and symmetry plane.

An S_3 element has six operators.

$\hat{S}_3^1$	$\hat{S}_3^2$	$\hat{S}_3^3$	$\hat{S}_3^4$	$\hat{S}_3^5$	$\hat{S}_3^6$
	$\hat{C}_3^2$	$\hat{\sigma}_h$	$\hat{C}_3^1$		$\hat{E}$

The equivalent operators listed below show that an S_3 element introduces only two distinct operators, $\hat{S}_3^1$ and $\hat{S}_3^5$. Thus an S_3 axis implies a C_3 axis and a horizontal symmetry plane σ_h. The *eclipsed* conformation of ethane has an S_3 axis, as shown in Fig. 9.3*a*.

An S_4 axis has four associated operations.

$\hat{S}_4^1$	$\hat{S}_4^2$	$\hat{S}_4^3$	$\hat{S}_4^4$
	$\hat{C}_2^1$		$\hat{E}$

Thus an S_4 axis implies a C_2 axis. This is illustrated by methane that has three equivalent S_4 axes at right angles to each other, as shown in Fig. 9.4. Note that $\hat{S}_n^{2n} = \hat{E}$ when n is odd, but $\hat{S}_n^n = \hat{E}$ when n is even.

It is always found that $\hat{S}_n^{n/2} = \hat{\imath}$ if n is even and $n/2$ is odd. The *staggered* configuration of ethane has an S_6 axis, as shown in Fig. 9.3*b*.

In general, S_n axes with n even contain n operations with neither a σ_h nor a C_n axis (but with a $C_{n/2}$ axis). In general, S_n axes with n odd contain a total of $2n$ operations, including σ_h and the operations generated by C_n.

Since a symmetry operation produces a nuclear configuration that is indistinguishable from the original one, the center of mass of a molecule has the same position before and after the operation. Thus the symmetry elements of a molecule all pass through the center of mass.

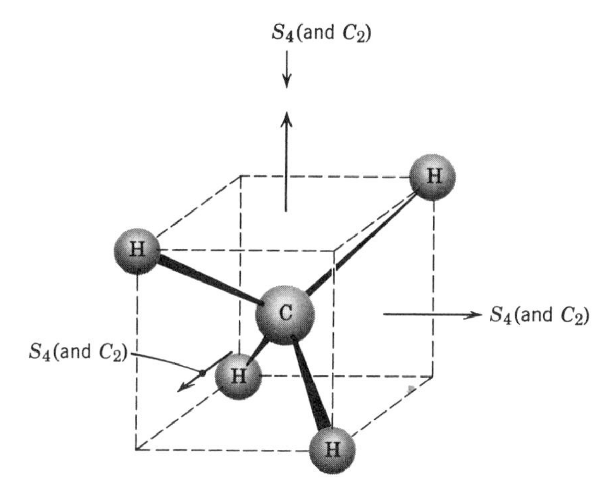

Fig. 9.4 Three equivalent S_4 rotation-reflection axes at right angles to one another are shown for the tetrahedral methane molecule, CH_4, which possesses T_d point-group symmetry. Any symmetrical tetrahedral molecule like methane also has four C_3 axes (at angles of 109° 28′ with one another) along the C—H bonds, three C_2 axes coincident with the S_4 axes, and six vertical planes (σ_d), each of which passes through the carbon and two hydrogens and relates the other two hydrogens to each other.

9.6 COMBINATION OF SYMMETRY OPERATIONS—POINT GROUPS

The set of symmetry operations of a molecule constitute a mathematical group. A set of operations belong to a mathematical group only when they are related to each other in a certain way. The four rules that define a group are the following.

1. If $\hat{A}$ represents a symmetry operation of the group and $\hat{B}$ represents another symmetry operation of the same group, the product $\hat{A}\hat{B} = \hat{F}$ also is an operation of the group. The product $\hat{A}\hat{B} = \hat{F}$ means that the operation $\hat{B}$ followed by the operation $\hat{A}$ is equivalent to an operation $\hat{F}$. In general, $\hat{A}\hat{B} \neq \hat{B}\hat{A}$, which means that the operation $\hat{B}$ followed by operation $\hat{A}$ is *not* necessarily equivalent to the operation $\hat{A}$ followed by operation $\hat{B}$. In other words, $\hat{A}$ *does not commute* with $\hat{B}$. If $\hat{A}\hat{B} = \hat{B}\hat{A}$, the multiplication is commutative. For a given molecule the various products can be summarized in a group multiplication table.
2. In each group there exists an identity operation $\hat{E}$ (corresponding to the $\hat{C}_1$ rotation of 360° about any given axis) such that for any other operation of the group (e.g., $\hat{A}$).

$$\hat{A}\hat{E} = \hat{E}\hat{A} = \hat{A} \tag{9.6}$$

3. For each operation $\hat{A}$ there exists in the group an inverse operation $\hat{A}^{-1}$ such that $\hat{A}^{-1}\hat{A} = \hat{A}\hat{A}^{-1} = \hat{E}$. The inverse operation $\hat{A}^{-1}$ is that which returns the object to its original position. Hence, the inverse of a $\hat{C}_2$, $\hat{\sigma}$, or $\hat{\imath}$ operation is

itself (viz., $\hat{C}_2\hat{C}_2 = \hat{E}$; $\hat{\sigma}\hat{\sigma} = \hat{E}$; $\hat{\imath}\hat{\imath} = \hat{E}$). The inverse of a $\hat{C}_3{}^1$ operation is a $\hat{C}_3{}^2$, while the inverse of an $\hat{S}_4{}^1$ operation is an $\hat{S}_4{}^3$.

4. The associative law of multiplication holds:

$$\hat{A}(\hat{B}\hat{C}) = (\hat{A}\hat{B})\hat{C} \tag{9.7}$$

The group of symmetry operations that leaves a molecule (or other finite objects such as a crystal) congruent with itself is called a *point group*, since all symmetry elements pass through a point. Each point group is represented by a *Schoenflies symbol* that designates sufficient symmetry elements to obtain the other symmetry elements and all *distinct* operations of the point group.

Consider H_2O (Fig. 9.2*a*), which has four symmetry operations: $\hat{E}$, $\hat{C}_2{}^1$, $\hat{\sigma}_v$, and $\hat{\sigma}'_v$. The operation of reflection in one vertical symmetry plane ($\hat{\sigma}_v$) followed by the operation of reflection in the other vertical symmetry plane ($\hat{\sigma}'_v$) is equivalent to the twofold operation, that is, $\hat{\sigma}'_v\hat{\sigma}_v = \hat{C}_2{}^1$. Similarly, the successive operations of $\hat{C}_2{}^1$ followed by $\hat{\sigma}_v$ yield the same result as the $\hat{\sigma}'_v$ operation (viz., $\hat{\sigma}_v\hat{C}_2{}^1 = \hat{\sigma}'_v$). Each of the four operations is its own inverse (e.g., $\hat{\sigma}_v\hat{\sigma}_v = \hat{E}$). The product operations for H_2O are summarized in Table 9.2 as a group multiplication table, which shows (*since no additional operations are generated*) that these four symmetry operations form a group and that the operations are commutative. The point group is designated by the Schoenflies symbol C_{2v}. The subscript "2" of this symbol signifies not only that the principal proper rotation axis (C_n) is a C_2, but also that there are *two vertical* symmetry planes, at right angles to each other, that contain the C_2 axis.

Table 9.2 Multiplication Table for the Group C_{2v}

		Operation $\hat{B}$			
		$\hat{E}$	$\hat{C}_2{}^1$	$\hat{\sigma}_v$	$\hat{\sigma}'_v$
Operation $\hat{A}$	$\hat{E}$	$\hat{E}$	$\hat{C}_2{}^1$	$\hat{\sigma}_v$	$\hat{\sigma}'_v$
	$\hat{C}_2{}^1$	$\hat{C}_2{}^1$	$\hat{E}$	$\hat{\sigma}'_v$	$\hat{\sigma}_v$
	$\hat{\sigma}_v$	$\hat{\sigma}_v$	$\hat{\sigma}'_v$	$\hat{E}$	$\hat{C}_2{}^1$
	$\hat{\sigma}'_v$	$\hat{\sigma}'_v$	$\hat{\sigma}_v$	$\hat{C}_2{}^1$	$\hat{E}$

The table contains the products $\hat{A}\hat{B}$ for the indicated operations. Note that each column and each row has each symmetry operation represented only once.

9.7 CLASSIFICATION OF SCHOENFLIES POINT GROUPS

In Table 9.3 a number of Schoenflies point groups are listed together with examples. There is, in principle, an infinite number of point groups. The external symmetries of crystals fall into only 32 point groups (Section 19.4).

In determining the Schoenflies point group of any molecule, it is convenient to carry out a systematic examination according to the following sequence of steps.

1. Determine whether a molecule possesses T_d point group symmetry, that of a regular tetrahedron, or O_h point group symmetry, that of a regular octahedron. In both cubic point groups there are *four* C_3 axes along the body diagonals of a cube. If the molecule does not belong to either of these point groups but contains more than one threefold axis (i.e., more than one axis of symmetry greater than a twofold), it is of a point group not given in this chapter.
2. Determine whether or not the molecule possesses any C_n proper axes of symmetry. If so, proceed to step 3. If not, look for a symmetry plane, in which case it belongs to the point group C_s, or a center of symmetry, in which case it has the point symmetry C_i. If no symmetry elements are located, the molecule then belongs to the point group C_1, which possesses only the identity element (and corresponding operation).
3. If at least one proper axis C_n is present, locate the principal axis C_n of highest symmetry if one exists (e.g., for the point group D_2 with three C_2 axes perpendicular to one another, there is no unique principal axis). Next determine if there is an improper axis S_{2n} collinear with the principal (or other) axis C_n. If an S_{2n} axis exists but no other elements of symmetry are present except possibly i, the molecule belongs to one of the S_n (n even) point groups. Otherwise, if no S_{2n} axes are found or if an S_{2n} axis is observed along with other symmetry elements, proceed to step 4.
4. Determine if the point group is dicyclic (i.e., D_n, D_{nh}, D_{nd}) by observing whether n twofold equally spaced axes lie in a plane perpendicular to the principal (or other) axis C_n. If so, proceed to step 5; if not, proceed to step 6.
5. To differentiate among the point groups D_n, D_{nh}, and D_{nd}, determine whether a horizontal mirror plane (σ_h) is perpendicular to the C_n axis. If a σ_h is present, the point group is D_{nh}. If not, determine whether there are *vertical* mirror planes (σ_d) containing the C_n axis that bisect pairs of horizontal C_2 axes. If σ_d planes are present, the point group is D_{nd}. If the molecule contains no mirror planes of any kind, the point group then is D_n.
6. If a molecule does not possess n horizontal C_2 axes perpendicular to the C_n axis, it must belong to a cyclic point group (viz., C_n, C_{nh}, or C_{nv}) containing only one proper rotation axis. If the C_n axis contains a horizontal mirror plane (σ_h), the point group is C_{nh}. If the C_n axis possesses n *vertical* symmetry planes intersecting the C_n axis, the molecule belongs to the point group C_{nv}. If the molecule contains neither horizontal nor vertical mirror planes, the point group is C_n.

9.8 SYMMETRY PROPERTIES OF WAVE FUNCTIONS

The presence of symmetry in a molecule indicates a simplification of the mathematical description of the molecule. The wave functions for the molecule must have all of the symmetry of the molecule. Another way of expressing this is to say that a molecular wave function must be an eigenfunction of the symmetry operators of the point group to which the molecule belongs.

An s orbital, being spherically symmetrical, has all possible point symmetry elements. A p_x orbital of Fig. 8.10 is symmetric with respect to reflection in the

Table 9.3 Common Schoenflies Point Groups with Examples

Schoenflies Symbol	Symmetry Elements	Molecular Configuration	Schoenflies Symbol	Symmetry Elements	Molecular Configuration
C_1	E	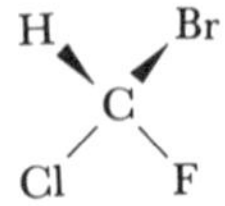 H, Br, C, Cl, F	D_{3h}	$E, C_3(S_3), 3C_2, \sigma_h, 3\sigma_v$	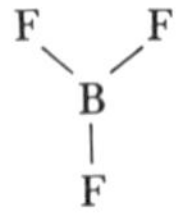 F, F, B, F
C_3	E, σ	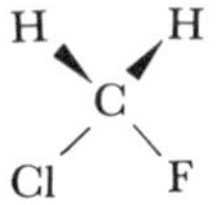 H, H, C, Cl, F	D_{4h}	$E, C_4(C_2, S_4), 4C_2, \sigma_h, 2\sigma_v, 2\sigma_d, i$	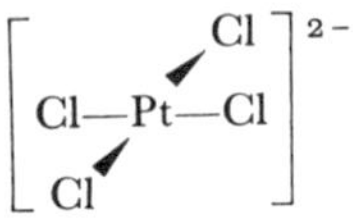 [Cl, Cl—Pt—Cl, Cl]$^{2-}$
C_i	E, i	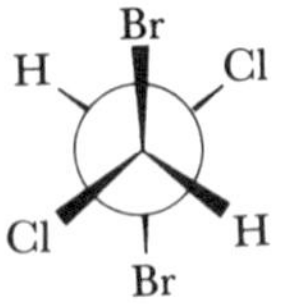 Br, H, Cl, Cl, H, Br	D_{5h}	$E, C_5(S_5), 5C_2, \sigma_h, 5\sigma_v$	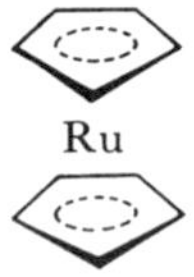 Ru
C_2	E, C_2	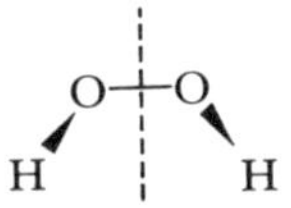 O—O, H, H	D_{6h}	$E, C_6(C_3, C_2, S_6, S_3), 6C_3, \sigma_h, 3\sigma_v, 3\sigma_d, i$	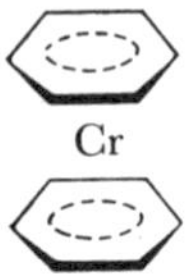 Cr
C_{2v}	$E, C_2, 2\sigma_v$	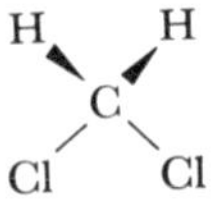 H, H, C, Cl, Cl	$D_{\infty h}$	$E, C_\infty(S_\infty), \infty C_3, \sigma_h, \infty\sigma, i$	H—C≡C—H
C_{3v}	$E, C_3, 3\sigma_v$	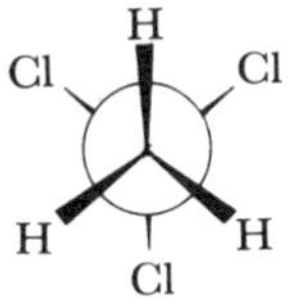 H, Cl, Cl, H, H, Cl	D_{2d}	$E, C_2(S_4), 2C_2, 2\sigma_d$	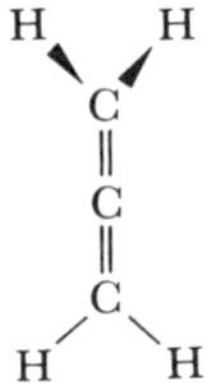 H, H, C=C=C, H, H
C_{4v}	$E, C_4(C_2), 4\sigma_v$	 O, F, F—Xe—F, F	D_{3d}	$E, C_3(S_6), 3C_2, 3\sigma_d, i$	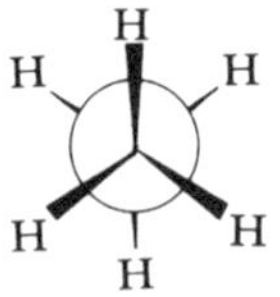 H, H, H, H, H, H
$C_{\infty v}$	$E, C_\infty, \infty\sigma_v$	 O≡C	D_{4d}	$E, C_4(S_8, C_2), 4C_2, 4\sigma_d$	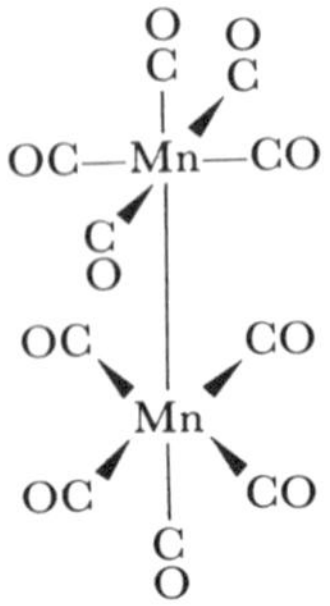 (OC)$_5$Mn—Mn(CO)$_5$
C_{2h}	E, C_2, σ_h, i	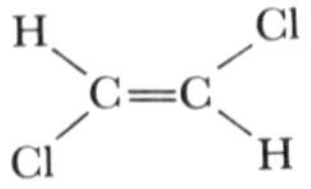 H, Cl, C=C, Cl, H			

Table 9.3—(*Continued*)

Schoenflies Symbol	Symmetry Elements	Molecular Configuration	Schoenflies Symbol	Symmetry Elements	Molecular Configuration
C_{3h}	$E, C_3(S_3), \sigma_h$	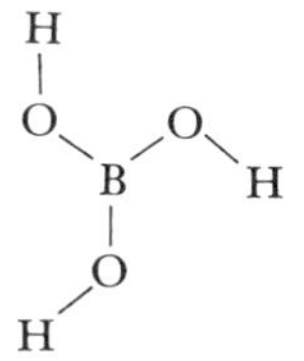 H–O–B(–O–H)–O–H	D_{5d}	$E, C_5(S_{10}), 5C_3, 5\sigma_d, i$	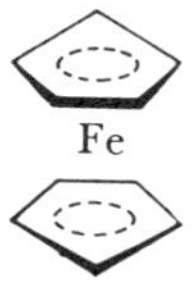 (pentagon) Fe (pentagon)
			T_d	$E, 3C_2(3S_2), 4C_3, 6\sigma_d$	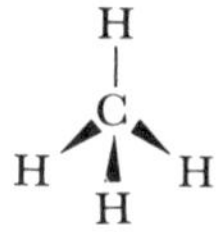 H–C(H)(H)–H
D_{2h}	$E, C_2, 2C_2, \sigma_h, 2\sigma_v, i$	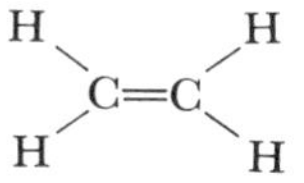 $H_2C{=}CH_2$	O_h	$E, 3C_4(3C_2, 3S_4), 4C_2(4S_6), 3\sigma_h, 6C_2, 6\sigma_d, i$	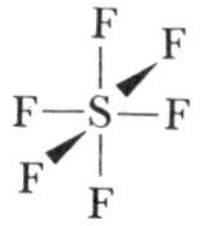 SF_6 (F–S–F, octahedral)

xz plane or the xy plane. However, when a p_x orbital is reflected in the yz plane, the wave function an equal distance on the other side of the plane has the same magnitude, *but the opposite sign.* Thus we say that the p_x wave function is *antisymmetric* with reflection in the yz plane. The p_x orbital is antisymmetric with respect to the $\hat{C}_2{}^z$ operation, the $\hat{C}_2{}^y$ operation, and the inversion operation $\hat{\imath}$, but it is symmetric with respect to any rotation about the x axis, a C_∞ axis. The d_{xz} orbital of Fig. 8.10 is symmetric with respect to $\hat{\sigma}^{xz}$, $\hat{\imath}$, and $\hat{C}_2{}^y$, and antisymmetric with respect to $\hat{\sigma}^{xy}$, $\hat{\sigma}^{yz}$, $\hat{C}_2{}^z$, $\hat{C}_2{}^x$, and $\hat{C}_4{}^y$.

9.9 SYMMETRY AND DIPOLE MOMENT

A dipole moment (Section 10.10) is a vector quantity that is not affected either in direction or in magnitude but any symmetry operation of the molecule. Therefore the dipole moment vector must be contained in each of the symmetry elements. Consequently, molecules that possess dipole moments belong only to the point groups C_n, C_s, and C_{nv}. The presence or absence of a dipole moment therefore tells something about the symmetry of a molecule. For example, carbon dioxide and water might have structures corresponding to a symmetrical linear molecule, to an unsymmetrical linear molecule, or to a bent molecule. The dipole moments recorded in Table 10.4 shows that carbon dioxide has zero moment; therefore, the molecule must be symmetrical and linear. If it were unsymmetrical or bent, there would have been a permanent dipole moment. On the other hand, water has a pronounced dipole moment and cannot have the symmetrical linear structure. A molecule with a center of symmetry cannot have a dipole moment.

9.10 SYMMETRY AND OPTICAL ACTIVITY

If a molecule and its mirror image cannot be superimposed, it is potentially optically active. Since a rotation followed by a reflection always converts a right-handed object to a left-handed object, an S_n axis guarantees that a molecule cannot exist in separate left- and right-handed forms.

All *improper* rotation axes (S_n), including a mirror plane ($\sigma = S_1$) and center of symmetry ($i = S_2$), convert a right-handed object into a left-handed object (i.e., produce a mirror image of the original object), whereas all *proper* rotation axes (C_n) leave a right-handed object unchanged in this respect. Hence, only molecules having only proper rotation elements of symmetry can be optically active.

In a molecule in which internal rotation can take place (e.g., ethane or H_2O_2) it is possible to have optically active conformations, but in a gas or solution these conformers are so rapidly interconverted that optical isomers cannot be resolved.

References

I. Bernal, W. C. Hamilton, and J. S. Ricci, *Symmetry*, W. H. Freeman and Co., San Francisco, 1972.

A. D. Boardman, D. E. O'Connor, and P. A. Young, *Symmetry and Its Applications in Science*, Wiley, New York, 1973.

F. A. Cotton, *Chemical Applications of Group Theory*, Wiley, New York, 1971.

L. H. Hall, *Group Theory and Symmetry in Chemistry*, McGraw-Hill Book Co., New York, 1969.

D. C. Harris and M. D. Bertolucci, *Symmetry and Spectroscopy*, Oxford University Press, 1978.

H. H. Jaffé and M. Orchin, *Symmetry in Chemistry*, Wiley, New York, 1965.

A. Vincent, *Molecular Symmetry and Group Theory*, Wiley, New York, 1977.

Problems

List the Schoenflies symbols and symmetry elements for each of the following molecules.

9.1 H_2S

S
H H

Ans. C_{2v}.

9.2 PCl_3

Cl Cl
P
Cl

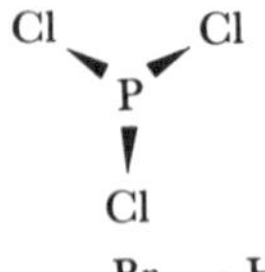

Ans. C_{3v}.

9.3 *trans*-$[CrBr_2(H_2O)_4]^+$ (ignore the H's)

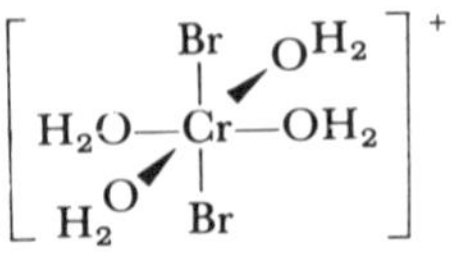

$[$ Br, OH_2, H_2O—Cr—OH_2, H_2O, Br $]^+$

Ans. D_{4h}.

9.4 FeF_6^{3-}

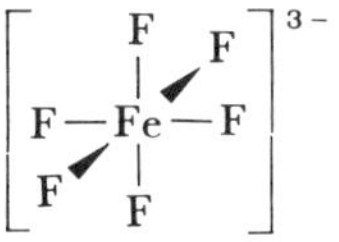

F, F, F—Fe—F, F, F; $3-$

Ans. O_h.

9.5 *gauche*-CH_2ClCH_2Cl

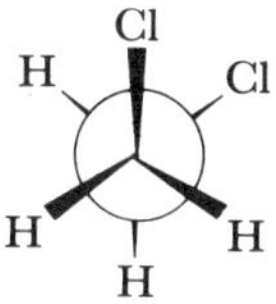

Cl, H, Cl, H, H, H

Ans. C_2.

9.6 $C_6H_3Br_3$ (1,3,5-tribromobenzene)

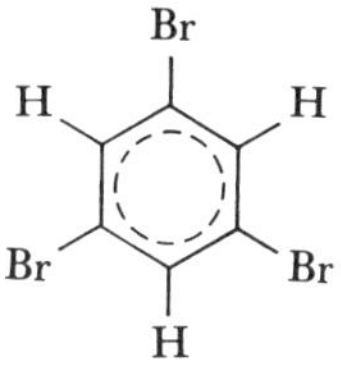

Br, H, H, Br, Br, H

Ans. D_{3h}.

9.7 $CHClBr(CH_3)$

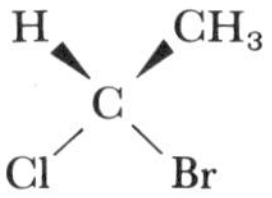

H, CH_3, C, Cl, Br

Ans. C_1.

9.8 IF_5

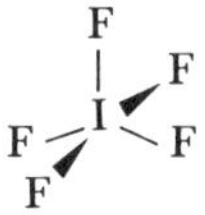

F, F, F—I—F, F

Ans. C_{4v}.

9.9 C_6H_{12} (cyclohexane)

Ans. D_{3d}.

9.10 B_2H_6

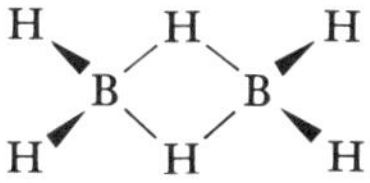

H, H, H, B, B, H, H, H

Ans. D_{2h}.

9.11 N_2

N≡N

Ans. $D_{\infty h}$.

9.12 $C_{10}H_8$ (naphthalene)

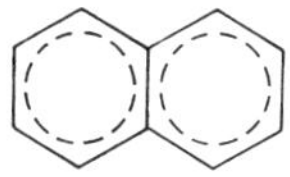

(planar)

Ans. D_{2h}.

9.13 C_5H_8 (spiropentane)

(triangles ⊥ to each other)

Ans. D_{2d}.

9.14 H_4C_4S (thiophene)

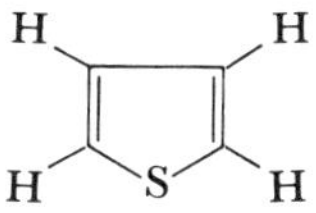

H, H, H, S, H

(planar)

Ans. C_{2v}.

9.15 $C_6H_4Cl_2$ (*p*-dichlorobenzene) *Ans.* D_{2h}.

9.16 CCl_4

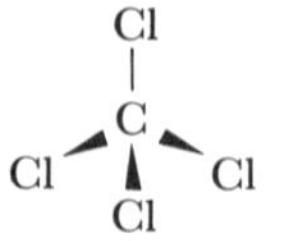

Ans. T_d.

9.17 PF_5 *Ans.* D_{3h}.

9.18 *trans*-CFClBrCFClBr

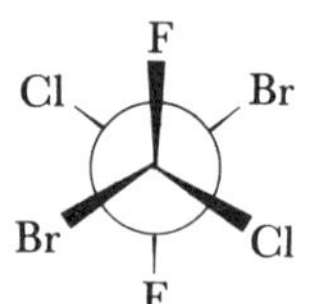

Ans. C_i.

9.19 HCN H—C≡N *Ans.* $C_{\infty v}$.

9.20 C_2F_6

Ans. D_{3d}.

9.21 Construct the operator multiplication table for the point group C_{2h}.

Ans.

	$\hat{E}$	$\hat{C}_2^1$	$\hat{\sigma}_h$	$\hat{\imath}$
$\hat{E}$	$\hat{E}$	$\hat{C}_2^1$	$\hat{\sigma}_h$	$\hat{\imath}$
$\hat{C}_2^1$	$\hat{C}_2^1$	$\hat{E}$	$\hat{\imath}$	$\hat{\sigma}_h$
$\hat{\sigma}_h$	$\hat{\sigma}_h$	$\hat{\imath}$	$\hat{E}$	$\hat{C}_2^1$
$\hat{\imath}$	$\hat{\imath}$	$\hat{\sigma}_h$	$\hat{C}_2^1$	$\hat{E}$

9.22 Write out the operations associated with the S_6 elements and their equivalents, if any. How many distinct operations are produced?

Ans.

$\hat{S}_6^1$	$\hat{S}_6^2$	$\hat{S}_6^3$	$\hat{S}_6^4$	$\hat{S}_6^5$	$\hat{S}_6^6$
	$\hat{C}_3^1$	$\hat{\imath}$	$\hat{C}_3^2$		$\hat{E}$

Thus an S_6 axis implies the existence of both a C_3 axis coincident with the S_6 axis and a center of symmetry (i). Only the S_6^1 and S_6^5 operations are *distinct* operations characteristic of only the S_6 axis.

9.23 $Cr(CO)_6$

9.24 C_8H_8 (cyclooctatetraene)

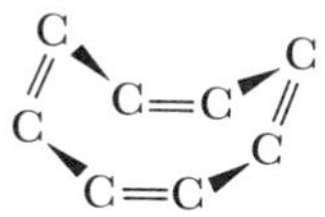

9.25 HCO_2H

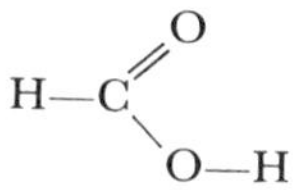

(planar)

9.26 $UO_2F_5^{3-}$

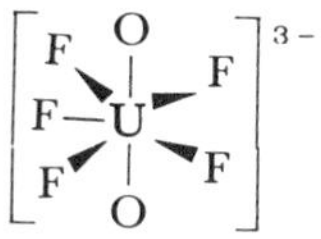

9.27 $C_{14}H_{10}$ (phenanthrene)

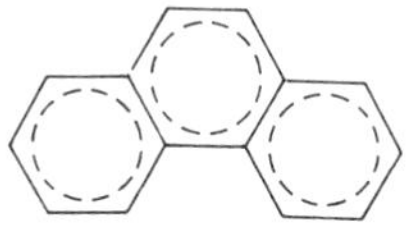

9.28 C_6Cl_6

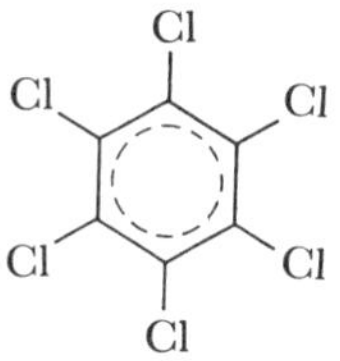

9.29 S_8

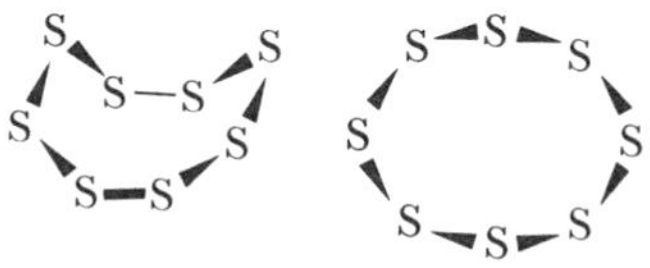

9.30 $Fe(CN)_6^{3-}$

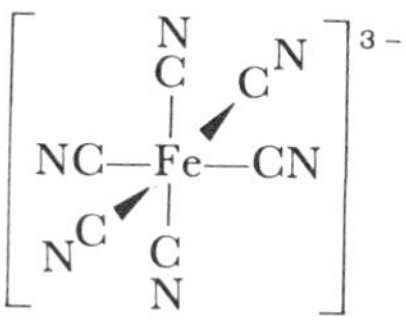

9.31 $[HNBCl]_3$

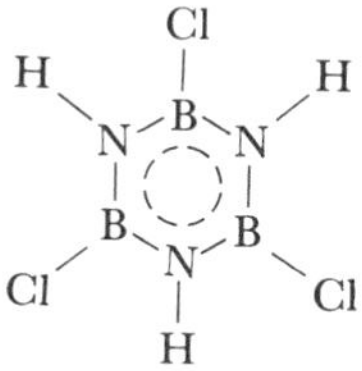

(planar)

9.32 $C_{10}H_{16}$ (adamantane)

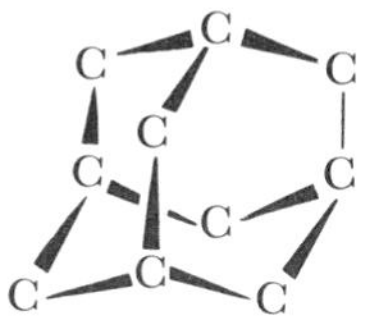

(ignoring the H's)

9.33 $C_6H_2O_2Cl_2$ (2,5-dichloroquinone) (planar)

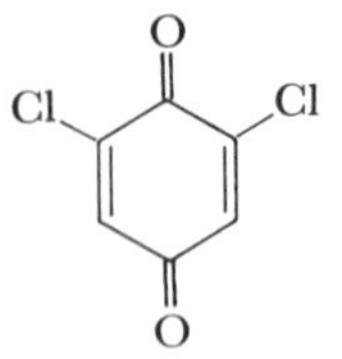

9.34 HOCl

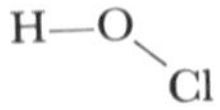

9.35 $[HNBH]_3$ 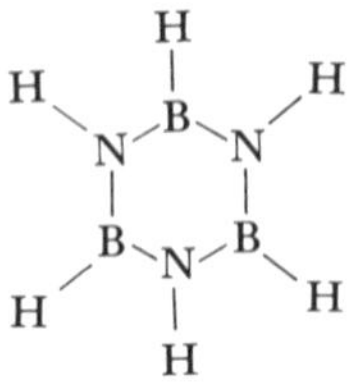 (planar)

9.36 $C_6H_3(C_6H_5)_3$ (1,3,5-triphenylbenzene) (not planar)

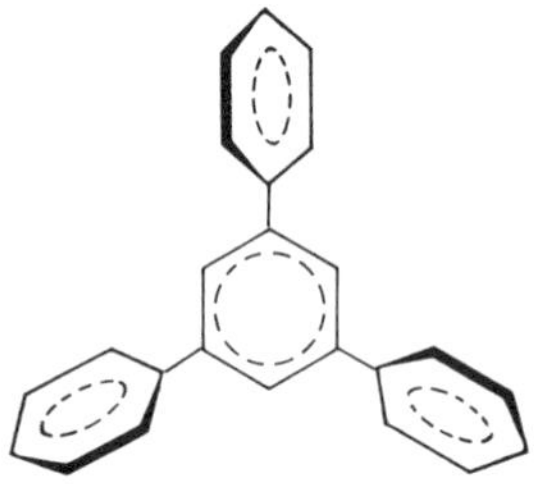

9.37 *cis*-CHCl=CHCl

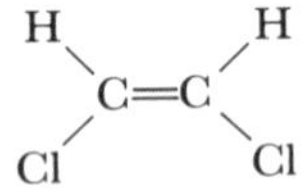

9.38 CH_3Cl (methyl chloride)

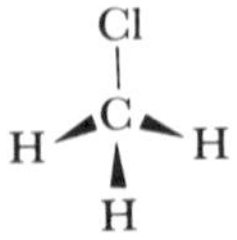

9.39 $Ni(CO)_4$

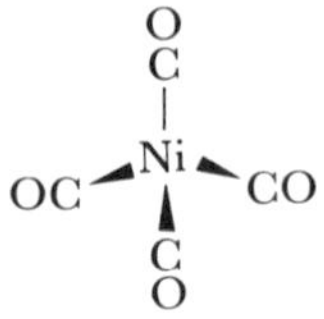

9.40 H_3CCH_2Br

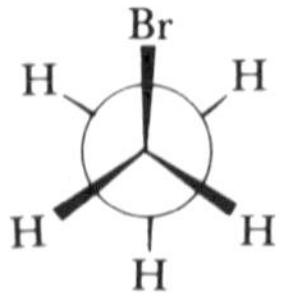

9.41 $PtBr_4(NH_3)_2$ [tetrabromodiammineplatinum(IV)] (ignore the H's)

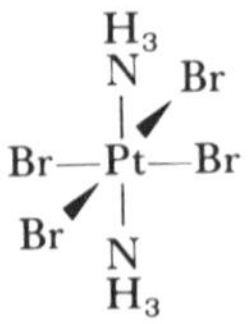

9.42 B_2Cl_4

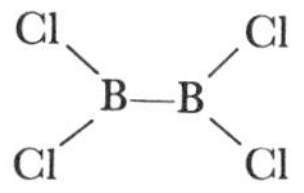

$$\begin{matrix} Cl & & & & Cl \\ & \diagdown & & \diagup & \\ & & B{-}B & & \\ & \diagup & & \diagdown & \\ Cl & & & & Cl \end{matrix}$$

9.43 Construct the operator multiplication table for the point group C_{3v}.

9.44 Write out the operations associated with the S_5 element and their equivalents, if any. How many distinct operations are produced?

CHAPTER 10

MOLECULAR ELECTRONIC STRUCTURE

Quantum mechanics has made it possible to understand the nature of chemical bonding and to predict the structures and properties of simple molecules. Our ideas about covalent bonds go back to 1916, when Lewis described the sharing of electron pairs between atoms. The pairs of electrons held jointly by two atoms were considered to be effective in completing a stable electronic configuration for each atom. This approach provided only a qualitative picture of chemical bonding. Today the electronic structures of simple molecules may be calculated with a high degree of accuracy, and energy levels, bond angles, bond distances, dipole moments, and other properties may be calculated. For molecules with a large number of electrons, approximations have to be introduced. However, even approximate calculations are very helpful in understanding molecular structure, chemical properties, and molecular spectra. The electronic structure of solids is discussed in Chapter 19.

10.1 DISSOCIATION ENERGIES OF DIATOMIC MOLECULES

The potential energy of a diatomic molecule as a function of internuclear distance R is shown in Fig. 10.1. The potential energy includes both the purely electronic energy E_{el} obtained by solving the Schrödinger equation for the molecule and the nuclear repulsion energy V_{NN}.

$$U = E_{el} + V_{NN} \tag{10.1}$$

As the nuclei are brought together, the energy decreases because of the attractive force of chemical bonding. But as the nuclei are pushed very close together, the potential energy increases rapidly because of internuclear repulsion. The internuclear distance R_e at the minimum of the potential energy curve is referred to as the equilibrium internuclear distance. The dissociation energy D^e measured from the minimum of the potential energy curve is referred to as the equilibrium dissociation energy. This is the dissociation energy that is obtained most directly from the solutions of the Schrödinger equation for a diatomic molecule, but we must be careful to distinguish it from the spectroscopic dissociation energy D^0. The spectroscopic dissociation energy D^0 is the energy required to dissociate a molecule in its ground vibrational state into two atoms. The equilibrium dissociation energy D^e and the spectroscopic dissociation energy D^0 are related by

$$D^e = D^0 + \tfrac{1}{2}h\omega_0 \tag{10.2}$$

where ω_0 is the ground state vibrational frequency.

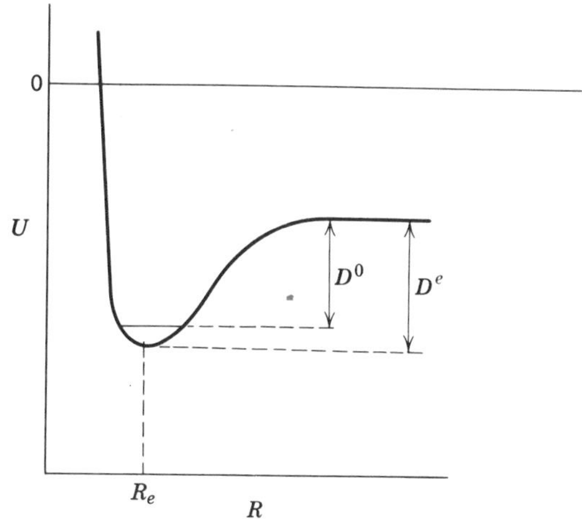

Fig. 10.1 Potential energy U for a diatomic molecule as a function of internuclear distance R. D^0 is the dissociation energy from the ground state and is often called the spectroscopic dissociation energy. D^e is the equilibrium dissociation energy, which is measured from the minimum of the potential energy curve. The potential energy is taken as zero for the completely separated electrons and nuclei.

A diatomic molecule can exist in one of a series of vibrational energy levels (see Section 8.10), of which only the zero point level is shown in Fig. 10.1. In accordance with the Heisenberg uncertainty principle (Section 8.4), a diatomic molecule has this amount of vibrational energy, even at the absolute zero of temperature.

Table 10.1 Dissociation Energies for $H_2^+(g)$ and $H_2(g)$ and Ionization Potentials for $H_2(g)$ and $H(g)$

	eV	cm^{-1}
	H_2^+	
D^0	2.650 79	21,380
$\frac{1}{2}\hbar\omega_0$	0.142	1,147
D^e	2.793	22,527
	H_2	
D^0	4.477 97[a]	36,117
$\frac{1}{2}\hbar\omega_0$	0.270 3	2,180
D^e	4.748 3	38,297
$IP(H_2)$	15.425 9	124,417
$IP(H)$	13.598 396	109,677.6

[a] This spectroscopic dissociation energy of H_2 is 432.057 kJ mol^{-1}, in agreement with $\Delta H_0^\circ =$ 432.074 kJ mol^{-1} calculated from Table A.2.

The two types of dissociation energies D^0 and D^e for H_2^+ and H_2 are given in Table 10.1. The values of dissociation energies are given in eV and cm^{-1} because both units are used, and each has certain advantages.* Spectroscopic measurements in the ultraviolet, visible, and infrared regions are usually expressed in terms of wavelengths or wave numbers. Thus experimental results are usually expressed in cm^{-1} because conversion to other units involves the use of physical constants whose values may change. The results of theoretical calculations are usually expressed in electron volts.

Before discussing the quantum mechanical calculation of the potential energy curves of H_2^+ and H_2, we must discuss the relations between these two molecules and certain aspects of their potential energy curves. The potential energy curves for the ground states of H_2 and H_2^+ are shown in Fig. 10.2. In this figures energies are measured with respect to two protons and two electrons, infinitely far from

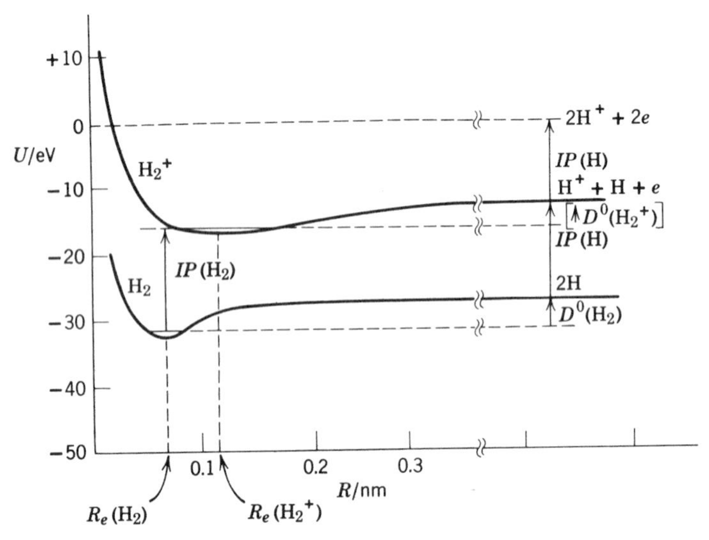

Fig. 10.2 Potential energy curves for the ground state of H_2 and H_2^+ with the zero point vibrational levels shown. Energies are measured with respect to two protons and two electrons, all at infinite distance.

---

* The relation between eV and cm^{-1} is given by

$$1\ \text{eV} = \frac{Ee}{hc} = \frac{(1\ \text{V})(1.602\ 189\ 2 \times 10^{-19}\ \text{C})}{(6.626\ 176 \times 10^{-34}\ \text{J s})(2.997\ 924\ 58 \times 10^{8}\ \text{m s}^{-1})(100\ \text{cm m}^{-1})}$$

$$= 8065.478\ \text{cm}^{-1}$$

The relation between eV and kJ mol^{-1} is given by

$$1\ \text{eV} = EeN_A = \frac{(1\ \text{V})(1.602\ 189\ 2 \times 10^{-19}\ \text{C})(6.022\ 045 \times 10^{23}\ \text{mol}^{-1})}{(10^{3}\ \text{J kJ}^{-1})}$$

$$= 96.484\ 55\ \text{kJ mol}^{-1}$$

each other. The ionization potential $IP(H_2)$ for hydrogen is the energy required to remove an electron to an infinite distance and form $H_2{}^+$.

$$H_2(g) = H_2{}^+(g) + e \qquad IP(H_2) = 15.4259 \text{ eV} \tag{10.3}$$

Thus the zero point levels of H_2 and $H_2{}^+$ are separated by 15.4259 eV, as shown in Fig. 10.2.

The potential energy curves for H_2 and $H_2{}^+$ at infinite internuclear distance are separated by the ionization potential of a hydrogen atom in its ground state. This ionization potential was calculated in Section 8.13.

$$H(g) = H^+(g) + e \qquad IP(H) = 13.598\,396 \text{ eV} \tag{10.4}$$

As can be seen from Fig. 10.2,

$$IP(H_2) + D^0(H_2{}^+) = D^0(H_2) + IP(H) \tag{10.5}$$

Since it is difficult to measure $D^0(H_2{}^+)$, this relation has been used to obtain the most accurate value for this quantity.*

$$\begin{aligned} D^0(H_2{}^+) &= D^0(H_2) + IP(H) - IP(H_2) \\ &= 36{,}117 \text{ cm}^{-1} + (13.598\,396 \text{ eV})(8065.478 \text{ cm}^{-1} \text{ eV}^{-1}) - 124{,}417 \text{ cm}^{-1} \\ &= 21{,}377 \text{ cm}^{-1} \end{aligned} \tag{10.6}$$

Now that we have placed these potential energy curves in a context, we want to talk about their calculation from quantum mechanics. We want to understand why a stable molecule $H_2{}^+$ is formed when two protons and an electron are brought together, and why an even more stable molecule H_2 is formed when two protons and two electrons are brought together.

10.2 THE BORN-OPPENHEIMER APPROXIMATION

Since electrons are much lighter than nuclei, they move much faster than nuclei in a molecule, and this makes it possible to write separate Schrödinger equations for electronic and nuclear motions. This approximation, which was introduced by Born and Oppenheimer, may be illustrated by considering the hydrogen molecule ion. The coordinates for $H_2{}^+$ are defined in Fig. 10.3. The molecular Hamiltonian for the hydrogen molecule ion is

$$\hat{H} = -\frac{\hbar^2}{2M}\nabla_a{}^2 - \frac{\hbar^2}{2M}\nabla_b{}^2 - \frac{\hbar^2}{2m}\nabla_{el}^2 - \frac{1}{4\pi\epsilon_0}\left(\frac{e^2}{r_a} + \frac{e^2}{r_b} - \frac{e^2}{r_{ab}}\right) \tag{10.7}$$

* G. Herzberg, *Science*, **177**, 123 (1972).

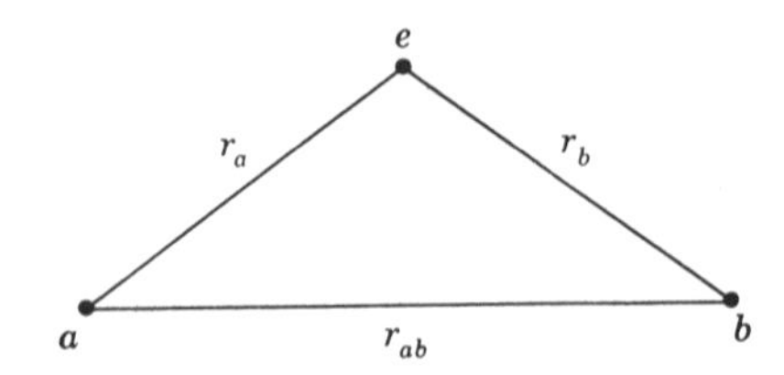

Fig. 10.3 Coordinates in H_2^+. Two protons are represented by a and b.

where M is the proton mass, m is the electron mass, and a and b designate the two protons. The first two terms are the kinetic energy operators of the nuclei, and the next term is the kinetic energy operator of the electron. The remaining terms represent electrostatic attractions between the electron and the nuclei and the electrostatic repulsion between the nuclei.

Born and Oppenheimer pointed out that since the electron moves so much more rapidly, we can consider the nuclei to be fixed and solve the electronic problem. The purely electronic Hamiltonian is

$$\hat{H}_{el} = -\frac{\hbar^2}{2m}\nabla_{el}^2 - \frac{1}{4\pi\epsilon_0}\left(\frac{e^2}{r_a} + \frac{e^2}{r_b}\right) \tag{10.8}$$

The Schrödinger equation for electronic motion is

$$\left(\hat{H}_{el} + \frac{e^2}{4\pi\epsilon_0 r_{ab}}\right)\psi_{el} = U\psi_{el} \tag{10.9}$$

The *potential* energy of proton-proton repulsion has to be included, but r_{ab} is fixed for a given calculation. The eigenvalue U contains the energy of nuclear repulsion. It may be shown that the omission of a constant term, such as $e^2/4\pi\epsilon_0 r_{ab}$, from the Hamiltonian does not affect the wave function ψ_{el}, but simply decreases the energy eigenvalue by the amount of the constant term. Thus equation 10.9 may be written

$$\hat{H}_{el}\psi_{el} = E_{el}\psi_{el} \tag{10.10}$$

where

$$U = E_{el} + \frac{e^2}{4\pi\epsilon_0 r_{ab}} \tag{10.11}$$

is the total energy of the hydrogen molecule, ignoring translational energy, vibrational energy, and rotational energy of the molecule as a whole.

In this chapter we will be dealing with Schrödinger equations of the form of equation 10.10. Once equation 10.10 has been solved for various internuclear distances, the wave function for nuclear motion may be calculated from the Schrödinger equation for nuclear motion only. The wave function for nuclear motion contains the information about vibrational and rotational motion.

10.3 THE HYDROGEN MOLECULE ION

The electronic Schrödinger equation for this simplest molecule, which is given by equation 10.10, can be solved exactly. The details of this calculation are not given here, but the results are. At each internuclear distance a series of values of E_{el}, and therefore of U, are obtained. For the ground electronic state, H_2^+ is dissociated into a proton and a ground state hydrogen atom. Thus, at $R = \infty$, $E_{el} = U = -\frac{1}{2}$ H, as shown in Fig. 10.4. At $R = 0$, $E_{el} = -\frac{1}{2}(2)^2 = -2$ H, because this is the energy of He^+ in its ground state (see Section 8.15). The minimum in the curve for U (the electronic energy including internuclear repulsion) is at 0.106 nm. This is in agreement with the actual value (see Table 10.2). The dissociation energy D^e measured from the minimum in the potential energy curve is 2.793 eV (269.483 kJ mol^{-1}).

Figure 10.5 shows the potential energy curves for the ground state of H_2^+ and a series of excited states. These energies may be calculated to any desired degree of accuracy. The potential energy curve labeled σ_u^* 1*s* does not lead to a stable molecule. A hydrogen molecule ion in this state immediately dissociates into a proton and a hydrogen atom.

Although the Schrödinger equation may be solved exactly for H_2^+, the addition of more electrons or nuclei requires the introduction of approximations. It is, therefore, of interest to see how a simple molecular orbital method works for H_2^+.

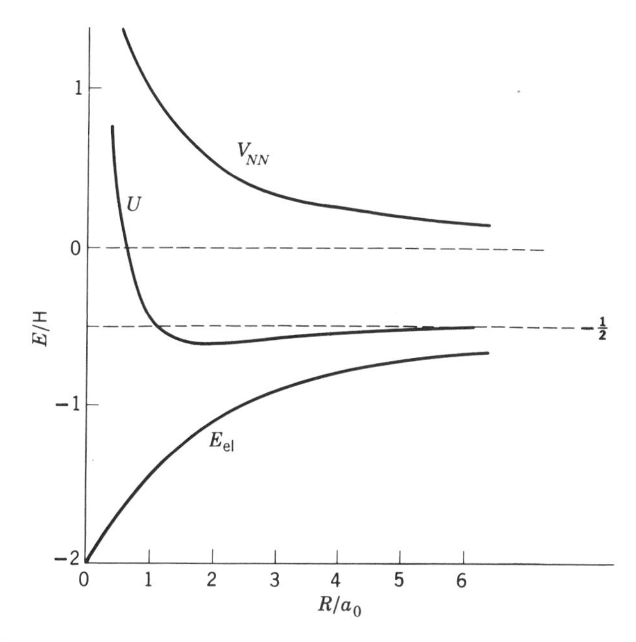

Fig. 10.4 Electronic energy of the H_2^+ ground state. Energy is plotted in Hartrees H. (From I. N. Levine, *Quantum Chemistry*, Second Edition. Copyright © 1974 by Allyn and Bacon, Inc., Boston. Reprinted with permission of the publisher.)

In the molecular orbital approach to H_2^+ we start out with the idea that if the two protons were far apart, either could have an electron in the 1*s* orbital. Therefore, we assume that as a first approximation, the electronic wave function for H_2^+ is

$$\psi = c_1(1s_a) + c_2(1s_b) \tag{10.12}$$

where $1s_a$ and $1s_b$ represent the normalized 1*s* wave functions associated with protons *a* and *b* and the constants c_1 and c_2 are to be evaluated by the variation method (Section 8.16). Such a wave function is referred to as a LCAO function, because it is a linear combination of atomic orbitals.

The variation energy *E* is given by

$$E = \frac{\int \psi^* \hat{H} \psi \, d\tau}{\int \psi^* \psi \, d\tau} = \frac{\int [c_1(1s_a) + c_2(1s_b)]\hat{H}[c_1(1s_a) + c_2(1s_b)] \, d\tau}{\int [c_1(1s_a) + c_2(1s_b)]^2 \, d\tau}$$

$$= \frac{c_1^2 H_{aa} + 2c_1c_2H_{ab} + c_2^2 H_{bb}}{c_1^2 S_{aa} + 2c_1c_2S_{ab} + c_2^2 S_{bb}} = \frac{c_1^2 H_{aa} + 2c_1c_2H_{ab} + c_2^2 H_{aa}}{c_1^2 + 2c_1c_2S + c_2^2} \tag{10.13}$$

where the following symbols have been used:

$$H_{aa} = \int (1s_a)\hat{H}(1s_a) \, d\tau = H_{bb} \tag{10.14}$$

$$H_{ab} = \int (1s_a)\hat{H}(1s_b) \, d\tau \tag{10.15}$$

$$H_{ba} = \int (1s_b)\hat{H}(1s_a) \, d\tau = H_{ab} \tag{10.16}$$

$$H_{bb} = \int (1s_b)\hat{H}(1s_b) \, d\tau = H_{aa} \tag{10.17}$$

$$S_{aa} = \int (1s_a)(1s_a) \, d\tau = 1 \tag{10.18}$$

$$S_{ab} = \int (1s_a)(1s_b) \, d\tau = S_{ba} = S \tag{10.19}$$

$$S_{ba} = \int (1s_b)(1s_a) \, d\tau = S_{ab} = S \tag{10.20}$$

$$S_{bb} = \int (1s_b)(1s_b) \, d\tau = 1 \tag{10.21}$$

The integrals represented by *S* are referred to as overlap integrals. According to the variation method, the best values of c_1 and c_2 in equation 10.12 are the ones that give the lowest value of *E*. Equation 10.13 is multiplied out and the derivatives taken with respect to c_1 and c_2. This yields

$$\frac{\partial E}{\partial c_1} = \frac{2c_1(H_{aa} - E) + 2c_2(H_{ab} - SE)}{c_1^2 + 2c_1c_2S + c_2^2} = 0 \tag{10.22}$$

$$\frac{\partial E}{\partial c_2} = \frac{2c_1(H_{ab} - SE) + 2c_2(H_{bb} - E)}{c_1^2 + 2c_1c_2S + c_2^2} = 0 \tag{10.23}$$

since these derivatives equal zero at the minimum energy. The values of c_1 and c_2 are obtained by solving the two simultaneous equations

$$c_1(H_{aa} - E) + c_2(H_{ab} - SE) = 0 \tag{10.24}$$

$$c_1(H_{ab} - SE) + c_2(H_{bb} - E) = 0 \tag{10.25}$$

There is a nontrivial solution for c_1 and c_2 from these simultaneous equations only if the determinant of the coefficients of c_1 and c_2 is equal to zero.

$$\begin{vmatrix} H_{aa} - E & H_{ab} - SE \\ H_{ab} - SE & H_{bb} - E \end{vmatrix} = 0 \tag{10.26}$$

This equation is referred to as a secular determinant. In this case it is a quadratic with two solutions E_g and E_u.

$$E_g = \frac{H_{aa} + H_{ab}}{1 + S} \tag{10.27}$$

$$E_u = \frac{H_{aa} - H_{ab}}{1 - S} \tag{10.28}$$

The integral H_{aa} is called the coulomb integral because the difference between H_{aa} and the energy of a single hydrogen atom is just that of the coulomb interaction of $(1s_a)^2$ with nucleus b. Since $(1s_a)^2$ represents a distribution of negative charge while the nucleus is positive, H_{aa} is a negative number. The integral H_{ab} is referred to as the resonance integral. When equation 10.27 or 10.28 is substituted back into equation 10.24, it is found that

$$c_1 = c_2 \qquad \text{or} \qquad c_1 = -c_2 \tag{10.29}$$

Substituting this relation into equation 10.12 and normalizing yields

$$\psi_g = \frac{1}{[2(1 + S)]^{1/2}}[(1s_a) + (1s_b)] \tag{10.30}$$

$$\psi_u = \frac{1}{[2(1 - S)]^{1/2}}[(1s_a) - (1s_b)] \tag{10.31}$$

Because of the symmetry of H_2^+ it is not surprising that the orbitals of the two protons are weighted equally. The first wave function ψ_g is symmetric; that is, inverting through the midpoint of the two nuclei does not change the magnitude or the sign of the wave function. Therefore this wave function is designated with a g, which stands for *gerade* (German for even). The second wave function ψ_u is antisymmetric and is designated by a u for *ungerade* (German for uneven).

The electron probability densities along the line of the two nuclei are given for ψ_g and ψ_u in Fig. 10.6 compared with the probability densities expected for two separate hydrogen atoms in the $1s$ state. The squares of the wave functions

$$\psi_g^2 = \frac{1}{2(1 + S)}[(1s_a)^2 + 2(1s_a)(1s_b) + (1s_b)^2] \tag{10.32}$$

$$\psi_u^2 = \frac{1}{2(1 - S)}[(1s_a)^2 - 2(1s_a)(1s_b) + (1s_b)^2] \tag{10.33}$$

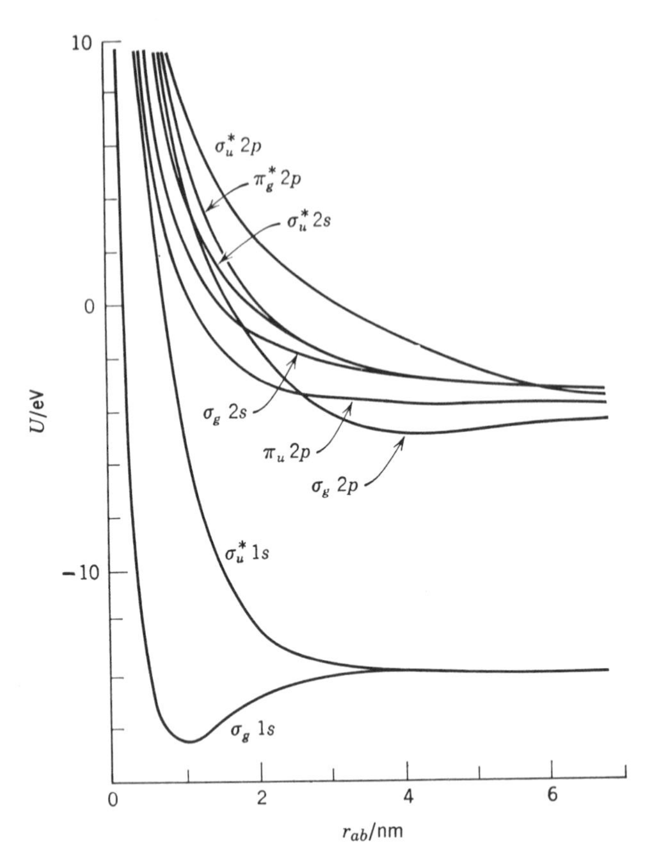

Fig. 10.5 Ground and excited states of H_2^+. The energy is measured relative to completely separated protons and one electron. (From J. C. Davis, *Advanced Physical Chemistry*, The Ronald Press Co., New York, 1965.)

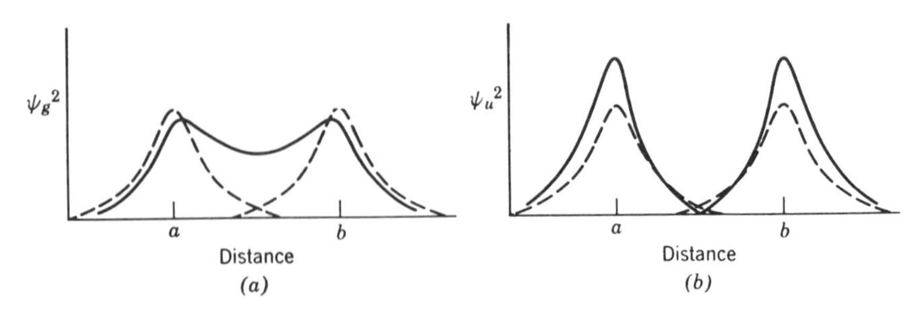

Fig. 10.6 Hydrogen molecule ion. (*a*) Probability density ψ_g^2 for the even wave function along a line passing through two nuclei. (*b*) Probability density ψ_u^2 for the uneven wave function. In both parts the dashed lines give the electron densities for two separate hydrogen atoms.

show that the electron density is increased between the nuclei with ψ_g and decreased with ψ_u. In fact, with ψ_u^2 the electron density is zero in a plane bisecting the line between the protons; that is, there is a node midway between the protons.

It is the buildup of electron density between the protons with ψ_g^2 that is responsible for the bonding. Since this orbital is symmetrical around the internuclear axis, it is referred to as a sigma orbital, and since it is even (*gerade*) and is made up of two $1s$ orbitals, it is designated $\sigma_g 1s$.

For the wave function ψ_u the molecule immediately dissociates, and so this molecular orbital is referred to as an *antibonding orbital.* This antibonding orbital is symmetrical about the internuclear axis and is referred to as a $\sigma_u^* 1s$ orbital, where the * reminds us that it is an antibonding orbital. The energy for ψ_u is evidently *higher* than that for ψ_g, and so H_{ab} must be negative.

The value of E_g calculated from equation 10.27 by carrying out the indicated integrations for different values of r_{ab} does show a minimum, and so this very simple molecular orbital theory does account for the bonding in H_2^+. However, it yields an internuclear distance of 0.123 nm (actual 0.106 nm) and a dissociation energy of 1.77 eV (actual 2.793 eV). The calculations can be improved by adding further terms to the wave functions and using the variation method to evaluate further adjustable parameters.

Further molecular orbitals may be constructed using $2s$, $2p$, and other orbitals. The angular momentum about the internuclear axis is given by $\lambda\hbar$, and the following symbols are used to represent λ.

λ	0	1	2
orbital	σ	π	δ

10.4 THE HYDROGEN MOLECULE

Using the Born-Oppenheimer approximation, the purely electronic Hamiltonian for the hydrogen molecule may be written

$$\hat{H} = -\frac{\hbar^2}{2m}(\nabla_1^2 + \nabla_2^2) + \frac{e^2}{4\pi\epsilon_0}\left(-\frac{1}{r_{a1}} - \frac{1}{r_{a2}} - \frac{1}{r_{b1}} - \frac{1}{r_{b2}} + \frac{1}{r_{12}}\right) \quad (10.34)$$

where the coordinates are defined in Fig. 10.7. As in the case of H_2^+, the electronic energy E_{el} may be calculated from equation 10.10 and the energy including the internuclear repulsion may be calculated using equation 10.11. The Schrödinger equation with this Hamiltonian cannot be solved exactly, but it may be solved to as high a degree of approximation as desired by use of the variation method. The potential energy curves for the ground and excited states are given in Fig. 10.8. The dissociation energy of H_2 (4.7483 eV) is greater than that of H_2^+ (2.793 eV), and the bond length in H_2 (0.0741 nm) is shorter than in H_2^+ (0.1060 nm).

Very precise ab initio calculations of the dissociation energies of H_2^+ and H_2 have been made by Kolos and Wolniewicz using 100-term wave functions.* The

* W. Kolos and L. Wolniewicz, *J. Chem. Phys.*, **41**, 3663 (1964); *ibid.*, **48**, 3672 (1968); *ibid.*, **49**, 404 (1968).

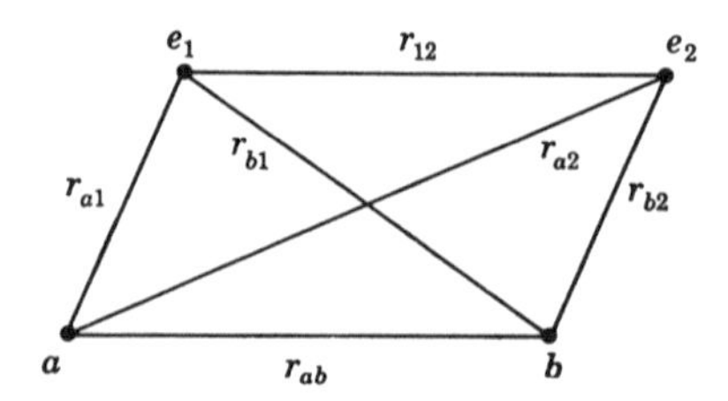

Fig. 10.7 Coordinates in the hydrogen molecule. The two protons are represented by a and b.

theoretical value of D^0 for H_2 is 36,117.8 cm^{-1} and the observed value* is 36,117.3 $\pm$ 1.0 cm^{-1}. The theoretical value of the internuclear distance for H_2 is 0.074 140 nm, compared with the experimental value from spectroscopic measurements of 0.074 139 nm.

In order to illustrate how simple molecular orbital theory is used with more than one electron, we will extend to H_2 the treatment of H_2^+ given in the preceding section.

If electron-electron repulsion is ignored, the probability $\psi(1)^2\,d\tau$ of finding electron 1 in a volume $d\tau$ is independent of the probability $\psi(2)^2\,d\tau$ of finding electron 2 in a volume $d\tau$. The probability of simultaneously finding electron 1 and electron 2 in volume $d\tau$ is the product of the individual probabilities or $\psi(1)^2\psi(2)^2\,d\tau^2$. Therefore it seems reasonable that the wave function that would describe the electron distribution of a hydrogen molecule without electron-electron repulsion is the product of the wave functions for the two electrons; $\psi = \psi(1)\psi(2)$. As a first approximation we can put two electrons into the H_2^+ LCAO-MO orbital (equation 10.30).†

$$\psi_{\text{MO}} = \psi_g(1)\psi_g(2) \tag{10.35}$$

$$\psi_{\text{MO}} = \frac{1}{2(1+S)}[(1s_a)(1) + (1s_b)(1)][(1s_a)(2) + (1s_b)(2)]$$

$$= \frac{1}{2(1+S)}[(1s_a)(1)(1s_b)(2) + (1s_a)(2)(1s_b)(1) + (1s_a)(1)(1s_a)(2) + (1s_b)(1)(1s_b)(2)] \tag{10.36}$$

The first two terms in this last expression correspond to configurations in which one electron is near one proton and the other is near the other. However, the third and fourth terms correspond to configurations in which both electrons 1 and 2 are near the same proton; that is, they correspond to the ionic structures $H_a^-H_b^+$ and

* G. Herzberg, *J. Mol. Spect.*, **33**, 147 (1970); *Science*, **177**, 123 (1972).

† As discussed in the case of the helium atom (Section 8.15), the total wave function consisting of a spatial part times a spin part must be antisymmetric with respect to the interchange of two electrons. However, the spin function is relatively unimportant in evaluating the various integrals encountered in the variation method, and so we will not include it here.

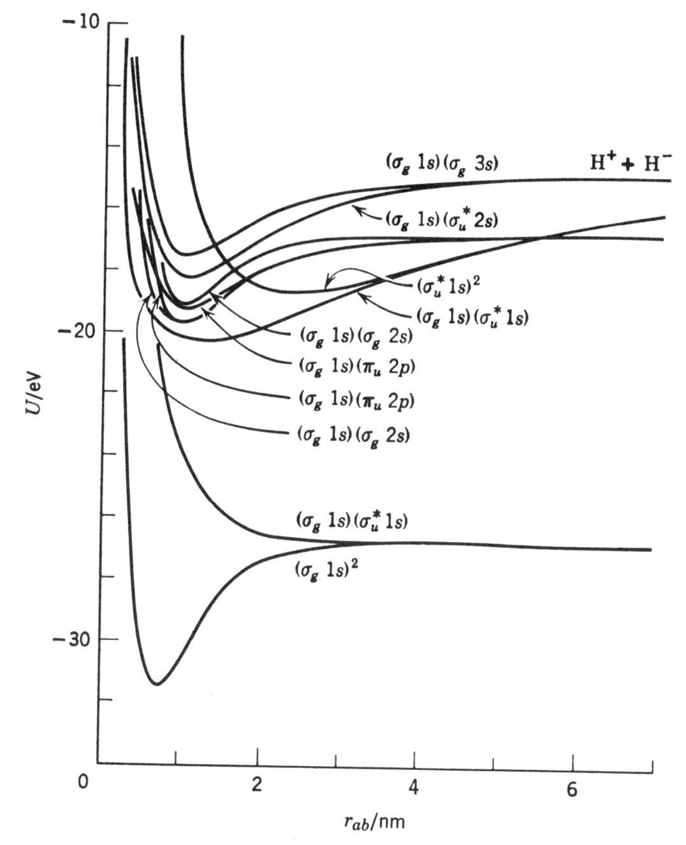

Fig. 10.8 Potential energy curves for ground and excited states of H_2. (From J. C. Davis, *Advanced Physical Chemistry*, The Ronald Press Co., New York, 1965.)

$H_a^+H_b^-$. The wave function ψ_{MO} leads to a dissociation energy for H_2 of 2.681 eV and an internuclear distance of 0.085 nm. Although this agreement with experiment is not very good, the simple LCAO-MO wave function given by equation 10.35 can be improved by introducing an effective nuclear charge and including excited H_2^+-like MOs in the trial function. When properties are calculated in this way, it is not possible to make simple interpretations of the terms of the wave function, but it does become possible to calculate the bond length and binding energy more accurately than they can be measured.

Another simple theory for chemical bonding in the valence-bond (VB) method. In this method complete atoms are brought together and allowed to interact. Whereas in the molecular-orbital theory all electrons are considered to belong to the whole molecule, in the valence-bond theory a pair of electrons is considered to belong to the pair of atoms that is bonded together.

The valence-bond theory leads to a different trial wave function for H_2. If two hydrogen atoms *a* and *b* are infinitely far apart, the system is described by the wave function

$$\psi = (1s_a)(1)(1s_b)(2) \tag{10.37}$$

where 1 designates the electron near proton *a* and 2 designates the electron near proton *b*. The square of the wave function ψ is the probability density function for the system that is equal to the product of two probability densities $[(1s_a)(1)]^2$ and $[(1s_b)(2)]^2$. Thus the probability of finding electron 1 in a given volume and electron 2 in another specified volume is equal to the product of two probabilities.

As the two atoms are brought closer together, the system might continue to be described by wave function 10.37, except that the electrons are indistinguishable. This indistinguishability of the electrons is taken into account by using as a wave function the sum of the two wave functions that are equally likely to be correct.

$$\psi_{\text{VB}} = (1s_a)(1)(1s_b)(2) + (1s_a)(2)(1s_b)(1) \tag{10.38}$$

This wave function is the basis of the Heitler-London treatment of 1927, which gave the first reasonable theoretical values for the dissociation energy of H_2. This wave function given by simple valence-bond theory differs from that for the simple molecular-orbital theory by lack of the "ionic terms." The simple molecular-orbital theory overemphasizes the ionic terms, and the simple-valence theory lacks these terms. The VB wave function can be improved by adding contributions from these ionic structures, and the MO wave function can be improved by providing for a smaller contribution by the ionic structures. Thus better results can be obtained with

$$\psi = c_1\psi_{\text{VB}} + c_2\psi_{\text{ionic}} \tag{10.39}$$

where

$$\psi_{\text{ionic}} = (1s_a)(1)(1s_a)(2) + (1s_b)(1)(1s_b)(2) \tag{10.40}$$

Thus with further improvement the distinction between the two methods disappears, and the same result is obtained from either starting point. Another way of improving the MO wave function, which again leads to the same result, is to "mix in" the excited-state wave function σ_u^*1s, which is given by

$$\psi'_{\text{MO}} = C[(1s_a)(1) - (1s_b)(1)][(1s_a)(2) - (1s_b)(2)] \tag{10.41}$$

to obtain the wave function

$$\psi = c_3\psi_{\text{MO}} + c_4\psi'_{\text{MO}} \tag{10.42}$$

By bringing in other functions and changing the effective nuclear charge in the wave functions, it is possible to get closer and closer to the experimental values for the dissociation energy and the equilibrium internuclear distance.

In the ground state of the hydrogen molecule the net electron spin for the molecule is zero. This is a *singlet* state and the energy level is not split in an electric or magnetic field. The ground states of most molecules containing an even number of electrons are singlet states. A hydrogen molecule with one electron in the bonding orbital and another with the same spin in the antibonding orbital is in a *triplet* state (Section 8.18).

The bond between two hydrogen atoms is exceptional in that the principal source of repulsion energy is due to the electrostatic repulsion of the protons. For a covalent bond not involving a hydrogen atom, each nucleus is shielded by a core of inner electrons. A repulsion arises when these cores begin to interpenetrate. If a third electron is added to the bond, giving H_2^-, it must be added to the antibonding orbital, as required by the Pauli exclusion principle. Since the antibonding orbital has a much higher energy, this leads to a high energy for H_2^- and instability with respect to $H_2 + e$.

10.5 ELECTRON CONFIGURATIONS OF HOMONUCLEAR DIATOMIC MOLECULES

From the quantum mechanical treatments of the hydrogen molecule ion and the hydrogen molecule we conclude that bonding occurs where there is overlap of atomic orbitals. In order to treat the bonding of diatomic molecules with more electrons, we must consider the possibility of forming bonds with other atomic orbitals. Not all combinations lead to bonding. First, it can be shown that only atomic orbitals of approximately the same energies can be combined. Second, the greater the overlap, the greater the strength of the bond. Overlap increases as atoms are brought closer together, but repulsion arises at close distances because of the electrostatic repulsions of the inner electron clouds and of the two positively charged nuclei.

As we discovered earlier, when two atomic orbitals are combined, two molecular orbitals are formed—a bonding orbital and an antibonding orbital. Antibonding orbitals are designated with an asterisk (*). Figure 10.9 shows the various possibilities for combining s and p orbitals. As we have seen, the combination of two $1s$ orbitals leads to a $\sigma_g 1s$ bonding orbital and a $\sigma_u^* 1s$ antibonding orbital. The $2s$ atomic orbitals combine in a similar way. The spherical node surrounding each nucleus in this case may be neglected because the orbital overlap is not significant this close to the nucleus.

An s orbital does not form a molecular orbital with a p_y or p_z orbital, that is, with a p orbital that is perpendicular to the internuclear axis. The overlap integral is equal to zero because the positive contribution due to one-half of the p orbital is exactly compensated by a negative contribution due to the other half of the p orbital.

There are two different ways of forming bonding orbitals from p orbitals. If the lobes of the p orbitals are pointed at each other along the internuclear axis, two σ orbitals are formed, one bonding and one antibonding. In contrast with the other LCAO-MOs the wave function for the bonding orbital has a negative sign. If the lobes of the p orbitals are perpendicular to the internuclear axis (p_y and p_z) they can overlap sideways to form π orbitals. The $\pi_u 2p$ orbital produces bonding because there is electron density that tends to draw the two nuclei together, even though it is not on the internuclear axis. There are two $\pi_u 2p$ bonding orbitals and two $\pi_g^* 2p$

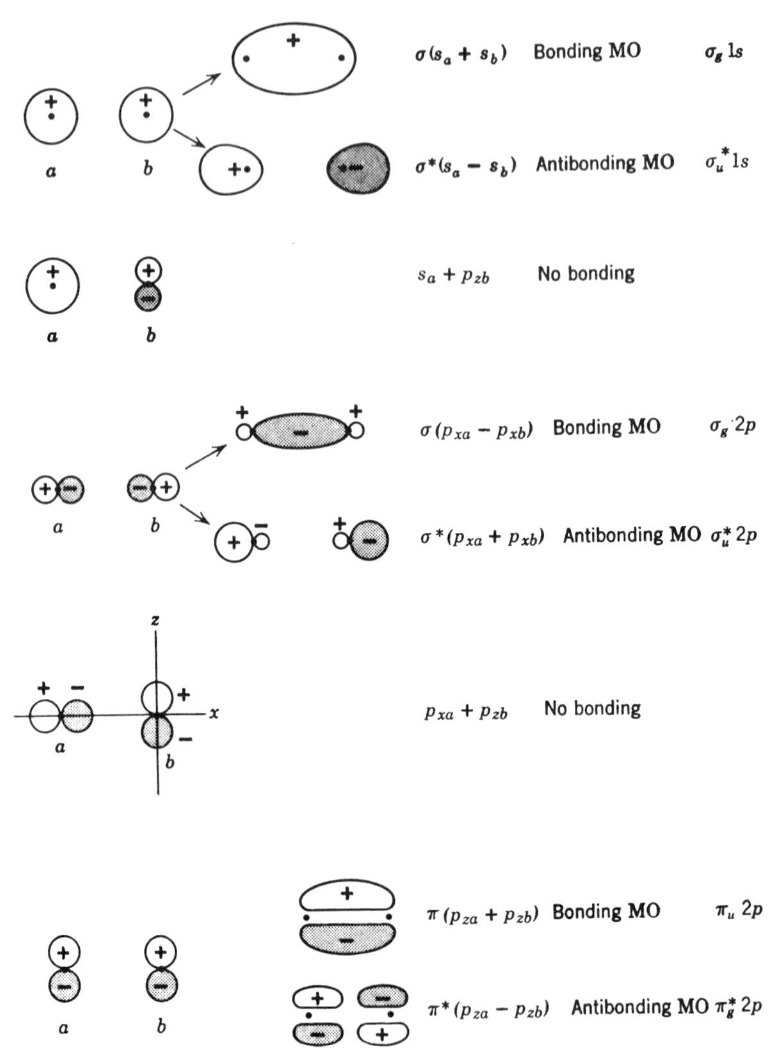

Fig. 10.9 Combinations of orbitals on nuclei *a* and *b* to form bonding and antibonding (indicated by *) orbitals. The MO wave functions and designations are given for the molecular orbitals. The *x* axis goes through the two nuclei. In the shaded region the wave function is negative.

antibonding orbitals because there are two p_y and two p_z atomic orbitals on the two nuclei. Thus six molecular orbitals are formed from the six $2p$ orbitals on the two nuclei, three bonding and three antibonding.

Orbitals of still higher energy can be formed from $3s$, $3p$, $3d$, and higher atomic orbitals, but we have carried the process far enough for the discussion of possible homonuclear diatomic molecules from He_2 to Ne_2. This simple molecular orbital theory helps us to understand whether stable molecules are formed and tells us something about the relative values of binding energies and bond lengths.

The electronic structure of molecules can be discussed in terms of the aufbau principle (Section 8.20) that is used to explain the periodic table. Following the Pauli principle that there may be only two electrons in an orbital, electrons are placed in orbitals starting with the lowest. In order to estimate the relative energies of various molecular orbitals of homonuclear diatomic molecules, it is convenient to use the correlation diagram given in Fig. 10.10. In constructing this diagram two separated atoms in the indicated electronic state are imagined to be pushed together until their nuclei coincide, in other words, until a *united atom* of twice the nuclear charge is formed. Figure 10.10 is based on the idea that the energies of the orbitals change smoothly in going from the separated atoms to the united atoms. The abscissa represents bond distances of homonuclear diatomic molecules. In carrying lines through to the united atom the principle is followed that molecular orbitals with a given angular momentum connect with atomic orbitals of the united atom with the same angular momentum $|m|$. The lines are also drawn so that the orbitals are either symmetric or antisymmetric with respect to a reflection plane halfway between the nuclei or through the nucleus of the united atom. The molecular orbitals $(1s_a + 1s_b)$, $(2s_a + 2s_b)$, $(2p_{za} - 2p_{zb})$, $(2p_{xa} + 2p_{xb})$, and $(2p_{ya} - 2p_{yb})$ are all symmetric and connect in the united atom with the symmetric atomic orbitals $1s$, $2s$, $3s$, $2p_x$, and $2p_y$, respectively. The molecular orbitals $(1s_a - 1s_b)$, $(2s_a - 2s_b)$, $(2p_{za} + 2p_{zb})$, $(2p_{xa} - 2p_{xb})$, and $(2p_{ya} + 2p_{yb})$ are antisymmetric and connect in the united atom with the antisymmetric atomic orbitals $2p_z$, $3p_z$, $4p_z$, $3d_{xz}$, and $3d_{yz}$, respectively. Such a correlation diagram is only approximate and is not a completely reliable guide to the order of energies of molecular orbitals. The net number of bonding electrons in a molecule equals the number of electrons in bonding orbitals minus the number of electrons in antibonding orbitals. The covalent bond order is the net number of bonding electrons divided by two. The bond orders and other information about homonuclear diatomic molecules from H_2^+ to F_2 are given in Table 10.2.

He_2^+

The third electron in this molecular ion is an antibonding orbital, but there is net bonding because there are two electrons in bonding orbitals. This ion is observed in electric arcs.

He_2

According to the simple molecular orbital theory, a pair of electrons would occupy the $\sigma_g 1s$ orbital and a pair would occupy the $\sigma_u^* 1s$ orbital. Since both the bonding and antibonding orbitals are filled, there is no decrease in energy as compared to two isolated helium atoms, and a stable He_2 molecule is not formed.

Li_2

The additional pair of electrons beyond those for He_2 go into the $\sigma_g 2s$ orbital. Thus there is a single bond between the two nuclei.

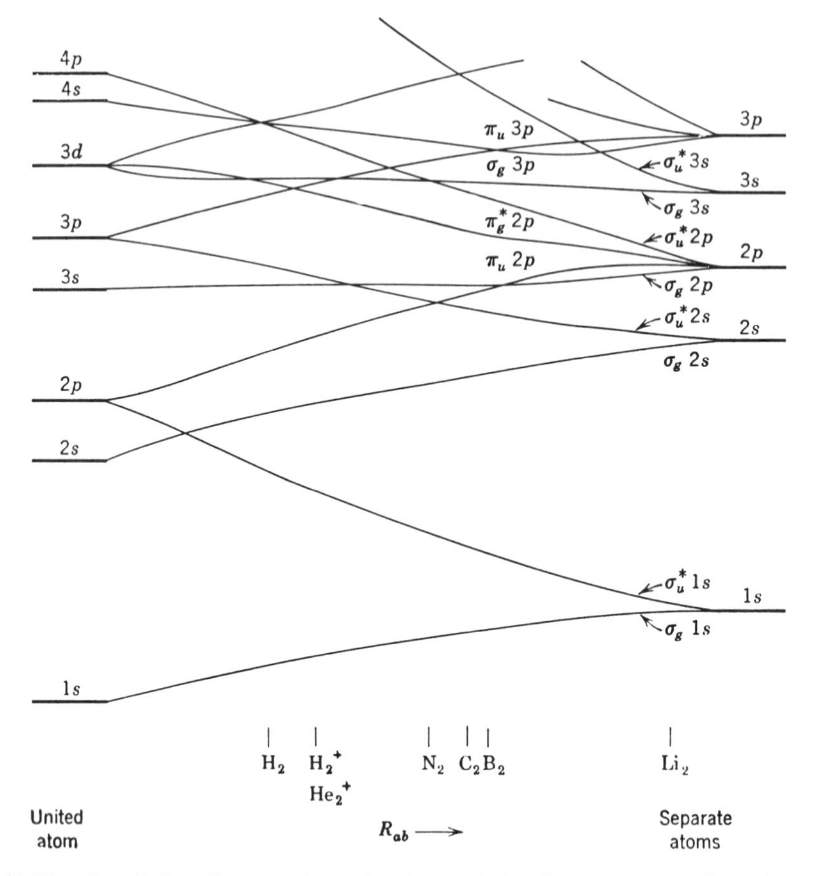

Fig. 10.10 Correlation diagram for molecular orbitals of homonuclear diatomic molecules. (From J. C. Davis, *Advanced Physical Chemistry*, The Ronald Press Co., New York, 1965.)

Be_2

The additional pair of electrons beyond Li_2 goes into the σ_u^*2s antibonding orbital. Thus, as in the case of He_2, there is no net stabilization as compared with isolated Be atoms.

B_2

According to Fig. 10.10, the additional pair of electrons beyond Be_2 might go into either the $\sigma_g 2p$ orbital or the $\pi_u 2p$ orbital, in either case producing a net stabilization. If the pair of electrons goes into the $\sigma_g 2p$ orbital, a molecule is formed in which all the electrons are paired (a singlet state). A stable B_2 molecule is formed, as indicated in Table 10.2, but spectroscopic measurements show that its ground state is a triplet (Section 8.18), so that the outer two electrons are in different $\pi 2p$ states. If there are several orbitals with the same energy, Hund's rules

Table 10.2 Ground States of Homonuclear Diatomic Molecules and Ions

Molecule	Number of Electrons	Configuration	Bond Order	R_{ab}/nm	D^e eV	D^e kJ mol^{-1}
H_2^+	1	$(\sigma 1s)$	$\frac{1}{2}$	0.106 0	2.793	269.483
H_2	2	$(\sigma 1s)^2$	1	0.074 12	4.748 3	458.135
He_2^+	3	$(\sigma 1s)^2(\sigma^* 1s)$	$\frac{1}{2}$	0.108 0	2.5	238
He_2	4	$(\sigma 1s)^2(\sigma^* 1s)^2$	0	—	—	–
Li_2	6	$[He_2](\sigma 2s)^2$	1	0.267 3	1.14	110.0
Be_2	8	$[He_2](\sigma 2s)^2(\sigma^* 2s)^2$	0	—	—	—
B_2	10	$[Be_2](\pi 2p)^2$	1	0.158 9	~3.0	~290
C_2	12	$[Be_2](\pi 2p)^4$	2	0.124 2	6.36	613.8
N_2^+	13	$[Be_2](\pi 2p)^4(\sigma 2p)$	$2\frac{1}{2}$	0.111 6	8.86	854.8
N_2	14	$[Be_2](\sigma 2p)^2(\pi 2p)^4$	3	0.109 4	9.902	955.42
O_2^+	15	$[N_2](\pi^* 2p)$	$2\frac{1}{2}$	1.112 27	6.77	653.1
O_2	16	$[N_2](\pi^* 2p)^2$	2	0.120 74	5.213	502.9
F_2	18	$[N_2](\pi^* 2p)^4$	1	0.143 5	1.34	118.8
Ne_2	20	$[N_2](\pi^* 2p)^4(\sigma^* 2p)^2$	0	—	—	—

(Section 8.20) require that the electrons spread themselves among several orbitals. Since there are two unpaired electrons, the ground state is a triplet.

C_2

The additional two electrons beyond B_2 fill the two half-vacant $\pi_u 2p$ bonding orbitals. Thus C_2 has four electrons in bonding orbitals that are not compensated by electrons in corresponding antibonding orbitals. Thus we would expect C_2 to have tighter binding and a smaller internuclear distance than B_2.

N_2

The additional pair of electrons beyond C_2 fills the remaining p bonding orbitals. Thus N_2 has a singlet ground state and a triple bond.

O_2

According to the simple molecular orbital theory, the additional pair of electrons beyond N_2 would go into the $\pi_u^* 2p$ orbitals. According to Hund's rule, one electron would go into each of these degenerate orbits, and their spins would be parallel. This is in accord with the facts, since molecular oxygen is paramagnetic (see Section 12.1) and its ground state is a triplet.

F_2

The additional two electrons beyond O_2 fill the $\pi_u^* 2p$ orbitals so that the ground state is a singlet. Since the electron pairs in two π_u^* antibonding orbitals approximately cancel the bonding due to the electron pairs in the two π_g orbitals, the bonding is weaker than in O_2, and the internuclear distance is greater.

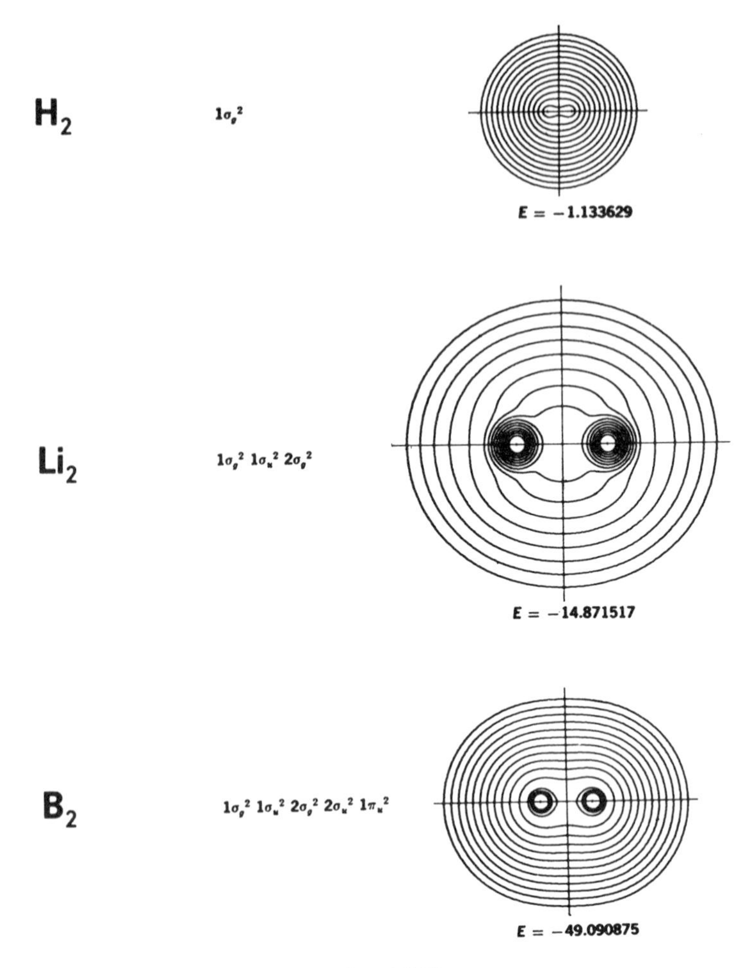

Fig. 10.11*a*

Fig. 10.11 Electron densities for covalent diatomic molecules. The charge density is plotted in electrons per cubic bohr. Adjacent contour lines differ by a factor of 2. The innermost contour line for each diagram is 1 electron (bohr)$^{-3}$, except for H_2, where it is 0.25 electron (bohr)$^{-3}$. The outermost contour line is 6.1×10^{-5} electron (bohr)$^{-3}$ in all cases. A bohr is 5.29×10^{-11} m. (From *Atomic and Molecular Structure: 4 Wall Charts* by Arnold C. Wahl. Copyright © 1970 by McGraw-Hill, Inc. Used with permission of McGraw-Hill Book Co.)

Ne_2

The σ_u^*2p orbital would be filled, and so the antibonding effects cancel the bonding effects, and there is no tendency to form a stable molecule.

Figure 10.11 displays contour diagrams for the electron densities of the stable diatomic molecules from H_2 to F_2. These electron densities were calculated by the Hartree-Fock method.

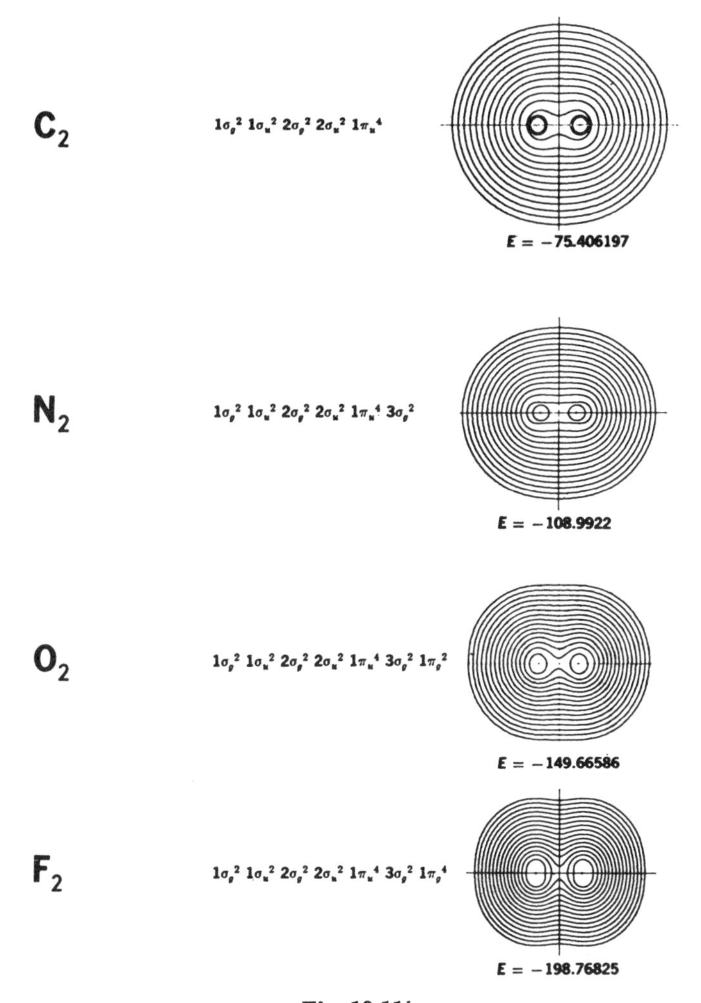

Fig. 10.11*b*

The treatment of diatomic molecules with different nuclei is complicated by the fact that the same atomic orbitals will not be involved on both atoms because of size, energy, and directional differences. In order for atomic orbitals on two atoms to be involved in bonding, they must have the same $\sigma, \pi, \ldots$ properties and energies that are not too different.

In heteronuclear diatomic molecules, bonding electrons are not shared equally between the two atoms. In an extreme case like Na^+Cl^-, an electron is transferred completely, and the electrostatic attraction of the two atoms makes the major contribution to the bonding, as discussed in Section 10.9. The difference in affinity of different atoms in a molecule for electrons is discussed in terms of electronegativity.

10.6 ELECTRONIC STRUCTURE OF POLYATOMIC MOLECULES

The wave function for a polyatomic molecule depends on all the bond distances and angles. In order to calculate the equilibrium configuration, it is necessary to calculate the wave function for a range of bond distances and angles. However, most molecular orbital calculations are made for the experimentally determined equilibrium configuration.

There are two types of quantum-mechanical calculations, ab initio and semiempirical. In ab initio calculations the correct Hamiltonian is used and a solution is sought without the use of experimental data. In semiempirical calculations an approximate Hamiltonian is used and parameters that can be adjusted to fit experimental data are incorporated.

10.7 ELECTRONEGATIVITY

There are several ways of estimating the tendency of an atom in a molecule to attract electrons to itself and become more negative. For example, this tendency in a molecule AB is measured by the relative stabilities of A^+B^- and A^-B^+. The energy difference between these structures is given by

$$\begin{aligned} E(\mathrm{A^+B^-}) - E(\mathrm{A^-B^+}) &= (IP_\mathrm{A} - EA_\mathrm{B}) - (IP_\mathrm{B} - EA_\mathrm{A}) \\ &= (IP_\mathrm{A} + EA_\mathrm{A}) - (IP_\mathrm{B} + EA_\mathrm{B}) \qquad (10.43) \end{aligned}$$

where IP is ionization potential and EA is electron affinity (Section 8.21). Mulliken defined the electronegativity difference between atoms A and B as $\frac{1}{2}[E(\mathrm{A^+B^-}) - E(\mathrm{A^-B^+})]$. Thus he defined the electronegativity of an atom as $\frac{1}{2}(IP + EA)$. The electronegativity of an atom depends on its valence state, and so the ionization potential and electron affinity used are not simply those of the ground state of the atom.

The Pauling electronegativity scale is based on his observation that the A—B bond energy exceeds the mean of the A—A and B—B bond energies if the A—B bond is polar. Pauling defined the electronegativity difference between A and B as $0.102(\Delta_\mathrm{AB}/\mathrm{kJ\ mol^{-1}})^{1/2}$, where

$$\Delta_\mathrm{AB} = E(\mathrm{A—B}) - \tfrac{1}{2}[E(\mathrm{A—A}) + E(\mathrm{B—B})]$$

The E's are the dissociation energies for the bonds in question. The electronegativity of H was set at 2.1. Values on the Mulliken scale in eV can be converted to the Pauling scale by dividing by 3.17. Exact agreement is not obtained, but the scales and in relatively good agreement. Fluorine is the most electronegative atom (4.0 on Pauling's scale), and cesium is the least electronegative atom (0.7 on Pauling's scale). The electronegativities of a number of elements are given in Fig. 10.12,

		H 2.1				
Li 1.0	Be 1.5	B 2.0	C 2.5	N 3.0	O 3.5	F 4.0
Na 0.9	Mg 1.2	Al 1.5	Si 1.8	P 2.1	S 2.5	Cl 3.0
K 0.8	Ca 1.0		Ge 1.7	As 2.0	Se 2.4	Br 2.8
Rb 0.8	Sr 1.0		Sn 1.7	Sb 1.8	Te 2.1	I 2.4
Cs 0.7	Ba 0.9					

Fig. 10.12 Dependence of electronegativity on position in the periodic table.

which shows that the electronegativity depends on the position of the element in the periodic table. As we go down the halogen column of the periodic table, the atoms become less electronegative because of the increasingly effective screening of the charge on the nucleus by inner electrons. The alkali-metal atoms have a great tendency to lose their outer electrons and therefore have a low electronegativity. Again the electronegativity decreases as we go down a column because of the increasingly effective screening of the charge on the nucleus by inner electrons.

By use of electronegativities it is possible to predict which bonds will be ionic and which bonds will be covalent. Two elements of very different electronegativity, like a halogen and an alkali metal, form an ionic bond because an electron is almost completely transferred to the atom of higher electronegativity. Two elements with nearly equal electronegativities form covalent bonds. For instance, carbon, which occupies an intermediate position in the electronegativity scale, forms covalent bonds with elements near it in the periodic table. If there is a considerable difference between the electronegativities of the two elements, the bond is polar (i.e., possessed of a high degree of ionic character), as in the case of sodium chloride. In the majority of chemical bonds the sharing of the electron pair is not exactly equal, so that the bond has some ionic character that results in a dipole moment (Section 10.10) for the bond.

10.8 HYDROGEN BONDS

A number of unusual structures such as HF_2^- and acetic acid dimer in the gas phase (see Fig. 10.13) are evidence for the formation of hydrogen bonds. The unusually high acid dissociation constant of salicylic acid, as compared with the meta and para isomers, is also evidence for a hydrogen bond. A hydrogen bond

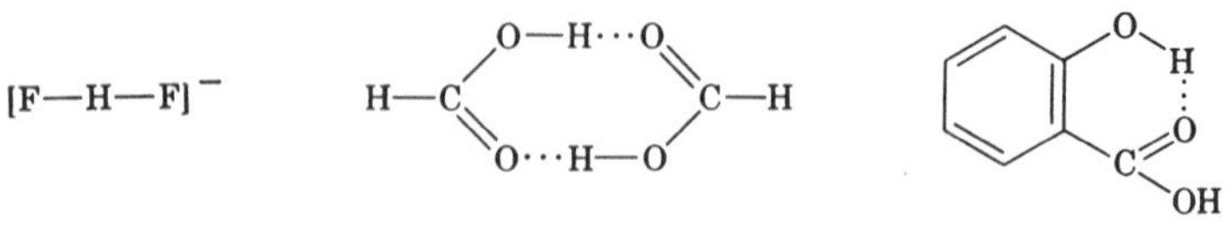

$[F—H—F]^-$

Formic acid dimer

Salicylic acid

Fig. 10.13 Examples of hydrogen bonding.

results when a proton may be shared between two electronegative atoms, such as F, O, or N, which are the right distance apart. The proton of the hydrogen bond is attracted by the high concentration of negative charge in the vicinity of these electronegative atoms. Fluorine forms very strong hydrogen bonds; oxygen, weaker ones; and nitrogen still weaker ones. The unusual properties of water are due to a large extent to the formation of hydrogen bonds involving the four lone pair electrons on oxygen. In ice there is a tetrahedral arrangement with each oxygen atom bonded to four hydrogen atoms. Hydrogen bonds are formed along the axis of each lone pair in ice, and their existence in liquid water is responsible for the high boiling point of water as compared with the boiling points of hydrides of other elements in the same column of the periodic table (H_2S, -62 °C; H_2Se, -42 °C; H_2Te, -4 °C). When water is vaporized, these hydrogen bonds are broken, but in formic and acetic acids the hydrogen bonds are strong enough for double molecules of the type illustrated in Fig. 10.13 to exist in the vapor. Benzoic and other carboxylic acids form dimers in certain nonpolar solvents, such as benzene and carbon tetrachloride.

Hydrogen bonds between N and O are responsible for the stability of the α helix formed by polypeptides. This structure is shown in Fig. 20.8. Helices of this structure are found in protein molecules.

10.9 IONIC BINDING

If an electron is completely transferred from one atom to another the molecule may be held together primarily by Coulomb attraction. For example, NaCl molecules exist in the gas phase, and the dissociation energy ($D^e = 4.29$ eV) is about what would be expected from ionic bonding. The energy of dissociation of NaCl(g) is equal to the energy required to move the ions from their equilibrium internuclear distance (0.236 nm) to infinity, and then transfer an electron from Cl^- to Na^+. The work required to separate the ions to infinite distance is 6.09 eV, the energy required to remove the electron from Cl^- is 3.61 eV, and the energy gained by transferring the electron to Na^+ is 5.14 eV. According to this simple model, the energy required to dissociate Na^+Cl^- into two neutral atoms is $6.09 + 3.61 - 5.14 = 4.56$ eV. The difference between this value and the experimental value is due to the neglect of the repulsion of Na^+ and Cl^- at close distances because of overlapping of their electron clouds.

10.10 DIPOLE MOMENT

When atoms of different electronegativity are bonded together, there is an excess of electron charge on the more electronegative atom and an excess of positive charge on the less electronegative atom. Thus a heteronuclear diatomic molecule may be an electric dipole and have a dipole moment $\boldsymbol{\mu}$. If a positive charge $+q$ is separated from a negative charge $-q$ by distance r, the dipole moment has the magnitude $\mu = qr$. The dipole moment $\boldsymbol{\mu}$ is a vector pointing from the negative charge to the positive charge with a magnitude of $|\boldsymbol{\mu}|$. The SI unit for dipole moment is C m. Dipole moments have often been expressed in debye units D after Peter Debye, who made so many contributions to the understanding of polar molecules. A debye unit D is 3.336×10^{-30} C m.*

The dipole moment $\boldsymbol{\mu}$ may be calculated for any arrangement of point charges q_i having a net charge of zero. If the vector from the origin to q_i is $\boldsymbol{r}_i$,

$$\boldsymbol{\mu} = \sum_i q_i \boldsymbol{r}_i \tag{10.44}$$

If the net charge of the system is zero ($\sum q_i = 0$) it is easy to show that the dipole moment $\boldsymbol{\mu}$ is independent of the choice of origin. The components of the dipole moment vector are

$$\mu_x = \sum_i q_i x_i \qquad \mu_y = \sum_i q_i y_i \qquad \mu_z = \sum_i q_i z_i \tag{10.45}$$

where x_i, y_i, and z_i are the coordinates of charge q_i.

10.11 DIELECTRIC CONSTANT

The capacitance C of a capacitor is defined as the ratio of the charge on one of the plates to the potential difference between the conductors. Thus the capacitance has the units of coulombs per volt; this unit is referred to as the farad F; that is, 1 F = 1 C/V.

If an insulating material, often referred to as a dielectric, is placed between the plates of the capacitor, the capacitance is increased because net charges appear on the surface of the dielectric, as shown in Fig. 10.14. As a result of these surface charges, the electric field within the dielectric is reduced below that which would have existed in the capacitor with the same charge on the plates in the absence of the dielectric. Since the potential difference between the plates is reduced by the

---

* When dipole moments are expressed in cgs units it is convenient to use the unit 10^{-18} esu cm, which is the original debye unit. The magnitude of the dipole moment for a dipole consisting of an electron separated from a unit positive charge by 10^{-8} cm is $(4.803 \times 10^{-10}\text{ esu})(10^{-8}\text{ cm}) = 4.803 \times 10^{-18}$ esu cm. The debye may be expressed in SI units as follows.

$$\frac{(10^{-18}\text{ esu cm})(1.602 \times 10^{-19}\text{ C})(10^{-2}\text{ m cm}^{-1})}{(4.803 \times 10^{-10}\text{ esu})} = 3.336 \times 10^{-30}\text{ C m}$$

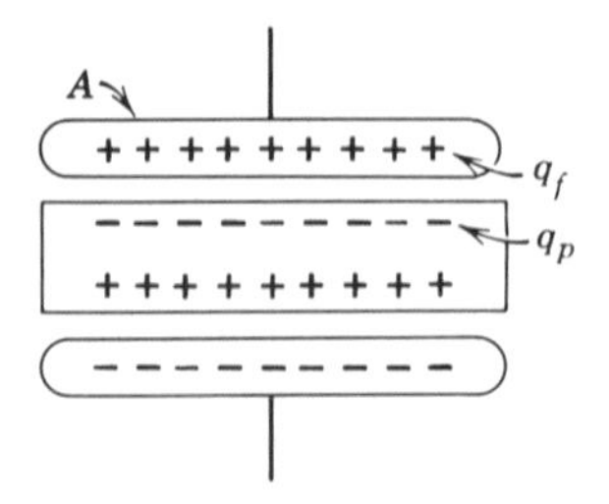

Fig. 10.14 Parallel-plate capacitor filled with a dielectric.

dielectric and the capacitance is equal to the ratio of the charge on one of the plates to the potential difference between the plates, the capacitance increases.

The dielectric constant κ of a dielectric is defined as the ratio of the capacitance C_d of the capacitor filled with dielectric to the capacitance C when the capacitor is evacuated.

$$\kappa = \frac{C_d}{C} \tag{10.46}$$

Thus the dielectric constant for a vacuum is unity. The magnitude of the dielectric constant of a substance depends on the temperature and, if an alternating electric field is used, the frequency. The dielectric constants of a few gases and liquids are given in Table 10.3. These values of the dielectric constant apply at frequencies sufficiently low that equilibrium is maintained as the electric field varies.

The electrical permittivity ϵ of a dielectric is given by $\epsilon = \kappa\epsilon_0$, where ϵ_0 is the permittivity of vacuum (see Section 5.1).

In Fig. 10.14 the free charge on the plates of the capacitor is represented by q_f and the polarization charge on the surface of the dielectric is represented by q_p. The surface charge density of the polarization charges is referred to as the electric polarization. The polarization vector $\boldsymbol{P}$ is perpendicular to the surface and points from the negative surface charges on the dielectric to the positive surface charges.

Table 10.3 Dielectric Constants κ of Gases and Liquids

Gas (1 atm)	At 0 °C	Liquid	At 20 °C
Hydrogen	1.000 272	Hexane	1.874
Argon	1.000 545	Benzene	2.283
Air (CO_2 free)	1.000 567	Toluene	2.387
Carbon dioxide	1.000 98	Chlorobenzene	5.94
Hydrogen chloride	1.004 6	Ammonia	15.5
Ammonia	1.007 2	Acetone	21.4
Water (steam at 110 °C)	1.012 6	Methanol	33.1
		Water	80

The magnitude of the polarization vector has the units C m^{-2} or C m m^{-3}, that is, electric dipole moment per unit volume. The magnitude of the polarization is proportional to the applied electric field.

Now we want to obtain the relation between the polarization $\boldsymbol{P}$, the dielectric constant κ, and the electric field $\boldsymbol{E}$. In order to do that we need to introduce the electric displacement vector $\boldsymbol{D}$, which is defined by

$$\boldsymbol{D} \equiv \epsilon_0 \boldsymbol{E} + \boldsymbol{P} \tag{10.47}$$

The electric displacement has the units C m^{-2}. The electric displacement vector is useful because it does not change when a dielectric is introduced into a capacitor and the same free charges remain on the plates of the capacitor. When there is no dielectric in the capacitor,

$$\boldsymbol{D} = \epsilon_0 \boldsymbol{E}_0 \tag{10.48}$$

where $\boldsymbol{E}_0$ is the electric field between the plates in this case. When a dielectric is inserted between the plates, $\boldsymbol{D}$ does not change and so

$$\boldsymbol{D} = \epsilon_0 \boldsymbol{E}_0 = \epsilon_0 \boldsymbol{E} + \boldsymbol{P} \tag{10.49}$$

Since the polarization of the medium opposes the charge on the capacitor plates, it is responsible for lowering the electric field by a factor equal to the dielectric constant. Thus $\boldsymbol{E} = \boldsymbol{E}_0/\kappa$, and

$$\boldsymbol{P} = \epsilon_0 \boldsymbol{E}(\kappa - 1) \tag{10.50}$$

When an electric field is applied to a sample, time is required for the establishment of the equilibrium polarization. Thus the dielectric constant is a frequency-dependent quantity. We will first consider a static field, or one that is changing so slowly that the polarization has its equilibrium value at all times.

10.12 POLARIZATION

Three contributions to the polarization must be considered: orientation polarization, electronic polarization, and vibrational polarization. Orientation polarization is due to the partial alignment of permanent dipoles. This is illustrated in Fig. 10.15. Molecules that have permanent dipole moments are not completely oriented in an electric field because of the disorienting effect of thermal motion. The extent to which dipoles can be oriented by an applied field was calculated by Debye* using the Boltzmann distribution law. The electric field acting on the molecule is represented by $\boldsymbol{E}_i$ and referred to as the internal field. The energy of a dipole in the field $\boldsymbol{E}_i$ is $-\boldsymbol{\mu}\cdot\boldsymbol{E}_i$, where $\boldsymbol{\mu}$ is the permanent dipole moment vector of the molecule and the dot refers to the dot product: $-\boldsymbol{\mu}\cdot\boldsymbol{E}_i = -\mu E_i \cos\theta$, where θ is the angle between the two vectors. If the energy of the dipole in the field is small compared

* P. Debye, *Polar Molecules*, Chemical Catalog Co., New York, 1929; or Dover Publications, Yew York.

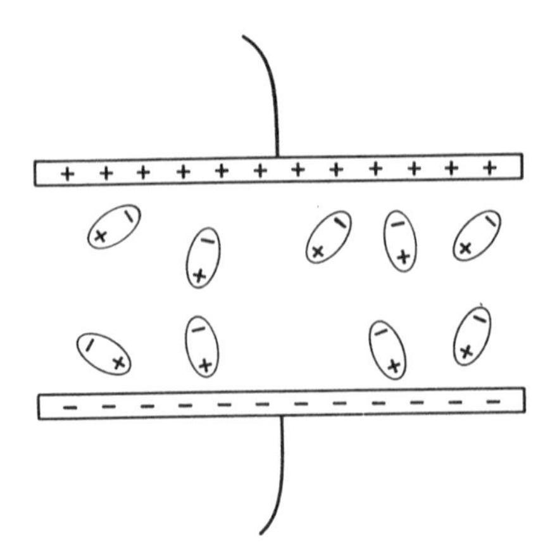

Fig. 10.15 Diagrammatic representation of a capacitor containing dipolar molecules. Although the molecules are all shown as partially oriented with the field, this is far from the true case, because of the disorienting effect of thermal motion.

with kT, it may be shown that the contribution per molecule in a gas to the mean moment in the direction of the field is given by $\mu^2 E_i/3kT$, where E_i is the magnitude of the internal field. As the temperature is increased, the thermal agitation becomes more vigorous and fewer of the permanent dipoles are oriented in the direction of the field.

The second contribution is from the field-induced displacement of the average positions of the electrons relative to the nuclei of the molecule. This electronic contribution $\alpha_e \boldsymbol{E}_i$ is practically independent of the temperature and is proportional to the electric field strength. The proportionality constant α_e is called the *mean molecular electronic polarizability*. The polarizability of a molecule is different in different directions in the molecule. The mean polarizability is the average taken over all possible orientations of the molecule with respect to the field. The mean polarizability is used because thermal motion ensures that all possible orientations are encountered. The polarizability α has the units of dipole moment divided by electric field strength, that is, $\mathrm{C\,m/V\,m^{-1}} = \mathrm{C^2\,m^2/J}$. The ease with which different molecules become polarized by distortion of the electron cloud differs greatly.

The third contribution $\alpha_v \boldsymbol{E}_i$ is from the deformation of the nuclear skeleton of the molecule by the electric field. The proportionality constant α_v is called the *mean molecular vibrational polarizability*. To a first approximation α_v is independent of temperature.

The magnitude $\boldsymbol{P}$ of the polarization, which includes all three contributions, is equal to the mean moment in the direction of the field multiplied by the number of molecules per unit volume.

$$\boldsymbol{P} = \frac{N}{V}\left(\frac{\mu^2}{3kT} + \alpha_e + \alpha_v\right)\boldsymbol{E}_i \tag{10.51}$$

There is a problem in knowing the average internal or local electric field strength $\boldsymbol{E}_i$. If the concentration of polar molecules is low it can be shown that $\boldsymbol{E}_i = \boldsymbol{E}(\kappa + 2)/3$, where $\boldsymbol{E}_i$ is the internal field and $\boldsymbol{E}$ the applied field. Eliminating $\boldsymbol{E}_i$ from equation 10.51 using this relation, eliminating $\boldsymbol{P}$ between this equation and equation 10.50, and introducing $N/V = N_A\rho/M$, we obtain

$$\mathscr{P} = \frac{\kappa - 1}{\kappa + 2}\frac{M}{\rho} = \frac{N_A}{3\epsilon_o}\left(\frac{\mu^2}{3kT} + \alpha_e + \alpha_v\right) \tag{10.52}$$

The quantity $\mathscr{P}$ is called the molar polarization, and ρ is the density. This equation is accurate for dilute gases and holds approximately for slightly polar liquids, but it does not apply to liquids of high dielectric constant because of the approximation of taking $\boldsymbol{E}_i = \boldsymbol{E}(\kappa + 2)/3$. Equation 10.52 neglects short-range molecular interactions; more complicated theories have been developed by Onsager and Kirkwood to take those effects into account.

10.13 DIPOLE MOMENTS OF GASEOUS MOLECULES

Since the orientation polarization depends on the absolute temperature, it is possible to calculate the polarizability $\alpha_e + \alpha_v$ and the dipole moment μ of gas molecules from measurements of the dielectric constant κ at a series of temperatures. In the gaseous state the molecules are so far apart that they do not induce dipole moments in each other.

To obtain the dipole moment of a gaseous molecule, the dielectric constants κ and gas densities are measured at several temperatures, and the corresponding values of the total molar polarization $\mathscr{P}$ are calculated. According to equation 10.52, a plot of $\mathscr{P}$ versus $1/T$ will have an intercept of $N_A(\alpha_e + \alpha_v)/3\epsilon_o$ and a slope of $N_A\mu^2/9k\epsilon_o$. Plots of molar polarization $\mathscr{P}$ versus reciprocal absolute temperature are illustrated in Fig. 10.16 for the chlorine-substituted compounds of methane as measured by Sanger.* It is seen that zero slopes are obtained for the symmetrical molecules CH_4 and CCl_4, indicating that these molecules have no permanent dipole moment. Because of the electronegative character of chlorine atoms, CH_3Cl, CH_2Cl_2, and $CHCl_3$ have permanent dipole moments with the chlorine atom the negative end of the dipole. The dipole moment of CH_3Cl is the largest of the three.

The dependence of the dielectric constant of a typical polar substance on frequency is shown in Fig. 10.17. At low frequencies all the terms on the right-hand side of equation 10.52 contribute. As the frequency is increased above the radio-frequency range, the dielectric constant decreases and the orientation polarization eventually becomes negligible because there is insufficient time for molecular orientation to occur before the field is reversed. In the region where the dielectric constant is labeled κ_∞, the polarizability is made up of vibrational, α_v, and

* R. Sanger, *Physik Z.*, **27**, 562 (1926).

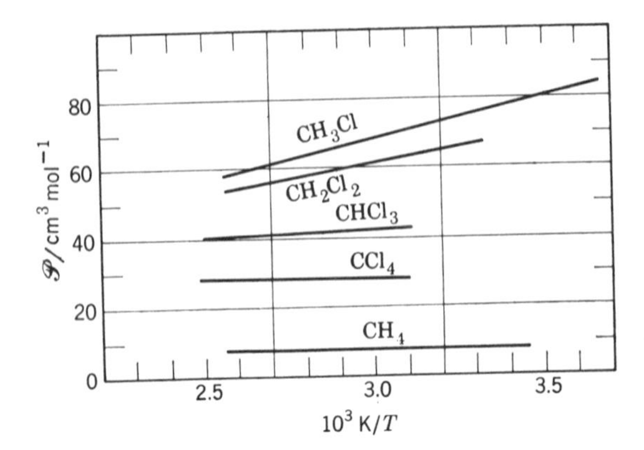

Fig. 10.16 Variation of molar polarization with temperature.

electronic, α_e, terms. The dielectric constant shows dispersion (Section 11.18) in the neighborhood of absorption lines in the infrared and ultraviolet.

Maxwell showed that the dielectric constant κ measured with a capacitor and the refractive index n of light are related by

$$\kappa = n^2 \tag{10.53}$$

provided the same frequency is used in both measurements.

The dipole moment of a solute in a nonpolar solvent may be determined from dielectric constant and density measurements on dilute solutions.* Refractive index measurements are used to evaluate the molecular electronic polarizability α_e. Dipole moments of a variety of substances are given in Table 10.4.

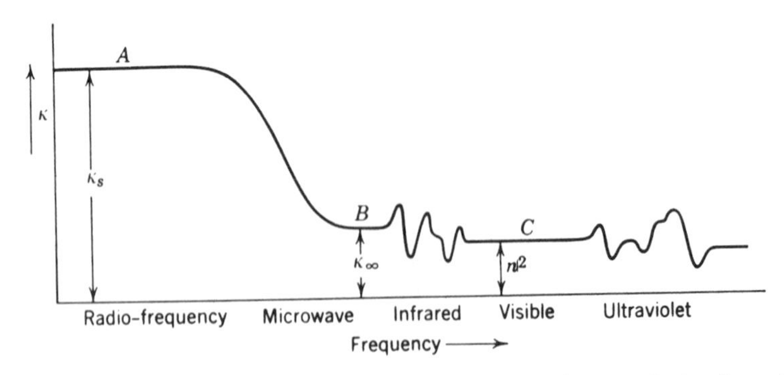

Fig. 10.17 Variation of dielectric constant with frequency for a typical polar substance.

* F. Daniels, J. W. Williams, P. Bender, R. A. Alberty, C. D. Cornwell, and J. E. Harriman, *Experimental Physical Chemistry*, McGraw-Hill Book Co., New York, 1970.

Table 10.4 Dipole Moments ($\mu/10^{-30}$ C m) of Gaseous Molecules

$AgClO_4$	15.7	C_6H_5Cl	5.17	SO_2	5.37
$C_6H_5NO_2$	13.21	C_6H_5OH	5.67	HCl	3.44
$(CH_3)_2CO$	9.3	C_2H_5OH	5.67	HBr	2.64
H_2O	6.14	$C_6H_5NH_2$	5.20	HI	1.00
CH_4	0	NH_3	4.87	N_2O	0.47
CH_3Cl	6.17	H_2S	3.67	CO	0.40
CH_3Br	4.84	H_2	0	CS_2	0
CH_3I	4.50	Cl_2	0	C_2H_4	0
CH_2Cl_2	5.30	CO_2	0	C_2H_6	0
$CHCl_3$	3.84	$C_6H_5CH_3$	1.33	C_6H_6	0
CCl_4	0				

10.14 VAN DER WAALS FORCES

The short-range attractive forces between molecules are named after van der Waals, who discussed these forces in gases and liquids (Section 1.18). These forces contribute to the nonideal behavior of gases and, at sufficiently high pressures and low temperatures, to condensation to the liquid state. The origin of these forces was explained in 1930 by London. The intermolecular attraction arises from the fluctuations of charge distribution in the two atoms or molecules that are close together. Since the electrons are moving, each molecule has an instantaneous dipole moment that is not zero. If the electron density fluctuations in the two atoms or molecules were unrelated there would be no net attraction between the molecules, since there would be repulsion as often as attraction. However, an instantaneous dipole in one atom or molecule induces an oppositely oriented dipole in the neighboring atom or molecule, and these instantaneous dipoles attract each other. This leads to a force of attraction that is referred to as a *dispersion force*.

The interaction energy is proportional to the square of the polarizability α (Section 10.12) and inversely proportional to r^6, where r is the distance between the two atoms or molecules. Thus the interaction is strong only at very short distances, and polarizable molecules attract each other more strongly than nonpolarizable molecules.

References

D. B. Cook, *Ab Initio Valence Calculations*, Wiley, New York, 1974.

C. A. Coulson, *The Shape and Structure of Molecules*, Clarendon Press, Oxford, 1973.

J. C. Davis, *Advanced Physical Chemistry*, The Ronald Press Co., New York, 1965.

W. H. Flygare, *Molecular Structure and Dynamics*, Prentice-Hall, Englewood Cliffs, N.J., 1978.

N. E. Hill, W. E. Vaughan, A. H. Price, and M. Davies, *Dielectric Properties and Molecular Behavior*, Van Nostrand Reinhold, London, 1969.

I. N. Levine, *Quantum Chemistry*, Allyn and Bacon, Boston, 1974.
F. L. Pilar, *Elementary Quantum Chemistry*, McGraw-Hill Book Co., New York, 1968.
L. Pauling, *The Nature of the Chemical Bond*, Cornell University Press, Ithaca, 1960.
G. C. Pimentel and A. L. McClellan, *The Hydrogen Bond*, W. H. Freeman and Co., San Francisco, 1960.
J. A. Pople and D. L. Beveridge, *Approximate Molecular Orbital Theory*, McGraw-Hill Book Co., New York, 1970.
W. G. Richards and J. A. Horsley, *Ab Initio Molecular Orbital Calculations for Chemists*, Clarendon Press, Oxford, 1970.
H. F. Schaefer, III, *The Electronic Structure of Atoms and Molecules*, Addison-Wesley Publ. Co., Reading, Mass., 1972.
E. Steiner, *The Determination and Interpretation of Molecular Wave Functions*, Cambridge University Press, Cambridge, England, 1976.
J. R. Van Wazer and I. Absar, *Electron Densities in Molecules and Molecular Orbitals*, Academic Press, New York, 1975.

Problems

10.1 Given the equilibrium dissociation energy D^e for N_2 in Table 10.2 and the fundamental vibration frequency 2331 cm^{-1}, calculate the spectroscopic dissociation energy D^0 in kJ mol^{-1}.
Ans. 942 kJ mol^{-1}.

10.2 Calculate the normalization factor shown in equation 10.31.

10.3 Derive the expression for the normalization constant c in

$$\psi = c[(1s_a) + (1s_b)]$$

Ans. $c = \pm 1/\sqrt{2(1 + S)}$, where S is the overlap integral.

10.4 The dipole moments of HCl, HBr, and HI are given in Table 10.4. Explain the relative magnitudes of these moments in terms of the electronegativity.
Ans. The electronegativities increase in going from I to Cl and so the negative charge on the halogen is greater in HCl than in HI and the dipole moment is larger.

10.5 Using data in Table A.1, calculate the electronegativities of Cl, Br, and I. The electronegativity of H is taken to be 2.1.
Ans. 3.1, 2.8, 2.3.

10.6 For KF(g) the dissociation constant D^e is 5.18 eV and the dipole moment is 28.7×10^{-30} C m. Estimate these values assuming that the bonding is entirely ionic. The ionization potential K(g) is 4.34 eV, and the electron affinity of F(g) is 3.40 eV. The equilibrium internuclear distance in KF(g) is 0.217 nm.
Ans. 5.70 eV. 34.7×10^{-30} C m.

10.7 Calculate the dipole moment HCl would have if it consisted of a proton and a chloride ion (considered to be a point charge) separated by 0.127 nm (the internuclear distance obtained from the infrared spectrum). The experimental value is 3.44×10^{-30} C m. How do you explain the difference?
Ans. 20.4×10^{-33} C m. The charges are not completely separated in HCl.

10.8 The molar polarization $\mathscr{P}$ of ammonia varies with temperature as follows.

t/°C	19.1	35.9	59.9	113.9	139.9	172.9
$\mathscr{P}$/cm³ mol⁻¹	57.57	55.01	51.22	44.99	42.51	39.59

Calculate the dipole moment of ammonia.
Ans. 5.28×10^{-30} C m.

10.9 Show that the SI units are the same on the two sides of equation 10.52.

10.10 At 1 atm the dielectric constant of NH_3 gas is 1.007 20 at 292.2 K and 1.003 24 at 446.0 K. Calculate the dipole moment μ and the polarizability α.

Ans. 5.28×10^{-30} C m. 2.36×10^{-40} C m/V m^{-1}.

10.11 Calculate the spectroscopic dissociation energy D^0 of O_2 from the data in Tables 10.2 and 11.5 and compare it with the dissociation energy for perfect gas O_2 at absolute zero.

10.12 Carry out the differentiations required to get from equation 10.13 to equations 10.24 and 10.25.

10.13 Figure 10.8 shows the minimum of the potential energy curve for H_2 in its ground state at -31.9 eV and the energy of two hydrogen atoms in their ground states as -27.2 eV, with respect to two protons and two electrons at infinite separations. Explain these values and the relationship between these values and the equilibrium dissociation energy of 4.7483 eV in Table 10.1.

10.14 By multiplying out terms show that the wave functions given in equations 10.36 and 10.39 are the same if c_1 and c_2 are defined appropriately.

10.15 Show that wave equations ψ_g and ψ_u in equations 10.30 and 10.31 are orthogonal.

10.16 Explain the relative enthalpies of formation (Table A.1) of HCl(g), HBr(g), and HI(g) in terms of electronegativity concept.

10.17 The dipole moments of CH_3Cl, CH_3Br, and CH_3I are given in Table 10.4. Explain the relative magnitudes of these moments in terms of the electronegativity.

10.18 Alternating positive and negative charges are arranged in a linear array with distances between successive charges of 100, 200, and 300 pm. Calculate the magnitude of the dipole moment. Show that the magnitude of the moment is independent of the choice of origin of the coordinate system

10.19 The dielectric constants of HCl(g) over a range of temperatures is given by:

t/°C	0	100	200
κ	1.0046	1.0025	1.0016

What is the dipole moment μ and the polarizability α?

10.20 Calculate the dipole moment of fluorobenzene, using a plot of $\mathscr{P}$ versus $1/T$, from the following molar polarizations for gaseous samples measured by K. B. McAlpine and C. P. Smyth [*J. Chem. Phys.*, **3**, 55 (1935)].

T/K	343.6	371.4	414.1	453.2	507.0
$\mathscr{P}$/cm^3 mol^{-1}	69.9	66.8	62.5	59.3	55.8

10.21 Show from equation 10.51 that the polarizability α has the units of dipole moment divided by electric field strength, that is, C m/V m^{-1}.

CHAPTER 11
MOLECULAR SPECTROSCOPY

Spectroscopy is the study of the interactions between matter and electromagnetic radiation. The spectra of atoms have been discussed briefly in connection with quantum mechanics. Atomic spectra involve transitions of electrons from one electronic energy level to another and are line spectra. Molecular spectra involve transitions between rotational and vibrational energy levels in addition to electronic transitions and are much more complicated than those of atoms. The study of rotational spectra yields moments of inertia and interatomic distances and angles. Vibrational spectra yield fundamental vibrational frequencies and force constants. Electronic spectra yield electronic energy levels and dissociation energies. Thus molecular spectroscopy is a powerful tool for learning about the structures and energy levels of molecules.

Additional topics in spectroscopy are discussed in other chapters, including nuclear magnetic resonance and electron spin resonance in Chapter 12 and fluorescence and phosphorescence in Chapter 17.

11.1 THE ELECTROMAGNETIC SPECTRUM

Electromagnetic radiation may be classified arbitrarily into regions according to the sources and detectors required. Typical limits of the various spectral regions are indicated in Table 11.1.

The chemical and physical effects of various types of radiation are quite different, and these differences can be understood in terms of the differing energies of the photons. The energy ϵ of a single photon is given by

$$\epsilon = h\nu \tag{11.1}$$

where h is Planck's constant (Section 8.1), 6.626×10^{-34} J s.

The energy of a mole of electrons that have been accelerated by 1 V is of the magnitude commonly encountered for chemical reactions.

$$(1.602\,19 \times 10^{-19}\ \text{J eV}^{-1})(6.022\,05 \times 10^{23}\ \text{mol}^{-1}) = 96{,}485\ \text{J mol}^{-1}\ \text{eV}^{-1}$$

This is a useful conversion factor between two commonly used energy units. Whereas kJ mol^{-1} is the basic energy unit in chemistry, wave numbers are used

Table 11.1 Energy of Electromagnetic Radiation

Description	Wavelength Range	Wave Number, cm^{-1}	Frequency, Hz	Energy	
				$kJ\ mol^{-1}$	eV
Radio frequency	3×10^3 m	3.33×10^{-6}	10^5	3.98×10^{-8}	4.12×10^{-10}
	0.30 m	0.0333	10^9	3.98×10^{-4}	4.12×10^{-6}
Microwave	0.0006 m	16.6	4.98×10^{11}	0.191	2.07×10^{-3}
Far infrared	(600 μm)				
	30 μm	333	10^{13}	3.98	0.0412
Near infrared	0.8 μm	1.25×10^4	3.75×10^{14}	149.8	1.55
Visible	(800 nm)				
	400 nm	2.5×10^4	7.5×10^{14}	299.2	3.10
Ultraviolet					
	150 nm	6.06×10^4	19.98×10^{14}	795	8.25
Vacuum ultraviolet	5 nm	2×10^6	6×10^{16}	2.39×10^6	247.8
X rays and γ rays	10^{-4} nm	10^{11}	3×10^{21}	1.19×10^9	1.24×10^7

in spectroscopy, and electron volts are used in physics. Useful conversion factors are given on the inside of the front cover.

Example 11.1 Calculate the energy in joules per quantum, joules per mole, and electron volts of photons of wavelength 300 nm.

$$h\nu = \frac{hc}{\lambda} = \frac{(6.62 \times 10^{-34}\ \mathrm{J\ s})(3 \times 10^8\ \mathrm{m\ s^{-1}})}{(300 \times 10^{-9}\ \mathrm{m})} = 6.62 \times 10^{-19}\ \mathrm{J}$$

$$N_A h\nu = (6.02 \times 10^{23}\ \mathrm{mol^{-1}})(6.62 \times 10^{-19}\ \mathrm{J}) = 3.98 \times 10^5\ \mathrm{J\ mol^{-1}}$$

$$= \frac{3.98 \times 10^5\ \mathrm{J\ mol^{-1}}}{96{,}485\ \mathrm{C\ mol^{-1}}}$$

$$= 4.13\ \mathrm{eV}$$

11.2 THE FUNDAMENTAL EQUATION

In absorption a photon of electromagnetic radiation is absorbed by a molecule that is in a particular quantum state characterized by a certain wave function and a definite energy E''. There is a transition to another quantum state, characterized by another wave function and a higher energy E'. In emission, a photon is emitted by a molecule that makes a transition from a higher energy state to a lower energy state. For either the absorption or emission process, the total energy must be conserved. This leads to the relation

$$h\nu = E' - E'' = \Delta E \tag{11.2}$$

where E' is the energy of the higher energy state, E'' is the energy of the lower energy state, and ν is the frequency of the radiation.

The types of molecular change that accompany the absorption of electromagnetic radiation may be described as follows. In the radio-frequency range the energy of one photon is very low, but changes of nuclear spin states of substances in a magnetic field may take place, as discussed in Chapter 12. In the microwave range there are changes in electron spin states for substances with unpaired electrons in a magnetic field (discussed in Chapter 12) and transitions between rotational energy levels of gas molecules. In the infrared region absorption causes changes in vibrational energy accompanied by changes in rotational energy. In the visible and ultraviolet regions absorption causes changes in the electronic energy of the molecule accompanied by changes in its vibrational and rotational energy. The changes in electronic energy involve the most loosely held, or so-called valence, electrons. In the far ultraviolet and X-ray regions the energies are sufficiently great to move inner electrons to higher levels and to ionize and dissociate molecules.

In spectroscopy we will not be concerned with translational energy, but only with the "internal" energy of a molecule. The internal energy E of a molecule in a certain state may be considered to be made up of contributions from rotational energy E_r, vibrational energy E_v, and electronic energy E_e.

$$E = E_r + E_v + E_e \tag{11.3}$$

The state of an isolated molecule may be specified by giving a series of quantum numbers to describe the rotational state, the vibrational state, and the electronic state. In the process of absorption or emission there is a change in this set of quantum numbers. For example, if a diatomic molecule in the ground electronic state and the zero vibrational state has some particular rotational energy, absorption of a photon may raise it to an excited electronic state, with several quanta of vibrational energy, and a different rotational energy. There are many transitions that might give rise to absorption or emission, but only those satisfying certain selection rules are allowed.

Electronic, vibrational, and rotational changes may all occur upon the absorption or emission of a single photon, and so equation 11.2 may be written

$$h\nu = (E'_\mathrm{e} - E''_\mathrm{e}) + (E'_\mathrm{v} - E''_\mathrm{v}) + (E'_\mathrm{r} - E''_\mathrm{r}) \tag{11.4}$$

where the upper state is represented by a single prime and the lower state by a double prime. A fact that considerably simplifies the interpretation of spectra is that usually

$$(E'_\mathrm{e} - E''_\mathrm{e}) \gg (E'_\mathrm{v} - E''_\mathrm{v}) \gg (E'_\mathrm{r} - E''_\mathrm{r}) \tag{11.5}$$

Consequently, rotational, vibrational, and electronic spectra occur in different regions of the spectrum.

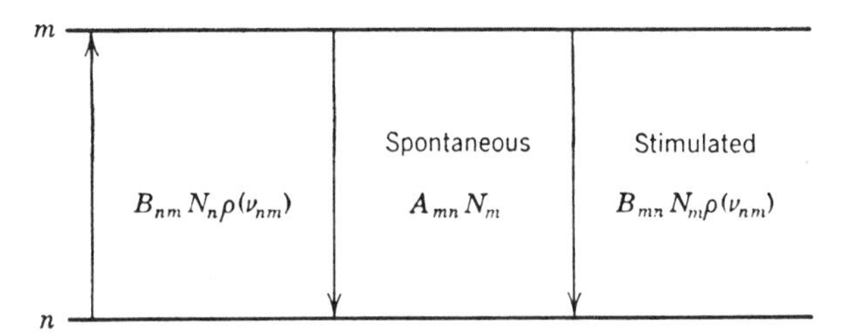

Fig. 11.1 Spontaneous and stimulated emission.

11.3 EINSTEIN COEFFICIENTS

Einstein formulated some of the most basic ideas about absorption and emission of radiation. He considered a system in which matter and radiation are in equilibrium in a closed cavity at temperature T. We consider transitions between two energy levels E_m and E_n where $E_m > E_n$, as shown in Fig. 11.1. The rate of absorption of radiation that raises a molecule from state n to state m is proportional to the density $\rho(\nu_{mn})$ of radiation of frequency ν_{mn} and to the number N_n of molecules in state n. Thus,

$$\text{Rate of absorption} = B_{nm}N_n\rho(\nu_{mn}) \tag{11.6}$$

where B_{nm} is the proportionality constant.

In addition, the radiation field *stimulates* the emission of photons of frequency ν_{mn} with transitions from state m to state n. This radiation has the same phase and direction as the incident radiation. We will see later (Section 17.12) that stimulated emission makes possible laser action. The rate of stimulated emission is given by

$$\text{Rate of stimulated emission} = B_{mn}N_m\rho(\nu_{mn}) \tag{11.7}$$

Finally, a molecule may emit spontaneously independent of the presence or absence of radiation.

$$\text{Rate of spontaneous emission} = A_{mn}N_m \tag{11.8}$$

At equilibrium the rates of absorption and emission are equal.

$$B_{nm}N_n\rho(\nu_{mn}) = B_{mn}N_m\rho(\nu_{mn}) + A_{mn}N_m \tag{11.9}$$

$$\frac{N_m}{N_n} = \frac{B_{nm}\rho(\nu_{mn})}{B_{mn}\rho(\nu_{mn}) + A_{mn}} \tag{11.10}$$

At equilibrium the numbers of molecules in the higher and lower energy states also satisfy the Boltzmann distribution law.

$$\frac{N_m}{N_n} = e^{-(E_m - E_n)/kT} = e^{-h\nu_{mn}/kT} \tag{11.11}$$

Eliminating N_{mn}/N_n between this equation and equation 11.10 yields

$$\rho(\nu_{mn}) = \frac{A_{mn}}{B_{nm}e^{h\nu_{mn}/kT} - B_{mn}} \tag{11.12}$$

This relation may be compared with the radiation law derived by Planck.

$$\rho(\nu_{mn}) = \frac{8\pi h\nu_{mn}^3}{c^3}\frac{1}{e^{h\nu_{mn}/kT} - 1} \tag{11.13}$$

where $\rho(\nu_{mn})\,d\nu$ is the energy per unit volume in the frequency range $d\nu$. In order for these two laws to be consistent,

$$B_{nm} = B_{mn} \tag{11.14}$$

$$A_{mn} = \left(\frac{8\pi h\nu_{mn}^3}{c^3}\right) B_{nm} \tag{11.15}$$

An expression for B_{nm} can be derived from quantum mechanics, which shows that B_{nm} is related to the absorption coefficient (Section 11.14) for a substance and also to the wave functions for states m and n. Thus the coefficient of spontaneous emission A_{mn} can be estimated from first principles or from absorbancy measurements. The coefficient for spontaneous emission A_{mn} determines the radiative lifetime of an excited state (Section 17.3). Because of the ν^3 dependence in equation 11.15, spontaneous emission is much more important in short wavelength radiation; it is almost unheard of in the microwave and radiofrequency regions.

The intensity of an absorption line is generally proportional to the number N_n of molecules in the lower state but, if there is a significant number in the higher state, the net absorption (ignoring spontaneous emission) is given by

$$B_{nm}\rho(\nu_{mn})N_n - B_{mn}\rho(\nu_{mn})N_m = B_{nm}\rho(\nu_{mn})(N_n - N_m) \tag{11.16}$$

since the stimulated emission is in the same direction as the incident radiation. If $N_n = N_m$, no absorption occurs; the transition is said to be saturated.

11.4 THE SCHRÖDINGER EQUATION FOR NUCLEAR MOTION OF A DIATOMIC MOLECULE

In the preceding chapter we discussed the form of the Schrödinger equation that yields the total electronic and nuclear wave functions for a molecule and its total energy. In that chapter we were primarily interested in the electronic wave function and the electronic energy levels. Since the electrons move so much more rapidly than the nuclei, the electronic wave function was calculated for fixed internuclear distances. In this chapter we will be interested in the wave functions for nuclear motion, because they yield the possible vibrational and rotational energy levels. We can solve the Schrödinger equation for nuclear motion using as the potential energy $U(R)$ the energy as a function of internuclear distance R that was calculated from the electronic Schrödinger equation. The Schrödinger equation for the nuclear motion of a diatomic molecule is

$$\left[-\frac{\hbar^2}{2\mu}\nabla^2 + U(R)\right]\psi_N = E\psi_N \tag{11.17}$$

where μ is the reduced mass (Section 8.12). This equation does not include the translational motion of the molecule. The origin of the coordinate system is taken at the center of mass of the molecule, as shown in Fig. 8.7. The wave function ψ_N for nuclear motion can be written as the product of a wave function, depending only on R and a wave function depending only on θ and ϕ. This first wave function is the vibrational wave function that contains the information about internal motions of the molecule, and the second is the rotational wave function that contains the information about angular motion of the molecule in the laboratory coordinate system.

11.5 ROTATIONAL SPECTRA OF DIATOMIC MOLECULES

As a first approximation to a rotating diatomic molecule, we consider a rigid rotor with the internuclear distance fixed at the equilibrium value R_e. As shown in Section 8.12, the rotational energy of a rigid diatomic molecule is

$$E_r = \frac{\hbar^2}{2I} J(J + 1) \qquad J = 0, 1, 2, \ldots \tag{11.18}$$

where I is the moment of inertia, $I = \mu R_e^2$, and μ is the reduced mass. The rotational levels are $(2J + 1)$-fold degenerate. The rotational energy levels for a rigid diatomic molecule, as calculated using equation 11.18, are shown in Fig. 11.2.

The quantum mechanical treatment of rotational transitions shows that a molecule has a pure rotational spectrum only if it possesses a permanent electric dipole moment. This is expected from the fact that a rotating dipole produces an oscillating electric field that can interact with the oscillating electric field of a light wave.

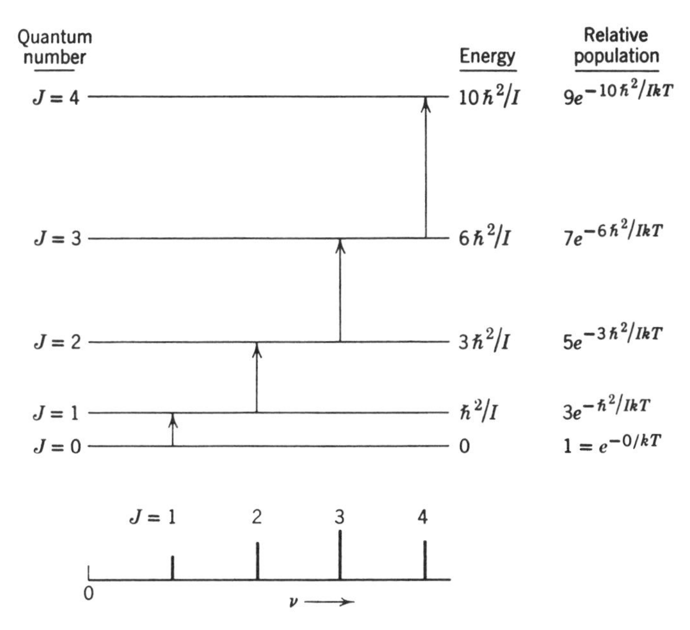

Fig. 11.2 Rotational levels for a rigid diatomic molecule and the absorption spectrum that results from $\Delta J = 1$. The energies and populations of the levels are indicated on the right. The transitions are labeled by the upper of the two J values involved.

Since homonuclear diatomic molecules such as H_2 and N_2 do not have dipole moments, they do not interact with the electromagnetic field of light and thus do not show pure rotational spectra, as do molecules such as HCl and CH_3Cl, which have dipole moments. Since the orientation of the dipole moment vector in a molecule must be unchanged by any symmetry operation of the molecule, the vector must lie along a symmetry axis. Only molecules belonging to point groups C_n, C_s, and C_{nv} can possess dipole moments (Section 9.9).

For electric dipole radiation only certain transitions are allowed between energy levels. For rotational transitions the selection rule is $\Delta J = \pm 1$. In absorption $\Delta J = +1$ and in emission $\Delta J = -1$.

In spectroscopy it is convenient to express the rotational part of the energy in terms of frequency or wave numbers and to represent this rotational term by $F(J)$ if it is expressed in terms of frequency or $\tilde{F}(J)$ if it is expressed in terms of wave numbers.* If units of frequency are being used, equation 11.18 may be written

$$F(J) = \frac{E_r}{h} = \frac{J(J+1)h}{8\pi^2 I} = J(J+1)B_e \tag{11.19}$$

where the rotational constant for the equilibrium configuration of the nuclei is represented by B_e.

$$B_e = \frac{h}{8\pi^2 I} \tag{11.20}$$

Or, if units of wave numbers are being used,

$$\tilde{F}(J) = \frac{E_r}{hc} = \frac{J(J+1)h}{8\pi^2 Ic} = J(J+1)\tilde{B}_e \tag{11.21}$$

where the rotational constant is written

$$\tilde{B}_e = \frac{h}{8\pi^2 Ic} \tag{11.22}$$

When SI units are used, $\tilde{B}_e$ comes out in m^{-1}, but this is conveniently converted to cm^{-1} by multiplying by 10^{-2} m cm^{-1}.

For a transition between two adjacent rotational levels the rotational term difference, which is represented by $\Delta_1 F(J + \frac{1}{2})$, is given by

$$\Delta_1 F(J+\tfrac{1}{2}) = F(J+1) - F(J) = \frac{h}{8\pi^2 I}[(J+1)(J+2) - J(J+1)]$$

$$= \frac{h}{8\pi^2 I}[2(J+1)] = 2(J+1)B_e \tag{11.23}$$

where J is the rotational quantum number for the lower level. The rotational term difference is expressed in terms of frequency units and is the frequency of the radiation absorbed or emitted. The rotational term difference may be obtained in

---

* Standard spectroscopic notation for diatomic molecules is described in F. A. Jenkins, *J. Optical Soc.*, **43**, 425 (1953).

wave numbers by dividing by the velocity of light. The frequencies of the successive lines in the rotational spectrum are given by $2B_e, 4B_e, 6B_e, \ldots$. Thus there is a series of equally spaced lines with separations of $2B_e$. A separate series of lines is found for each isotopically different species of a given molecule, because the moments of inertia of isotopically substituted molecules are different.

Example 11.2 Calculate the reduced mass μ for $H^{35}Cl$. Atomic masses of some isotopes are given in the Appendix.

$$\mu = \frac{(1.007\,825 \times 10^{-3}\ \text{kg mol}^{-1})(34.968\,85 \times 10^{-3}\ \text{kg mol}^{-1})}{[(1.007\,825 + 34.968\,85) \times 10^{-3}\ \text{kg mol}^{-1}](6.022\,045 \times 10^{23}\ \text{mol}^{-1})}$$

$$= 1.626\,68 \times 10^{-27}\ \text{kg}$$

Example 11.3 The pure rotational spectrum of $H^{35}Cl$ has lines at 21.18, 42.36, 63.54, 84.72, and 105.91 cm^{-1}. What is the moment of inertia and the equilibrium internuclear distance?

The spacing between lines is 21.18 cm^{-1}, and so $\tilde{B}_e = (10.59\ \text{cm}^{-1})(10^2\ \text{cm m}^{-1}) = 1059\ \text{m}^{-1}$.

$$I = \frac{h}{8\pi^2 c\tilde{B}_e} = \frac{6.626 \times 10^{-34}\ \text{J s}}{8\pi^2(2.998 \times 10^8\ \text{m s}^{-1})(1059\ \text{m}^{-1})} = 2.643 \times 10^{-47}\ \text{kg m}^2$$

Since the reduced mass of $H^{35}Cl$ is $1.626\,68 \times 10^{-27}$ kg,

$$I = \mu R_e^2$$

$$R_e = \sqrt{\frac{I}{\mu}} = \sqrt{\frac{2.643 \times 10^{-47}\ \text{kg m}^2}{1.626\,68 \times 10^{-27}\ \text{kg}}} = 1.275 \times 10^{-10}\ \text{m} = 0.1275\ \text{nm}$$

The relative intensities of rotational lines depend on the relative populations of the levels in the initial state. The populations of the levels at thermal equilibrium are given by the Boltzmann distribution (Section 13.3).

$$N_i = g_i e^{-\epsilon_i/kT} \Big/ \sum_i g_i e^{-\epsilon_i/kT} \tag{11.24}$$

where g_i is the degeneracy (Section 8.7) of the ith level. As discussed earlier, the component of the angular momentum in a particular direction is equal to $m\hbar$, where m may have values of $J, (J-1), \ldots, 0, \ldots, -J$, where J is the rotational quantum number. Thus there are in all $2J + 1$ different possible states with quantum number J. In the absence of an external electric or magnetic field the energies are identical for these various sublevels, and so the Jth energy level is said to have a degeneracy of $2J + 1$.

The rotational energy in the absence of an external electric or magnetic field is given by $\epsilon_i = hJ(J+1)B_e$ so that, using equation 11.24, the population of the Jth rotational level is given by

$$N_J = K(2J+1)e^{-[hJ(J+1)B_e]/kT} \tag{11.25}$$

According to this equation, the number of molecules in level J increases with J at low J values, goes through a maximum and then, because of the exponential term, decreases as J is further increased. The lines in the spectrum at the bottom of Fig. 11.2 have been labeled with the rotational quantum number J of the upper of the two states involved, and the relative lengths of these lines are in proportion to the relative populations of these states.

For molecules with larger moments of inertia I the rotational energies are smaller, in fact, small compared with kT. The quantum numbers may become quite large before $e^{-\epsilon_J/kT}$ becomes appreciably different from unity. For small quantum numbers populations are given by the degeneracies, since $e^{-\epsilon_J/kT} \approx 1$ for $\epsilon_J \ll kT$.

The lines in the rotational spectrum of a diatomic molecule are split when the molecules being studied are placed in an external electric field. This rotational Stark effect results from the interaction of the molecular dipole moment with the electric field.

Although homonuclear diatomic molecules do not have permanent electric dipole moments and do not exhibit pure rotational spectra, they do show rotational Raman spectra (Section 11.11), and their electronic spectra show rotational fine structure.

11.6 ROTATIONAL SPECTRA OF POLYATOMIC MOLECULES

For the treatment of its pure rotational spectrum we may consider a polyatomic molecule to be a rigid framework with fixed bond lengths and angles equal to their mean values. For a polyatomic molecule the moment of inertia about a particular axis is simply the sum of the moments due to the various nuclei about that axis.

$$I = \sum_i m_i R_i^2 \tag{11.26}$$

where R_i is the perpendicular distance of the nucleus of mass m_i from the axis. The symmetry axes of a molecule all pass through the center of mass of the molecule.

Molecules may be classified according to their ellipsoid of inertia, constructed as follows. Lines are drawn from the center of mass of the molecule in various directions with length proportional to $1/\sqrt{I_\alpha}$, where I_α is the moment of inertia about that line as an axis. The three mutually perpendicular principal axes of the ellipsoid of inertia are designated a, b, and c. The principal moments of inertia about these axes are labeled so that $I_a \leqslant I_b \leqslant I_c$.

The symmetry axes of a molecule help in locating its principal axes. Any symmetry operation of a molecule must apply to its ellipsoid of inertia.

The principal moments of inertia are used to classify molecules, as shown in Table 11.2. If all three principal moments of inertia are equal, the molecule is a *spherical* top. If two principal moments are equal, the molecule is a *symmetric* top. A molecule is a *prolate* top (cigar shaped) if the two larger moments are equal. The molecule is an *oblate* top (discus shaped) if the two smaller moments are equal. The molecule is an *asymmetric* top if all three principal moments are unequal.

Table 11.2 Classification of Polyatomic Molecules According to Their Moments of Inertia

Moments of Inertia	Type of Rotor	Examples
$I_a = I_b \quad I_c = 0$	Linear	HCN
$I_a = I_b = I_c$	Spherical top	CH_4, SF_6, UF_6
$I_a < I_b = I_c$	Prolate symmetric top	CH_3Cl
$I_a = I_b < I_c$	Oblate symmetric top	C_6H_6
$I_a \neq I_b \neq I_c$	Asymmetric top	CH_2Cl_2, H_2O

In order for a rotational transition to absorb or emit electromagnetic radiation, a molecule must have a properly oriented permanent dipole moment. Spherical top molecules have no observable rotational spectrum, since they cannot have a dipole moment. Some linear molecules, symmetric tops, and asymmetric tops also have zero dipole moment by symmetry and so no rotational spectrum.

In classical mechanics the rotational energy of a rotator with one degree of freedom is $(I\omega)^2/2I$, where ω is the angular velocity in radians per second and I is the moment of inertia. For a molecule with three degrees of freedom the classical energy of rotation is

$$E_r = \frac{P_a^2}{2I_a} + \frac{P_b^2}{2I_b} + \frac{P_c^2}{2I_c} \tag{11.27}$$

in which the components of the total angular momentum about the three principal axes are given by

$$P_a = I_a\omega_a \tag{11.28a}$$

$$P_b = I_b\omega_b \tag{11.28b}$$

$$P_c = I_c\omega_c \tag{11.28c}$$

The total angular momentum is given by

$$P^2 = P_a^2 + P_b^2 + P_c^2 \tag{11.29}$$

Before considering the quantization of rotational energy we will consider the expressions for the rotational energy for the four classes of molecules:

1. Spherical tops. For a spherical top $I_a = I_b = I_c = I$, the momental ellipsoid is a sphere, and equation 11.27 becomes

$$E_r = \frac{(P_a^2 + P_b^2 + P_c^2)}{2I} = \frac{P^2}{2I} \tag{11.30}$$

where the second form has been obtained by introducing equation 11.29.

2. Linear molecule. For a linear molecule, $I_a = 0$ and $I_b = I_c$. Thus equation 11.27 becomes

$$E_r = \frac{(P_b^2 + P_c^2)}{2I_b} = \frac{P^2}{2I_b} \tag{11.31}$$

3. Symmetric top. For a prolate symmetric top, $I_b = I_c$, and equation 11.27 becomes

$$E_r = \frac{P^2}{2I_b} + \frac{P_a^2}{2}\left(\frac{1}{I_a} - \frac{1}{I_b}\right) \tag{11.32}$$

For an oblate symmetric top, $I_a = I_b$, and equation 11.27 becomes

$$E_r = \frac{P^2}{2I_b} + \frac{P_c^2}{2}\left(\frac{1}{I_c} - \frac{1}{I_b}\right) \tag{11.33}$$

4. Asymmetric top. Since the three moments are unequal, equation 11.27 cannot be simplified.

The rotational motion of a molecule is quantized and, as we have seen in Section 8.11, the square of the total angular momentum P^2 is given by

$$P^2 = J(J+1)\hbar^2 \qquad J = 0, 1, 2, \ldots \tag{11.34}$$

and the square of the component P_z^2 along the unique symmetry axis is given by

$$P_z^2 = M^2\hbar^2 \qquad M = 0, \pm 1, \pm 2, \ldots, \pm J \tag{11.35}$$

The square of the axial component of the total angular momentum P_k may be written

$$P_k^2 = K^2\hbar^2 \qquad K = 0, \pm 1, \pm 2, \ldots, \pm J \tag{11.36}$$

For a prolate symmetric top, introducing equation 11.34 into equation 11.32 and replacing P_a^2 with K^2h^2 yields

$$E_r = J(J+1)\frac{h^2}{8\pi^2 I_b} + K^2\left(\frac{1}{I_a} - \frac{1}{I_b}\right)\frac{h^2}{8\pi^2} \tag{11.37}$$

Since this equation is used to calculate frequencies, it is usually written

$$F(J) = J(J+1)B + K^2(A - B) \tag{11.38}$$

where the rotational constants A and B are given by

$$A = \frac{h}{8\pi^2 I_a} \tag{11.39}$$

$$B = \frac{h}{8\pi^2 I_b} \tag{11.40}$$

For an oblate top the corresponding equation is

$$F(J) = J(J+1)B + K^2(C - B) \tag{11.41}$$

For a linear polyatomic molecule the equation for the rotational term $F(J)$ is the same as that given earlier for a diatomic molecule.

Asymmetric top molecules have three different moments of inertia. There is no simple formula for the rotational levels in terms of J and K.

11.7 MICROWAVE SPECTROSCOPY

Microwave radiation is produced by special electronic oscillators called klystrons. Monochromatic radiation is produced, and the frequency may be varied continuously over wide ranges. The usual experimental arrangement is shown in Fig. 11.3. Microwave radiation is transmitted down a wave guide that contains the gas being studied. The intensity of the radiation at the other end of the wave guide is measured by use of a crystal diode detector and amplifier. The oscillator frequency is swept over a range, and the transmitted intensity is presented on an oscilloscope or a recorder as a function of frequency.

According to the Heisenberg uncertainty principle, the accuracy with which an energy level may be determined is inversely proportional to the time the molecule is in this level. Hence, to obtain sharp rotational lines of a gas, the pressure must be maintained sufficiently low so that the average time between collisions is long compared to the period of a rotation. Usually it is necessary to determine microwave spectra at pressures below 0.1 torr to reduce the line broadening effects of collisions.

Microwave lines at low pressures are very sharp, and their frequencies may often be determined to the order of 1 ppm. Therefore, the moments of inertia of a molecule may be determined with great accuracy. As a matter of fact, this method has made possible the most precise evaluations of bond lengths and bond angles. The spectrum of a polyatomic molecule gives at most three principal moments of inertia; since usually more than three bond lengths and angles are involved, isotopically different molecules must be studied. In effect, a number of simultaneous equations are solved for the internuclear distances and angles.

The lines in the microwave spectrum are split if the molecules being studied are

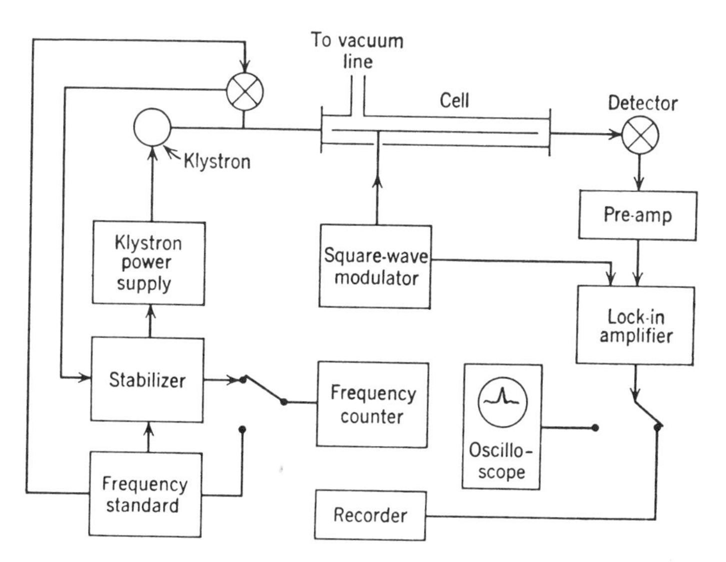

Fig. 11.3 Block diagram of a Stark-modulated microwave spectrometer.

in an electric field. This so-called Stark effect is due to the interaction of the dipole moment of the gaseous molecule and the electric field. Since the splitting is proportional to the permanent dipole moment, the magnitude of the dipole moment may be derived from the spectrum.

11.8 VIBRATIONAL SPECTRA OF DIATOMIC MOLECULES

The Schrödinger equation for the vibrational motion of a diatomic molecule includes a potential energy function $V(R)$ that is different for each molecule. In the preceding chapter we discussed the shape of this function for H_2^+ and H_2. Since, in general, this function is not known and we are particularly interested in its behavior near the potential energy minimum, we expand it as a Taylor series about $R = R_e$ to obtain

$$V(R) = V(R_e) + \left(\frac{dV}{dR}\right)_{R_e}(R - R_e) + \frac{1}{2}\left(\frac{d^2V}{dR^2}\right)_{R_e}(R - R_e)^2 + \frac{1}{3!}\left(\frac{d^3V}{dR^2}\right)_{R_e}(R - R_e)^3 + \cdots \tag{11.43}$$

Since we chose the zero of energy at the minimum of the curve, $V(R_e) = 0$, and $(dV/dR)_{R_e} = 0$ at this point since it is a minimum, we have

$$V(R) = \frac{1}{2}\left(\frac{d^2V}{dR^2}\right)_{R_e}(R - R_e)^2 + \frac{1}{3!}\left(\frac{d^2V}{dR^2}\right)_{R_e} + \cdots$$
$$= \tfrac{1}{2}k_e(R - R_e)^2 + \cdots \tag{11.44}$$

where k_e is the force constant. Neglect of terms higher than $(R - R_e)^2$ in this expression for the potential energy of a molecule as a function of distance amounts to replacing the potential energy curve with a parabola. This reduces the Schrödinger equation for the vibration of a diatomic molecule to that for a simple harmonic oscillator. In Section 8.10 we found that the energy levels for a simple harmonic oscillator are given by

$$E_v = (v + \tfrac{1}{2})hc\tilde{\omega}_e \qquad v = 0, 1, 2, \ldots \tag{11.45}$$

where v is the vibrational quantum number and $\tilde{\omega}_e$ is the fundamental vibrational frequency expressed in wave numbers.*

In spectroscopy it is convenient to express the vibrational part of the energy in terms of frequency or wave numbers and to represent this vibrational term by $G(v)$ if it is expressed in terms of frequency or $\tilde{G}(v)$ if it is expressed in terms of wave numbers. Thus,

$$\tilde{G}(v) = \frac{E_v}{hc} = (v + \tfrac{1}{2})\tilde{\omega}_e \tag{11.46}$$

The fundamental vibrational frequency ν for a diatomic molecule considered classically is given by equation 8.60, with m replaced by the reduced mass of the

* According to standard spectroscopic notation [F. A. Jenkins, *J. Opt. Soc.*, **43**, 425 (1953)], $\tilde{\omega}_e$ is used for vibrational frequencies expressed in wave numbers.

diatomic molecule. This new equation may be derived as follows. In a diatomic molecule the masses of the two atoms are represented by m_1 and m_2, and their distance from the center of mass are represented by R_1 and R_2. The internuclear distance R is equal to $R_1 + R_2$. The center of mass is located so that

$$R_1 m_1 = R_2 m_2 \tag{11.47}$$

Combining this equation with $R = R_1 + R_2$ yields

$$R_1 = \frac{m_2}{m_1 + m_2} R \tag{11.48}$$

$$R_2 = \frac{m_1}{m_1 + m_2} R \tag{11.49}$$

The force tending to restore the diatomic molecule to its equilibrium internuclear distance is equal to $m_1 a_1 = m_2 a_2$, where a_1 is the acceleration of nucleus 1 and a_2 is the acceleration of nucleus 2. By use of equations 11.48 and 11.49 we have

$$a_1 = \frac{d^2 R_1}{dt^2} = \frac{m_2}{m_1 + m_2} \frac{d^2 R}{dt^2} \tag{11.50}$$

$$a_2 = \frac{d^2 R_2}{dt^2} = \frac{m_1}{m_1 + m_2} \frac{d^2 R}{dt^2} \tag{11.51}$$

Thus

$$\text{Force} = \frac{m_1 m_2}{m_1 + m_2} \frac{d^2 R}{dt^2} = \mu \frac{d^2 R}{dt^2} = \mu \frac{d^2(R - R_e)}{dt^2} \tag{11.52}$$

where R_e is the equilibrium internuclear distance and μ is the reduced mass. Equating the force of attraction or repulsion $-k(R - R_e)$ to that given in equation 11.52, we obtain

$$\frac{d^2(R - R_e)}{dt^2} = -\frac{k}{\mu}(R - R_e) \tag{11.53}$$

If the nuclei are closed together than R_e, there is repulsion. If the nuclei are further apart than R_e, there is attraction. The attraction is due to electronic bonding and the repulsion is due to the overlapping of inner electronic shells.

By comparing equations 11.53 and 8.58, we see that we can write

$$\nu = \frac{1}{2\pi} \left(\frac{k}{\mu}\right)^{1/2} \tag{11.54}$$

Thus the fundamental vibration frequency for a diatomic molecule depends on the force constant and the reduced mass of the molecule.

Example 11.4 Calculate the force constant for $H^{35}Cl$ from the fact that the fundamental vibration frequency is $8.667 \times 10^{13}\ s^{-1}$ and the reduced mass is 1.627×10^{-27} kg.

$$k = (2\pi 8.667 \times 10^{13}\ s^{-1})^2 (1.627 \times 10^{-27}\ kg) = 483\ kg\ s^{-1} = 483\ N\ m^{-1}$$

The possible energy levels for a simple harmonic oscillator were shown in Section 8.10.

Vibrational transitions can be involved in the emission or absorption of electromagnetic radiation only if the molecule has a dipole moment that varies with internuclear distance. A fluctuating dipole moment offers a mechanism for interaction between the molecule and electromagnetic radiation. Homonuclear diatomic molecules such as H_2, N_2, etc., have zero dipole moment for all bond lengths and therefore do not show vibrational spectra. In general, heteronuclear diatomic molecules do have dipole moments that depend on internuclear distance, and so they exhibit vibrational spectra.

The quantum mechanical treatment of transitions between vibrational levels shows that for the simple harmonic oscillator with a dipole moment that changes linearly with the internuclear distance the selection rule is

$$\Delta v = \pm 1$$

where the + applies to absorption and the − to emission. Thus, for a molecule in its ground state, a single vibrational line would be expected in the absorption spectrum, if the molecule were a simple harmonic oscillator. However, overtone lines that correspond to $\Delta v = \pm 2, \pm 3, \ldots$ are observed with diminishing intensities. These transitions are permitted quantum mechanically by the higher terms in the Taylor series, which were neglected in equation 11.44 or by a higher order $d\mu/dR$ term in the expansion of dipole moment μ as a function of distance R. The vibrational frequencies for diatomic molecules are of the order of 1000 cm^{-1}, with higher values for molecules with hydrogen atoms or strong bonds and lower values for molecules with heavy atoms or weak bonds (see Table 11.5).

To a first approximation the vibrational energy of a diatomic molecule may be represented by equation 11.46, but this equation does not explain why overtone frequencies are not exact integral multiples of the fundamental frequency. Equation 11.45 can be improved by adding an anharmonicity term to give

$$\tilde{G}(v) = \tilde{\omega}_e(v + \tfrac{1}{2}) - \tilde{\omega}_e\tilde{x}_e(v + \tfrac{1}{2})^2 + \cdots \qquad (11.55)$$

where $\tilde{\omega}_e$ is the hypothetical vibrational wave number for a harmonic oscillator with force constant k and $\tilde{\omega}_e\tilde{x}_e$ is an anharmonicity constant. Transitions from the $v = 0$ level to higher levels give rise to the vibrational absorption spectrum.

The vibrational absorption spectrum of HCl is represented schematically in Fig. 11.4. The strongest absorption band is at 3.46 μm; there is a much weaker band at 1.76 μm and a very much weaker one at 1.198 μm. Since the equilibrium ratio of $H^{35}Cl$ molecules having $v = 1$ to those with $v = 0$ is 8.9×10^{-7} at room temperature (see equation 13.2), the molecules with $v = 1$ do not produce a significant absorption.

The coefficients $\tilde{\omega}_e$ and $\tilde{\omega}_e\tilde{x}_e$ in the power series for the vibrational energy may be determined in the way indicated in Table 11.3. The second column gives the observed wave numbers of the successive band centers. The next column gives the differences between the successive lines in wave numbers. This so-called first difference is equal to the vibrational term difference between states $v + 1$ and v and is represented by $\Delta\tilde{G}(v + \tfrac{1}{2})$.

$$\Delta\tilde{G}(v + \tfrac{1}{2}) \equiv \tilde{G}(v + 1) - \tilde{G}(v) \qquad (11.56)$$

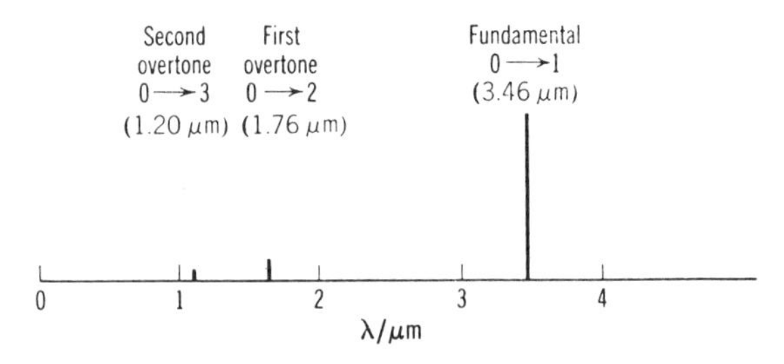

Fig. 11.4 Vibrational absorption spectrum of HCl. The relative intensities of the lines fall off five times as fast as indicated.

Inserting equation 11.55 and excluding higher terms yields

$$\Delta\tilde{G}(v + \tfrac{1}{2}) = \tilde{\omega}_e - 2\tilde{\omega}_e\tilde{x}_e(v + 1) \tag{11.57}$$

where v is the lower of the two levels.

The last column of Table 11.3 gives second differences $\Delta^2\tilde{G}(v + 1)$, which are defined by

$$\Delta^2\tilde{G}(v + 1) \equiv \Delta\tilde{G}(v + \tfrac{3}{2}) - \Delta\tilde{G}(v + \tfrac{1}{2}) \tag{11.58}$$

Substituting equation 11.57 yields

$$\Delta^2\tilde{G}(v + 1) \equiv -2\tilde{\omega}_e\tilde{x}_e \tag{11.59}$$

The second differences in Table 11.3 indicate that the anharmonicity constant $\tilde{\omega}_e\tilde{x}_e$ is equal to 103.0 cm^{-1}/2 = 51.5 cm^{-1}.

Table 11.3 Band Centers for $H^{35}Cl$ Vibrational Bands[a]

Transition	$\tilde{G}(v)/\text{cm}^{-1}$	$\Delta\tilde{G}(v + \frac{1}{2})/\text{cm}^{-1}$	$\Delta^2\tilde{G}(v + 1)/\text{cm}^{-1}$
$v = 0 \rightarrow 0$	0		
		2885.98	
$v = 0 \rightarrow 1$	2885.98		−103.98
		2782.00	
$v = 0 \rightarrow 2$	5667.98		−103.20
		2678.80	
$v = 0 \rightarrow 3$	8346.78		−102.77
		2576.03	
$v = 0 \rightarrow 4$	10,992.81		−102.65
		2473.38	
$v = 0 \rightarrow 5$	13,396.19		

[a] D. H. Rank, D. P. Eastman, B. S. Rao, and T. A. Wiggins, *J. Opt. Soc. Am.*, **52**, 1 (1962).

11.9 VIBRATION-ROTATION SPECTRA OF DIATOMIC MOLECULES

The use of spectrometers with high resolution to study the vibrational spectra of diatomic molecules yields a great deal of fine structure. This is illustrated by Fig. 11.5, which shows the fundamental vibrational band for gaseous HCl. These absorption lines are caused by transitions in which both v and J change.

A diatomic molecule in a particular vibrational level may be in one of several rotational levels. The populations of the various rotational levels at equilibrium are determined by the Boltzmann distribution as discussed in Section 13.3. When the molecule absorbs a quantum of radiation that is sufficiently large to raise it to a higher vibrational level, the molecule has to go to a different rotational sublevel of the higher vibrational state because the selection rule for the rotational quantum number J is $\Delta J = \pm 1$. The possible transitions are shown in Fig. 11.6, where the lower group of rotational levels all belong to the $v'' = 0$ vibrational level and the upper group of rotational levels all belong to the $v' = 1$ vibrational level. Vibration-rotation absorption transitions with $\Delta J = +1$ form the R branch of a vibration-rotation band, and transitions with $\Delta J = -1$ form the P branch.*

The relative intensities of the lines in Fig. 11.6 are proportional to the relative

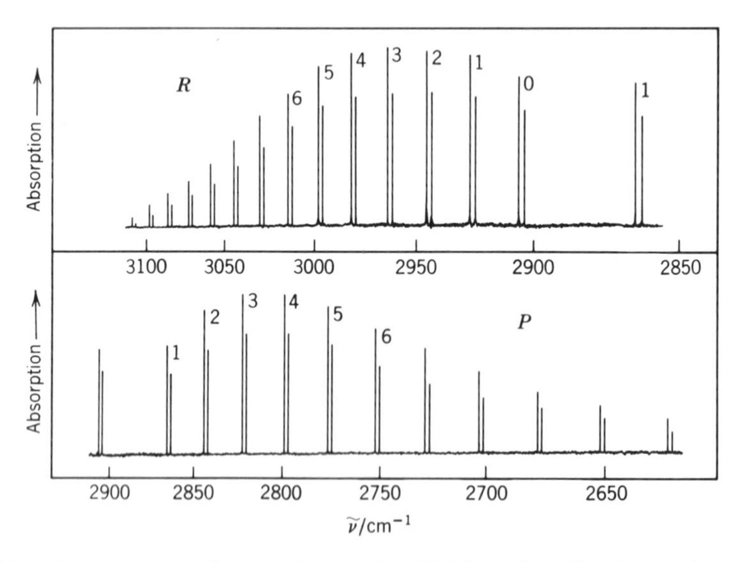

Fig. 11.5 Fundamental vibrational band for HCl ($v = 0 \rightarrow 1$). The double peaks are due to the presence of $H^{35}Cl$ (75% abundance) and $H^{37}Cl$ (25% abundance). (From A. R. H. Cole, *Tables of Wavenumbers for the Calibration of Infrared Spectrometers: Second Edition*, Pergamon Press on behalf of IUPAC, 1977.)

---

* A few diatomic molecules with electronic angular momentum in the ground state, such as NO, can also possess a Q branch for which $\Delta J = 0$.

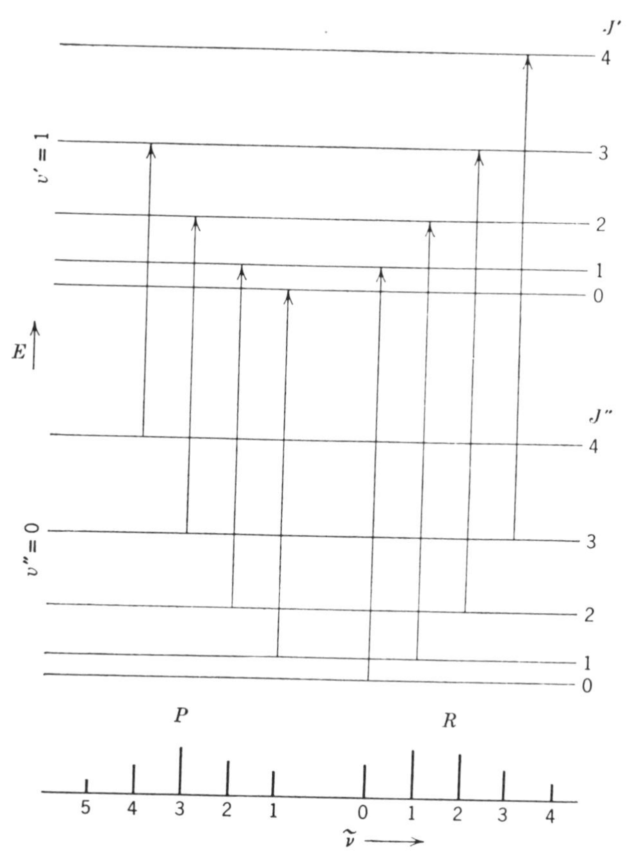

Fig. 11.6 Vibrational and rotational energy levels for a diatomic molecule and the transitions observed in the vibration-rotation spectrum. In the spectrum shown at the bottom the lines are indexed according to the rotational quantum number of the initial state. The relative heights of the spectral lines indicate relative intensities of absorption.

populations of the initial levels. As explained earlier, the relative populations of rotational levels are proportional to the product of the $(2J + 1)$-fold degeneracy of the levels and the Boltzmann factor. Thus the intensity of the lines goes through a maximum on either side of the center of the band, and the variation in intensity follows Fig. 11.2. The maxima in the P and R branches correspond closely to the rotational levels that have the greatest populations of molecules.

The wave numbers for the individual lines are obtained by use of the expression for the energy of a state of the diatomic molecule. For lines in the R branch $(v'', J'' \rightarrow v', J' = J'' + 1)$,

$$\tilde{\nu}_R = \tilde{\omega}_e(v'' - v') + 2\tilde{B}(J'' + 1) \qquad J'' = 0, 1, 2, \ldots \qquad (11.60)$$

According to this approximate equation, we expect a series of equally spaced lines on the wave-number scale at wave numbers higher than $\tilde{\omega}_e(v'' - v') + 2\tilde{B}$.

The wave numbers for the lines of the P branch (v'', $J'' \rightarrow v'$, $J' = J'' - 1$) are obtained in a similar way.

$$\tilde{\nu}_P = \tilde{\omega}_e(v'' - v') - 2\tilde{B}J'' \qquad J'' = 1, 2, 3, \ldots \tag{11.61}$$

According to this equation, we expect a series of equally spaced lines on the wave-number scale at lower wave numbers than $\tilde{\omega}_e(v'' - v')$. Note that according to these equations there is no line at $\tilde{\omega}_e$, in agreement with the spectrum in Fig. 11.5.

Figure 11.5 shows that the spacing between lines changes with J. In the P branch the spacing increases with increasing J, and in the R branch the spacing decreases with increasing J.

The following more complete analysis of this spectrum takes into account the fact that the rotational constant $\tilde{B}$ has different values in the two vibrational states. This analysis is based on the following equation for the energy of any rotation-vibration state.

$$E(v, J)/hc = \tilde{\omega}_e(v + \tfrac{1}{2}) - \tilde{\omega}_e\tilde{x}_e(v + \tfrac{1}{2})^2 + \tilde{B}_v J(J + 1) \tag{11.62}$$

where

$$\tilde{B}_v = \tilde{B}_e - \tilde{\alpha}_e(v + \tfrac{1}{2}) \tag{11.63}$$

where $\tilde{\alpha}_e$ is the vibration-rotation coupling constant expressed in wave numbers. For the R branch $J' = J'' + 1$, so that use of equation 11.2 yields

$$\begin{aligned}\tilde{\nu}_R &= \tilde{\nu}_0 + \tilde{B}'_v(J + 1)(J + 2) - \tilde{B}''_v J(J + 1) \\ &= \tilde{\nu}_0 + (3\tilde{B}'_v - \tilde{B}''_v)J + (\tilde{B}'_v - \tilde{B}''_v)J^2 + 2\tilde{B}'_v \\ &= \tilde{\nu}_0 + (B'_v + B''_v)(J + 1) + (B'_v - B''_v)(J + 1)^2\end{aligned} \tag{11.64}$$

This equation shows why the successive lines in the R branch draw together, as seen in Fig. 11.5. As we will see shortly, $\tilde{B}'_v < \tilde{B}''_v$, so that the coefficient of the $(J + 1)^2$ term is negative. In equation 11.64,

$$\tilde{\nu}_0 = \tilde{\omega}_e - 2\tilde{\omega}_e\tilde{x}_e(v + 1) \tag{11.65}$$

Similarly, for the P branch, where $J' = J'' - 1$,

$$\tilde{\nu}_P = \tilde{\nu}_0 + \tilde{B}'_v J(J - 1) - \tilde{B}''_v J(J + 1) = \tilde{\nu}_0 - (\tilde{B}'_v + \tilde{B}''_v)J + (\tilde{B}'_v - \tilde{B}''_v)J^2 \tag{11.66}$$

Since the coefficient of the J^2 term is negative, it causes the frequencies of lines in the P branch to decrease more rapidly so that the lines draw apart, as illustrated in Fig. 11.5.

The data for the first several lines of the P and R branches of the fundamental vibration-rotation band of $H^{35}Cl$ given in Table 11.4 may be analyzed to obtain $\tilde{B}'_v$ and $\tilde{B}''_v$. The rotational constant $\tilde{B}'_v$ for the upper state can be calculated by taking differences between the R and P wave numbers for the same lower state rotational levels.

$$\tilde{\nu}_R(J) - \tilde{\nu}_P(J) = 2\tilde{B}'_v(2J + 1) \tag{11.67}$$

The data of Table 11.4 yield $\tilde{B}'_v = 10.12\ \text{cm}^{-1}$.

The rotational constant for the lower state can be calculated from

$$\tilde{\nu}_R(J) - \tilde{\nu}_P(J + 2) = 2\tilde{B}''_v(2J + 3) \tag{11.68}$$

The data of Table 11.4 yield $B''_v = 10.44\ \text{cm}^{-1}$. It is to be expected that the lower

Table 11.4 Wave Numbers (cm^{-1}) for Lines in the Fundamental Vibration-Rotation Band of $H^{35}Cl$

J	$\tilde{\nu}_P(J)$	$\tilde{\nu}_R(J)$
0		2906.25
1	2865.09	2925.78
2	2843.56	2944.89
3	2821.49	2963.24
4	2798.78	
	$\tilde{\nu}_R(J) - \tilde{\nu}_P(J)$	B'_v
1	60.69	10.11
2	101.33	10.13
3	141.75	10.12
	$\tilde{\nu}_R(J) - \tilde{\nu}_P(J+2)$	$\tilde{B}''_v$
0	62.69	10.45
1	104.29	10.43
2	146.11	10.44

Table 11.5 Constants of Diatomic Molecules in Their Ground States[a]

Molecule	$N_A\mu/10^{-3}$ kg mol^{-1}	R_e/nm	$\tilde{\omega}_e$/cm^{-1}	D^0/eV
Br_2	39.958	$0.228\ 3_6$	323.2	1.971
CH	0.930 024	0.111 98	2861.6	3.47
Cl_2	17.489 42	0.198 8	564.9	2.475
CO	6.858 41	0.112 82	2170.21	11.108
H_2	0.504 066	$0.074\ 16_6$	4395.24	4.476_3
H_2^+	0.503 928	0.106	2297	2.648_1
HCl	0.979 889	0.127 460	2989.74	4.430
HBr	0.995 58	0.141 38	2649.67	3.75_4
HI	1.000 187	$0.160\ 4_1$	2309.53	3.056_4
KCl	18.599	0.279	280	4.42
LiH	0.881 506	$0.159\ 53_5$	1405.649	2.5
Na_2	11.498 22	$0.307\ 8_6$	159.23	0.73
NO	7.468 81	0.115 08	1904.03	6.487
O_2	8.000 00	$0.120\ 739_8$	1580.361	5.080
OH	0.948 38	0.097 06	3735.21	4.35
N_2	7.001 535	0.109 4	2359.61	9.756
I_2	63.452 20	0.266 6	214.57	1.5417

[a] From G. Herzberg, *Molecular Spectra and Molecular Structure*, D. Van Nostrand Co., Princeton, N.J., 1950. The reduced masses are for the molecules with the most abundant isotopic species. The lowered position of the last figure of some of the numbers indicates that this figure is uncertain.

vibrational state would have a slightly smaller mean internuclear distance and therefore a larger rotational constant than the higher vibrational state. To calculate $\tilde{B}_e$ for the molecule at its equilibrium configuration, we use equation 11.63, which yields

$$10.44 \text{ cm}^{-1} = \tilde{B}_e - \tilde{\alpha}_e/2 \tag{11.69a}$$

$$10.12 \text{ cm}^{-1} = \tilde{B}_e - \tfrac{3}{2}\tilde{\alpha}_e \tag{11.69b}$$

Thus $\tilde{B}_e = 10.59 \text{ cm}^{-1}$ and $\tilde{\alpha}_e = 0.32 \text{ cm}^{-1}$. Using this more complete analysis, we obtain the same value for the rotational constant $\tilde{B}_e$ as from the pure rotational spectrum in Example 11.3.

The fundamental vibration frequencies $\tilde{\omega}_e$, spectroscopic dissociation energies D^0, and other constants for some diatomic molecules are given in Table 11.5. (The difference between D^0 and D^e is shown in Fig. 10.1.)

11.10 VIBRATIONAL SPECTRA OF POLYATOMIC MOLECULES

In order to describe the positions of N nuclei in a molecule it is necessary to specify $3N$ coordinates. Three of these may be used to specify the position, and therefore the translational motion, of the center of mass of the molecule. Another three may be used to specify the angular position of a nonlinear molecule with respect to some external coordinate system. For example, the orientation of a water molecule may be described by giving two angles to locate the direction of the C_2 axis (Section 9.3) and the angle of the plane of the molecule. For a linear molecule only two angles need to be specified. Thus $3N$-6 coordinates remain for a nonlinear molecule to describe vibrations, and $3N$-5 coordinates remain for a linear molecule. These are commonly called internal degrees of freedom, and they include vibrations and internal rotations. If there is an appreciable potential energy barrier for an internal rotation about some bond, there will be an oscillation about a mean position. In ethylene, $CH_2{=}CH_2$, there is a large potential barrier for internal rotation so that there are only small oscillations about the C=C bond. In some cases, like $CH_3{-}CH_3$, the potential energy barrier is small enough that the internal rotation is said to be "free" at room temperature.

The vibratory motion of a polyatomic molecule may be very complicated, but all of the possible motions may be described in terms of simple motions, called the *normal modes of vibration*. Some modes are degenerate. The number of normal modes times the degeneracy of each normal mode is equal to the number of internal degrees of freedom that we have just calculated.

In a normal mode of vibration, the nuclei move in phase (i.e., the nuclei pass through the extremes of their motion simultaneously). The motions of the nuclei in a normal mode are such that the center of mass does not move, and the molecule does not rotate as a whole. This means that different atoms move different distances. Each normal mode has a characteristic vibration frequency. Any vibratory motion of a molecule may be expressed as a sum of displacements in the various normal modes.

For a linear triatomic molecule like CO_2, $3N - 5 = 4$, and so there are four vibrational degrees of freedom. Two are symmetrical and asymmetrical stretching vibrations. These normal modes are shown in Fig. 11.7. Two are bending motions that differ only in that they are in mutually perpendicular planes and are therefore degenerate. It is noted in this figure that the stretching vibrations have higher frequencies than the bending vibrations. This is often true and is a result of the fact that a larger force is required to stretch a bond between two atoms than to deform a bond angle.

In the figure the displacement vectors show the direction and relative amplitude of vibration of each nucleus. The symmetry of a molecule restricts the possible forms of normal vibrations. The normal modes of vibration of a molecule are either symmetric or antisymmetric with respect to all symmetry elements in the molecule.

Of the three normal mode vibrations for CO_2 the symmetric stretch is not active in the infrared, but the other vibrations are. Since CO_2 is symmetrical in its equilibrium state, it does not have a dipole moment, and the symmetrical stretching vibration does not create one. The asymmetric stretch and bending vibrations produce a changing dipole moment.

For a nonlinear triatomic molecule like H_2O, $3N - 6 = 3$, and so there are three normal modes of vibration, all of which are infrared active. These normal modes are shown in Fig. 11.7*b*.

The frequencies, in cm^{-1}, of the strongest bands for H_2O vapor are summarized in Table 11.6. The weaker bands in the spectrum are the overtones and combinations shown in the table. The vibrations are not strictly harmonic, and so the

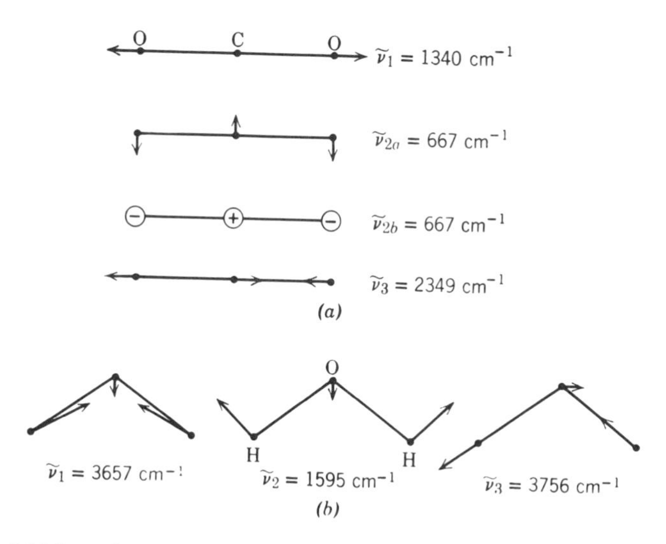

Fig. 11.7 (*a*) Normal modes of vibration of symmetrical linear triatomic molecule CO_2. (*b*) Normal modes of vibration of nonlinear triatomic molecule H_2O. The magnitudes of the oxygen vibrations have been increased relative to those of hydrogen.

overtones are not exact multiples, and the combinations are not exact sums as shown in the table.

Table 11.6 Infrared Bands of H_2O Vapor

$\tilde{\nu}/\text{cm}^{-1}$	Intensity	Interpretation
1595.0	Very strong	$\tilde{\nu}_2$
3151.4	Medium	$2\tilde{\nu}_2$
3651.7	Strong	$\tilde{\nu}_1$
3755.8	Very strong	$\tilde{\nu}_3$
5332.0	Medium	$\tilde{\nu}_2 + \tilde{\nu}_3$
6874	Weak	$2\tilde{\nu}_2 + \tilde{\nu}_3$

The use of normal coordinates in describing molecular vibrations allows the separation of the Schrödinger equation into as many separate equations as there are normal modes, if the potential is quadratic. The eigenvalue for each of the separated equations is

$$E_i = (v_i + \tfrac{1}{2})h\nu_i \tag{11.70}$$

and so the total vibrational energy of the molecule is

$$E_v = \sum_i (v_i + \tfrac{1}{2})h\nu_i \tag{11.71}$$

The vibrational spectra of polyatomic molecules are very useful in identifying them and serve also as a criterion of purity. For such practical applications the infrared spectra for a large number of compounds have been cataloged and are used like "fingerprints." Groups of atoms within the molecule have quite characteristic absorption bands. The wavelengths at which a certain group absorbs vary slightly, depending on the structure of the rest of the molecule.

The infrared spectrum may be considered to be made up of several regions.*

3700–2500 cm^{-1}—hydrogen stretching vibrations. These vibrations occur at high frequencies because of the low mass of the hydrogen atom. If an OH group is not involved in hydrogen bonding (Section 10.8), it usually has a frequency in the vicinity of 3600–3700 cm^{-1}. Hydrogen bonding causes this frequency to drop by 300 to 1000 cm^{-1} or more. The NH absorption falls in the 3300–3400 cm^{-1} range, and the CH absorption falls in the 2850–3000 cm^{-1} range. As the atom to which hydrogen is attached becomes heavier, the frequencies decrease; for SH, PH, and SiH they are approximately 2500, 2400 and 2200 cm^{-1}.

2500–2000 cm^{-1}—triple-bond region. These bonds have high frequencies because of the large force constants. The C≡C group usually causes absorption between 2050 and 2300 cm^{-1}, but this absorption may be weak or absent because of the symmetry of the molecule. The C≡N group absorbs near 2200–2300 cm^{-1}.

* R. P. Bauman, *Absorption Spectroscopy*, Wiley, New York, 1962, pp. 336–347.

2000–1600 cm^{-1}—double-bond region. Absorption bands of substituted aromatic compounds fall in this range and are a good indicator of the position of the substitution. Carbonyl groups, C=O, of ketones, aldehydes, acids, amides, and carbonates usually show strong absorption in the vicinity of 1700 cm^{-1}. Olefins, C=C, may show absorption in the vicinity of 1650 cm^{-1}. The bending of the C—N—H bond also occurs in this region.

1500–1700 cm^{-1}—single-bond stretch and bend region. This region is not diagnostic for particular functional groups, but it is a useful "fingerprint" region, since it shows differences between similar molecules. Organic compounds usually show peaks in the region between 1300 and 1475 cm^{-1} because of the bending motions of hydrogens. Out-of-plane bending motions of olefinic and aromatic CH groups usually occur between 700 and 1000 cm^{-1}.

11.11 RAMAN SPECTRA

When light passes through a liquid or gas, a small fraction of the light is scattered. A perfect crystalline solid would not scatter light, because scattering by one unit of the crystal would interfere destructively with scattering from another symmetrically located unit. The mechanism of the scattering of light involves the polarization of the molecules or atoms by the electric field of the light. The electric field of the light induces a rapidly fluctuating dipole in the atoms or molecules in its path. As discussed in Section 20.7, the fluctuating dipole leads to the emission of electromagnetic waves in various directions at the same frequency as the incident radiation, and this radiation is seen as the scattered light. This scattering, referred to as Rayleigh scattering, may be viewed as the *elastic* scattering of a photon by a molecule.

In 1928 Raman discovered experimentally that there is sometimes also present in the scattered light weak radiation of frequencies not present in the incident light. The frequency differences between the weak lines and the exciting line are characteristic of the scattering substance and are independent of the frequency of the exciting line. Raman scattering may be viewed as *inelastic* scattering of a photon by a molecule. It is different from fluorescence or phosphorescence (Section 17.3) in that the sample does not need to have an absorption band at the wavelength of the incident light. In other words, any wavelength may be used to study the Raman effect.

Since Raman lines are weak, long exposures have been required with conventional light sources, but the development of lasers has made available powerful monochromatic light sources for the study of Raman scattering. The advantages of these sources are their high intensity of light collimated in a particular direction, the small line width, and the polarization of the beam.

The experimental arrangement for observing Raman spectra is indicated in Fig. 11.8. Light scattered by the sample is focused on the slit of the spectrophotometer. Intensity is measured as a function of wavelength by use of a photomultiplier tube. The Raman spectrum of liquid CCl_4 obtained in this way is shown in Fig. 11.9.

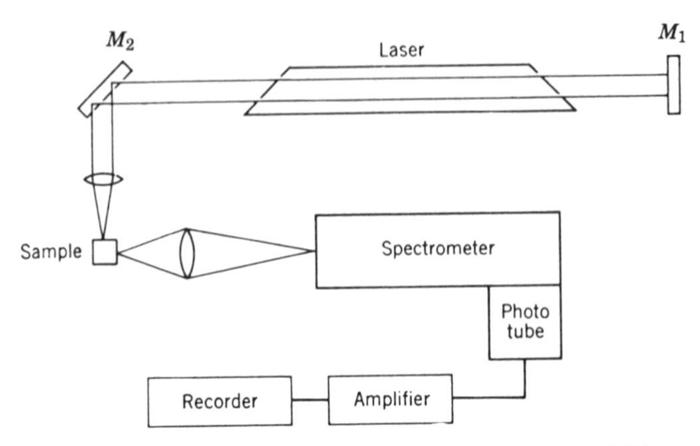

Fig. 11.8 Apparatus for obtaining Raman spectra. Mirrors M_1 and M_2 are required to obtain laser action.

The development of lasers has made it possible to use high-incident intensities and to obtain Raman spectra at wavelengths near absorption lines of the substance being irradiated. Near an absorption line Raman scattering is greatly enhanced, so that Raman spectra may be obtained at low concentrations. This is called the resonance Raman effect. The particular chromophore responsible for the near

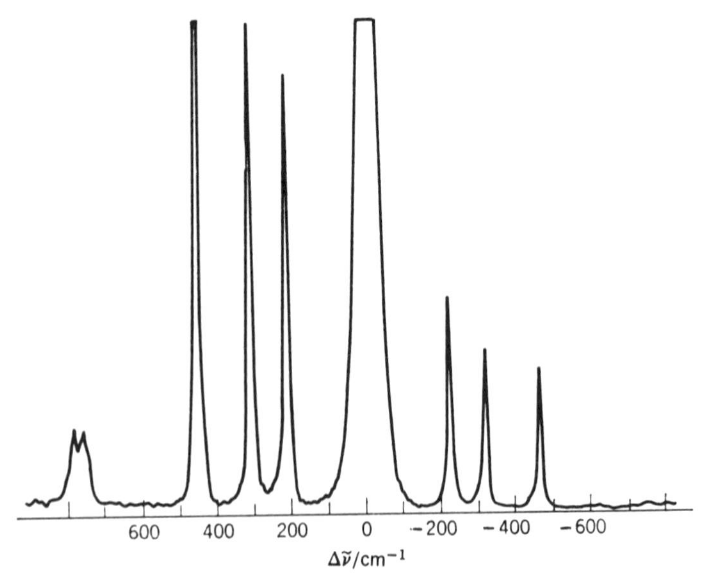

Fig. 11.9 Raman spectrum of liquid CCl_4 obtained using a He-Ne laser. The Raman frequencies are measured from the parent line at 632.8 nm.

resonant electronic transition will dominate the resonance Raman spectrum, because only vibrations of that chromophore are enhanced.

The change in frequency between the incident and scattered light can be viewed as an exchange of energy between the incident photon and the scattering molecule. The photon, of energy $h\nu$ insufficient to cause a transition to an excited electronic state and be absorbed, induces a forced oscillation in the molecule, which is in its ground electronic state and in a low vibration and rotational level. If the molecule is shifted from a level with energy E'' to another energy E', the scattered light has a frequency of ν' and, from conservation of energy,

$$h\nu + E'' = h\nu' + E' \tag{11.72}$$

Generally, the scattered light has a lower frequency (called a Stokes line), because energy is lost to the molecule. When the photon interacts with a molecule in a higher vibrational or rotational level, however, the molecule may give energy to the photon so that a line of higher frequency (anti-Stokes line) is found in the scattered radiation. There will be both Stokes and anti-Stokes lines for any permitted transition, but the anti-Stokes lines are weaker because of the relatively small number of molecules in higher energy states, as given by the Boltzmann distribution.

Equation 11.72 may be rearranged to

$$E' - E'' = h(\nu - \nu') = h\nu_{\text{Raman}} = hc\tilde{\nu}_{\text{Raman}} \tag{11.73}$$

which shows that the shift in frequency is a measure of the difference in energy between two levels. The shifts in frequencies $\nu - \nu'$ are called Raman shifts ν_{Raman} or $\tilde{\nu}_{\text{Raman}}$. These shifts fall in the range 100 to 4000 cm^{-1} for vibrational changes, with smaller values for rotational changes.

The selection rules for the Raman spectrum depend on the polarizability of a molecule. The electric field of the light wave may distort the electron distribution of a molecule to induce a dipole moment. When a spherically symmetric atom is placed in an electric field, a dipole moment $\boldsymbol{\mu}$ is induced in the same direction as the electric field $\boldsymbol{E}$.

$$\boldsymbol{\mu} = \alpha \boldsymbol{E} \tag{11.74}$$

The constant of proportionality is the polarizability α (Section 10.12). However, for a molecule that does not have spherical symmetry, each component (μ_x, μ_y, μ_z) of the induced dipole moment $\boldsymbol{\mu}$ can depend on each component (E_x, E_y, E_z) of the electric field $\boldsymbol{E}$. Therefore, the molecular polarizability $\boldsymbol{\alpha}$ is a tensor, and the induced dipole-moment vector is not necessarily in the same direction as the applied electric field. This can be expressed by the matrix equation

$$\begin{pmatrix} \mu_x \\ \mu_y \\ \mu_z \end{pmatrix} = \begin{pmatrix} \alpha_{xx} & \alpha_{xy} & \alpha_{xz} \\ \alpha_{yx} & \alpha_{yy} & \alpha_{yz} \\ \alpha_{zx} & \alpha_{zy} & \alpha_{zz} \end{pmatrix} \begin{pmatrix} E_x \\ E_y \\ E_z \end{pmatrix} \tag{11.75}$$

This is equivalent to the following set of algebraic equations:

$$\mu_x = \alpha_{xx}E_x + \alpha_{xy}E_y + \alpha_{xz}E_z \tag{11.76a}$$

$$\mu_y = \alpha_{yx}E_x + \alpha_{yy}E_y + \alpha_{yz}E_z \tag{11.76b}$$

$$\mu_z = \alpha_{zx}E_x + \alpha_{zy}E_y + \alpha_{zz}E_z \tag{11.76c}$$

Only six of the nine components of the polarizability are independent, since it can be shown that $\alpha_{xy} = \alpha_{yx}$, $\alpha_{xz} = \alpha_{zx}$, and $\alpha_{yz} = \alpha_{zy}$.

In order for a particular vibration of a molecule to appear in the Raman spectrum, it is necessary for the particular vibration to change the polarizability of the molecule. The reason for this selection rule is readily understood. If there is a change in polarizability attending the vibration there will be a variation in the induced moment at the frequency of the vibration. Since the polarizability of a homonuclear diatomic molecule does change during a vibration, these molecules have vibrational Raman lines, although they have no IR spectrum. For the vibrational Raman effect,

$$\Delta v = \pm 1 \tag{11.77}$$

Rotational transitions may also be observed by use of the Raman effect if the polarizability of a molecule depends on its orientation. This is true for linear molecules, even those that because of symmetry do not show infrared or microwave spectra. The moments of inertia of homonuclear diatomic molecules can be obtained from rotational Raman spectra, and this makes it possible to determine their internuclear distances very accurately. For the Raman effect of linear molecules,

$$\Delta J = 0, \pm 2 \tag{11.78}$$

The data obtained from Raman and infrared spectra are often complementary. A detailed consideration of the selection rules for the Raman effect shows that if a molecule has a center of symmetry (Section 9.2), a given vibrational transition cannot appear in both the IR and Raman spectra.

11.12 ELECTRONIC SPECTRA OF DIATOMIC MOLECULES

Transitions between electronic levels of molecules lead to absorption and emission in the visible and ultraviolet parts of the spectrum. For some molecules the changes in energy accompanying changes in electronic structures are so great that absorption occurs only in the vacuum ultraviolet. The transitions between electronic states may be accompanied by transitions between rotational and vibrational states, and so the spectra are very complicated.

All molecules show electronic spectra because the change from one electronic state to another always provides the opportunity for interaction with electromagnetic radiation. Homonuclear diatomic molecules, which do not have rotation or vibration-rotation spectra, do show vibrational and rotational structure in their electronic spectra.

An important principle for the interpretation of electronic spectra was stated by Franck and Condon. They pointed out that since electrons move much more rapidly than nuclei, it is a good approximation to assume that, in an electronic transition, the nuclei do not change their positions. Therefore an electronic transition may be represented by a vertical line on a plot of potential energy versus internuclear distance.

Figure 11.10*a* shows the potential energy curves for the ground state and first excited state of a diatomic molecule *XY*. The molecule has a different electronic structure in the excited state, and the potential energy of the excited state is higher. The equilibrium internuclear separation is usually greater in the excited state and the bonding is weaker, as shown by the smaller dissociation energy for the excited state. When the molecule in the ground state absorbs radiation, it is most likely to be at its equilibrium internuclear separation, as discussed in Section 8.10 for the ground state of a harmonic oscillator. This is not true for the higher vibrational levels where the classical situation (in which the highest probability is at the turning points) is approached more and more closely as the vibrational energy is increased. According to the Franck-Condon principle, the most probable transition is $v = 0 \rightarrow 2$, because the excited molecule in the $v = 2$ state does not have much internuclear kinetic energy and has a high probability density at the equilibrium internuclear distance of the vibrational zero point level of the ground electronic state. Transitions to other vibrational levels of the excited state occur with lower probabilities, as indicated by the schematic absorption spectrum shown in Fig. 11.10*b*.

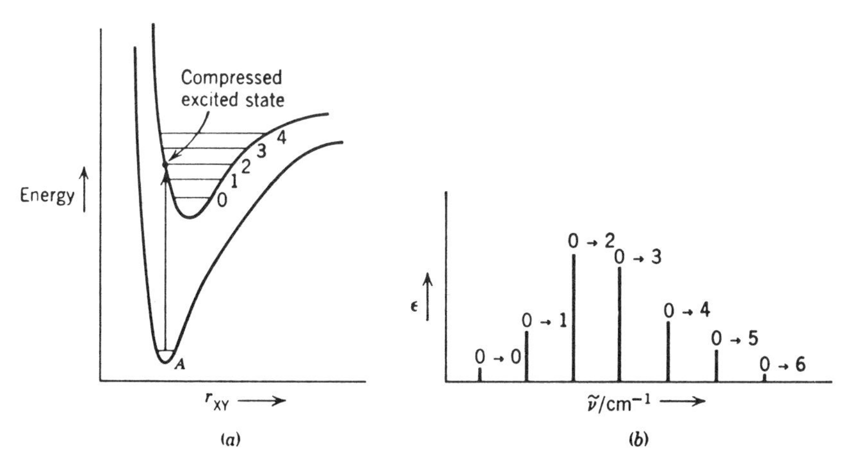

Fig. 11.10 (*a*) Potential energy curves for the ground state and first excited state of a diatomic molecule *XY*. (*b*) Absorption spectrum of diatomic molecule *XY*. (From N. J. Turro, *Molecular Photochemistry*. Copyright © 1965 by W. A. Benjamin, Inc., Menlo Park, Calif.)

If the minimum potential energy of the excited state is at a much greater internuclear distance, the absorption of radiation by the ground state will produce excited molecules with enough energy to dissociate, as shown in Fig. 11.11*a*. Since variable amounts of energy may lead to dissociation, continuous absorption is produced as shown by the spectrum in Fig. 11.11*b*. There is some probability of excitation to vibrational levels with $v = 3$, 4, and 5, and so some lines also appear in the absorption spectrum.

Dissociation may also be produced if a molecule is excited to an antibonding excited state [see, for example, curve $(\sigma_g 1s)(\sigma_u^* 1s)$ of Fig. 10.8].

The dissociation energy for a molecule can frequently be determined with great accuracy from its electronic spectrum. The dissociation energy is simply calculated from the wavelength that separates the discontinuous part of the spectrum from the continuous part of the spectrum resulting from the dissociation of the molecule into two parts with various amounts of kinetic energy.

Since dissociation of a diatomic molecule may take place into excited atoms, it is necessary to know the energy states of the atoms formed in order to calculate the dissociation energy from the limit of continuous emission or absorption. The energies approached by the various potential energy curves at infinite internuclear distance correspond to the different energies of excitation of the two separated atoms. Since the atoms may be produced in excited electronic states (as implied by Fig. 11.11*b*), the excitation energy of the atoms must be subtracted to obtain the energy of dissociation into ground state atoms.

The dissociation energy of O_2 has been determined very accurately from its

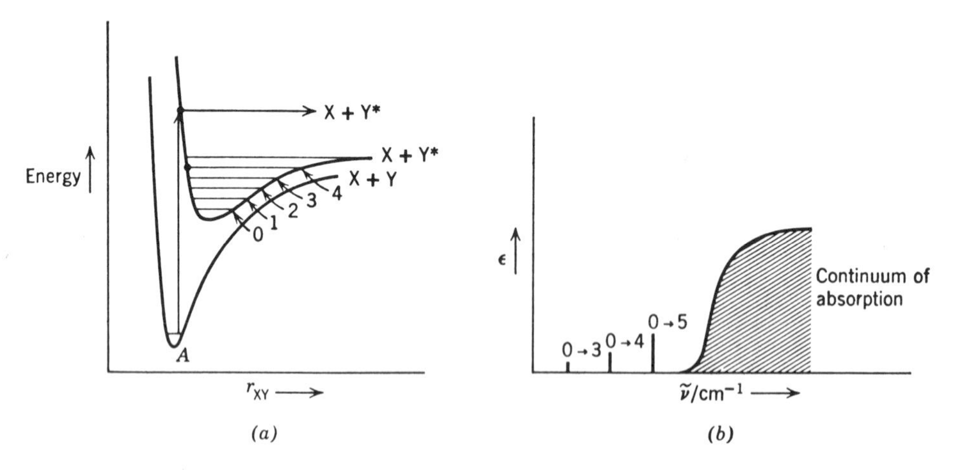

Fig. 11.11 (*a*) Potential energy curves for the ground state and first excited state of a diatomic molecule *XY*. The excited state has a larger equilibrium internuclear distance than the one in the preceding figure. (*b*) Absorption spectrum of diatomic molecule *XY*. (Reproduced from *Molecular Photochemistry*, 1965, written by N. J. Turro, with permission of publishers, Addison-Wesley/W. A. Benjamin Inc., Advanced Book Program, Reading, Mass., U.S.A.)

ultraviolet spectrum. Continuous absorption begins at 1759 nm (56,877 cm^{-1}). One of the oxygen atoms produced is in an excited state with an energy of 15,868 cm^{-1} above the ground state and the other is in the ground electronic state. Therefore the dissociation energy into two atoms in their ground states is 56,877 − 15,868 = 41,009 cm^{-1}, or 5.08 eV.

The dissociation generally takes place from the lowest vibrational level of the ground electronic state. However, for heavy diatomic molecules such as I_2, the molecule at room temperature occupies several of the lower vibrational levels of the ground electronic state. In this situation the dissociation may take place from vibrationally excited states.

The fluorescence and phosphorescence spectra of diatomic molecules are discussed in Sections 17.3 and 17.4.

11.13 SPECTROPHOTOMETERS

When polychromatic radiation is passed into a substance, some of the radiation may be absorbed and the rest is either transmitted or scattered. The fraction transmitted may be determined as a function of frequency by use of a spectrophotometer. The design of a spectrophotometer is shown schematically in Fig. 11.12.

The principal parts of a spectrophotometer are the source of electromagnetic radiation, the monochromator, the cell compartment, the photoelectric detector, and a device for indicating the output from the detector (electric meter, potentiometer, or recording potentiometer). The cell compartment contains an optical absorption cell filled with the solution to be studied and another absorption cell filled with a reference solution, usually pure solvent. The ratio of the intensity I of transmitted light for the solution of the intensity I_0 for the solvent is called the *transmittance*.

The transmittance I/I_0 can be determined at different wavelengths, and the absorption spectrum can be mapped. With some spectrophotometers such a plot is recorded automatically. The positions and intensities of the absorption bands and lines serve for identification and for criteria of purity; the percentage transmissions serve for quantitative analyses of the concentration of material present.

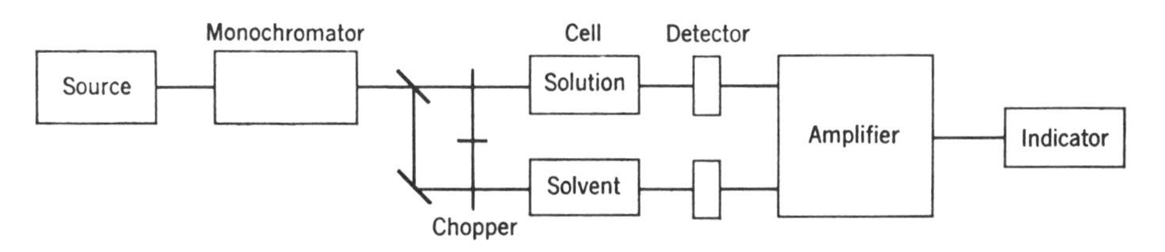

Fig. 11.12 A spectrophotometer. The source and monochromator may be replaced by a tunable laser.

11.14 LAMBERT-BEER LAW

The probability that a photon will be absorbed is usually directly proportional to the concentration of absorbing molecules and to the thickness of the sample for a very thin sample. This probability is expressed mathematically by the equation

$$\frac{dI}{I} = -kc\,dx \tag{11.79}$$

where I is the intensity of light of a particular wavelength, that is, the number of photons per unit area per unit time, and dI is the change in light intensity produced by absorption in a thin layer of thickness dx and concentration c. Distance x is measured through the cell in the direction of the beam of light that is being absorbed.

The intensity of a beam of light after passing through length l of solution is related to the incident intensity I_0 be equation 11.81, which is obtained by integrating equation 11.79 between the limits I_0 when $x = 0$ and I when $x = l$.

$$\int_{I_0}^{I} \frac{dI}{I} = -kc \int_0^l dx \tag{11.80}$$

$$\ln \frac{I}{I_0} = 2.303 \log \frac{I}{I_0} = -kcl \tag{11.81}$$

Since it is convenient to use logarithms to the base 10, the Lambert-Beer law is used in the form

$$\log \frac{I_0}{I} = A = \epsilon cl \tag{11.82}$$

where $\epsilon = k/2.303$ is referred to as the molar absorption coefficient and c is amount-of-substance concentration.

The quantity $\log (I_0/I)$ is referred to as the absorbance A. It can be seen from equation 11.82 that the absorbance is directly proportional to the concentration c and to the path length l. The proportionality constant is characteristic of the solute and depends on the wavelength of the light, the solvent, and the temperature. Since the molar absorption coefficient ϵ depends on wavelength, the Lambert-Beer law is obeyed at each wavelength. If the radiation is not monochromatic the Lambert-Beer law may not be obeyed. The apparent absorption coefficient for a substance that associates or dissociates will change with concentration because of the changing ratio of concentrations of the absorbing species.

Figure 11.13 shows the absorption spectra of benzene and paraxylene.*

---

* P. E. Stevenson, *J. Chem. Ed.*, **41**, 234 (1964).

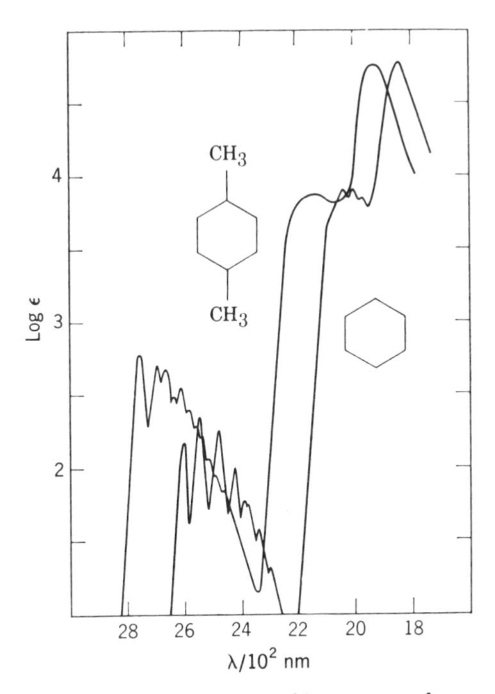

Fig. 11.13 Absorption spectra of benzene and paraxylene.

Example 11.5 The percentage transmittance of an aqueous solution of disodium fumarate at 250 nm and 25 °C is 19.2% for a 5×10^{-4} mol L^{-1} solution in a 1-cm cell. Calculate the absorbance A and the molar absorption coefficient ϵ.

$$A = \log\left(\frac{I_0}{I}\right) = \log\left(\frac{100}{19.2}\right) = 0.717$$

$$\epsilon = \frac{A}{lc} = \frac{0.717}{(1\ \text{cm})(5 \times 10^{-4}\ \text{mol L}^{-1})} = 1.43 \times 10^3\ \text{L mol}^{-1}\ \text{cm}^{-1}$$

What will be the percentage transmittance of a 1.75×10^{-5} mol L^{-1} solution in a 10-cm cell?

$$\log\left(\frac{I_0}{I}\right) = (1.43 \times 10^3\ \text{L mol}^{-1}\ \text{cm}^{-1})(10\ \text{cm})(1.75 \times 10^{-5}\ \text{mol L}^{-1}) = 0.251$$

$$\frac{I_0}{I} = 1.782 \quad \text{and} \quad \frac{100I}{I_0} = 56.1\%$$

For mixtures of independently absorbing substances the absorbance is given by the equation

$$\log\left(\frac{I_0}{I}\right) = A = (\epsilon_1 c_1 + \epsilon_2 c_2 + \cdots)l \tag{11.83}$$

where $c_1, c_2, \ldots$ are the concentrations of the substances having absorption coefficients of $\epsilon_1, \epsilon_2, \ldots$. A mixture of n components may be analyzed by measuring A at

n wavelengths at which the absorption coefficients are known for each substance, provided that these coefficients are sufficiently different. The concentrations of the several substances may then be obtained by solving the n simultaneous linear equations.

When a sample is irradiated continuously, its absorption coefficient remains constant in the absence of chemical reaction, and this indicates that excited molecules are continuously deactivated so that they do not accumulate. Usually the excitation energy is simply degraded to thermal energy in molecular collisions, but a chemical reaction may occur and change the composition and absorption spectrum of the sample (cf. Chapter 17). An excited molecule may also emit a quantum of radiation. Such emission is referred to as fluorescence or phosphorescence, depending on the type of excited state (Sections 17.3 and 17.4).

11.15 ELECTRONIC SPECTRA OF POLYATOMIC MOLECULES

Electronic transitions lead to absorption spectra in the visible and ultraviolet regions. For organic compounds this involves promotion of electrons in n (nonbonding), σ, and π orbitals in the ground state to σ^* and π^* antibonding orbitals in the excited state. Since n electrons do not form bonds, there are no antibonding orbitals associated with them.

The energy required for a $\sigma^* \leftarrow \sigma$ transition is generally so high that the absorption occurs in the vacuum ultraviolet. Thus saturated hydrocarbons and other compounds in which all valence shell electrons are involved in single bonds do not absorb in the ordinary ultraviolet or visible regions.

Ultraviolet absorptions due to $\sigma^* \leftarrow n$ transitions are shown by compounds that contain nonbonding electrons on oxygen, nitrogen, sulfur, or halogen atoms. For example, methyl alcohol vapor shows an absorption maximum at 183 nm with a molar absorption coefficient of 150 L mol^{-1} cm^{-1}.

Groups like C=C, C=O, —N=N—, and —N=O, which cause absorption at wavelengths longer than 180 nm, are called *chromophores*. The positions and intensities of the absorptions caused by these groups are rather characteristic. For example, C=C double bonds generally produce an absorption maximum at 180 to 190 nm with a molar absorption coefficient of about 10^4 L mol^{-1} cm^{-1}. Ketone and aldehyde C=O groups usually have absorption maxima at 270 to 290 nm with molar absorption coefficients of 10–30 L mol^{-1} cm^{-1}. The most intense electronic absorption spectra are produced by certain dyes that may have molar absorption coefficients as high as 10^5 L mol^{-1} cm^{-1} in solution.

Unsaturated molecules can show $\pi^* \leftarrow n$ and $\pi^* \leftarrow \pi$ transitions. One of the best understood cases is the carbonyl absorption of aldehydes and ketones. The stronger absorption band around 180 nm is due to the $\pi^* \leftarrow \pi$ transition, and the weaker absorption band around 285 nm is due to the $\pi^* \leftarrow n$ transition. The structure of the rest of the molecule affects the strength of the absorption and the wavelength of the maximum absorption, but a series of compounds with the same chromophore will generally show about the same ultraviolet absorption spectrum. When chromophoric groups are separated by two or more single bonds, their effects are usually additive but, if they are a part of a conjugated* system, large non-additive effects are seen because the π electron system is spread over at least four atomic centers. The absorption band is generally shifted 15 to 45 nm to longer wavelengths. As further unsaturated groups are added to the conjugated system, there are further shifts to a longer wavelength because the energy required for the $\pi^* \leftarrow \pi$ transition is less, and there is an increase in molar absorption coefficient. Single-ring aromatics absorb in the vicinity of 250 nm, naphthalenes in the vicinity of 300 nm, and anthracenes and phenanthrenes in the vicinity of 360 nm.

11.16 FREE-ELECTRON MODEL

For molecules with conjugated systems of double bonds [i.e., $R(CH{=}CH)_nR'$] it is found that the electronic absorption bands shift to longer wavelengths as the number of conjugated double bonds is increased. Approximate quantitative calculations of the absorption frequencies may be made on the basis of the free-electron model for the π electrons of these molecules. The energy for the lowest electronic transition is that required to raise an electron from the highest filled level to the lowest unfilled level. In a system of conjugated double bonds each carbon atom has three σ bonds that lie in a plane, and each sigma bond involves one outer electron of that carbon atom. Above and below this plane are the π orbital systems (Fig. 10.9). Each carbon atom contributes one electron to this π system, but these electrons are free to move the entire length of the series of π orbitals and are not localized at a given carbon atom. In the free-electron model it is assumed that the π system is a region of uniform potential and that the potential energy rises sharply to infinity at the ends of the system (i.e., a square-well potential). Thus the energy levels E available to the π electrons would be expected to be those calculated for the particle restricted to movement in one direction (Section 8.6).

$$E = \frac{n^2h^2}{8m_ea^2} \tag{11.84}$$

The length of the box a is usually taken to be the length of the chain between terminal carbon atoms plus a bond length or two.

The π electrons (one for each carbon atom) are assigned to orbitals so that there are two (one with spin $+\frac{1}{2}$ and the other with spin $-\frac{1}{2}$) in each level, starting with the lowest. For a completely conjugated hydrocarbon the number of π electrons is even, and the quantum number of the highest filled level will be $n = N/2$, where N is the number of π electrons (the number of carbon atoms involved). In absorption an electron from the highest filled level is excited to the next higher level with

* In conjugated molecules double and single bonds alternate.

quantum number $n' = N/2 + 1$. The difference in energy of these two levels is

$$\Delta E = \frac{h^2}{8m_e a^2}(n'^2 - n^2) = \frac{h^2}{8m_e a^2}\left[\left(\frac{N}{2}+1\right)^2 - \left(\frac{N}{2}\right)^2\right] = \frac{h^2}{8m_e a^2}(N+1) \qquad (11.85)$$

The absorption frequency in wave numbers is given by

$$\tilde{\nu} = \frac{\Delta E}{hc} = \frac{h(N+1)}{8cm_e a^2} \qquad (11.86)$$

Example 11.6 Calculate the lowest absorption frequency for octatetraene (C_8H_{10}) that contains a series of four conjugated double bonds. The length of the π bond system is about 0.95 nm.

$$\tilde{\nu} = \frac{n(N+1)}{8cm_e a^2} = \frac{(6.62 \times 10^{-34}\ \text{J s})(9)(10^{-2}\ \text{m cm}^{-1})}{8(3 \times 10^8\ \text{m s}^{-1})(9.110 \times 10^{-31}\ \text{kg})(0.95 \times 10^{-9}\ \text{m})^2}$$

$$= 30{,}200\ \text{cm}^{-1}$$

The observed absorption band is at 33,100 cm^{-1}.

11.17 PHOTOELECTRON SPECTROSCOPY

When a gas M is irradiated with ultraviolet or X-ray photons, the photoionization process can be represented by

$$M + h\nu \rightarrow M^+ + e^- \qquad (11.87)$$

The kinetic energy $E(e^-)$ of the emitted electron is given by

$$E(e^-) = h\nu + E(M) - E(M^+) \qquad (11.88)$$

where $E(M)$ is the energy of M in its initial electronic and vibrational state (usually the ground state) and $E(M^+)$ is the energy of the ion in the electronic and vibrational state in which it is produced. Thus measurement of the energy of electrons emitted gives information about the different electronic and vibrational states of M^+.

The most commonly used source of ultraviolet photons is a helium gas discharge tube; the most intense line has a wavelength of 58.4 nm and an energy of 21.22 eV. The spectrum of energy of emitted electrons may be measured with a focusing deflection analyzer, utilizing either magnetic or electrostatic fields. An example of an electron energy spectrum obtained by sweeping the appropriate field strength is shown in Fig. 11.14 for hydrogen gas. The most energetic electrons emitted by the sample (line labeled 0) come from the production of hydrogen molecule ions in their ground state.

$$H_2(v = 0) + h\nu = H_2^+(v = 0) + e^- \qquad (11.89)$$

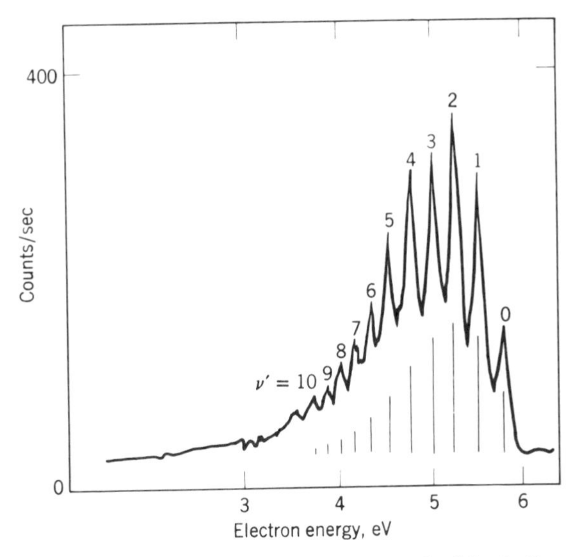

Fig. 11.14 Photoelectron spectrum of hydrogen gas excited by helium resonance radiation (58.4 nm or 21.22 eV). [From D. W. Turner and D. P. May, *J. Chem. Phys.*, **45**, 471 (1966).]

The electrons emitted in this peak have an energy of 5.77 eV so that the energy difference $E(M^+) - E(M)$ for the $v = 0 \rightarrow 0$ transition is $21.22 - 5.77 = 15.45$ eV, which is the adiabatic ionization potential for hydrogen gas (cf. Section 10.1).

The other lines in the spectrum result when H_2^+ is produced in higher vibrational states. Since the minimum of the potential energy curve for H_2^+ is at a somewhat larger internuclear distance than for H_2 (Section 10.3), the Franck-Condon principle leads us to expect that the 0–0 transition will not be the most probable. The energy difference for the strongest band (in this case $v = 0 \rightarrow 2$) is referred to as the vertical ionization potential; this ionization potential is $21.22 - 5.24 = 15.98$ eV.

The photoelectron spectra of molecules with more electrons are, of course, considerably more complicated and yield information about the dissociation of inner electrons as well as valence electrons. Photoelectron spectroscopy offers the most direct method for determining ionization potentials and provides a great deal of information about molecular electronic structure. The spectra, because of the dissociation of inner shell electrons of the same atom, appear at different energies, depending on its chemical environment. These "chemical shifts" offer the possibility of obtaining structural information. For example, the nitrogen 1*s* spectrum of $Na^+N_3^-$ shows two peaks with relative intensities of 2:1, indicating that the three nitrogen atoms are not chemically equivalent.

11.18 CIRCULAR BIREFRINGENCE AND CIRCULAR DICHROISM*

The dependence of refractive index and absorption coefficient on wavelength can also be measured with circularly polarized light. Circularly polarized light is light in which the electric vector rotates as the light beam advances. If the electric vector rotates clockwise as observed facing the light source the light is said to be right-handed circularly polarized light. If it rotates counterclockwise, it is left-handed circularly polarized light. When left and right circularly polarized beams of equal intensity are combined, they yield plane polarized light. In plane polarized light the electric vector remains in a plane. A separated beam of circularly polarized light may be obtained by passing plane polarized light through a quarter wave plate oriented at 45° to the direction of the electric vector of the polarized light. Since the quarter wave plate may be inclined to the right or to the left, both right and left circularly polarized light may be obtained in this way.†

Figure 11.15 gives the variation of refractive index (dispersion curve) and the variation of the absorption coefficient (absorption curve) for an optically active material measured with left and right circularly polarized light. The difference in refractive index for the two components is referred to as *circular birefringence,* and the difference in absorption is referred to as *circular dichroism.* The difference curves are shown in the lower part of Fig. 11.15. The plot of Δn versus λ is referred to as the rotatory dispersion curve, and the plot of $\Delta\epsilon$ versus λ is referred to as the circular dichroism spectrum. When a rotatory dispersion curve rises on going to shorter λ, it is referred to as normal, because that is the normal pattern for a refractive index curve. When an absorption band causes the effects shown in Fig. 11.15 the whole phenomenon is referred to as a Cotton effect. In contrast with ordinary dispersion, a strong absorption band may or may not give a large effect on the rotatory dispersion and a weak absorption band may give a large effect on the rotatory dispersion.

Substances that show circular birefringence and circular dichroism are said to be optically active. They can be divided into two classes: one in which optical activity is found only in the crystal form, for example, quartz, and the other in which it is found in the solid, gaseous, and liquid states of the pure substance or in solutions. Optical activity arises in the former group due to a right- or left-hand spiral structure in the crystal and disappears when this structure is melted. Substances in the latter category are optically active because of the asymmetry of the molecule itself. For a molecule whose mirror image is not superimposable on itself, left and right circularly polarized light have different refractive indices and correspondingly different absorption coefficients. This may happen for any molecule having only proper rotation elements of symmetry (Section 9.3). A molecule possessing any improper rotation axis (S_n), including a mirror plane or a center of symmetry, cannot be optically active.

In order to clarify what is observed experimentally, let us consider Fig. 11.16. This figure shows the vector sum of the intensities of left and right circularly polarized light at one instant only. If plane polarized light (Fig. 11.16*a*) is passed through a sample having

---

* D. J. Caldwell and H. Eyring, *The Theory of Optical Activity,* Wiley, New York, 1971; C. Djerassi, *Optical Rotatory Dispersion,* McGraw-Hill Book Co., New York, 1959; J. G. Foss, *J. Chem. Ed.,* **40,** 592 (1963); G. Snatzke, *Optical Rotatory Dispersion and Circular Dichroism in Organic Chemistry,* Sadtler Research Labs., Inc., Philadelphia, Pa., 1967; L. Velluz, M. Legrand, and M. Grosjean, *Optical Circular Dichroism,* Academic Press, New York, 1965.

† D. Halliday and R. Resnick, *Physics,* Part 2, Wiley, New York, 1978, p. 1081.

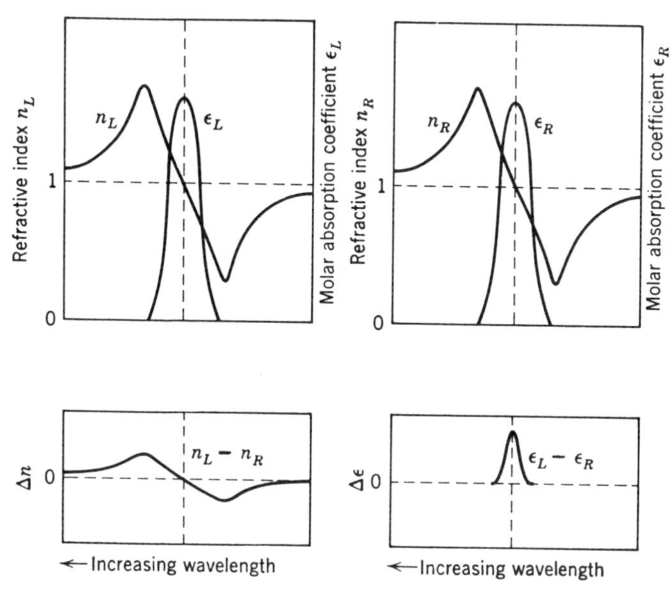

Fig. 11.15 Variation of refractive index n and molar absorption coefficient ϵ in the neighborhood of a single absorption line of an optically active substance, as measured with left- and right-handed circularly polarized light. The difference curves are referred to as the rotatory dispersion curve (Δn) and the circular dichroism spectrum ($\Delta\epsilon$). [From J. G. Foss, *J. Chem. Ed.*, **40**, 592 (1963).]

different refractive indices n_L and n_R for left- and right-handed circularly polarized light, there will be a rotation of the plane of polarization due to the different velocities of the left and right circularly polarized components, as shown in Fig. 11.16*b*. If the refractive indices of the medium are different for the two circularly polarized components, the angles of the two electric vectors measured with respect to the plane of the plane polarized light will be different after passage through a layer of the medium. It can be seen from the figure that the sum of the two electric vectors does not lie in the original plane, so that the plane has been rotated. The angle δ_n (in radians) through which the resultant electric vector is rotated is given by

$$\delta_n = \frac{\pi}{\lambda}(n_R - n_L)l \tag{11.90}$$

where n_R and n_L are the refractive indices for the right and left circularly polarized rays, l is the length of the medium, and λ is the wavelength in vacuum. Thus optical rotation may rise from a difference in the refractive indices for the two circularly polarized components.

The rotation of plane-polarized light is measured with a polarimeter that consists of a light source, linear polarizer, sample, and analyzer (another linear polarizer). The rotation of the plane of polarization by the sample is measured by rotating the analyzer. If a substance rotates the plane of polarized light to the right, or clockwise, as viewed looking toward the light source, it is said to be dextrorotatory; if the rotation is counterclockwise, the substance is levorotatory. The magnitude of the rotation α is directly proportional to

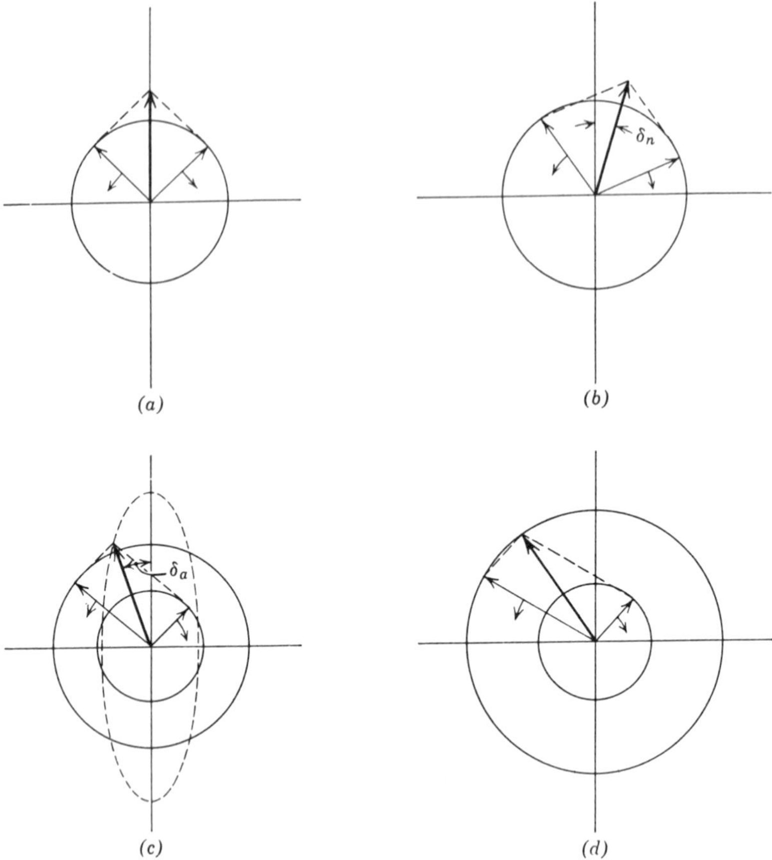

Fig. 11.16 Circular birefringence and circular dichroism. (*a*) Representation of plane polarized light as the sum of two components of circularly polarized light rotating in opposite directions. (*b*) Rotation of plane polarized light by an angle δ_n due to different velocities of propagation of left and right circularly polarized light. (*c*) Production of elliptically polarized light from plane polarized light by different absorption coefficients for left and right circularly polarized light. (*d*) Effects on plane polarized light of both differences in refractive indices and absorption coefficients for the two circularly polarized components. In (*c*) and (*d*) the emerging beam is elliptically polarized.

the length l of the sample and the concentration c of the optically active molecules, and so it is convenient to calculate a *specific rotation* $[\alpha]$ that is defined by

$$[\alpha] = \frac{\alpha}{cl} \tag{11.91}$$

where l is the path length in decimeters and c is the concentration in grams per cubic centimeter of solution. For a pure substance, c equals ρ, the density of the pure substance. The specific rotation varies with the wavelength, temperature, and solvent, so these variables must be specified.

In Fig. 11.16*c* we consider the effect of greater absorption of one of the circularly polarized components than the other. In this figure the electric vectors of the left and right circularly polarized components are shown as having rotated through equal angles in opposite directions, but the intensity of the right circularly polarized component has been reduced by absorption. The locus traced out by the tip of the resultant vector is an ellipse for this case, and so the resultant beam is elliptically polarized light, in contrast with the situation in Fig. 11.16*b*, which produced plane-polarized light. Although partial absorption of one of the components produces elliptically polarized light, the major axis of the ellipse still lies along the vertical direction, that is, it is not rotated.

In general, the effects described in Fig. 11.16*b* and 11.16*c* occur together, and the resultant is described in Fig. 11.16*d*. The emerging beam is elliptically polarized, and the major axis is rotated. The ellipticity is determined by the difference in absorbance, and the optical rotation is determined by the difference in refractive indices.

A molecule may have several optically active absorption bands and therefore several Cotton effects. Both the classical electromagnetic and quantum mechanical theories for optical rotation predict a variation of $[\alpha]$ with wavelength of the form

$$[\alpha] = \sum_{i=1}^{n} \frac{k_i}{\lambda^2 - \lambda_{oi}^2} \tag{11.92}$$

where the k_i and λ_{oi} are constants, and λ is the wavelength of the incident monochromatic light. Formally the λ_{oi} are the wavelengths of certain absorption bands in the molecule—those that are optically active. This equation applies only at some distance from an absorption band; the rotation does not become infinite. Not all absorption bands in an optically active molecule are necessarily optically active. If the chromophore is far removed from the asymmetric center, the difference in absorption of left and right circularly polarized light may be negligible.

The rotatory dispersion curve and circular dichroism spectrum are used in determining the structure, configuration, and conformation of complicated, optically active molecules such as steroids. Another region of extensive interest is that of proteins and synthetic polypeptides. Here information about gross conformational changes can be obtained, since optical rotation is very sensitive to the configuration and conformation of molecules.

The Faraday effect is the rotation of plane-polarized light by an otherwise optically inactive medium placed in a magnetic field. The Kerr electro-optic effect is the rotation of plane-polarized light by an otherwise optically isotropic substance placed in an electric field.

References

C. N. Banwell, *Fundamentals of Molecular Spectroscopy*, McGraw-Hill Book Co., New York, 2nd ed., 1972.

E. F. H. Brittain, W. O. George, and C. H. J. Wells, *Introduction to Molecular Spectroscopy*, Academic Press, New York, 1970.

R. Chang, *Basic Principles of Spectroscopy*, McGraw-Hill Book Co., New York, 1971.

J. C. Davis, Jr., *Advanced Physical Chemistry*, Ronald Press, New York, 1965.

H. B. Dunford, *Elements of Diatomic Molecular Spectra*, Addison-Wesley Publ. Co., Reading, Mass., 1968.

W. H. Flygare, *Molecular Structure and Dynamics*, Prentice-Hall, Englewood Cliffs, N.J., 1978.

T. R. Gilson and P. J. Hendra, *Laser Raman Spectroscopy*, Wiley, New York, 1970.
W. Gordy, W. V. Smith, and R. F. Trambarulo, *Microwave Spectroscopy*, Wiley, New York, 1963.
W. A. Guillory, *Introduction to Molecular Structure and Spectroscopy*, Allyn and Bacon, Boston, 1977.
M. D. Harmony, *Introduction to Molecular Energies and Spectra*, Holt, Rinehart and Winston, Inc., New York, 1972.
D. C. Harris and M. D. Bertolucci, *Symmetry and Spectroscopy*, Oxford University Press, Oxford, 1978.
G. Herzberg, *Infrared and Raman Spectra of Polyatomic Molecules*, D. Van Nostrand Co., Princeton, N.J., 1950.
G. Herzberg, *Molecular Spectra and Molecular Structure*, D. Van Nostrand Co., Princeton, N.J., 1950.
G. Herzberg, *The Spectra and Structures of Simple Free Radicals*, Cornell University Press Ithaca, 1971.
I. N. Levine, *Molecular Spectroscopy*, Wiley, New York, 1975.
D. A. Long, *Raman Spectroscopy*, McGraw-Hill Book Co., New York, 1977.
J. I. Steinfeld, *Molecules and Radiation*, MIT Press, Cambridge, Mass., 1977.
H. A. Szymanski, *Raman Spectroscopy*, Plenum Press, New York, 1967.
C. H. Townes and A. L. Schawlow, *Microwave Spectroscopy*, McGraw-Hill Book Co., New York, 1955.
D. H. Whiffen, *Spectroscopy*, Longman, London, 1972.
E. B. Wilson, J. C. Decius, and P. C. Cross, *Molecular Vibrations*, McGraw-Hill Book Co., New York, 1955.

Problems

11.1 Since the energy of a molecular quantum state is divided by kT in the Boltzmann distribution, it is of interest to calculate the temperature at which kT is equal to the energy of photons of different wavelength. Calculate the temperature at which kT is equal to the energy of photons of wavelength 10^3 cm, 10^{-1} cm, 10^{-3} cm, 10^{-5} cm.

Ans. 0.001 44, 14.4, 1440, 144,000 K.

11.2 Most chemical reactions require activation energies ranging between 40 and 400 kJ mol^{-1}. What are the equivalents of 40 and 400 kJ mol^{-1} in terms of (*a*) nm, (*b*) wave numbers, and (*c*) electron volts?

Ans. (*a*) 2991, 299.1 nm. (*b*) 3343, 33,430 cm^{-1}. (*c*) 0.415, 4.15 eV.

11.3 Calculate the reduced mass and the moment of inertia of $D^{35}Cl$, given that $R_e = 0.1275$ nm. *Ans.* 3.162×10^{-27} kg, 5.141×10^{-47} kg m^2.

11.4 Calculate the frequency in wave numbers and the wavelength in cm of the first rotational transition ($J = 0 \rightarrow 1$) for $D^{35}Cl$. *Ans.* 10.89 cm^{-1}, 0.091 83 cm.

11.5 The moment of inertia of $^{12}C^{16}O$ is 18.75×10^{-47} kg m^2. Calculate the frequencies in wave numbers and the wavelengths in centimeters for the first four lines in the pure rotational spectrum.

Ans. 2.986, 5.972, 8.957, 11.943 cm^{-1}; 0.3349, 0.1675, 0.1116, 0.0837 cm.

11.6 The separation of the pure rotation lines in the spectrum of CO is 3.86 cm^{-1}. Calculate the equilibrium internuclear separation. *Ans.* 113 pm.

11.7 The moment of inertia of $^{16}O{=}^{12}C{=}^{16}O$ is 7.167×10^{-46} kg m^2. (*a*) Calculate the CO bond length, R_{CO}, in CO_2. (*b*) Assuming that isotopic substitution does not alter R_{CO}, calculate the moments of inertia of

$$1.\ ^{18}O{=}^{12}C{=}^{18}O \quad \text{and} \quad 2.\ ^{16}O{=}^{13}C{=}^{16}O$$

Ans. (*a*) 0.1162 nm. (*b*) 8.071×10^{-46}, 7.167×10^{-46} kg m^2.

11.8 What are the reduced mass and moment of inertia of $H^{37}Cl$? Given: $R_e =$ 0.1275 nm. *Ans.* $1.629\,14 \times 10^{-27}$ kg. 2.648×10^{-47} kg m^2.

11.9 Calculate the zero-point energies of (*a*) H_2, and (*b*) Cl_2 in kilojoules per mole. The fundamental vibration frequencies are to be found in Table 11.5.

Ans. (*a*) 26.288, (*b*) 3.381 kJ mol^{-1}.

11.10 (*a*) What vibrational frequency in wave numbers corresponds to a thermal energy of kT at 25 °C? (*b*) What is the wavelength of this radiation?

Ans. (*a*) 207 cm^{-1}. (*b*) 4.83×10^{-3} cm.

11.11 Given the following fundamental vibration frequencies:

$H^{35}Cl$	2989 cm^{-1}	H^2D	3817 cm^{-1}
$^2D^{35}Cl$	2144 cm^{-1}	$^2D^2D$	2990 cm^{-1}

Calculate $\Delta H°$ for the reaction

$$H^{35}Cl(v = 0) + {}^2D^2D(v = 0) = {}^2D^{35}Cl(v = 0) + H^2D(v = 0)$$

Ans. 108 J mol^{-1}.

11.12 (*a*) Calculate the wavelength in μm that corresponds with the fundamental vibration frequency of CO. The necessary data are in Table 11.5. What wavelengths correspond with the (*b*) first, and (*c*) second overtones of the fundamental vibration frequency? *Ans.* (*a*) 4.61, (*b*) 2.31, (*c*) 1.54 μm.

11.13 Calculate the wavelengths in (*a*) wave numbers, and (*b*) micrometers of the center two lines in the vibration-rotation spectrum of HBr for the fundamental vibration. The necessary data are to be found in Table 11.5.

Ans. (*a*) 2632.7, 2666.6 cm^{-1}, (*b*) 3.80, 3.75 μm.

11.14 The fundamental vibration frequency of $H^{35}Cl$ is 8.667×10^{13} s^{-1}. What would be the separation between the infrared absorption lines for $H^{35}Cl$ and $H^{37}Cl$ if the force constants of the bonds are assumed to be the same? *Ans.* 26 nm.

11.15 The spectroscopic dissociation energy of $H_2(g)$ is 4.4763 eV, and the fundamental vibrational frequency is 4395.24 cm^{-1}. What is the spectroscopic dissociation energy of $D_2(g)$ if it has the same force constant? *Ans.* 4.5560 eV

11.16 How many normal modes of vibration are there for (*a*) SO_2(bent), (*b*) H_2O_2(bent), (*c*) HC≡CH(linear), and (*d*) C_6H_6? *Ans.* (*a*) 3, (*b*) 6, (*c*) 7, (*d*) 30.

11.17 List the numbers of translational, rotational, and vibrational degrees of freedom for (*a*) Ne, (*b*) N_2, (*c*) CO_2, and (*d*) CH_2O.

Ans. (*a*) 3, 0, 0, (*b*) 3, 2, 1, (*c*) 3, 2, 4, (*d*) 3, 3, 6.

11.18 When CCl_4 is irradiated with the 435.8 nm mercury line, Raman lines are obtained at 439.9, 444.6, and 450.7 nm. Calculate the Raman frequencies of CCl_4 (expressed in wave numbers). Also calculate the wavelengths (expressed in μm) in the infrared at which absorption might be expected.

Ans. $\tilde{\nu}_{Raman}$/cm^{-1}	214	312	454	759
λ/μm	46.8	32.0	22.0	13.2

11.19 The first several Raman frequencies of $^{14}N_2$ are 19.908, 27.857, 35.812, 43.762, 51.721, and 59.662 cm^{-1}. These lines are due to pure rotational transitions with J = 1, 2, 3, 4, 5, and 6. The spacing between the lines is $4B_e$. What is the internuclear distance?
Ans. 110 pm.

11.20 According to the hypothesis of Franck, the molecules of the halogens dissociate into one normal atom and one excited atom. The wavelength of the convergence limit in the spectrum of iodine is 499.5 nm. (*a*) What is the energy of dissociation of iodine into one normal and one excited atom? (*b*) The lowest excitation energy of the iodine atom is 0.94 eV. What is the energy corresponding to this excitation? (*c*) Compute the heat of dissociation of the iodine molecule into two normal atoms, and compare it with the value obtained from thermochemical data, 144.4 kJ mol^{-1}.
Ans. (*a*) 239.5, (*b*) 90.7, (*c*) 148.8 kJ mol^{-1}.

11.21 A solution of a dye containing 0.1 g L^{-1} transmits 80% of the light at 435.6 nm in a glass cell 1 cm thick. (*a*) What percent of light will be absorbed by a solution containing 2 g/100 cm^3 in a cell 1 cm thick? (*b*) What concentration will be required to absorb 50% of the light? (*c*) What percent of the light will be transmitted by a solution of the dye containing 1 g/100 cm^3 in a cell 5 cm thick? (*d*) What thickness should the cell be in order to absorb 90% of the light with solution of this concentration?
Ans. (*a*) 36.0%, (*b*) 3.11 g/100 cm^3, (*c*) 32.8%, (*d*) 10.3 cm.

11.22 The absorption coefficient α for a solid is defined by $I = I_0 e^{-ax}$, where x is the thickness of the sample. The absorption coefficients for NaCl and KBr at a wavelength of 28 μm are 14 and 0.25 cm^{-1}. Calculate the percentage of this infrared radiation transmitted by 0.5 cm thicknesses of these crystals. *Ans.* NaCl, 0.09%; KBr, 88.2%.

11.23 The following absorption data are obtained for solutions of oxyhemoglobin in pH 7 buffer at 575 nm in a 1-cm cell:

g/100 cm^3	Transmission, %
0.03	53.5
0.05	35.1
0.10	12.3

The molar mass of hemoglobin is 64.0 kg mol^{-1}. (*a*) Is Beer's law obeyed? What is the molar absorption coefficient? (*b*) Calculate the percent transmission for a solution containing 0.01 g/100 cm^3. *Ans.* (*a*) Yes. 5.81 × 10^4 L mol^{-1} cm^{-1}. (*b*) 81%.

11.24 The protein metmyoglobin and imidazole form a complex in solution. The molar absorption coefficients in L mol^{-1} cm^{-1} of the metmyoglobin (Mb) and the complex (C) are as follows.

λ	$\dfrac{\epsilon_{Mb}}{10^3\ L\ mol^{-1}\ cm^{-1}}$	$\dfrac{\epsilon_C}{10^3\ L\ mol^{-1}\ cm^{-1}}$
500 nm	9.42	6.88
630 nm	3.58	1.30

An equilibrium mixture in a cell of 1-cm path length has an absorbance of 0.435 at 500 nm and 0.121 at 650 nm. What are the concentrations of metmyoglobin and complex?
Ans. 2.17 × 10^{-5}, 3.37 × 10^{-5} mol L^{-1}.

11.25 (*a*) Calculate the energy levels for n = 1 and n = 2 for an electron in a potential well of width 0.5 nm with infinite barriers on either side. The energies should be expressed in J and in kJ mol^{-1}. (*b*) If an electron makes a transition from n = 2 to n = 1, what will be the wavelength of the radiation emitted?
Ans. (*a*) 2.41 × 10^{-19}, 9.64 × 10^{-19} J; 145.2, 580.7 kJ mol^{-1}, (*b*) 275 nm.

11.26 When α-D-mannose ($[\alpha]_D^{20} = +29.3°$) is dissolved in water, the optical rotation decreases as β-D-mannose is formed until at equilibrium $[\alpha]_D^{20} = +14.2°$. This process is referred to as mutarotation. As expected, when β-D-mannose ($[\alpha]_D^{20} = -17.0°$) is dissolved in water, the optical rotation increases until $[\alpha]_D^{20} = +14.2°$ is obtained. Calculate the percentage of α form in the equilibrium mixture. *Ans.* 67%.

11.27 What is the wavelength of light that has energy equal to the energy of electrons that have been accelerated by a potential of (*a*) 400 V, and (*b*) 3 V?

11.28 The internuclear distance in CO is 0.112 82 nm. Calculate (*a*) the reduced mass, and (*b*) the moment of inertia.

11.29 Calculate the frequencies in cm^{-1} and the wavelengths in micrometers for the pure rotational lines in the spectrum of $H^{35}Cl$ corresponding to the following changes in rotational quantum number: $0 \rightarrow 1$, $1 \rightarrow 2$, $2 \rightarrow 3$, and $8 \rightarrow 9$.

11.30 Assuming that the internuclear distance is 0.0742 nm for (*a*) H_2, (*b*) HD, (*c*) HT and (*d*) D_2, calculate the moments of inertia of these molecules.

11.31 Calculate the energy difference between the $J = 0$ and $J = 1$ rotational levels for OH radicals, in kilojoules per mole. Calculate the wavelength in centimeteres at which this transition will appear. The equilibrium internuclear distance and reduced mass are given in Table 11.5.

11.32 The far infrared spectrum of HI consists of a series of equally spaced lines with $\Delta\tilde{\nu} = 12.8\ cm^{-1}$. What is (*a*) the moment of inertia, and (*b*) the internuclear distance?

11.33 Show that for large J the frequency of radiation absorbed in exciting a rotational transition is approximately equal to the classical frequency of rotation of the molecule in its initial or final state.

11.34 Calculate the rotational constant B and the frequencies of the first three lines in the rotational spectrum of $^{16}O^{12}C^{32}S$ given that the O–C distance is 116.47 pm and the C–S distance is 155.76 pm and that the molecule is linear. Atomic masses of isotopes are given in the Appendix. The moment of inertia of a linear molecule ABC is given by

$$I = \frac{m_A m_B r_{AB}^2 + m_B m_C r_{BC}^2 + m_A m_C (r_{AB} + r_{BC})^2}{m_A + m_B + m_C}$$

11.35 Calculate the zero-point energies of gaseous HCl and KCl in electron volts. The fundamental vibration frequencies are to be found in Table 11.5.

11.36 How many cm^{-1} correspond to 1 kJ mol^{-1}?

11.37 Calculate the energy difference between the ground and first vibrational levels for H_2 in (*a*) joules per molecule, (*b*) kilojoules per mole, and (*c*) electron volts. The fundamental vibration frequency is given in Table 11.5.

11.38 Calculate the wavelengths in (*a*) wave numbers, and (*b*) μm of the center four lines in the infrared spectrum of HI at the first harmonic of the fundamental vibration frequency. The necessary data are to be found in Table 11.5.

11.39 Gaseous HBr has an absorption band centered at about 2645 cm^{-1} consisting of a series of lines approximately equally spaced with an interval of 16.9 cm^{-1}. For gaseous DBr estimate the frequency in wave numbers of the band center and the interval between lines.

11.40 The fundamental vibration frequency of $H^{35}Cl$ is $8.667 \times 10^{13}\ s^{-1}$. Calculate the fundamental vibration frequency of $D^{35}Cl$ on the assumption that the force constants of the bonds are the same. About what wavelength in micrometers will the fundamental infrared absorption of $D^{35}Cl$ be grouped? The observed value is 4.8 μm.

11.41 The fundamental ($v = 0 \rightarrow 1$) absorption of gaseous $H^{81}Br$ occurs at 2650 cm^{-1}. Calculate the force constant.

11.42 Using the Boltzmann distribution (Section 13.3), calculate the ratio of the population of the first vibrational excited state to the population of the ground state for $H^{35}Cl(\tilde{\omega}_0 = 2990\ cm^{-1})$ and $^{127}I_2(\tilde{\omega}_0 = 213\ cm^{-1})$.

11.43 List the numbers of translational, rotational, and vibrational degrees of freedom of Cl_2, H_2O, and C_2H_2.

11.44 List the numbers of translational, rotational, and vibrational degrees of freedom of NNO (a linear molecule) and NH_3.

11.45 The rotational Raman spectrum of hydrogen gas is measured using a 488-nm laser. Stokes lines are observed at 355, 588, 815, and 1033 cm^{-1}. Since these transitions are of the type $J \rightarrow J + 2$, it may be shown that the wave numbers of these lines are given by

$$\tilde{\nu} = 4\tilde{B}_e(J + \tfrac{3}{2})$$

where J is the rotational quantum number of the initial state (0, 1, 2, 3, respectively, for the above lines) and $\tilde{B}_e$ is given by equation 11.22. What is R_e? [L. C. Hoskins, *J. Chem. Ed.*, **54**, 642 (1977).]

11.46 The dissociation energies of $HCl(g)$, $H_2(g)$, and $Cl_2(g)$ into normal atoms have been determined spectroscopically and are 4.431, 4.476, and 2.476 eV, respectively. Calculate the enthalpy of formation of $HCl(g)$ at 0 K in kilojoules per mole from these data.

11.47 The limit of continuous absorption for Br_2 gas occurs at 19,750 cm^{-1}. The dissociation that occurs is

$$Br_2(\text{normal}) = Br(\text{normal}) + Br(\text{excited})$$

The transition of a normal bromine atom to an excited one corresponds to a wave number of 3685 cm^{-1}.

$$Br(\text{normal}) = Br(\text{excited})$$

Calculate the energy increase for the process

$$Br_2(\text{normal}) = 2Br(\text{normal})$$

in (*a*) cm^{-1}, and (*b*) electron volts.

11.48 When a 1.9-cm absorption cell was used, the transmittance of 436-nm light by bromine in carbon tetrachloride solution was found to be as follows.

c/mol L^{-1}	0.005 46	0.003 50	0.002 10	0.001 25	0.000 66
I/I_0	0.010	0.050	0.160	0.343	0.570

Calculate the molar absorption coefficient. What percentage of the incident light would be transmitted by 2 cm of solution containing 1.55×10^{-3} mol L^{-1} bromine in carbon tetrachloride?

11.49 Commercial chlorine from electrolysis contains small amounts of chlorinated organic impurities. The concentrations of impurities may be calculated from infrared absorption spectra of liquid Cl_2. Calculate the concentration of $CHCl_3$ in grams per milliliter in a sample of liquid Cl_2 if the transmittance at $\tilde{\nu} = 1216\ cm^{-1}$ is 45% for a 5-cm cell. At this wavelength liquid Cl_2 does not absorb, and the absorption coefficient for $CHCl_3$ dissolved in liquid Cl_2 is $900 \pm 80\ cm^{-1}\ (g\ cm^{-3})^{-1}$.

11.50 To test the validity of Beer's law in the determination of vitamin A, solutions of known concentration were prepared and treated by a standard procedure with antimony trichloride in chloroform to produce a blue color. The percent transmission of the incident filtered light for each concentration, expressed in micrograms per milliliter, was as follows.

Concentration, μg mL^{-1}	1.0	2.0	3.0	4.0	5.0
Transmission, %	66.8	44.7	29.2	19.9	13.3

Plot these data so as to test Beer's law. A solution, when treated in the standard manner with antimony trichloride, transmitted 35% of the incident light in the same cell. What was the concentration of vitamin A in the solution?

11.51 The protein metmyoglobin and the azide ion (N_3^-) for a complex. The molar absorption coefficients (L mol^{-1} cm^{-1}) of the metmyoglobin (Mb) and of the complex (C) in a buffer are as follows:

λ	$\dfrac{\epsilon_{Mb}}{10^4 \text{ L mol}^{-1}\text{ cm}^{-1}}$	$\dfrac{\epsilon_{C}}{10^4 \text{ L mol}^{-1}\text{ cm}^{-1}}$
490 nm	0.850	0.744
540 nm	0.586	1.028

An equilibrium mixture in a 1-cm cell gave an absorbance of 0.656 at 490 nm and of 0.716 at 540 nm. (*a*) What are the concentrations of metmyoglobin and complex? (*b*) Since the total azide concentration is 1.048×10^{-4} mol L^{-1}, what is the equilibrium constant for

$$\text{Mb} + \text{N}_3^- = \text{C}$$

11.52 A dye having a conjugated series of double bonds has an electronic absorption band at 300 nm. Calculate the length of the π bond system assuming that the transition is from $n = 1$ to $n = 2$ for an electron in a one-dimensional box.

11.53 The electronic spectrum of a molecule

$$R\text{—(CH=CH)}_k\text{—}R'$$

can be described by considering the transitions of $2k\pi$ electrons among the levels in a one-dimensional box of length kl, where l is the length of the unit —CH=CH—. (*a*) Ignore electron-electron interactions and assign electrons to orbitals in accord with the Pauli principle. Obtain an expression for the lowest energy transition as a function of k. (*b*) If $l = 0.28$ nm, what is the wavelength of the lowest energy transition for $k = 2$, 5, and 10?

11.54 The most prominent line in the photoelectron spectrum of H_2 is due to the transition

$$\text{H}_2(v = 0) + h\nu \rightarrow \text{H}_2^+(v = 2) + e^-$$

If helium resonance radiation with an energy of 21.22 eV is used, what will be the electron energy, assuming that H_2^+ is a harmonic oscillator with a fundamental vibration frequency of 4395 cm^{-1}. The 0–0 ionization potential is 15.45 V.

11.55 The specific rotation of *l*-leucine is $[\alpha]_D^{25} = -14.0°$. If the specific rotation of a mixture of *d* and *l* forms is $[\alpha]_D^{25} = +2.3°$, calculate the fraction of *l* form in the mixture.

11.56 When α-D-glucose ($[\alpha]_D^{20} = +112.2°$) is dissolved in water, the optical rotation decreases as β-D-glucose is formed until at equilibrium $[\alpha]_D^{20} = +52.7°$. As expected, when β-D-glucose ($[\alpha]_D^{20} = +18.7°$) is dissolved in water, the optical rotation increases until $[\alpha]_D^{20} = +52.7°$ is obtained. Calculate the percentage of the β form in the equilibrium mixture.

CHAPTER 12

MAGNETIC RESONANCE SPECTROSCOPY

Magnetic resonance spectroscopy differs from other kinds of spectroscopy in that a magnetic field is used to produce energy level separations. For magnetic fields that can be routinely produced in the laboratory, the transitions between energy levels for nuclei that are magnetic dipoles occur in the radiofrequency range, and the transitions between energy levels for the spins of unpaired electrons occur in the microwave range. Nuclear magnetic resonance (NMR) and electron spin resonance (ESR) yield such valuable structural information that they have become indispensable in chemistry.

In order to relate these magnetic phenomena to the bulk magnetic properties of matter, we begin the chapter with a discussion of magnetic susceptibility.

12.1 MAGNETIC SUSCEPTIBILITY

The principal measure of the strength of a magnetic field is the magnetic flux density vector $\boldsymbol{B}$. Since magnetic fields are produced by currents, it is possible to connect the strength of the field of any magnet, including a single magnetic dipole, with basic mechanical and electrical units. The SI unit of the magnetic flux density is the tesla T, which is equal to a newton per ampere-meter.*

In the presence of a magnetic field, magnetic dipoles within a material become partially oriented; the magnetic dipole moment per unit volume is referred to as the magnetization $\boldsymbol{M}$. The units of the magnetic dipole moment are A m^2, and the unit of the magnitude of the magnetization vector $\boldsymbol{M}$ is A m^{-1}. Since the magnetic moments in a diamagnetic material oppose the magnetizing field, $\boldsymbol{M}$ is negative in a diamagnetic material.

It is convenient to introduce another vector quantity, the magnetic field intensity $\boldsymbol{H}$, which is defined by

$$\boldsymbol{H} \equiv \frac{\boldsymbol{B}}{\mu_0} - \boldsymbol{M} \tag{12.1}$$

where μ_0 is the permeability of vacuum. The permeability of vacuum has the value $4\pi \times 10^{-7}$ N A^{-2} exactly. (Note that $\epsilon_0\mu_0 = 1/c^2$, where ϵ_0 is the permittivity of

* The gauss G is the unit for magnetic flux density in the cgs gaussian system of units; 1 T = 10^4 G.

vacuum and c is the speed of light in vacuum.) The magnetic field intensity $\boldsymbol{H}$ is a vector, and its magnitude is expressed in A m^{-1}. The physical significance of $\boldsymbol{H}$ is better seen by writing equation 12.1 as

$$\boldsymbol{B} = \mu_0(\boldsymbol{H} + \boldsymbol{M}) \tag{12.2}$$

The magnetic flux density $\boldsymbol{B}$, which determines the magnetic force on a moving charged particle, is determined by $\mu_0\boldsymbol{H}$ arising from the electric currents producing the field and $\mu_0\boldsymbol{M}$ arising from the magnetic moments induced by the magnetizing field.

For isotropic substances the magnetic susceptibility χ is defined by

$$\chi = \frac{M}{H} \tag{12.3}$$

where M and H are the magnitudes of the magnetization and magnetic field intensity vectors. Thus the magnetic susceptibility is a dimensionless quantity.

In contrast with polarization by an electric field, the magnetic moment may be in the direction of the applied field (χ is positive) or in the opposite direction (χ is negative). For a *diamagnetic* substance χ is negative, small, independent of the magnetic field intensity, and independent of temperature. For a *paramagnetic* substance χ is positive, small, independent of the magnetic field intensity, and decreases with increasing temperature. For a ferromagnetic substance χ is positive, large, dependent on the magnetic field and temperature, and dependent on previous history. For an antiferromagnetic substance χ is small and positive, but shows nonlinearity and hysteresis effects and has a transition temperature known as the Néel point.

Paramagnetism results from the orientation of permanent magnetic dipoles in a substance. These permanent magnetic dipoles are due to the spin of unpaired electrons or to the angular momentum of electrons in orbitals of atoms or molecules. Electrons in orbitals with $l = 1, 2, 3, \ldots$ have angular momentum and therefore produce a magnetic dipole moment. Nuclei with magnetic moments produce a paramagnetic effect, but this effect is only about a millionth as large as paramagnetism due to orbital moments or unpaired electrons.

In the absence of a magnetic field the magnetic dipoles that produce paramagnetism are not oriented. In the presence of a field the orientation of permanent magnetic dipoles by the field is opposed by the disorganizing effect of thermal motion, as in the alignment of electric dipoles by an electric field (Section 10.12). In most substances the magnetic effects of electron spin and electron orbital motions cancel because electrons are paired in filled shells. Many rare earth and transition metal ions are paramagnetic because they have unpaired electrons. Free radicals have an odd number of electrons and are therefore paramagnetic. The most familiar substance that is paramagnetic is molecular oxygen. As we have seen earlier (Section 10.5), it has two unpaired electrons. This property of oxygen gas makes it possible to determine its partial pressure in a gas stream with a little torsion balance in the field of a magnet.

The magnetic moment μ of a paramagnetic molecule or ion is commonly measured in units of Bohr magnetons μ_B.

$$\mu_B = \frac{e\hbar}{2m_e} = \frac{(1.602\,189\,2 \times 10^{-19}\ \text{C})(6.626\,176 \times 10^{-34}\ \text{J s})}{4\pi(9.109\,534 \times 10^{-31}\ \text{kg})}$$

$$= 9.274\,078 \times 10^{-24}\ \text{J T}^{-1} \qquad (12.4)$$

This is the magnetic moment in the direction of the magnetic field for an electron in the first Bohr orbit of a hydrogen atom.

Diamagnetism results from the induction of microscopic currents in a sample by the external magnetic field. The magnetic dipoles produced in this way are aligned in a direction opposite of that of the external field. Since the induced magnetic dipoles oppose the field, a diamagnetic substance experiences a force in the direction of the weaker part of an inhomogeneous magnetic field. A paramagnetic substance is attracted by the poles of a magnet with an inhomogeneous magnetic field. The diamagnetic effect is produced in all substances, but it is only about a hundredth or a thousandth as strong as the paramagnetic effect when the latter exists.

The metals iron, cobalt, nickel, gadolinium, dysprosium, and certain of their alloys and compounds are ferromagnetic below a certain critical temperature for each substance. The origin of ferromagnetism was a puzzle that was explained by quantum mechanics. The question is: Why do so many electrons in incomplete shells have their spins aligned, and why do they remain aligned even after the applied magnetic field is removed? The explanation is that the lowest energy state for certain solids is one in which spins are parallel instead of opposed, as they are for the two electrons in a hydrogen molecule, for example. The requirements of certain atomic distances and certain radii of the orbitals for d electrons limit this phenomena to just a few elements. Ferromagnetic substances show hysteresis in their magnetic properties. This means that the magnetic moment depends on the magnetic history of the sample; the change in moment with increasing field is not retraced when the field is reduced.

The atoms in an antiferromagnetic substance are arranged so that the magnetic moments of nearest neighbors are opposed.

12.2 PRINCIPLES OF MAGNETIC RESONANCE

In the preceding section it was pointed out that since electrons and certain nuclei have both charge and spin angular momentum, they behave like little magnets. The energy of a magnetic dipole $\boldsymbol{\mu}$ in a magnetic field $\boldsymbol{B}$ is $-\boldsymbol{\mu}\cdot\boldsymbol{B}$. Since the angular momentum $\boldsymbol{I}$ that accompanies $\boldsymbol{\mu}$ ($\boldsymbol{\mu} = \gamma\hbar\boldsymbol{I}$) can take on only certain orientations, the magnetic dipole energies are quantized. In the presence of a magnetic field these magnetic dipoles give rise to a series of energy levels that correspond with the orientations of the angular momentum that are allowed quantum mechanically. Transitions between these levels can be produced by use of an oscillating magnetic field of electromagnetic radiation with a frequency that corresponds with separation of the energy levels. This phenomenon is referred to as nuclear magnetic resonance (NMR) when nuclei are involved and electron spin resonance (ESR) when electrons are involved. Because of the similarities of these methods, we will discuss the basic equations for both in this section.

NMR and ESR are quantum mechanical effects but, as usual, it is helpful to look at the classical analog. In classical mechanics the application of a torque to an angular momentum causes it to precess. An example is the motion of a top in a gravitational field. The motion of the angular momentum $\boldsymbol{I}$ under the influence of the torque $\boldsymbol{\mu} \times \boldsymbol{B}$ exerted on $\boldsymbol{\mu}$ by the field is another. If an oscillating magnetic field is then applied with a frequency equal to the frequency of the precession, the precessing magnetic dipole can absorb energy from the oscillating magnetic field, and the system is said to be in resonance.

All nuclei with odd mass numbers have the property of spin, and those nuclei with even mass number that have an odd number of protons also have spin. The spin I of a nucleus with an odd mass number is always an odd integral multiple of $\frac{1}{2}$, and the spin of a nucleus with an even mass number, but an odd number of protons, is 1, 2, 3, ... The spins of a number of nuclei are summarized in Table 12.1. The commonly occurring nuclei ^{12}C and ^{16}O do not have spin.

The proton and neutron each have spin $\frac{1}{2}$, and the spin of a nucleus (which is always integral or half integral) may be thought of as the resultant of the spins of the protons and neutrons that comprise the nucleus. A deuterium nucleus, which contains one proton and one neutron, would be expected to have a spin of 1 or 0, depending on whether the spins of the proton and neutron were aligned parallel (↑↑) or antiparallel (↑↓). The deuteron is found to have a spin of 1 in its nuclear ground state, and so the proton and neutron are in parallel alignment.

For a nucleus that has spin the magnitude of the intrinsic angular momentum $\boldsymbol{I}$ is $\sqrt{I(I+1)}\hbar$, where I is the spin. This relation may be compared with that for the spin angular momentum for an electron in Section 8.14. The difference is that the spin of an electron is $\frac{1}{2}$, but the spin of a nucleus may have one of a range of half integer or integer values. In a magnetic field a nucleus of spin I may have any one of $2I + 1$ orientations with respect to the direction of the field. The possible orientations of angular momentum vectors were shown earlier for spin $\frac{1}{2}$ in Fig. 8.11 and for spin 1 and 2 in Fig. 8.6. In the case of nuclei the $2I + 1$ orientations correspond with the possible values of the nuclear spin quantum number m_I of $-I, -(I-1), \ldots, (I-1), I$. The component of the angular momentum in the field direction is given by $m_I\hbar$. Thus the largest possible value of the angular momentum in the field direction is $I\hbar$ (see Figs. 8.6 and 8.11).

A nucleus with spin also has an associated magnetic dipole moment $\boldsymbol{\mu}$ that is proportional to the angular momentum $\boldsymbol{I}$.

$$\boldsymbol{\mu} = g_N \frac{e}{2m_p} \boldsymbol{I} \tag{12.5}$$

where g_N is the nuclear g factor and m_p is the mass of the proton.* The magnetic moment and angular momentum behave as if they are parallel or antiparallel vectors, and so the nuclear g factor may have a negative value. The electronic g factor has been calculated theoretically, but accurate values for the nuclear g factor have to be obtained from experiment. Values of g_N for a number of nuclei are given in Table 12.1 along with values of I and isotopic abundances. Nuclei with even numbers of both protons and neutrons have zero spin because there is a pairing of nucleonic angular momenta.

The magnitude of the magnetic moment of a nucleus is given by

$$|\mu| = |g_N| \frac{e}{2m_p} \sqrt{I(I+1)}\hbar = |g_N|\mu_N\sqrt{I(I+1)} \tag{12.6}$$

where the nuclear magneton μ_N is defined by

$$\mu_N = \frac{e\hbar}{2m_p} \tag{12.7}$$

The magnetic moment may have a component along the direction of the magnetic field of $\mu_z = g_N\mu_N m_I$.

Example 12.1 Calculate the value of the nuclear magneton.

$$\mu_N = \frac{e\hbar}{2m_p} = \frac{(1.602\,189\,2 \times 10^{-19}\ \text{C})(6.626\,176 \times 10^{-34}\ \text{J s})}{4\pi(1.672\,648\,5 \times 10^{-27}\ \text{kg})}$$
$$= 5.050\,824 \times 10^{-27}\ \text{J T}^{-1}$$

The nuclear magneton is $\frac{1}{1836}$ of the Bohr magneton.

The energy eigenvalues for a nuclear magnetic dipole in a magnetic flux density $\boldsymbol{B}$ are

$$E_{\text{mag}} = -g_N\mu_N B m_I \qquad m_I = I, I-1, \ldots, -I \tag{12.8}$$

The selection rule for a magnetic dipole transition is $\Delta m_I = \pm 1$. Therefore the frequency of an NMR transition is

$$\nu = \frac{g_N\mu_N B}{h} \tag{12.9}$$

(If g_N is negative, as it is for ^{15}N and ^{17}O, a negative sign has to be inserted in this equation.) Since the energy levels are equally spaced, there is a single NMR frequency. Figure 12.1*a* shows how the energy levels of a proton depend on the magnetic flux density. The proton is in the lower spin level when its magnetic moment is parallel with the field; the component of spin in the z direction is $+\frac{1}{2}$. In the upper level the magnetic moment is antiparallel, and $m_I = -\frac{1}{2}$.

The frequencies for a 1.4094-T field are shown for various nuclei in Table 12.1.

---

* The gyromagnetic ratio γ is the ratio of the maximum expectation value of the component of the magnetic moment and the angular momentum in the direction of the magnetic field $\gamma = \mu/I\hbar$.

Table 12.1 Nuclear Magnetic Properties

Nucleus	% Abundance	Spin I	g_N	Frequency in MHz at 1.4094 T
^1H	99.99	$\frac{1}{2}$	5.585	60.000
^2D	0.01	1	0.857	9.2104
^7Li	92.5	$\frac{3}{2}$	2.171	23.317
^{13}C	1.11	$\frac{1}{2}$	1.405	15.087
^{14}N	99.6	1	0.403	4.3343
^{15}N	0.4	$\frac{1}{2}$	−0.567	6.0798
^{17}O	0.04	$\frac{5}{2}$	−0.757	8.134
^{19}F	100	$\frac{1}{2}$	5.257	56.446
^{23}Na	100	$\frac{3}{2}$	1.478	15.871

Example 12.2 Calculate the magnetic field strength required to give a precessional frequency for protons of 60 MHz.

$$B = \frac{h\nu}{g_N\mu_N} = \frac{(6.6262 \times 10^{-34}\ \mathrm{J\ s})(60 \times 10^6\ \mathrm{s^{-1}})}{(5.585)(5.0508 \times 10^{-27}\ \mathrm{J\ T^{-1}})} = 1.4094\ \mathrm{T}$$

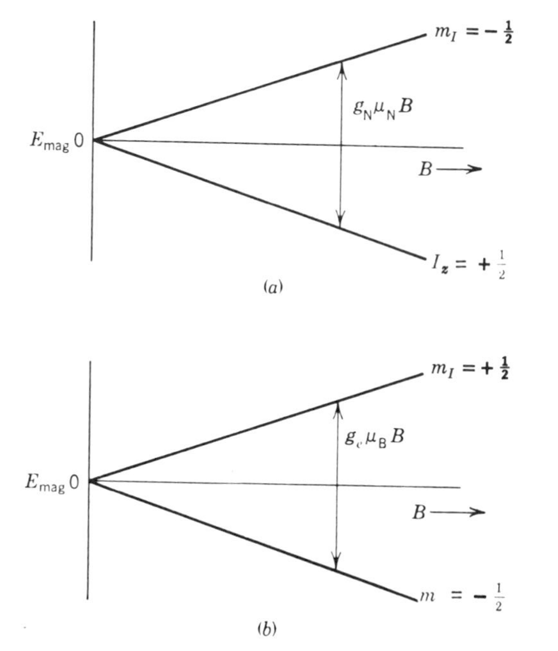

Fig. 12.1 (*a*) Energy levels of a proton in a magnetic field. (*b*) Energy levels of an electron in a magnetic field. The vertical scales are different in (*a*) and (*b*).

The basic equations for electron spin resonance (ESR) follow the same pattern as for nuclear magnetic resonance. The magnetic energy of an electron in a magnetic field is

$$E_{\text{mag}} = g_e \mu_B m_s B \tag{12.10}$$

where g_e is the g factor for the electron (2.002 322 for a free electron), μ_B is the Bohr magneton (Section 12.1), and m_s is the quantum number corresponding to the z component of the electron spin. The two energy levels of a single electron in a magnetic field are shown in Fig. 12.1*b*. Because of the negative charge of the electron, the magnetic moment $\boldsymbol{\mu}_e$ of an electron is in the direction opposite to its spin angular momentum, and the electron spin quantum number is $-\frac{1}{2}$ in the lower level in contrast with the situation with nuclei. For a transition from $m_s = -\frac{1}{2}$ to $m_s = +\frac{1}{2}$,

$$\Delta E = h\nu = g_e \mu_B B \tag{12.11}$$

Example 12.3 Calculate the magnetic field strength required to give a precessional frequency of 9500 MHz for a free electron.

$$B = \frac{h\nu}{g_e \mu_B} = \frac{(6.6262 \times 10^{-34}\ \text{J s})(9500 \times 10^6\ \text{s}^{-1})}{(2.0023)(9.2741 \times 10^{-24}\ \text{J T}^{-1})} = 0.3390\ \text{T}$$

12.3 HIGH RESOLUTION NMR SPECTROMETER

An NMR spectrometer provides the magnetic field to produce the energy levels, the radio frequency to excite transitions, and a radiofrequency receiver to detect emitted radiation. The resolution of an NMR spectrometer depends on the strength and homogeneity of the magnetic field and the constancy of the radiofrequency radiation. High resolution is required for most chemical applications.

The high resolution spectra are usually obtained with magnetic field strengths of the order of 1 T. A field of 1.4094 T corresponds with a precession frequency of 60 MHz for protons, as shown in Example 12.2. For several reasons it is advantageous to use the highest possible fields, but there are great practical difficulties in obtaining sufficiently stable and homogeneous magnetic fields for high resolution NMR work above about 7 T. Since chemical shifts (Section 12.6) are in the range of parts per million, the field needs to be constant to a part in 10^8 or 10^9. The fixed frequency radiofrequency sources that are used are crystal controlled and are stable to a part in 10^9.

Different methods for studying nuclear magnetic resonance were developed independently by Purcell and Bloch in 1946. A diagram for a crossed-coil NMR spectrometer is shown in Fig. 12.2. The sample to be investigated is placed in a tube between the poles of the electromagnet. The direction of the constant magnetic field B_0 is taken as the z direction. The strength of this magnetic field is set at a value just below that required for resonance, and then a secondary field is applied by use of coils around the faces of the pole pieces. The current for these

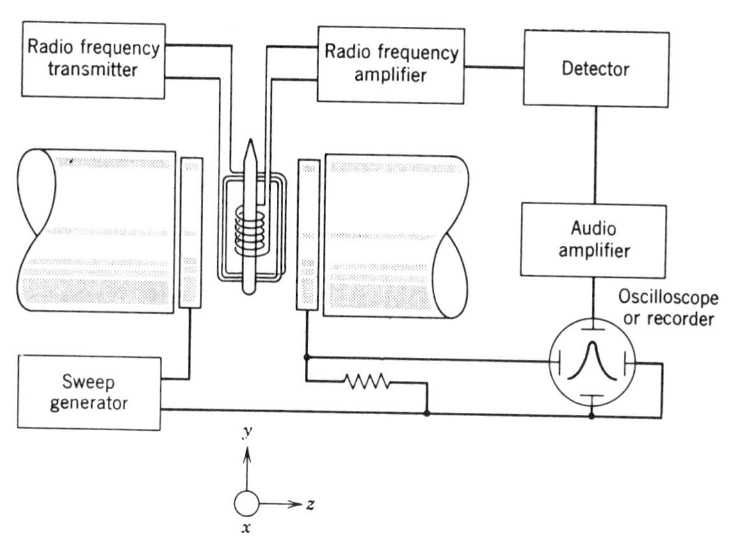

Fig. 12.2 Block diagram of a high resolution nuclear magnetic resonance spectrometer. The axis of the transmitter coil is the x axis, and that of the receiver coil is the y axis.

coils is supplied by a sweep generator so that the magnetic field strength at the sample can be swept through the resonance condition. A sweep generator produces a steadily increasing voltage that drops to zero when a certain voltage is reached and then repeats the process. The sweep generator also controls the sweep of the oscilloscope or chart recorder.

The magnetic nuclei in the sample precess about the z direction (the axis of the electromagnet). Transitions between the possible orientations of the nuclear spin are produced by exposing the sample to radio waves of a constant frequency, which are transmitted by a coil around the sample tube with its axis perpendicular to the z direction. The axis of the transmitter coil is the x axis, and that of the receiver coil is the y axis, which is the axis of the sample tube. When the radiofrequency and the magnetic field strength correspond with a resonance for a certain set of nuclei, energy is absorbed from the transmitter. This energy is radiated by the precessing nuclei and picked up by the detector coil, which is oriented so that it does not otherwise receive energy from the transmitter. If the signal from the radiofrequency receiver is applied to the vertical amplifier of the oscilloscope, the absorption line is traced out as shown. Very frequently the limiting factor in high resolution NMR spectroscopy is the inhomogeneity of the magnetic field. An improvement in resolution can be obtained by spinning the sample tube so that the nuclei experience a field that is averaged over the volume of the sample. Even if the field were perfect, an NMR absorption line would have a finite width because of the interaction of a magnetic moment with its surroundings.

In Fourier transform NMR spectroscopy the sample is irradiated with a whole range of radiofrequencies (so called "white" radiation). To avoid saturating the system, as described in the next section, the irradiation is carried out in very short pulses. After a pulse the nuclei reemit their energy. This reemitted energy is a composite of the resonance frequencies of all the nuclei in the sample. If there are two nuclei and no coupling, two frequencies ν_A and ν_X will be emitted. The two frequencies will produce a "beat" pattern in the detector, and from the beat pattern it is possible to calculate ν_A and ν_X. This process is referred to as a Fourier transform, and a small computer is required to analyze the beat pattern if there are several frequencies. This method is important because of the increase in sensitivity in essentially looking at all the resonances after each pulse instead of one at a time in the conventional type of NMR experiment described above. Thus smaller samples can be used and less abundant isotopic species (like ^{13}C) may be studied.

The NMR absorption lines of solids are broad, but they give important information about the distance between nuclei with magnetic moments and about their orientations in the crystal. The nuclei interact by a direct dipole-dipole coupling between their magnetic moments, and this causes very broad absorption lines.

The NMR absorption lines in liquids are very much narrower than in solids because the dipole-dipole coupling with nuclei in other molecules is averaged to zero by the rapid random rotations of molecules in a liquid.

12.4 THERMAL EQUILIBRIUM

The difference in energy between nuclear or electron spin states in magnetic fields that can be produced in the laboratory is very small at room temperature and, therefore, the populations in the various states are nearly equal. According to the Boltzmann distribution (Section 13.3), more spins will be in the low-energy state, shown in Fig. 12.1, and

$$\frac{N_h}{N_l} = e^{-\Delta E/kT} = e^{-g_N \mu_N B/kT} \tag{12.12}$$

where N_h is the number of spins in the high-energy state and N_l is the number of spins in the low-energy state. At ordinary temperatures, $g_N\mu_N B \ll kT$ so that the exponential may be approximated by a power series, and to a sufficiently good approximation.

$$\frac{N_l}{N_h} = 1 + \frac{g_N \mu_N B}{kT} \tag{12.13}$$

Thus the excess population in the low-energy state is extremely small. Only the spins in the low-energy state can absorb radiation. That radiation has to have the right frequency to be absorbed. Since the size of the excess population in the low-energy state is proportional to B, the intensity of the NMR or ESR signal increases as the magnetic field strength increases. If the system is irradiated with too high a power the excess spins at the lower energy will be used up, the higher- and lower-energy states will come to have equal populations, and no resonance signal will be obtained. This phenomenon, referred to as saturation, can be understood from equation 11.16.

Example 12.4 What is the ratio of the number of proton spins in the lower state to the number in the higher state in a magnetic field of 1 T at room temperature?

$$\frac{N_l}{N_h} = 1 + \frac{g_N \mu_N B}{kT} = 1 + \frac{(5.585)(5.05 \times 10^{-27}\,\text{J T}^{-1})(1\,\text{T})}{(1.38 \times 10^{-23}\,\text{J K}^{-1})(298\,\text{K})}$$

$$= 1 + 6.86 \times 10^{-6}$$

12.5 RELAXATION PROCESSES

Figure 12.3*a* represents the precession of spin $\frac{1}{2}$ nuclei about the magnetic flux density vector B_0. The precessing vectors represent the small excess number of nuclei in the low-energy level. At equilibrium these vectors are uniformly distributed over the conical surface and lead to a net magnetization M_z in the direction of the field, but no magnetization perpendicular to the z axis. This equilibrium situation may be perturbed in two ways: (1) the magnetic flux density B_0 may be suddenly increased, or (2) a B_1 field rotating at the same angular velocity and in the same direction as the nuclei may be applied.

If the magnetic flux density B_0 is suddenly doubled, we expect the net magnetization M_z to increase, but this does not happen instantaneously. The spin system approaches the new equilibrium situation represented in Fig. 12.3*b* exponentially. If the excess number of nuclei in the low-energy level is represented by n and the initial number is n_0, the approach to the new equilibrium number n_{eq} is described by

$$n - n_{ep} = (n_0 - n_{eq})e^{-t/T_1} \tag{12.14}$$

where t is time and T_1 is the relaxation time. (This equation is derived for a reversible first-order chemical reaction in Section 15.8.) In this case the relaxation time is referred to as the longitudinal or spin-lattice relaxation time T_1. It is the characteristic time for exchange of energy between the spin system and the thermal surroundings (the lattice). In solids T_1 may be as long as 1000 s because of the absence of mechanisms for interchanging nuclear spin energy with the rest of the system. In liquids T_1 is typically in the range 0.5 to 50 s. The relaxation process is faster in liquids because of the fluctuating magnetic fields caused by other nuclear magnets in the system that are undergoing random Brownian motion. The local time-varying magnetic field at a nucleus has components of the frequency necessary to cause transitions between the low- and high-energy levels.

The spin system described by Fig. 12.3*a* may also be perturbed by applying a rotating B_1 field. If the angular velocity and direction coincide with that of the precessing nuclei, the precessing spin vectors tend to bunch up, as shown in Fig. 12.3*c*. Since the distribution of spins about the cone is not uniform, a M_{xy} magnetization is produced. The time constant for the spontaneous decay of M_{xy} after removal of B_1 is referred to as the transverse relaxation time T_2. Since the energy of the spins depends only on M_z, a T_2 relaxation does not involve energy transfer between the spins and the surroundings. Since the T_2 relaxation amounts to loss of phase coherence of the precessing spins, it can be regarded as a randomization process and T_2 as the characteristic time for the transfer of entropy from the surroundings to the spins. In liquids the fluctuating magnetic fields caused by the Brownian motion of the nuclear magnets also cause a relaxation of the M_{xy} magnetization, and so T_1 and T_2 are usually about equal in liquids. In solids T_2 is often of the order of microseconds.

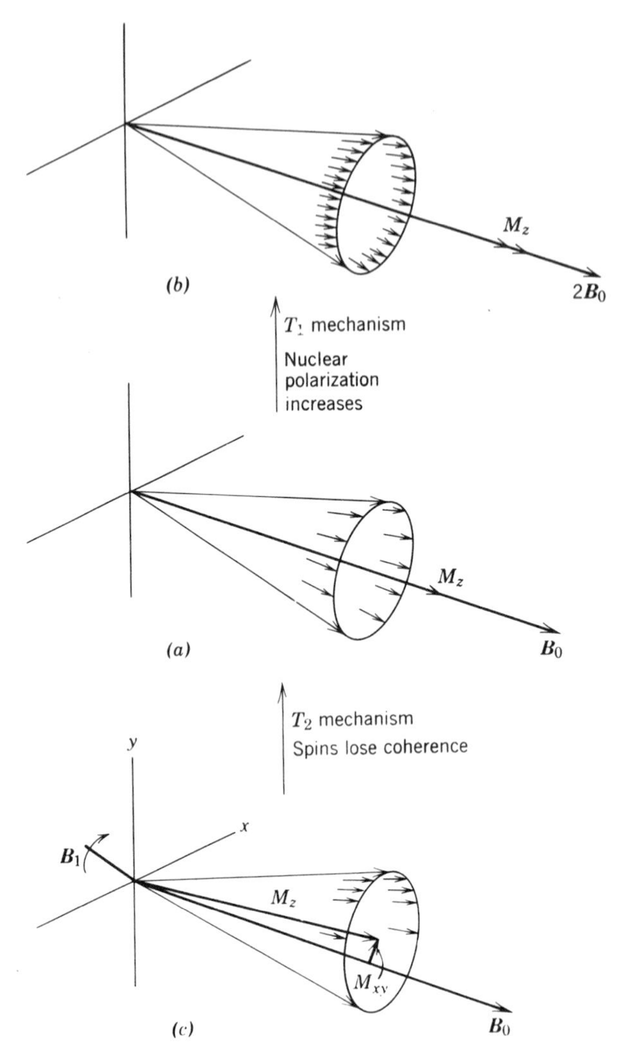

Fig. 12.3 Vectors representing the orientation of spin-$\frac{1}{2}$ nuclei (*a*) at equilibrium, (*b*) at equilibrium at twice the field, and (*c*) after application of a B_1 oscillating field. (From J. W. Akitt, *NMR and Chemistry*, Chapman and Hall Ltd., London, 1973.)

12.6 THE CHEMICAL SHIFT

When an atom or molecule is placed in a magnetic field, the magnetic field at the nucleus is slightly different from the applied field. In the presence of a magnetic field the motion of the orbital electrons is altered so as to produce a secondary magnetic field at the nucleus that opposes the main field. This screening effect, which is called diamagnetic screening because it is related to the mechanism that causes diamagnetism, is proportional to the magnetic flux density B_0. Because of this screening, identical nuclei in chemically different environments resonate at different values of the *applied* magnetic field. For example, Fig. 12.4 shows a sketch of the proton magnetic resonance spectrum of CH_3CH_2OH under conditions of low resolution. The areas under the three peaks are proportional to the numbers of protons of the three different types. As illustrated in Fig. 12.4, the secondary field opposes the applied field, so that the effective field at a proton is less than the applied field. The more a proton is shielded by the surrounding electrons, the higher the applied field required to produce resonance at a given frequency. The local field depends on the chemical environment, so the effect is referred to as a *chemical shift*. In ethyl alcohol the protons in CH_3 are shielded to a greater extent than those in CH_2, which are shielded to a greater extent than the proton in OH. The three protons in CH_3— are in identical chemical environments because of the rapid rotation about the CH_3—CH_2 bond. This rotation also assures the equivalence of the two CH_2 protons.

The secondary field is proportional to the applied field but in the opposite direction, and so the effective magnetic field B_{eff} at the nucleus is expressed by

$$B_{eff} = (1 - \sigma)B \tag{12.15}$$

where the screening constant σ

$$\sigma = \frac{B - B_{eff}}{B} \approx \frac{B - B_{eff}}{B_{eff}} \tag{12.16}$$

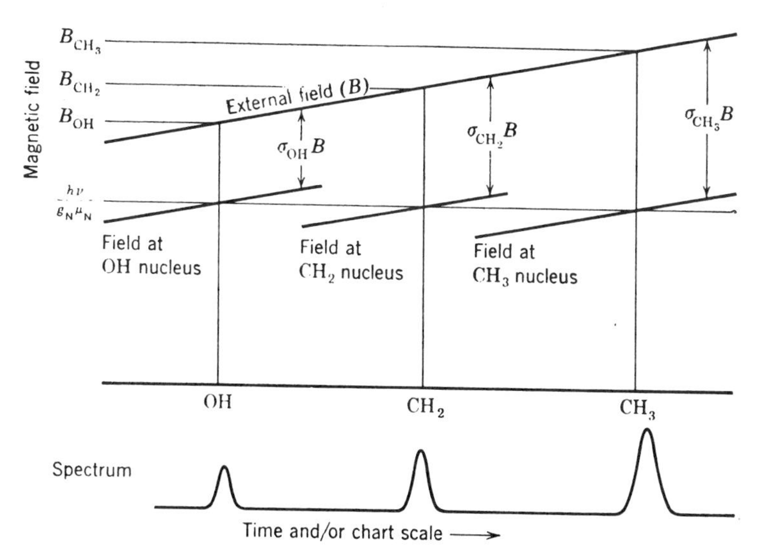

Fig. 12.4 Low-resolution proton-magnetic-resonance spectrum of CH_3CH_2OH. As the magnetic field is increased, protons in different chemical environments come into resonance. As indicated, the magnetic field strengths at the various protons are different from the strength of the external field. (Adapted from J. C. Davis, *Advanced Physical Chemistry*, The Ronald Press, New York, 1965.)

is of the order of 10^{-6}. Thus a very slightly higher field is required to produce resonance of a proton when it is in a molecule as compared to when it is isolated. Although chemical shifts are not large, the nuclear resonance lines obtained from liquids are so narrow that very small changes in local fields may be detected. Since it is not convenient to use an isolated nucleus as a reference, chemical shifts are measured relative to σ for some standard compound in the same solvent. If there are nuclei in two chemical environments the chemical shift difference can be calculated using equation 12.17.

$$\sigma_1 - \sigma_2 = \frac{B_2 - B_1}{B_{\text{eff}}} \tag{12.17}$$

Chemical shifts may be expressed in frequency units by multiplying $\sigma_1 - \sigma_2$ by the resonance frequency, but it is generally more convenient to use $\sigma_1 - \sigma_2$ in parts per million because this applies to experiments at any frequency.

For proton NMR tetramethylsilane $Si(CH_3)_4$ is used as a reference standard. A small amount of the standard is added to the liquid sample. The shielding constant σ for tetramethylsilane (about 31×10^{-6}) is larger than that of protons in almost any other organic compound. To express chemical shifts relative to tetramethylsilane (TMS) and in a form independent of the spectrometer frequency, a δ scale is defined, so that

$$\delta_i = \frac{B_{0,\text{TMS}} - B_{0,i}}{B_{0,\text{TMS}}} \times 10^6 \text{ ppm} \tag{12.18}$$

On this scale $Si(CH_3)_4$ protons have a δ_i of 0 and —CHO protons have a δ_i of approximately 10 ppm.

Chemical shifts for protons in various environments, expressed as δ values, are given in Table 12.2.

Table 12.2 Chemical Shifts of ^1H

Compound	Shift (δ) (ppm)	Compound	Shift (δ) (ppm)
Methyl protons		Olefinic protons	
$(CH_3)_4Si$	0.00	$(CH_3)_2C{=}CH_2$	4.6
$(CH_3)_4C$	0.92	Cyclohexene	5.57
CH_3CH_2OH	1.17		
CH_3COCH_3	2.07	Acetylenic protons	
CH_3OH	3.38	$HOCH_2C{\equiv}CH$	2.33
Methylene protons		Aromatic protons	
Cyclopropane	0.22	Benzene	7.27
Cyclohexane	1.44	Naphthalene	7.73
CH_3CH_2OH	3.59		
		Aldehydic protons	
Methine protons		CH_3CHO	9.72
$(CH_3)_2CHOH$	3.95	C_6H_5CHO	9.96

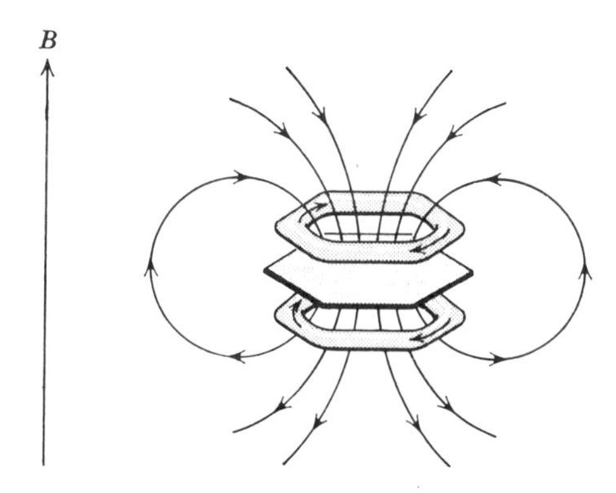

Fig. 12.5 Magnetic field produced by electron circulation in the π orbitals of the benzene ring. At the protons, the induced field is in the same direction as the applied field.

Proton chemical shifts in organic compounds can be correlated with the electronegativities of neighboring groups, types of carbon bonding, and hydrogen bonding. In H—C—X, the chemical shift for the proton depends on the electronegativity of X. The greater the electronegativity (Section 10.7) of X, the more it will draw electrons away from H. The more electrons are drawn away, the lower the magnetic field required for resonance, and the larger the δ value. The electronegativities of the halogens are F > Cl > Br > I, and the proton chemical shifts δ for protons in the methyl halides are $CH_3F > CH_3Cl > CH_3Br > CH_3I$. Protons bonded to aromatic rings have resonances at lower magnetic field strengths than aliphatic protons. When a benzene ring is oriented perpendicular to the magnetic field, the circulation of electrons in the π orbitals induces a field that is in the same direction as the applied field at the protons, as shown in Fig. 12.5. Therefore aromatic protons resonate at a lower field than they otherwise would. This effect is reduced by molecular tumbling because, when the benzene ring is oriented parallel to the field, there is no such effect.

Chemical shifts for ^{19}F and ^{11}B are much larger than those for protons, because they are surrounded by a larger number of electrons capable of having currents induced in them.

12.7 SPIN-SPIN SPLITTING

Figure 12.6, which shows the high-resolution, proton-magnetic-resonance spectrum for CH_3CH_2OH, shows that the spectrum is more complicated than indicated in Fig. 12.4. With higher resolution the lines in the spectrum are split into multiplets. This splitting is called spin-spin splitting, and it results from the effects of nuclear spins in the same molecule. In the pattern for ethyl alcohol, the absorption line due to methyl (CH_3) protons is split into three components because the neighboring

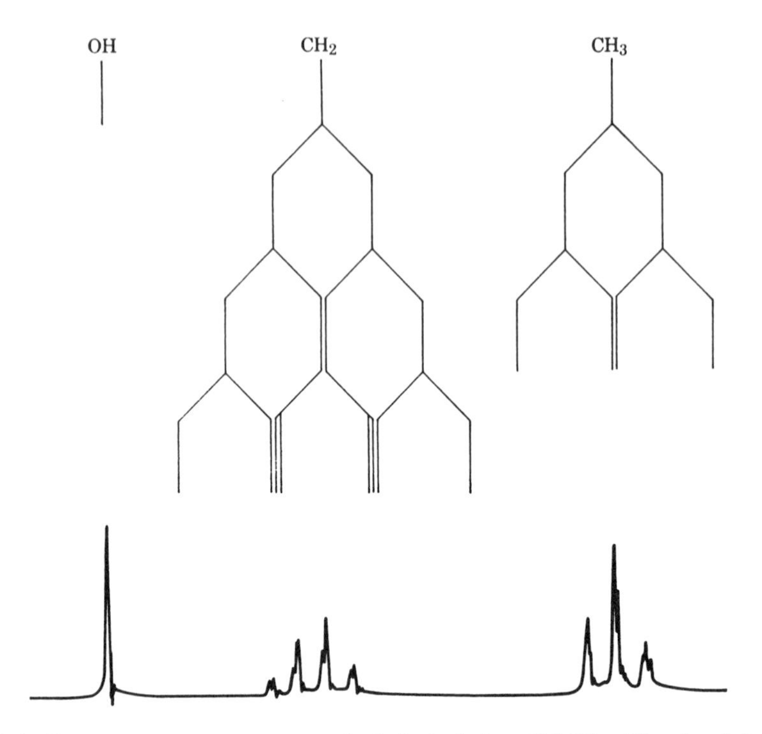

Fig. 12.6 Proton resonance spectrum of ethyl alcohol at 40 MHz. The signal from the radiofrequency receiver is plotted vertically, and the magnetic field strength is plotted horizontally. The field increases linearly from left to right.

methylene group (CH_2) contains two protons, each with spin $\frac{1}{2}$. We can think of the first methylene proton splitting the methyl proton resonance into a doublet, as shown in the figure, and then the second methylene proton splitting the doublet into a triplet with the center line twice as intense as the other two. The two methylene protons produce the same spin-spin splitting because there is rapid internal rotation around the C—C bond.

The absorption line due to methylene protons is split into four components by the three protons of the neighboring CH_3 group. The reason for the relative intensities of 1:3:3:1 is evident from the diagram.

In this spectrum the proton in the hydroxyl group does not cause further splitting because it undergoes chemical exchange so rapidly with protons in other molecules that it does not produce a splitting effect. This rapid chemical exchange occurs only when acid catalyzed, but only traces of acid are required to eliminate the splitting.

We can generalize these results by saying that if there are n equivalent protons that interact with the proton being studied, the absorption will be split into $n + 1$ lines, and their relative intensities will be proportional to the coefficients of the binomial expansion of $(1 + x)^n$. These coefficients are given by Pascal's triangle.

0 proton	1
1 proton	1 1
2	1 2 1
3	1 3 3 1
4	1 4 6 4 1
5	1 5 10 10 5 1
6	1 6 15 20 15 6 1

Since the spin-spin splitting is due to neighboring groups, it gives important additional structural information.

The direct dipolar coupling of two nuclear spins does not explain the observed spin-spin coupling effects, because rapid molecular tumbling in liquids averages it to zero. Nuclear spins in molecules are coupled through intervening electrons. The major effect results from the coupling of the spin of the first nucleus with electrons through the so-called contact interaction, first proposed by Fermi. This effect, which tends to align the spin of orbital electrons antiparallel to the nuclear spin, can only occur when the electron has a significant probability density at the nucleus. The electrons that become partially aligned in this way affect the magnetic field at the second nucleus. In addition, nuclear spin induces a current in the valence electrons by the action of the magnetic field of the nucleus on the orbital magnetic moment of the electrons.

The strength of spin-spin splitting is measured by the coupling constant J which, in the case we have been discussing, is simply the separation of peaks in a multiplet, usually expressed as a frequency by multiplying the σ value for the splitting by the resonance frequency. A few spin-spin coupling constants are summarized in Table 12.3. Spin-spin coupling constants are independent of the magnitude of the applied magnetic field; this is in contrast to chemical shifts, which are directly proportional to the applied field.

For the purpose of discussing different types of spin systems, it is convenient to

Table 12.3 Proton Spin-Spin Coupling Constants

Structure	J (Hz)	Structure	J (Hz)
$>\mathrm{C}<$ (H, H on same C)	−20 to +6	H—C=C—H (H on different sides)	12 to 19
—C(H)—C(H)—	5.5 to 7.5	benzene ring	o 6 to 9 m 0.5 to 4 p 0 to 2.5
—C(H)=C(H)—	7 to 10	cyclohexane (ax, eq)	ax, ax 9 to 14 ax, eq 2 to 4 eq, eq 2.5 to 4

use letters of the alphabet to represent nuclei. A spin system with two nuclei may be represented by AX, AB, or A_2 depending on whether the difference in chemical shifts is large compared with the coupling constant, whether they are of the same order of magnitude, or whether the nuclei are equivalent.

Example 12.5 Sketch the proton resonance spectrum for a compound, represented by AMX, with three protons with rather different chemical shifts ($\nu_0\delta_A = 100$, $\nu_0\delta_M = 200$, and $\nu_0\delta_X = 700$ Hz). The three protons are coupled with $J_{AM} = 9$, $J_{AX} = 7$, and $J_{MX} = 3$ Hz.

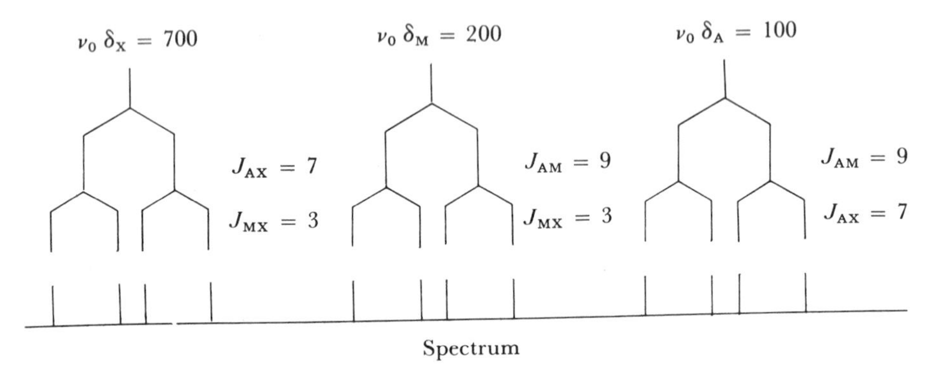

12.8 SECOND-ORDER EFFECTS

The interpretation of NMR spectra in the preceding section applies when the chemical shifts are large compared with the coupling constant. When the chemical shifts are of the order of the coupling constants or smaller, the interpretation of the spectrum becomes more complicated because second-order effects have to be taken into account. It is the ratio of the chemical shift separation in frequency units to the coupling constant J that is important. In order to make this comparison the chemical shift δ needs to be expressed in Hz by multiplying by the spectrometer operating frequency ν_0 to obtain $\nu_0\delta$. We will consider only one example of second-order effects: those encountered in an AB system with two spin $\frac{1}{2}$ nuclei. As shown in Fig. 12.7, the spectrum consists of two doublets if the chemical shift $\nu_0\delta$ is large compared with the coupling constant J. As the chemical shift is reduced, the doublets approach each other, but the question is "How can the two doublets collapse to give a single line when the coupling constant is reduced to zero?" In order to answer this question it is necessary to consider the four possible energy states for AB with two spin $\frac{1}{2}$ nuclei. The wave function for the two possible wave functions for each nucleus are represented by α for $I = +\frac{1}{2}$ and β for $I = -\frac{1}{2}$. There are four possible wave functions: $\alpha(A)\alpha(B)$, $\alpha(A)\beta(B)$, $\beta(A)\alpha(B)$, and $\beta(A)\beta(B)$. (See Section 8.18.) The spin states $\alpha(A)\beta(B)$ and $\beta(A)\alpha(B)$ have the same component of the spin angular momentum in the direction of the field, and so these states have to be combined to form mixed wave functions of the form shown in Table 12.4 with constants C_1 and C_2. The energy levels corresponding with the four wave functions* are shown in Table 12.4.

The four possible transitions between the four energy levels are shown in Table 12.5.

* I. N. Levine, *Molecular Spectroscopy*, Wiley-Interscience, New York, 1975.

Table 12.4 Wave Functions and Energy Levels for an AB System with Two Spin $\frac{1}{2}$ Nuclei[a]

Number	Wave Function	Energy Level (Hz)
1	$\alpha\alpha$	$\frac{1}{2}(\nu_A + \nu_B) + \frac{1}{4}J$
2	$C_1(\alpha\beta) + C_2(\beta\alpha)$	$\frac{1}{2}[(\nu_0\delta)^2 + J^2]^{1/2} - \frac{1}{4}J$
3	$-C_2(\alpha\beta) + C_1(\beta\alpha)$	$-\frac{1}{2}[(\nu_0\delta)^2 + J^2]^{1/2} - \frac{1}{4}J$
4	$\beta\beta$	$-\frac{1}{2}(\nu_A + \nu_B) + \frac{1}{4}J$

[a] From J. W. Akitt, *NMR and Chemistry*, Chapman and Hall, London, 1973.

Only those transitions can take place for which the change in the z component of the spin is ± 1. For α the energy eigenvalue is $\frac{1}{2}\hbar$, and for β it is $-\frac{1}{2}\hbar$. The transition energies, which may be calculated from Table 12.4, are given in Table 12.5 with respect to the mean frequency $(\nu_A + \nu_B)/2$. The transitions between the four energy levels and the corresponding four-line spectrum are shown in Fig. 12.8. The separation of the two lines in each doublet is the chemical shift J_{AB}, but the separation of the two doublets is no longer equal to the chemical shift $\nu_0\delta$. The separation between lines a and c (or b and d or doublet centers) is now given by $[(\nu_0\delta)^2 + J^2]^{1/2}$. It is convenient to calculate the chemical shift from the a–d and b–c distances from

$$\nu_0\delta = [(a - d)(b - c)]^{1/2} \tag{12.19}$$

which may be verified by substituting the energy differences between these pairs of lines.

As shown in Fig. 12.7, the relative intensities of the lines change as the chemical shift is reduced. In the limit as the chemical shift approaches zero, the spectrum approaches the single-line spectrum characteristics of A_2. In other words, no spin-spin couplings are observed within a group of magnetically equivalent nuclei.

It is important to make a distinction between nuclei that are chemically equivalent and nuclei that are magnetically equivalent. This distinction is illustrated by CH_2F_2 and $CH_2{=}CF_2$, which at first both look like A_2X_2 systems. Fluorine, represented by X, has a spin of $\frac{1}{2}$. In both compounds the protons are chemically equivalent and have the same chemical shift; they are equivalent by symmetry operations. The proton resonance spectra of these compounds are shown in Fig. 12.9. The proton spectrum of CH_2F_2 is the expected 1:2:1 triplet. The proton spectrum of $CH_2{=}CF_2$ is more complicated because the coupling of H and F, which are *cis* (on the same side) is different from the coupling of

Table 12.5 Transition Energies and Line Intensities for an AB System with Two Spin $\frac{1}{2}$ Nuclei[a]

	Transition	Energy (Hz)	Relative Intensity
a	$3 \rightarrow 1$	$+\frac{1}{2}J + \frac{1}{2}[(\nu_0\delta)^2 + J^2]^{1/2}$	$1 - J/[(\nu_0\delta)^2 + J^2]^{1/2}$
b	$4 \rightarrow 2$	$-\frac{1}{2}J + \frac{1}{2}[(\nu_0\delta)^2 + J^2]^{1/2}$	$1 + J/[(\nu_0\delta)^2 + J^2]^{1/2}$
c	$2 \rightarrow 1$	$+\frac{1}{2}J - \frac{1}{2}[(\nu_0\delta)^2 + J^2]^{1/2}$	$1 + J/[(\nu_0\delta)^2 + J^2]^{1/2}$
d	$4 \rightarrow 3$	$-\frac{1}{2}J - \frac{1}{2}[(\nu_0\delta)^2 + J^2]^{1/2}$	$1 - J/[(\nu_0\delta)^2 + J^2]^{1/2}$

[a] From J. W. Akitt, *NMR and Chemistry*, Chapman and Hall, London, 1973.

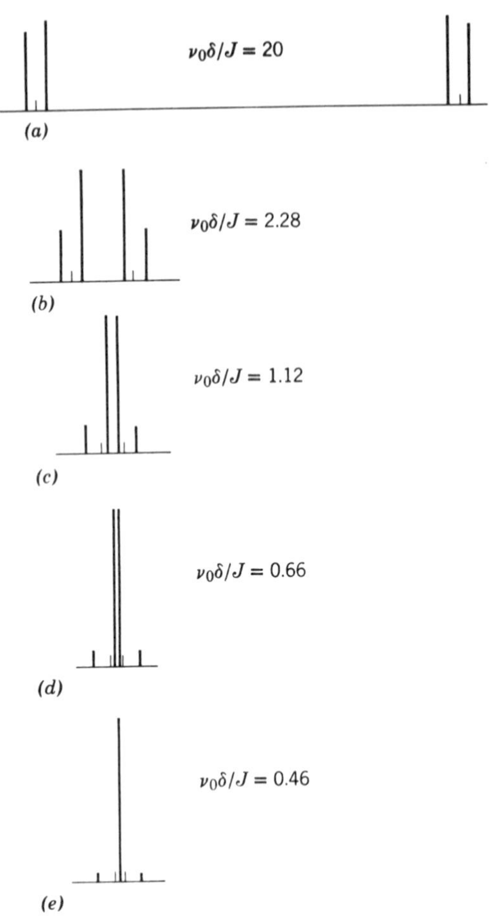

Fig. 12.7 Spectra for an AB system with two spin-$\frac{1}{2}$ nuclei for various ratios of $\nu_0\delta/J$ and constant coupling constant. The A and B chemical shifts are indicated by the small markers under the baseline. (From J. W. Akitt, *NMR and Chemistry*, Chapman and Hall Ltd., London, 1973.)

H and F, which are *trans* (on opposite sides). Thus the two protons in $CH_2{=}CF_2$ are not magnetically equivalent. In order for nuclei to be magnetically equivalent, it is necessary for them to be chemically equivalent *and* to couple to all other nuclei in the system in exactly the same way. If the ethanol molecule CH_3CH_2OH were perfectly rigid the methyl protons would not be equivalent. However, because of rapid rotation about the C—C bond, the electronic environments of the three methyl protons are magnetically equivalent.

The separation of peaks due to the chemical shift is directly proportional to the field strength, but the separation due to spin-spin splitting J is independent of field strength. Thus, at higher field strengths, the spin-spin splittings do not cause as much complication in the interpretation of the NMR pattern.

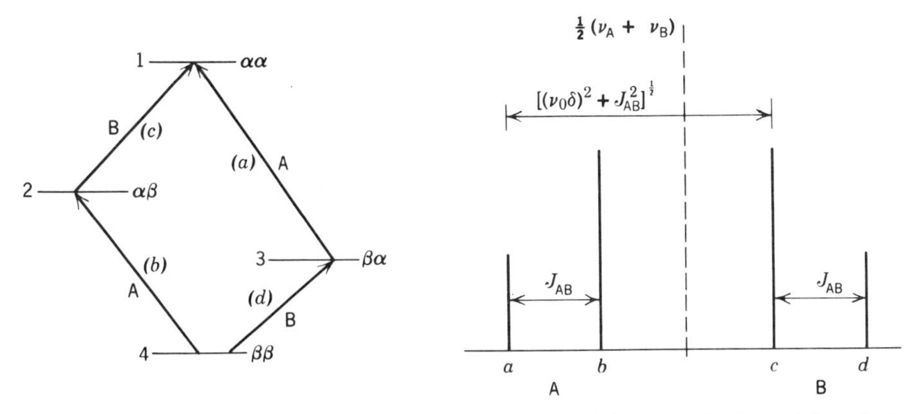

Fig. 12.8 Energy level diagram for an AB system with two spin-½ nuclei and the resulting spectrum. (From J. W. Akitt, *NMR and Chemistry*, Chapman and Hall Ltd., London, 1973.)

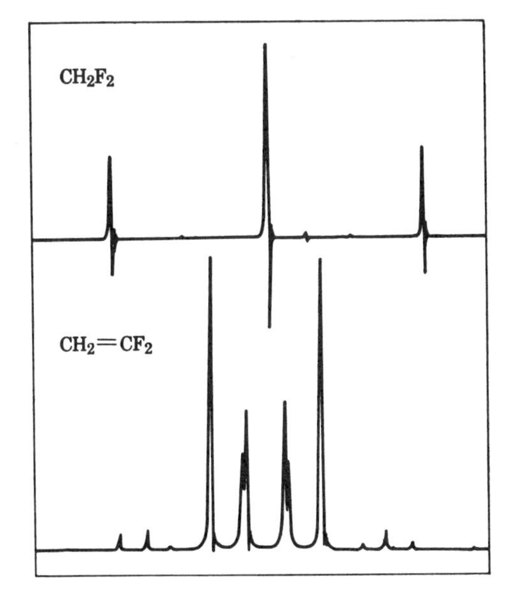

Fig. 12.9 Proton-resonance spectra of CH_2F_2 and $CH_2{=}CF_2$ at 60 MHz. [From E. D. Becker, *J. Chem. Ed.*, **42**, 591 (1965).]

12.9 ELECTRON SPIN RESONANCE

The basic theory of electron spin resonance (ESR) has been described in Section 12.2. An electron resonance experiment is similar to an NMR experiment but, since the gyromagnetic ratio of the electron is about 10^3-fold larger than for nuclei, the frequencies required fall in the microwave range instead of the radiofrequency

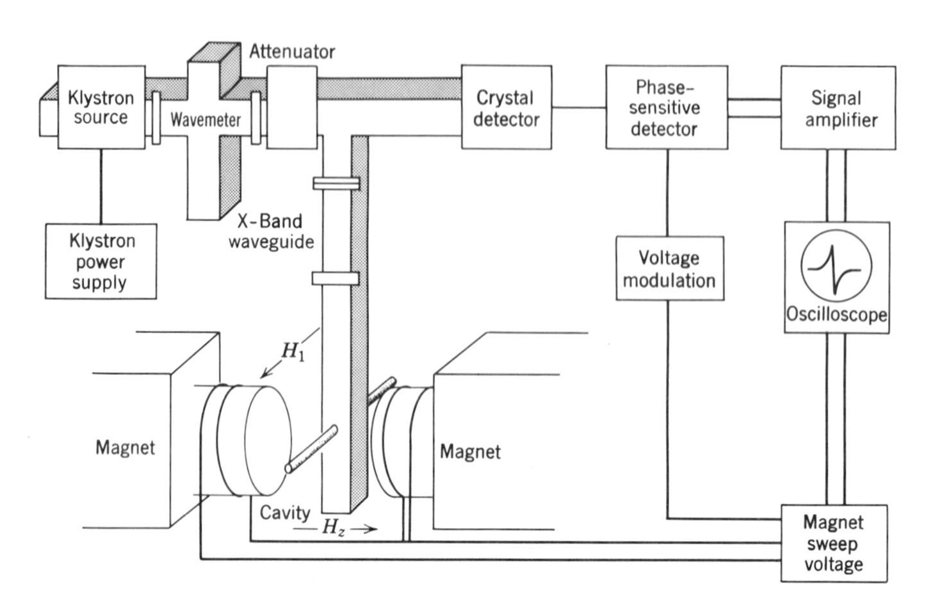

Fig. 12.10 Block diagram of simple ESR spectrometer. (From A. W. Guillory, *Introduction to Molecular Structure and Spectroscopy*. Copyright © 1977 by Allyn and Bacon, Inc., Boston. Reprinted with permission of the publisher.)

range when magnetic fields of convenient laboratory strength are used. Usually a frequency of about 10 GHz ($\lambda = 3$ cm) is used with a magnetic field of 0.3 to 0.4 T.

Substances that show ESR spectra include free radicals, odd electron molecules, triplet states of organic molecules, and paramagnetic transition metal ions and their complexes. Any paramagnetic substance can be studied but, as we have seen, most substances are not paramagnetic, because electron spins are usually paired.

The block diagram for a simple ESR spectrometer is shown in Fig. 12.10. As in the NMR spectrometer, the frequency is held constant, and the magnetic field is swept through resonance. Microwave radiation from a klystron passes down a waveguide to a resonant cavity containing the sample. When there are transitions between electron spin levels in the sample, energy is absorbed from the microwave radiation, and less microwave energy is received at the crystal detector. By use of field modulation and a phase sensitive detector the derivative of the absorption line is recorded on the oscilloscope or strip chart recorder. The shape of the derivative curve is illustrated on the face of the oscilloscope in Fig. 12.10.

12.10 HYPERFINE COUPLING

ESR spectroscopy is especially useful for chemistry because the electron magnetic moment interacts with other magnetic moments in the molecule including protons and other nuclei listed in Table 12.1. The splitting of absorption lines that results is called hyperfine splitting instead of spin-spin splitting, as in NMR.

As a simple example of hyperfine splitting, let us consider a free radical with two protons that affect the electron energy levels in a magnetic field to different extents. Figure 12.11*a* shows the effects of the two protons on the possible energy levels for the electron. In the presence of a magnetic field an unpaired electron has two energy levels, $m_s = \frac{1}{2}$ and $m_s = -\frac{1}{2}$. The two protons each split these levels so that the unpaired electron has eight energy levels. In electron spin resonance transitions the electron spin flips, but the nuclear spins do not. Thus, in absorbing energy in an ESR transition, the electron goes from an energy level in the lower group ($m_s = -\frac{1}{2}$) to the corresponding level in the upper group ($m_s = +\frac{1}{2}$). As the magnetic field strength is increased, the four possible transitions one-by-one come into resonance, and four ESR lines are obtained. Since the four nuclear-spin states ($\alpha_1\alpha_2$, $\alpha_1\beta_2$, $\beta_1\alpha_2$, and $\beta_1\beta_2$) are equally probable, the four lines are of equal intensity. The two hyperfine splittings a_1 and a_2 may be calculated from the spectrum as shown.

If the two protons have the same coupling constant to the unpaired electron, the spectrum has three lines, and the center line is twice as intense because it represents two possible transitions, as shown in Fig. 12.11*b*. Two or more protons having the same coupling to an unpaired electron are said to be equivalent, and they usually occupy symmetrically equivalent positions in the molecule.

Figure 12.12 shows a simple way of looking at these two spectra. We can describe the effect of two proton spins in the molecule by saying that the first proton splits the original single line due to electron spin into a doublet, and the second proton further splits the two lines into a quadruplet. If the two hyperfine splittings are different, four absorption lines are obtained. If the two hyperfine splittings are the same (the protons are equivalent), three lines are obtained.

Thus, in general, if n equivalent protons interact with an unpaired electron there will be $n + 1$ lines, and their relative intensities are given by Pascal's triangle (Section 12.7). The ESR spectrum of benzene negative ion ($C_6H_6^-$) in Fig. 12.13*a* illustrates this very nicely. The ESR spectrum of naphthalene negative ion ($C_{10}H_8^-$) shown in Fig. 12.13*b* is more complicated because there are two types of protons, labeled A and B in the following structural formula:

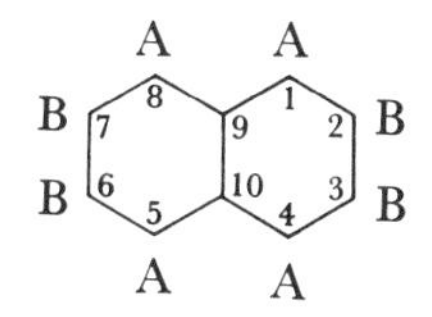

The four A protons alone would give rise to a five-line spectrum. Each of these lines is further split into five lines by the B protons, to give the 25-line spectrum shown in the figure.

The splitting by other magnetic nuclei is similar, except that, of course, nuclei of spin 1, like ^{14}N, cause splitting into triplets of equal intensity, since the quantum number for the ^{14}N nucleus may be +1, 0, or −1.

The hyperfine coupling between the proton and electron of the hydrogen atom is 0.0507 T. This coupling (due largely to Fermi's contact interaction, discussed in

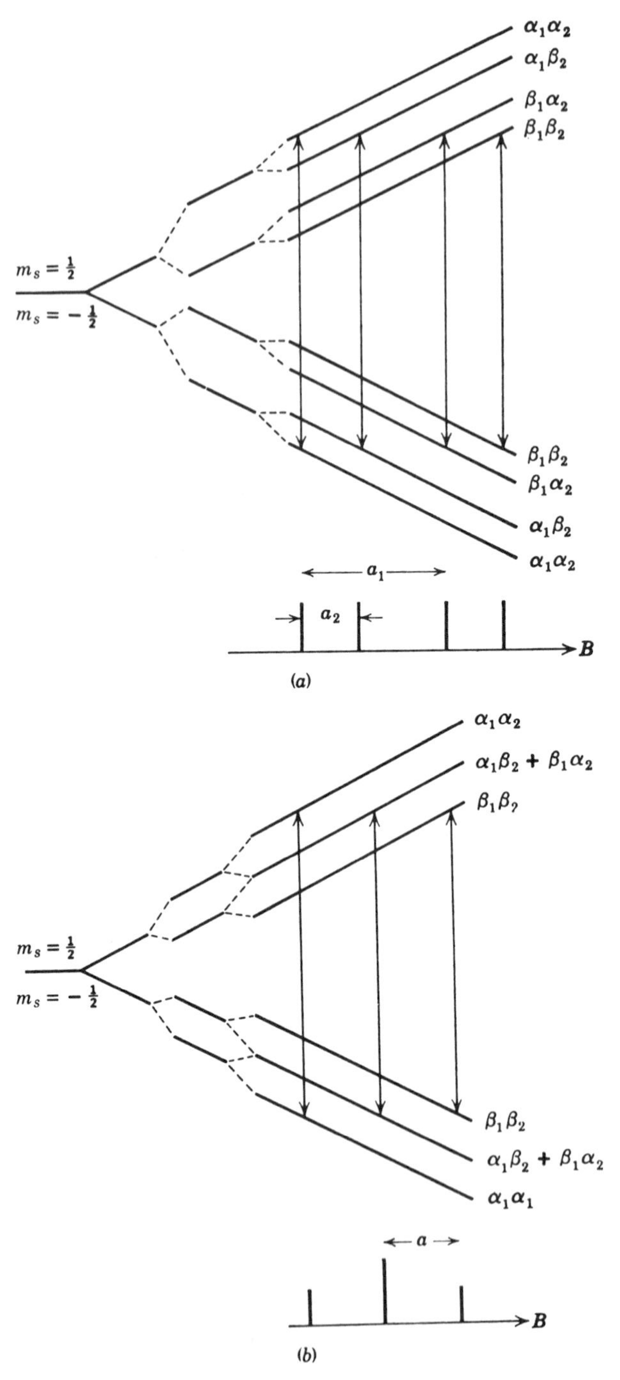

Fig. 12.11 (*a*) Hyperfine energy levels and ESR spectrum for an unpaired electron in the presence of two nonequivalent protons. (*b*) Hyperfine energy levels and ESR spectrum for an unpaired electron in the presence of two equivalent protons. The spin states of the protons are represented by $\alpha_1\alpha_2$, $\alpha_1\beta_2$, $\beta_1\alpha_2$, and $\beta_1\beta_2$.

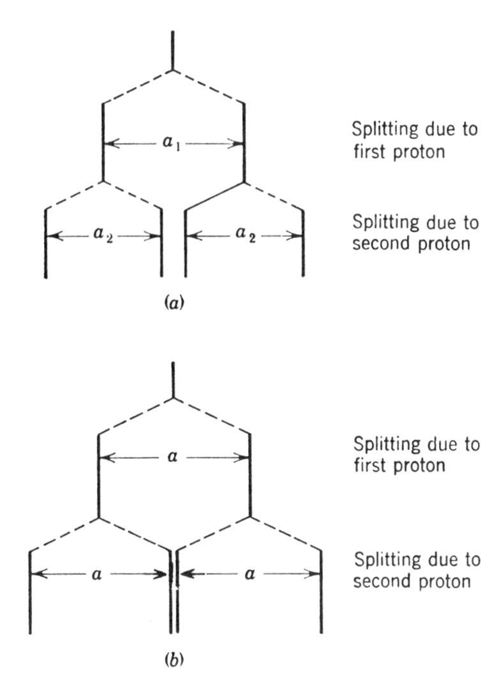

Fig. 12.12 (*a*) Hyperfine splitting due to two nonequivalent protons. (*b*) Hyperfine splitting due to two equivalent protons.

Section 12.7) is proportional to the probability of finding the 1*s* electron at the nucleus. The coupling between unpaired electrons and protons in organic radicals are much less than this (up to about 3×10^{-3} T) because the unpaired electron occupies orbitals that extend over the whole molecule, and the probability that the unpaired electron is at a particular hydrogen nucleus is quite small.

12.11 ESR APPLICATIONS

ESR spectroscopy has been very useful in determining the structure of organic and inorganic free radicals. Free radicals may be produced chemically, photochemically, or by use of high-energy radiation. If a radical has a very short life a flow system or continuous radiation may have to be used to maintain a sufficiently high concentration for detection. Actually, a concentration of only about 10^{-10} mol L^{-1} is required to obtain a spectrum under favorable conditions.

ESR spectra may be determined for unstable radicals trapped in glasses, frozen rare gas matrices, or in crystals. If the radicals are regularly oriented with respect to crystal axes the directional effects of magnetic interactions may be studied.

A molecule in a triplet state has a total electron spin of one ($S = 1$). In this case there are three sublevels that have spin angular momentum S_z about a chosen axis of +1, 0, or −1. A triplet molecule has an even number of electrons, two of them unpaired, while a radical with spin $\frac{1}{2}$ has an odd number of electrons. In order for

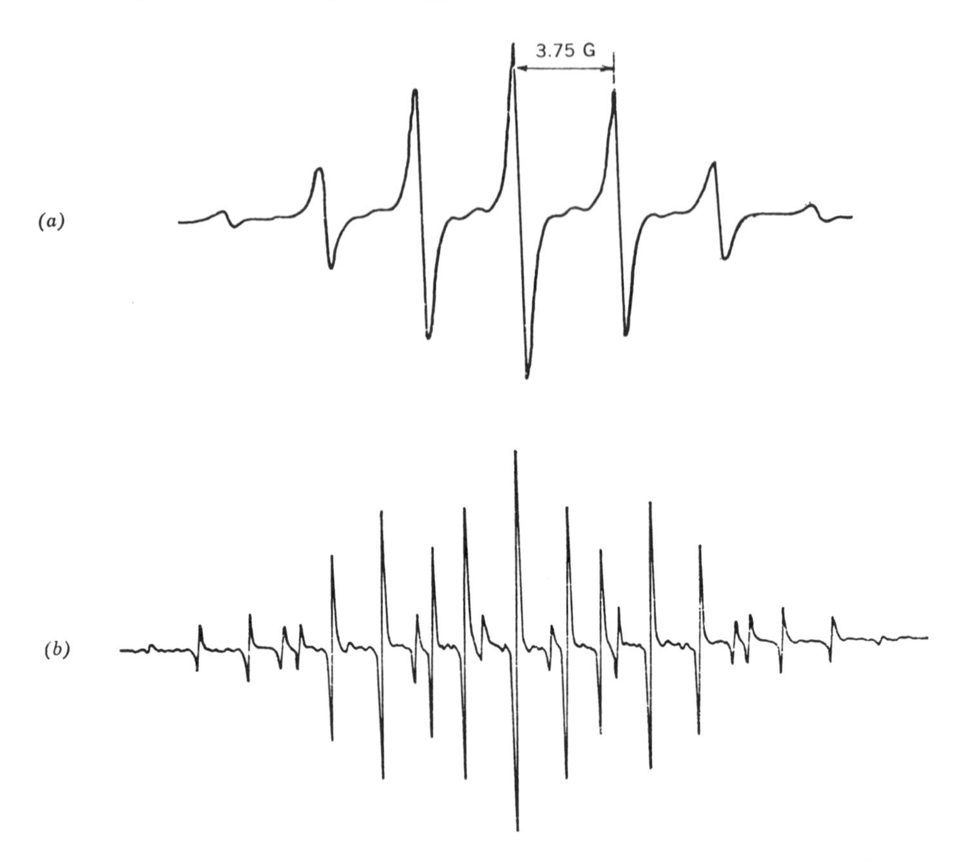

Fig. 12.13 (*a*) ESR spectrum of benzene negative ion. (*b*) ESR spectrum of naphthalene negative ion. (From A. Carrington and A. D. McLachlan, *Introduction to Magnetic Resonance*, Harper and Row, New York, 1967.)

a molecule to have a triplet state, the unpaired electrons must interact; a molecule with two unpaired electrons a great distance apart is a diradical, not a triplet.

The information about electron densities in molecules that is obtained from ESR spectra has been useful for checking molecular-orbital and valence-bond calculations.

Groups with unpaired electrons (spin labels) may be attached to proteins to provide a very sensitive detector of changes in protein structures.

References

J. W. Akitt, *NMR and Chemistry*, Chandler and Hall, London, 1973.

C. N. Banwell, *Fundamentals of Molecular Spectroscopy*, McGraw-Hill Book Co., New York, 1972.

E. F. H. Brittain, W. O. George, and C. H. J. Wells, *Introduction to Molecular Spectroscopy*, Academic Press, New York, 1970.

A. Carrington and A. D. McLachlan, *Introduction to Magnetic Resonance*, Harper and Row, New York, 1967.
R. Chang, *Basic Principles of Spectroscopy*, McGraw-Hill Book Co., New York, 1971.
W. A. Guillory, *Introduction to Molecular Structure and Spectroscopy*, Allyn and Bacon, Inc., Boston, 1977
O. Howarth, *Theory of Spectroscopy*, Wiley, New York, 1973.
I. N. Levine, *Molecular Spectroscopy*, Wiley-Interscience, New York, 1975.
W. W. Paudler, *Nuclear Magnetic Resonance*, Allyn and Bacon, Inc., Boston, 1971.
J. A. Pople, W. G. Schneider, and H. J. Bernstein, *High Resolution Nuclear Magnetic Resonance* McGraw-Hill Book Co., New York, 1959.
J. E. Wertz and J. R. Bolton, *Electron Spin Resonance*, McGraw-Hill Book Co., New York, 1972

Problems

12.1 Calculate the magnetic flux density to give a precessional frequency for fluorine of 60 MHz. *Ans.* 1.4973 T.

12.2 Frequencies used in nuclear magnetic resonance spectra are of the order of 60 MHz. Calculate the corresponding energy in kilojoules per mole.
Ans. 2.394×10^{-5} kJ mol^{-1}.

12.3 What magnetic flux density is required for proton magnetic resonance at 220 MHz?
Ans. 5.167 T.

12.4 The gyromagnetic ratio γ_N for a nucleus is defined by

$$\boldsymbol{\mu}_N = \gamma_N \hbar \boldsymbol{I}$$

What is the value of γ_N for H? *Ans.* 2.675×10^8 s^{-1} T^{-1}.

12.5 What is the ratio of the number of ^{31}P spins in the lower state to the number in the upper state in a magnetic field of 1.724 T at room temperature? (Given: $g_N = 2.263$.)
Ans. $1 + 4.79 \times 10^{-6}$.

12.6 In a magnetic field of 2 T, what fraction of the protons have their spin lined up with the field at room temperature? *Ans.* 0.500 003 43.

12.7 Using information from Tables 12.2 and 12.3, sketch the spectrum you would expect for ethyl acetate ($CH_3CO_2CH_2CH_3$). *Ans.*

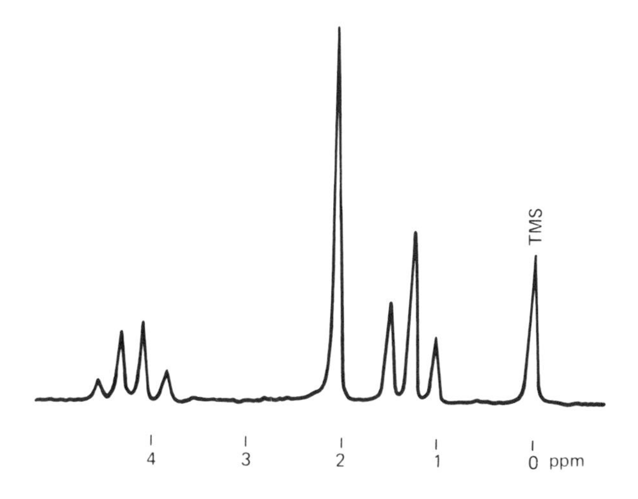

12.8 Are protons *a* and *b* magnetically equivalent in

NO_2

H_c H_d

H_a H_b

F

Ans. They are not magnetically equivalent (although they are chemically equivalent) because the coupling $H_a - H_c$ is different from the coupling $H_a - H_d$.

12.9 At room temperature the chemical shift of cyclohexane protons is an average of the chemical shifts of the axial and equatorial protons. Explain.

12.10 The proton resonance pattern of 2,3-dibromothiophene shows an AB-type spectrum with lines at 405.22, 410.85, 425.07, and 430.84 Hz measured from tetramethylsilane at 1.41 T [K. F. Kuhlmann and C. L. Braun, *J. Chem. Ed.*, **46**, 750 (1969)]. (*a*) What is the coupling constant *J*? (*b*) What is the difference in the chemical shifts of the A and B hydrogens? (*c*) At what frequencies would the lines be found at 2 T?

Ans. (*a*) 5.70 Hz. (*b*) 0.318 ppm. (*c*) 576.18, 581.88, 603.84, and 609.54 Hz.

12.11 Calculate the precessional frequency of electrons in a 1.5-T field.

Ans. 42,000 MHz.

12.12 Line separations in ESR may be expressed in G or MHz. Show how the conversion factor 1 T = 2.80×10^4 MHz is obtained.

12.13 Sketch the ESR spectrum expected for *p*-benzosemiquinone radical ion

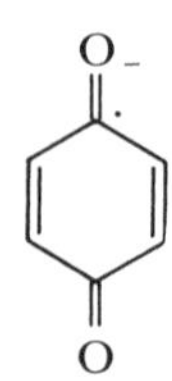

The four hydrogens are magnetically equivalent.

Ans. Five lines with relative intensities of 1, 4, 6, 4, 1,

12.14 An unpaired electron in the presence of two protons gives the following four-line ESR spectrum: $\Delta B/10^{-4}$ T = 0, 1, 3, 4. What are the two coupling constants in T and in MHz?

Ans. $a_1 = 3 \times 10^{-4}$ T or 8.40 MHz, $a_2 = 1 \times 10^{-4}$ T or 2.80 MHz.

12.15 Sketch the ESR spectrum for an unpaired electron in the presence of three protons for the following cases: (*a*) the protons are not equivalent, (*b*) the protons are equivalent, and (*c*) two protons are equivalent and the third is different.

Ans. (*a*) Eight lines arranged in pairs.

(*b*) Four equally spaced lines with intensities 1:3:3:1. (*c*) Six lines.

12.16 Sketch the ESR spectrum of $CH_3\cdot$.

Ans. Four lines equally spaced 2.3×10^{-3} T apart with intensities of 1:3:3:1.

12.17 Using data from Table 12.1, calculate the magnetic field strengths at which (*a*) ^{13}C, and (*b*) ^{19}F will precess at 5 MHz.

12.18 What is the magnitude of the magnetic moment of the proton?

12.19 Calculate the magnetic flux density *B* required for ^{19}F resonance at 20 MHz.

12.20 What is the ^{13}C resonance at 1 T?

12.21 A sample containing protons is placed in a magnetic field with a strength of 1 T. Calculate the fraction of the protons that at 25 °C will have their spins lined up with the field.

12.22 At a magnetic flux density of 1.41 T, the frequency separation between protons in benzene and protons in tetramethylsilane is 436.2 Hz. What is the chemical shift?

12.23 Sketch the expected nuclear-magnetic-resonance spectra for the protons in *n*-propyl, isopropyl, and *tert*-butyl alcohols, showing the number of components in each band. (Note that only protons on adjacent carbon atoms need to be considered in predicting the splitting.)

12.24 Describe the proton and deuteron NMR spectra of HD.

12.25 On the basis of the spin-spin coupling constants in Table 12.3, describe the spectrum expected for

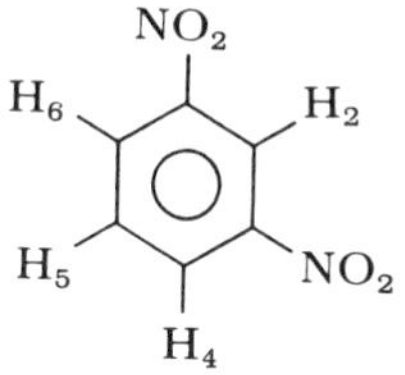

NO_2

H_6 H_2

H_5 NO_2

H_4

The protons are labeled to assist with labeling the spectrum.

12.26 Calculate the resonance frequency for electrons at 0.33 T.

12.27 Sketch the ESR spectrum of $CD_3\cdot$, giving the relative intensities of the various lines.

12.28 Sketch the ESR spectrum for the ethyl radical with the indicated hyperfine splittings.

$$\cdot CH_2\text{—}CH_3$$
$$2.238 \times 10^{-3}\ \text{T} \qquad 2.687 \times 10^{-3}\ \text{T}$$

12.29 Sketch the spectrum for an unpaired electron in the presence of ^{14}N and ^{1}H.

CHAPTER 13

STATISTICAL MECHANICS

The equilibrium properties of matter may be considered from two points of view: the macroscopic and the microscopic. From the macroscopic point of view represented by thermodynamics, the behavior of large numbers of molecules is described in terms of pressure, volume, temperature, composition, and exchanges of heat and work. The relationships between various types of equilibrium measurements derived in thermodynamics are not based on any theory of the structure of matter or of the mechanisms by which changes occur.

From the microscopic point of view represented by the applications of classical mechanics in kinetic theory and by quantum mechanics, the behavior of individual molecules is described in terms of their structures and the mechanisms by which they interact with other molecules. Ideally we would like to be able to predict the thermodynamic behavior of substances using our knowledge about individual molecules obtained from spectroscopic measurements and from theoretical calculations of wave functions.

Statistical mechanics provides the needed bridge between microscopic mechanics (classical and quantum) and macroscopic thermodynamics. The classical aspects of this science were developed during the latter part of the nineteenth century by Boltzmann in Austria, Maxwell in England, and Gibbs in the United States. For simple systems, such as perfect gases, the calculations are not very difficult, and values of thermodynamic properties so obtained are often more accurate than the ones measured directly. For more complicated systems, especially those involving strong interactions between molecules, the theory is much more difficult and is the subject of current research.

Statistical mechanics provides insight into the laws of thermodynamics, and through it we will see heat, work, temperature, irreversible processes, and state functions in a new light.

13.1 MICROSTATES AND RANDOMNESS

A microstate of a system is a particular spatial configuration and distribution of energy among particles that is described in full detail. For a given thermodynamic state (i.e., a certain number of particles at a specified temperature, pressure, and volume) there are many possible microstates.

Although it will take a whole chapter to introduce statistical mechanics, the basic idea is quite simple. Large groups of molecules in a given thermodynamic state can exist in many, many microstates and will tend to move toward equilibrium because that is the thermodynamic state with the largest number of microstates.

As an example, let us consider the distribution of energy among atoms in a crystal. By considering crystals with small numbers of atoms we can actually count the possible microstates and see the origin of the Boltzmann distribution. It is assumed that the atoms in the crystal are perfect one-dimensional harmonic oscillators so that an atom may have energies ϵ_v given by $vh\nu$, where v is the vibrational quantum number, h is Planck's constant, and ν is the vibrational frequency. This amounts to taking the zero point energy as the reference and measuring energy with respect to the zero point level.

Figure 13.1 shows the different ways of distributing three quanta of energy among the three atoms of the crystal. The atoms may be distinguished because they have different locations, and they are labeled *a*, *b*, and *c*. Each little diagram represents a microstate of the system. Thus our crystal can be in any one of 10 different microstates, each of which has exactly the same energy, three quanta of a particular frequency. There does not seem to be any reason why any of these 10 states should be favored over others. Thus it is assumed that they are all equally probable, and that at any given instant the crystal would be as likely to be in one microstate as another. We assume that there is a mechanism for quanta to move from one atom to another, and therefore we expect that over a long period of time the crystal spends equal amounts of time in each of the 10 microstates consistent with an overall energy of $3h\nu$ for the entire three-atom crystal.

The microstates of the three-atom crystal are of three different types: I, II, and III. We will refer to them as three different distributions. In the first distribution one atom has two quanta, another has one, and the third is in the ground state: there are six of these microstates. In the second distribution one atom has three quanta and the other two are in the ground state: there are three of these microstates. In the third distribution each atom has one quantum of energy: there is only one microstate with this distribution.

The number W of microstates in a distribution can be calculated using the relation

$$W = \frac{N!}{N_1!\, N_2!\, N_3! \cdots N_n!} \tag{13.1}$$

where $N! = 1 \cdot 2 \cdot 3 \cdots N$, N_1 is the number of atoms in state 1 (e.g., having zero quanta), N_2 is the number of atoms in state 2 (e.g., having one quantum), N_3 is the number of atoms in state 3 (e.g., having two quanta), etc., and N is the total number of atoms. Remember that $0! = 1$. We do not really need equation 13.1 at this point, because we can write out all the possible microstates, as in Fig. 13.1, but this relation will be useful later when the numbers are larger.

With Fig. 13.1 we can answer the question: "On the average how many atoms of the crystal will be in the $v = 0$ level, how many in the $v = 1$ level, etc.?"

In considering crystals we are not interested in *which* atoms have a certain

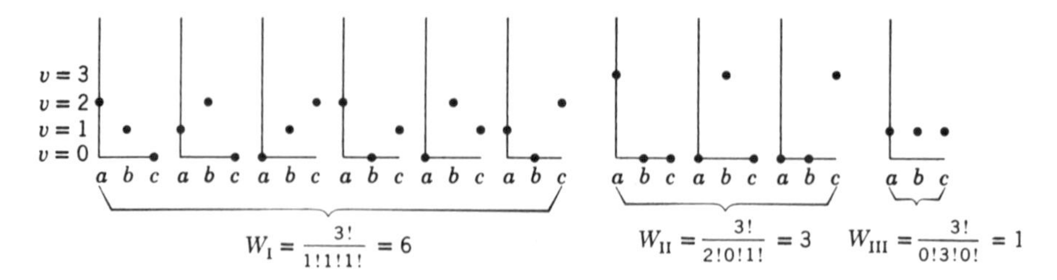

Fig. 13.1 Different ways of distributing three quanta of energy between atoms *a*, *b*, and *c* in an idealized crystal. When there are three quanta of energy, there are 10 possible ways of distributing them, and so we say there are 10 microstates.

number of quanta, but *how many* atoms have 1, 2, 3, . . . quanta. The numbers of atoms in various levels are shown in Fig. 13.1 and are given in the second column of Table 13.1. The average occupation numbers N_i are obtained by dividing by the total number of microstates. We can compare the average occupation numbers N_i with what we would expect from the Boltzmann distribution, which gives the correct answer when the number of atoms is very large. We will derive the Boltzmann distribution in Section 13.3, but we will use it here to calculate the occupation numbers N_i. The Boltzmann distribution is

$$N_i = N \frac{e^{-\epsilon_i/kT}}{\sum_i e^{-\epsilon_i/kT}} \tag{13.2}$$

where N_i is the average occupation number of level i, $N = \sum N_i$, and ϵ_i is the energy of level i. Since for a crystal of harmonic oscillators

$$\epsilon_i = v_i h\nu \tag{13.3}$$

then

$$N_i = N \frac{e^{-v_i h\nu/kT}}{\sum_i e^{-v_i h\nu/kT}} \tag{13.4}$$

Table 13.1 Average Occupation N_i of Energy Levels in the Three-Atom Crystal

Level (v)	Number of Atoms in Level	N_i[a]	N_i (Boltzmann Distribution 13.7)
0	12	1.2	1.932
1	9	0.9	0.711
2	6	0.6	0.261
3	3	0.3	0.096
Total		3	3

[a] Number of atoms in level ÷10.

We have to realize that when we analyzed the crystal by counting microstates, we implicitly assumed something about the temperature. In this example we have taken the energy of the crystal to be $Nh\nu$, where N is the number of oscillators in the crystal. Since we know from thermodynamics that the energy of the crystal is NkT (see Section 1.13), we are assuming

$$Nh\nu = NkT \tag{13.5}$$

so that equation 13.4 becomes

$$N_i = N \frac{e^{-v_i}}{\sum_i e^{-v_i}} \tag{13.6}$$

For the three-atom crystal,

$$N_i = \frac{3}{\sum_{i=0}^{3} e^{-v_i}} e^{-v_i} = \frac{3}{1.553} e^{-v_i} \tag{13.7}$$

since $\sum_{i=0}^{3} e^{-v_i} = 1.553$. We would not expect the Boltzmann distribution to give the average occupation N_i of energy levels for this small a crystal, but it gets better and better as we consider larger and larger crystals.

13.2 ENSEMBLES

Gibbs introduced the concept of an ensemble of systems as a means of calculating thermodynamic properties from molecular properties. When this has been done for mechanical properties like pressure, energy, and volume, the nonmechanical thermodynamic properties like entropy and Gibbs energy can be calculated by using the equations of thermodynamics.

We consider a system that has N molecules or atoms in a volume V and is in thermal equilibrium in a heat bath at temperature T. From a macroscopic point of view the system may be described by a few parameters—for example, the pressure, volume, and the temperature. From a microscopic point of view the system may be in one of a very large number of quantum states that are consistent with the thermodynamic properties. In order to calculate a mechanical property (say the pressure) of the system from the microscopic point of view, it would appear to be necessary to know which quantum state the system is in. But since this is impossible, another approach had to be found.

In statistical mechanics the mechanical properties of a system are obtained by calculating the value of the mechanical property for each of the quantum states of the system that is consistent with the thermodynamic state of the system. The average of the values of this mechanical property over all the possible quantum states is postulated to correspond with the thermodynamic property. This averaging process is carried out by means of the ensemble.

An ensemble of systems is the collection of a very large number of systems, each constructed to be a replica on a thermodynamic level of the system of interest.

In a *canonical ensemble** the individual systems have N and V fixed. Such an ensemble is illustrated in Fig. 13.2. The temperature is fixed by immersing the individual system in a heat bath of the desired temperature. Since the systems may exchange energy with the heat bath, they do not have the same energy. After the $\mathscr{A}$ individual systems have come to equilibrium with the heat bath they may be taken from the heat bath and each surrounded by insulation. The total energy of this ensemble is now constant because it is an isolated system. We will represent the total energy of the ensemble containing $\mathscr{A}N$ molecules by $\mathscr{E}$.

An individual system of the canonical ensemble may have an energy corresponding with any one of the energy eigenvalues for the system. The possible energy eigenvalues may be represented by $E_1, E_2, E_3, \ldots$, with the lowest energy represented by E_1 and with any particular energy E_i repeated according to its degeneracy $\Omega(E_i)$.

The state of the entire ensemble may be specified by giving the numbers a_1, a_2, a_3, ... of the systems in states 1, 2, 3, ...

State number	1	2	3	...
Energy	E_1	E_2	E_3	...
Occupation number	a_1	a_2	a_3	...

The set of occupation numbers $a_1, a_2, a_3, \ldots$ is called a distribution.

Any set of occupation numbers for an ensemble has to satisfy two conditions.

$$\mathscr{A} = \sum_i a_i \tag{13.8}$$

$$\mathscr{E} = \sum_i a_i E_i \tag{13.9}$$

The number of systems in an ensemble is fixed, and the total energy of the ensemble is fixed.

There are many possible distributions $a_1, a_2, a_3, \ldots$ that satisfy equations 13.8 and 13.9. According to the *principle of equal a priori probabilities* every distribution

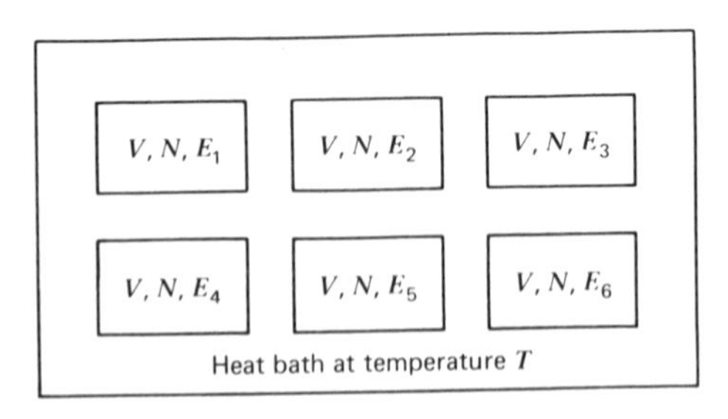

Fig. 13.2 Canonical ensemble of $\mathscr{A}$ systems showing only six of them.

* The other principal types of ensembles are the microcanonical and grand canonical. In the microcanonical ensemble the systems are isolated systems each having the same N, V, and E. In the grand canonical ensemble the systems each have the same V, T, and μ.

that satisfies the two conditions must be given equal weight in calculating an average property for the ensemble.

In order to count the numbers of distributions of different types we again make use of equation 13.1 for the number of ways W that distinguishable objects can arrange in groups so that a_1 are in the first group, a_2 in the second, and so on.

$$W = \frac{\mathscr{A}!}{a_1!\, a_2!\, a_3! \cdots} \tag{13.10}$$

In a particular distribution, $a_i/\mathscr{A}$ of systems in the ensemble are in the ith state, which has energy E_i. The overall probability p_i that a system is in this quantum state is obtained by averaging $a_i/\mathscr{A}$ over all of the possible distributions. This averaging process is simplified by letting $\mathscr{A}$ go to infinity. As $\mathscr{A}$ gets larger and larger one distribution tends more and more to predominate. In the limit $\mathscr{A} \to \infty$, we will represent the most probable distribution by $a_1^*, a_2^*, a_3^*, \ldots$. In this case the probability p_i that a system in the ensemble is in the ith quantum state is given by

$$p_i = a_i^*/\mathscr{A} \tag{13.11}$$

If we can calculate the probabilities p_i we can calculate the mechanical properties of the system represented by the ensemble. For example, the internal energy U of the system under consideration is the average value of the energy, which is given by

$$U = \sum_i p_i E_i \tag{13.12}$$

13.3 DERIVATION OF THE BOLTZMANN DISTRIBUTION

In order to calculate the needed probabilities we must calculate the distribution $a_1^*, a_2^*, a_3^*, \ldots$ that maximizes W.

When we change the a_i to $a_i + da_i$, $\ln W$ changes by

$$d \ln W = \sum_i (\partial \ln W/\partial a_i)\, da_i \tag{13.13}$$

since W depends on all the a_i. At the maximum, $d \ln W$ vanishes, but we cannot simply solve the problem by setting all the $\partial \ln W/\partial a_i = 0$ because equations 13.8 and 13.9 also have to be satisfied. Since $\mathscr{A}$ and $\mathscr{E}$ are constants, these equations require that

$$\sum_i da_i = 0 \tag{13.14}$$

$$\sum_i E_i\, da_i = 0 \tag{13.15}$$

In Lagrange's method of undetermined multipliers the additional constraints are each multiplied by a constant (in this case α and $-\beta$, respectively) and then added to the main variation equation (13.13).

$$d \ln W = \sum_i (\partial \ln W/\partial a_i)\, da_i + \alpha \sum_i da_i - \beta \sum_i E_i\, da_i$$

$$= \sum_i [(\partial \ln W/\partial a_i) + \alpha - \beta E_i]\, da_i \qquad (13.16)$$

Now the variations da_i are treated as independent, and the constants α and β are evaluated later. Since the da_i are independent, $d \ln W = 0$ only if

$$(\partial \ln W/\partial a_i) + \alpha - \beta E_i = 0 \qquad (13.17)^*$$

for every i.

Since the a_i are all large we use Stirling's approximation to eliminate the factorials in equation 13.10. Stirling's approximation

$$\ln N! = N \ln N - N \qquad (13.18)$$

gets more and more accurate as N increases. Using equation 13.18, equation 13.10 may be written

$$\ln W = \mathscr{A} \ln \mathscr{A} - \mathscr{A} - \sum_j a_j \ln a_j + \sum_j a_j$$

$$= \mathscr{A} \ln \mathscr{A} - \sum_j a_j \ln a_j \qquad (13.19)$$

since $\mathscr{A} = \sum_j a_j$. Differentiating with respect to a_i gives

$$\frac{\partial \ln W}{\partial a_i} = -\sum_i \frac{\partial(a_j \ln a_j)}{\partial a_i} = -\sum_i (\ln a_i + 1) \qquad (13.20)$$

Since a_j is independent of a_i unless $j = i$, $\partial a_j/\partial a_i = 0$ except when $j = i$, $\partial a_i/\partial a_i = 1$. Neglecting the unity in comparison with $\ln a_i$, equation 13.17 becomes

$$-\ln a_i + \alpha - \beta E_i = 0 \qquad (13.31)$$

The occupation numbers a_i satisfying this equation are the most probable occupation numbers, and will be designated by a_i^*. Thus

$$a_i^* = e^{\alpha} e^{-\beta E_i} \qquad (13.22)$$

Since $\mathscr{A} = \sum_i a_i^*$,

$$\mathscr{A} = e^{-\alpha} \sum_i e^{-\beta E_i} = e^{-\alpha} Q \qquad (13.23)$$

where

$$Q = \sum_i e^{-\beta E_i} \qquad (13.24)$$

---

* Strictly, the set of a_i that satisfies equation 13.17 just determines an *extremum* of $\ln W$; that is, $\ln W$ may be a maximum or minimum. It may be shown that the solution to equation 13.17 corresponds to a maximum of $\ln W$.

is the *canonical ensemble partition function.* Thus equation 13.22 becomes

$$a_i^* = \frac{\mathscr{A}}{Q} e^{-\beta E_i} \tag{13.25}$$

which is the Boltzmann distribution that we have been seeking.*

It is sometimes convenient to write the partition function in terms of allowed energy levels E_j instead of the energies E_i of quantum states.

$$Q = \sum_{j(\text{levels})} \Omega_j e^{-\beta E_j} \tag{13.26}$$

where Ω_j is the degeneracy (Section 8.7) of the jth level.

The exponential factor $e^{-\beta E_i}$ is called the Boltzmann factor. β has to be a positive quantity so that it will be possible to normalize as shown above. The reason behind the exponential decrease with E_i is that as one system gets more energy, the number of states available to the other $\mathscr{N} - 1$ systems is substantially reduced because of the constraint of constant energy.

We will be able to evaluate β in the next section.

13.4 INTERNAL ENERGY AND ENTROPY

As pointed out at the end of Section 13.1, the value of a mechanical property for a system can be obtained by summing p_i times the mechanical property of state i over all of the states of the ensemble. For the internal energy U,

$$U = \sum_i p_i E_i \tag{13.27}$$

In the symbols of the preceding section $p_i = a_i^*/\mathscr{A}$ and $U = \mathscr{E}/\mathscr{A}$. Therefore, substituting from equation 13.25, we obtain

$$U = \frac{1}{Q} \sum_i E_i e^{-\beta E_i} \tag{13.28}$$

Since the numerator is the negative derivative of the denominator with respect to β,

$$\sum_i E_i e^{-\beta E_i} = -\left(\frac{\partial}{\partial \beta} \sum_i e^{-\beta E_i}\right)_{V,N} \tag{13.29}$$

we obtain

$$U = -\frac{1}{Q}\left(\frac{\partial Q}{\partial \beta}\right)_{V,N} = -\left(\frac{\partial \ln Q}{\partial \beta}\right)_{V,N} \tag{13.30}$$

---

* In Section 13.1 we have already used a Boltzmann distribution

$$N_i = \frac{N e^{-\epsilon_i/kT}}{\sum_i e^{-\epsilon_i/kT}}$$

where N_i is the number of atoms in state i and ϵ_i is the energy of state i.

The differentiation is carried out at constant volume because the energy levels E_i depend on the volume (see equation 8.36).

In order to evaluate β we will compare the relations of statistical mechanics with the combined first and second laws of thermodynamics for a closed system with only pressure-volume work, that is, $dU = T\,dS - P\,dV$.

In statistical mechanical terms, if V and T are changed reversibly at constant N for our system, the differential of U may be calculated from equation 13.27.

$$dU = \sum_i E_i\,dp_i + \sum_i p_i\,dE_i \tag{13.31}$$

The first term is due to changes that do not affect the energy levels but involve a change in the populations of the various energy levels; thus it comes from thermal changes. The second term does involve a change in energy levels, and this suggests that it corresponds with work done on the system.

Now we want to substitute the expression for E_i from the Boltzmann distribution into the first term of equation 13.31. Replacing $a_i^*/\mathscr{A}$ in equation 13.25 by p_i and taking the logarithm of that equation,

$$E_i = -\frac{1}{\beta}(\ln p_i + \ln Q) \tag{13.32}$$

Substituting this expression into the first term of equation 13.31 and recognizing in the second that the energy states E_i at constant N_i can only be changed by changing the volume, we obtain

$$dU = -\frac{1}{\beta}\sum_i(\ln p_i + \ln Q)\,dp_i + \sum_i p_i\left(\frac{\partial E_i}{\partial V}\right)_N dV \tag{13.33}$$

This equation may be simplified by noting that $\ln Q \sum dp_i = 0$, since $\sum dp_i = 0$ from $\sum p_i = 1$. The first term may be rewritten using

$$d\left(\sum_i p_i \ln p_i\right) = \sum_i dp_i + \sum_i \ln p_i\,dp_i = \sum_i \ln p_i\,dp_i \tag{13.34}$$

so that equation 13.33 becomes

$$dU = -\frac{1}{\beta}\,d\left(\sum \ln p_i\,dp_i\right) - P\,dV \tag{13.35}$$

where the pressure P_i of state i is

$$P_i = -\left(\frac{\partial E_i}{\partial V}\right)_N \tag{13.36}$$

This relation comes from the fact that $-P_i\,dV = dE_i$ is the work that has to be done on the system to change the energy of the ith state by dE_i and

$$P = \sum_i p_i P_i \tag{13.37}$$

Comparing equation 13.35 with $dU = T\,dS - P\,dV$ indicates that

$$T\,dS = -\frac{1}{\beta}\,d\left(\sum_i p_i \ln p_i\right) \tag{13.38}$$

Thus β is inversely proportional to the temperature T introduced in the second law.

$$\beta = \frac{1}{kT} \tag{13.39}$$

where the proportionality constant k depends on the size of the unit of temperature. The Boltzmann constant ($k = R/N_A$) is used because this brings statistical mechanics into agreement with the temperature scale of the second law.

According to equation 13.38, the entropy is given by

$$S = -k \sum_i p_i \ln p_i \qquad (13.40)^*$$

Thus the entropy depends only on the probabilities p_i of the various states accessible to a system. Since the probabilities are fractions, the entropy is necessarily a positive quantity. If there is only one possible state (e.g., a perfect crystal at absolute zero), $S = 0$.

13.5 EXPRESSION OF THERMODYNAMIC PROPERTIES IN TERMS OF THE PARTITION FUNCTION Q

With the identification $\beta = 1/kT$, equation 13.30 for the internal energy U may be written

$$U = kT^2\left(\frac{\partial \ln Q}{\partial T}\right)_{V,N} \tag{13.41}$$

The expression for the entropy in terms of the partition function may be derived from the expression for the entropy in terms of the probabilities of the various states (equation 13.40). Substituting $\beta = 1/kT$ in equation 13.32, solving for $\ln p_i$, and substituting it into equation 13.30 yields

$$S = -k \sum_i p_i\left(-\frac{E_i}{kT} - \ln Q\right) = \frac{1}{T}\sum_i p_i E_i + k \ln Q \sum_i p_i \tag{13.42}$$

---

* This expression may be used to derive the relation $S = k \ln \Omega$ used in Section 2.24 as follows: if all of the members of the ensemble have the same energy E_i, they all have the same degeneracy, $\Omega_j = \Omega$. Since every state of a member of the ensemble has equal a priori probability, $p_i = 1/\Omega$. Therefore the summation of equation 13.40 consists of Ω identical terms

$$S = -k\Omega\left(\frac{1}{\Omega}\ln\frac{1}{\Omega}\right) = k \ln \Omega$$

There should be a constant of integration in equation 13.40, which turns out to be equal to zero.

Since $\sum_i p_i E_i = U$ (equation 13.27) and $\sum_i p_i = 1$,

$$S = k \ln Q + \frac{U}{T} \tag{13.43}$$

Since $A = U - TS$,

$$A = -kT \ln Q \tag{13.44}$$

Since pressure $P = -(\partial A/\partial V)_{T,N}$,

$$P = kT\left(\frac{\partial \ln Q}{\partial V}\right)_{T,N} \tag{13.45}$$

Since $H = U + PV$

$$H = kT^2\left(\frac{\partial \ln Q}{\partial T}\right)_{V,N} + PV \tag{13.46}$$

Since $G = H - TS$

$$G = -kT \ln Q + PV \tag{13.47}$$

Now we need to discuss the calculation of Q so that these expressions for thermodynamic properties can be used. Up to this point the theory is perfectly general and applies to any collection of molecules.

13.6 EVALUATION OF THE PARTITION FUNCTION

For an ideal gas, the energy of the system is given by the sum of the energies of the individual molecules. The evaluation of the partition function is very much more complicated for systems in which the molecules interact strongly. If it is possible to label the independent particles $a, b, c, \ldots$ (i.e., there are localized subsystems), the possible energy eigenvalues for the whole system are given by

$$E_i = \epsilon_{ai} + \epsilon_{bi} + \cdots \tag{13.48}$$

where ϵ_{ai} is the energy of molecule a in one of its possible states. In calculating the canonical ensemble partition function all of the possible energy eigenvalues for the systems are to be used. The following example shows that all possible combinations of energies of individual molecules are included if the canonical ensemble partition function is written as the product of the molecular partition functions q of the individual particles.

$$Q = q_a q_b \cdots \tag{13.49}$$

where

$$q_a = \sum_i e^{-\epsilon_{ai}/kT} \tag{13.50}^*$$

$$q_b = \sum_i e^{-\epsilon_{bi}/kT} \tag{13.51}$$

* Molecular partition functions were used in Section 4.9.

Since each term $e^{-\epsilon_i/kT}$ of the partition function for a molecule has a value between zero and unity, the partition function q may be interpreted approximately as the number of quantum states available to the molecule at temperature T.

Example 13.1 Verify that the product of summations indicated in equation 13.43 does indeed yield the indicated summation. Taking the first two terms in the molecular partition functions,

$$q_a q_b = (e^{-\epsilon_{a1}/kT} + e^{-\epsilon_{a2}/kT})(e^{-\epsilon_{b1}/kT} + e^{-\epsilon_{b2}/kT})$$
$$= e^{-(\epsilon_{a1}+\epsilon_{b1})/kT} + e^{-(\epsilon_{a1}+\epsilon_{b2})/kT} + e^{-(\epsilon_{a2}+\epsilon_{b1})/kT} + e^{-(\epsilon_{a2}+\epsilon_{b2})/kT}$$

Thus all possible combinations of molecular energy states are obtained.

Molecular partition functions may also be written in terms of summations over levels j by including the degeneracies g_j.

$$q = \sum_j g_j e^{-\epsilon_j/kT} \tag{13.52}$$

If molecules $a, b, \ldots$ are identical their molecular partition functions are identical so that for N independent molecules the canonical ensemble partition function is given by

$$Q = q^N \quad \text{(localized subsystems)} \tag{13.53}$$

This relation is limited to localized subsystems (molecules) because we have assumed that molecules can be identified as $a, b, \ldots$. If molecules are in a crystal they are localized and may be identified by position.

If molecules are in a gas they are not localized and, according to quantum mechanics, they are indistinguishable. In order to see how equation 13.53 has to be modified for gases, let us consider a gas in which one molecule has quantum numbers n_1', n_2', n_3' and another has quantum number n_1'', n_2'', n_3''. In deriving equation 13.53 we have assumed that the state of the system is different when the two molecules are interchanged so that the first has quantum numbers n_1'', n_2'', n_3'' and the second has quantum numbers n_1', n_2', n_3'. However, this is not so, because the molecules are identical and the state of the system is the same in the two cases. Fortunately it is easy to make a correction because, in a gas, the number of microstates is far greater than the number of molecules so that it is very unlikely that there will be more than one molecule in a quantum state. Therefore, a correction to equation 13.53 can be made by dividing by $N!$, the number of ways of permuting N molecules.

$$Q = \frac{q^N}{N!} \quad \text{(nonlocalized subsystems)} \tag{13.54}$$

The molecular partition function q may be calculated from molecular properties calculated from quantum mechanics or deduced from spectroscopy. As discussed in Chapter 11 on spectroscopy, a molecule in a gas has various types of energy: translational, rotational, vibrational, and electronic. To a fairly high degree of approximation it may be considered that the total energy of a molecule in a particular state is the sum of these various types of energy:

$$\epsilon_i = \epsilon_{ti} + \epsilon_{ri} + \epsilon_{vi} + \epsilon_{ei} \tag{13.55}$$

Writing the energy as a sum of terms is an approximation because the rotational, vibrational, and electronic energy levels are not completely independent. The rotational energy levels depend on the vibrational state, and the vibrational energy levels depend on the rotational state. In general, the rotational- and vibrational-energy level schemes depend on the electronic state. At normal temperatures, however, only the ground electronic state, or a group of electronic states with nearly the same energy, need be considered in calculating the molecular partition function. Substituting equation 13.55 into equation 13.50 and using $e^{a+b+c} = e^a e^b e^c$, we obtain

$$q = \sum_i e^{-\epsilon_{ti}/kT} \sum_i e^{-\epsilon_{ri}/kT} \sum_i e^{-\epsilon_{vi}/kT} \sum_i e^{-\epsilon_{ei}/kT} \tag{13.56}$$

or

$$q = q_t q_r q_v q_e \tag{13.57}$$

where

$$q_t = \sum_i e^{-\epsilon_{ti}/kT} \tag{13.58}$$

etc. To this degree of approximation the canonical ensemble partition function is given by

$$Q = \frac{1}{N!}(q_t q_r q_v q_e)^N = Q_t Q_r Q_v Q_e \tag{13.59}$$

or

$$\ln Q = \ln Q_t + \ln Q_r + \ln Q_v + \ln Q_e \tag{13.60}$$

where

$$Q_t = \frac{q_t{}^N}{N!} \tag{13.61}$$

$$Q_r = q_r{}^N \tag{13.62}$$

$$Q_v = q_v{}^N \tag{13.63}$$

$$Q_e = q_e{}^N \tag{13.64}$$

The $N!$ is included in the translational factor for the reasons described in connection with equation 13.54.

Since the canonical ensemble partition function can be written as the product of translational, rotational, vibrational, and electronic factors (equation 13.59), the thermodynamic properties are sums of translational, rotational, vibrational, and electronic terms. For example, substitution of equation 13.60 into equation 13.41 yields

$$U = kT^2\left(\frac{\partial \ln Q_t}{\partial T}\right)_{V,N} + kT^2\left(\frac{\partial \ln Q_r}{\partial T}\right)_{V,N} + kT^2\left(\frac{\partial \ln Q_v}{\partial T}\right)_{V,N} + kT^2\left(\frac{\partial \ln Q_e}{\partial T}\right)_{V,N} \tag{13.65}$$

$$U = U_t + U_r + U_v + U_e \tag{13.66}$$

Similar equations can be written for the other thermodynamic properties.

13.7 TRANSLATIONAL PARTITION FUNCTION FOR A PERFECT GAS

A molecule in a box can be in one of a number of possible energy levels. The quantum mechanical energy levels for a molecule of perfect gas in a container with dimensions $a \times b \times c$ are those given in equation 8.36. Substituting the equation for the energies of the allowed quantum states of a molecule of mass m into equation 13.50,

$$\begin{aligned} q_t &= \sum_{n_1=1}^{\infty}\sum_{n_2=1}^{\infty}\sum_{n_3=1}^{\infty} \exp\left[-\frac{h^2}{8mkT}\left(\frac{n_1^2}{a^2}+\frac{n_2^2}{b^2}+\frac{n_3^2}{c^2}\right)\right] \\ &= \sum_{n=1}^{\infty} \exp\left[-\frac{h^2n^2}{8ma^2kT}\right]\sum_{n=1}^{\infty} \exp\left[-\frac{h^2n^2}{8mb^2kT}\right]\sum_{n=1}^{\infty} \exp\left[-\frac{h^2n^2}{8mc^2kT}\right] \end{aligned} \tag{13.67}$$

In order to evaluate these sums, we note that for macroscopic containers the exponents are very small, unless T is very small. Since the successive terms in the summations differ by only small amounts, we may replace the summations by integrations, considering the quantum number n as a continuous variable that is not restricted to integer values.

$$\int_0^{\infty} e^{-h^2x^2/8ma^2kT}\,dx = \frac{(2\pi mkT)^{1/2}a}{h} \tag{13.68}$$

since

$$\int_0^{\infty} e^{-\alpha x^2}\,dx = \frac{1}{2}\left(\frac{\pi}{\alpha}\right)^{1/2} \tag{13.69}$$

Substituting integrals obtained in this way in equation 13.67 and noting that $abc = V$, we obtain the following expression for the translational partition function.

$$q_t = \left(\frac{2\pi mkT}{h^2}\right)^{3/2} V \tag{13.70}$$

Thus the translational partition function for a molecule is proportional to the volume V in which it moves.

Example 13.2 What is the molecular partition function for translational motion for a hydrogen atom at 3000 K in a volume of 0.246 170 5 m³? (This is the volume that would contain 1 mol of perfect gas at this temperature and 1 atm pressure.)

Using equation 13.70

$$q_t = \left[\frac{2\pi(1.0080 \times 10^{-3}\ \text{kg mol}^{-1})(1.380\,662 \times 10^{-23}\ \text{J K}^{-1})(3000\ \text{K})}{(6.626\,176 \times 10^{-34}\ \text{J s})^2}\right]^{3/2}$$

$$\times\ (0.246\,170\,5\ \text{m}^3)$$

$$= 7.691\,15 \times 10^{30}$$

Note that the molecular partition function is a dimensionless quantity and may be interpreted approximately as the number of energy levels accessible to the single hydrogen atom in this volume.

13.8 TRANSLATIONAL CONTRIBUTIONS TO THE THERMODYNAMIC PROPERTIES OF PERFECT GASES

In order to calculate the translational contributions to the thermodynamic properties of perfect gases, we need to calculate the canonical ensemble partition function for translation, which is given by equation 13.61. Substituting equation 13.70 into equation 13.61 yields

$$Q_t = \left(\frac{2\pi mkT}{h^2}\right)^{3N/2}\frac{V^N}{N!} = \left(\frac{2\pi mkT}{h^2}\right)^{3N/2}\frac{V^N e^N}{N^N} \tag{13.71}$$

where the last form is obtained using Stirling's approximation $N! = N^N/e^N$.

This equation may be used to calculate the translational contributions to the thermodynamic properties for a perfect gas using equations 13.41, 13.43, 13.44, 13.45, 13.46, and 13.47.

$$P = kT\left(\frac{\partial \ln Q_t}{\partial V}\right)_{T,N} = \frac{NkT}{V} \tag{13.72}$$

$$U_t = kT^2\left(\frac{\partial \ln Q_t}{\partial T}\right)_{V,N} = \tfrac{3}{2}NkT \tag{13.73}$$

$$(C_V)_t = \left(\frac{\partial U_t}{\partial T}\right)_V = \tfrac{3}{2}Nk \tag{13.74}$$

$$H_t = U_t + PV = \tfrac{3}{2}NkT + NkT = \tfrac{5}{2}NkT \tag{13.75}$$

$$(C_P)_t = \left(\frac{\partial H_t}{\partial T}\right)_P = \tfrac{5}{2}Nk \tag{13.76}$$

$$S_t = k \ln Q_t + \frac{U_t}{T} = \tfrac{5}{2}Nk + Nk \ln\left[\left(\frac{2\pi mkT}{h^2}\right)^{3/2} \frac{V}{N}\right] \tag{13.77}$$

$$A_t = -kT \ln Q_t = -NkT - NkT \ln\left[\left(\frac{2\pi mkT}{h^2}\right)^{3/2} \frac{V}{N}\right] \tag{13.78}$$

$$G_t = NkT - kT \ln Q_t = -NkT \ln\left[\left(\frac{2\pi mkT}{h^2}\right)^{3/2} \frac{V}{N}\right] \tag{13.79}$$

These equations may be used to calculate molar thermodynamic properties by letting N equal Avogadro's constant N_A. To calculate thermodynamic quantities for the standard state, the value of V corresponding to the standard state is used. It is therefore convenient to rewrite these equations for calculating molar thermodynamic properties at the standard pressure of $P°$. To do this N_A has been substituted for N in Nk and, inside the logarithmic term, the quantity V/N_A has been replaced by $kT/P°$. If the standard pressure is 1 atm, $P° = 101{,}325$ Pa when the calculation is carried out in SI units. Rewriting the above equations yields

$$P = \frac{RT}{V} \tag{13.80}$$

$$U_t° = \tfrac{3}{2}RT \tag{13.81}$$

$$(C_V°)_t = \tfrac{3}{2}R \tag{13.82}$$

$$H_t° = \tfrac{5}{2}RT \tag{13.83}$$

$$(C_P°)_t = \tfrac{5}{2}R \tag{13.84}$$

$$S_t° = R\left\{\tfrac{5}{2} + \ln\left[\left(\frac{2\pi mkT}{h^2}\right)^{3/2} \frac{kT}{P°}\right]\right\} \tag{13.85}$$

$$A_t° = -RT\left\{1 + \ln\left[\left(\frac{2\pi mkT}{h^2}\right)^{3/2} \frac{kT}{P°}\right]\right\} \tag{13.86}$$

$$G_t° = -RT \ln\left[\left(\frac{2\pi mkT}{h^2}\right)^{3/2} \frac{kT}{P°}\right] \tag{13.87}$$

The thermodynamic properties $U_t°$, $H_t°$, $A_t°$, and $G_t°$ calculated with these equations approach zero as the thermodynamic temperature approaches zero. Thus we may say that these thermodynamic properties are with respect to a standard state of a hypothetical perfect gas at absolute zero. Since a different standard state is used in chemical thermodynamics we will have to wait until Section 13.13 to compare the results of equations 13.81, 13.83, 13.86, and 13.87 with standard thermodynamic tables.

The value of $S_t°$ approaches $-\infty$ as $T \to 0$, but, of course, the gas condenses before this happens and the entropy of a perfect crystal approaches zero as the thermodynamic temperature approaches zero (Section 2.22).

Example 13.3 What are the translational contributions to C_P°, H°, S°, and G° for $O_2(g)$ at 298.15 K?

$$m = \frac{M}{N_A} = \frac{31.9988 \times 10^{-3}\ \text{kg mol}^{-1}}{6.022\,045 \times 10^{23}\ \text{mol}^{-1}} = 5.313\,61 \times 10^{-26}\ \text{kg}$$

$$(C_P^\circ)_t = \tfrac{5}{2}R = \tfrac{5}{2}(8.314\,41\ \text{J K}^{-1}\ \text{mol}^{-1}) = 20.786\ \text{J K}^{-1}\ \text{mol}^{-1}$$

$$H_t^\circ = \tfrac{5}{2}RT = \tfrac{5}{2}(8.314\,41\ \text{J K}^{-1}\ \text{mol}^{-1})(298.15\ \text{K}) = 6.197\ \text{kJ mol}^{-1}$$

$$\left(\frac{2\pi mkT}{h^2}\right)^{3/2}\frac{kT}{P^\circ}$$

$$= \left[\frac{2\pi(5.313\,61 \times 10^{-26}\ \text{kg})(1.380\,662 \times 10^{-23}\ \text{J K}^{-1})(298.15\ \text{K})}{(6.626\,176 \times 10^{-34}\ \text{J s})^2}\right]^{3/2}$$

$$\times \frac{(1.380\,662 \times 10^{-23}\ \text{J K}^{-1})(298.15\ \text{K})}{101\,325\ \text{N m}^{-2}} = 7.114\,66 \times 10^6$$

$$S_t^\circ = (8.314\,41\ \text{J K}^{-1}\ \text{mol}^{-1})(2.5 + \ln 7.114\,66 \times 10^6) = 151.968\ \text{J K}^{-1}\ \text{mol}^{-1}$$

$$G_t^\circ = -(8.314\,41\ \text{J K}^{-1}\ \text{mol}^{-1})(298.15\ \text{K}) \ln 7.114\,66 \times 10^6 = -39.112\ \text{kJ mol}^{-1}$$

When this calculation is repeated at a series of additional temperatures the following results are obtained.*

T/K	$(C_P^\circ)_t$ J K^{-1} mol^{-1}	S_t° J K^{-1} mol^{-1}	H_t° kJ mol^{-1}	G_t° kJ mol^{-1}
298.150	20.786	151.968	6.197	−39.112
500.000	20.786	162.715	10.393	−70.964
1000.000	20.786	177.122	20.876	−156.336
2000.000	20.786	191.530	41.572	−341.488
3000.000	20.786	199.958	62.358	−537.517

Note that H_t° is directly proportional to T.

The translational contribution to the entropy is a linear function of ln T, and the translation contribution to the Gibbs energy is a linear function of $T \ln T$.

The values of $(C_P^\circ)_t$ and S_t° are less than C_P° and S° in Table A.2 because of additional contributions due to rotation and vibration. The values of H_t° and G_t° cannot be compared with ΔH_f° and ΔG_f° in Table A.2 because of the difference in standard states.

Example 13.4 What are the translational contributions to C_P°, S°, H°, and G° for $O(g)$ at 298.15 K?

The translational contributions to C_P° and H° are the same as for $O_2(g)$, or any

* The tables of calculated values in this chapter have been computed using an IBM 5100 Computer with APL programs.

other perfect gas. Because the mass of the atom is one-half that of the diatomic molecule, the translational entropy for the atomic gas is smaller by $-\frac{3}{2}R \ln 2 = -8.645\ \text{J K}^{-1}\ \text{mol}^{-1}$, and the translational Gibbs energy is more positive by $\frac{3}{2}RT \ln 2 = 2.577\ \text{kJ mol}^{-1}$. Thus at 298.15 K

$$(C_P^\circ)_t = 20.786\ \text{J K}^{-1}\ \text{mol}^{-1}$$

$$S_t^\circ = 143.323\ \text{J K}^{-1}\ \text{mol}^{-1}$$

$$H_t^\circ = 6.197\ \text{kJ mol}^{-1}$$

$$G_t^\circ = -36.535\ \text{kJ mol}^{-1}$$

When this calculation is repeated at a series of additional temperatures the following results are obtained.

T/K	$(C_P^\circ)_t$ J K^{-1} mol^{-1}	S_t° J K^{-1} mol^{-1}	H_t° kJ mol^{-1}	G_t° kJ mol^{-1}
298.150	20.786	143.323	6.197	−36.535
500.000	20.786	154.070	10.393	−66.642
1000.000	20.786	168.478	20.786	−147.692
2000.000	20.786	182.886	41.572	−324.199
3000.000	20.786	191.314	62.358	−511.583

13.9 ROTATIONAL CONTRIBUTIONS TO THE THERMODYNAMIC PROPERTIES OF PERFECT GASES

The expression for the rotational energy of a rigid diatomic molecule has already been discussed in Section 11.5.

$$\epsilon_r = \frac{J(J+1)h^2}{8\pi^2 I} \tag{13.88}$$

Here J is the rotational quantum number, and I is the moment of inertia of the diatomic molecule. Although the rotational energy depends only on J, the state of a rigid rotator is specified by the quantum number J and an additional quantum number M, where M can have integral values between $-J$ and $+J$. Thus there are $2J + 1$ values of M for each value of J; in other words, the rotational levels are $(2J + 1)$-fold degenerate. The rotational partition function is thus

$$q_r = \sum_{J=0}^{\infty} (2J+1)e^{-J(J+1)h^2/8\pi^2 IkT} \tag{13.89}$$

For molecules with large moments of inertia, the energy levels are ordinarily so close together that they may be considered to be continuous above 10 to 100 K, so that the summation may be replaced by an integration.

$$q_r = \int_0^\infty (2J + 1)e^{-J(J+1)h^2/8\pi^2IkT}\, dJ \tag{13.90}$$

Since $(2J + 1)\, dJ$ is the differential of $J(J + 1) = J^2 + J$, this equation may be integrated by multiplying and dividing through by $8\pi^2IkT/h^2$ to obtain

$$q_r = \frac{8\pi^2IkT}{h^2}\int_0^\infty e^{-u}\, du = \frac{8\pi^2IkT}{h^2} = \frac{T}{\Theta_r} \tag{13.91}$$

where Θ_r is the characteristic rotational temperature defined by

$$\Theta_r = \frac{h^2}{8\pi^2Ik} \tag{13.92}$$

The calculation of the rotational partition function for a homonuclear diatomic molecule (like H_2) has to take into account the fact that rotation by 180° interchanges two equivalent nuclei. Since the new orientation is indistinguishable from the first, we have to divide by 2 so that the indistinguishable orientations are counted only once. For a heteronuclear diatomic molecule (like HCl), a 180° rotation produces a distinguishable orientation. The effect of symmetry is taken into account by introducing a symmetry number that has the value 2 for a homonuclear molecule and the value 1 for a heteronuclear molecule. In general then

$$q_r = \frac{T}{\sigma\Theta_r} \tag{13.93}$$

Example 13.5 What is the characteristic rotational temperature Θ_r for $H_2(g)$ at 3000 K and what is the value of the molecular partition function for rotation? (The moment of inertia is 4.6054×10^{-47} kg m².)

$$\Theta_r = \frac{h^2}{8\pi^2Ik} = \frac{(6.626\,18 \times 10^{-34}\text{ J s})^2}{8\pi^2(4.6054 \times 10^{-48}\text{ kg m}^2)(1.380\,66 \times 10^{-23}\text{ J K}^{-1})} = 87.547\text{ K}$$

$$q_r = \frac{T}{\sigma\Theta_r} = \frac{3000\text{ K}}{(2)(87.547\text{ K})} = 17.134$$

In order to calculate the rotational contributions to the thermodynamic properties of perfect gases, we need to calculate the canonical ensemble partition function Q_r. Since $Q_r = q_r^N$, equations 13.41 and 13.43 lead to the following expressions for the rotational contributions to the molar thermodynamic quantities for diatomic molecules.

$$U_r^\circ = kT^2\left(\frac{\partial \ln Q_r}{\partial T}\right)_{V,N} = RT \tag{13.94}$$

$$(C_V^\circ)_r = R \tag{13.95}$$

$$H_r^\circ = RT \tag{13.96}$$

$$(C_P^\circ)_r = R \tag{13.97}$$

$$S_r^\circ = k \ln Q_r + U_r/T = R \ln \left(\frac{eT}{\sigma\Theta_r}\right) \tag{13.98}$$

$$G_r^\circ = H_r - TS_r = -RT \ln \left(\frac{T}{\sigma\Theta_r}\right) \tag{13.99}$$

The expression for C_V° and C_P° confirms the expectation from the classical principle of equipartition (Section 1.12) that the two components of rotational motion of a rigid linear molecule each contribute $R/2$ to C_V° or C_P°

Example 13.6 What are the rotational contributions to C_P°, S°, H°, and G° for $O_2(g)$ at 298.15 K? The moment of inertia for $O_2(g)$ is $I = 1.9373 \times 10^{-46}$ kg m^2.

$$\Theta_r = \frac{h^2}{8\pi^2 Ik} = \frac{(6.626\ 18 \times 10^{-34}\ \text{J s})^2}{8\pi^2(1.9373 \times 10^{-46}\ \text{kg m}^2)(1.380\ 66 \times 10^{-23}\ \text{J K}^{-1})} = 2.079\ \text{K}$$

$$(C_P^\circ)_r = R = 8.314\ \text{J K}^{-1}\ \text{mol}^{-1}$$

$$S_r^\circ = R \ln \left(\frac{eT}{\sigma\Theta_r}\right) = (8.314\ 41\ \text{J K}^{-1}\ \text{mol}^{-1}) \ln \left[\frac{(2.7183)(298.15\ \text{K})}{(2)(2.079\ \text{K})}\right]$$

$$= 43.838\ \text{J K}^{-1}\ \text{mol}^{-1}$$

$$H_r^\circ = RT = (8.314\ 41\ \text{J K}^{-1}\ \text{mol}^{-1})(298.15\ \text{K}) = 2.479\ \text{kJ mol}^{-1}$$

$$G_r^\circ = -RT \ln \left(\frac{T}{\sigma\Theta_r}\right) = -(8.314\ 41\ \text{J K}^{-1}\ \text{mol}^{-1})(298.15\ \text{K}) \ln \left(\frac{298.15\ \text{K}}{2 \times 2.079\ \text{K}}\right)$$

$$= -10.591\ \text{kJ mol}^{-1}.$$

When this calculation is repeated at a series of additional temperatures the following results are obtained.

T/K	$(C_P^\circ)_r$ J K^{-1} mol^{-1}	S_r° J K^{-1} mol^{-1}	H_r° kJ mol^{-1}	G_r° kJ mol^{-1}
298.150	8.314	43.838	2.479	−10.591
500.000	8.314	48.137	4.157	−19.911
1000.000	8.314	53.900	8.314	−45.586
2000.000	8.314	59.663	16.629	−102.697
3000.000	8.314	63.034	24.943	−164.160

Equations 13.94 to 13.99 may also be used in calculations on linear polyatomic molecules like CO_2, C_2H_2, N_2O, and COS. The symmetry number is 2 for symmetrical linear molecules like CO_2 and C_2H_2, and it is 1 for unsymmetrical linear molecules like N_2O and COS.

A linear molecule with a small moment of inertia I has a small rotational partition function and a small rotational contribution to $S°$ and $G°$. Molecules with larger amounts of inertia have larger rotational contributions to $S°$ and $G°$. The reason for this is that when the moment of inertia is large, the rotational energy levels are close together and more states are populated.

For polyatomic molecules in general the calculation of rotational contributions has to take into account the fact that the molecule may have three different moments of inertia. As shown in Section 11.6 the three moments of inertia for a spherical top molecule, like CH_4, are equal: $I_a = I_b = I_c$. Two moments are equal for a symmetric top: $I_a < I_b = I_c$ for a prolate top like CH_3Cl, and $I_a = I_b < I_c$ for an oblate top like C_6H_6. For an asymmetric top, like H_2O, the three moments of inertia are all different.

For temperatures above the rotational temperatures it may be shown* that the rotational partition function for a polyatomic molecule is given by

$$q_r = \frac{\pi^{1/2}}{\sigma}\left(\frac{T^3}{\Theta_a\Theta_b\Theta_c}\right)^{1/2} \tag{13.100}$$

where the characteristic rotational temperatures are given by

$$\Theta_a = \frac{h^2}{8\pi^2 I_a k} \tag{13.101a}$$

$$\Theta_b = \frac{h^2}{8\pi^2 I_b k} \tag{13.101b}$$

$$\Theta_c = \frac{h^2}{8\pi^2 I_c k} \tag{13.101c}$$

and σ is the symmetry number. The symmetry number can be obtaincd by counting the number of indistinguishable orientations a molecule can have when acted on by symmetry operations. For NH_3, CH_4, and C_6H_6, σ is 3, 12, and 12, respectively.

Thus the rotational contributions to the thermodynamic properties of nonlinear polyatomic molecules are

$$U_r^\circ = kT^2\left(\frac{\partial \ln q_r^N}{\partial T}\right)_{V,N} = \tfrac{3}{2}RT \tag{13.102}$$

$$(C_V^\circ)_r = \tfrac{3}{2}R \tag{13.103}$$

$$H_r^\circ = \tfrac{3}{2}RT \tag{14.104}$$

$$(C_P^\circ)_r = \tfrac{3}{2}R \tag{13.105}$$

$$S_r^\circ = k \ln q_r^N + U_r^\circ/T = R \ln \left[\frac{\pi^{1/2}}{\sigma}\left(\frac{T^3 e^3}{\Theta_a\Theta_b\Theta_c}\right)^{1/2}\right] \tag{13.106}$$

$$G_r^\circ = H_r^\circ - TS_r^\circ = -RT \ln \left[\frac{\pi^{1/2}}{\sigma}\left(\frac{T^3}{\Theta_a\Theta_b\Theta_c}\right)^{1/2}\right] \tag{13.107}$$

---

* D. McQuarrie, *Statistical Mechanics*, Harper & Row, New York, 1976, p. 133.

Nonlinear polyatomic molecules have three degrees of rotational freedom rather than the two for linear molecules; therefore, the rotational contribution to C_P° is $3R/2$, and the enthalpy is $3RT/2$ relative to the hypothetical perfect gas at absolute zero.

Example 13.7 For $H_2O(g)$, $I_aI_bI_c = 5.7658 \times 10^{-141}$ kg^3 m^6 and $\sigma = 2$. What is the value of the rotational partition function for $H_2O(g)$ and what are the rotational contributions to C_P°, S°, H°, and G° at 3000 K?

$$\Theta_a\Theta_b\Theta_c = \left(\frac{h^2}{8\pi^2 k}\right)^3 \frac{1}{I_aI_bI_c} = \left[\frac{(6.626\,176 \times 10^{-34}\text{ J s})^2}{8\pi^2(1.380\,662 \times 10^{-23}\text{ J K}^{-1})}\right]^3$$

$$\times \frac{1}{5.7658 \times 10^{-141}\text{ kg}^3\text{ m}^6} = 1.133\,15 \times 10^4$$

$$q_r = \frac{\pi^{1/2}}{2}\left(\frac{3000^2}{1.133\,15 \times 10^4}\right)^{1/2} = 13\,680.0$$

$$(C_P^\circ)_r = \tfrac{3}{2}R = \tfrac{3}{2}(8.314\,41\text{ J K}^{-1}\text{ mol}^{-1}) = 12.472\text{ J K}^{-1}\text{ mol}^{-1}$$

$$S_r^\circ = (8.314\,41\text{ J K}^{-1}\text{ mol}^{-1})\ln\left[\frac{\pi^{1/2}}{2}\left(\frac{3000^3 e^3}{1.133\,15 \times 10^4}\right)^{1/2}\right]$$

$$= 72.511\text{ J K}^{-1}\text{ mol}^{-1}$$

$$H_r^\circ = \tfrac{3}{2}RT = \tfrac{3}{2}(8.314\,41\text{ J K}^{-1}\text{ mol}^{-1})(3000\text{ K}) = 37.415\text{ kJ mol}^{-1}$$

$$G_r^\circ = -(8.314\,41\text{ J K}^{-1}\text{ mol}^{-1})(3000\text{ K})\ln\left[\frac{\pi^{1/2}}{2}\left(\frac{3000^3}{1.133\,15 \times 10^4}\right)^{1/2}\right]$$

$$= -180.118\text{ kJ mol}^{-1}.$$

When this calculation is repeated at a series of temperatures the following results are obtained.

T/K	$(C_P^\circ)_r$ J K^{-1} mol^{-1}	S_r° J K^{-1} mol^{-1}	H_r° kJ mol^{-1}	G_r° kJ mol^{-1}
298.150	12.472	43.717	3.718	−9.316
500.000	12.472	50.165	6.236	−18.847
1000.000	12.472	58.809	12.472	−46.338
2000.000	12.472	67.454	24.943	−109.965
3000.000	12.472	72.511	37.415	−180.118

13.10 VIBRATIONAL CONTRIBUTIONS TO THE THERMODYNAMIC PROPERTIES OF PERFECT GASES

The vibrational levels of diatomic molecules have been discussed in Section 11.8 on spectroscopy. If there is excitation to very high vibrational levels, the anharmonicity has to be taken into account in calculating the partition function, but, for many purposes, it is sufficient to use the harmonic oscillator approximation because only lower levels are occupied. For the purpose of calculating the partition function, vibrational energy is usually measured from the ground state ($v = 0$) rather than from the bottom of the potential energy curve. This simplifies the equations, because the *zero* point energy ($h\omega_0/2$, where ω_0 is the fundamental vibration frequency) does not appear.

In the harmonic oscillator approximation

$$\epsilon_v = vh\omega \qquad v = 0, 1, 2, \ldots$$

so that

$$q_v = \sum_{v=0}^{\infty} e^{-vh\omega/kT} = 1 + x + x^2 + x^3 + \cdots \tag{13.108}$$

where $x = e^{-h\omega/kT}$. When $x < 1$,

$$\frac{1}{1 - x} = 1 + x + x^2 + x^3 + \cdots \tag{13.109}$$

Since $e^{-h\omega/kT} < 1$,

$$q_v = \frac{1}{1 - e^{-h\omega/kT}} = \frac{1}{1 - e^{-\Theta_v/T}} \tag{13.110}$$

where Θ_v is the characteristic vibrational temperature; $\Theta_v = h\omega/k = hc\tilde{\omega}/k$. The characteristic temperature of a vibration Θ_v is the temperature at which the energy of the vibrational quantum is equal to kT.

Since $Q_v = q_v{}^N$, the vibrational contributions to the various thermodynamic quantities for a diatomic molecules are at follows:

$$U_v^\circ = RT^2\left(\frac{\partial \ln q_v}{\partial T}\right)_{N,V} = RT\,\frac{(\Theta_v/T)}{e^{\Theta_v/T} - 1} \tag{13.111}$$

$$(C_V^\circ)_v = \frac{\partial U_v^\circ}{\partial T} = R\left(\frac{\Theta_v}{T}\right)^2 \frac{e^{\Theta_v/T}}{(e^{\Theta_v/T} - 1)^2} \tag{13.112}$$

$$H_v^\circ = RT\,\frac{(\Theta_v/T)}{e^{\Theta_v/T} - 1} \tag{13.113}$$

$$(C_p^\circ)_v = R\left(\frac{\Theta_v}{T}\right)^2 \frac{e^{\Theta_v/T}}{(e^{\Theta_v/T} - 1)^2} \tag{13.114}$$

$$S_v^\circ = R\left[\frac{(\Theta_v/T)}{e^{\Theta_v/T} - 1} - \ln(1 - e^{-\Theta_v/T})\right] \tag{13.115}$$

$$G_v^\circ = RT\ln(1 - e^{-\Theta_v/T}) \tag{13.116}$$

Equation 13.112 explains why vibrational motion is not fully excited at ordinary temperatures. As mentioned in Section 1.12, classical mechanics was not able to explain this fact. As the temperature increases the vibrational contribution to the heat capacity approaches R in accordance with the principle of equipartition (Section 1.12).

Example 13.8 What are the vibrational contributions to C_P°, S°, H°, and G° for $O_2(g)$ at 298.15 K? The vibrational frequency is 1580.246 cm^{-1}.

$$\Theta_v = \frac{hc}{k} = \frac{(6.26\,176 \times 10^{-34}\,\text{J s})(2.997\,925 \times 10^9\,\text{m s}^{-1})(1.580\,246 \times 10^5\,\text{m}^{-1})}{1.380\,662 \times 10^{-23}\,\text{J K}^{-1}}$$

$$= 2273.64\ \text{K}$$

$$(C_P^\circ)_v = (8.314\,41\,\text{J K}^{-1}\,\text{mol}^{-1})\left(\frac{2273.64}{298.15}\right)^2 \frac{e^{2273.64/298.15}}{(e^{2273.64/298.15} - 1)^2}$$

$$= 0.236\,\text{J K}^{-1}\,\text{mol}^{-1}$$

$$S_v^\circ = (8.314\,41\,\text{J K}^{-1}\,\text{mol}^{-1})\left[\frac{(2273.64)/(298.15)}{e^{2273.64/298.15} - 1} - \ln\left(1 - e^{-2273.64/298.15}\right)\right]$$

$$= 0.035\,\text{J K}^{-1}\,\text{mol}^{-1}$$

$$H_v^\circ = (8.314\,41\,\text{J K}^{-1}\,\text{mol}^{-1})(298.15\,\text{K})\,\frac{(2273.64\,\text{K})/(298.15\,\text{K})}{e^{2273.64/298.15} - 1}$$

$$= 0.009\,\text{kJ mol}^{-1}$$

$$G_v^\circ = (8.314\,41\,\text{J K}^{-1}\,\text{mol}^{-1})(298.15\,\text{K})\ln\left(1 - e^{-2273.64/298.15}\right)$$
$$= -0.001\,\text{kJ mol}^{-1}.$$

When this calculation is repeated at a series of additional temperatures the following results are obtained.

T/K	$(C_P^\circ)_v$ J K^{-1} mol^{-1}	S_v° J K^{-1} mol^{-1}	H_v° kJ mol^{-1}	G_v° kJ mol^{-1}
298.150	0.236	0.035	0.009	−0.001
500.000	1.861	0.493	0.202	−0.044
1000.000	5.498	3.072	2.169	−0.903
2000.000	7.474	7.862	8.930	−6.434
3000.000	7.928	10.816	16.674	−15.773

The small value of the vibrational contribution at 298.15 K indicates that $O_2(g)$ has very little vibrational energy at this temperature. The classical vibrational heat capacity is $R = 8.314\,\text{J K}^{-1}\,\text{mol}^{-1}$, and this value of $(C_p^\circ)_v$ is approached at 3000 K.

For polyatomic molecules there are $3N - 5$ normal modes of vibration for linear molecules and $3N - 6$ normal modes for nonlinear molecules, as we have seen in Section 11.10. Each normal mode makes an independent contribution to the thermodynamic properties. If a normal mode is degenerate, the contribution is multiplied by the degeneracy. Therefore, the vibrational contributions to the thermodynamic properties for polyatomic molecules are obtained by summing terms of the types shown in equations 13.111 to 13.116.

13.11 ELECTRONIC CONTRIBUTIONS TO THE THERMODYNAMIC PROPERTIES OF PERFECT GASES

The electronic partition function q_e is given by

$$q_e = \sum_i e^{-\epsilon_i/kT} \tag{13.117}$$

where the summation is carried over all electonic states. This partition function is more generally written as a sum over levels instead of ovr states. Thus, if the degeneracy of the ground state is g_0 and the degeneracy of the first excited state is g_1,

$$q_e = g_0 + g_1 e^{-\epsilon_1/kT} + \cdots \qquad (13.118)*$$

where the energies of the excited states are measured with respect to the energy of the ground state. If the energies of the excited states are all large with respect to kT, only the degeneracy g_0 of the ground state has to be taken into account in calculating thermodynamic properties. The electronic contribution to the canonical ensemble partition function Q_e is $q_e{}^N$ (see equation 13.64). Thus the degeneracy g_0 of the ground state of an atom adds $R \ln g_0$ to the entropy, $-RT \ln g_0$ to the Helmholtz and Gibbs energies, but nothing to C_P° and H°.

If an atom has low-lying electronic levels, its partition function has to be calculated by summing equation 13.118. In this case the contribution to S° is

$$S_e^\circ = R \ln (g_0 + g_1 e^{-\epsilon_1/kT} + \cdots) \tag{13.119}$$

ignoring a term in $d \ln q_e/dT$. The contribution to G° is

$$G_e^\circ = -RT \ln (g_0 + g_1 e^{-\epsilon_1/kT} + \cdots) \tag{13.120}$$

When electronic levels have energies greater than about 10^4 cm^{-1} their contributions to the thermodynamic properties at $T < 3000$ K are negligible. We will not

---

* S. J. Strickler, *J. Chem. Ed.*, **43**, 364 (1966), discusses the fact that the electronic partition function q_e for the hydrogen atom at 25 °C calculated from

$$q_e = \sum_{n=1}^{\infty} n^2 e^{-\epsilon_n/kT}$$

is infinite. In this equation n^2 is the degeneracy, and ϵ_n is the energy of the orbital with quantum number n. This is contrary to experience, since it would require the population of the ground state to be zero. The problem is that there is an infinite number of terms in the partition function with an energy less than 13.60 eV. The paradox is resolved by considering the size of the orbital of a hydrogen atom for a very large quantum number n. For a finite container the maximum n is finite, and the value of q_e is not detectably different from unity. In fact, the cutoff probably comes from neighboring molecules that limit the size of a hydrogen atom to about $(V/N)^{1/3}$.

deal with the electronic contributions to H° and C_P° because differentiation of q_e once or twice with respect to T yields very complicated expressions.

Example 13.9 What are the electronic contributions to S° and G° for O(g) at 298.15 K to the degree of completeness we have used here? The degeneracy in the ground state is 5, the first excited state has an energy of 158.2 cm^{-1} with respect to the ground state and has a degeneracy of 3, and the second excited has an energy of 226.5 cm^{-1} with no degeneracy.

$$\Theta_{e1} = \frac{hc\tilde{\mu}}{k} = \frac{(6.626\,176 \times 10^{-34}\,\text{J s})(2.997\,925 \times 10^{8}\,\text{m s}^{-1})(1.582 \times 10^{4}\,\text{m}^{-1})}{1.380\,662 \times 10^{-23}\,\text{J K}^{-1}}$$

$$= 228.1\text{ K}$$

$$\Theta_{e2} = 325.9\text{ K}$$

$$q_e = 5 + 3e^{-228.1/298.15} + e^{-325.1/298.15} = 6.7311$$

$$S_e^\circ = R \ln q_e = (8.314\,41\,\text{J K}^{-1}\,\text{mol}^{-1}) \ln 6.7311 = 15.853\,\text{J K}^{-1}\,\text{mol}^{-1}$$

$$G_e^\circ = -TS_e^\circ = -(298.15\text{ K})(15.853\,\text{J K}^{-1}\,\text{mol}^{-1}) = -4.726\text{ kJ mol}^{-1}$$

When this calculation is repeated at an additional series of temperatures the following results are obtained.

T/K	$(C_P^\circ)_e^*$ J K^{-1} mol^{-1}	S_e° J K^{-1} mol^{-1}	H_e° kJ mol^{-1}	G_e° kJ mol^{-1}
298.150	0	15.853	0	−4.727
500.000	0	16.666	0	−8.333
1000.000	0	17.403	0	−17.403
2000.000	0	17.819	0	−35.638
3000.000	0	17.965	0	−53.895

As the temperature increases the electronic contribution to the entropy approaches $S_e = R \ln (g_0 + g_1 + g_2 + \ldots)$.

In calculating the electronic partition function for a diatomic molecule it is convenient to measure energy with respect to the atoms in a perfect gas state at absolute zero. When this is done the calculated thermodynamic properties of the molecule and its constituent atoms are relative to the same reference state. On this basis the energy of the electronic ground state of the molecule is $-D^\circ$, where D° is the spectroscopic dissociation energy (Section 10.1). The energy of the molecule in its first excited state is then $-D^\circ + \epsilon_1$, where ϵ_1 is the energy with respect to the ground state. Equation 13.118 may therefore be written

$$q_e = g_0\, e^{D^\circ/kT} + g_1 e^{(D^\circ - \epsilon_1)/kT} + \cdots$$

$$= e^{D^\circ/kT}(g_0 + g_1 e^{-\epsilon_1/kT} + \cdots) \tag{13.121}$$

* We have chosen to neglect the electronic contributions to C_P° and H°, but they are not zero.

In this case the contributions to the thermodynamic functions are

$$(C_P^\circ)_e = 0 \tag{13.122}$$

ignoring the term in equation 13.21 in parentheses,

$$H_e^\circ = -D^\circ \tag{13.123}$$

ignoring the term in equation 13.121 in parentheses,

$$S_e^\circ = R \ln (g_0 + g_1 e^{-\epsilon_1/kT} + \cdots) \tag{13.124}$$

ignoring a term involving $d \ln q_e/dT$, and

$$G_e^\circ = -RT \ln (g_0 + g_1 e^{-\epsilon_1/kT} + \cdots) \tag{13.125}$$

Example 13.10 What are the electronic contributions to C_P°, H°, S°, and G° for $O_2(g)$ at 298.15 K? The electronic ground state is a triplet, and the spectroscopic dissociation energy is 491.888 kJ mol^{-1}. The energy of the first excited electronic state is so high that it does not have to be considered.

$$(C_P^\circ)_e = 0$$

$$H_e^\circ = -D = -491.888 \text{ kJ mol}^{-1}$$

$$S_e^\circ = (8.314\,41 \text{ J K}^{-1} \text{ mol}^{-1}) \ln 3 = 9.134 \text{ J K}^{-1} \text{ mol}^{-1}$$

$$\begin{aligned} G_e^\circ &= -491.888 \text{ kJ mol}^{-1} - (8.314\,41 \times 10^{-3} \text{ kJ K}^{-1} \text{ mol}^{-1})(298.15 \text{ K}) \ln 3 \\ &= -494.611 \text{ kJ mol}^{-1} \end{aligned}$$

The electronic contributions to thermodynamic quantities at a series of temperatures are as follows.

T/K	$(C_P^\circ)_e$ J K^{-1} mol^{-1}	S_e° J K^{-1} mol^{-1}	H_e° kJ mol^{-1}	G_e° kJ mol^{-1}
298.150	0	9.134	−491.888	−494.611
500.000	0	9.134	−491.888	−496.455
1000.000	0	9.134	−491.888	−501.022
2000.000	0	9.134	−491.888	−510.157
3000.000	0	9.134	−497.888	−519.291

13.12 THERMODYNAMIC PROPERTIES OF PERFECT GASES

Now that we have shown how to calculate the translational, rotational, vibrational, and electronic contributions to the thermodynamic properties, we can obtain the properties by summing up the contributions, as shown, for example, in equation 13.66.

Example 13.11 What are the values of C_P°, S°, H°, and G° for $O_2(g)$ at 298.15 K?

These values are obtained by adding the translational, rotational, vibrational, and electronic contributions.

$$C_P^\circ = (C_P^\circ)_t + (C_P^\circ)_r + (C_P^\circ)_v + (C_P^\circ)_e = (20.786 + 8.314 + 0.236 + 0)\ \text{J K}^{-1}\ \text{mol}^{-1}$$
$$= 29.336\ \text{J K}^{-1}\ \text{mol}^{-1}$$

$$S^\circ = S_t^\circ + S_r^\circ + S_v^\circ + S_e^\circ = (151.968 + 43.838 + 0.035 + 9.134)\ \text{J K}^{-1}\ \text{mol}^{-1}$$
$$= 204.976\ \text{J K}^{-1}\ \text{mol}^{-1}$$

$$H^\circ = H_t^\circ + H_r^\circ + H_v^\circ + H_e^\circ = (6.197 + 2.479 + 0.009 - 491.888)\ \text{kJ mol}^{-1}$$
$$= -483.203\ \text{kJ mol}^{-1}$$

$$G^\circ = G_t^\circ + G_r^\circ + G_v^\circ + G_e^\circ = (-39.112 - 10.591 - 0.001 - 494.611)\ \text{kJ mol}^{-1}$$
$$= -544.316\ \text{kJ mol}^{-1}$$

Since the value of C_P° calculated here is in good agreement with the value in Table A.2 (29.372 J K^{-1} mol^{-1}), the electronic contribution that we ignored is small at this temperature. The value of S° is also in good agreement with the value in Table A.2 (205.033 J K^{-1} mol^{-1}). The values of H° and G° calculated here cannot be compared with ΔH_f° and ΔG_f° because of the difference in standard states, but we will calculate ΔH_f° and ΔG_f° in Example 13.13.

Example 13.12 What are the values of C_P°, S°, H°, and G° for O(g) at 298.15 K?

Atoms have only translational and electronic contributions, and therefore we sum the contributions calculated in Examples 13.4 and 13.9.

$$C_P^\circ = (C_P^\circ)_t + (C_P^\circ)_e = 20.786\ \text{J K}^{-1}\ \text{mol}^{-1} \qquad \text{since } (C_P^\circ)_e \text{ is taken here as zero}$$

$$S^\circ = S_t^\circ + S_e^\circ = (143.323 + 15.853)\ \text{J K}^{-1}\ \text{mol}^{-1} = 159.177\ \text{J K}^{-1}\ \text{mol}^{-1}$$

$$H^\circ = H_t^\circ + H_e^\circ = 6.197\ \text{kJ mol}^{-1} \qquad \text{since } H_e^\circ \text{ is taken here as zero}$$

$$G^\circ = G_t^\circ + G_e^\circ = (-36.535 - 4.727)\ \text{kJ mol}^{-1} = -41.261\ \text{kJ mol}^{-1}$$

Molecular parameters of a number of gases are given in Table 13.1. The dissociation energies D° are with respect to the corresponding atoms in the ground state. For diatomic molecules these dissociation energies have been obtained from spectra. For polyatomic molecules the values of D° have been obtained from thermochemical measurements and theoretical calculations of heat capacities of gases. In Table 13.1 values of Θ_e only up to about 10,000 K are given.

Table 13.1 Molecular Parameters of Gases[a]

	M/g mol^{-1}	D°/kJ mol^{-1}	σ	Θ_r/K[b]	Θ_v/K[c]	g_0	Θ_e/K[c]
H	1.0080	0				2	
C	12.001	0				1	23.6(3)
							62.6(5)
N	14.008	0				4	

Table 13.1—(*Continued*)

	M/g mol^{-1}	$D°$/kJ mol^{-1}	σ	Θ_r/K[b]	Θ_v/K[c]	g_0	Θ_e/K[c]
O	15.994	0				5	228.1(3) 325.9
Cl	35.453	0				2	1269.53(2)
I	126.9045	0				4	10 939.3(2)
H_2	2.016	432.073	2	87.547	6338.3	1	
N_2	28.0134	941.4	2	2.875 05	3392.01	1	
O_2	31.9988	491.888	2	2.079	2273.64	3	
Cl_2	70.906	239.216	2	0.3456	807.3	1	
I_2	253.82	148.81	2	0.053 76	308.65	1	
HCl	36.465	427.772	1	15.2344	4301.38	1	
HI	127.918	294.67	1	9.369	3322.24	1	
CO	28.010 55	1070.11	1	2.7771	3121.48	1	
NO	30.008	627.7	1	2.4520	2738.87	2	174.2(2)
CO_2	44.009 95	1596.23	2	0.561 67	960.10(2) 1932.09 3380.14	1	
NO_2	46.008	928.3	2	4.243 01	1088.9 1953.6 2396.3	2	
H_2O	18.016	917.773	2	11,331.5	2294.27 5261.71 5403.78	1	
NH_3	17.0361	1157.77	3	1876.0	1367 2341(2) 4800 4955(2)	1	
CH_4	16.043	1640.57	12	435.6	1957(3) 2207.1(2) 4196.2 4343.3(3)	1	
N_2O_4	92.016	1909.82	4	6.5793×10^{-3}	72 374 554 619 691 971 1079 1184 1814 1975 2460 2515	1	

[a] These values have all been obtained from D. R. Stull and H. Prophet, *JANAF Thermochemical Tables*, 2nd ed., NSRDS-NBS 37, 1971 and supplements (see footnote to Table A.2).

[b] The values of Θ_r for nonlinear polyatomic molecules are values of $\Theta_a\Theta_b\Theta_c$.

[c] The values in parentheses are degeneracies. Otherwise the levels have a degeneracy of unity.

The calculated values of the thermodynamic properties of the 21 perfect gases listed in Table 13.1 are shown in Table 13.2. These values have been calculated with the equations and approximations discussed in this chapter. The values of C_P° and S° may be compared with the values from the JANAF Thermochemical Tables (Table A.2), which are more accurate because they take more effects into account.

Table 13.2 Standard Thermodynamic Properties of Gases Calculated from Molecular Parameters[a]

	T/K	C_P° J K^{-1} mol^{-1}	S° J K^{-1} mol^{-1}	H° kJ mol^{-1}	G° kJ mol^{-1}
H(g)	298.150	20.786	114.608	6.197	−27.973
	500.000	20.786	125.354	10.393	−52.284
	1000.000	20.786	139.762	20.786	−118.976
	2000.000	20.786	154.170	41.572	−266.768
	3000.000	20.786	162.598	62.358	−425.435
C(g)	298.150	20.786	156.853	6.197	−40.563
	500.000	20.786	168.063	10.393	−73.638
	1000.000	20.786	182.818	20.786	−162.032
	2000.000	20.786	197.401	41.572	−353.231
	3000.000	20.786	205.888	62.358	−555.307
N(g)	298.150	20.786	153.192	6.197	−39.477
	500.000	20.786	163.938	10.393	−71.576
	1000.000	20.786	178.346	20.786	−157.560
	2000.000	20.786	192.754	41.572	−343.936
	3000.000	20.786	201.182	62.358	−541.188
O(g)	298.150	20.786	159.177	6.197	−41.261
	500.000	201786	170.736	10.393	−74.975
	1000.000	20.786	185.881	20.786	−165.095
	2000.000	20.786	200.705	41.572	−359.837
	3000.000	20.786	209.279	62.358	−565.478
Cl(g)	298.150	20.786	159.126	6.197	−41.246
	500.000	20.786	170.388	10.393	−74.801
	1000.000	20.786	186.223	20.786	−165.437
	2000.000	20.786	202.108	41.572	−362.644
	3000.000	20.786	211.188	62.358	−571.207
I(g)	298.150	20.786	180.677	6.197	−47.671
	500.000	20.786	191.423	10.393	−85.319
	1000.000	20.786	205.831	20.786	−185.045
	2000.000	20.786	220.257	41.572	−398.941
	3000.000	20.786	228.775	62.358	−623.966

[a] The reference state is the hypothetical perfect gas of atoms at absolute zero.

Table 13.2—(*Continued*)

	T/K	C_P° J K^{-1} mol^{-1}	S° J K^{-1} mol^{-1}	H° kJ mol^{-1}	G° kJ mol^{-1}
$H_2(g)$	298.150	29.100	130.229	−423.397	−462.225
	500.000	29.105	145.275	−417.523	−490.160
	1000.000	29.693	165.553	−402.879	−568.433
	2000.000	32.926	187.130	−371.559	−745.819
	3000.000	34.907	200.903	−337.524	−940.232
$N_2(g)$	298.150	29.113	191.453	−932.723	−989.805
	500.000	29.534	206.570	−926.818	−1030.103
	1000.000	32.547	227.934	−911.318	−1139.252
	2000.000	35.679	251.691	−876.864	−1380.246
	3000.000	36.583	266.360	−840.654	−1639.735
$O_2(g)$	298.150	29.336	204.976	−483.202	−544.316
	500.000	30.961	220.479	−477.135	−587.375
	1000.000	34.598	243.229	−460.618	−703.847
	2000.000	36.574	268.010	−424.757	−960.776
	3000.000	37.028	282.942	−387.913	−1236.740
$Cl_2(g)$	298.150	33.767	222.809	−230.060	−296.491
	500.000	35.822	240.851	−222.999	−343.424
	1000.000	36.978	266.159	−204.711	−470.870
	2000.000	37.303	291.927	−167.517	−751.372
	3000.000	37.365	307.067	−130.177	−1051.377
$I_2(g)$	298.150	36.710	260.412	−138.720	−216.362
	500.000	37.156	279.525	−131.255	−271.017
	1000.000	37.349	305.362	−112.612	−417.974
	2000.000	37.398	331.271	−75.230	−737.772
	3000.000	37.408	346.437	−37.827	−1077.137
$HCl(g)$	298.150	29.101	186.639	−419.096	−474.742
	500.000	29.213	201.699	−413.215	−514.065
	1000.000	31.242	222.460	−398.180	−620.640
	2000.000	34.834	245.411	−364.860	−855.682
	3000.000	36.126	259.822	−329.276	−1108.740
$HI(g)$	298.150	29.115	206.335	−285.993	−347.512
	500.000	29.579	221.462	−280.084	−390.815
	1000.000	32.663	242.889	−264.536	−507.425
	2000.000	35.740	266.710	−229.993	−763.413
	3000.000	36.615	281.398	−193.738	−1037.933
$CO(g)$	298.150	29.126	197.505	−1061.433	−1120.319
	500.000	29.733	212.665	−1055.509	−1161.842
	1000.000	33.010	234.290	−1039.812	−1274.103
	2000.000	35.914	258.298	−1005.011	−1521.607
	3000.000	36.704	273.038	−968.631	−1787.745

Table 13.2—(*Continued*)

	T/K	C_P° J K^{-1} mol^{-1}	S° J K^{-1} mol^{-1}	H° kJ mol^{-1}	G° kJ mol^{-1}
NO(g)	298.150	29.172	208.852	−619.021	−681.291
	500.000	30.152	224.871	−613.054	−725.490
	1000.000	33.709	247.575	−597.026	−844.610
	2000.000	36.861	272.276	−561.735	−1106.288
	3000.000	26.861	287.225	−525.133	−1386.808
$CO_2(g)$	298.150	37.124	213.665	−1586.864	−1650.569
	500.000	44.599	234.767	−1578.561	−1695.944
	1000.000	54.122	269.109	−1553.514	−1822.623
	2000.000	59.699	308.833	−1495.889	−2113.555
	3000.000	61.109	333.358	−1435.357	−2435.430
$NO_2(g)$	298.150	36.973	239.918	−918.114	−989.645
	500.000	43.206	260.522	−910.015	−1040.276
	1000.000	52.166	293.773	− 885.770	−1179.543
	2000.000	56.441	331.672	−830.855	−1494.198
	3000.000	57.394	354.774	−773.847	−1838.168
H_2O	298.150	33.482	188.553	−907.849	−964.066
	500.000	35.119	206.197	−900.946	−1004.044
	1000.000	41.025	232.302	−881.945	−1114.247
	2000.000	50.208	263.946	−835.756	−1363.648
	3000.000	54.042	285.143	−783.349	−1638.779
$NH_3(g)$	298.150	35.478	192.464	−1147.652	−1205.035
	500.000	41.447	212.158	−1139.911	−1245.990
	1000.000	55.648	245.439	−1115.494	−1360.933
	2000.000	71.576	289.810	−1050.557	−1630.177
	3000.000	77.300	320.098	−975.677	−1935.960
$CH_4(g)$	298.150	35.338	186.075	−1630.563	−1686.041
	500.000	45.668	206.570	−1622.741	−1725.756
	1000.000	71.360	246.688	−1592.785	−1839.472
	2000.000	94.257	304.805	−1507.624	−2117.234
	3000.000	101.323	344.601	−1409.238	−2443.042
$N_2O_4(g)$	298.150	77.201	304.229	−1893.429	−1984.135
	500.000	97.161	349.272	−1875.670	−2050.306
	1000.000	119.194	424.913	−1820.473	−2245.387
	2000.000	129.027	511.545	−1694.941	−2718.032
	3000.000	131.198	564.356	−1564.622	−3257.689

The values of H° and G° are with respect to perfect gases of the consistent atoms at absolute zero. Therefore, these values of H° and G° cannot be compared with the values of ΔH_f° and ΔG_f° in Table A.2 because of the difference in standard state. However, they can be used to calculate ΔH° and ΔG° for chemical reactions because they have all been calculated for the same standard state—the hypothetical standard state of the perfect gas of the constituent atoms at absolute zero.

13.13 CHANGES IN THERMODYNAMIC PROPERTIES FOR CHEMICAL REACTIONS OF PERFECT GASES

The tables that have been developed here for the thermodynamic properties of individual gases may be used to calculate ΔC_P°, ΔS°, ΔH°, and ΔG° for chemical reactions. As mentioned earlier the values of H° and G° are not the customary enthalpies and Gibbs energies of formation, but they may be used in the same way because they are based on the same defined standard state.

Example 13.13 What are the values of ΔC_P°, ΔS°, ΔH°, and ΔG° for $\frac{1}{2}O_2(g) = O(g)$ calculated statistically mechanically and what is the equilibrium constant at 298.15 K?

Using values calculated in Example 13.11 and Example 13.12,

$$\Delta C_P^\circ = 20.786 - \tfrac{1}{2}(29.336) = 6.118 \text{ J K}^{-1} \text{ mol}^{-1}$$

$$\Delta S^\circ = 159.177 - \tfrac{1}{2}(204.976) = 56.689 \text{ J K}^{-1} \text{ mol}^{-1}$$

$$\Delta H^\circ = 6.197 - \tfrac{1}{2}(-483.203) = 247.799 \text{ kJ mol}^{-1}$$

$$\Delta G^\circ = -41.261 - \tfrac{1}{2}(-544.316) = 230.897 \text{ kJ mol}^{-1}$$

$$K = e^{-\Delta G^\circ/RT} = e^{-230{,}897/(8.31441)(298.15)} = 3.53 \times 10^{-41}$$

Table 13.3 gives ΔC_P°, ΔS°, ΔH°, and ΔG° for the dissociation of homonuclear diatomic molecules calculated from the values in the preceeding table. These reactions are written in the form $\frac{1}{2}X_2 = X$ so that ΔH° and ΔG° are equal to ΔH_f° and ΔG_f° for the atoms. The last column gives the equilibrium constant at each temperature. These equilibrium constants are numerically equal to the pressure in atmospheres of atoms in equilibrium with the diatomic gas at 1 atm. In all cases, ΔH° is positive, and therefore the equilibrium constant increases as the temperature increases. It might be noted that as a given dissociation reaction becomes spontaneous as the temperature is raised, the heat capacity at constant pressure of the diatomic molecule approaches its classical value, that is, the vibrational motion becomes completely active.

The Gibbs energy of formation of $H(g)$ at 298.15 K given in Table 13.3 is 203.139 kJ mol^{-1} compared with 203.296 kJ mol^{-1} in JANAF Thermochemical Tables. The difference in ΔG_f° is due to the incompleteness of the calculation given here for $H_2(g)$. At low temperatures $H_2(g)$ can exist in two forms, ortho with spins parallel and para with spins opposite.

Table 13.3 also shows the values from the current calculation of ΔC_P°, ΔS°, ΔH_f°, and ΔG_f° for the formation of NH_3, NO, NO_2, H_2O, HI, and HCl. The values of ΔG_f° agree with the values in JANAF Thermochemical Tables to about 1 kJ mol^{-1} at 298.15 K and 2 kJ mol^{-1} at 3000 K. The differences are due to the neglect in the present calculation of anharmonicity, centrifugal distortion, higher electronic levels, and contributions of electronic levels to C_P° and ΔH°.

Table 13.3 Changes in Standard Thermodynamic Properties for Chemical Reactions of Gases

T/K	ΔC_P° J K^{-1} mol^{-1}	ΔS° J K^{-1} mol^{-1}	ΔH° kJ mol^{-1}	ΔG° kJ mol^{-1}	K
		$\frac{1}{2}H_2(g) = H(g)$			
298.15	6.235	49.493	217.896	203.139	2.58×10^{-36}
500.00	6.234	52.717	219.154	192.796	7.23×10^{-21}
1000.00	5.940	56.985	222.226	165.240	2.34×10^{-9}
2000.00	4.323	60.605	227.352	106.142	1.69×10^{-3}
3000.00	3.333	62.146	231.120	44.681	1.67×10^{-1}
		$\frac{1}{2}N_2(g) = N(g)$			
298.15	6.230	57.465	472.559	455.426	
500.00	6.019	60.653	473.802	443.475	4.69×10^{-47}
1000.00	4.513	64.379	476.445	412.066	2.99×10^{-22}
2000.00	2.947	66.909	480.004	346.187	9.09×10^{-10}
3000.00	2.495	68.002	482.685	278.680	1.41×10^{-5}
		$\frac{1}{2}O_2(g) = O(g)$			
298.15	6.118	56.689	247.799	230.897	3.53×10^{-41}
500.00	5.305	60.496	248.961	218.713	1.42×10^{-23}
1000.00	3.487	64.266	251.095	186.829	1.74×10^{-10}
2000.00	2.499	66.700	253.950	120.551	7.11×10^{-4}
3000.00	2.272	67.807	256.314	52.892	1.20×10^{-1}
		$\frac{1}{2}Cl_2(g) = Cl(g)$			
298.15	3.902	47.722	121.227	106.999	1.80×10^{-19}
500.00	2.875	49.962	121.892	96.911	7.51×10^{-9}
1000.00	2.297	53.143	123.141	69.998	2.21×10^{-4}
2000.00	2.135	56.144	125.331	13.042	4.56×10^{-1}
3000.00	2.104	57.655	127.447	−45.519	6.20
		$\frac{1}{2}I_2(g) = I(g)$			
298.15	2.431	50.471	75.558	60.510	2.51×10^{-11}
500.00	2.208	51.661	76.020	50.190	5.71×10^{-6}
1000.00	2.111	53.150	77.092	23.942	5.62×10^{-2}
2000.00	2.087	54.621	79.187	−30.055	6.09
3000.00	2.082	55.556	81.271	−85.398	3.07×10^{1}
		$\frac{1}{2}N_2(g) + \frac{3}{2}H_2(g) = NH_3(g)$			
298.15	−22.729	−98.606	−46.195	−16.796	8.76×10^{2}
500.00	−16.977	−109.039	−50.218	4.302	3.55×10^{-1}
1000.00	−5.165	−116.858	−55.516	61.342	6.25×10^{-4}
2000.00	4.347	−116.729	−54.785	178.674	2.16×10^{-5}
3000.00	6.648	−114.437	−49.054	294.256	7.53×10^{-6}

Table 13.3—(*Continued*)

T/K	ΔC_P° J K^{-1} mol^{-1}	ΔS° J K^{-1} mol^{-1}	ΔH° kJ mol^{-1}	ΔG° kJ mol^{-1}	K
		$\frac{1}{2}N_2(g) + \frac{1}{2}O_2(g) = NO(g)$			
298.15	−0.052	10.638	88.942	85.770	9.41×10^{-16}
500.00	−0.096	11.346	88.922	83.249	2.01×10^{-9}
1000.00	0.136	11.994	88.942	76.948	9.56×10^{-5}
2000.00	0.102	12.426	89.075	64.223	2.10×10^{-2}
3000.00	0.055	12.574	89.151	51.429	1.27×10^{-1}
		$\frac{1}{2}N_2(g) + O_2(g) = NO_2(g)$			
298.15	−6.919	−60.784	31.451	49.573	2.07×10^{-9}
500.00	−2.523	−63.242	30.529	62.151	3.22×10^{-7}
1000.00	1.295	−63.423	40.507	93.930	1.24×10^{-5}
2000.00	2.027	−62.183	32.334	156.700	8.08×10^{-5}
3000.00	2.075	−61.349	34.393	218.440	1.57×10^{-4}
		$H_2(g) + \frac{1}{2}O_2(g) = H_2O(g)$			
298.15	−10.287	−44.163	−242.851	−229.683	1.73×10^{40}
500.00	−9.467	−49.318	−244.856	−220.197	1.01×10^{23}
1000.00	−5.967	−54.866	−248.757	−193.891	1.34×10^{10}
2000.00	−1.005	−57.188	−251.818	−137.442	3.89×10^{3}
3000.00	0.621	−57.231	−251.868	−80.176	2.49×10
		$\frac{1}{2}H_2(g) + \frac{1}{2}Cl(g) = HCl(g)$			
298.15	−2.333	10.120	−92.367	−95.385	5.14×10^{16}
500.00	−3.250	8.636	−92.955	−97.273	1.45×10^{10}
1000.00	−2.093	6.604	−94.385	−100.989	1.88×10^{5}
2000.00	−0.280	5.883	−95.321	−107.086	6.26×10^{2}
3000.00	−0.010	5.837	−95.435	−112.936	9.25×10
		$\frac{1}{2}H_2(g) + \frac{1}{2}I_2(g) = HI(g)$			
298.15	−3.790	11.014	−4.935	−8.219	2.75×10
500.00	−3.551	9.062	−5.695	−10.226	1.17×10
1000.00	−0.858	7.431	−6.790	−14.222	5.53
2000.00	0.578	7.510	−6.598	−21.618	3.67
3000.00	0.458	7.729	−6.063	−29.249	3.23
		$CO(g) + H_2O(g) = CO_2(g) + H_2(g)$			
298.15	3.616	−42.164	−40.980	−28.408	9.48×10^{4}
500.00	8.852	−38.819	−39.628	−20.219	1.29×10^{2}
1000.00	9.781	−31.930	−34.636	−2.706	1.38
2000.00	6.503	−26.281	−26.681	25.881	2.11×10^{-1}
3000.00	5.207	−23.921	−20.901	50.861	1.30×10^{-1}
		$CO(g) + 3H_2(g) = CH_4(g) + H_2O(g)$			
298.15	−47.608	−213.563	−206.788	−143.114	1.18×10^{25}
500.00	−36.260	−235.723	−215.340	−97.479	1.53×10^{10}
1000.00	−9.704	−251.960	−226.280	25.681	4.56×10^{-2}
2000.00	9.773	−250.936	−223.691	278.180	5.43×10^{-8}
3000.00	13.941	−246.001	−211.383	526.620	6.77×10^{-10}

Table 13.3—(*Continued*)

T/K	ΔC_P° J K^{-1} mol^{-1}	ΔS° J K^{-1} mol^{-1}	ΔH° kJ mol^{-1}	ΔG° kJ mol^{-1}	K
		$CO(g) + \frac{1}{2}O_2(g) = CO_2(g)$			
298.15	−6.671	−86.327	−283.830	−258.092	1.64×10^{45}
500.00	−0.615	−88.137	−284.484	−240.415	1.31×10^{25}
1000.00	3.813	−86.796	−283.392	−196.597	1.86×10^{10}
2000.00	5.498	−83.470	−278.500	−111.561	8.20×10^{2}
3000.00	5.891	−81.151	−272.769	−29.315	3.24
		$NH_3(g) + \frac{7}{4}O_2(g) = NO_2(g) + \frac{3}{2}H_2(g)$			
298.15	0.380	−28.423	−286.630	−278.156	5.38×10^{48}
500.00	0.255	−28.180	−286.536	−272.446	2.90×10^{28}
1000.00	−2.492	−28.863	−287.112	−258.249	3.09×10^{13}
2000.00	−3.828	−31.236	−290.608	−228.136	9.08×10^{5}
3000.00	−3.641	−32.758	−294.355	−196.081	2.59×10^{3}
		$NH_3(g) + \frac{5}{4}O_2(g) = NO(g) + \frac{3}{2}H_2O(g)$			
298.15	7.247	42.999	−229.139	−241.959	2.45×10^{42}
500.00	2.681	46.409	−228.143	−251.348	1.81×10^{26}
1000.00	−3.650	46.554	−228.677	−275.230	2.38×10^{14}
2000.00	−5.753	43.373	−233.867	−320.614	2.36×10^{8}
3000.00	−5.661	41.165	−239.597	−363.091	2.10×10^{6}
		$N_2O_4(g) = 2NO_2(g)$			
298.15	−3.254	175.608	57.202	4.844	1.42×10^{-1}
500.00	−10.749	171.772	55.640	−30.246	1.44×10^{3}
1000.00	−14.862	162.633	48.933	−113.700	8.69×10^{5}
2000.00	−16.145	151.798	33.231	−270.365	1.15×10^{7}
3000.00	−16.409	145.192	16.928	−418.646	1.95×10^{7}

Values of ΔH_f° and ΔG_f° are obtained for all the substances in Table 13.1, except for $C(g)$, $CO(g)$, $CO_2(g)$, and $CH_4(g)$. Since the standard state for carbon is graphite and we have not introduced experimental values of the thermodynamic properties of $C(s)$, we have not been able here to calculate the standard ΔH_f° and ΔG_f° for gases involving carbon.

The reaction for the formation of NH_3 from its elements is an example of a reaction that is thermodynamically spontaneous at room temperature, but does not occur at an appreciable rate. As the temperature is raised to increase the rate of formation of NH_3, the equlibrium constant becomes less favourable. Therefore, it has been important to find catalysts for $\frac{1}{2}N_2(g) + \frac{3}{2}H_2(g) = NH_3(g)$ so that the reaction can be carried out a lower temperatures.

It is of interest to compare the reactions for the formation of HCl and HI. These reactions have approximately the same ΔS° and both evolve heat, but ΔH° is much more negative for the formation of HCl. Therefore, the equilibrium constant for the formation of HCl is much larger than for HI at room temperature.

The equilibrium constant for $\frac{1}{2}H_2(g) + \frac{1}{2}I_2(g) = HI(g)$ only changes from 27.5 at 298.15 K to 3.23 at 3000 K. In contrast the equilibrium cnostant for the formation of HCl changes by a factor of 5×10^{14} over this temperature range.

It is interesting to realize that CH_4 can be formed spontaneously from CO and H_2 (or CO_2 and H_2) at temperatures below about 700 K, but not at higher temperatures.

Oxygen and nitrogen do not react spontaneously at room temperature, but in high-temperature processes NO is formed. Since the $\Delta H°$ for the oxidation of nitrogen to NO_2 is less positive than for NO, the equilibrium constant for this latter reaction does not change so rapidly with temperature. If a mixture of N_2 and O_2 is heated, NO_2 predominates at low temperatures, but NO predominates at high temperatures.

It is interesting to see that CO "burns" with O_2 to form CO_2 almost completely at temperatures below about 3500 K, but not above. As the temperature is raised, $\Delta S°$ becomes more important in determining the equilibrium constant for the reaction. In thermodynamic terms we can say that $CO + \frac{1}{2}O_2$ has a higher entropy than CO_2, and in molecular terms we can say that the increasing vibrational and rotational energy of CO_2 at very high temperatures causes it to dissociate.

The next to the last two reactions in Table 13.3 are of interest because they are the reactions by which nitric acid is produced from NH_3.

13.16 HEAT CAPACITIES OF SOLIDS

In order to make a simple calculation of the heat capacity of a monatomic solid we can imagine that each atom oscillates about its equilibrium lattice point with a small amplitude. As mentioned in Section 1.13, according to classical theory each atom would contribute R for each of its three vibrational degrees of freedom, so that the molar heat capacity at constant volume would be $3R = 25 \text{ J K}^{-1} \text{ mol}^{-1}$. This is observed at high enough temperatures for all atomic solids. However, classical theory was not able to explain the decrease of C_V to zero as absolute zero is approached.

In 1907 Einstein calculated the heat capacity of an atomic crystal on the basis of the assumption that each of the $3N$ independent harmonic oscillators have the same frequency ν_E. Using the vibrational partition function (equation 13.89) that we have already derived for a diatomic gas, he was able to show that the heat capacity for such an idealized solid is given by

$$C_V = 3Nk\left(\frac{\Theta_E}{T}\right)^2 \frac{e^{-\Theta_E/T}}{(1 - e^{-\Theta_E/T})^2} \tag{13.127}$$

where Θ_E is the Einstein temperature defined by $\Theta_E = h\nu_E/k$. This equation with one adjustable parameter Θ_E represents C_V quite well at higher temperatures, but fails as the temperature approaches absolute zero. An important feature of equation 13.127 is that the heat capacity curves for all atomic solids should superimpose if plotted versus T/Θ_E. As $T \to \infty$, $C_V \to 3R$, and as $T \to 0$, $C_V \to 0$.

The deficiency that the Einstein theory used a single vibrational frequency was remedied by Debye, who provided for a distribution of vibration frequencies. A crystal containing N atoms can be treated like a large polyatomic molecule in that the vibrational motion can be decomposed into a set of independent harmonic oscillations by introducing normal coordinates (Section 11.10). The normal frequencies of a crystal range from very low values to high values of about 10^{13} Hz. The low-frequency modes are important at low temperatures because the energy of an oscillator is proportional to the frequency, and the lower-energy vibrational modes are populated at low temperatures. Debye calculated the frequency distribution by treating the crystal as a continuous elastic body. He showed that the heat capacity is given by

$$C_V = 9Nk\left(\frac{T}{\Theta_D}\right)^3 \int_0^{\Theta_D/T} \frac{x^4 e^x}{(e^x - 1)^2}\, dx \tag{13.128}$$

where the Debye temperature $\Theta_D = h\nu/kT$ and x is a dummy variable. As the temperature approaches absolute zero, this equation reduces to

$$C_V = \frac{12\pi^4}{5} Nk\left(\frac{T}{\Theta_D}\right)^3 \tag{13.129}$$

which represents the experimental results rather well.

References

F. C. Andrews, *Equilibrium Statistical Mechanics*, Wiley, New York, 1975.

S. M. Blinder, *Advanced Physical Chemistry*, The Macmillan Co., London, 1969.

N. Davidson, *Statistical Mechanics*, McGraw-Hill Book Co., New York, 1962.

T. L. Hill, *An Introduction to Statistical Thermodynamics*, Addison-Wesley Publishing Co., Reading, Mass., 1960.

W. Kauzmann, *Thermodynamics and Statistics: With Applications to Gases*, W. A. Benjamin, Inc., New York, 1967.

J. H. Knox, *Molecular Thermodynamics*, Wiley-Interscience, New York, 1971.

J. E. Mayer and M. G. Mayer, *Statistical Mechanics*, Wiley, New York, 1976.

B. J. McClelland, *Statistical Thermodynamics*, Wiley, New York, 1973.

D. S. McQuarrie, *Statistical Thermodynamics*, Harper and Row, New York, 1973.

D. S. McQuarrie, *Statistical Mechanics*, Harper and Row, New York, 1976.

D. Rapp, *Statistical Mechanics*, Holt, Rinehart and Winston, Inc., New York, 1972.

Problems

13.1 (*a*) How many different ways can two distinguishable balls be placed in two boxes? (*b*) How many different ways can two distinguishable balls be placed in three boxes? (*c*) What are the answers to (*a*) and (*b*) if the balls are indistinguishable?

Ans. (*a*) 4, (*b*) 9, (*c*) 3, 6.

13.2 (*a*) How many different ways can four distinguishable balls be placed in two boxes? (*b*) How many different ways are there if the balls are indistinguishable?

Ans. (*a*) 16, (*b*) 5.

13.3 What are the average occupation numbers N_i for an idealized five-atom crystal with five quanta of energy? As in the example discussed in the chapter, we assume that the atoms in the crystal are perfect one-dimensional harmonic oscillators.

Ans.

v	0	1	2	3	4	5
N_i	3.169	1.166	0.429	0.158	0.058	0.020

13.4 How many microstates are there for a three-atom crystal of one-dimensional harmonic oscillators if only two quanta of energy are available? On the average how many atoms will have zero, one, or two quanta? *Ans.* 6, 1.5, 1, 0.5.

13.5 Using the Boltzmann distribution calculate the ratio of populations at 25 °C of energy levels separated by (*a*) 1000 cm^{-1}, and (*b*) 10 kJ mol^{-1}.
Ans. (*a*) 0.0080, (*b*) 0.0177.

13.6 Calculate the temperature at which 10% of the molecules in a system will be in the first excited electronic state if this state is 400 kJ mol^{-1} above the ground state.
Ans. 22,000 K.

13.7 Calculate the fraction of hydrogen atoms that at equilibrium at 1000 °C would have $n = 2$. *Ans.* 4.23×10^{-41}.

13.8 The difference of 4.60 J K^{-1} mol^{-1} between the third law entropy of CO and the statistical mechanical value can be attributed to the randomness of orientation of CO molecules in its crystals at absolute zero. If half of the molecules are oriented CO and half OC, calculate the entropy of a crystal at absolute zero using equation 13.1 and Stirling's approximation. *Ans.* 5.76 J K^{-1} mol^{-1}.

13.9 Calculate the translational partition function for $H_2(g)$ at 1000 K and 1 atm.
Ans. 1.396×10^{30}.

13.10 Calculate the entropy of one mole of H-atom gas at 1000 K and (*a*) 1 atm, and (*b*) 1000 atm. *Ans.* 139.762 J K^{-1} mol^{-1}, (*b*) 82.327 J K^{-1} mol^{-1}.

13.11 Claculate the entropy of neon at 25 °C and 1 atm. *Ans.* 146.218 J K^{-1} mol^{-1}.

13.12 Calculate $S°$ and $C_P°$ for argon ($M = 39.948$ g mol^{-1}) at 25 °C and 1 atm.
Ans. 154.735, 20.786 J K^{-1} mol^{-1}.

13.13 Ignoring the vibrational motion of $H_2(g)$ and any electronic excitation of $H(g)$ or $H_2(g)$, what value of $\Delta H_f°$ does statistical mechanics yield for $H(g)$ at 25 °C?
Ans. 214.089 kJ mol^{-1}.

13.14 Calculate the translational partition functions for H, H_2, and H_3 at 1000 K and 1 atm. What are the rotational partition functions of H_2 and H_3 (linear) at 1000 K? The internuclear distances in H_3 are 0.094 nm.
Ans. 4.94×10^{29}, 1.40×10^{30}, 2.57×10^{30}, 5.72, 36.8.

13.15 Calculate the ratio of the number of HBr molecules in state $v = 2$, $J = 5$ to the number in state $v = 1$, $J = 2$ at 1000 K. Assume that all of the molecules are in their electronic ground states. ($\Theta_v = 3700$ K, $\Theta_r = 12.1$ K.) *Ans.* 0.0407.

13.16 The ground state of $Cl(g)$ is twofold degenerate. The first excited state is 875.4 cm^{-1} higher in energy and is twofold degenerate. What is the value of the electronic pertition function at 25 °C? At 1000 K? *Ans,* 2.029, 2.568.

13.17 Derive the expression for the vibrational contribution to the internal energy

$$U = \frac{RTx}{e^x - 1}$$

where $x = h\nu/kT$.

13.18 Calculate the entropy of nitrogen gas at 25 °C and 1 atm pressure. The equilibrium separation of atoms is 0.1095 nm and the vibrational wave number is 2330.7 cm^{-1}.
Ans. 191.414 J K^{-1} mol^{-1}.

13.19 Using

$$\left(\frac{\partial G}{\partial T}\right)_P = -S$$

and the contribution of vibration to the Gibbs energy for a diatomic molecule in a perfect gas

$$G = RT \ln (1 - e^{-x})$$

derive the expression for the corresponding contribution to the entropy.

Ans. $R\left[\frac{x}{e^x - 1} - \ln (1 - e^{-x})\right]$.

13.20 Calculate the statistical mechanical values of C_P°, S°, H°, and G° for $H(g)$ at 3000 K. *Ans.* 20.786, 156.835 J K^{-1} mol^{-1}. 62.358, -408.146 kJ mol^{-1}.

13.21 Calculate the statistical mechanical values of C_P°, S°, H°, and G° for $H_2(g)$ at 3000 K. *Ans.* 34.907, 200.903 J K^{-1} mol^{-1}. -337.524, -940.232 kJ mol^{-1}.

13.22 What are the value of ΔC_P°, ΔH°, ΔS°, and ΔG° for $H_2(g) = 2H(g)$ at 3000 K calculated in the preceeding two problems? What is the value of K_P? What is the degree of dissociation at 1 atm?

Ans. 6.665, 124.293 J K^{-1} mol^{-1}. 462.240, 89.362 kJ mol^{-1}. 2.78×10^{-2}. 0.083.

13.23 Calculate the equilibrium constant for the isotope exchange reaction

$$D + H_2 = H + DH$$

at 25 °C. Assume that the equilibrium distance and force constants of H_2 and DH are the same. *Ans.* 1.73.

13.24 (*a*) How many different ways can three distinguishable balls be placed in two boxes? (*b*) How many different ways are there if the balls are indistinguishable?

13.25 Calculate the average occupation N_i of energy levels of a five-atom crystal having three vibrational quanta.

13.26 A crystal contains N atoms, each of which can have energy $+u$ or energy $-u$ (this is the type of situation one finds when a paramagnetic substance is placed in a magnetic field). (*a*) Suppose $N = 3$. Write down all the possible microstates that the system could be in. (*b*) Determine the values of $\mathcal{N}$, E_i, N_i, and p_i for the ensemble of microstates in the answer to (*a*). (*c*) Show that $q^N = Q$ for this system. (*d*) Find S and U for this system. (*e*) Find S and U for the system for arbitrary $N \gg 1$.

13.27 The energies of the $n = 2$ and $n = 1$ orbitals of the hydrogen atom are 27,420 and 109,678 cm^{-1}, respectively. What are the relative populations in these levels at (*a*) 25 °C, and (*b*) 2000 °C?

13.28 Calculate the ratio of populations at 25 °C of energy levels separated by (*a*) 1 eV and (*b*) 10 eV. (*c*) Calculate the ratios at 1000 °C.

13.29 Show that

$$\sum \sum g_i g_j e^{-\beta\epsilon_i} e^{-\beta\epsilon_j} = \left(\sum g_i e^{-\beta\epsilon_i}\right)\left(\sum g_j e^{-\beta\epsilon_j}\right)$$

13.30 Compare the translational partition function of $I(g)$ at 1000 K and 1 atm with that for $H(g)$ calculated in Problem 13.14.

13.31 Calculate the entropies of 1 mol of each of the atomic gases H, N, and C at 25 °C and 1 atm. Compare these values with those in Table A.1.

13.32 Calculate the molar entropy of helium in the perfect gas state at 25 °C and 1 atm pressure.

13.33 Calculate the rotational energy level of $H^{35}Cl$ that has the highest occupation number at (*a*) 25 °C, and (*b*) 500 K.

13.34 Calculate q_r and q_v for H_2 and NO at 3000 K. What are the differences?

13.35 What fraction of HCl molecules is in the state $v = 2$, $J = 7$ at 500 °C?

13.36 A molecule exists in singlet and triplet forms with the singlet having a higher energy by 4.11×10^{-21} J per molecule. The singlet level has a degeneracy of 1 and the triplet level has a degeneracy of 3. (*a*) Ignoring higher levels, what is the electronic partition function? (*b*) What is the ratio of the concentration of triplets to singlet molecules at 298 K?

13.37 What are the rotational contributions to C_P°, S°, H°, and G° for $NH_3(g)$ at 25 °C?

13.38 What are the vibrational contributions to C_P°, S°, H°, and G° for $NH_3(g)$ at 25 °C?

13.39 Calculate the entropy of nitrogen gas at 2000 K and 1 atm pressure. (Compare Problem 13.18.)

13.40 Calculate the entropy for chlorine gas at 25 °C and 1 atm pressure.

13.41 Calculate the equilibrium constant K_P for

$$I_2(g) = 2I(g)$$

at 1000 K. [For experimental determinations of this equilibrium constant see M. L. Perlman and G. K. Rollefson, *J. Chem. Phys.*, **9**, 362 (1941).]

13.42 Using statistical mechanics, calculate the equilibrium constant K_P for the following reaction at 2000 K.

$$N_2(g) + O_2(g) = 2NO(g)$$

The experimental value in Table 4.2 is 4.08×10^{-4}.

13.43 Calculate K_P for

$$Cl_2(g) = 2Cl(g)$$

at 2000 K.

13.44 What are the electronic contributions to S° and G° for $I(g)$ at 298.15 K and 3000 K?

13.45 Calculate the equilibrium constant at 25 °C for the reaction

$$H_2 + D_2 = 2HD$$

It may be assumed that the equilibrium distance and the force constant k are the same for all three molecular species, so that the additional vibrational frequencies required may be calculated from $2\pi\nu = (k/\mu)^{1/2}$. Because of the zero point vibration, ΔU_0 for this reaction is given by

$$\Delta U_0 = \tfrac{1}{2} N_A h (2\nu_{HD} - \nu_{H_2} - \nu_{D_2})$$

13.46 Calculate C_P for CO_2 at 1000 K. Compare the actual contributions to C_P from the various normal modes with the classical expectations.

13.47 Calculate C_P for hydrogen gas at 298 and 2000 K. This calculation is discussed in some detail by C. Marzzacco and M. Waldman, *J. Chem. Ed.*, **50**, 444 (1973).

13.48 It may be shown (W. Kauzmann, *Thermodynamics and Statistics: With Application to Gases*, W. A. Benjamin, Inc., New York, 1967, p. 249) that the second virial coefficient in

$$\frac{PV}{RT} = 1 + \frac{B}{V} + \frac{C}{V^2} + \cdots$$

is given by

$$B = 2\pi N_A \int_0^\infty (1 - e^{-U/kT}) r^2 \, dr$$

where r is intermolecular distance, k is the Boltzmann constant (R/N_A), and U is the potential energy of interaction. Derive the expression for B for a gas of rigid spheres of diameter d. The molecules do not interact unless they touch, and so $U = 0$ if $r > d$. The molecules cannot penetrate each other, and so $U = \infty$ if $r < d$.

PART THREE

CHEMICAL DYNAMICS

After considering equilibrium properties in Parts One and Two we now turn to a consideration of the rates at which physical and chemical processes occur. This is more difficult, because rates depend on the mechanisms by which processes occur and so a microscopic instead of macroscopic approach must be used. The consideration of rate processes has been postponed until now because quantum mechanics and statistical mechanics are needed for the interpretation or prediction of rate constants.

The kinetic theory of gases uses a simple model to calculate the rates of certain processes. The probabilities of molecular speeds and the values of average speeds depend on the molar mass and temperature for noninteracting gas molecules. The frequency of collisions and the transport properties (viscosity, diffusion, and heat conduction) for gases of rigid spherical molecules may be calculated. However, the behavior of real gases is more complicated because of molecular interactions.

Gas-phase kinetics and liquid-phase kinetics are discussed in separate chapters because of the very different theoretical bases for understanding. Rates of reactions in the gas phase are interpreted in terms of the kinetic theory of gases and the statistical mechanics of isolated molecules. In the liquid phase molecules are in close proximity, and this makes the application of molecular theories much more difficult. In both gas-phase kinetics and liquid-phase kinetics the determination of the rate law (i.e., the dependence of rate on various concentrations) gives an indication of the mechanism of the reaction. The factors that determine the rates of unimolecular, bimolecular, and termolecular reactions are considered. The chapter on kinetics in the gas phase closes with a consideration of various types of chain reactions and of heterogeneous reactions.

In the liquid phase many reactions are diffusion controlled; that is, their rate constants are determined by the rates with which the reactants can diffuse together. Relaxation methods are useful for studying very fast reactions in the liquid phase, and the theory of diffusion-controlled reactions yields an upper limit for the rate constants of bimolecular reactions. Catalysis is important in the liquid phase, and the most remarkable catalysts are the enzymes. The steady-state rate laws for enzymatic reactions tell us a great deal about the mechanism, and this is illustrated for the effects of inhibitors, hydrogen ion, and coenzymes.

The energy of light may be used to provide the activation energy for a chemical reaction. Since excited electronic states have different chemical properties from ground states, reactivities of photo-activated molecules may be quite different. To understand photochemistry, we draw on quantum mechanics. Once a molecule has absorbed a quantum of light, many different intramolecular processes may occur before that molecule dissociates or reacts with another molecule. It is important to understand these other intramolecular processes, including fluorescence and phosphorescence, which compete with chemical reactions. There are several ways in which activated molecules may transfer energy to other molecules. The quantum yields for various chemical reactions and for photosynthesis are discussed. The development of lasers has opened up new possibilities in photochemistry.

Part Three ends with a consideration of irreversible processes in solution: viscosity, electrical conductivity, and diffusion. These time-dependent processes yield useful information about the properties of solutions.

CHAPTER 14

KINETIC THEORY OF GASES

Kinetic theory helps us to understand certain thermodynamic properties in molecular terms. But kinetic theory goes beyond this and helps us to understand the rates of various processes. In this chapter the distribution of molecular velocities in a gas will be derived and used to calculate various types of average velocities and the rate of escape from a small hole. By introduction of the concept of collision cross section it is possible to calculate the frequency of molecular collisions and of the rates of transfer of mass, energy, and momentum in a gas composed of spherical molecules.

14.1 MODEL OF A PERFECT GAS

Elementary kinetic theory is based on a series of assumptions about gas molecules. It assumes that a volume of gas contains many molecules, but that the molecules fill a very small fraction of the volume. It assumes that molecules do not attract or repel each other and therefore travel in straight lines between collisions. It is also based on the assumption that collisions between molecules and with the wall are perfectly elastic. An elastic collision is one in which there is no change in total kinetic energy; thus the molecules do not absorb energy into internal motions, and the wall does not absorb energy from the gas. The molecules of real gases attract and repel each other in a way that we will describe later, and chemical reactions are examples of inelastic collisions.

According to kinetic theory, the pressure exerted by a gas on the walls of the container is equal to the average force per unit area exerted by the impacts of many molecules over a period of time.

Since the velocity of a molecule has both direction and magnitude it may be represented by a vector $\boldsymbol{v}$. The magnitude of this vector is represented by the scalar v, and we will refer to v as the *speed* of the molecule. The velocity $\boldsymbol{v}$ may be considered to be made up of components in the x, y, and z directions, as shown in Fig. 14.1. The magnitudes v_x, v_y, and v_z of these components are related to the speed v by

$$v^2 = v_x{}^2 + v_y{}^2 + v_z{}^2 \tag{14.1}$$

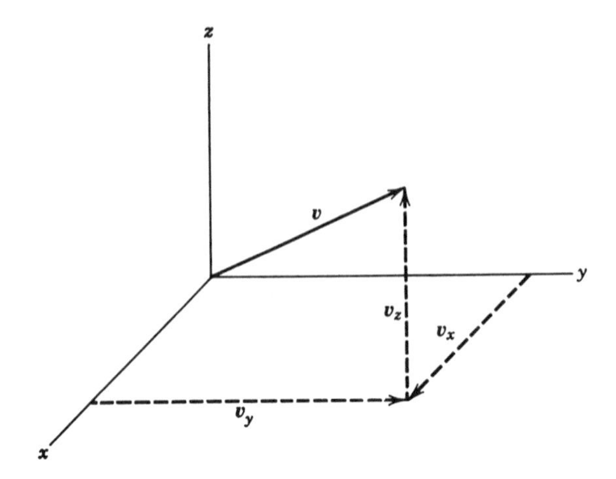

Fig. 14.1 Vector components $\boldsymbol{v}_x$, $\boldsymbol{v}_y$, and $\boldsymbol{v}_z$ of the vector velocity $\boldsymbol{v}$. The magnitude of the speed v and the magnitudes of the three components v_x, v_y, v_z are scalars.

A similar equation applies to each of the N molecules in the system.

$$\begin{aligned} v_1{}^2 &= v_{x1}^2 + v_{y1}^2 + v_{z1}^2 \\ v_2{}^2 &= v_{x2}^2 + v_{y2}^2 + v_{z2}^2 \\ v_3{}^2 &= v_{x3}^2 + v_{y3}^2 + v_{z3}^2 \\ &\vdots \end{aligned} \tag{14.2}$$

Thus

$$\sum_{i=1}^{N} v_i{}^2 = \sum_{i=1}^{N} v_{xi}^2 + \sum_{i=1}^{N} v_{yi}^2 + \sum_{i=1}^{N} v_{zi}^2 \tag{14.3}$$

The mean square speed in the gas sample is

$$\langle v^2 \rangle = \frac{1}{N} \sum_{i=1}^{N} v_i{}^2 \tag{14.4}$$

and the mean square speed components are similarly defined. Thus equation 14.3 may be written

$$\langle v^2 \rangle = \langle v_x{}^2 \rangle + \langle v_y{}^2 \rangle + \langle v_z{}^2 \rangle \tag{14.5}$$

Since molecular motion is random, and the orientation of the coordinate system is chosen arbitrarily,

$$\langle v_x{}^2 \rangle = \langle v_y{}^2 \rangle = \langle v_z{}^2 \rangle \tag{14.6}$$

so that

$$\langle v^2 \rangle = 3\langle v_x{}^2 \rangle \tag{14.7}$$

In order to calculate the pressure exerted by a gas, let us consider N molecules of mass m in a rectangular container with sides of length a, b, and c, as illustrated in Fig. 14.2. Molecule number 1 will move back and forth, colliding with face A with a time interval Δt between collisions of

$$\Delta t = \frac{2a}{v_{x1}} \tag{14.8}$$

where v_{x1} is the component of the velocity in the x direction.

The time-averaged pressure exerted by molecule 1 can be calculated using Newton's second law: force is equal to mass times acceleration. Thus the contribution of molecule number 1 to the force on side A in direction x is $m(dv_{x1}/dt) = d(mv_{x1})/dt$, the rate of transfer of momentum. Since molecule number 1 approaches the surface with momentum mv_{x1} and bounces off with momentum $-mv_{x1}$ in the x direction, the transfer of momentum to side A in the x direction is $2mv_{1x}$, and the average rate of change of momentum or force F is

$$F = \frac{2mv_{x1}}{\Delta t} = \frac{mv_{x1}^2}{a} \tag{14.9}$$

using equation 14.8. Since the area of side A is bc, the force per unit area or pressure produced by molecule 1 is

$$P_1 = \frac{mv_{x1}^2}{abc} = \frac{mv_{x1}^2}{V} \tag{14.10}$$

where V is the volume of the container.

The total pressure is the sum of the pressures exerted by the N molecules.

$$P = P_1 + P_2 + \cdots = \sum_{i=1}^{N} P_i = \sum_{i=1}^{N} \frac{mv_{xi}^2}{V} = \frac{Nm\langle v_x{}^2\rangle}{V} \tag{14.11}$$

where $\langle v_x{}^2\rangle$ is the mean square speed in the x direction.

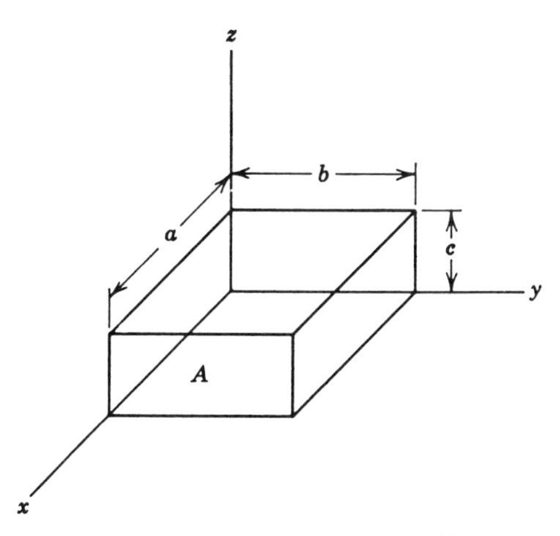

Fig. 14.2 Rectangular container with sides of length a, b, and c.

The expression for the total pressure may be written in terms of the mean square speed $\langle v^2 \rangle$ by use of equation 14.7.

$$PV = \tfrac{1}{3}Nm\langle v^2 \rangle = \tfrac{2}{3}(\tfrac{1}{2}Nm\langle v^2 \rangle) = \tfrac{2}{3}U_t \tag{14.12}$$

where the translational energy U_t of our idealized gas is the total kinetic energy $\tfrac{1}{2}Nm\langle v^2 \rangle$. If this equation is applied to a mole of molecules, $PV = RT$ and

$$U_t = \tfrac{3}{2}RT \tag{14.13}$$

or, for a single molecule,

$$\epsilon_t = \tfrac{3}{2}kT \tag{14.14}$$

Thus we arrive at the conclusion that the kinetic energy of a perfect gas is directly proportional to the thermodynamic temperature. The thermodynamic temperature is a measure of the average kinetic energy of gas molecules. At 300 K the kinetic energy is $\tfrac{3}{2}(8.314\ \text{J K}^{-1}\ \text{mol}^{-1})(300\ \text{K}) = 3741\ \text{J mol}^{-1}$. At 1273 K it is 15.9 kJ mol^{-1}. It is important to note that for a perfect gas the translational kinetic energy is independent of the volume or pressure and the molar mass or type of molecule and depends only on the temperature. Thus a helium atom has the same translational kinetic energy, on the average, as a heavy hydrocarbon molecule at the same temperature.

Example 14.1 What are the mean-square and root-mean-square speeds of hydrogen molecules at 0 °C?

Considering 1 mol of hydrogen, equations 14.12 and 14.13 can be combined to obtain

$$\langle v^2 \rangle = \frac{3RT}{M} = \frac{3(8.314\ \text{J K}^{-1}\ \text{mol}^{-1})(273.15\ \text{K})}{2.016 \times 10^{-3}\ \text{kg mol}^{-1}} = 3.379 \times 10^6\ \text{m}^2\ \text{s}^{-2}$$

$$\langle v^2 \rangle^{1/2} = 1.84 \times 10^3\ \text{m s}^{-1}$$

The root-mean-square speed of a hydrogen molecule at 0 °C is 4110 mph, but at ordinary pressures it travels only an exceedingly short distance before colliding with another molecule and changing direction.

14.2 DISTRIBUTION OF MOLECULAR VELOCITIES

To illustrate what we mean by velocity distribution, let us consider the distribution of velocities in the x direction for a sample of gas containing a very large number N of molecules. The distribution of velocity components in the x direction, v_x, at any instant may be described by dividing the span of velocities into equal intervals Δv_x and determining the numbers of molecules whose v_x values fall in the different intervals. It is convenient to plot as the ordinate, not the fraction of molecules in a given velocity interval, but the fraction divided by the width of the interval, as

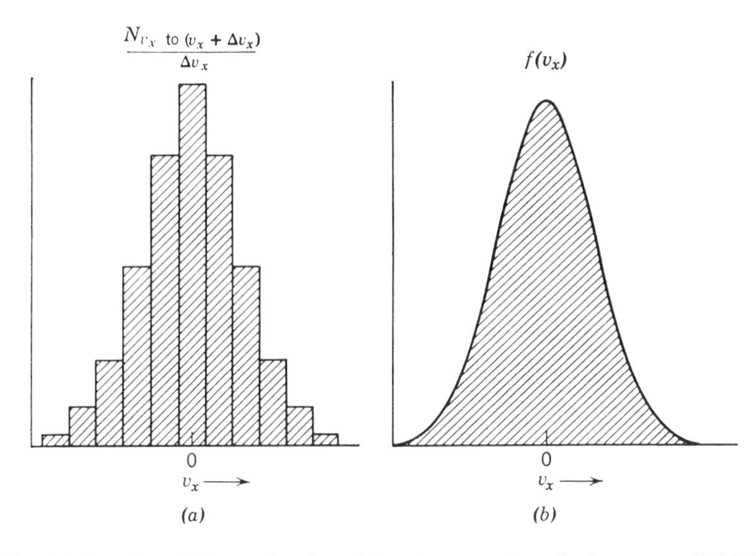

Fig. 14.3 (*a*) Fraction of the molecules with v_x in a certain velocity range divided by the width of the velocity range plotted against average velocity. (*b*) Continuous distribution curve for velocities in the *x* direction.

shown in Fig. 14.3*a*. When this is done, the *area* of each rectangle in the bar graph gives the *fraction* of the molecules in the specified interval.

The accuracy of this representation can be increased by reducing the interval Δv_x. When this is done, the bar graph approaches a smooth curve. In the limit of smaller and smaller intervals the fraction of molecules divided by the velocity interval is represented by $f(v_x)$, the *probability density* function for molecular velocities. The *fraction* of molecules having velocities within the range v_x to $v_x + dv_x$ is then $f(v_x)\, dv_x$, which is the *probability* that the velocity is in this range. The dv_x appears in the probability $f(v_x)\, dv_x$ because the fraction of molecules in the velocity range of width dv_x is proportional to the width of the range dv_x. A plot of the probability density $f(v_x)$ versus v_x such as that shown in Fig. 14.3*b* completely describes the distribution of molecular velocities in a single direction. The fraction of molecules with *x* component of velocity between any two given values is equal to the area under the curve between these two values of v_x. The mathematical forms of the velocity distributions in the three directions are the same because these are arbitrarily chosen directions, and the gas is isotropic.

In deriving the equation for $f(v_x)$ we will use a method first used by Maxwell in 1859. In order to do that we need to consider the probability density $F(v_x, v_y, v_z)$ for velocity components v_x, v_y, and v_z. The quantity $F(v_x, v_y, v_z)$ is defined so that $F(v_x, v_y, v_z)\, dv_x\, dv_y\, dv_z$ is the fraction of the molecules that simultaneously have v_x between v_x and $v_x + dv_x$, v_y between $v_y + dv_y$, and v_z between v_z and $v_z + dv_z$. Maxwell arrived at the functional form of $F(v_x, v_y, v_z)$ by realizing that the velocity components are *independent* and that the choice of directions of the coordinate axes is *arbitrary*. Since the velocity components in the three directions are independent, the probability density $F(v_x, v_y, v_z)$ is simply the product of the probability densities in the three directions.

$$F(v_x, v_y, v_z) = f(v_x)f(v_y)f(v_z) \tag{14.15}$$

Furthermore, since the velocity distribution must be independent of the orientation of the x, y, and z axes in space, the probability density $F(v_x, v_y, v_z)$ can depend only on the magnitude v of the velocity vector. Thus $F(v_x, v_y, v_z) = g(v)$, and equation 14.15 becomes

$$g(v) = f(v_x)f(v_y)f(v_z) \tag{14.16}$$

Differentiating both sides of this equation with respect to v_x yields

$$\frac{\partial}{\partial v_x}[g(v)]_{v_y,v_z} = \left(\frac{dg(v)}{dv}\right)\left(\frac{\partial v}{\partial v_x}\right)_{v_y,v_z} = f(v_y)f(v_z)\frac{df(v_x)}{dv_x} \tag{14.17}$$

Differentiating equation 14.1 yields $\partial v/\partial v_x = v_x/v$. Substituting this relation in the preceding equation,

$$\frac{dg(v)}{v\,dv} = f(v_y)f(v_z)\frac{df(v_x)}{v_x\,dv_x} \tag{14.18}$$

Dividing both sides by $g(v)$ yields

$$\frac{dg(v)}{g(v)v\,dv} = \frac{df(v_x)}{f(v_x)v_x\,dv_x} \tag{14.19}$$

This process may also be carried out with v_y and v_z so that we obtain

$$\frac{df(v_x)}{f(v_x)v_x\,dv_x} = \frac{df(v_y)}{f(v_y)v_y\,dv_y} = \frac{df(v_z)}{f(v_z)v_z\,dv_z} \tag{14.20}$$

Since a function of v_x alone is equal to a function of v_y alone, which is equal to a function of v_z alone, each must separately equal a constant, which we will represent by $-\lambda$. Thus

$$\frac{df(v_x)}{f(v_x)} = -\lambda v_x\,dv_x \tag{14.21}$$

$$d[\ln f(v_x)] = -\frac{\lambda}{2}d(v_x^2) \tag{14.22}$$

Integration yields

$$f(v_x) = Ae^{-\lambda v_x^2/2} \tag{14.23}$$

The constant λ is a positive quantity; otherwise $f(v_x)$ would increase without limit. The value of the constant A may be determined from the fact that the probability that v_x is between $-\infty$ and $+\infty$ is unity.

$$1 = \int_{-\infty}^{+\infty} f(v_x)\,dv_x = A\int_{-\infty}^{+\infty} e^{-\lambda v_x^2/2}\,dv_x \tag{14.24}$$

Since

$$\int_{-\infty}^{\infty} e^{-\alpha x^2}\,dx = \left(\frac{\pi}{\alpha}\right)^{1/2} \tag{14.25}$$

$A = (\lambda/2\pi)^{1/2}$, and so the normalized velocity distribution in the x direction is given by

$$f(v_x) = \left(\frac{\lambda}{2\pi}\right)^{1/2} e^{-\lambda v_x^2/2} \tag{14.26}$$

In order to determine the value of λ we can make use of the fact that we already have an expression for the translational kinetic energy of a perfect gas; according to equation 14.14, $\epsilon_t = 3kT/2$. Since this is the translational kinetic energy for motion in three dimensions, the average translational kinetic energy per molecule for motion in the x direction is

$$\epsilon_x = \tfrac{1}{2}m\langle v_x^2\rangle = \tfrac{1}{2}kT \tag{14.27}$$

where the Boltzmann constant k is equal to the perfect gas constant divided by the Avogadro constant N_A.

$$k = \frac{R}{N_A} = \frac{8.314\ \text{J K}^{-1}\ \text{mol}^{-1}}{6.022 \times 10^{23}\ \text{mol}^{-1}}$$

$$= 1.381 \times 10^{-23}\ \text{J K}^{-1}. \tag{14.28}$$

Thus we may think of the Boltzmann constant as the gas constant per molecule or as the conversion factor between the temperature scale and the molecular energy scale.

The average square of the velocity in the x direction may be calculated using equation 14.26 in

$$\langle v_x^2\rangle = \int_{-\infty}^{\infty} v_x^2 f(v_x)\, dv_x = \frac{1}{\lambda} \tag{14.29}$$

since

$$\int_{-\infty}^{\infty} x^2 e^{-\alpha x^2}\, dx = \frac{\pi^{1/2}}{2\alpha^{3/2}} \tag{14.30}$$

Thus $\lambda = m/kT$. This yields the following probability $f(v_x)\, dv_x$ that a molecule will have a velocity between v_x and $v_x + dv_x$.

$$f(v_x)\, dv_x = \frac{\exp(-mv_x^2/2kT)\, dv_x}{(2\pi kT/m)^{1/2}} \tag{14.31}^*$$

---

* This expression for the velocity distribution in the x direction may also be derived by substituting the expression for the kinetic energy of a molecule in the x direction ($\epsilon_x = \tfrac{1}{2}mv_x^2$) into the Boltzmann distribution (Section 13.3).

$$N_i = Ae^{-\varepsilon_i/kT}$$

and normalizing to determine the value of A. Another simple example of the application of the Boltzmann distribution is the so-called barometric formula for the pressure P in an isothermal atmosphere.

$$P = P_0 e^{-mgh/kT}$$

where P_0 is the pressure at height $h = 0$, and mgh is the potential energy of a single molecule at height h. Here m is the mass of a molecule and g is the acceleration of gravity.

A plot of $f(v_x)$ versus v_x has the form shown by Fig. 14.3*b*. Since the probability that a molecule will have a velocity between v_x and $v_x + dv_x$ is $f(v_x)\,dv_x$, it can be seen that the most probable velocity in the x direction is zero. The probabilities of velocities in the positive and negative directions are, of course, equal, the gas as a whole being stationary. The probabilities of very high velocities in either the positive or negative x direction are very small.

The distribution functions for $f(v_y)$ and $f(v_z)$ have the same functional form as $f(v_x)$.

14.3 PROBABILITY OF MOLECULAR SPEEDS

Now that we have calculated the probability densities in the x, y, and z directions we can calculate the fraction of molecules that simultaneously have component velocities in the ranges v_x to $v_x + dv_x$, v_y to $v_y + dv_y$, and v_z to $v_z + dv_z$. Using equations 14.15 and 14.31, this fraction is

$$F(v_x, v_y, v_z)\,dv_x\,dv_y\,dv_z = \left(\frac{m}{2\pi kT}\right)^{3/2} e^{-m(v_x{}^2 + v_y{}^2 + v_z{}^2)/2kT}\,dv_x\,dv_y\,dv_z \quad (14.32)$$

since $e^a e^b e^c = e^{a+b+c}$. However, for most purposes we are interested in the fraction of the molecules having a speed in the range v to $v + dv$, independent of the direction. Imagine that the origins of the velocity vectors of all the molecules in a sample of gas are brought to the origin of a coordinate system: we are interested in the molecules having speed vectors ending in a spherical shell of radius v around the origin with thickness dv. The volume of this shell is $4\pi v^2\,dv$, and so the fraction of the molecules with velocities v in the range v to $v + dv$ is given by

$$f(v)\,dv = \left(\frac{m}{2\pi kT}\right)^{3/2} e^{-mv^2/2kT} 4\pi v^2\,dv \quad (14.33)$$

This equation, which was first derived by Maxwell in 1860, is of fundamental importance to the kinetic theory of gases.

A plot of $f(v)$ versus the magnitude of the molecular speed v is shown in Fig. 14.4 for oxygen at 100, 300, 500, and 1000 K. The probability that a molecule has its speed between any two values is given by the area under the curve between these two values of the speed. In contrast with Fig. 14.3*b* for the velocity components in a given direction, it can be seen that the probability of a zero speed is zero. Figure 14.3*b* is based on equation 14.31, which does not have the term v^2 contained in equation 14.33. Small speeds are favored by the exponential factor in equation 14.33, but high speeds are favored by the v^2 factor. The plot of $f(v)$ versus v is approximately parabolic near the origin. At higher speeds the probability decreases toward zero because the exponential term decreases much more rapidly than v^2 increases. Thus very few molecules have very high or very low speeds. The fraction

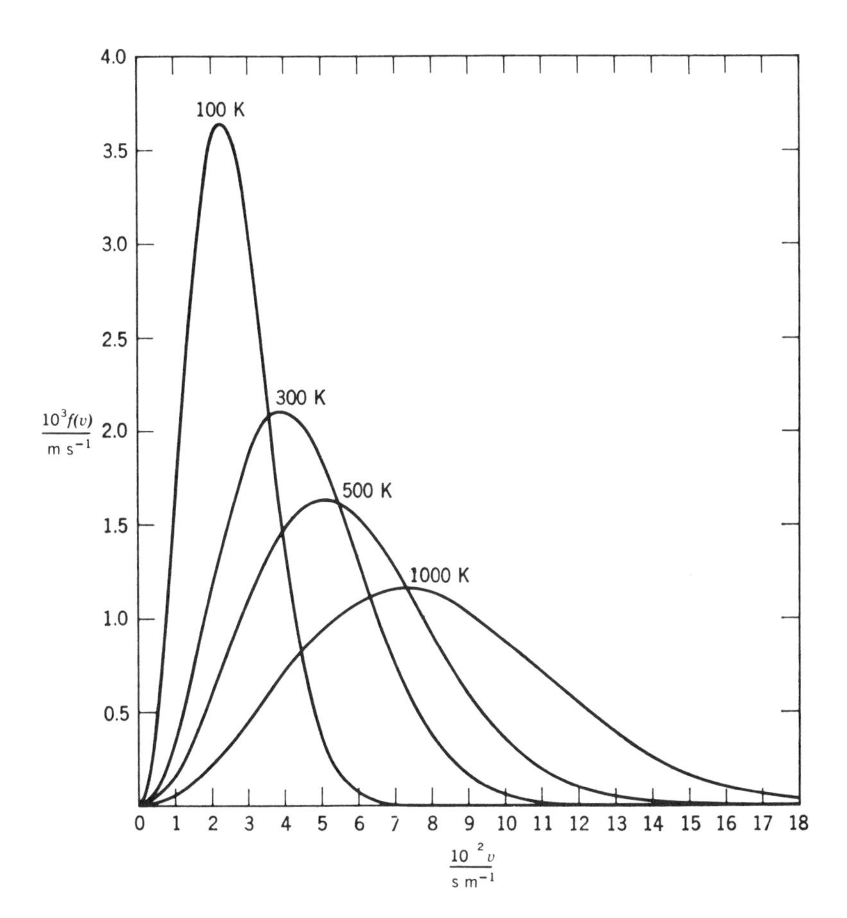

Fig. 14.4 Probability density of various speeds v for oxygen at 100, 300, 500, and 1000 K, calculated using equation 14.33.

of the molecules having speeds greater than 10 times the most probable speed is 9×10^{-42} at any temperature. The Avogadro constant times this fraction is a very small number. As the temperature is increased, the most probable speed (which is given by the maximum in the plot) moves to higher speeds and the distribution of speeds also becomes broader, but the relative shape is independent of temperature.

14.4 TYPES OF AVERAGE SPEEDS

Since there is a distribution of speeds, various types of averages may be considered. The most probable speed v_p is obtained by setting $df(v)/dv$ equal to zero.

$$\frac{df(v)}{dv} = \left(\frac{m}{2\pi kT}\right)^{3/2} e^{-mv^2/2kT}\left[8\pi v + 4\pi v^2\left(\frac{-mv}{kT}\right)\right] \tag{14.34}$$

This derivative vanishes at $v = v_p$, where

$$v_p = \left(\frac{2kT}{m}\right)^{1/2} = \left(\frac{2RT}{M}\right)^{1/2} \tag{14.35}$$

The mean speed $\langle v \rangle$ is obtained by summing all the speeds and dividing by the total number of molecules N.

$$\langle v \rangle = \frac{1}{N} \sum_{i=1}^{N} v_i \tag{14.36}$$

We may also think of the mean speed $\langle v \rangle$ as the summation of each possible speed times the probability $P(v)$ of that speed.

$$\langle v \rangle = \sum vP(v) \tag{14.37}$$

As discussed earlier, the probability that a molecule has a speed between v and $v + dv$ is represented by $f(v)\, dv$, so that

$$\langle v \rangle = \int_0^\infty vf(v)\, dv \tag{14.38}$$

We may use an integration because the number of molecules in a macroscopic sample is so large that the speed distribution may be taken to be continuous. Substituting the Maxwell distribution,

$$\langle v \rangle = 4\pi \left(\frac{m}{2\pi kT}\right)^{3/2} \int_0^\infty e^{-mv^2/2kT} v^3\, dv \tag{14.39}$$

Substituting $mv^2/2kT = x^2$ and $dv = 2kTx\, dx/mv$ and using

$$\int_0^\infty x^3 e^{-x^2}\, dx = \tfrac{1}{2} \tag{14.40}$$

we obtain

$$\langle v \rangle = \left(\frac{8kT}{\pi m}\right)^{1/2} = \left(\frac{8RT}{\pi M}\right)^{1/2} \tag{14.41}$$

The root-mean-square speed $\langle v^2 \rangle^{1/2}$ is defined by

$$\langle v^2 \rangle^{1/2} = \left(\frac{1}{N} \sum_{i=1}^{N} v_i^2\right)^{1/2} \tag{14.42}$$

The symbol $\sum v_i^2$ indicates the sum of the squares of the speeds $v_1, v_2, v_3, \ldots, v_N$ of all the N individual molecules. Since the speed distribution is continuous, the root-mean-square speed is obtained by multiplying each speed squared by the probability of that speed, integrating over all speeds, and taking the square root.

$$\langle v^2 \rangle^{1/2} = \left[\int_0^\infty v^2 f(v)\, dv\right]^{1/2} \tag{14.43}$$

Substituting the Maxwell distribution and using the relation

$$\int_0^\infty x^4 e^{-x^2}\, dx = \tfrac{3}{8}\sqrt{\pi} \tag{14.44}$$

we obtain a root-mean-square speed of

$$\langle v^2\rangle^{1/2} = \left(\frac{3kT}{m}\right)^{1/2} = \left(\frac{3RT}{M}\right)^{1/2} \tag{14.45}$$

in agreement with the simple calculation in Example 14.1. It can be seen that at a given temperature each of these average speeds is inversely proportional to the square root of the molar mass. Lighter molecules move more rapidly, so that their average kinetic energies are exactly equal to those for the heavier molecules. The speed of sound in a gas is of the same order of magnitude as these molecular speeds.*

Example 14.2 Calculate the most probable speed v_p and the arithmetic mean speed $\langle v\rangle$ for hydrogen molecules at 0 °C.

$$v_p = \left(\frac{2RT}{M}\right)^{1/2} = \left[\frac{(2)(8.314\ \text{J K}^{-1}\ \text{mol}^{-1})(273\ \text{K})}{(2.016 \times 10^{-3}\ \text{kg mol}^{-1})}\right]^{1/2}$$

$$= 1.50 \times 10^3\ \text{m s}^{-1}$$

$$\langle v\rangle = \left(\frac{8RT}{\pi M}\right)^{1/2} = \left[\frac{(8)(8.314\ \text{J K}^{-1}\ \text{mol}^{-1})(273\ \text{K})}{(3.1416)(2.016 \times 10^{-3}\ \text{kg mol}^{-1})}\right]^{1/2}$$

$$= 1.69 \times 10^3\ \text{m s}^{-1}$$

These values may be compared with the root-mean-square speed calculated in Example 14.1.

The distribution of molecular velocities may be determined experimentally using the molecular beam method. A beam of molecules moving in a vacuum is collimated by slits and passed through a velocity selector that allows only those molecules in a narrow range of velocity to reach the detector. A velocity selector may be constructed from a system of toothed disks on a shaft that is rotated at an accurately controlled speed with slots appropriately arranged. A typical molecular beam apparatus for studying velocity distributions as well as molecular collision cross sections (Section 14.6) is shown in Fig. 14.5. Through the use of such an apparatus the Maxwell-Boltzmann velocity distribution has been checked experimentally.

---

* The speed of the pressure wave depends on the speed of the gas molecules themselves. Since the sound wave causes adiabatic heating, it is not surprising that the speed of sound in a gas depends on the heat capacity. A detailed analysis (R. D. Present, *Kinetic Theory of Gases*, McGraw-Hill Book Co., New York, 1958) shows that the speed of sound in a gas is $(C_P RT/C_V M)^{1/2}$.

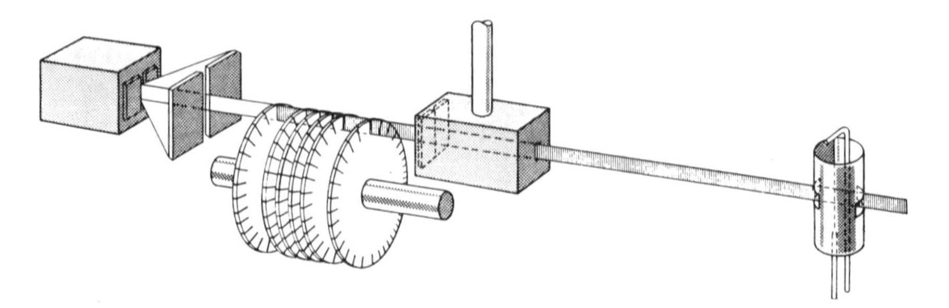

Fig. 14.5 Molecular beam apparatus. Starting at the left, the parts are: effusion source, collimating slit, slotted-disk-type velocity selector, scattering chamber, and detector.

14.5 COLLISIONS WITH A WALL OR OPENING

The number of collisions per unit time per unit area of wall may be derived by considering Fig. 14.6, which depicts the volume containing all the molecules with certain components v_x, v_y, and v_z that will strike area A in time δt. The volume of the cylinder is equal to its height, $v_z\ \delta t$, times the area of the end A. Thus the number of molecules in the cylinder is $\rho A v_z\ \delta t$, where ρ is the concentration of molecules. The probability that the molecules will have the required velocity components is given by equation 14.15. The number of collisions is therefore

$$\rho(Av_z\ \delta t)F(v_x, v_y, v_z)\ dv_x\ dv_y\ dv_z = \rho(Av_z\ \delta t)f(v_x)f(v_y)f(v_z)\ dv_x\ dv_y\ dv_z \quad (14.46)$$

where the probability densities are given by equations like equation 14.31. We must now integrate over all values of the components of the velocities to obtain the total number of collisions. The number z of collisions per unit area of wall per unit time is

$$z = \rho \int_0^{\infty} \int_{-\infty}^{+\infty} \int_{-\infty}^{-\infty} v_z f(v_x)f(v_y)f(v_z)\ dv_x\ dv_y\ dv_z \quad (14.47)$$

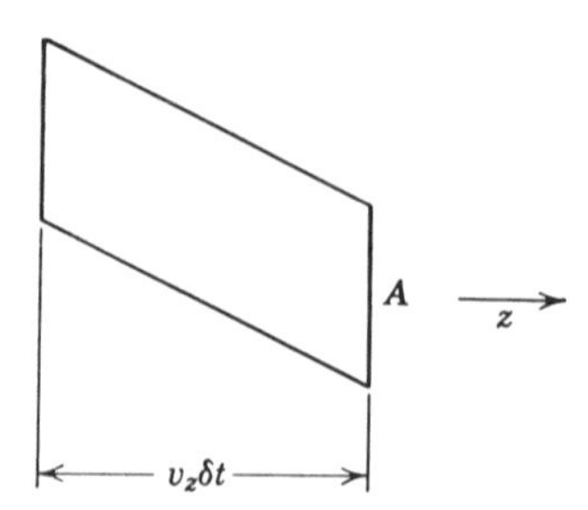

Fig. 14.6 Oblique cylinder containing all the molecules with certain components v_x, v_y, v_z, which will strike area A in time δt.

The integration for the z component is only for positive velocities, since we are only interested in molecules that are moving toward the surface. Since the probability densities in equation 14.31 have been normalized, the integrations over v_x and v_y simply yield unity, so that

$$z = \rho\left(\frac{m}{2\pi kT}\right)^{1/2}\int_0^{\infty} v_z e^{-mv_z^2/2kT}\, dv_z = \rho\sqrt{\frac{kT}{2\pi m}} = \rho\sqrt{\frac{RT}{2\pi M}} = \frac{\rho\langle v\rangle}{4} \tag{14.48}$$

This is also the number of molecules per square meter per second that would escape through a small hole in the vessel containing the gas. We refer to a hole that is small compared with the mean free path (Section 14.6), so that the spatial and velocity distributions of the molecules inside the vessel are not appreciably affected by the effusion loss. If the pressure on the other side of the hole is so low that no molecules return, equation 14.48 gives the flux (molecules m^{-2} s^{-1}) out of the hole. The rates of flow of two gases through a small aperture are inversely proportional to the square root of the molar mass. The proportionality is exact if the cross section of the aperture is small compared with the average distance through which a molecule travels before colliding with another molecule.

Equation 14.48 is the basis for the Knudsen method for measuring the vapor pressures of solids. The mass w of molecules striking an opening per unit area per unit time is $w = zM/N_A$. Introducing the perfect gas law in the form $P = \rho RT/N_A$, where ρ is the concentration of molecules, and equations 14.41 and 14.48, we obtain

$$P = w\left(\frac{2\pi RT}{M}\right)^{1/2} \tag{14.49}$$

where P is pressure in newtons per square meter and w is the rate of effusion through the hole in kilograms per square meter per second. A sufficiently large surface area of the solid must be exposed to maintain the saturation vapor pressure.

Example 14.3 The vapor pressure of solid beryllium was measured by R. B. Holden, R. Speiser, and H. L. Johnston [*J. Am. Chem. Soc.*, **70**, 3897 (1948)] using a Knudsen cell. The effusion hole was 0.318 cm in diameter, and they found a weight loss of 9.54 mg in 60.1 min at a temperature of 1457 K. What is the vapor pressure?

$$P = w\sqrt{\frac{2\pi RT}{M}} = \frac{(9.54 \times 10^{-6}\ \text{kg})}{\pi(0.159 \times 10^{-2}\ \text{m})^2(60 \times 60.1\ \text{s})}$$

$$\times \sqrt{\frac{2\pi(8.314\ \text{J K}^{-1}\ \text{mol}^{-1})(1457\ \text{K})}{(9.013 \times 10^{-3}\ \text{kg mol}^{-1})}} = 0.968\ \text{Pa}$$

Since an atmosphere is 101,325 Pa, the vapor pressure is 9.55×10^{-6} atm.

14.6 COLLISION DIAMETER AND MEAN FREE PATH

The preceding results have been obtained without any consideration of collisions between molecules. However, since the mean free path and the transport properties (diffusion, heat conductivity, and viscosity) are determined by molecular collisions, it is necessary to introduce the concept of molecular size. To a first approximation, we may assume that the molecules are rigid, noninteracting spheres with diameter σ. Figure 14.7 shows that for molecules with a diameter of σ, the effective area of the target is $\pi\sigma^2$. The quantity $\pi\sigma^2$ is called the *collision cross section* for the rigid spherical molecule, because it is the cross-sectional area of an imaginary sphere surrounding the molecule into which the center of another molecule cannot penetrate.

The mean free path is the average distance traversed by a molecule between collisions. If the number of molecules per unit volume is ρ, the number of targets in a volume perpendicular to the x direction and of thickness dx is $A\rho\, dx$, where A is the area of the plane. The total target area in this volume is $A\pi\sigma^2\rho\, dx$. The probability that a molecule moving in the x direction will collide with a molecule in this volume is equal to the fraction of the area covered by molecules.

$$\text{Probability} = \pi\sigma^2\rho\, dx \tag{14.50}$$

To obtain an approximate expression for the mean free path, let us consider a beam of molecules moving in the x direction. The number of molecules in the beam at distance x that have not collided with another molecule is represented by $n(x)$. The number of molecules colliding in distance dx is $n(x)\pi\sigma^2\rho\, dx$. Thus the attenuation of the beam in distance dx is

$$n(x + dx) - n(x) = -n(x)\pi\sigma^2\rho\, dx \tag{14.51}$$

or

$$\frac{dn}{dx} = -\pi\sigma^2\rho n(x) \tag{14.52}$$

If there were n_0 molecules in the beam at $x = 0$, equation 14.52 may be integrated.

$$\int_{n_0}^{n(x)} \frac{dn}{n(x)} = -\pi\sigma^2\rho \int_0^x dx \qquad n(x) = n_0 e^{-\pi\sigma^2\rho x} \tag{14.53}$$

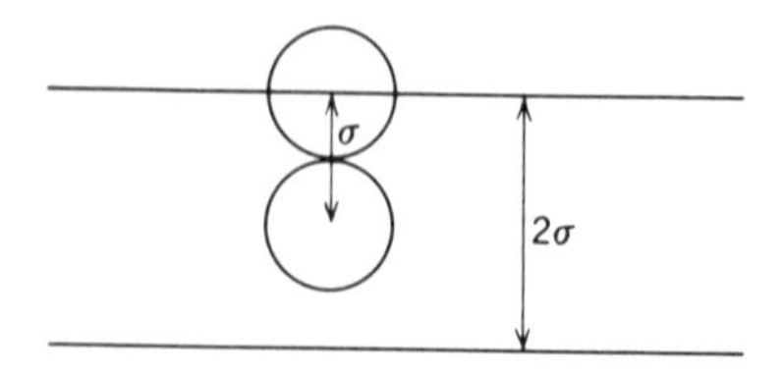

Fig. 14.7 A rigid spherical molecule collides with like molecules striking an area $\pi\sigma^2$ perpendicular to the path of the approaching molecule.

Thus, according to this simple model, the intensity of the beam decreases exponentially. The use of a molecular beam apparatus such as that illustrated in Fig. 14.5 allows the determination of the collision diameter for a particular pair of molecules by use of equation 14.53.

The mean free path is obtained by multiplying the distance x by dn, the number of molecules colliding between x and $x + dx$, integrating over all values of x, and dividing by the number of molecules in the beam.

$$l = \frac{1}{n_0}\int_0^\infty x\,dn = -\frac{1}{n_0}\int_0^\infty x\frac{dn}{dx}\,dx = \frac{1}{\pi\sigma^2\rho} \qquad \text{(approximate)} \qquad (14.54a)$$

$$l = \frac{1}{2^{1/2}\pi\sigma^2\rho} \qquad \text{(exact)} \qquad (14.54b)$$

The error of $2^{1/2}$ comes from the assumption that the target molecules are stationary.

Assuming that the collision diameter σ is independent of temperature, the temperature and pressure dependence of the mean free path may be obtained by substituting the perfect gas law in the form $\rho = P/kT$.

$$l = \frac{kT}{2^{1/2}\pi\sigma^2 P} \qquad (14.55)$$

Thus, at constant temperature, the mean free path is inversely proportional to the pressure.

Example 14.4 For oxygen at 25 °C the collision diameter is 0.361 nm. What is the mean free path at (*a*) 1 atm pressure, and (*b*) 0.1 Pa.

$$(a)\ \rho = \frac{N}{V} = \frac{PN_A}{RT} = \frac{(1\ \text{atm})(6.022\times10^{23}\ \text{mol}^{-1})(10^3\ \text{L m}^{-3})}{(0.082\,051\ \text{L atm K}^{-1}\ \text{mol}^{-1})(298\ \text{K})} = 2.46\times10^{25}\ \text{m}^{-3}$$

$$l = [(1.414)(3.14)(3.61\times10^{-10}\ \text{m})^2(2.46\times10^{25}\ \text{m}^{-3})]^{-1} = 7.02\times10^{-8}\ \text{m}$$

$$(b)\ \rho = \frac{PN_A}{RT} = \frac{(0.1\ \text{Pa})(6.022\times10^{23}\ \text{mol}^{-1})}{(8.314\ \text{J K}^{-1}\ \text{mol}^{-1})(298\ \text{K})} = 2.43\times10^{19}\ \text{m}^{-3}$$

$$l = [(1.414)(3.14)(3.61\times10^{-10}\ \text{m})^2(2.43\times10^{19}\ \text{m}^{-3})]^{-1} = 0.071\ \text{m} = 7.1\ \text{cm}$$

At pressures so low that the mean free path becomes comparable with the dimensions of the containing vessel, the flow properties of the gas become markedly different.

In this section we have assumed that collisions are elastic, but they may also be inelastic (energy is absorbed into internal motions) or chemical reaction may occur.

14.7 MOLECULAR COLLISIONS IN A GAS

In deriving the expression for the number of collisions between two types of gas molecules it is necessary to use the average relative velocity $\langle v_r\rangle$, which is defined by

$$\langle v_r\rangle = (\langle v_1^2\rangle + \langle v_2^2\rangle)^{1/2} = \left(\frac{8kT}{\mu\pi}\right)^{1/2} \qquad (14.56)$$

where μ is the reduced mass (Section 8.12). One molecule of type 1 will strike

$$Z_1 = \rho_2 \pi \sigma^2 \langle v_r \rangle \tag{14.57}$$

molecules of type 2 per unit time. To calculate the number of collisions Z_{12} of molecules of type 1 with molecules of type 2 per unit time per unit volume of gas, we simply have to multiply Z_1 by the number density ρ_1 of molecules of type 1.

$$Z_{12} = \rho_1 \rho_2 \pi \sigma^2 \langle v_r \rangle \tag{14.58}$$

If the collisions are between like molecules,

$$Z_{11} = \frac{1}{2^{1/2}} \rho^2 \pi \sigma^2 \langle v \rangle \tag{14.59}$$

where a factor of $\frac{1}{2}$ has been introduced so that each collision is not counted twice and $\langle v_r \rangle$ has been replaced by $2^{1/2}\langle v \rangle$, where $\langle v \rangle = (8kT/\pi m)^{1/2}$. Similarly, a single molecule moving through like molecules undergoes

$$Z_1 = 2^{1/2} \rho \pi \sigma^2 \langle v \rangle \tag{14.60}$$

collisions per unit time.

Example 14.5 For oxygen at 25 °C, calculate the number Z_1 of collisions per second per molecule, the number Z_{11} of collisions per cubic meter per second, and the number of moles of collisions per liter per second at a pressure of 1 atm. The collision diameter of oxygen in 0.361 nm or 3.61×10^{-10} m, as determined in a manner to be described shortly.

$$\langle v \rangle = \left(\frac{8RT}{\pi M}\right)^{1/2} = \left[\frac{(8)(8.314 \text{ J K}^{-1} \text{ mol}^{-1})(298 \text{ K})}{\pi(32 \times 10^{-3} \text{ kg mol}^{-1})}\right]^{1/2} = 444 \text{ m s}^{-1}$$

From Example 14.4, $\rho = 2.46 \times 10^{25}$ m^{-3}.

$$\begin{aligned} Z_1 &= \sqrt{2}\pi\sigma^2\langle v \rangle \rho \\ &= (1.414)(3.14)(3.61 \times 10^{-10} \text{ m})^2(444 \text{ m s}^{-1})(2.46 \times 10^{25} \text{ mol m}^{-3}) \\ &= 6.32 \times 10^9 \text{ s}^{-1} \end{aligned}$$

which is the number of collisions per molecule per second.

$$\begin{aligned} Z_{11} &= \frac{1}{2^{1/2}} \rho^2 \pi \sigma^2 \langle v \rangle \\ &= (0.707)(2.46 \times 10^{25} \text{ m}^{-3})^2(3.14)(3.61 \times 10^{-10} \text{ m})^2(444 \text{ m s}^{-1}) \\ &= 7.77 \times 10^{34} \text{ m}^{-3} \text{ s}^{-1} = \frac{(7.77 \times 10^{34} \text{ m}^{-3} \text{ s}^{-1})(10^{-3} \text{ m}^3 \text{ L}^{-1})}{6.022 \times 10^{23} \text{ mol}^{-1}} \\ &= 1.29 \times 10^8 \text{ mol}^{-1} \text{ L}^{-1} \text{ s}^{-1} \end{aligned}$$

which is the number of moles of collisions per liter per second.

Actual chemical reaction rates are usually far smaller than the collision rates, indicating that the collisions must involve certain energies and orientations to lead to reaction.

14.8 MOLECULAR INTERACTIONS

Collisions between gas molecules are more complicated than indicated in the preceding section because of intermolecular attractive and repulsive forces. There is attraction at longer distances and repulsion at shorter distances. The attractive force usually varies inversely with the seventh power of the intermolecular distance (Section 10.14) for large separations. The repulsive force varies inversely with a higher power of the intermolecular distance.

The interaction of two molecules or atoms may be represented by a plot of potential energy versus intermolecular distance, as shown in Fig. 14.8. The potential energy V is the negative of the work required to bring one of the molecules from infinity (where the potential energy is defined as zero) to distance r. The force F, which one molecule exerts on the other, is the negative derivative of the potential energy.

$$F = -\frac{dV}{dr} \tag{14.61}$$

To the right of the minimum dV/dr is positive so that the force is negative; this means that the molecules are attracted toward each other. To the left of the

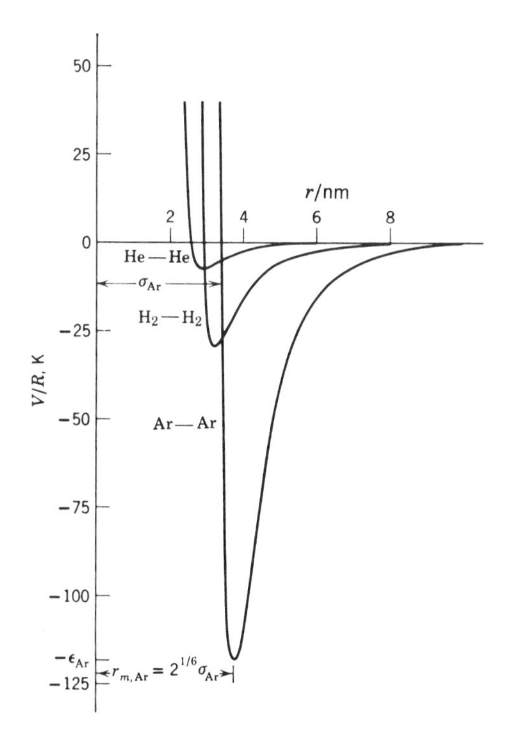

Fig. 14.8 Potential energy expressed as V/R for the interaction between two molecules.

minimum dV/dr is negative so that the force is positive; this means that the molecules repel each other.

The functional form that is often used to represent the intermolecular potential energy was first introduced by Lennard-Jones.

$$V = 4\epsilon\left[\left(\frac{\sigma}{r}\right)^{12} - \left(\frac{\sigma}{r}\right)^{6}\right] \tag{14.62}$$

In this so-called "6–12 equation" the inverse sixth power dependence comes from the theory for the van der Waals attraction (Section 10.14), but the twelfth power dependence for the repulsive energy is chosen empirically. The significance of the parameters ϵ and σ is shown in Fig. 14.8. The depth of the potential well is ϵ. The distance between atoms or molecules at which $V = 0$ is σ.

The values of parameters ϵ and σ may be obtained from scattering experiments and from studies of transport properties (Section 14.11).

Example 14.6 By differentiation of the expression for the Lennard-Jones potential, show that the distance r_m at the minimum where $dV/dr = 0$ is $r_m = 2^{1/6}\sigma$. Substituting this relationship into equation 14.62, show that the Lennard-Jones potential may also conveniently be given by

$$V = \epsilon\left[\left(\frac{r_m}{r}\right)^{12} - 2\left(\frac{r_m}{r}\right)^{6}\right]$$

Differentiating equation 14.62 and setting the derivative equal to zero, we obtain

$$\frac{dV}{dr} = 4\epsilon\left[-\frac{12\sigma^{12}}{r^{13}} + \frac{6\sigma^{6}}{r^{7}}\right] = 0$$

If we represent the value of r at the minimum by r_m,

$$r_m = 2^{1/6}\sigma$$

Substituting this in equation 14.62 yields the desired expression for V.

For molecules, in contrast with atoms, the intermolecular forces depend on the relative orientation of the molecules, in addition to the distance from each other.

The effects of intermolecular attractions and repulsions on elastic molecular collisions are illustrated in Fig. 14.9 for two molecules that are spherically symmetrical. In an actual intermolecular collision both molecules would recoil, but it is more convenient to consider molecular collisions in a coordinate system fixed at the center of mass of the colliding particles that allows us to consider the collision as if it were one molecule scattered by a fixed center. This system is used in Fig. 14.9. Collisions are shown for different values of the impact parameter b. This is the distance between centers when the molecules are at their closest on hypothetical trajectories that would occur if there were no interaction. For large values of the impact parameter the molecules attract each other along the whole trajectory, and the deflection is negative by definition (although it is not possible experimentally to distinguish positive from negative deflections). As the value of the impact

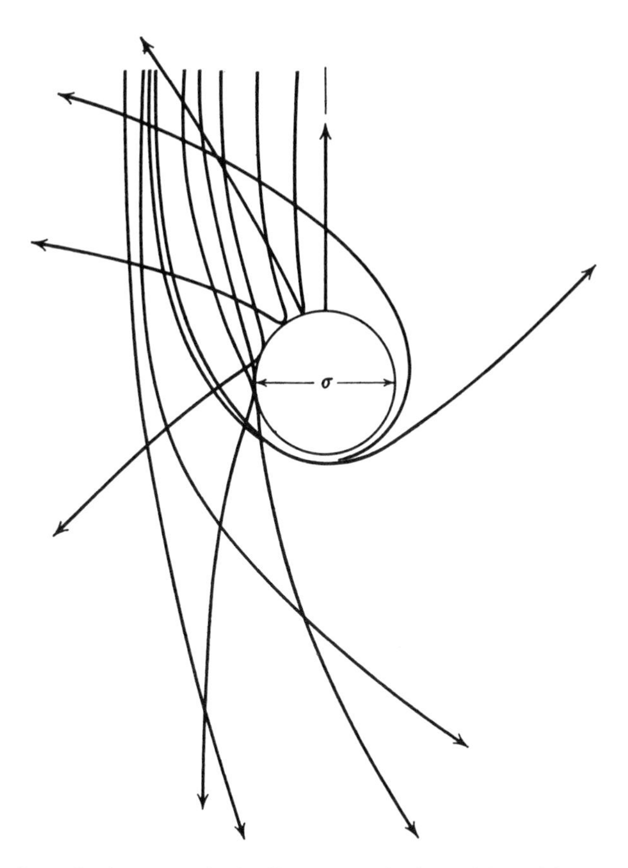

Fig. 14.9 Trajectories for two spherically symmetrical particles with one relative kinetic energy and several values of the impact parameter b. The particles are assumed to interact according to the Lennard-Jones potential, given by equation 14.62. The collisions are represented in relative coordinates. These trajectories are for low-velocity collisions. The two molecules are assumed to approach with a relative kinetic energy one-tenth as large as the energy parameter ϵ of the Lennard-Jones potential. (Reproduced from *Thermal Properties of Matter, Vol. 1: Kinetic Theory of Gases*, 1966, by W. Kauzmann, with permission of publishers, Addison-Wesley/W. A. Benjamin Inc., Advanced Book Program, Reading, Mass., U.S.A.)

parameter decreases, the deflection becomes more and more negative, as shown in the diagram, until the repulsive force begins to be felt. The greatest negative deflection θ_r is known as the rainbow angle.* The deflection at the rainbow angle is greater the greater the depth of the potential well. As the impact parameter is further reduced, the repulsive force becomes dominant and there are large positive deflections. For a head-on collision ($b = 0$) the deflection is 180°. If there are many

---

* The rainbow angle receives its name from the similarity with the refraction of light by water drops that produce rainbows.

collisions at random values of the impact parameter, theory and experiment show that there is a concentration of trajectories at angles just smaller than the rainbow angle. The rainbow scattering phenomena is less prominent for higher relative velocities because at higher relative velocities the particles spend less time close to each other, and there is less time for the attractive and repulsive forces to produce a deflection. For very low relative energies "orbiting" occurs; that is, θ_r becomes infinite at a certain impact parameter for which the attractive and repulsive forces exactly balance.

The classical cross section for molecules interacting according to the Lennard-Jones potential goes to infinity as the relative velocity goes to zero. This happens because the potential does not become zero except for infinite separation of the molecules. Of course, it does not make sense to have an infinite cross section, and the explanation of this unphysical prediction was provided by quantum mechanics. When collision cross sections are calculated quantum mechanically, infinite values are not obtained.

14.9 TRANSPORT PHENOMENA IN GASES

If a gas is not uniform with respect to composition, temperature, and velocity, transport processes occur until the gas does become uniform. The transport of matter in the absence of bulk flow is referred to as *diffusion*. The transport of heat from regions of high temperature to regions of lower temperature without convection is referred to as *thermal* conduction, and the transfer of momentum from a region of higher velocity to a region of lower velocity gives rise to the phenomenon of *viscous* flow. In each case the rate of flow is proportional to the rate of change of some property with distance, a so-called gradient.

The flux of component i in the z direction due to diffusion is proportional to the concentration gradient dc_i/dz.

$$J_{iz} = -D\frac{dc_i}{dz} \tag{14.63}$$

The proportionality constant is the diffusion coefficient D. The flux J_{iz} is expressed in terms of quantity per unit area per unit time. If SI units are used, J_{iz} has the units mol m^{-2} s^{-1}, dc_i/dz has the units of mol m^{-4}, and D has the units of m^2 s^{-1}. The negative sign comes from the fact that if c_i increases in the positive z direction, dc_i/dz is positive, but the flux is in the negative z direction because the flow is in the direction of lower concentrations.

The diffusion coefficient for the diffusion of one gas into another may be determined by use of a cell such as that shown schematically in Fig. 14.10*a*. The heavier gas is placed in chamber A and the lighter in chamber B. The sliding partition is withdrawn for a definite interval of time. From the average composition of one chamber or the other, after a time interval, D may be calculated.

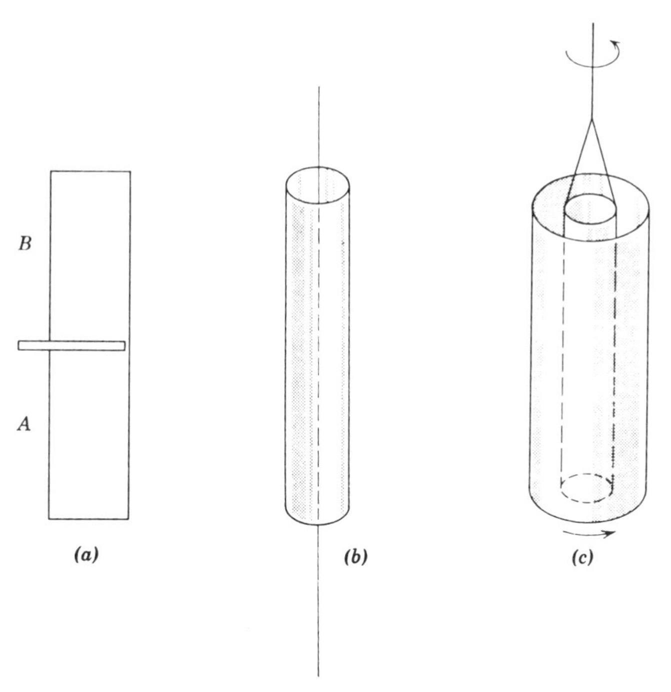

Fig. 14.10 Schematic diagrams of apparatus for the measurements of (*a*) the diffusion coefficient D, (*b*) the coefficient of thermal conductivity λ, and (*c*) the coefficient of viscosity η of gases.

The transport of heat is due to a gradient in temperature. Thus the flux of energy q_z in the z direction due to the temperature gradient in that direction is given by

$$q_z = -\lambda \frac{dT}{dz} \tag{14.64}$$

where the proportionality constant λ is the coefficient of thermal conductivity. When q_z has the units of $\mathrm{J\ m^{-2}\ s^{-1}}$ and dT/dz has the units of $\mathrm{K\ m^{-1}}$, λ has the units of $\mathrm{J\ m^{-1}\ s^{-1}\ K^{-1}}$. The negative sign in equation 14.64 indicates that if dT/dz is positive, the flow of heat is in the negative z direction, which is the direction of lower temperature.

The determination of the coefficient of thermal conductivity by the hot-wire method is illustrated schematically in Fig. 14.10*b*. The outer cylinder is kept at a constant temperature by a thermostatically controlled bath. The tube is filled with the gas under investigation, and the fine wire at the axis of the tube is heated electrically. When a steady state is achieved, the temperature of the wire is measured by determining its electrical resistance. The coefficient of thermal conductivity is calculated from the temperature of wire and wall, the heat dissipation, and the dimensions of the apparatus.

Thermal diffusion is the flux of material due to a temperature gradient of dT/dz.

The fact that the thermal diffusion coefficients depend on mass makes it possible to separate isotopes by use of this effect.

14.10 COEFFICIENT OF VISCOSITY

Viscosity is a measure of the resistance that a fluid offers to an applied shearing force. Consider what happens to the fluid between parallel planes, illustrated in Fig. 14.11, when the top plane is moved in the y direction at a constant speed relative to the bottom plane while maintaining a constant distance between the planes (coordinate z). The planes are considered to be very large, so that edge effects may be ignored. The layer of fluid immediately adjacent to the moving plane moves with the velocity of this plane. The layer next to the stationary plane is stationary; in between the velocity usually changes linearly with distance, as shown. The velocity gradient (i.e., the rate of change of velocity with respect to distance measured *perpendicular* to the direction of flow) is represented by dv_y/dz. The coefficient of viscosity η is defined by the equation

$$F = -\eta \frac{dv_y}{dz} \tag{14.65}$$

Here F is the force per unit area required to move one plane relative to the other. The negative sign comes from the fact that, if F is in the $+y$ direction, the velocity v_y decreases in successive layers away from the moving plane and dv_y/dz is negative. If F has the units of kg m s^{-2}/m^2 and dv_y/dz has the units of m s^{-1}/m, the coefficient of viscosity η has the units of kg m^{-1} s^{-1}. The SI unit of viscosity is the Pascal second. The Pascal (abbreviated Pa) is the name of the SI unit of pressure, that is, 1 N/m^2. Since 1 N = 1 kg m s^{-2}, 1 Pa s = 1 kg m^{-1} s^{-1}. A fluid has a viscosity of 1 Pa s if a force of 1 N is required to move a plane of 1 m^2 at a velocity of 1 m s^{-1} with respect to a plane surface a meter away and parallel with it.*

Although the coefficient of viscosity is conveniently defined in terms of this hypothetical experiment, it is easier to measure it by determining the rate of flow

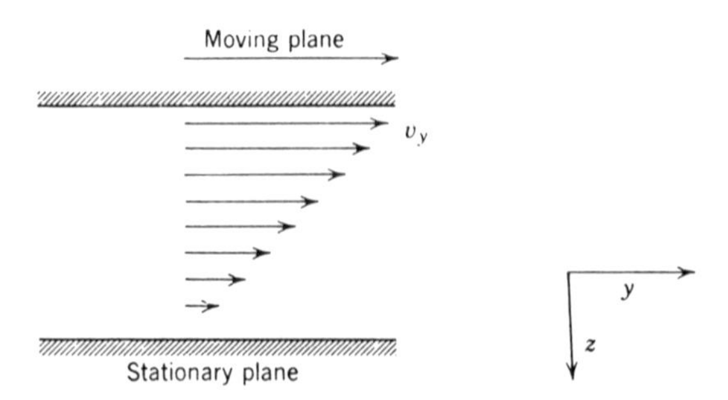

Fig. 14.11 Velocity gradient in a fluid due to a shearing action.

* The cgs unit of viscosity is the poise that is 1 g s^{-1} cm^{-1}. 0.1 Pa s = 1 poise.

through a tube, the torque on a disk that is rotated in the fluid, or other experimental arrangement. An experimental arrangement is illustrated in Fig. 14.10*c*. The outer cylinder is rotated at a constant velocity by an electric motor. The inner coaxial cylinder is suspended on a torsion wire. A torque is transmitted to the inner cylinder by the fluid, and this torque is calculated from the angular twist of the torsion wire.

14.11 COEFFICIENT OF VISCOSITY OF A PERFECT GAS OF RIGID SPHERES

Viscosity results from a transfer of momentum from one layer of a moving fluid to the next. Molecules from a faster moving layer that move down into a slower moving layer have more momentum in the direction of flow than molecules in this layer. Thus they tend to speed up the flow in this layer. Molecules from a slower moving layer that move up into a faster moving layer have a lower momentum in the direction of flow than molecules in this layer, and therefore tend to slow down the flow in this layer. This is the origin of the linear velocity gradient perpendicular to the direction of flow. The transfer of momentum is readily calculated for a perfect gas composed of rigid spheres.

The rigorous theory for rigid spherical molecules yields

$$\eta = \frac{5}{16} \frac{(\pi mkT)^{1/2}}{\pi\sigma^2} \tag{14.66}$$

for a dilute gas. According to equation 14.66 the coefficient of viscosity of a gas is independent of pressure. This is confirmed for real gases in the range of approximately 10^{-2} to 10 atm. When the pressure is so low that the mean free path becomes comparable with the dimensions of the apparatus, the collisions of molecules are primarily with the wall, and equation 14.66 is not applicable. At high pressures intermolecular interactions lead to higher viscosities.

Equation 14.66 is useful for calculating collision diameters. Some collision diameters calculated from gas viscosities by use of this equation are summarised in Table 14.1. Molecules are not rigid spheres, and so these values are not to be taken literally.

Table 14.1 Rigid-Sphere Molecular Diameters (Calculated from Gas Viscosity Measurements at 0 °C and 1 atm)[a]

Gas	$\eta/10^{-5}$ Pa s	σ/nm
A	2.099	0.364
Ne	2.967	0.258
N_2	1.663	0.375
O_2	1.918	0.361
CH_4	1.030	0.414
CO_2	1.366	0.363

[a] J. O. Hirschfelder, C. F. Curtiss, and R. B. Bird, *The Molecular Theory of Gases and Liquids*, John Wiley & Sons, New York, 1954.

Equation 14.66 also indicates that η should be proportional to $T^{1/2}$, but a somewhat larger exponent is obtained for real gases, partly because the cross-sectional diameter becomes smaller at high temperatures due to the increased penetration of the potential energy barrier by gas molecules of higher velocities. In contrast with the viscosity of perfect gases, the viscosity of liquids decreases with increasing temperature, as described in Section 18.1.

The kinetic theory of gases also provides the means for calculating the coefficients of diffusion and thermal conductivity for perfect gases. The treatments for D and λ are similar to that of η. The theoretical results* for gases of rigid spherical molecules are

$$D = \frac{3}{8}\frac{(\pi mkT)^{1/2}}{\pi\sigma^2\rho m} = \frac{12\lambda}{25\rho c_v} = \frac{6}{5}\cdot\frac{\eta}{\rho} \tag{14.67}$$

$$\lambda = \frac{25}{32}\frac{(\pi mkT)^{1/2}c_v}{\pi\sigma^2 m} = \frac{5}{2}\cdot\frac{c_v\eta}{m} \tag{14.68}$$

where m is the mass of a single molecule, $\rho = P/kT$, and c_v is the heat capacity per molecule (C_V/N_A). It is interesting to note that the coefficient of thermal conductivity is also independent of pressure for rigid spherical molecules. Real gases show deviations from these relations because they are not made up of rigid spheres. More accurate relations may be derived by taking more realistic molecular interactions into account.

References

R. B. Bird, W. E. Stewart, and E. N. Lightfoot, *Transport Phenomena*, Wiley, New York, 1960.

S. Chapman and T. G. Cowling, *The Mathematical Theory of Non-Uniform Gases*, 3rd ed., Cambridge University Press, 1970.

H. J. M. Hanley, *Transport Phenomena in Fluids*, Marcel Dekker, New York, 1969.

J. O. Hirschfelder, C. F. Curtiss, and R. B. Bird, *The Molecular Theory of Gases and Liquids*, Wiley, New York, 1954.

J. H. Jeans, *Introduction to the Kinetic Theory of Gases*, Cambridge University Press, 1940.

W. Kauzmann, *Kinetic Theory of Gases*, W. A. Benjamin, Inc., New York, 1966.

J. E. Mayer and M. G. Mayer, *Statistical Mechanics*, Wiley, New York, 1977, Chapter 2.

D. A. McQuarrie, *Statistical Mechanics*, Harper and Row, New York, 1976, Chapter 16.

R. D. Present, *Kinetic Theory of Gases*, McGraw-Hill Book Co., New York, 1958.

Problems

14.1 If the diameter of a gas molecule is 0.4 nm and each is imagined to be in a separate cube, what is the length of the side of the cube in molecular diameters at 0 °C and pressures of (*a*) 1 atm, and (*b*) 1 Pa.

Ans. (*a*) 8.3 molecular diameters, (*b*) 389 molecular diameters.

14.2 Assuming that the atmosphere is isothermal at 0 °C and that the average molar mass of air is 29 g mol^{-1}, calculate the atmospheric pressure at 20,000 ft above sea level, using the Boltzmann distribution.

Ans. 0.466 atm.

---

* J. O. Hirschfelder, C. F. Curtiss, and R. B. Bird, *The Molecular Theory of Gases and Liquids*, John Wiley & Sons, New York, 1965, p. 14.

14.3 Calculate the pressure at an altitude of 500 miles above the earth assuming that the atmosphere is isothermal with a temperature of 0 °C and that the average molar mass is 29 g mol^{-1}, independent of height. *Ans.* 1.7×10^{-44} atm.

14.4 Plot the probability density $f(v)$ of various molecular speeds versus speed for oxygen at 25 °C.

14.5 Calculate the mean speed and the root-mean-square speed for the following set of molecules: 10 molecules moving at 5×10^2 m s^{-1}, 20 molecules moving 10×10^2 m s^{-1}, and 5 molecules moving 15×10^2 m s^{-1}. *Ans.* 9.28×10^2, 9.82×10^2 m s^{-1}.

14.6 Calculate the ratio of the root-mean-square speed to the mean speed to the most probably speed. *Ans.* 1.225:1.128:1.000.

14.7 Calculate the velocity of sound in nitrogen gas at 25 °C. (See Section 14.4.) *Ans.* 352 m s^{-1}.

14.8 Calculate the number of collisions per square centimeter per second of oxygen molecules with a wall at a pressure of 1 atm and 25 °C. *Ans.* 2.73×10^{23}.

14.9 A Knudsen cell containing crystalline benzoic acid (M = 122 g mol^{-1}) is carefully weighed and placed in an evacuated chamber thermostated at 70 °C for 1 hr. The circular hole through which effusion occurs is 0.60 mm in diameter. Calculate the sublimation pressure of benzoic acid at 70 °C in Pa from the fact that the weight loss is 56.7 mg. *Ans.* 21.3 Pa.

14.10 Large vacuum chambers have been built for testing space vehicles at 10^{-6} Pa. Calculate (*a*) the mean-free path of nitrogen at this pressure, and (*b*) the number of molecular impacts per square meter of wall per second at 25 °C. $\sigma_{N_2} = 0.375$ nm. *Ans.* (*a*) 6590 m, (*b*) 2.88×10^{16} m^{-2} s^{-1}.

14.11 Calculate the mean free path of nitrogen at 1 atm and 25 °C. *Ans.* 65 nm.

14.12 (*a*) Calculate the mean free path for hydrogen gas (σ = 0.247 nm) at 1 atm and 0.1 Pa at 25 °C. (*b*) Repeat the calculation for chlorine gas (σ = 0.496 nm). *Ans.* (*a*) 1.50×10^{-7}, 0.152, (*b*) 3.72×10^{-8}, 0.037 m.

14.13 Consider an atomic beam of potassium passing through a scattering gas of Ar contained in a cell of 1 cm length at 0 °C. Assuming a collision cross section of 6×10^{-18} m^2 for potassium-argon collisions, calculate the pressure of argon required to produce an attenuation of the beam of 25%. *Ans.* 1.81×10^{-5} Pa.

14.14 Oxygen is contained in a vessel at 250 Pa pressure and 25 °C. Calculate (*a*) the number of collisions between molecules per second per cubic meter, and (*b*) the mean free path. $\sigma_{O_2} = 0.361$ nm. *Ans.* (*a*) 4.75×10^{29} s^{-1} m^{-3}, (*b*) 2.86×10^{-5} m.

14.15 The pressure in interplanetary space is estimated to be of the order of 10^{-14} Pa. Calculate (*a*) the average number of molecules per cubic centimeter, (*b*) the number of collisions per second per molecule, and (*c*) the mean free path in miles. Assume that only hydrogen atoms are present and that the temperature is 1000 K. Assume σ = 0.2 nm. *Ans.* (*a*) 0.724 cm^{-3}. (*b*) 5.92×10^{-10} s^{-1}. (*c*) 4.83×10^9 miles.

14.16 The Lennard-Jones parameters for nitrogen are ϵ/k = 95.1 K and σ = 0.37 nm. Plot the potential energy (expressed as V/R in K) for the interaction of two molecules of nitrogen.

14.17 For Ne the parameters of the Lennard-Jones 6-12 potential are ϵ/k = 35.6 K and σ = 0.275 nm. Plot V in J mol^{-1} versus r and calculate the distance r_m where $dV/dr = 0$. *Ans.* 0.309 nm.

14.18 For rigid sphere molecules the coefficient of viscosity is given by equation 14.66. Given that for N_2 σ = 0.375 nm, what is the coefficient of viscosity of N_2 at 0 °C? *Ans.* 1.66×10^{-5} Pa s.

14.19 Calculate the mean kinetic energy in electron volts of a molecule in a gas at 300 K.

14.20 At sea level and 25 °C, what is the change in barometric pressure in (*a*) centimeters of mercury, and (*b*) inches of mercury per 1000-ft change in altitude?

14.21 What is the mean atmospheric pressure in Denver, Colorado, which is a mile high, assuming an isothermal atmosphere at 25 °C? Air may be taken as 20% O_2 and 80% N_2.

14.22 What is the ratio of the number of molecules having twice the most probable speed to the number having the most probable speed?

14.23 Calculate the root-mean-square, mean, and most probable speeds for oxygen molecules at 25 °C.

14.24 Suppose that a gas contains 10 molecules having an instantaneous speed of 2×10^2 m s^{-1}, 30 molecules with a speed of 4×10^2 m s^{-1}, and 15 molecules with a speed of 6×10^2 m s^{-1}. Calculate v, v_p, and $\langle v^2 \rangle^{1/2}$.

14.25 Calculate the root-mean-square speed of oxygen molecules having a kinetic energy of 10 kJ mol^{-1}. At what temperature would this be the root-mean-square speed?

14.26 Calculate the velocity of sound in (*a*) He, and (*b*) N_2 at 25 °C.

14.27 The surface of a metal has 10^{15} atoms cm^{-2} that can react with oxygen. Assuming that one oxygen atom reacts with each metal atom and that all oxygen molecules stick, calculate the time for sufficient oxygen to be transferred to the surface by collisions of oxygen molecules if the partial pressure of oxygen is 10^{-7} Pa and the temperature is 25 °C.

14.28 The vapor pressure of water at 25 °C is 3160 Pa. (*a*) If every water molecule that strikes the surface of liquid water sticks, what is the rate of evaporation of molecules from a square centimeter of surface. (*b*) Using this result, find the rate of evaporation in g cm^{-2} min^{-1} of water into perfectly dry air.

14.29 R. B. Holden, R. Speiser, and H. L. Johnston [*J. Am. Chem. Soc.*, **70**, 3897 (1948)] found the rate of loss of weight of a Knudsen effusion cell containing finely divided beryllium to be 19.8×10^{-7} g cm^{-2} s^{-1} at 1320 K and 1210×10^{-7} g cm^{-2} s^{-1} at 1537 K. Calculate ΔH_{sub} for this temperature range.

14.30 The vapor pressure of naphthalene (M = 128.16 g mol^{-1}) is 17.7 Pa at 30 °C. Calculate the weight loss in a period of 2 hr of a Knudsen cell filled with naphthalene and having a round hole 0.50 mm in diameter.

14.31 (*a*) At a pressure of 10^{-8} Pa, how many molecules are there per cubic centimeter at 0 °C? (*b*) What is the mean free path for oxygen molecules at this pressure? σ_{O_2} = 0.361 nm.

14.32 A molecular beam of CsCl is passed through a 1-cm path of *cis*-dichloroethylene at 0 °C and a pressure of 3.9×10^{-3} Pa, and the attenuation of the CsCl beam is found to be 25%. In another experiment with *trans*-dichloroethylene as the scattering gas at the same temperature and pressure an attenuation of 13.4% is observed. Calculate the collision cross sections for the collision of CsCl with the (*a*) *cis*, and (*b*) *trans* compounds. (*c*) Why would you expect the interaction with the *cis* compound to be greater? [H. Schumacher, R. B. Bernstein, and E. W. Rothe, *J. Chem. Phys.*, **33**, 584 (1960).]

14.33 The collision cross section for collisions of Cs atoms with CH_4 molecules at 0 °C is $\pi\sigma^2 = 7 \times 10^{-18}$ m^2. Calculate the pressure of methane that would attenuate a Cs atomic beam 10% over a path of 2 cm.

14.34 What is the average time between collisions of a molecule of N_2 at 1 atm and 25 °C?

14.35 (*a*) Calculate the number of collisions per second undergone by a single nitrogen molecule in nitrogen at 1 atm pressure and 25 °C. (*b*) What is the number of collisions per cubic centimeter per second? What is the effect on the number of collisions (*c*) of doubling the absolute temperature at constant pressure, and (*d*) of doubling the pressure at constant temperature?

14.36 Calculate the number of acetaldehyde (CH_3CHO) molecules colliding per milliliter per second at 800 K and 1 atm. The molecular diameter may be taken as 0.50 nm.

14.37 The Lennard-Jones parameters for argon are $\epsilon/k = 122$ K and $\sigma = 0.34$ nm. Plot the potential energy (expressed as V/R in K) for the interaction of two molecules of argon.

14.38 For methane the parameters for the Lennard-Jones 6-12 potential are $\epsilon/k = 14.82$ K and $\sigma = 0.382$ nm. Plot V versus r and calculate r_m.

14.39 The coefficient of viscosity of helium is 1.88×10^{-5} Pa s at 0 °C. Calculate (*a*) the collision diameter, and (*b*) the diffusion coefficient at 1 atm.

14.40 The van der Waals constant b of *n*-heptane is 0.2654 L mol^{-1}. Estimate the coefficient of viscosity of this gas at 25 °C.

CHAPTER 15

KINETICS: GAS PHASE AND BASIC RELATIONSHIPS

Chemical kinetics is concerned with the rates of reactions and the mechanisms by which reactions occur. A mechanism is a series of elementary reactions that accounts for the overall reaction. To learn about the mechanism of a reaction, the changes in rate due to variations of concentrations of reactants, products, and catalysts are studied. Important information may also be obtained from studies of the effect on rate of changing the temperature, pressure, solvent, electrolyte concentration, or isotopic composition. Since a mechanism is a hypothesis to explain experimental facts, it is not necessarily unique, even if it accounts for all of the facts; however, some otherwise plausible mechanisms may be definitely excluded using kinetic data.

15.1 REACTION RATE

In terms of the symbols introduced in Section 1.20 a general chemical reaction

$$\nu_1 A_1 + \nu_2 A_2 = \nu_3 A_3 + \nu_4 A_4 \tag{15.1}$$

may be represented by

$$\sum \nu_i A_i = 0 \tag{15.2}$$

by giving the stoichiometric coefficients of the products positive signs and the stoichiometric coefficients of the reactants negative signs. The extent of reaction $\xi(xi)$ is defined by $n_i = n_{i0} + \nu_i \xi$, where n_{i0} is the initial number of moles of a reactant or product and n_i is the number of moles at some later time; thus $dn_i = \nu_i\, d\xi$, where ν_i is dimensionless and ξ has the unit mole. The reaction rate may be defined as the rate of change of the extent of reaction.

$$\frac{d\xi}{dt} = \frac{1}{\nu_i}\frac{dn_i}{dt} \tag{15.3}$$

If there are intermediates the reaction rates for various reactants may not be equal; however, if the intermediates are present only at very low concentrations, the rates calculated from the numbers n_i of moles of different reactants will be equal. The reaction rate defined in this way may be used even when there are volume changes during the reaction or two or more phases are present.

If the volume is constant during the reaction the reaction rate generally is expressed in terms of the rate of change of the concentration of one of the reactants or products.

$$\frac{1}{\nu_i}\frac{d(A_i)}{dt} = \frac{1}{V}\frac{d\xi}{dt} \tag{15.4}$$

where (A_i) is the concentration of A_i and V is the volume in which the reaction is occurring.

In practice reaction rates are frequently expressed in terms of rate of change of concentration, or pressure, of one of the reactants or products. In this chapter we will refer to $d(A_i)/dt$ as a reaction rate with the realization that this is not satisfactory if there are intermediates at appreciable concentrations or a volume change in the reaction.

If the stoichiometric coefficients are not all unity it is obviously very important to specify the concentration of a particular reactant or product. For example, for the reaction

$$A + 2B = AB_2 \qquad -\frac{d(A)}{dt} = \frac{1}{2}\left\{-\frac{d(B)}{dt}\right\} = \frac{d(AB_2)}{dt} \tag{15.5}$$

15.2 RATE EQUATION

A rate equation gives the dependence of the rate upon concentrations of reactants, products, and catalysts. For many reactions the rate is proportional to integer powers of the concentrations of reactants if the volume of the reaction mixture is essentially constant and the concentrations of intermediates are negligible.

$$\frac{d(A_i)}{dt} = k(A_1)^{n_1}(A_2)^{n_2}\cdots \tag{15.6}$$

If $n_1 = 1$, the reaction is said to be first order in A_1, and if $n_1 = 2$ the reaction is said to be second order in A_1. For a rate equation of this simple form the total reaction order is the sum $\sum n_i$ of the exponents.* The proportionality factor k is referred to as the rate constant and, according to equation 15.6, it has the units $C^{1-\Sigma n_i}$ time^{-1}. If the reaction is first order, k is usually given in s^{-1} or min^{-1}. If the reaction is second order overall, k is usually given in L mol^{-1} s^{-1}, cm^3 mol^{-1} s^{-1}, or cm^3 s^{-1}. The last set of units is obtained if concentrations are expressed in molecules/cm^3.

---

* If reactants are mixed in stoichiometric proportions, only the total order $\sum n_i$ is obtained. For example, if for the reaction $A + 2B = AB_2$ the rate equation is $-d(A)/dt = k(A)(B)$, the rate equation can be written as follows

$$-\frac{d(A)}{dt} = k(A)\{(B)_0 - 2[(A)_0 - (A)]\}$$

If $(B)_0 = 2(A)_0$, this becomes

$$-\frac{d(A)}{dt} = k(A)\{2(A)_0 - 2[(A)_0 - (A)]\} = 2k(A)^2$$

It is important to note that the exponents n_i in the rate equation are not the stoichiometric coefficients ν_i of the balanced chemical equation, but must be determined in rate experiments.

Not all rate equations have the simple form of equation 15.6; a rate equation may be a complicated function of the concentrations of reactants, products, catalysts, and inhibitors. If a reaction can go by two paths, for example, a catalyzed path and an uncatalyzed path, the rate equation will consist of two additive terms, one for each path (see Section 16.7). A complete rate equation for a reversible reaction contains the equilibrium constant expression for the reaction; it contains positive and negative terms and, when the rate is set equal to zero, the equilibrium constant expression is obtained. However, many reactions go so far toward completion that only the rate equation of the forward reaction is of interest. Before discussing further the relation between rate equation and equilibrium, we need to discuss the commonly occurring rate equations and the determination of the rate constant k.

15.3 FIRST-ORDER REACTIONS

The rate equation for a first-order reaction

$$-\frac{d(\mathrm{A})}{dt} = k(\mathrm{A}) \tag{15.7}$$

may be integrated after it is written in the form

$$-\frac{d(\mathrm{A})}{(\mathrm{A})} = k\,dt \tag{15.8}$$

If the concentration of A is $(\mathrm{A})_1$ at t_1 and $(\mathrm{A})_2$ at t_2,

$$-\int_{(\mathrm{A})_1}^{(\mathrm{A})_2} \frac{d(\mathrm{A})}{(\mathrm{A})} = k\int_{t_1}^{t_2} dt \tag{15.9}$$

$$\ln\frac{(\mathrm{A})_1}{(\mathrm{A})_2} = k(t_2 - t_1) \tag{15.10}$$

An especially useful form of this equation is obtained if t_1 is taken to be zero, and the initial concentration is represented by $(\mathrm{A})_0$.

$$\ln\frac{(\mathrm{A})_0}{(\mathrm{A})} = kt \tag{15.11}$$

or

$$(\mathrm{A}) = (\mathrm{A})_0 e^{-kt} \tag{15.12}$$

or

$$\ln(\mathrm{A}) = \ln(\mathrm{A})_0 - kt \tag{15.13}$$

The last form indicates that the rate constant k may be calculated from a plot of ln (A) versus t; the slope of such a plot is $-k$.

The calculation of first-order rate constants may be illustrated for the decomposition of nitrogen pentoxide.* Nitrogen pentoxide decomposes completely in the gas phase, or when dissolved in inert solvents, at a rate that is conveniently measured at room temperature. The reaction is strictly first order (except at very low pressures), and the end products are oxygen and a mixture of nitrogen tetroxide and nitrogen dioxide. The following equation represents the overall reaction:

$$\begin{array}{c} 2N_2O_5 \rightarrow 2N_2O_4 + O_2 \\ \quad\quad \rightleftharpoons \\ \quad\quad 4NO_2 \end{array} \tag{15.14}$$

For every molecule of oxygen produced, two molecules of nitrogen pentoxide have decomposed. It will be shown later (Section 15.10), however, that this reaction is much more complicated, having several intermediate steps. It has been suggested that in chemical kinetics the only simple reactions are the ones that have not been thoroughly studied.

When a solution of nitrogen pentoxide in carbon tetrachloride decomposes, the nitrogen tetroxide and nitrogen dioxide remain in solution while the oxygen, having a low solubility in this solvent, escapes and may be measured in a gas buret. The reaction vessel is carefully thermostated, and it is agitated to prevent supersaturation of the oxygen.

Experimental data for the decomposition of nitrogen pentoxide dissolved in carbon tetrachloride at 45 °C are plotted in Fig. 15.1. In Fig. 15.1*b* the straight line produced by plotting the logarithm of the concentration against time shows that the reaction is first order and follows strictly the relation given by equation 15.13. The value of the first-order rate constant, $6.22 \times 10^{-4}\ s^{-1}$, is obtained from the slope of the line in Fig. 15.1*b*.

It is evident from equations 15.10 and 15.11 that to determine the rate constant for a first-order reaction it is only necessary to determine the *ratio* of the concentrations at two times. Quantities proportional to concentration may be substituted for concentrations in these equations, since the proportionality constants cancel.

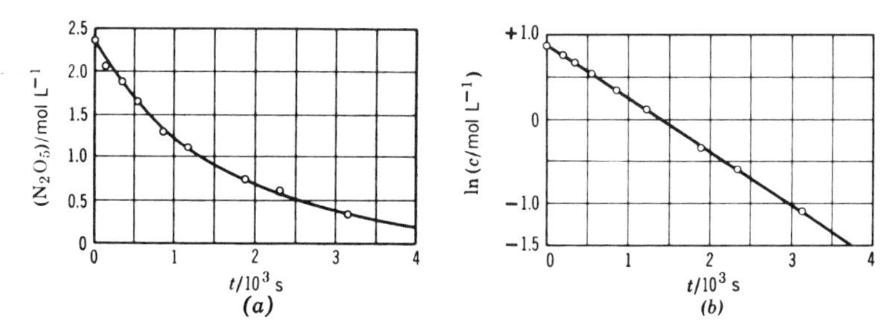

Fig. 15.1 First-order decomposition of N_2O_5 at 45 °C.

* F. Daniels and E. H. Johnston, *J. Am. Chem. Soc.*, **43**, 53 (1921); H. Eyring and F. Daniels, *J. Am. Chem. Soc.*, **52**, 1472 (1930).

The half-life $t_{1/2}$ of a reaction is the time required for half of the reactant to disappear. For a first-order reaction the half-life is independent of the initial concentration. Thus 50% of the substance remains after one half-life, 25% remains after two half-lives, 12.5% after three, etc. The relation between the half-life and the rate constant is obtained from equation 15.11.

$$k = \frac{1}{t_{1/2}} \ln \frac{1}{\frac{1}{2}} = \frac{0.693}{t_{1/2}} \tag{15.15}$$

15.4 SECOND-ORDER REACTIONS

The simplest example of a second-order reaction is one in which the rate is proportional to the square of the concentration of a reactant. The rate law may be integrated after arranging it in the form

$$-\frac{d(\mathrm{A})}{(\mathrm{A})^2} = k\,dt \tag{15.16}$$

If the concentration is $(\mathrm{A})_0$ at $t = 0$ and (A) at time t, integration yields

$$kt = \frac{1}{(\mathrm{A})} - \frac{1}{(\mathrm{A})_0} \tag{15.17}$$

Thus a plot of $1/(\mathrm{A})$ versus t is linear for such a second-order reaction, and the slope is equal to the second-order rate constant. This integrated rate equation also applies if the rate is given by $k(\mathrm{A})(\mathrm{B})$, the stoichiometry is represented by $\mathrm{A} + \mathrm{B} =$ products, and A and B are initially at the same concentration. As may be seen from equation 15.17, the half-life for such a second-order reaction is given by

$$t_{1/2} = \frac{1}{k(\mathrm{A})_0} \tag{15.18}$$

If the rate is given by $k(\mathrm{A})(\mathrm{B})$, the stoichiometry is given by

$$a\mathrm{A} + b\mathrm{B} \rightarrow \text{products}$$

and the reactants are not in stoichiometric proportions [i.e., $b(\mathrm{A})_0 \neq a(\mathrm{B})_0$], then the integrated rate equation is

$$kt = \frac{a}{[b(\mathrm{A})_0 - a(\mathrm{B})_0]} \ln \frac{(\mathrm{A})(\mathrm{B})_0}{(\mathrm{A})_0(\mathrm{B})} \tag{15.19}$$

If the stoichiometric coefficients of A and B are unity, equation 15.19 becomes

$$kt = \frac{1}{[(\mathrm{A})_0 - (\mathrm{B})_0]} \ln \frac{(\mathrm{A})(\mathrm{B})_0}{(\mathrm{A})_0(\mathrm{B})} \tag{15.20}$$

15.5 THIRD-ORDER REACTIONS

The integration of a third-order rate equation is readily accomplished for the case where the reactants are present in stoichiometric proportions and the total order of the reaction is third.

Integrating

$$\int_{(A)_0}^{(A)} \frac{d(A)}{(A)^3} = -k \int_0^t dt \tag{15.21}$$

yields

$$kt = \frac{1}{2}\left[\frac{1}{(A)^2} - \frac{1}{(A)_0^2}\right] \tag{15.22}$$

15.6 ZERO-ORDER REACTIONS

These are reactions in which the rate is unaffected by changes in the concentrations of one or more reactants because it is determined by some limiting factor other than concentration, such as the amount of light absorbed in a photochemical reaction or the amount of catalyst in a catalytic reaction. Then

$$-\frac{d(A)}{dt} = k \tag{15.23}$$

A catalytic reaction might be first order in catalyst and zero order in reactant.

Integration of equation 15.23 yields

$$kt = (A)_0 - (A) \tag{15.24}$$

The value of k calculated in this way may be a function of the intensity of light or the concentration of catalyst.

15.7 RESTRICTIONS OF THERMODYNAMICS ON RATE EQUATIONS*

If the rate equation for the forward reaction is known, thermodynamics allows us to say something about possible forms for the rate equation of the reverse reaction. For example, for the reaction

$$A + 2B = C \tag{15.25}$$

suppose that the rate equation for the forward reaction (determined before there is any significant accumulation of C) is

$$-\left[\frac{d(A)}{dt}\right]_f = k_f(A)(B) \tag{15.26}$$

* K. Denbigh, *The Principles of Chemical Equilibrium*, 3rd ed., Cambridge University Press, Cambridge, 1971, p. 442.

The rate equation for the reverse reaction is given by

$$\left[\frac{d(\mathrm{A})}{dt}\right]_r = k_r(\mathrm{A})^\alpha(\mathrm{B})^\beta(\mathrm{C})^\gamma \tag{15.27}$$

The net rate is given by

$$-\frac{d(\mathrm{A})}{dt} = -\left[\frac{d(\mathrm{A})}{dt}\right]_f - \left[\frac{d(\mathrm{A})}{dt}\right]_r = k_f(\mathrm{A})(\mathrm{B}) - k_r(\mathrm{A})^\alpha(\mathrm{B})^\beta(\mathrm{C})^\gamma \tag{15.28}$$

At equilibrium the net rate is zero, and so

$$\frac{(\mathrm{A})_{\mathrm{eq}}^\alpha(\mathrm{B})_{\mathrm{eq}}^\beta(\mathrm{C})_{\mathrm{eq}}^\gamma}{(\mathrm{A})_{\mathrm{eq}}(\mathrm{B})_{\mathrm{eq}}} = \frac{k_f}{k_r} \tag{15.29}$$

If the equilibrium expression is written in the usual way for reaction 15.25,

$$\frac{(\mathrm{C})_{\mathrm{eq}}}{(\mathrm{A})_{\mathrm{eq}}(\mathrm{B})_{\mathrm{eq}}^2} = K \tag{15.30}$$

we conclude that $\alpha = 0$, $\beta = -1$, and $\gamma = 1$, so that the rate equation for the reverse reaction is

$$\left[\frac{d(\mathrm{A})}{dt}\right]_r = \frac{k_r(\mathrm{C})}{(\mathrm{B})} \tag{15.31}$$

However, the balanced chemical equation may be written in other ways, which leads us to admit the possibility of other possible rate equations for the reverse reaction. For example, if the equilibrium expression is written

$$\frac{(\mathrm{C})_{\mathrm{eq}}^2}{(\mathrm{A})_{\mathrm{eq}}^2(\mathrm{B})_{\mathrm{eq}}^4} = K \tag{15.32}$$

the rate equation expected for the reverse reaction would be

$$\left[\frac{d(\mathrm{A})}{dt}\right]_r = k_r(\mathrm{A})^{-1}(\mathrm{B})^{-3}(\mathrm{C})^2 \tag{15.33}$$

Thus the form of the rate equation for the forward reaction implies something about the form of the rate equation for the reverse reaction.

15.8 REVERSIBLE FIRST-ORDER REACTIONS

As the simplest example of a reversible reaction let us consider

$$\mathrm{A} \underset{k_2}{\overset{k_1}{\rightleftharpoons}} \mathrm{B} \tag{15.34}$$

The rate law for this reversible elementary reaction is

$$\frac{d(\mathrm{A})}{dt} = -k_1(\mathrm{A}) + k_2(\mathrm{B}) \tag{15.35}$$

If initially only A is present,

$$\frac{d(\mathrm{A})}{dt} = -k_1(\mathrm{A}) + k_2[(\mathrm{A})_0 - (\mathrm{A})] = k_2(\mathrm{A})_0 - (k_1 + k_2)(\mathrm{A})$$

$$= -(k_1 + k_2)\left[(\mathrm{A}) - \frac{k_2}{k_1 + k_2}(\mathrm{A})_0\right] = -(k_1 + k_2)[(\mathrm{A}) - (\mathrm{A})_{\mathrm{eq}}] \quad (15.36)$$

where the expression for $(\mathrm{A})_{\mathrm{eq}}$ is obtained as follows.

$$\frac{(\mathrm{B})_{\mathrm{eq}}}{(\mathrm{A})_{\mathrm{eq}}} = \frac{(\mathrm{A})_0 - (\mathrm{A})_{\mathrm{eq}}}{(\mathrm{A})_{\mathrm{eq}}} = \frac{k_1}{k_2} \quad (15.37)$$

and

$$(\mathrm{A})_{\mathrm{eq}} = \frac{k_2}{k_1 + k_2}(\mathrm{A})_0 \quad (15.38)$$

Integrating equation 15.36 yields

$$-\int_{(\mathrm{A})_0}^{(\mathrm{A})} \frac{d(\mathrm{A})}{(\mathrm{A}) - (\mathrm{A})_{\mathrm{eq}}} = (k_1 + k_2)\int_0^t dt \quad (15.39)$$

$$\ln \frac{(\mathrm{A})_0 - (\mathrm{A})_{\mathrm{eq}}}{(\mathrm{A}) - (\mathrm{A})_{\mathrm{eq}}} = (k_1 + k_2)t \quad (15.40)$$

For such a reaction the concentrations of A and B as functions of time are illustrated in Fig. 15.2. The concentration of A will be halfway to its equilibrium value in a time of $0.693/(k_1 + k_2)$.

Thus a plot of $-\ln[(\mathrm{A}) - (\mathrm{A})_{\mathrm{eq}}]$ versus time is linear, and $(k_1 + k_2)$ may be calculated from the slope. It should be especially noted that the rate of approach to equilibrium in this reaction is determined by the sum of the rate constants of the forward and reverse reactions, not by the rate constant for the forward reaction. Since the ratio k_1/k_2 may be calculated from the equilibrium concentrations by use of equation 15.37, the values of k_1 and k_2 may be obtained.

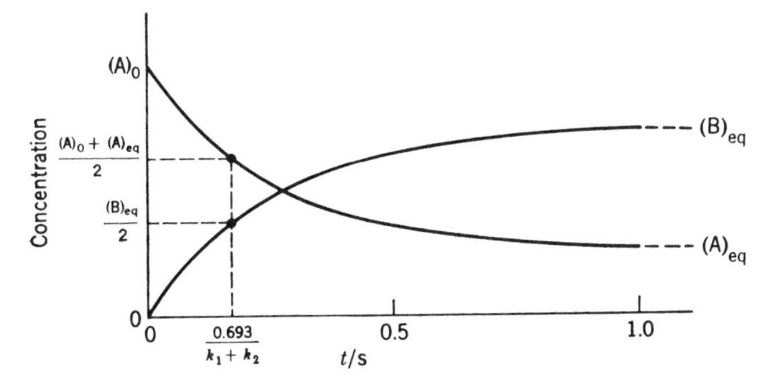

Fig. 15.2 Reversible first-order reaction starting with A at concentration $(\mathrm{A})_0$. The values of the rate constants are $k_1 = 3\ \mathrm{s}^{-1}$ and $k_2 = 1\ \mathrm{s}^{-1}$.

15.9 CONSECUTIVE FIRST-ORDER REACTIONS

Consecutive reactions occur when the product of a reaction undergoes further reaction. Two consecutive first-order reactions may be represented by

$$A \xrightarrow{k_1} B \xrightarrow{k_2} C \tag{15.41}$$

To determine the way in which the concentrations of the compounds in such a mechanism depend on time, the rate equations are first written down for each substance. It is then necessary to obtain the solution of these simultaneous differential equations. For the foregoing reactions the rate equations are as follows.

$$\frac{d(A)}{dt} = -k_1(A) \tag{15.42}$$

$$\frac{d(B)}{dt} = k_1(A) - k_2(B) \tag{15.43}$$

$$\frac{d(C)}{dt} = k_2(B) \tag{15.44}$$

It will be assumed that, at $t = 0$, $(A) = (A)_0$, $(B) = 0$, and $(C) = 0$. The rate equation for A is readily integrated to obtain

$$(A) = (A)_0 e^{-k_1 t} \tag{15.45}$$

Substitution of this expression into equation 15.43 yields

$$\frac{d(B)}{dt} = k_1(A)_0 e^{-k_1 t} - k_2(B) \tag{15.46}$$

which may be integrated to obtain

$$(B) = \frac{k_1(A)_0}{(k_2 - k_1)}(e^{-k_1 t} - e^{-k_2 t}) \tag{15.47}$$

Because of conservation of the number of moles, $(A)_0 = (A) + (B) + (C)$ at any time, and so the concentration of C is given by

$$(C) = (A)_0 - (A) - (B) = (A)_0\left[1 + \frac{1}{k_1 - k_2}(k_2 e^{-k_1 t} - k_1 e^{-k_2 t})\right] \tag{15.48}$$

The concentrations of A, B, and C are shown in Fig. 15.3 for $(A)_0 = 1$ mol L^{-1}, $k_1 = 0.1$ s^{-1}, and $k_2 = 0.05$ s^{-1}.

If the course of the reaction were followed by analyzing for A, curve A would be obtained; if it were followed by measuring the concentration of the end product C, curve C would result; and, finally, if only the intermediate product B were determined, it would be found that its concentration would rise to a maximum and then fall off, as shown by curve B. The actual rate of production of C is seen to be quite complicated, and the existence of an induction period or time lag at the beginning of the reaction is evident.

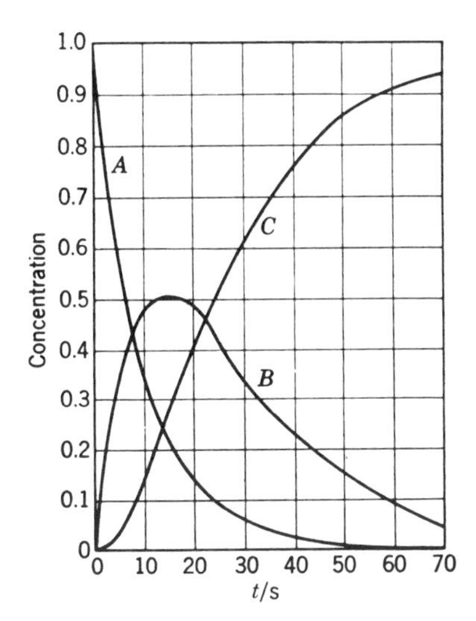

Fig. 15.3 Consecutive first-order reactions: $A \xrightarrow{k_1} B \xrightarrow{k_2} C$; $k_1 = 0.10\,s^{-1}$, $k_2 = 0.05\,s^{-1}$.

Many examples of consecutive first-order reactions are found among the disintegration reactions of the radioactive nuclei. These radioactive reactions can be expressed with exactness by simple first-order equations.

Complex systems of interconnected first-order reactions may be treated mathematically to find the path the system will take to equilibrium.*

15.10 MECHANISMS OF CHEMICAL REACTIONS

Most chemical reactions proceed through a series of steps. Even when the rate law is simple, a series of steps may be involved. One of the objectives of kinetic studies is to identify the intermediate steps because only in this way can we understand how the reaction occurs. The individual steps are referred to as *elementary reactions*. The series of elementary reactions that account for the overall reaction is called the *mechanism* of the reaction. In discussing a mechanism we refer to the *molecularity* of the steps. The molecularity is the number of reactant molecules in an elementary step. The individual steps of a mechanism are referred to as *unimolecular*, *bimolecular*, and *termolecular*, depending on whether one, two, or three molecules are involved as reactants. For elementary reactions the molecularity (uni-, bi-, and ter-) and the order (first, second, and third) are the same, but these names are not synonyms at the level of the overall rate law. For example, a unimolecular step in a mechanism

---

* J. Wei, *Advances in Catalysis*, **13**, 203 (1962); R. Aris, *Elementary Chemical Reactor Analysis*, Prentice-Hall, Englewood Cliffs, N.J., 1969, p. 104.

is first order, but a first-order reaction is not necessarily unimolecular, as we will see in this section.

The discovery of the mechanism of a reaction cannot be described briefly because many types of empirical and theoretical ideas are involved. Kinetic data, however, are very useful for helping to determine a mechanism. Information about mechanism may also be obtained by use of isotopes to determine the paths of various atoms through the reaction and by use of spectroscopy to identify intermediates. The mechanism must yield the observed rate law and account satisfactorily for the net chemical change.

The rate equations may be integrated for a few mechanisms involving reversible, consecutive, and parallel steps of first and second orders. The rate equations for each reactant and intermediate are written down and solved simultaneously. For more complicated mechanisms, however, a closed solution of the differential equations may not exist. If no closed solution is obtainable, one can use computers to obtain numerical solutions or introduce experimentally suitable approximations to simplify the mathematics. Valuable information can be obtained by considering the rate equations for a complex reaction even if the equations cannot be solved explicitly. Through them it is often possible to understand why there are sometimes induction periods, why complex reactions may approach first order or second order.

By the use of approximations it may be possible to obtain the overall rate laws for quite complicated mechanisms. As an illustration, consider the decomposition of N_2O_5 that follows the mechanism*

$$N_2O_5 \underset{k_2}{\overset{k_1}{\rightleftharpoons}} NO_2 + NO_3 \tag{15.49}$$

$$NO_2 + NO_3 \xrightarrow{k_3} NO + O_2 + NO_2 \tag{15.50}$$

$$NO + NO_3 \xrightarrow{k_4} 2NO_2 \tag{15.51}$$

The net chemical reaction

$$2N_2O_5 = 4NO_2 + O_2$$

is made up of step 15.49 occurring twice and steps 15.50 and 15.51 each occurring once. The rate equations for the intermediates NO_3 and NO are

$$\frac{d(NO_3)}{dt} = k_1(N_2O_5) - (k_2 + k_3)(NO_2)(NO_3) - k_4(NO)(NO_3) \tag{15.52}$$

$$\frac{d(NO)}{dt} = k_3(NO_2)(NO_3) - k_4(NO)(NO_3) \tag{15.53}$$

Since the concentrations of the reaction intermediates NO and NO_3 are never very great during the reaction,

$$\frac{d(NO)}{dt} \ll -\frac{d(N_2O_5)}{dt} \quad \text{and} \quad \frac{d(NO_3)}{dt} \ll -\frac{d(N_2O_5)}{dt}$$

* H. S. Johnston, *Gas Phase Reaction Rate Theory*, The Ronald Press Co., New York, 1966, Chapt. 1.

It is therefore useful to use the steady-state approximation and set the derivatives in equations 15.52 and 15.53 equal to zero. This leads to two simultaneous equations that may be solved to give the following expressions for the steady-state concentrations of NO and NO_3.

$$(NO) = \left(\frac{k_3}{k_4}\right)(NO_2) \tag{15.54}$$

$$(NO_3) = \frac{k_1(N_2O_5)}{(k_2 + 2k_3)(NO_2)} \tag{15.55}$$

The rate equation for the decomposition of N_2O_5 is

$$\frac{d(N_2O_5)}{dt} = -k_1(N_2O_5) + k_2(NO_2)(NO_3) \tag{15.56}$$

Substituting equation 15.55 yields the predicted rate equation

$$\frac{d(N_2O_5)}{dt} = -\frac{2k_1k_3}{k_2 + 2k_3}(N_2O_5) \tag{15.57}$$

so that the overall reaction behaves in a first-order manner, as shown experimentally in Fig. 15.1*b*.

A more drastic assumption that is sometimes useful is to assume that a reaction has a rate-determining step. If one of the steps is rate determining, all the steps in the mechanism up to that point remain at equilibrium. Thus the concentrations of all of the substances involved in these prior steps may be calculated from the equilibrium expressions.

15.11 PRINCIPLE OF DETAILED BALANCING

In addition to the ways in which thermodynamics restricts the values of rate constants for forward and reverse reactions, there is another restriction that does not come from thermodynamics. This is the principle of detailed balancing that says that at equilibrium the forward rate of *each* reaction in a mechanism is equal to the reverse rate of that same reaction in the mechanism.* The application of this principle becomes important when a reaction has more than one path.

For the mechanism

$$\begin{array}{ccc} A & \underset{k_2}{\overset{k_1}{\rightleftharpoons}} & B \\ {\scriptstyle k_5}\nwarrow\searrow{\scriptstyle k_6} & & {\scriptstyle k_4}\swarrow\nearrow{\scriptstyle k_3} \\ & C & \end{array} \tag{15.58}$$

writing down the equations for $d(A)/dt$, $d(B)/dt$, and $d(C)/dt$, letting these rates

* L. Onsager, *Phys. Rev.*, **37**, 405 (1931).

be equal to zero, and solving the three simultaneous equations for $(\mathrm{A})_{\mathrm{eq}}$, $(\mathrm{B})_{\mathrm{eq}}$, and $(\mathrm{C})_{\mathrm{eq}}$ yields the following equations:

$$\frac{(\mathrm{B})_{\mathrm{eq}}}{(\mathrm{A})_{\mathrm{eq}}} = \frac{k_1k_4 + k_1k_5 + k_4k_6}{k_2k_4 + k_2k_5 + k_3k_5} \tag{15.59}$$

$$\frac{(\mathrm{C})_{\mathrm{eq}}}{(\mathrm{A})_{\mathrm{eq}}} = \frac{k_1k_3 + k_2k_6 + k_3k_6}{k_2k_4 + k_2k_5 + k_3k_5} \tag{15.60}$$

which do not look much like the correct equilibrium expressions, which are:

$$\frac{(\mathrm{B})_{\mathrm{eq}}}{(\mathrm{A})_{\mathrm{eq}}} = \frac{k_1}{k_2} \tag{15.61}$$

$$\frac{(\mathrm{C})_{\mathrm{eq}}}{(\mathrm{B})_{\mathrm{eq}}} = \frac{k_3}{k_4} \tag{15.62}$$

$$\frac{(\mathrm{A})_{\mathrm{eq}}}{(\mathrm{C})_{\mathrm{eq}}} = \frac{k_5}{k_6} \tag{15.63}$$

Multiplying the left-hand sides of these equations yields unity, so that

$$\frac{k_1k_3k_5}{k_2k_4k_6} = 1 \tag{15.64}$$

and the six equilibrium constants of mechanism 15.58 are not independent. This relation is not required by thermodynamics but by the principle of detailed balancing. Substitution of equation 15.64 into equation 15.59 yields equation 15.61.

Thus the principle of detailed balancing does not allow equilibrium to be maintained by the following cyclic mechanism.

$$\begin{array}{ccc} \mathrm{A} & \longrightarrow & \mathrm{B} \\ & \nwarrow \quad \swarrow & \\ & \mathrm{C} & \end{array} \tag{15.65}$$

If a rate equation has a sum of terms for the forward reaction, indicating multiple paths, the principle of detailed balancing requires that *each* term for the forward reaction be balanced by a thermodynamically appropriate term for the reverse reaction at equilibrium.

15.12 EFFECT OF TEMPERATURE

If the temperature range is not too great, the dependence of rate constants on temperature can usually be represented by an empirical equation proposed by Arrhenius in 1889.

$$k = Ae^{-E_a/RT} \tag{15.66}$$

where A is the *preexponential factor* and E_a is the *activation energy*. The preexponential factor A has the same units as the rate constant, and for a first-order reaction the

Table 15.1 Rate Constants for the Decomposition of Gaseous Nitrogen Pentoxide at Different Temperatures

$t/°C$	$k/10^{-5}\ s^{-1}$	Half-Life[a]
65	487	2.49 m
55	150	7.64 m
45	49.8	25.2 m
35	13.5	89.5 m
25	3.46	5.78 h
0	0.0787	10.9 d

[a] h, hour; d, day; m, minute.

usual unit is s^{-1}. Since this is the unit of a frequency, A is sometimes referred to as a frequency factor. Equation 15.66 may be written in logarithmic form.

$$\ln k = \ln A - \frac{E_a}{RT} \tag{15.67}$$

According to this equation, a straight line should be obtained when the logarithm of the rate constant is plotted against the reciprocal of the absolute temperature. Differentiating equation 15.67 with respect to temperature,

$$\frac{d \ln k}{dT} = \frac{E_a}{RT^2} \tag{15.68}$$

and integrating between limits,

$$\ln \frac{k_2}{k_1} = \frac{E_a}{R}\left(\frac{T_2 - T_1}{T_1 T_2}\right) \tag{15.69}$$

The rate constants for the decomposition of gaseous nitrogen pentoxide* at different temperatures are given in the second column of Table 15.1 in s^{-1}. The values of $\ln k$ are plotted against $1/T$ in Fig. 15.4.

The slope of the line in Fig. 15.4 is $-12{,}400$ K, and E_a has a value of $-R(\text{slope}) =$ 103 kJ mol^{-1}. Equation 15.67 becomes

$$\ln k = 31.31 - \frac{103 \times 10^3\ \text{J mol}^{-1}}{(8.314\ \text{J K}^{-1}\ \text{mol}^{-1})T} \tag{15.70}$$

and equation 15.66 becomes

$$k = (4.35 \times 10^{13}\ s^{-1}) \exp\left[-\frac{103 \times 10^3\ \text{J mol}^{-1}}{(8.314\ \text{J K}^{-1}\ \text{mol}^{-1})T}\right] \tag{15.71}$$

It should be realized that for any given frequency factor there is a fairly narrow range of activation energies that will give reaction rates in the range measurable by

* F. Daniels and E. H. Johnston, *J. Am. Chem. Soc.*, **43**, 53 (1921).

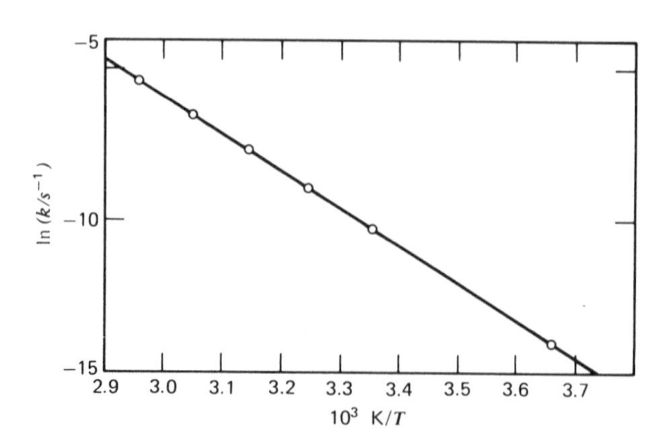

Fig. 15.4 Plot of $\ln k$ versus $1/T$ for the decomposition of N_2O_5 from which the Arrhenius activation energy E_a may be calculated.

conventional techniques; that is, with half-lives from 1 min to 10 days. For example, if $A = 10^{13}\ s^{-1}$ and the temperature is 298 K, reactions with activation energies less than about 80 kJ mol^{-1} will be too fast to study with ordinary methods, and reactions with activation energies greater than about 100 kJ mol^{-1} will be too slow.

15.13 COLLISION THEORY OF BIMOLECULAR REACTIONS OF GASES

We now consider a series of theories that relate the rate constant for a reaction to the properties of the molecules involved. The simplest and oldest of these theories is the collision theory for bimolecular reactions. According to this theory the reaction rate is the product of the collision frequency and the fraction of the collisions with sufficient energy for reaction.

As shown in Section 14.7, the frequency of collisions between molecules of type 1 and molecules of type 2 is

$$Z_{12} = \rho_1 \rho_2 \pi \sigma^2 \langle v_r \rangle \tag{15.72}$$

where $\sigma = r_A + r_B$ is the collision diameter, $\langle v_r \rangle$ is the average relative velocity of the two types of molecules, and ρ_1 and ρ_2 are the numbers of molecules per unit volume. The collision radii of A and B are represented by r_A and r_B. In a pure gas the frequency of collision is

$$Z_{11} = \frac{1}{2^{1/2}} \rho^2 \pi \sigma^2 \langle v \rangle \tag{15.73}$$

The average relative velocities of molecules of types 1 and 2 are given by equation 14.56, which is

$$\langle v_r \rangle = \left(\frac{8kT}{\pi\mu}\right)^{1/2} \tag{15.74}$$

and the relative average velocity of two molecules of type 1 is given by $2^{1/2}\langle v \rangle$, where

$$\langle v \rangle = \left(\frac{8kT}{\pi m}\right)^{1/2} \tag{15.75}$$

In order for reaction to occur the energy of a collision must equal or exceed some critical value. The effective energy is not the total kinetic energy of the two colliding molecules but, instead, the kinetic energy corresponding to the component of the relative velocity of the two molecules along the line of their centers at the moment of collision. This is the energy with which the two molecules are pressed together. Assuming that the kinetic energy of this component of the relative velocity must be greater than some minimum value E_0, the threshold energy per mole, it can be shown that the fraction of all collisions with a line-of-centers component of the kinetic energy greater than E_0 is given by the Boltzmann factor $e^{-E_0/RT}$.*

The rate of reaction of molecules of type 1 with molecules of type 2 is therefore given by

$$-\frac{d\rho_1}{dt} = \pi\sigma^2\left(\frac{8RT}{\pi\mu N_A}\right)^{1/2} e^{-E_0/RT}\rho_1\rho_2 \tag{15.76}$$

so that the second-order rate constant is given by

$$k = \pi\sigma^2\left(\frac{8RT}{\pi\mu N_A}\right)^{1/2} e^{-E_0/RT} \tag{15.77}$$

Using SI units, the value of k calculated from this equation has the units of $m^3\ s^{-1}$. It needs to be multiplied by Avogadro's constant to obtain the value in $(mol\ m^{-3})^{-1}\ s^{-1}$ and further by $10^3\ L\ m^{-3}$ to obtain the value in $L\ mol^{-1}\ s^{-1}$ or by $10^6\ cm^3\ m^{-3}$ to obtain the value in $cm^3\ mol^{-1}\ s^{-1}$. The value of the preexponential factor for a typical bimolecular reaction is of the order of $10^{14}\ cm^3\ mol^{-1}\ s^{-1}$. Rate constants of bimolecular gas reactions may also be expressed in $cm^3\ s^{-1}$; the preexponential factor for a typical bimolecular gas reaction is of the order of

$$\frac{10^{14}\ cm^3\ mol^{-1}\ s^{-1}}{6 \times 10^{23}\ mol^{-1}} \approx 10^{-10}\ cm^3\ s^{-1}$$

A collision that is sufficiently energetic to supply the activation energy may still fail to produce a reaction if the colliding molecules are not oriented in such a way that they can react with each other. Thus a steric factor p may be introduced into equation 15.77. Steric factors tend to be in the range 1 to 10^{-4}, with smaller steric

* For a much more complete discussion of bimolecular reactions, see W. C. Gardiner, Jr., *Rates and Mechanisms of Chemical Reactions*, W. A. Benjamin, Inc., New York, 1969.

factors being found for reactions with complex reactants. However, collision theory does not predict the values of the steric factor or of the minimum energy required for reaction. Neither does it explain why certain reactions are unimolecular and others are termolecular.

Example 15.1 Using collision theory, calculate the preexponential factor for the rate constant for a "typical" reaction with a reduced mass μ of 30×10^{-3} kg mol$^{-1}/N_A$ and a collision diameter σ of 5×10^{-10} m at 298 K.

$$k = \pi\sigma^2\left(\frac{8RT}{\pi\mu N_A}\right)^{1/2} = (3.14)(5 \times 10^{-10}\ \text{m})^2\left[\frac{8(8.314\ \text{J K}^{-1}\ \text{mol}^{-1})(298\ \text{K})}{(3.14)(30 \times 10^{-3}\ \text{kg mol}^{-1})}\right]^{1/2}$$

$$= 3.60 \times 10^{-16}\ \text{m}^3\ \text{s}^{-1}$$

$$= (3.60 \times 10^{-16}\ \text{m}^3\ \text{s}^{-1})(6.022 \times 10^{23}\ \text{mol}^{-1})(10^6\ \text{cm}^3\ \text{m}^{-3})$$

$$= 2.17 \times 10^{14}\ \text{cm}^3\ \text{mol}^{-1}\ \text{s}^{-1}$$

15.14 POTENTIAL ENERGY SURFACES

We may also look at a bimolecular reaction in terms of the potential energy diagram for various configurations of the system. Consider the reaction A + BC → AB + C. If A, B, and C are atoms the potential energy depends on the A–B, B–C, and A–C distances (or the A–B and B–C distances and the ABC angle), so that four dimensions would be required to plot the potential energy against the three independent parameters. However, if A, B, and C are confined to a line (line-of-centers collision), the potential energy of the system can be plotted as a function of the A–B and B–C distances. This is done by means of a contour diagram in Fig. 15.5. Each point on this surface represents the potential energy of a particular linear configuration of A–B–C. In the reaction the AB distance is initially very large and the system is represented by a point to the right of ①. The BC internuclear distance oscillates, as discussed in Section 11.8. As A approaches BC, the system follows the dashed line path and the potential energy of the system increases until the saddle point ‡ is reached. This is the high energy point along the reaction path from A + BC to AB + C. The configuration at ‡ plays a special role in the theory discussed in the next section, and it is called the activated complex.

If the potential energy surface for various configurations of atoms involved in a reaction can be obtained theoretically, the expected reaction rate can be calculated using a computer to obtain the probabilities of reaction when reactants approach each other with different relative velocities.

The potential energy surface for the hydrogen atom–hydrogen molecule reaction

$$H_A + H_BH_C \rightarrow H_AH_B + H_C \qquad (15.78)$$

has been calculated, and the theoretically expected kinetics have been studied in some detail and compared with the reaction rate constants obtained experimentally

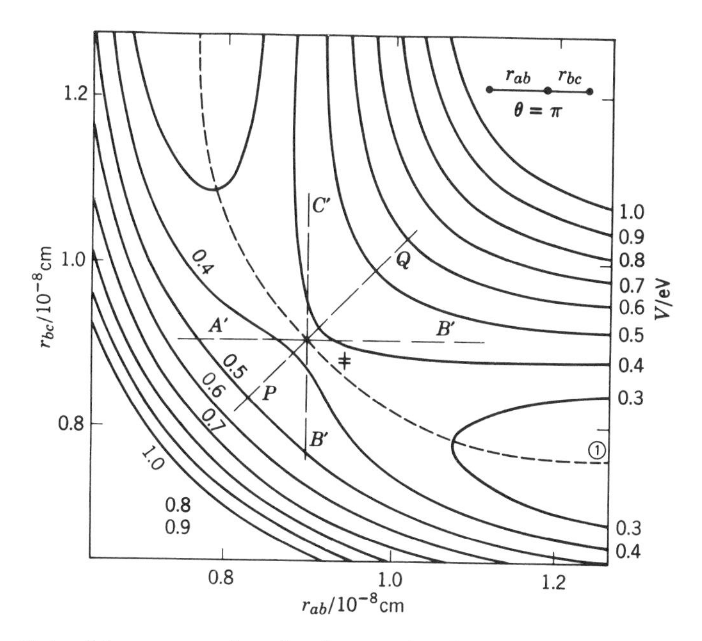

Fig. 15.5 Potential energy surface for the reaction $H_A + H_BH_C \rightarrow H_AH_B + H_C$ for a linear approach and departure. [Based on R. N. Porter and M. Karplus, *J. Chem. Phys.*, **40**, 1105 (1964).]

from rate studies of the ortho-para hydrogen conversion occurring by this mechanism. It is difficult to calculate sufficiently good potential energy surfaces that the barrier height can be obtained accurately. The energy of a hydrogen molecule and a hydrogen atom is about 50 eV (4810 kJ mol^{-1}) less than that of three isolated hydrogen atoms, but the energy barrier is only about 0.4 eV (38 kJ mol^{-1}). Thus the energies of the possible three-atom systems have to be calculated with great accuracy to yield a very accurate energy barrier by difference.

A computer may be used to calculate the probability that H_BH_C is converted into H_AH_B. H_A and H_BH_C may approach with different energies. The movement of nuclei during a collision may be calculated classically. The experimental Arrhenius parameters are given by

$$k = (5.4 \times 10^{13}\ \text{mol}^{-1}\ \text{cm}^3\ \text{s}^{-1}) \exp\left[-\frac{34{,}400\ \text{J mol}^{-1}}{(8.314\ \text{J K}^{-1}\ \text{mol}^{-1})\,T}\right] \quad (15.79)$$

Theoretical calculations* yield

$$k = (4.3 \times 10^{13}\ \text{mol}^{-1}\ \text{cm}^3\ \text{s}^{-1}) \exp\left[-\frac{31{,}100\ \text{J mol}^{-1}}{(8.314\ \text{J K}^{-1}\ \text{mol}^{-1})\,T}\right] \quad (15.80)$$

* R. N. Porter and M. Karplus, *J. Chem. Phys.*, **40**, 1105 (1964).

15.15 ACTIVATED COMPLEX THEORY FOR BIMOLECULAR REACTIONS

This theory* postulates that the activated complex (represented by $\ddagger$ in Fig. 15.5) is in equilibrium with the reactants. This equilibrium is treated by the methods of thermodynamics or statistical mechanics, even though the activated complex is not an ordinary molecule but is in the process of flying apart. Then the reaction rate is obtained by multiplying the concentration of the activated complex by the decomposition frequency.

The variation of energy along the reaction coordinate is shown in Fig. 15.6. The reaction coordinate is the dashed line in Fig. 15.5.

Thus we can think of the activation process in terms of going over a mountain pass from one valley to the next. The difference in elevation of the floors of the two valleys is analogous to the difference in internal energy U between reactants and products. The height of the mountain pass measured from one of the valleys is analogous to the activation energy. The greater the activation energy, the fewer are the collisions involving sufficient energy to cause reaction at a given temperature, and the slower is the reaction.

The solid line in Fig. 15.6 represents the lowest energy path over the classical surface. Since reactants, products, and the activated complex have vibrational energy, their vibrational levels (Section 8.10) are indicated. (The rotational energies are small enough to neglect here.)

According to the ACT (activated complex theory) the rate of the reaction

$$A + B \rightarrow \text{products} \tag{15.81}$$

is the product of the decomposition frequency $\nu^\ddagger$ for the activated complex $AB^\ddagger$ and the concentration of the activated complex.

$$-\frac{d(A)}{dt} = \nu^\ddagger c_{AB^\ddagger} \tag{15.82}$$

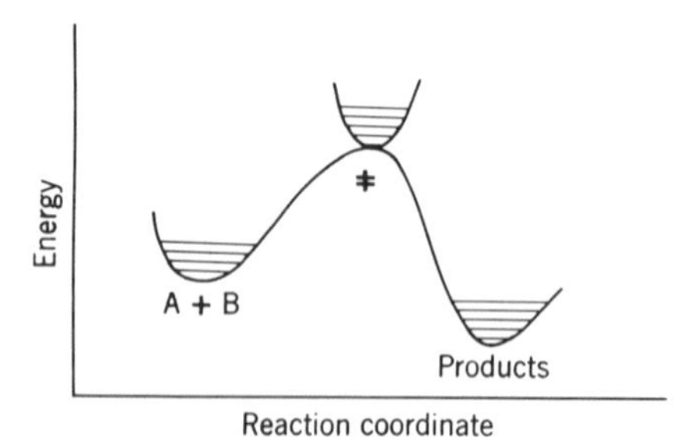

Fig. 15.6 Variation of energy along the reaction coordinate.

* H. Eyring, *J. Chem. Phys.*, **3**, 107 (1935); M. G. Evans and M. Polanyi, *Trans. Faraday Soc.*, **31**, 875 (1935).

Since the activated complex is assumed to be in equilibrium with the reactants,

$$K^{\ddagger} = \frac{c^{\circ} c_{AB^{\ddagger}}}{c_A c_B} \tag{15.83}$$

where the reference concentration (for example, mol L^{-1}) c° is required to keep $K^{\ddagger}$ dimensionless (Section 4.2). Thus equation 15.82 may be written

$$-\frac{d(A)}{dt} = \frac{\nu^{\ddagger} K^{\ddagger} c_A c_B}{c^{\circ}} \tag{15.84}$$

The equilibrium constant $K^{\ddagger}$ may be expressed in terms of the partition functions of A, B, and AB‡ and the difference $\Delta\epsilon_0$ between the energies of the activated complex and reactants at absolute zero (see Section 13.14).

$$K^{\ddagger} = \frac{c^{\circ} N_A V q_{AB^{\ddagger}}}{q_A q_B} e^{-\Delta\epsilon_0/kT} \tag{15.85}$$

where $q_{AB^{\ddagger}}$, q_A, and q_B are molecular partition functions. The magnitudes of these various partition functions have been discussed in Chapter 13. In order to eliminate the V from equation 15.85 it is convenient to use partition functions $q'_{AB^{\ddagger}} = q_{AB^{\ddagger}}/V$, $q'_A = q_A/V$, and $q'_B = q_B/V$, that is, partition functions without the volume factor, so that

$$K^{\ddagger} = \frac{c^{\circ} N_A q'_{AB^{\ddagger}}}{q'_A q'_B} e^{-\Delta\epsilon_0/kT} \tag{15.86}$$

The partition function for the activated complex is the product of the partition functions for the various degrees of freedom, except that a special treatment is required for the vibrational mode corresponding to the decomposition. The vibrational partition function $q_{v^{\ddagger}}$ for a normal mode is

$$q_{v^{\ddagger}} = (1 - e^{-h\nu^{\ddagger}/kT})^{-1} \tag{15.87}$$

However, if the vibration frequency is low, $h\nu^{\ddagger} \ll kT$, so that using the series expansion $e^{-x} = 1 - x + \cdots$ yields

$$q_{v^{\ddagger}} = \frac{kT}{h\nu^{\ddagger}} = \frac{RT}{N_A h\nu^{\ddagger}} \tag{15.88}$$

Thus the molecular partition function for AB‡ may be written

$$q'_{AB^{\ddagger}} = \frac{RT}{N_A h\nu^{\ddagger}} q''_{AB^{\ddagger}} \tag{15.89}$$

where $q''_{AB^{\ddagger}}$ is the molecular partition function with one vibrational factor missing. Substituting equations 15.86 and 15.89 into 15.84 yields

$$-\frac{d(A)}{dt} = \frac{RT}{h} \frac{q''_{AB^{\ddagger}}}{q'_A q'_B} e^{-E^{\ddagger}/RT} c_A c_B \tag{15.90}$$

where $E^{\ddagger} = N_A \Delta\epsilon_0$. Since the rate constant k is defined by

$$-\frac{d(A)}{dt} = kc_A c_B \tag{15.91}*$$

$$k = \frac{RT}{h} \frac{q''_{AB^{\ddagger}}}{q'_A q'_B} e^{-E^{\ddagger}/RT} \tag{15.92}$$

Since the partition functions used here have the units of m^{-3}, the rate constant calculated using SI units has the units of $(\text{mol m}^{-3})^{-1}\ s^{-1}$.

In order to calculate k it is necessary to make assumptions about the vibrational frequencies and moments of inertia of $AB^{\ddagger}$.

The right side of equation 15.92 is frequently multiplied by a transmission coefficient κ, which is the probability that a molecule once across the barrier will go on and not return. The transmission coefficient κ is usually taken as unity, but it is zero in a few cases. For example, in the recombinations of atoms $A + A = A_2$, A_2 contains all the exothermic energy of this reaction and will dissociate on its first vibration unless some energy is lost to a third body on collision. Thus atomic recombination reactions are always termolecular, as described in the next section.

The experimental Arrhenius parameters for a number of bimolecular reactions are given in Table 15.2. The preexponential factors may be compared with the 10^{13}–10^{15} $\text{mol}^{-1}\ \text{cm}^3\ \text{s}^{-1}$ calculated from simple collision theory (Section 15.13). The last column gives preexponential factors calculated using activated complex theory using *estimated* frequencies and geometrical parameters for the activated complex. The results show that activated complex theory is a considerable advance over simple collision theory.

Table 15.2 Preexponential Factors for Some Bimolecular Reactions Calculated by ACT Compared with Experimental Values[a]

Reaction	$A_{exp}/\text{mol}^{-1}\ \text{cm}^3\ \text{s}^{-1}$	$A_{calc}/\text{mol}^{-1}\ \text{cm}^3\ \text{s}^{-1}$
$H + H_2 \rightarrow H_2 + H$	5.4×10^{13}	7.4×10^{13}
$Br + H_2 \rightarrow HBr + H$	3×10^{13}	1×10^{14}
$H + CH_4 \rightarrow H_2 + CH_3$	1×10^{13}	2×10^{13}
$H + C_2H_6 \rightarrow H_2 + C_2H_5$	3×10^{12}	1×10^{13}
$CH_3 + H_2 \rightarrow CH_4 + H$	2×10^{12}	1×10^{12}
$CH_3 + CH_3COCH_3 \rightarrow CH_4 + CH_3COCH_2$	4×10^{11}	1×10^{11}
$CD_3 + CH_4 \rightarrow CD_3H + CH_3$	1×10^{11}	2×10^{11}
$2ClO \rightarrow Cl_2 + O_2$	6×10^{10}	1×10^{11}

[a] From J. Nicholas, *Chemical Kinetics*, John Wiley & Sons, New York, 1976.

* This rate equation indicates that the activated complex contains one molecule of A and one molecule of B. The generalization to more complicated rate equations is that the order of the rate law indicates the number of molecules in the activated complex.

Example 15.2 Using activated complex theory, calculate the preexponential factor for the rate constant for the reaction

$$H + H_2 \rightarrow H_2 + H$$

at 500 K. Assume a linear activated complex with the nuclei each separated by 0.94×10^{-10} m. The vibrational partition functions may be ignored because their values are so close to unity. (The experimental value is 5.4×10^{13} mol^{-1} cm^3 s^{-1}.)

From equation 15.92 the preexponential factor is given by

$$\frac{RT}{h}\frac{q''_{H_3{}^\ddagger}}{q'_H q'_{H_2}}$$

where q' and q'' are molecular partition functions without the volume factor, and q'' lacks one of the vibrational factors.

The translational partition function for a hydrogen atom at 3000 K was calculated in Example 13.2 to be 7.69×10^{30}. At 500 K the translational partition function without the volume factor is

$$q'_{tH} = \frac{(7.69 \times 10^{30})}{(0.246 \text{ m}^3)}\left(\frac{500}{3000}\right)^{3/2} = 2.13 \times 10^{30} \text{ m}^{-3}$$

The translational partition function for H_2 and H_3 are larger by factors of $2^{3/2}$ and $3^{3/2}$, so that

$$q'_{tH_2} = 2^{3/2}(2.13 \times 10^{30} \text{ m}^{-3}) = 6.02 \times 10^{30} \text{ m}^{-3}$$

$$q'_{tH_3} = 3^{3/2}(2.13 \times 10^{30} \text{ m}^{-3}) = 1.11 \times 10^{31} \text{ m}^{-3}$$

The rotational partition function for H_2 at 3000 K was calculated in Example 13.4 to be 17.14. At 500 K the rotational partition function is

$$q_{rH_2} = 17.14\frac{500}{3000} = 2.86$$

Since $H_3{}^\ddagger$ is linear with nuclei separated by 0.9×10^{-10} m, the moment of inertia is

$$I = \mu R^2 = \frac{m_1 m_2}{m_1 + m_2}R^2 = \frac{(1.0078 \times 10^{-3} \text{ kg mol}^{-1})^2(1.8 \times 10^{-10})^2}{(2.0156 \times 10^{-3} \text{ kg mol}^{-1})(6.022 \times 10^{23} \text{ mol}^{-1})}$$

$$= 2.71 \times 10^{-47} \text{ kg m}^2$$

(Note that the central hydrogen atom is on the axis of rotation and does not contribute to the moment of inertia.)

$$q_{rH_3{}^\ddagger} = \frac{8\pi^2 IkT}{2h^2} = \frac{8\pi^2(2.71 \times 10^{-47} \text{ kg m}^2)(1.381 \times 10^{-23} \text{ J K}^{-1})(500 \text{ K})}{2(6.62 \times 10^{-34} \text{ J s})^2} = 16.68$$

Thus the preexponential factor is calculated to be

$$\frac{RT}{h}\frac{q''_{H_3{}^\ddagger}}{q'_H q'_{H_2}} = \frac{(8.314 \text{ J K}^{-1} \text{ mol}^{-1})(500 \text{ K})}{(6.626 \times 10^{-34} \text{ J s})}\frac{(1.11 \times 10^{31} \text{ m}^{-3})(16.86)}{(2.13 \times 10^{30} \text{ m}^{-3})(6.02 \times 10^{30} \text{ m}^{-3})(2.86)}$$

$$= 3.20 \times 10^7 (\text{mol m}^{-3})^{-1} \text{ s}^{-1} = (3.20 \times 10^7 \text{ mol}^{-1} \text{ m}^3 \text{ s}^{-1})(10^6 \text{ cm}^3 \text{ m}^{-3})$$

$$= 3.20 \times 10^{13} \text{ cm}^3 \text{ mol}^{-1} \text{ s}^{-1}$$

This may be compared with the experimental value of 5.4×10^{13} cm^3 mol^{-1} s^{-1} in Table 15.2.

15.16 THERMODYNAMIC FORMULATION OF ACTIVATED COMPLEX THEORY

According to equation 15.84, the rate constant for a bimolecular reaction is given by

$$k = \frac{RT}{N_A h}\frac{K^\ddagger}{c^\circ} \tag{15.93}$$

The equilibrium constant $K^\ddagger$ for the formation of the activated complex from the reactants A and B may be expressed in terms of the standard Gibbs energy of activation $\Delta G^{\ddagger\circ}$ or in terms of the standard entropy of activation $\Delta S^{\ddagger\circ}$ and the standard enthalpy of activation $\Delta H^{\ddagger\circ}$.

$$k = \frac{RT}{N_A h c^\circ} e^{-\Delta G^{\ddagger\circ}/RT} \tag{15.94a}$$

$$= \frac{RT}{N_A h c^\circ} e^{\Delta S^{\ddagger\circ}/R} e^{-\Delta H^{\ddagger\circ}/RT} \tag{15.94b}$$

for a bimolecular reaction. In general,

$$k = \frac{RT}{N_A h} (c^\circ)^{\Delta n^\ddagger} e^{\Delta S^{\ddagger\circ}/R} e^{-\Delta H^{\ddagger\circ}/RT} \tag{15.94c}$$

where $\Delta n^\ddagger$ is the change in number of moles going from reactants to activated complex; $\Delta n^\ddagger = 0$ for a unimolecular reaction, -1 for a bimolecular reaction, and -2 for a termolecular reaction of gases. The standard states for these thermodynamic functions are unit concentrations.

Equation 15.94*c* has approximately the form of the Arrhenius equation. In order to write equation 15.95*c* in the Arrhenius form we need to find the relationship between the Arrhenius activation energy and $\Delta H^{\ddagger\circ}$. The Arrhenius activation energy E_a is given by

$$E_a = RT^2 \frac{d \ln k}{dT} \tag{15.95}$$

Substituting equation 15.93,

$$E_a = RT^2 \frac{d}{dT} \ln \left(\frac{RT}{N_A h}\frac{K^\ddagger}{c_0}\right) = RT + RT^2 \frac{d \ln K^\ddagger}{dT} = RT + \Delta U^{\ddagger\circ} \tag{15.96}$$

Since $\Delta U^{\ddagger\circ} = \Delta H^{\ddagger\circ} - \Delta(PV)^\ddagger$,

$$E_a = \Delta H^{\ddagger\circ} + RT - \Delta(PV)^\ddagger \tag{15.97}$$

For a reaction with perfect gases, $\Delta(PV)^\ddagger = \Delta n^\ddagger RT$, where $\Delta n^\ddagger$ is defined above. Thus

$$\begin{aligned} E_a &= \Delta H^{\ddagger\circ} - (\Delta n^\ddagger - 1)RT \\ &= \Delta H^{\ddagger\circ} + mRT \end{aligned} \tag{15.98}$$

where m is the order of the reaction.

Substituting this relation into 15.94c yields

$$k = e^m (c^0)^{1-m} \frac{RT}{N_A h} e^{\Delta S^{\ddagger\circ}/R} e^{-E_a/RT} \tag{15.99}$$

where the factor $(c^0)^{1-m}$ has been used so that the equation will be applicable to a reaction of order m, rather than being restricted to second order as assumed in equation 15.93.* Thus the preexponential factor A in the Arrhenius equation is given by

$$A = e^m (c^0)^{1-m} \frac{RT}{N_A h} e^{\Delta S^{\ddagger\circ}/R} \tag{15.100}$$

Thus the standard entropy of activation may be calculated from the preexponential factor in the Arrhenius equation.

For many unimolecular, bond-breaking gas reactions $\Delta S^{\ddagger\circ} = 0$ because the activated complex is so much like the original reactants and there is very little change in configuration in going from the reactants to the activated complex. In this case $e^{\Delta S^{\ddagger\circ}/R} = 1$ and, at room temperatures,

$$A = \frac{eRT}{N_A h} = \frac{(2.718)(8.31\ \text{J K}^{-1}\ \text{mol}^{-1})(300\ \text{K})}{(6.02 \times 10^{23}\ \text{mol}^{-1})(6.63 \times 10^{-34}\ \text{J s})} = 1.7 \times 10^{13}\ \text{s}^{-1} \tag{15.101}$$

for a unimolecular reaction.

If the activation of the molecule involves a rearrangement of atoms or a change in configuration, there will be a change in entropy and $e^{\Delta S^{\ddagger\circ}/R}$ is not unity. The values of $\Delta S^{\ddagger\circ}$ are rarely large enough to give a value of more than 10^2 or less than 10^{-2} to the term $e^{\Delta S^{\ddagger\circ}/R}$, and so frequency factors $e(RT/N_A h)e^{\Delta S^{\ddagger\circ}/R}$ may range from about 10^{11} to 10^{15}. If there is an increase in rotational and vibrational freedom in the activated complex, $\Delta S^{\ddagger\circ}$ is positive, and the preexponential factor is larger than $1.7 \times 10^{13}\ \text{s}^{-1}$. If there is a decrease in rotational and vibrational freedom in the activated complex, $\Delta S^{\ddagger\circ}$ is negative, and the preexponential factor is smaller than $1.7 \times 10^{13}\ \text{s}^{-1}$.

Example 15.3 What is the entropy of activation $\Delta S^{\ddagger\circ}$ for the reaction

$$H + CH_4 = H_2 + CH_3$$

which has a preexponential factor of $10^{13}\ \text{cm}^3\ \text{s}^{-1}\ \text{mol}^{-1}$ in the neighborhood of 500 K?

$$A = \frac{RTe^2}{N_A hc^0} e^{\Delta S^{\ddagger\circ}/R}$$

$$\Delta S^{\ddagger\circ} = R \ln \frac{AN_A hc^0}{RTe^2} = (8.314\ \text{J K}^{-1}\ \text{mol}^{-1})$$

$$\times \ln \frac{(10^{13}\ \text{cm}^3\ \text{s}^{-1}\ \text{mol}^{-1})(10^{-2}\ \text{m cm}^{-1})^3(6.02 \times 10^{23}\ \text{mol}^{-1}) \times (6.63 \times 10^{-34}\ \text{J s})(10^{-3}\ \text{mol m}^{-3})}{(8.314\ \text{J K}^{-1}\ \text{mol}^{-1})(500\ \text{K})(2.72)^2}$$

$$= -189\ \text{J K}^{-1}\ \text{mol}^{-1}$$

* P. J. Robinson, *J. Chem. Ed.*, **55**, 509 (1978).

The negative value of $\Delta S^{\ddagger\circ}$ indicates that the activated complex has a more organized structure than the reactants.

15.17 TERMOLECULAR REACTIONS

The combination of atoms or small radicals has to take place in the presence of a third body M which carries away the energy that is liberated; otherwise the product molecule would immediately fly apart again.

$$A + A + M \rightarrow A_2 + M \tag{15.102}$$

$$A + B + M \rightarrow AB + M \tag{15.103}$$

These third-order reactions generally have preexponential factors in the range 10^{14}–10^{16} $mol^{-2}\ cm^6\ s^{-1}$ and *negative* activation energies. There is a wide range of efficiencies for third bodies, and rate constants may change up to 10^3 for different third bodies.

The negative activation energies may be explained by the mechanism

$$A + M \underset{k_{-1}}{\overset{k_1}{\rightleftharpoons}} A..M \tag{15.104}$$

$$A..M + A \overset{k_2}{\rightleftharpoons} A_2 + M \tag{15.105}$$

where A..M is a short-lived intermediate. The steady-state rate equation is

$$\frac{d(A_2)}{dt} = \frac{k_1 k_2 (A)^2 (M)}{k_{-1} + k_2 (A)} \tag{15.106}$$

If the intermediate is very short lived, $k_{-1} \gg k_2(A)$, and equation 15.106 becomes

$$\frac{d(A_2)}{dt} = K_1 k_2 (A)^2 (M) \tag{15.107}$$

where $K_1 = k_1/k_{-1}$. If k_2 follows the Arrhenius equation with activation energy E_2, the activation energy for the termolecular reaction is given by

$$E_a = RT^2 \frac{d \ln K_1}{dT} + RT^2 \frac{d \ln k_2}{dT} = \Delta H_1^\circ + E_2 \tag{15.108}$$

Since ΔH_1° is quite negative for the combination of atoms or small radicals, the overall activation energy E_a for the termolecular reaction is negative.

15.18 UNIMOLECULAR REACTIONS

A large number of gaseous isomerization and dissociation reactions follow first-order kinetics at about atmospheric pressure. At first sight this is puzzling, because one would think that since the activation energy is provided in bimolecular collisions, the reactions would be second order. In addition, it is found that as the reactant

Table 15.3 Arrhenius Parameters for k_∞ for Unimolecular Gas Reactions

Reactant	Product	log (A/s^{-1})	E_a/kJ mol^{-1}
cis-2-butene	*trans*-2-butene	13.8	263
cyclopropane	propylene	15.2	272
N_2O_5	$NO_2 + NO_3$	13.6	103
C_2H_6	$2CH_3$	16.5	368
CO_2	$CO + O$	11.3	460

pressure decreases, the first-order rate constant decreases, and at very low pressures the reaction becomes second order. The Arrhenius parameters for a few first-order gas isomerization or decomposition reactions at their high-pressure limits are given in Table 15.3. In 1922 Lindemann proposed a mechanism for unimolecular reactions that explains why they are second order at low pressures but become first order as the pressure is increased. He pointed out that when a molecule is energized by a bimolecular collision, there may be a time lag before decomposition or isomerization, and that during this time lag the energized molecule may lose its extra energy in a second bimolecular collision. Since such reactions are usually studied by diluting the reacting gas A with an excess of inert gas M, the energizing and deenergizing collisions are mostly collisions of A and M, as indicated in the following mechanism.

$$A + M \xrightarrow{k_e} A^* + M \tag{15.109}$$

$$A^* + M \xrightarrow{k_{de}} A + M \tag{15.110}$$

$$A^* \xrightarrow{k_{uni}} \text{products} \tag{15.111}$$

Here A* represents an energized A molecule. Any collision of A* with M is assumed to be deenergizing, and the bimolecular rate constant for this process is k_{de}. Some M molecules are more effective than others in activating A upon collision and inactivating A* upon collision. The decomposition of A* is unimolecular with a rate constant of k_{uni}.

When a molecule absorbs a large amount of energy in a collision, the translational, rotational, and vibrational energy is increased. Electronic excitation is involved only in those rare cases where there are low-lying electronic levels. It is primarily the vibrational energy that plays a role in tearing the molecule apart. This energy can be transferred from one type of vibrational motion to another. A molecule with a large amount of vibrational energy may vibrate for a short time and then dissociate when the energy in one particular vibrational mode becomes so great that the molecule flies apart by breaking a bond that was stretched excessively in the vibration.

Since A* is never present at a very high concentration, the rate law for this mechanism may be derived assuming A* is in a steady state (Section 15.10).

$$\frac{d(A^*)}{dt} = k_e(A)(M) - [k_{de}(M) + k_{uni}](A^*) = 0 \tag{15.112}$$

Thus

$$-\frac{d(A)}{dt} = k_{uni}(A^*) = \frac{k_{uni}k_e(A)(M)}{k_{de}(M) + k_{uni}} = k(A) \tag{15.113}$$

This equation can be simplified to two limiting forms, depending on the relative magnitudes of the two denominator terms. At sufficiently low pressures $k_{de}(M) \ll k_{uni}$ and

$$-\frac{d(A)}{dt} = k_e(A)(M) \tag{15.114}$$

Under these conditions the reaction is second order, and all A* molecules decompose to products before they can be deenergized in a collision with a second M molecule. At sufficiently high pressures $k_{de}(M) \gg k_{uni}$ and

$$-\frac{d(A)}{dt} = \frac{k_{uni}k_e}{d_{de}}(A) = k_\infty(A) \tag{15.115}$$

Under these conditions the reaction is first order in A. The concentration of A* is at its equilibrium value, $(A^*) = (A)k_e/k_{de}$, at the high-pressure limit, and so the rate of reaction is determined by the equilibrium constant for the production of A* and by k_{uni}.

Experimental data on unimolecular reactions deviate significantly from equation 15.113, but the basic idea of the theory is correct, and modern extensions of the theory are able to account for the experimental results quite satisfactorily.

15.19 RADICAL REACTIONS

Free radicals are reactive molecules (or atoms) with unpaired electrons. This term is not applied to stable species like Fe^{3+} and O_2 in spite of the fact that their paramagnetic behavior (Section 12.1) demonstrates that they have unpaired electrons. At very high temperatures organic molecules may be partially dissociated into free radicals. Alkyl free radicals in the gas phase may be prepared by the thermal decomposition of metal alkyls. For example, methyl radicals $CH_3\cdot$ may be obtained from

$$Pb(CH_3)_4 \rightarrow Pb + 4CH_3\cdot$$

When this reaction is carried out by flowing gaseous $Pb(CH_3)_4$ down a heated glass tube, a lead mirror is produced on the inside of the tube. If a gas containing free radicals is passed over such a mirror, the mirror is removed. The removal of the lead mirror of methyl radicals is just the reverse of this reaction. During flow

Table 15.4 Arrhenius Parameters for Elementary Radical Reactions

	$\log_{10}$ (A/s^{-1})	E_a/kJ mol^{-1}
Molecular fission		
$C_2H_6 \rightarrow 2CH_3$	17.4	384
$N_2O_5 \rightarrow NO_2 + NO_3$	14.8	88
Radical fission		
$CH_3CO \rightarrow CH_3 + CO$	15	43
$C_2H_5 \rightarrow C_2H_4 + H$	13	167
	$\log_{10}$ (A/mol^{-1} cm^3 s^{-1})	
Radical-molecule addition		
$H + C_2H_4 \rightarrow C_2H_5$	11.0	~0
$CH_3 + C_2H_4 \rightarrow C_3H_7$	11.0	29
Exchange reactions		
$CH_3 + H_2 \rightarrow CH_4 + H$	11.7	44
$Cl + H_2 \rightarrow HCl + H$	13.9	23
$H + HCl \rightarrow H_2 + Cl$	13.6	19
Radical combination		
$2CH_3 \rightarrow C_2H_6$	13.3	~0
	$\log_{10}$ (A/mol^{-2} cm^6 s^{-1})	
$H + H + H_2 \rightarrow H_2 + H_2$	16.0	~0
$I + I + Ar \rightarrow I_2 + Ar$	15.8	−5

down a tube the radicals disappear by colliding with the wall or by combining with each other. Free radicals may also be produced by the absorption of light of sufficiently short wavelength or by passing a gas through an electrical discharge.

Since a radical has an unpaired electron, its reaction with a molecule having paired electrons gives rise to another radical. In this way the reactive center is maintained and can give rise to a chain of reactions. We may ask why such a reaction ever stops. Sometimes, as a matter of fact, the chain reaction does not stop until all the material is consumed. At other times, however, the chain is broken when one of the activated molecules in it collides with the wall of the containing vessel or with another radical to form a spin-paired molecule. The length of the chain (i.e., the number of molecules reacting per molecule activated) is determined by the relative rates of the chain-propagating and the chain-breaking reactions.

Many gas-phase reactions proceed by a series of elementary radical reactions. Mechanisms involving radicals may be nonchain, straight chain, and branched chain. Branched-chain reactions produce explosions if they are highly exothermic; they are discussed in Section 15.21. The rate parameters for several types of elementary radical reactions are given in Table 15.4. Although the rate constants for radical fission reactions may have very large preexponential factors, they may be slow, except at very high temperatures, because of high activation energies. The activation energy of a radical fission reaction may be lower when rearrangement to a molecular product compensates for some of the energy required to break a bond. Radical-molecule additions are important in polymerizations. These reactions are exothermic and may have low activation energies. Bimolecular radical exchange reactions are very common. The heat of reaction may be positive or negative, depending on whether the newly formed bond has a higher or lower dissociation energy than the bond broken. Radical combination reactions may be bimolecular if the energy liberated may be absorbed in the product. But, as discussed in Section 15.17, the combination of atoms or small radicals requires a third body to carry away the energy liberated.

The decomposition of nitrogen pentoxide discussed in Section 15.10 is an example of a complicated nonchain radical mechanism.

15.20 UNBRANCHED CHAIN REACTIONS

The reaction of gaseous bromine with hydrogen

$$H_2 + Br_2 = 2HBr \tag{15.116}$$

is an example of an unbranched chain reaction.

In the temperature range 200 to 300 °C the rate of reaction of H_2 and Br_2 is given by

$$\frac{d(\mathrm{HBr})}{dt} = \frac{k(\mathrm{H_2})(\mathrm{Br_2})^{1/2}}{1 + k'(\mathrm{HBr})/(\mathrm{Br_2})} \tag{15.117}$$

If no HBr is initially present, the initial velocity $d(\mathrm{HBr})/dt$ is directly proportional to $(\mathrm{H_2})(\mathrm{Br_2})^{1/2}$. If sufficient HBr is added initially so that $k'(\mathrm{HBr})/(\mathrm{Br_2}) \gg 1$, the initial velocity is directly proportional to $(\mathrm{H_2})(\mathrm{Br_2})^{3/2}/(\mathrm{HBr})$, and it is seen that doubling the initial HBr concentration cuts the initial velocity in half.

In 1919 Christiansen, Herzfeld, and Polanyi independently proposed a radical chain mechanism to explain this rate law. At that time there was no direct evidence of gaseous free radicals. They proposed the following chain reaction.

Initiation $$Br_2 + M \xrightarrow{k_1} 2Br + M \tag{15.118}$$

Propagation $$Br + H_2 \xrightarrow{k_2} HBr + H \tag{15.119}$$

$$H + Br_2 \xrightarrow{k_3} HBr + Br \tag{15.120}$$

Termination $$2Br + M \xrightarrow{k_4} Br_2 + M \tag{15.121}$$

Inhibition $$H + HBr \xrightarrow{k_5} H_2 + Br \tag{15.122}$$

The reaction is initiated by the splitting of bromine instead of hydrogen, because the bond in bromine is much weaker. Step 2 is endothermic and has a high activation energy, so that it is slow relative to the others. Step 5 is the reverse of step 2 and is fast.

The chain reaction rapidly gets into a steady state in which $d(\mathrm{Br})/dt$ and $d(\mathrm{H})/dt$ are both approximately zero. Using the expressions for these derivatives obtained from the mechanism and setting them equal to zero yields the following steady-state rate equation.

$$\frac{d(\mathrm{HBr})}{dt} = \frac{2k_2(k_1/k_4)^{1/2}(\mathrm{H_2})(\mathrm{Br_2})^{1/2}}{1 + k_5(\mathrm{HBr})/k_3(\mathrm{Br_2})} \tag{15.123}$$

Thus the empirical rate constant k in equation 15.117 may be identified with $2k_2(k_1/k_4)^{1/2}$, where the equilibrium constant k_1/k_4 for the dissociation of bromine may be calculated using statistical mechanics.

Reaction 15.116 may be carried out photochemically at lower temperatures where the thermal chain reaction is slow, because the concentration of atomic bromine is very low. The experimental rate equation is

$$\frac{d(\mathrm{HBr})}{dt} = \frac{k''I_a^{1/2}(\mathrm{H_2})}{(\mathrm{M})^{1/2}[1 + k'(\mathrm{HBr})/(\mathrm{Br_2})]} \tag{15.124}$$

where I_a is the intensity of radiation absorbed and M is any other gas that may be present. The mechanism of the photochemical reaction is the same as the thermal reaction except that the first step is replaced with

$$\mathrm{Br_2} + h\nu \rightarrow 2\mathrm{Br} \tag{15.125}$$

The steady-state rate equation is

$$\frac{d(\mathrm{HBr})}{dt} = \frac{2k_2I_a^{1/2}(\mathrm{H_2})}{k_4^{1/2}(\mathrm{M})^{1/2}[1 + k_5(\mathrm{HBr})/k_3(\mathrm{Br_2})]} \tag{15.126}$$

Note that in the absence of light, the equilibrium concentration of Br is

$$(\mathrm{Br}) = \left[\frac{k_1}{k_4}(\mathrm{Br_2})\right]^{1/2} \tag{15.127}$$

and under photochemical conditions

$$(\mathrm{Br}) = \left[\frac{I}{k_4(\mathrm{M})}\right]^{1/2} \tag{15.128}$$

The hydrogen-chlorine thermal reaction is faster than the hydrogen-bromine thermal reaction because the propagation step

$$\mathrm{Cl} + \mathrm{H_2} \rightarrow \mathrm{HCl} + \mathrm{H}$$

is only endothermic by 1.2 kcal mol^{-1} and is much faster than reaction 15.119. Since the reverse of this step is slower than for bromine, there is no inhibitory term in the denominator of the rate law. The hydrogen-iodine thermal reaction is very slow below 800 K because the propagation step

$$\mathrm{I} + \mathrm{H_2} \rightarrow \mathrm{HI} + \mathrm{H}$$

has an activation energy of 140 kJ mol^{-1}. At lower temperatures the hydrogen-iodine reaction goes by a nonchain mechanism.*

* J. Nicholas, *Chemical Kinetics*, John Wiley & Sons, 1976, p. 120.

15.21 BRANCHED CHAIN REACTIONS

If a step in a chain reaction produces two or more radicals from one, there is a possibility of a rapid increase in rate and, for an exothermic reaction, an explosion. The reaction of hydrogen with oxygen can be explosive above about 700 K. The ranges of temperature and pressure within which there are explosions are shown In Fig. 15.7. For example, at about 550 °C stoichiometric hydrogen-oxygen mixtures react very slowly at pressures below 100 Pa. As the pressure is increased, the reaction rate increases slowly, but at a pressure of about 100 Pa, depending on the volume of the vessel, there is a sudden explosion. On the other hand, if the gases are at a considerably higher pressure, the rate is again quite low. Hinshelwood found that, if hydrogen at 2.6×10^4 Pa and oxygen at 1.3×10^4 Pa are placed in a 300-cm^3 quartz vessel at 550 °C, the rate of reaction is quite slow and becomes slower if the pressure is further reduced to 1.33×10^4 Pa. If the pressure is reduced to 1.30×10^4 Pa, however, an explosion occurs. Finally, as the total pressure is increased above the explosion zone, the reaction rate increases until it becomes so fast that the reaction mixture may be said to explode. The fact that the exact limits depend on the vessel surface and the vessel diameter indicates that radical chains may be terminated by reaction at the wall. If the vessel surface is coated with potassium chloride, radicals disappear when they strike the wall; if the vessel surface is coated with boric oxide, radicals are not destroyed so rapidly by collisions with the wall.

The first explosion limit can be understood in terms of the following mechanism:

$$\text{Initiation} \qquad H_2 + O_2 \xrightarrow{\text{wall}} 2OH \qquad (15.129)$$

$$\text{Propagation} \qquad OH + H_2 \xrightarrow{k_2} H_2O + H \qquad (15.130)$$

$$\text{Branching} \qquad H + O_2 \xrightarrow{k_3} OH + O \qquad (15.131)$$

$$O + H_2 \xrightarrow{k_4} OH + H \qquad (15.132)$$

$$\text{Termination} \qquad H \xrightarrow{k_5} \text{wall} \qquad (15.133)$$

The propagation reaction is exothermic and fast. The third and fourth reactions are called branching reactions because two radicals are formed from one. If the rate of branching is greater than the rate of termination, the number of radicals increases exponentially with time, and an explosion results. Reaction 15.131 is endothermic and slow below 700 K. The conditions of the first explosion limit are governed by the relative rates for branching $2k_3(\text{H})(\text{O}_2)$ and termination $k_5(\text{H})$. As the concentration of oxygen is increased, the rate of branching becomes greater than the rate of termination and an explosion occurs.

When experiments are done at successively higher pressures, a second limit is reached above which there is no explosion. This is due to the fact that a new termination step has become sufficiently important to prevent the exponential increase in the number of radicals. In order for a new termination step to become

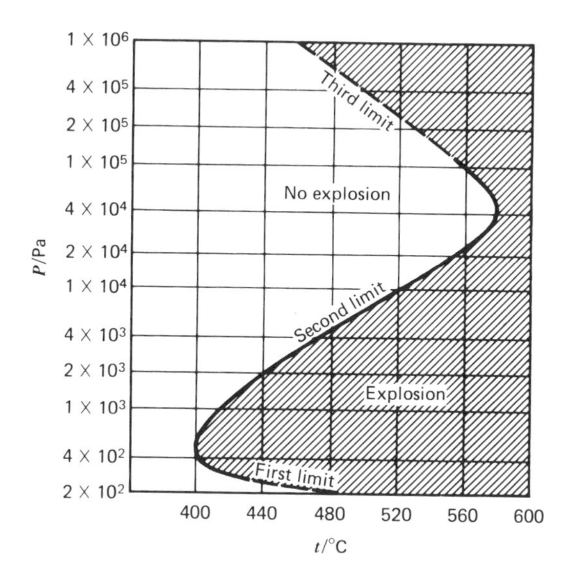

Fig. 15.7 Explosion limits of a stoichiometric oxygen-hydrogen mixture.

more important as the pressure is increased, the new termination step must be higher order than the branching reaction. Thus, to explain the second limit, the following reaction must be added to the previous reactions.

$$\text{Termination} \qquad H + O_2 + M \xrightarrow{k_6} HO_2 + M \tag{15.134}$$

In a stoichiometric mixture of oxygen and hydrogen, M may be hydrogen or oxygen, but these two gases have different efficiencies in this reaction. The HO_3 radical is relatively unreactive and does not produce another radical before it is quenched on the wall.

The third explosion limit results from the fact that the following reaction diminishes termination.

$$\text{Propagation} \qquad HO_2 + H_2 \rightarrow H_2O + OH \tag{15.135}$$

15.22 MOLECULAR BEAM EXPERIMENTS

This method provides the opportunity to learn a great deal more about the reactivity of molecules than can be learned from the usual measurements of reaction rates. In ordinary experiments reactant molecules approach each other on trajectories with random angles and impact parameters and with relative energies that range around those having the highest probabilities at the reaction temperature. Colliding molecules may also have different amounts of vibrational, rotational, and electronic energies. Reaction rate constants are averages over these various angles of approach, relative energies, and the like.

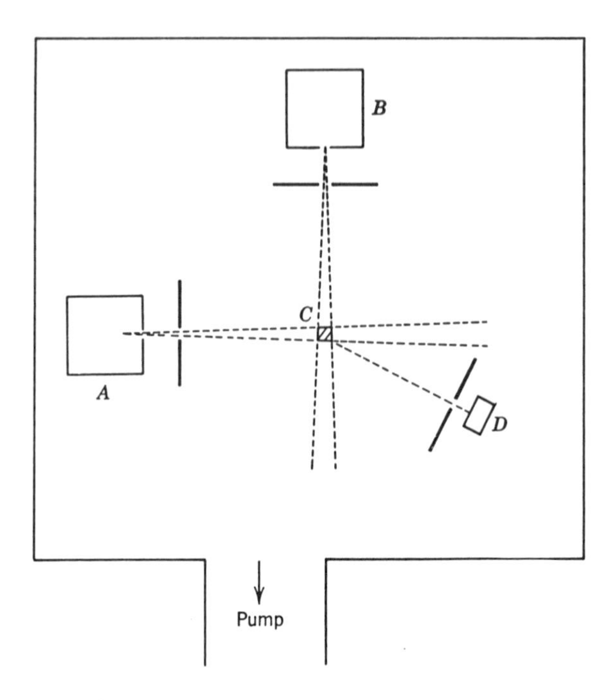

Fig. 15.8 Schematic diagram for a molecular beam apparatus for studying the reaction of molecules from source *A* with molecules from source *B*. Products are detected at *D*.

In order to get a microscopic view of a bimolecular gas chemical reaction, we would need to carry out the reaction in such a way that reactant molecules are in known quantum states and the quantum states of product molecules are identified. In the simplest type of apparatus, illustrated schematically in Fig. 15.8, *A* and *B* are sources of beams of the two reactants that collide in region *C*. These collisions occur in a chamber evacuated with a high-speed pump so that the only collisions are between the molecules from *A* and *B*. Product molecules and elastically scattered reactant molecules are detected at *D*. The effect of changing the angle of approach may be studied by moving *A* or *B*, and the effect of the relative velocity of the reactants may be studied by use of velocity selectors (see Fig. 14.5) on the beams as they leave *A* and *B*. The orientation of molecules as they collide is also important; the effect of orientation on rate may be studied for molecules with dipole moments (Section 10.10), because molecules may be oriented by use of an electric field.

15.23 HETEROGENEOUS REACTIONS

A homogeneous reaction is one occurring in a single phase and a heterogeneous reaction is one occurring in more than one phase; for example, in the gas phase and on the surface of a solid catalyst. Heterogeneous reactions are carried out on a large scale in industry. The development of catalysts is especially important for slow reactions that evolve heat. Although the equilibrium constant may be favorable at room temperature, the rate may be so low that the reaction cannot be used practically. When the temperature is raised, the rate is increased, but at high temperatures the equilibrium constant may be unfavorable, as is readily surmised from Le Châtelier's principle. A reaction of this type is $N_2 + 3H_2 = 2NH_3$, which is the basis for the Haber process for the manufacture of ammonia. To carry out this reaction at temperatures where the equilibrium constant is sufficiently favorable, special iron catalysts are used. Vanadium and other metals are added to make the catalyst more effective and to protect it from too rapid inactivation by traces of impurities in the gases. Substances that enhance the activity of catalysts are called *promoters*, and substances that inhibit catalytic activity are called *poisons*, Since only a fraction of the surface of the catalyst may be involved, it is easy to see how relatively small amounts of promoters and poisons may be effective. In the manufacture of sulfuric acid by the contact process, the presence of a very minute amount of arsenic completely destroys the catalytic activity of the platinum catalyst by forming platinum arsenide at the surface.

Enormous quantities of solid catalysts are used in the petroleum industry for "cracking," which increases the yield of gasoline from petroleum, and for "reforming," which causes the rearrangement of the molecular structures and raises the octane rating of gasoline. To facilitate the regeneration of these catalysts, they may be circulated as fine particles in the gas stream.

The catalytic activity of the walls of the containing vessel is a factor in many gaseous reactions. For surface reactions the temperature coefficient is usually small, because the activation energies are low and the slowest process is the diffusion of the products away from the walls. A 10 K rise at about 300 K increases the diffusion rate by only about 3%, whereas the rate of a chemical reaction often increases about 300%, as already explained. Since homogeneous reactions generally have higher activation energies that the corresponding heterogeneous reactions, homogeneous reactions are favored at higher temperatures and heterogeneous reactions at lower temperatures.

The mechanisms of surface-catalyzed reactions involve a sequence of steps: (1) adsorption of the reacting gases on the surface, (2) reaction on the surface, and (3) desorption of the products. Under the influence of forces at the surface, reactions occur at much higher rates than in the gas phase. The nature of the surface determines the products that are obtained, different catalysts yielding different products with the same reacting gases.

Surface-catalyzed reactions often obey a rate equation that may be derived from the Langmuir adsorption isotherm (Section 7.12). The rate of appearance of product dx/dt formed by the surface reaction is proportional to the fraction of the surface occupied by reactant molecules as given by equation 7.53.

$$\frac{dx}{dt} = k'\theta = \frac{k'(k/r)P}{1 + (k/r)P} \tag{15.136}$$

It can be seen from equation 15.136 that at low pressures, where $1 \gg (k/r)P$, the rate is directly proportional to the pressure.

$$\frac{dx}{dt} = k'\left(\frac{k}{r}\right)P \qquad (15.137)$$

It can also be seen from equation 15.136 that at high pressures, where $1 \ll (k/r)P$, the rate is independent of the pressure of the reactant: $(dx/dt) = k'$. Under these conditions the surface is saturated, and further increasing the pressure cannot increase the number of reactant molecules adsorbed. When ammonia is decomposed on a tungsten filament, for example, the rate of decomposition is independent of the pressure, over a wide range of pressure.

It is often found that surface-catalyzed reactions are slowed down by an accumulation of product molecules. This is a result of the adsorption of product molecules on the active sites, which may be taken into account quantitatively in the theory.

One of the experimental problems in studying gas reactions is that they may be catalyzed by the surface of the reaction vessel. If we want to study a homogeneous gas reaction, therefore, it is important to find conditions under which the surface-catalyzed reaction (heterogeneous reaction) is negligible. In general, a catalyst lowers the activation energy, and so the rate of a surface-catalyzed reaction does not increase as rapidly with increasing temperature as the homogeneous reaction. Consequently, at sufficiently high temperatures the homogeneous reaction is bound to be faster.

References

S. W. Benson, *Thermochemical Kinetics*, Wiley, New York, 1976.

S. W. Benson and H. E. O'Neal, *Kinetic Data on Gas Phase Unimolecular Reactions*, NSRDS-NBS-21, National Bureau of Standards, U.S. Dept. of Commerce, Washington, D.C., 1970.

J. N. Bradley, *Shock Waves in Chemistry and Physics*, Wiley, New York, 1962.

F. S. Dainton, *Chain Reactions: An Introduction*, 2nd ed., Barnes and Noble, New York, 1966.

W. H. Flygare, *Molecular Structure and Dynamics*, Prentice-Hall, Inc., Englewood Cliffs, N.J., 1978.

W. C. Gardiner, Jr., *Rates and Mechanisms of Chemical Reactions*, W. A. Benjamin, Inc., New York, 1969.

S. Glasstone, K. J. Laidler, and H. Eyring, *The Theory of Rate Processes*, McGraw-Hill Book Co., New York, 1941.

E. F. Greene and J. P. Toennies, *Chemical Reactions in Shock Waves*, Academic Press, New York, 1964.

G. G. Hammes, *Principles of Chemical Kinetics*, Academic Press, New York, 1978.

G. G. Hammes (ed.), *Investigation of Rates and Mechanisms of Reactions*, 3rd ed. (Vol. 6, pt. 2 of A. Weissberger, *Techniques of Chemistry*), Wiley, New York, 1974.

H. S. Johnston, *Gas Phase Reaction Rate Theory*, The Ronald Press Co., New York, 1966.

E. L. King, *How Chemical Reactions Occur*, W. A. Benjamin, Inc., New York, 1963.

K. J. Laidler, *Chemical Kinetics*, McGraw-Hill Book Co., New York, 1965.

R. D. Levine and R. B. Bernstein, *Molecular Reaction Dynamics*, Oxford University Press, New York, 1974.

E. S. Lewis (ed.), *Investigation of Rates and Mechanisms of Reactions*, 3rd ed. (Vol. 6, pt. 1 of A. Weissberger, *Techniques of Chemistry*), Wiley, New York, 1974.

J. Nicholas, *Chemical Kinetics*, Wiley, New York, 1976.

M. F. R. Mulcahy, *Gas Kinetics*, Wiley, New York, 1973.

A. F. Trotman-Dickenson and G. S. Milne, *Tables of Bimolecular Gas Phase Reactions*, NSRDS-NBS-9, National Bureau of Standards, U.S. Dept. of Commerce, Washington, D.C., 1967.

R. E. Weston, Jr., and H. A. Schwartz, *Chemical Kinetics*, Prentice-Hall, Inc., Englewood Cliffs, N.J., 1972.

Problems

15.1 The half-life of a first-order chemical reaction A $\rightarrow$ B is 10 min. What percent of A remains after 1 hr?

Ans. 1.56%.

15.2 The following data were obtained on the rate of hydrolysis of 17% sucrose in 0.099 mol L^{-1} HCl aqueous solutions at 35 °C.

t/min	9.82	59.60	93.18	142.9	294.8	589.4
Sucrose remaining, %	96.5	80.3	71.0	59.1	32.8	11.1

What is the order of the reaction with respect to sucrose and the value of the rate constant k?

Ans. First order, $6.20 \times 10^{-5}\ s^{-1}$.

15.3 Methyl acetate is hydrolyzed in approximately 1 mol L^{-1} HCl at 25 °C. Aliquots of equal volume are removed at intervals and titrated with a solution of NaOH. Calculate the first-order rate constant from the following experimental data.

t/s	339	1242	2745	4546	∞
v/cm^3	26.34	27.80	29.70	31.81	39.81

Ans. $1.26 \times 10^{-4}\ s^{-1}$,

15.4 Prove that in a first-order reaction, where $dn/dt = -kn$, the average life, that is the average life expectancy of the molecules, is equal to $1/k$.

15.5 It is found that the decomposition of HI to $H_2 + I_2$ at 508 °C has a half-life of 135 min when the initial pressure of HI is 0.1 atm and 13.5 min when the pressure is 1 atm. (*a*) Show that this proves that the reaction is second order. (*b*) What is the value of the rate constant in L $mol^{-1}\ s^{-1}$? (*c*) What is the value of the rate constant in $atm^{-1}\ s^{-1}$?

Ans. (*a*) $t_{1/2} = 1/k(A)_0$. (*b*) 7.91×10^{-2} L $mol^{-1}\ s^{-1}$. (*c*) $1.24 \times 10^{-3}\ atm^{-1}\ s^{-1}$.

15.6 The reaction between propionaldehyde and hydrocyanic acid has been studied at 25 °C by W. J. Svirbely and J. F. Roth [*J. Am. Chem. Soc.*, **75**, 3106 (1953)]. In a certain aqueous solution at 25 °C the concentrations at various times were as follows.

t/min	2.78	5.33	8.17	15.23	19.80	∞
(HCN)/mol L^{-1}	0.0990	0.0906	0.0830	0.0706	0.0653	0.0424
(C_3H_7CHO)/mol L^{-1}	0.0566	0.0482	0.0406	0.0282	0.0229	0.0000

What is the order of the reaction and the value of the rate constant k?

Ans. Second order, 0.675 L $mol^{-1}\ min^{-1}$.

15.7 The reaction $CH_3CH_2NO_2 + OH^- \rightarrow H_2O + CH_3CHNO_2^-$ is of second order, and k at 0 °C is 39.1 L $mol^{-1}\ min^{-1}$. An aqueous solution is made 0.004 molar in nitroethane and 0.005 molar in NaOH. How long will it take for 90% of the nitroethane to react?

Ans. 26.3 min.

15.8 Hydrogen peroxide reacts with thiosulfate ion in slightly acid solution as follows.

$$H_2O_2 + 2S_2O_3^{2-} + 2H^+ \rightarrow 2H_2O + S_4O_6^{2-}$$

This reaction rate is independent of the hydrogen-ion concentration in the pH range 4 to 6. The following data were obtained at 25 °C and pH 5.0.
Initial concentrations: $(H_2O_2) = 0.036\ 80$ mol L^{-1}; $(S_2O_3^{2-}) = 0.020\ 40$ mol L^{-1}

t/min	16	36	43	52
$(S_2O_3^{2-})/10^{-3}$ mol L^{-1}	10.30	5.18	4.16	3.13

(*a*) What is the order of the reaction? (*b*) What is the rate constant?
Ans. (*a*) Second order. (*b*) 0.59 L mol^{-1} min^{-1}.

15.9 A solution of A is mixed with an equal volume of a solution of B containing the same number of moles, and the reaction A + B = C occurs. At the end of 1 hr A is 75% reacted. How much of A will be left unreacted at the end of 2 hr if the reaction is (*a*) first order in A and zero order in B; (*b*) first order in both A and B; and (*c*) zero order in both A and B?
Ans. (*a*) 6.25, (*b*) 14.3, (*c*) 0%.

15.10 Derive the integrated rate equation for a reaction of ½ order. Derive the expression for the half-life of such a reaction.

$$Ans.\ (A)_0^{1/2} - (A)^{1/2} = \frac{k}{2}t;\ t_{1/2} = \frac{\sqrt{2}}{k}(\sqrt{2} - 1)(A)_0^{1/2}.$$

15.11 For the reaction 2A = B + C the rate law for the forward reaction is

$$-\frac{d(A)}{dt} = k(A)$$

Give two possible rate laws for the reverse reaction. *Ans.* $k(B)(C)/(A)$; $k(B)^{1/2}(C)^{1/2}$.

15.12 The following table gives kinetic data [Y. T. Chia and R. E. Connick, *J. Phys. Chem.*, **63**, 1518 (1959)] for the following reaction at 25 °C.

$$OCl^- + I^- = OI^- + Cl^-$$

(OCl^-)	(I^-)	(OH^-)	$\frac{d(IO^-)}{dt}\Big/10^{-4}$
	mol L^{-1}		mol L^{-1} s^{-1}
0.0017	0.0017	1.00	1.75
0.0034	0.0017	1.00	3.50
0.0017	0.0034	1.00	3.50
0.0017	0.0017	0.5	3.50

What is the rate law for the reaction and what is the value of the rate constant?

$$Ans.\ \frac{d(OI^-)}{dt} = \frac{(60\ s^{-1})(I^-)(OCl^-)}{(OH^-)}.$$

15.13 When an optically active substance is isomerized, the optical rotation decreases from that of the original isomer to zero in a first-order manner. In a given case the half-time for this process is found to be 10 min. Calculate the rate constant for the conversion of one isomer to another. *Ans.* 5.78×10^{-4} s^{-1}.

15.14 Parallel first-order reactions that compete for the reactant are often encountered because many products may be thermodynamically possible. For the reactions

$$A \xrightarrow{k_1} B \qquad A \xrightarrow{k_2} C$$

show that B and C appear with the same half-life even though k_1 and k_2 have different values.

Ans. $(\mathrm{A}) = (\mathrm{A})_0 e^{-(k_1+k_2)t}$. $(\mathrm{B}) = \dfrac{k_1(\mathrm{A})_0}{k_1 + k_2}[1 - e^{-(k_1+k_2)t}]$.

$(\mathrm{C}) = \dfrac{k_2(\mathrm{A})_0}{k_1 + k_2}[1 - e^{-(k_1+k_2)t}]$.

15.15 The reaction $2NO + O_2 \rightarrow 2NO_2$ is third order. Assuming that a small amount of NO_3 exists in rapid reversible equilibrium with NO and O_2 and that the rate-determining step is the slow bimolecular reaction $NO_3 + NO \rightarrow 2NO_2$, derive the rate equation for this mechanism. *Ans.* $d(\mathrm{NO_2})/dt = k(\mathrm{NO})^2(\mathrm{O_2})$.

15.16 Set up the rate expressions for the following mechanism.

$$\mathrm{A} \underset{k_2}{\overset{k_1}{\rightleftharpoons}} \mathrm{B} \qquad \mathrm{B} + \mathrm{C} \xrightarrow{k_3} \mathrm{D}$$

If the concentration of B is small compared with the concentrations of A, C, and D, the steady-state approximation may be used to derive the rate law. Show that this reaction may follow the first-order equation at high pressures and the second-order equation at low pressures. *Ans.* $d(\mathrm{D})/dt = k_1k_3(\mathrm{A})(\mathrm{C})/[k_2 + k_2(\mathrm{C})]$.

15.17 The reaction $NO_2Cl = NO_2 + \frac{1}{2}Cl_2$ is first order and appears to follow the mechanism

$$\mathrm{NO_2Cl} \xrightarrow{k_1} \mathrm{NO_2} + \mathrm{Cl} \qquad \mathrm{NO_2Cl} + \mathrm{Cl} \xrightarrow{k_2} \mathrm{NO_2} + \mathrm{Cl_2}$$

(*a*) Assuming a steady state for the chlorine atom concentration, show that the empirical first-order rate constant can be identified with $2k_1$. (*b*) The following data were obtained by H. F. Cordes and H. S. Johnston [*J. Am. Chem. Soc.*, **76**, 4264 (1954)] at 180 °C. In a single experiment the reaction is first order, and the empirical rate constant is represented by k. Show that the reaction is second order at these low gas pressures and calculate the second-order rate constant.

$c/10^{-8}$ mol cm^{-3}	5	10	15	20
$k/10^{-4}$ s^{-1}	1.7	3.4	5.2	6.9

Ans. 3.4×10^3 cm^3 mol^{-1} s^{-1}.

15.18 Suppose the transformation of A to B occurs by both a reversible first-order reaction and a reversible second-order reaction involving hydrogen ion.

$$\mathrm{A} \underset{k_2}{\overset{k_1}{\rightleftharpoons}} \mathrm{B} \qquad \mathrm{A} + \mathrm{H^+} \underset{k_4}{\overset{k_3}{\rightleftharpoons}} \mathrm{B} + \mathrm{H^+}$$

What is the relationship between these four rate constants? *Ans.* $k_1k_4 = k_2k_3$.

15.19 If a first-order reaction has an activation energy of 104,600 J mol^{-1} and, in the equation $k = Ae^{-E_a/RT}$, A has a value of 5×10^{13} s^{-1}, at what temperature will the reaction have a half-life of (*a*) 1 min, and (*b*) 30 days? *Ans.* (*a*) 76 °C, (*b*) −3 °C.

15.20 Isopropenyl allyl ether in the vapor state isomerizes to allyl acetone according to a first-order rate equation. The following equation gives the influence of temperature on the rate constant (in s^{-1}).

$$k = 5.4 \times 10^{11} e^{-123{,}000/RT}$$

where the activation energy is expressed in J mol^{-1}. At 150 °C, how long will it take to build up a partial pressure of 0.395 atm of allyl acetone, starting with 1 atm of isopropenyl allyl ether [L. Stein and G. W. Murphy, *J. Am. Chem. Soc.*, **74**, 1041 (1952)]?

Ans. 1420 s.

15.21 The hydrolysis of $(CH_2)_6C(Cl)(CH_3)$ in 80% ethanol follows the first-order rate equation. The values of the specific reaction-rate constants, as determined by H. C. Brown and M. Borkowski [*J. Am. Chem. Soc.*, **74**, 1896 (1952)], are as follows.

t/°C	0	25	35	45
k/s^{-1}	1.06×10^{-5}	3.19×10^{-4}	9.86×10^{-4}	2.92×10^{-3}

(*a*) Plot log k against $1/T$; (*b*) calculate the activation energy; (*c*) calculate the preexponential factor. *Ans.* (*b*) 88.1 kJ mol^{-1}, (*c*) 8.4×10^{11} s^{-1}.

15.22 (*a*) The viscosity of water changes about 2% per degree at room temperature. What is the activation energy for this process? (*b*) The activation energy for a reaction is 62.8 kJ mol^{-1}. Calculate k_{35}/k_{25}. *Ans.* (*a*) 15.1 kJ mol^{-1}. (*b*) 2.27.

15.23 The preexponential factor for the termolecular reaction

$$2NO + O_2 \rightarrow 2NO_2$$

is 10^9 cm^6 mol^{-2} s^{-1}. What is the value in L^2 mol^{-2} s^{-1}? *Ans.* 10^3 L^2 mol^{-2} s^{-1}.

15.24 For the two parallel reactions $A \xrightarrow{k_1} B$ and $A \xrightarrow{k_2} C$, show that the activation energy E' for the disappearance of A is given in terms of the activation energies E_1 and E_2 for the two paths by

$$E' = \frac{k_1E_1 + k_2E_2}{k_1 + k_2}$$

15.25 For the mechanism

$$A + B \underset{k_2}{\overset{k_1}{\rightleftharpoons}} C \qquad C \xrightarrow{k_3} D$$

(*a*) Derive the rate law using the steady-state approximation to eliminate the concentration of C. (*b*) Assuming that $k_3 \ll k_2$, express the preexponential factor A and E_a for the apparent second-order rate constant in terms of A_1, A_2, and A_3 and E_{a1}, E_{a2}, and E_{a3} for the three steps. *Ans.* (*a*) $\dfrac{d(D)}{dt} = \dfrac{k_1k_3(A)(B)}{k_2 + k_3}$. (*b*) $E_a = E_{a1} + E_{a3} - E_{a2}$. $A = \dfrac{A_1A_3}{A_2}$.

15.26 The thermal decomposition of gaseous acetaldehyde is a second-order reaction. The value of E_a is 190,400 J mol^{-1}, and the molecular diameter of the acetaldehyde molecule is 5×10^{-8} cm. (*a*) Calculate the number of molecules colliding per cm^3 per second at 800 K and 1 atm pressure. (*b*) Calculate k in L mol^{-1} s^{-1}.

Ans. (*a*) 2.9×10^{28} cm^{-3} s^{-1}, (*b*) 0.077 L mol^{-1} s^{-1}.

15.27 Show that the activated complex theory yields the simple collision theory result when it is applied to the reaction of two rigid spherical molecules.

15.28 F. W. Schuler and G. W. Murphy [*J. Am. Chem. Soc.*, **72**, 3155 (1950)] studied the thermal rearrangement of vinyl allyl ether to allyl acetaldehyde in the range 150 to 200 °C and found that

$$k = 5 \times 10^{11} e^{-128{,}000/RT}$$

where k is in s^{-1} and the activation energy is in J mol^{-1}. Calculate (*a*) the enthalpy of activation, and (*b*) the entropy of activation, and (*c*) give an interpretation of the latter.

Ans. (*a*) 124.3 kJ mol^{-1}, (*b*) -32.6 J K^{-1} mol^{-1},
(*c*) The activated complex may be an improbable ring structure.

15.29 The vapor-phase decomposition of di-*t*-butyl peroxide is first order in the range 110 to 280 °C and follows the equation

$$k = 3.2 \times 10^{16} e^{-163,600/RT}$$

where k is in s^{-1} and E_a is in J mol^{-1}. Calculate (*a*) $\Delta H^{\ddagger\circ}$, and (*b*) $\Delta S^{\ddagger\circ}$.

Ans. (*a*) 159.9 kJ mol^{-1}. (*b*) 59.4 J K^{-1} mol^{-1}.

15.30 The apparent activation energy for the recombination of iodine atoms in argon is -5.9 kJ mol^{-1}. This negative temperature coefficient may result from the following mechanism.

$$I + M = IM \qquad K = \frac{(IM)}{(I)(M)} \qquad IM + I \underset{k_{-1}}{\overset{k_1}{\rightleftharpoons}} I_2 + M$$

Assuming that the first step remains at equilibrium, derive the rate equation that includes both the forward and reverse reactions. Show that the reverse reaction is bimolecular and the equilibrium constant expression for the dissociation of iodine is independent of the concentration of the third body.

15.31 Derive the steady-state rate equation for the following mechanism for a termolecular reaction.

$$A + A \underset{k_{-1}}{\overset{k_1}{\rightleftharpoons}} A_2^* \qquad A_2^* + M \xrightarrow{k_2} A_2 + M$$

Ans. $\dfrac{d(A_2)}{dt} = \dfrac{k_1 k_2 (M)(A)^2}{k_{-1} + k_2(M)}$.

15.32 For the mechanism

$$H_2 + X_2 \underset{k_{-1}}{\overset{k_1}{\rightleftharpoons}} 2HX \qquad X + H_2 \underset{k_{-2}}{\overset{k_2}{\rightleftharpoons}} HX + H$$

$$X_2 \underset{k_{-3}}{\overset{k_3}{\rightleftharpoons}} 2X \qquad H + X_2 \underset{k_{-4}}{\overset{k_4}{\rightleftharpoons}} HX + X$$

show that the steady-state rate law is

$$\frac{d(HX)}{dt} = 2k_1(H_2)(X_2)\left[1 - \frac{(HX)^2}{K(H_2)(X_2)}\right]\left[1 + \frac{\dfrac{k_3}{k_1}\sqrt{\dfrac{2k_2}{k_{-2}(X_2)}}}{1 + \dfrac{k_{-3}(HX)}{k_4(X_2)}}\right]$$

15.33 For the gas reaction

$$O + O_2 + M \underset{k'}{\overset{k}{\rightleftharpoons}} O_3 + M$$

When M = O_2, Benson and Axworthy [*J. Chem. Phys.*, **26**, 1718 (1957)] obtained

$$k = 6.0 \times 10^7 e^{2.5/RT} \text{ L}^2 \text{ mol}^{-2} \text{ s}^{-1}$$

where the activation energy is in kJ mol^{-1}. Calculate the values of the parameters in the Arrhenius equation for the reverse reaction.

Ans. $A = 1.1 \times 10^{13}$ L mol^{-1} s^{-1}. $E_a = 104$ kJ mol^{-1}.

15.34 The kinetics of the hydrolysis of an ester is studied by titrating the acid produced. A sample is withdrawn and titrated with alkali. The volumes required at various times are

t/min	0	27	60	∞
V/mL	0	18.1	26.0	29.7

(*a*) Prove this reaction is first order. (*b*) Calculate the half-life.

15.35 The following values of percent transmission are obtained with a spectrophotometer at a series of times during the decomposition of a substance absorbing light at a particular wavelength. Calculate k, $t_{1/2}$, and τ assuming the reaction is first order.

t	Percent Transmission
5 min	14.1
10 min	57.1
∞	100

Beer's law: $\log 100/T = abc$, where T = percent transmission, a = absorbancy index, b = cell thickness, and c = concentration.

15.36 The hydrolysis of 1-chloro-1-methylcycloundecane in 80% ethanol has been studied by H. C. Brown and M. Borkowiski [*J. Am. Chem. Soc.*, **74**, 1894 (1952)] at 25 °C. The extent of hydrolysis was measured by titrating the acid formed after measured intervals of time with a solution of NaOH. The data are as follows.

t/hr	0	1.0	3.0	5.0	9.0	12	∞
x/cm^3	0.035	0.295	0.715	1.055	1.505	1.725	2.197

(*a*) What is the order of the reaction? (*b*) What is the value of the rate constant? (*c*) What fraction of the 1-chloro-1-methylcycloundecane will be left unhydrolyzed after 8 hr?

15.37 From the following data on the rate of the rearrangement of 1-cyclohexenyl allyl malonitrile at 135.7 °C, determine graphically the first-order rate constant. Check the graphical evaluation by calculating k from data at two different times.

t/min	0	5	10	20	30	45
% rearranged	19.8	34.2	46.7	64.7	77.0	86.3

15.38 The reaction

$$SO_2Cl_2 = SO_2 + Cl_2$$

is first order with a rate constant of $2.2 \times 10^{-5}\ \text{s}^{-1}$ at 320 °C. What percentage of SO_2Cl_2 is decomposed after being heated at 320 °C for 2 hr?

15.39 A solution of ethyl acetate and sodium hydroxide was prepared that contained (at $t = 0$) 5×10^{-3} mol L^{-1} ethyl acetate and 8×10^{-3} mol L^{-1} sodium hydroxide. After 400 s at 25 °C a 25-mL aliquot was found to neutralize 33.3 mL of 5×10^{-3} mol L^{-1} hydrochloric acid. (*a*) Calculate the rate constant for this second-order reaction. (*b*) At what time would you expect 20.0 mL of hydrochloric acid to be required?

15.40 The second-order rate constant for an alkaline hydrolysis of ethyl formate in 85% ethanol (aqueous) at 29.86 °C is 4.53 L mol^{-1} s^{-1} [H. M. Humphreys and L. P. Hammett, *J. Am. Chem. Soc.*, **78**, 521 (1956)]. (*a*) If the reactants are both present at 0.001 mol L^{-1}, what will be the half-life of the reaction? (*b*) If the concentration of one of the reactants is doubled and of the other is cut in half, how long will it take for half the reactant present at the lower concentration to react?

15.41 The second-order rate constant for a gas reaction is 1×10^3 L mol^{-1} s^{-1} at 25 °C. Calculate the value of the rate constant when the rate equation is written in terms of pressure in atmospheres.

15.42 Equal molar quantities of A and B are added to a liter of a suitable solvent. At the end of 500 s half of A has reacted according to the reaction A + B = C. How much of A will be reacted at the end of 800 s if the reaction is (*a*) zero order with respect to both A and B; (*b*) first order with respect to A and zero order with respect to B; (*c*) first order with respect to both A and B?

15.43 It is more difficult than commonly realized to differentiate between a first- and a second-order reaction by the shape of the plot of percentage reaction versus time. Make these plots for first- and second-order reactions ($a = b$) having the same half-life. By what percent do they differ at (*a*) $t = t_{1/2}/2$, and (*b*) $t = 2t_{1/2}$?

15.44 Show that the interconversion of ortho- and parahydrogen will be $\frac{3}{2}$ order, as obtained experimentally in the range 600 to 750 °C, if the rate-determining step is that between atoms and molecules of hydrogen.

$$\mathrm{H} + \underset{\uparrow\downarrow}{\text{para-H}_2} \underset{k_2}{\overset{k_1}{\rightleftharpoons}} \underset{\uparrow\uparrow}{\text{ortho-H}_2} + \mathrm{H}$$

where the arrows represent the directions of the nuclear spins (cf. Section 12.1).

15.45 The reaction between selenious acid and iodide ion in acid solution is

$$H_2SeO_3 + 6I^- + 4H^+ = Se(s) + 2I_3^- + 3H_2O$$

The initial reaction rates were measured at 0 °C at a variety of concentrations, as indicated in the following table, in moles per liter. These initial rates were evaluated from plots of H_2SeO_3 versus time. Determine the form of the rate law. (*Note.* This rate law holds only as long as insignificant quantities of I_3^- are present.)

$(H_2SeO_3)/10^{-4}$ mol L^{-1}	$(H^+)/10^{-2}$ mol L^{-1}	$(I^-)/10^{-2}$ mol L^{-1}	Initial Rate$/10^{-7}$ mol L^{-1} s^{-1}
0.712	2.06	3.0	4.05
2.40	2.06	3.0	14.6
7.20	2.06	3.0	44.6
0.712	2.06	1.8	0.93
0.712	2.06	3.0	4.05
0.712	2.06	9.0	102
0.712	2.06	15.0	508
0.712	2.06	3.0	4.05
0.712	5.18	3.0	28.0
0.712	12.5	3.0	173.0

15.46 The initial rate of the reaction

$$BrO_3^- + 3SO_3^{2-} = Br^- + 3SO_4^{2-}$$

was found by F. S. Williamson and E. L. King [*J. Am. Chem. Soc.*, **79**, 5397 (1953)] to be given by $k(BrO_3^-)(SO_3^{2-})(H^+)$. Give one of the thermodynamically possible rate laws for the reverse reaction.

15.47 For a reversible first-order reaction

$$A \underset{k_2}{\overset{k_1}{\rightleftharpoons}} B$$

$k_1 = 10^{-2}$ s^{-1} and $(B)_{eq}/(A)_{eq} = 4$. If $(A)_0 = 0.01$ mol L^{-1} and $(B)_0 = 0$, what will be the concentration of B after 30 s?

15.48 The reaction $2NO + O_2 \rightarrow 2NO_2$ is third order and $dc_{NO_2}/dt = kc_{NO}^2 c_{O_2}$. The rate constant k has a value of 7.1×10^9 mol^{-2} cm^6 s^{-1} at 25 °C. Air blown through a certain hot chamber and cooled quickly at 25 °C and 1 atm contains 1% by volume of nitric oxide, NO, and 20% of oxygen. (*a*) How long will it take for 90% of this NO to be converted into nitrogen dioxide, NO_2 (or N_2O_4)? (*b*) If the gases are blown through at the rate of 5000 ft^3 min^{-1}, how large a chamber must be constructed to obtain this 90% conversion?

15.49 Derive the integrated rate equation for an nth-order reaction

$$-\frac{d(\mathrm{A})}{dt} = k(\mathrm{A})^n, \qquad n > 1$$

Derive the expression for the half-life.

15.50 For the consecutive first-order reactions

$$\mathrm{A} \xrightarrow[k_1 = 0.15\,\mathrm{s}^{-1}]{} \mathrm{B} \xrightarrow[k_2 = 1.1\,\mathrm{s}^{-1}]{} \mathrm{C}$$

plot curves that give the concentrations of A, B, and C as a function of time.

15.51 The equations for (B) and (C) in Section 15.9 give an indeterminate result if $k_1 = k_2$. Rederive the equations, giving (B) and (C) as functions of time for the special case that

$$\mathrm{A} \xrightarrow{k_1} \mathrm{B} \xrightarrow{k_2} \mathrm{C}$$

15.52 Ozone is decomposed by the catalytic chain

$$\mathrm{NO} + \mathrm{O_3} \xrightarrow{k_1} \mathrm{NO_2} + \mathrm{O_2} \qquad \mathrm{NO_2} + \mathrm{O} \xrightarrow{k_2} \mathrm{NO} + \mathrm{O_2}$$

What is the steady-state rate law for the formation of O_2?

15.53 For the decomposition of ozone

$$2\mathrm{O_3} = 3\mathrm{O_2}$$

the rate law is

$$-\frac{d(\mathrm{O_3})}{dt} = k\frac{(\mathrm{O_3})^2}{(\mathrm{O_2})}$$

Devise a mechanism to explain this rate law. (*Hint*. The first step might be expected to be the production of an oxygen atom.)

15.54 The following mechanism has been proposed for the thermal decomposition of ethyleneoxide [*J. Chem. Phys.*, **31**, 506 (1959)].

Initiation: $\mathrm{CH_2{-}CH_2}$ (epoxide ring, bridged by O) $\xrightarrow{k_1}$ $\mathrm{CH_2{-}CH\cdot}$ (ring, bridged by O) $+ \mathrm{H}$

Propagation: $\mathrm{CH_2{-}CH\cdot}$ (ring, bridged by O) $\xrightarrow{k_2}$ $\mathrm{CH_3\cdot} + \mathrm{CO}$

$\mathrm{CH_2{-}CH_2}$ (ring, bridged by O) $+ \mathrm{CH_3\cdot}$ $\xrightarrow{k_3}$ $\mathrm{CH_2{-}CH\cdot}$ (ring, bridged by O) $+ \mathrm{CH_4}$

Termination: $\mathrm{CH_2{-}CH\cdot}$ (ring, bridged by O) $+ \mathrm{CH_3\cdot}$ $\xrightarrow{k_4}$ stable products

Assuming the radicals are in a steady state, show that the decomposition is first order in ethyleneoxide concentration.

15.55 The following rate constants were obtained by Wiig for the first-order decomposition of acetone dicarboxylic acid in aqueous solution.

$t/°\mathrm{C}$	0	20	40	60
$k/10^{-5}\ \mathrm{s}^{-1}$	2.46	47.5	576	5480

(*a*) Calculate the energy of activation. (*b*) Calculate the preexponential factor A. (*c*) What is the half-life of this reaction at 80 °C?

15.56 Although the thermal decomposition of ethyl bromide is complex, the overall rate is first order and the rate constant is given by the expression $k = (3.8 \times 10^{14}\ \text{s}^{-1})e^{-229,000/RT}$, where $R = 8.314\ \text{J K}^{-1}\ \text{mol}^{-1}$. Estimate the temperature at which (*a*) ethyl bromide decomposes at the rate of 1% per second, and (*b*) the decomposition is 70% complete in 1 hr.

15.57 Given that the first-order rate constant for the overall decomposition of N_2O_5 is $k = (4.3 \times 10^{13}\ \text{s}^{-1})e^{-103,000/RT}\ \text{s}^{-1}$, calculate (*a*) the half-life at $-10\ °\text{C}$; and (*b*) the time required for 90% reaction at 50 °C.

15.58 At 700 K, what is the half-life of SiH_4? The equation for the first-order rate constant for decomposition is $k = (2 \times 10^{13}\ \text{s}^{-1})e^{-216,000/RT}$, where $R = 8.314\ \text{J K}^{-1}\ \text{mol}^{-1}$.

15.59 Suppose that a substance X decomposes into A and B in parallel paths with rate constants given by

$$k_A = (10^{15}\ \text{s}^{-1})e^{-126,000/RT} \qquad k_B = (10^{13}\ \text{s}^{-1})e^{-83,700/RT}$$

where the activation energies are given in J mol^{-1}. (*a*) At what temperature will the two products be formed at the same rate? (*b*) At what temperature will A be formed 10 times as fast as B? (*c*) At what temperature will A be formed 0.1 as fast as B? (*d*) State a generalization concerning the effect of temperature on the relative rates of reactions with different activation energies.

15.60 For the reaction

$$O + NO + M \rightarrow NO_2 + M$$

$k_{1000\,K} = 6 \times 10^9\ \text{L}^2\ \text{mol}^{-2}\ \text{s}^{-1}$ and $k_{300\,K} = 3 \times 10^{10}\ \text{L}^2\ \text{mol}^{-2}\ \text{s}^{-1}$. Calculate the parameters in the Arrhenius equation.

15.61 In the reaction

$$N_2 + O_2 \underset{k_2}{\overset{k_1}{\rightleftharpoons}} 2NO$$

$$-\frac{d(N_2)}{dt} = k_1\left(p'_{N_2} - \frac{p_{NO}}{2}\right)\left(p'_{N_2} - \frac{p_{NO}}{2}\right) - k_2 p_{NO_2}$$

where p'_{N_2} is the original pressure of N_2, p'_{O_2} is the original pressure of O_2, and p_{NO} is the pressure of NO formed. Values of the equilibrium constant K are given in Table 4.2. The rate constant for the reverse reaction is given in atm^{-1} s^{-1} by

$$k_2 = 1 \times 10^9 e^{-293,000/RT}$$

(*a*) Calculate k_1 at 2400 and 1900 K. (*b*) Calculate the time required for NO at 0.02 atm to undergo 10% decomposition at 2400 and 1900 K, using k_2 and neglecting k_1.

15.62 For the reaction

$$CO_2(+H_2O) \underset{k_2}{\overset{k_1}{\rightleftharpoons}} H_2CO_3$$

(the parentheses indicate that H_2O is not included in the equilibrium constant expression or in the rate equation)

$$\Delta H° = 4730\ \text{J mol}^{-1} \qquad \Delta S° = -33.5\ \text{J K}^{-1}\ \text{mol}^{-1}$$

At 25 °C $\qquad k_1 = 0.0375\ \text{s}^{-1}$

At 0 °C $\qquad k_1 = 0.0021\ \text{s}^{-1}$

Calculate (*a*) the activation energy for the forward reaction, (*b*) the activation energy for the reverse reaction, (*c*) k_{-1} at 25 °C, and (*d*) k_{-1} at 0 °C, assuming that $\Delta H°$ is independent of temperature in this range.

15.63 (*a*) Calculate the second-order rate constant for collisions of dimethyl ether molecules with each other at 777 K. It is assumed that the molecules are spherical and have a radius of 0.25 nm. If every collision was effective in producing decomposition, what would be the half-life of the reaction (*b*) at 1 atm pressure, and (*c*) at a pressure of 0.13 Pa?

15.64 Estimate the preexponential factor for the reaction

$$2CH_3 \rightarrow C_2H_6$$

using collision theory. The molecular diameter of CH_4 obtained from gas viscosity measurements at 0 °C is 0.414 nm (Table 14.1). The experimental value of A in Table 15.4 is $10^{13.3}$ cm³ mol⁻¹ s⁻¹.

15.65 Estimate the preexponential factor in the neighborhood of 500 K for the reaction

$$Cl + H_2 \rightarrow HCl + H$$

assuming the activated complex is linear. The H—H and H—Cl bond distances in the activated complex may be assumed to be 0.092 and 0.145 nm, respectively, and the vibrational partition functions may be taken as unity. The experimental value from Table 15.4 is $10^{13.9}$ cm³ mol⁻¹ s⁻¹.

15.66 Interpret $\Delta S^{\ddagger\circ}$ calculated from the preexponential factors for the following bimolecular elementary reactions. In each case the temperature may be taken as room temperature.

	A/cm³ mol⁻¹ s⁻¹
$H + H_2 \rightarrow H_2 + H$	$10^{-14.7}$
$CH_3 + CH_4 \rightarrow CH_4 + CH_3$	$10^{-11.8}$
$C_6H_5 + H_2 \rightarrow C_6H_6 + H$	$10^{-10.7}$

15.67 For the first-order gaseous decomposition of propylene oxide $\Delta H^{\ddagger\circ} = 238.1$ kJ mol⁻¹ and $\Delta S^{\ddagger\circ} = 25$ J K⁻¹ mol⁻¹ in the neighborhood of 285 °C. Calculate (*a*) the frequency factor A, and (*b*) the first-order rate constant at 285 °C.

15.68 The first-order rate constant for the thermal decomposition of $C_2H_5Br(g)$ is given by

$$k = (3.8 \times 10^{14}\ \text{s}^{-1})e^{-230,000/RT}$$

where the activation energy is in J mol⁻¹. Calculate (*a*) $\Delta H^{\ddagger\circ}$, and (*b*) $\Delta S^{\ddagger\circ}$ at 5000 °C.

15.69 A reaction A + B + C → D follows the mechanism

$$A + B \rightleftharpoons AB \qquad AB + C \rightarrow D$$

in which the first step remains essentially in equilibrium. Show that the dependence of rate on temperature is given by

$$k = Ae^{-(E_a + \Delta H)/RT}$$

where ΔH is the enthalpy change for the first reaction.

CHAPTER 16
KINETICS: LIQUID PHASE

The kinetics of reactions in the liquid phase cannot be interpreted in terms of kinetic theory and statistical mechanics of gases. The interpretation of rates in the liquid phase is necessarily more complicated from a molecular viewpoint because of the much greater interaction between molecules. However, bimolecular reactions in solution cannot occur more rapidly than the reactant molecules can diffuse together, and this rate may be calculated from measured diffusion coefficients. A number of reactions in the liquid phase occur at diffusion-controlled rates, and special methods have been developed for studying the rates of these reactions.

Acids and bases catalyze many reactions, and so we will discuss two types of mechanisms for acid catalysis. Enzymes catalyze the reactions in living things and provide the mechanisms by which rates of these reactions are controlled. These reactions provide interesting examples of solution kinetics.

The last section of the chapter is about the kinetics of reactions occurring at the surface of an electrode. These rates, which are important in corrosion, in batteries, and in electroanalytical methods, are extremely sensitive to electrical potential differences.

16.1 CHARACTERISTICS OF REACTION RATES IN THE LIQUID PHASE

In the liquid phase the motion of an individual molecule is very much hindered by its immediate neighbors. No matter in which direction a molecule moves, it bumps into another molecule and tends to bounce back to a position near its original position. Therefore most of the motion of a molecule in a liquid is restricted to the volume in a solvent *cage* defined by its nearest neighbors. Typically a molecule collides many times with its nearest neighbors before it undergoes a displacement so large that it finds itself surrounded by a new set of neighbors.

In order for molecules A and B to react they must, in general, share the same cage; when they are in the same cage they are referred to as an *encounter pair*. We can distinguish between two types of bimolecular reactions in solution by consideration of the following simple mechanism.

$$A + B \underset{k_{-1}}{\overset{k_1}{\rightleftharpoons}} \{AB\} \xrightarrow{k_2} \text{Products} \tag{16.1}$$

where {AB} is the encounter pair. By use of the steady-state approximation (Section 15.10) it is readily shown that

$$-\frac{d(A)}{dt} = \frac{k_1 k_2}{k_{-1} + k_2}(A)(B) \tag{16.2}$$

If $k_2 \gg k_{-1}$, the reaction rate is determined by the rate $k_1(A)(B)$ with which the reactants diffuse together, and so we refer to such a reaction as a *diffusion-controlled reaction.* In a diffusion-controlled reaction reactant molecules that have encountered each other collide enough times so that reaction is highly likely before they can diffuse away from each other. For aqueous solutions it has been estimated that the cage lifetime for a pair of noninteracting molecules is of the order 10^{-12} to 10^{-8} s, during which time they may undergo 10 to 10^5 collisions with each other.

If $k_2 \ll k_{-1}$, the reaction rate is

$$-\frac{d(A)}{dt} = k_2 K_{AB}(A)(B) \tag{16.3}$$

where $K_{AB} = k_1/k_{-1}$ is the equilibrium constant for the formation of the encounter pair. This is an *activation-controlled reaction* because the reaction rate is largely determined by the activation energy for k_2.

16.2 DIFFUSION CONTROLLED REACTIONS IN LIQUIDS

The maximum rate with which reactants can diffuse together in liquids may be calculated using the macroscopic theory of diffusion and experimentally determined diffusion coefficients of the reactants. The elementary theory of diffusion-controlled reactions was developed in 1917 by Smoluchowski in connection with his theoretical study of the coagulation of colloidal gold.

The diffusion coefficient D is defined in terms of Fick's first law, which is given in Section 18.9. The diffusion coefficients of low molar mass solutes in aqueous solution at 25 °C are of the order of 10^{-9} m² s⁻¹. Experimental methods for determining diffusion coefficients in dilute aqueous solutions are discussed in Section 18.11. The diffusion coefficients for ions may be calculated from their ionic mobilities (Section 18.7).

Smoluchowski considered spherical particles with radii R_1 and R_2 that could be considered to react when they diffused within a distance $R_{12} = R_1 + R_2$ of each other.

We may imagine one reactant molecule stationary and serving as a sink. Since $C = 0$ at distance R_{12}, a spherically symmetrical concentration gradient is set up. The flux through this concentration gradient is calculated and is expressed as a second-order rate constant k_a for association by

$$k_a = 4\pi N_A(D_1 + D_2)R_{12}f \tag{16.4}$$

where f is an electrostatic factor. The electrostatic factor f is different from unity if the reactants are ions. It is larger than unity if the reactants have opposite charges and attract each other, and it is smaller than unity if the reactants have the same charge and repel each other. One ion can be visualized as moving in the electric field created by the other ion.

Because of the electrostatic factor f, the effective reaction radius ($R_{\text{eff}} = R_{12}f$) is substantially increased for the reaction of oppositely charged ions and substantially decreased for the reaction of two ions with the same sign. If the ionic strength is so low that ion atmospheres may be neglected, f is given by

$$f = \frac{z_1 z_2 e^2}{4\pi\epsilon_0 \kappa k T R_{12}} \left[\exp\left(\frac{z_1 z_2 e^2}{4\pi\epsilon_0 \kappa k T R_{12}}\right) - 1\right]^{-1} \tag{16.5}$$

where the charges on the ions are $z_1 e$ and $z_2 e$, κ is the dielectric constant, and ϵ_0 is the permittivity of free space. More details on the derivation of the equations for diffusion-controlled reactions are given by Amdur and Hammes.*

The temperature coefficients for diffusion-controlled reactions in water are small, because they correspond with the temperature coefficient of the viscosity of liquid water ($E_a = 17.4$ kJ mol^{-1} at 25 °C).

Example 16.1 A typical diffusion coefficient for a small molecule in aqueous solution at 25 °C is 5×10^{-9} m^2 s^{-1}. If the reaction radius is 0.4 nm, what value is expected for the second-order rate constant for a diffusion controlled reaction of neutral molecules?

$$k = 4\pi N_A(D_1 + D_2)R_{12} = 4\pi(6.022 \times 10^{23}\ \text{mol}^{-1})(10^{-8}\ \text{m}^2\ \text{s}^{-1})(0.4 \times 10^{-9}\ \text{m})$$
$$= 3.0 \times 10^7\ \text{m}^3\ \text{mol}^{-1}\ \text{s}^{-1} = (3.0 \times 10^7\ \text{m}^3\ \text{mol}^{-1}\ \text{s}^{-1})(10^3\ \text{L m}^{-3})$$
$$= 3.0 \times 10^{10}\ (\text{mol L}^{-1})^{-1}\ \text{s}^{-1}$$

Example 16.2 What is the electrostatic factor f in water at 25 °C if the reaction radius R_{12} is 0.2 nm for opposite unit charges? For like unit charges? The dielectric constant κ of water is 80. For opposite charges,

$$\frac{z_1 z_2 e^2}{4\pi\epsilon_0 \kappa k T R_{12}} = -\frac{(1.602 \times 10^{-19}\ \text{C})^2(8.988 \times 10^9\ \text{N C}^{-2}\ \text{m}^2)}{80(1.3807 \times 10^{-23}\ \text{J K}^{-1})(298.15\ \text{K})(0.2 \times 10^{-9}\ \text{m})} = -3.502$$

$$f = -3.502[e^{-3.502} - 1]^{-1} = 3.61$$

Thus a diffusion-controlled reaction is expected to be 3.61 times faster in this case than for uncharged particles.

For ions of the same charge,

$$f = 3.502[e^{3.502} - 1]^{-1} = 0.11$$

Thus a diffusion-controlled reaction is expected to be 0.11 as fast for singly charged particles with the same sign as for neutral particles.

---

* I. Amdur and G. G. Hammes, *Chemical Kinetics*, McGraw-Hill Book Co., New York, 1966.

16.3 EXPERIMENTAL METHODS FOR STUDYING FAST REACTIONS IN SOLUTIONS

If a reaction occurs so rapidly that appreciable reaction occurs during the process of mixing the reactants, a flow method may be used. An early example was the study of the reaction of hemoglobin and oxygen.* A hemoglobin solution was forced into one arm of a Y-mixer and a solution of oxygen in a buffer into the other. In this way it is possible to mix liquids in about 10^{-3} s. In the stopped-flow method, reagents are forced into the mixer, the flow is brought to a sudden stop, and observations are made of the extent of reaction. In the continuous flow method the solutions are mixed and forced down the tube at a steady rate; the extent of reaction is constant at any given distance down the tube, but increases with distance from the mixing chamber.

Some reactions occur in much less than 10^{-3} s, and so their kinetics may not be studied by mixing methods. The time range has been extended down to about 10^{-9} s by the use of relaxation methods developed by M. Eigen and co-workers† in Göttingen, Germany. A solution in equilibrium is perturbed by rapidly changing one of the independent variables (usually temperature or pressure) on which the equilibrium depends. The change of the system to the new equilibrium is then followed by use of a rapidly responding physical method, for example, light absorption or electrical conductivity.

Equilibria may be shifted by changing the temperature (if $\Delta H \neq 0$) or by changing the pressure (if $\Delta V \neq 0$). A solution may be heated in a microsecond (10^{-6} s) by use of a pulsed laser or by discharging a large electrical capacitor through a special conductivity cell containing the sample. Equilibria may also be shifted by reducing the pressure suddenly by allowing high-pressure gas to escape through a rupture disk. Figure 16.1 is a schematic diagram of a temperature-jump apparatus in which an increase in temperature in a small volume of solution is produced by passing a large current for about 1 μs. If there is a single reaction the return to equilibrium at the new higher temperature is represented by

$$\Delta C = \Delta C_0 e^{-t/\tau} \tag{16.6}$$

where τ is the relaxation time and ΔC_0 is the displacement of the concentration of one of the reactants from its equilibrium value at $t = 0$. If several reactions are involved in the return to equilibrium, ΔC is expressed by a sum of exponential terms with different relaxation times. The relaxation time τ is the time required for ΔC to drop to $1/e$ of its initial value ΔC_0.

In another form of relaxation method an independent parameter is varied in

---

* H. Hartridge and F. J. W. Roughton, *Proc. Roy. Soc.*, **A, 104**, 395 (1923).

† M. Eigen and L. De Maeyer, in G. G. Hammes, ed., *Investigation of Rates and Mechanisms of Reactions*, 3rd ed., Wiley, New York, 1974, p. 63; G. G. Hammes, *Principles of Chemical Kinetics*, Academic Press, New York, 1978; D. N. Hague, *Fast Reactions*, Wiley-Interscience, New York, 1971; D. N. Hague, *Fast Reactions*, Wiley-Interscience, New York, 1971.

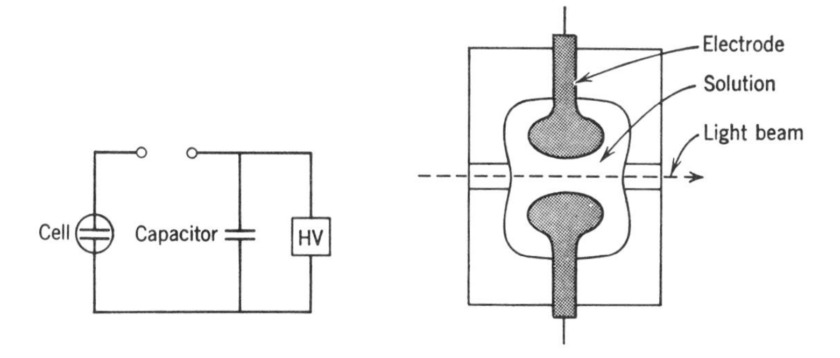

Fig. 16.1 Schematic diagram of a temperature-jump apparatus.

a sinusoidal manner and the phase lag in the response of the chemical system is measured. In a sound wave both the pressure and temperature vary and, if the frequency is in the neighborhood of $1/\tau$, there will be enhanced sound absorption.

The line width in various types of spectroscopic experiments gives information about the rates of molecular processes. For example, nuclear magnetic resonance (NMR) spectroscopy may be used to determine the rate of a chemical reaction. Electron spin resonance (ESR) spectroscopy may also be used if the chemical environment of an unpaired electron is affected by a reaction.

16.4 RELAXATION TIME FOR A SIMPLE REACTION

When a single-step reaction is displaced slightly from equilibrium, the return to equilibrium is first order in the displacement from equilibrium. For example, consider the reaction

$$\mathrm{A} + \mathrm{B} \underset{k_{-1}}{\overset{k_1}{\rightleftharpoons}} \mathrm{C} \tag{16.7}$$

for which the rate equation is

$$\frac{d(\mathrm{C})}{dt} = k_1(\mathrm{A})(\mathrm{B}) - k_{-1}(\mathrm{C}) \tag{16.8}$$

At equilibrium

$$0 = k_1(\mathrm{A})_{\mathrm{eq}}(\mathrm{B})_{\mathrm{eq}} - k_{-1}(\mathrm{C})_{\mathrm{eq}} \tag{16.9}$$

Equation 16.8 may be written in terms of displacements $\Delta(\mathrm{C})$ from the final equilibrium concentrations by introducing

$$(\mathrm{A}) = (\mathrm{A})_{\mathrm{eq}} - \Delta(\mathrm{C}) \tag{16.10}$$

$$(\mathrm{B}) = (\mathrm{B})_{\mathrm{eq}} - \Delta(\mathrm{C}) \tag{16.11}$$

$$(\mathrm{C}) = (\mathrm{C})_{\mathrm{eq}} + \Delta(\mathrm{C}) \tag{16.12}$$

The fact that the displacements from equilibrium are the same for the three reactants, except for sign, comes from the stoichiometry of reaction 16.7. Substituting equations 16.10 to 16.12 into equation 16.8 yields

$$\frac{d\Delta(\mathrm{C})}{dt} = k_1[(\mathrm{A})_{\mathrm{eq}} - \Delta(\mathrm{C})][(\mathrm{B})_{\mathrm{eq}} - \Delta(\mathrm{C})] - k_{-1}[(\mathrm{C})_{\mathrm{eq}} + \Delta(\mathrm{C})] \quad (16.13)$$

If the displacement from equilibrium $\Delta(\mathrm{C})$ is small,

$$\frac{d\Delta(\mathrm{C})}{dt} = -\{k_1[(\mathrm{A})_{\mathrm{eq}} + (\mathrm{B})_{\mathrm{eq}}] + k_{-1}\}\,\Delta(\mathrm{C}) = -\frac{\Delta(\mathrm{C})}{\tau} \quad (16.14)$$

where equation 16.9 has been used and the term in $[\Delta(\mathrm{C})]^2$ has been neglected because $\Delta(\mathrm{C})$ is small. Thus the rate of approach to equilibrium is proportional to the displacement from equilibrium $\Delta(\mathrm{C})$. It is customary to use the relaxation time τ (equation 16.6) to characterize the rate of return to equilibrium. From equation 16.14 we see that for this example

$$\tau = \{k_{-1} + k_1[(\mathrm{A})_{\mathrm{eq}} + (\mathrm{B})_{\mathrm{eq}}]\}^{-1} \quad (16.15)$$

Thus k_{-1} and k_1 may be obtained as slope and intercept of a plot τ^{-1} versus $(\mathrm{A})_{\mathrm{eq}} + (\mathrm{B})_{\mathrm{eq}}$.

Example 16.3 When a sample of pure water in a small conductivity cell is heated suddenly with a pulse of microwave radiation, equilibrium in the water dissociation reaction does not exist at the new higher temperature until additional dissociation occurs. It is found that the relaxation time for the return to equilibrium at 25 °C is 36 μs. Calculate k_1 and k_{-1}.

$$\mathrm{H^+ + OH^-} \underset{k_{-1}}{\overset{k_1}{\rightleftharpoons}} \mathrm{H_2O}$$

$$\tau = \frac{1}{k_{-1} + k_1[(\mathrm{H^+}) + (\mathrm{OH^-})]}$$

$$K = \frac{(\mathrm{H^+})(\mathrm{OH^-})}{(\mathrm{H_2O})} = \frac{k_{-1}}{k_1} = \frac{10^{-14}}{55.5} = 1.8 \times 10^{-16}\ \mathrm{mol\ L^{-1}}$$

Eliminating k_{-1}, we have

$$\tau = \frac{1}{k_1[K + (\mathrm{H^+}) + (\mathrm{OH^-})]} = \frac{1}{k_1[(1.8 \times 10^{-16}) + (2 \times 10^{-7})]} = 36 \times 10^{-6}\ \mathrm{s}$$

$$k_1 = 1.4 \times 10^{11}\ (\mathrm{mol\ L^{-1}})^{-1}\ \mathrm{s^{-1}}$$

$$k_{-1} = Kk_1 = (1.8 \times 10^{-16}\ \mathrm{mol\ L^{-1}})(1.4 \times 10^{11}\ \mathrm{L\ mol^{-1}\ s^{-1}}) = 2.5 \times 10^{-5}\ \mathrm{s^{-1}}$$

If the reaction being studied involves two steps and the concentration of the intermediate is significant, there will be two independent rate equations. If the reactions are both near to equilibrium, these equations may be linearized, and the two linear differential equations will yield two relaxation times. The return to equilibrium will then be given by the sum of two exponential terms. In general, the number of exponential terms is equal to the number of independent reactions.*

16.5 RATE CONSTANTS FOR ELEMENTARY REACTIONS IN WATER

The rate constant of 1.4×10^{11} (mol $L^{-1})^{-1}$ s^{-1} for the combination of hydrogen ions and hydroxyl ions in water at 25 °C is the largest second-order rate constant known for aqueous solutions. The rate for this reaction would be expected to be large because of the anomalously high mobilities for these ions in aqueous solutions (Section 18.4). However, the reaction is even faster because of an interesting additional effect. Substitution of 1.4×10^{11} L mol^{-1} s^{-1} in equation 16.4, along with the proper electrostatic factor, yields $R_{12} = 0.75$ nm. Since this reaction radius is equal to about three O—H bond distances, the proton apparently "tunnels" to the hydroxyl ion once it is close. Tunneling is a quantum mechanical effect in which a particle can penetrate a region even though it does not have the energy to penetrate it classically.

The rate constants k_a for some diffusion-controlled association reactions of H^+ and OH^- are given in Table 16.1, as measured by Eigen and co-workers. Since the association reactions are diffusion controlled, the difference in acid dissociation constants for various weak acids can be attributed primarily to the rate of dissociation of protons. The reaction of hydroxyl ions with various weak acids is also approximately diffusion controlled because the rate constants are in accord with equation 16.4.

Table 16.1 Rate Constants at 25 °C for Elementary Reactions in Dilute Aqueous Solutions (k_a applies to the association reaction and k_d to the dissociation reaction)

	$k_a/(\text{mol L}^{-1})^{-1}\ \text{s}^{-1}$	k_d/s^{-1}
$H^+ + OH^- \rightleftharpoons H_2O$	1.4×10^{11}	2.5×10^{-5}
$D^+ + OD^- \rightleftharpoons D_2O$	8.4×10^{10}	2.5×10^{-6}
$H^+ + F^- \rightleftharpoons HF$	1.0×10^{11}	7×10^{7}
$H^+ + CH_3CO^- \rightleftharpoons CH_3CO_2H$	4.5×10^{10}	7.8×10^{5}
$H^+ + C_6H_5CO_2^- \rightleftharpoons C_6H_5CO_2H$	3.5×10^{10}	2.2×10^{6}
$H^+ + NH_3 \rightleftharpoons NH_4^+$	4.3×10^{10}	24.6
$H^+ + C_3N_2H_4 \rightleftharpoons C_3N_2H_5^+$ (imidazole)	1.8×10^{10}	1.1×10^{3}
$OH^- + NH_4^+ \rightleftharpoons NH_3 + H_2O$	3.4×10^{10}	6×10^{5}
$OH^- + C_3N_2H_5^+ \rightleftharpoons C_3N_2H_4 + H_2O$ (imidazole)	2.5×10^{10}	2.5×10^{3}
$OH^- + {}^+H_3NCH_2CO_2^- \rightleftharpoons H_2NCH_2CO_2^- + H_2O$ (glycine)	1.4×10^{10}	8.4×10^{5}

* G. G. Hammes, *Principles of Chemical Kinetics*, Academic Press, New York, 1978, p. 192.

16.6 KINETICS OF THE HYDRATION OF CO_2 AND DEHYDRATION OF H_2CO_3

In water CO_2 exists largely as dissolved CO_2 (see Section 6.3) instead of as H_2CO_3. The half-life for the hydration-dehydration reaction depends on the pH, and at neutral pH values is long enough that it may be studied with simple equipment. It is the slowness of this reaction that accounts for the fading of the endpoint when carbonate ion is titrated with acid. The half-time for this uncatalyzed reaction in aqueous buffers is of special interest because it is too long to account for the process of elimination of CO_2 in our lungs. In living things this reaction is catalyzed by the enzyme carbonic anhydrase so that CO_2 may be hydrated, and H_2CO_3 dehydrated, more rapidly.

In order to discuss the kinetics of the uncatalyzed hydration of CO_2, it is necessary to consider the following reactions.

$$
\begin{aligned}
CO_2 + H_2O &\underset{k_{-1} = 13.7\ \text{s}^{-1}}{\overset{k_1 = 0.0375\ \text{s}^{-1}}{\rightleftharpoons}} H_2CO_3 \\
&\underset{k_{-2} = 4.7 \times 10^{10}\ (\text{mol L}^{-1})^{-1}\ \text{s}^{-1}}{\overset{k_2 = 8 \times 10^{6}\ \text{s}^{-1}}{\rightleftharpoons}} H^+ + HCO_3^- \\
OH^- + CO_2 &\underset{k_{-3} = 1.9 \times 10^{-4}\ \text{s}^{-1}}{\overset{k_3 = 8.5 \times 10^{5}\ (\text{mol L}^{-1})^{-1}\ \text{s}^{-1}}{\rightleftharpoons}} HCO_3^-
\end{aligned}
\tag{16.16}
$$

The rate constants are for 25 °C.

The rate equation for CO_2 is

$$\frac{d(CO_2)}{dt} = -k_1(CO_2) + k_{-1}(H_2CO_3) - k_3(OH^-)(CO_2) + k_{-3}(HCO_3^-) \tag{16.17}$$

The reaction $H^+ + HCO_3^- \rightleftharpoons H_2CO_3$ occurs so rapidly that it remains in equilibrium in an experiment in which we follow the hydration of CO_2. Since HCO_3^- and H_2CO_3 remain in equilibrium, we may eliminate H_2CO_3 by use of

$$\frac{(H^+)(HCO_3^-)}{(H_2CO_3)} = K_{H_2CO_3} = 1.70 \times 10^{-4} \tag{16.18}$$

to obtain

$$\frac{d(CO_2)}{dt} = -[k_1 + k_3(OH^-)](CO_2) + [k_{-3} + k_{-1}(H^+)/K_{H_2CO_3}](HCO_3^-) \tag{16.19}$$

Since the total concentration of dissolved CO_2 is $(CO_2) + (H_2CO_3) + (HCO_3^-) \approx (CO_2) + (HCO_3^-)$, this equation has the form

$$\frac{d(A)}{dt} = -k_f(A) + k_r(B) \tag{16.20}$$

Table 16.2 Half-Lives for the Hydration-Dehydration Reaction for CO_2—HCO_3^- at 25 °C

pH	$t_{1/2}$
3	0.0086 s
4	0.086
5	0.82
6	5.9
7	14.9
8	14.8
9	5.7

which corresponds to a reversible first-order reaction. As shown in Section 15.8, the half-life of such a reaction is given by

$$t_{1/2} = \frac{0.693}{k_f + k_r} \tag{16.21}$$

Applying this relation to the hydration of CO_2,

$$\begin{aligned} t_{1/2} &= \frac{0.693}{k_1 + k_3K_w/(\mathrm{H^+}) + k_{-3} + k_{-1}(\mathrm{H^+})/K_{\mathrm{H_2CO_3}}} \\ &= \frac{0.693}{0.0375 + \dfrac{8.5 \times 10^{-11}}{(\mathrm{H^+})} + 8.06 \times 10^4(\mathrm{H^+})} \end{aligned} \tag{16.22}$$

Values of the half-life calculated with this equation are given in Table 16.2. The reaction

$$\mathrm{OH^-} + \mathrm{HCO_3^-} \underset{1.3 \times 10^3\ \mathrm{s^{-1}}}{\overset{6 \times 10^9 (\mathrm{mol\ L^{-1}})^{-1}\ \mathrm{s^{-1}}}{\rightleftharpoons}} \mathrm{CO_3^{2-}} + \mathrm{H_2O} \tag{16.23}$$

has to be considered at higher pH values.

16.7 ACID AND BASE CATALYSIS

Acids and bases catalyze many reactions in which they are not consumed. Suppose the rate of disappearance of a substance S (often called the substrate of the catalytic reaction) is first order in S; $-d(\mathrm{S})/dt = k(\mathrm{S})$. The first-order rate constant k for the reaction in a buffer solution may be a linear function of $(\mathrm{H^+})$, $(\mathrm{OH^-})$, (HA), and $(\mathrm{A^-})$, where HA is the weak acid in the buffer and A^- is the corresponding anion.

$$k = k_0 + k_{\mathrm{H^+}}(\mathrm{H^+}) + k_{\mathrm{OH^-}}(\mathrm{OH^-}) + k_{\mathrm{HA}}(\mathrm{HA}) + k_{\mathrm{A^-}}(\mathrm{A^-}) \tag{16.24}$$

In this expression k_0 is the first-order rate constant at sufficiently low concentrations of all of the catalytic species H^+, OH^-, HA, and A^-. The so-called catalytic coefficients k_{H^+}, k_{OH^-}, k_{HA}, and k_{A^-} may be calculated from experiments with different concentrations of these species. If only the term $k_{H^+}(H^+)$ is important, the reaction is said to be subject to specific hydrogen-ion catalysis. If the term $k_{HA}(HA)$ is important, the reaction is said to be subject to *general acid catalysis* and, if the term $k_{A^-}(A^-)$ is important, the reaction is said to be subject to *general base catalysis*.

By considering two types of catalytic mechanisms, we can see how different types of terms arise in equation 16.24. In the first mechanism a proton is transferred from an acid AH to the substrate S, and then the acid form of the substrate reacts with a water molecule to form the product P.

$$\begin{gathered} S + AH^+ \underset{k_{-1}}{\overset{k_1}{\rightleftharpoons}} SH^+ + A \\ SH^+ + H_2O \xrightarrow{k_2} P + H_3O^+ \end{gathered} \tag{16.25}$$

Assuming that SH^+ is in a steady state, then

$$\frac{d(SH^+)}{dt} = 0 = k_1(S)(AH^+) - [k_{-1}(A) + k_2](SH^+) \qquad (16.26)^*$$

The rate of appearance of product is given by

$$\frac{d(P)}{dt} = k_2(SH^+) = \frac{k_1k_2(S)(AH^+)}{k_{-1}(A) + k_2} \tag{16.27}$$

where the second form is obtained by solving equation 16.26 for (SH^+). If $k_2 \gg k_{-1}(A)$,

$$\frac{d(P)}{dt} = k_1(A)(AH^+) \tag{16.28}$$

and the reaction is said to be general acid catalyzed. However, if $k_2 \ll k_{-1}(A)$,

$$\frac{d(P)}{dt} = \frac{k_1k_2(S)(AH^+)}{k_{-1}(A)} = \frac{k_1k_2}{k_{-1}K}(S)(H^+) \tag{16.29}$$

where the second form is obtained by inserting $K = (A)(H^+)/(AH^+)$. In this case the reaction is specifically hydrogen-ion catalyzed.

In the second mechanism the acid form of the substrate reacts with a base A instead of a water molecule.

$$S + AH^+ \underset{k_{-1}}{\overset{k_1}{\rightleftharpoons}} SH^+ + A \qquad SH^+ + A \xrightarrow{k_2} P + AH^+ \tag{16.30}$$

The steady-state treatment of this mechanism leads to

$$\frac{d(P)}{dt} = k_2(SH^+)(A) = \frac{k_1k_2(S)(AH^+)}{k_{-1} + k_2} \tag{16.31}$$

which is an example of general acid catalysis.

---

* Note (H_2O) is not written after k_2 because in dilute aqueous solutions (H_2O) cannot be appreciably changed, and so k_2 represents a first order rate constant.

16.8 ENZYME CATALYSIS

The most amazing catalysts are the enzymes, which catalyze the multitudinous reactions in living organisms and also provide the means for controlling reaction rates in such systems. Enzymes are proteins; this means that they are copolymers of aminoacids with specific aminoacid sequences and definite three-dimensional structures. Proteins provide various functional groups at the catalytic site that can become associated with a substrate molecule and thereby catalyze a reaction. Some enzymes catalyze a single reaction. An example is fumarase, which catalyzes the hydration of fumarate to L-malate.

$$\begin{array}{c} H \quad\;\; CO_2^- \\ \diagdown \;\diagup \\ C \\ \| \\ C \\ \diagup \;\diagdown \\ {}^-O_2C \quad\;\; H \end{array} \quad + H_2O = \quad \begin{array}{c} H \;\; H \;\; CO_2^- \\ \diagdown \;|\; \diagup \\ C \\ | \\ C \\ \diagup \;|\; \diagdown \\ {}^-O_2C \;\; OH \;\; H \end{array} \tag{16.32}$$

The reactants in an enzyme reaction are generally referred to as substrates. This reaction may be represented by S $=$ P, since the concentration of water is constant. Other enzymes catalyze a class of reactions of a given type like ester hydrolysis. To operate some enzymes require particular metal ions or coenzymes.

Since enzymes are very effective catalysts, they are usually used in laboratory experiments at concentrations much lower than the concentration of the substrate. Generally, the reaction rate that is measured in the laboratory is a steady-state rate. This rate for most reactions is found to be directly proportional to the concentration of the enzyme. When the concentration of a substrate is varied, it is usually found that the initial rate is first order in substrate at low substrate concentration and approaches zero order in substrate as the substrate concentration is increased; this is illustrated in Fig. 16.2*a*. When solutions of enzyme and substrate are mixed, there is initially a very fast reaction of the enzyme with the substrate

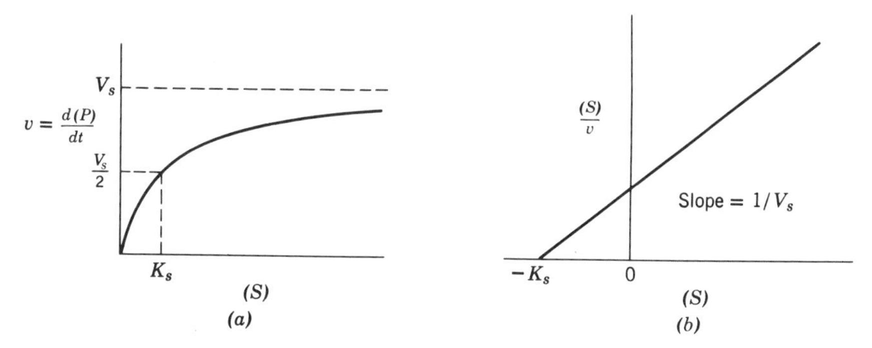

Fig. 16.2 (*a*) Plot of initial steady-state velocity v of appearance of product P at different initial concentrations of substrate S. (*b*) Plot of $(S)/v$ versus (S).

that can only be observed using special methods (cf. relaxation methods in Section 16.3). The nature of the steady-state reaction can be understood by considering the simplest type of mechanism for the overall reactions S = P.

$$\mathrm{E + S} \underset{k_2}{\overset{k_1}{\rightleftharpoons}} \mathrm{X} \underset{k_4}{\overset{k_3}{\rightleftharpoons}} \mathrm{E + P} \tag{16.33}$$

where E is the enzymatic site and X is an intermediate, often referred to as the enzyme-substrate complex. The rate equations for this mechanism are

$$\frac{d(\mathrm{X})}{dt} = k_1(\mathrm{E})(\mathrm{S}) - (k_2 + k_3)(\mathrm{X}) + k_4(\mathrm{E})(\mathrm{P}) \tag{16.34}$$

$$\frac{d(\mathrm{P})}{dt} = k_3(\mathrm{X}) - k_4(\mathrm{E})(\mathrm{P}) \tag{16.35}$$

These two rate equations cannot be solved to obtain analytic expressions for (E), (S), (X), and (P) as functions of time, but these concentrations may be calculated by use of a computer for specific values of the four rate constants.

Since enzymatic reactions are generally studied with enzyme concentrations (strictly speaking molar concentrations of enzymatic sites) much lower than the concentrations of substrates, it is a good approximation to assume that the enzymatic reaction is in a steady state in which $d(\mathrm{X})/dt = 0$. Since $(\mathrm{E}) + (\mathrm{X}) = (\mathrm{E})_0$, where $(\mathrm{E})_0$ is the initial concentration of enzymatic sites, equation 16.34 may be written

$$\frac{d(\mathrm{X})}{dt} = k_1[(\mathrm{E})_0 - (\mathrm{X})](\mathrm{S}) - (k_2 + k_3)(\mathrm{X}) + k_4[(\mathrm{E})_0 - (\mathrm{X})](\mathrm{P}) = 0 \tag{16.36}$$

Eliminating (X) between equations 16.35 and 16.36, we obtain

$$v = \frac{d(\mathrm{P})}{dt} = \frac{\dfrac{V_\mathrm{S}}{K_\mathrm{S}}(\mathrm{S}) - \dfrac{V_\mathrm{P}}{K_\mathrm{S}}(\mathrm{P})}{1 + \dfrac{(\mathrm{S})}{K_\mathrm{S}} + \dfrac{(\mathrm{P})}{K_\mathrm{P}}} \tag{16.37}$$

where v is the steady-state velocity and

$$V_\mathrm{S} = k_3(\mathrm{E})_0 \qquad V_\mathrm{P} = k_2(\mathrm{E})_0$$

$$K_\mathrm{S} = \frac{k_2 + k_3}{k_1} \qquad K_\mathrm{P} = \frac{k_2 + k_3}{k_4}$$

Equation 16.37 is written in this way because V_S, V_P, K_S, and K_P represent the collections of kinetic parameters that may be determined experimentally. If only S is added to the enzyme solution and the initial rate is determined before an appreciable amount of product has been formed, equation 16.37 becomes

$$v = \frac{V_\mathrm{S}}{1 + \dfrac{K_\mathrm{S}}{(\mathrm{S})}} \tag{16.38}$$

If (S) $\ll K_S$, the steady-state velocity is directly proportional to (S), so that the reaction is first order in substrate as shown in Fig. 16.2*a*. If (S) $\gg K_S$, the reaction is zero order in S. In both cases the reaction is first order with respect to the total concentration of enzymatic sites.

The quantity $V_S = k_3(E)_0$ is called the maximum velocity for the enzymatic reaction, and k_3 is called the turnover number for the forward reaction. This latter quantity is greater than $10^6\ s^{-1}$ for catalase, which catalyzes the decomposition of H_2O_2 to $H_2O + \frac{1}{2}O_2$, and is about $100\ s^{-1}$ for chymotrypsin, which catalyzes the hydrolysis of a number of esters and amides.

The quantity $K_S = (k_2 + k_3)/k_1$ is called the Michaelis constant for the substrate, and it is evident from equation 16.38 that it is equal to the concentration of substrate required to give half the maximum velocity. The values of V_S and K_S may be obtained from the slope and intercept of linear plots of $(S)/v$ versus (S), or v versus $v/(S)$, or $1/v$ versus $1/(S)$. The first type of plot, which is shown in Fig. 16.2*b*, comes from writing equation 16.38 in the form

$$\frac{(S)}{v} = \frac{K_S}{V_S} + \frac{(S)}{V_S} \tag{16.39}$$

The way in which V_S and K_S depend on pH, salt concentration, coenzyme concentration, etc., gives further information about the enzymatic mechanism.

At equilibrium the two terms in the numerator of equation 16.37 are equal and so

$$K = \frac{(P)_{eq}}{(S)_{eq}} = \frac{V_S K_P}{V_P K_S} \tag{16.40}$$

Thus, the kinetic parameters for the forward and reverse reactions are not independent, but are related through the equilibrium constant K.

Compounds that are structurally related to the substrate or product often combine with the catalytic site of the enzyme and cause inhibition; that is, the enzyme-catalyzed reaction is slowed down by the inhibitor. Since the substrate and the inhibitor compete for the same site, the effect of the inhibitor may be reduced by raising the substrate concentration. This type of competitive inhibition may be represented by the mechanism

$$E + S \rightleftharpoons ES \rightarrow E + P \qquad E + I \rightleftharpoons EI \qquad K_I = \frac{(E)(I)}{(EI)} \tag{16.41}$$

where I in the inhibitor. The steady-state rate law is

$$v = \frac{V_S}{1 + \frac{K_S}{(S)}\left[1 + \frac{(I)}{K_I}\right]} \tag{16.42}$$

where K_I is the dissociation constant of EI into E and I. Substances that bind to the enzyme, although not at the active site, may not interfere with the binding of the substrate at the active site, but may alter K_S and V_S. Such inhibitors are referred to as noncompetitive.

In general, an enzymatic reaction has an optimum pH; the maximum velocity

V_S decreases as the pH is raised or lowered from the optimum pH. In the neutral pH range the effects are generally reversible, but proteins are irreversibly denatured at extreme pH values. Reversible effects of pH on V_S may be attributable to the ionization of the enzyme-substrate complex. If the enzyme-substrate complex exists in three states with different numbers of protons, and if only the intermediate form breaks down to give product, the expression for the effect of pH on the maximum velocity may be derived from

$$\begin{array}{c} \text{ES} \\ K_{b\text{ES}} \updownarrow \\ \text{HES} \xrightarrow{k} \text{enzyme} + \text{product} \\ K_{a\text{ES}} \updownarrow \\ \text{H}_2\text{ES} \end{array} \qquad (16.43)$$

where $K_{a\text{ES}}$ and $K_{b\text{ES}}$ are acid dissociation constants, and k is the rate constant for the rate-determining step. Since

$$\begin{aligned} (\text{E})_0 &= (\text{ES}) + (\text{HES}) + (\text{H}_2\text{ES}) \\ &= (\text{HES})[1 + (\text{H}^+)/K_{a\text{ES}} + K_{b\text{ES}}/(\text{H}^+)] \end{aligned} \qquad (16.44)$$

$$V_S = k(\text{HES}) = \frac{k(\text{E})_0}{1 + (\text{H}^+)/K_{a\text{ES}} + K_{b\text{ES}}/(\text{H}^+)} \qquad (16.45)$$

If this is the mechanism a plot of V_S versus pH is a symmetrical bell-shaped curve.

Since a protein has many dissociable acid groups, it is perhaps surprising that the experimental results are sometimes as simple as they are. There is a simple explanation of why two acid groups in the catalytic site might have the total effect on the kinetics that is represented by equation 16.45. If the catalytic function involves both an acidic function and a basic function, we would expect H_2ES to be inactive because the basic site is occupied by a proton and ES to be inactive because it cannot donate a proton to the substrate. Only HES can yield product because it has one group that can donate a proton and another that can accept a proton.

Mechanisms like the ones we have been discussing can give a variety of complicated effects; that is, the rate law may have a very complicated form. But, in general, the steady-state rate is somewhere between zero order and first order in substrate concentration. However, there are some enzymes for which the steady-state rate varies with a higher power of the substrate concentration. In other words, curves analogous to the sigmoid oxygen-binding curve for hemoglobin (Section 6.7) are obtained. This has been found to be especially true of enzymes of importance in the regulation of metabolic pathways. These cooperative effects are encountered with multisite enzymes, not single-site enzymes, because the cooperative effect involves an increased affinity of a second site for a substrate when a first site is occupied. As in the case of hemoglobin, this interaction involves a structural change. According to the Monod-Changeaux-Wyman model, the multisite enzyme can exist in at least two states. In each of the two states the conformations of all the subunits are assumed to be the same. The binding of substrate shifts the equilibrium toward one or the other of these two states. If the effector drives the equilibrium in the direction that produces an enhanced rate of reaction, the effector is called an *activator*. If it causes a reduction in rate it is called an inhibitor. As we have seen in the case of hemoglobin, the effect is multiplied by the fact that one effector molecule affects several catalytic sites on the molecule. The fact that enzymatic activities may be affected by various substances present in the cell provides a mechanism for the control of the rates of reactions in living things so that metabolic intermediates do not accumulate.

16.9 COENZYME REACTIONS

In the living cell the oxidation of alcohol to aldehyde involves the coenzyme nicotinamide adenine dinucleotide. The overall reaction that is catalyzed by alcohol dehydrogenase is

$$\text{alcohol} + \text{NAD}^+ = \text{aldehyde} + \text{NADH} + \text{H}^+ \tag{16.46}$$

where NAD^+ is nicotinamide adenine dinucleotide in the oxidized form and NADH is the reduced form. In the cell there is a means of oxidizing NADH to NAD^+ so that it can participate over and over again in the oxidation of alcohol. Since nicotinamide adenine dinucleotide is recycled in this way, it is called a coenzyme. The concentration of the oxidized coenzyme in the cell depends on the amount present and the rate with which it is oxidized in some other reaction. The concentration of oxidized coenzyme affects the rate of oxidation of alcohol to aldehyde.

Under certain conditions the horse liver alcohol dehydrogenase reaction, equation 16.46, behaves kinetically in accord with the mechanism

$$\text{E} + \text{NAD}^+ \underset{k_{-1}}{\overset{k_1}{\rightleftharpoons}} \text{ENAD}^+ \tag{16.47}$$

$$\text{ENAD}^+ + \text{alc} \underset{k_{-2}}{\overset{k_2}{\rightleftharpoons}} \text{ald} + \text{ENADH} + \text{H}^+ \tag{16.48}$$

$$\text{ENADH} \underset{k_{-3}}{\overset{k_3}{\rightleftharpoons}} \text{E} + \text{NADH} \tag{16.49}$$

The initial steady-state velocities may be measured in both the forward (v_f) and reverse (v_r) directions. Rearrangement of the steady-state rate equations gives

$$\frac{(\text{E})_0}{v_f} = \frac{1}{k_3} + \frac{1}{k_1(\text{NAD}^+)} + \frac{1}{k_2(\text{alc})} + \frac{k_{-1}}{k_1 k_2(\text{NAD}^+)(\text{alc})} \tag{16.50a}$$

$$\frac{(\text{E})_0}{v_r} = \frac{1}{k_{-1}} + \frac{1}{k_{-3}(\text{NADH})} + \frac{1}{k_{-2}(\text{ald})} + \frac{k_3}{k_{-2}k_{-3}(\text{NADH})(\text{ald})} \tag{16.50b}$$

If the steady-state velocities for the forward and reverse reactions are found to follow these equations, this is strong evidence that the mechanism is of the ordered type that has been assumed in reactions 16.47 to 16.49. Thus all six rate constants in mechanism 16.47, 16.48, and 16.49 may be calculated from the dependence of steady-state velocities on substrate concentration. They must satisfy the equilibrium relation

$$K = \frac{(\text{aldehyde})(\text{NADH})}{(\text{alcohol})(\text{NAD}^+)} = \frac{k_1 k_2 k_3}{k_{-1} k_{-2} k_{-3}} \tag{16.51}$$

16.10 ELECTRODE KINETICS

Electrode kinetics is concerned with the rates of electrochemical reactions at the interface between a metallic conductor and a solution. The rate of an electrode reaction is very dependent on the electric potential of the electrode. In discussing working cells it is frequently convenient to relate their behavior when a finite current is passing to the behavior of the reversible cell in terms of the *overpotential*. The overpotential is defined as the difference between the measured potential under working conditions and the thermodynamic potential. For a Galvanic cell the measured potential is lower than the reversible one, and in an electrolysis cell the potential measured during electrolysis is higher than the thermodynamic potential. For a Galvanic cell the potential difference is lowered because of the energy required to surmount physical or chemical barriers in the cell at a certain rate of discharge. For an electrolysis cell an extra potential difference must be applied to overcome the various energy barriers to the discharge of ions and the movement of ions throughout the cell.

We may identify three contributions to the overpotential: activation, concentration, and ohmic overpotentials. The activation overpotential arises from the activation energy required for rate-determining chemical processes in the reaction at the surface of an electrode. The Gibbs energy of activation $\Delta G^{\ddagger\circ}$ for an electron-transfer reaction is a function of the electrode-solution potential difference. In an electrolysis cell, application of a potential lowers the activation barrier and speeds the electron-transfer reaction. The concentration overpotential arises from the decrease in concentration of reactants in the immediate vicinity of an electrode where they are being consumed. Concentration polarization can be reduced by vigorous stirring. Ohmic overpotential is simply the IR drop (current times resistance) due to the internal resistance of a cell. This type of overpotential can be reduced by using electrolytes of high conductivity and keeping the distance between electrodes short.

In 1905 Tafel discovered that the overpotential E_η is a logarithmic function of the current density (current per unit area) i.

$$E_\eta = a \log \left(\frac{i}{i_0}\right) \tag{16.52}$$

where a is an empirical constant and i_0, another empirical constant, is the exchange current density. Theoretical derivations of the Tafel equation* show that i_0 is the current

* J. Albery, *Electrode Kinetics*, Clarendon Press, Oxford, 1975; A. McDougall, *Fuel Cells*, Wiley, New York, 1976.

Table 16.3 Tafel Constants for the Evolution of Hydrogen Gas from 1 mol L^{-1} HCl at 25 °C

Electrode	$-\log (i_0/\text{A cm}^{-2})$	a/V
Pt	2.6	0.028
Ag	4	0.130
Hg	12	0.119
Fe	6	0.130

density within the cell in both directions at equilibrium. At equilibrium the net current is zero because of the canceling of the two electron transfers in opposite directions.

Alternatively, the Tafel equation may be written

$$i = i_0 e^{E_\eta/2.303a} \tag{16.53}$$

This emphasizes the exponential dependence of current density on the overpotential. The Tafel equation is generally only applicable for overpotentials greater than about 0.05 V, but an equation covering the full range may be derived theoretically. The values of the Tafel constants for the evolution of hydrogen gas from various metal electrodes are given in Table 16.3.

References

R. P. Bell, *The Proton in Chemistry*, Cornell University Press, Ithaca, 1959.

P. D. Boyer, *The Enzyme*, Vol. II, *Kinetics and Mechanism*, 3rd ed., Academic Press, New York, 1970.

A. Cornish-Bowden, *Principles of Enzyme Kinetics*, Butterworths, London, 1976.

A. Fersht, *Enzyme Structure and Mechanism*, W. H. Freeman and Co., San Francisco, 1977.

W. C. Gardiner, Jr., *Rates and Mechanisms of Chemical Reactions*, W. A. Benjamin, Inc., New York, 1969.

H. Gutfreund, *Enzymes: Physical Principles*, Wiley-Interscience, New York, 1972.

G. G. Hammes, *Principles of Chemical Kinetics*, Academic Press, New York, 1978.

G. G. Hammes, ed., *Investigation of Rates and Mechanisms of Reactions*, 3rd ed., Wiley, New York, 1974.

I. H. Segal, *Enzyme Kinetics*, Wiley-Interscience, New York, 1975.

E. Zeffren and P. L. Hall, *The Study of Enzyme Mechanisms*, Wiley-Interscience, New York, 1973.

Problems

16.1 Show that if A and B can be represented by spheres of the same radius that react when they touch, the second-order rate constant is given by

$$k_a = \frac{8 \times 10^3 R}{3\eta} \text{ L mol}^{-1}\text{ s}^{-1}$$

where R is in J K^{-1} mol^{-1}. To obtain this result the diffusion coefficient is expressed in terms of the radius of a spherical particle by use of equation 20.12. For water at 25 °C, $\eta = 8.95 \times 10^{-4}$ kg m^{-1} s^{-1}. Calculate k at 25 °C. *Ans.* 7.4×10^9 L mol^{-1} s^{-1}.

16.2 The diffusion coefficient D of an ion is related to its ionic mobility u by

$$D = \frac{uRT}{zF}$$

where D = diffusion coefficient in $m^2\ s^{-1}$
u = ion mobility in $m^2\ V^{-1}\ s^{-1}$
R = $8.314\ J\ K^{-1}\ mol^{-1}$
z = number of charges on ion
F = Faraday constant = 96,500 C mol^{-1}

The ionic mobilities of H^+ and OH^- are $3.63 \times 10^{-7}\ m^2\ V^{-1}\ s^{-1}$ and $2.05 \times 10^{-7}\ m^2\ V^{-1}\ s^{-1}$ at 25 °C. What is the rate constant for the reaction

$$H^+ + OH^- \rightarrow H_2O$$

The reaction radius is 0.75 nm, because once the proton is this close the reaction can proceed very rapidly by quantum mechanical tunneling. The electrostatic factor f is 1.70.
Ans. $1.4 \times 10^{11}\ L\ mol^{-1}\ s^{-1}$.

16.3 For acetic acid at 25 °C

$$CH_3CO_2H \underset{4.5 \times 10^{10}\ L\ mol^{-1}\ s^{-1}}{\overset{7.8 \times 10^5\ s^{-1}}{\rightleftharpoons}} CH_3CO_2^- + H^+$$

what is the relaxation time τ for a 0.1 mol L^{-1} solution? *Ans.* 8.5 ns.

16.4 Derive the relation between the relaxation time τ and the rate constants for the reaction $A \underset{k_2}{\overset{k_1}{\rightleftharpoons}} B$, which is subjected to a small displacement from equilibrium.
Ans. $\tau = (k_1 + k_2)^{-1}$.

16.5 Calculate the first-order rate constants for the dissociation of the following weak acids: acetic acid, acid form of imidazole $C_3N_2H_5^+$, NH_4^+. The corresponding acid dissociation constants are 1.75×10^{-5}, 1.2×10^{-7}, and 5.71×10^{-10}, respectively. The second-order rate constants for the formation of the acid forms from a proton plus the base are 4.5×10^{10}, 1.5×10^{10}, and $4.3 \times 10^{10}\ L\ mol^{-1}\ s^{-1}$, respectively.
Ans. $7.9 = 10^5$, 1.8×10^3, $25\ s^{-1}$.

16.6 The mutarotation of glucose is first order in glucose concentration and is catalyzed by acids (A) and bases (B). The first-order rate constant may be expressed by an equation of the type that is encountered in reactions with parallel paths.

$$k = k_0 + k_{H^+}(H^+) + k_A(A) + k_B(B)$$

where k_0 is the first-order rate constant in the absence of acids and bases other than water. The following data were obtained by J. N. Brönsted and E. A. Guggenheim [*J. Am. Chem. Soc.*, **49**, 2554 (1927)] at 18 °C in a medium containing 0.02 mol L^{-1} sodium acetate and various concentrations of acetic acid.

(CH_3CO_2H)/mol L^{-1}	0.020	0.105	0.199
$k/10^{-4}\ min^{-1}$	1.36	1.40	1.46

Calculate k_0 and k_A. The term involving k_{H^+} is negligible under these conditions.
Ans. $k_0 = 1.35 \times 10^{-4}\ min^{-1}$, $k_A = 5.5 \times 10^{-5}\ L\ mol^{-1}\ min^{-1}$.

16.7 The mutarotation of glucose is catalyzed by acids and bases and is first order in the concentration of glucose. When perchloric acid is used as a catalyst, the concentration of hydrogen ions may be taken to be equal to the concentration of perchloric acid, and the catalysis by perchlorate ion may be ignored since it is such a weak base. The following first-order constants were obtained by J. N. Brönsted and E. A. Guggenheim [*J. Am. Chem. Soc.*, **49**, 2554 (1927)] at 18 °C.

$(HClO_4)$/mol L^{-1}	0.0010	0.0048	0.0099	0.0192	0.0300	0.0400
$k/10^{-4}$ min^{-1}	1.25	1.38	1.53	1.90	2.15	2.59

Calculate the values of the constants in the equation $k = k_0 + k_{H^+}(H^+)$.

Ans. $k_0 = 1.21 \times 10^{-4}$ min^{-1}, $k_{H^+} = 3.4 \times 10^{-3}$ L mol^{-1} min^{-1}.

16.8 Suppose an enzyme has a turnover number of 10^4 min^{-1} and a molar mass of 60,000 g mol^{-1}. How many moles of substrate can be turned over per hour per gram of enzyme if the substrate concentration is twice the Michaelis constant? It is assumed that the substrate concentration is maintained constant by a preceding enzymatic reaction and that products do not accumulate and inhibit the reaction.

Ans. 6.7 mol hr^{-1} g^{-1}.

16.9 The kinetics of the fumarase reaction

$$\text{fumarate} + H_2O = \text{L-malate}$$

is studied at 25 °C using an 0.01 ionic strength buffer of pH 7. The rate of the reaction is obtained using a recording ultraviolet spectrometer to measure the fumarate concentration. The following rates of the forward reaction are obtained using a fumarase concentration of 5×10^{-10} mol L^{-1}.

$(F)/10^{-6}$ mol L^{-1}	$v_F/10^{-7}$ mol L^{-1} s^{-1}
2	2.2
40	5.9

The following rates of the reverse reaction are obtained using a fumarase concentration of 5×10^{-10} mol L^{-1}.

$(M)/10^{-6}$ mol L^{-1}	$v_M/10^{-7}$ mol L^{-1} s^{-1}
5	1.3
100	3.6

(*a*) Calculate the Michaelis constants and turnover numbers for the two substrates. In practice many more concentrations would be studied. (*b*) Calculate the four rate constants in the mechanism

$$E + F \underset{k_{-1}}{\overset{k_1}{\rightleftharpoons}} X \underset{k_{-2}}{\overset{k_2}{\rightleftharpoons}} E + M$$

where E represents the catalytic site. There are four catalytic sites per fumarase molecule. (*c*) Calculate K_{eq} for the reaction catalyzed. The concentration of H_2O is omitted in the expression for the equilibrium constant because its concentration cannot be varied in dilute aqueous solutions.

Ans. (*a*) $V_F = 6.5 \times 10^{-7}$ mol L^{-1} s^{-1}, $K_F = 3.9 \times 10^{-6}$ mol L^{-1},
$V_M = 4.0 \times 10^{-7}$ mol L^{-1} s^{-1}, $K_M = 1.03 \times 10^{-5}$ mol L^{-1},
$k_2 = 3.3 \times 10^2$ s^{-1}, $k_{-1} = 2.0 \times 10^2$ s^{-1}.
(*b*) $k_1 = 1.4 \times 10^8$ L mol^{-1} s^{-1}, $k_{-2} = 5.1 \times 10^7$ L mol^{-1} s^{-1}. (*c*) 4.4.

16.10 At pH 7 the measured Michaelis constant and maximum velocity for the enzymatic conversion of fumarate to L-malate

$$\text{fumarate} + H_2O = \text{L-malate}$$

are 4.0×10^{-6} mol L^{-1} and $(1.3 \times 10^3\ s^{-1})(E)_0$, where $(E)_0$ is the total molar concentration of the enzyme. The Michaelis constant and maximum velocity for the reverse reaction are 10×10^{-6} mol L^{-1} and $(0.80 \times 10^3\ s^{-1})(E)_0$. What is the equilibrium constant for this hydration reaction? (The activity of water is set equal to unity since it is in excess.) *Ans.* 4.1.

16.11 At 25 °C and pH 8 the maximum initial velocities V and Michaelis constants K of the fumarase reaction $F + H_2O = M$ are

$$V_F = (0.2 \times 10^3 \text{ s}^{-1})(E)_0 \qquad V_M = (0.6 \times 10^3 \text{ s}^{-1})(E)_0$$
$$K_F = 7 \times 10^{-6} \text{ mol L}^{-1} \qquad K_M = 100 \times 10^{-6} \text{ mol L}^{-1}$$

where $(E)_0$ is the total molar concentration of enzymatic sites. Calculate the values of the four rate constants in the mechanism

$$\text{E} + \text{F} \underset{k_2}{\overset{k_1}{\rightleftharpoons}} \text{EX} \underset{k_4}{\overset{k_3}{\rightleftharpoons}} \text{E} + \text{M}$$

and the equilibrium constant $(M)_{eq}/(F)_{eq}$.

Ans. $k_1 = 1.1 \times 10^8 \text{ L mol}^{-1} \text{ s}^{-1}$, $k_2 = 0.6 \times 10^3 \text{ s}^{-1}$, $k_3 = 0.2 \times 10^3 \text{ s}^{-1}$, $k_4 = 8 \times 10^6 \text{ L mol}^{-1} \text{ s}^{-1}$, 4.6.

16.12 Derive the steady-state rate equation for the mechanism

$$\text{E} + \text{S} \underset{k_2}{\overset{k_1}{\rightleftharpoons}} \text{X} \xrightarrow{k_3} \text{E} + \text{P} \qquad \text{E} + \text{I} \underset{k_5}{\overset{k_4}{\rightleftharpoons}} \text{EI}$$

for the case that $(S) \gg (E)_0$ and $(I) \gg (E)_0$. *Ans.* Equation 16.42.

16.13 The following initial velocities were determined spectrophotometrically for solutions of sodium succinate to which a constant amount of succinoxidase was added. The velocities are given as the change in absorbancy at 250 nm in 10 s. Calculate V, K_S, and K_I for malonate.

	$\frac{A \times 10^3}{10 \text{ s}}$	
(Succinate) 10^{-3} mol L^{-1}	No Inhibitor	15×10^{-6} mol L^{-1} malonate
10	16.7	14.9
2	14.2	10.0
1	11.3	7.7
0.5	8.8	4.9
0.33	7.1	—

Ans. 15.1. 0.38×10^{-3} mol L^{-1}, 8.6×10^{-6} mol L^{-1}.

16.14 The maximum initial velocities for an enzymatic reaction are determined at a series of pH values.

pH	6.0	6.5	7.0	7.5	8.0	8.5	9.0
V	11	30	74	129	147	108	53

Calculate the values of the parameters V', K_a, and K_b in

$$V = \frac{V'}{1 + (H^+)/K_a + K_b/(H^+)}$$

Hint. A plot of V versus pH may be constructed and the hydrogen ion concentration at the midpoint on the acid side referred to as $(H^+)_a$ and the hydrogen ion concentration at the midpoint on the basic side is referred to as $(H^+)_b$. Then

$$K_a = (H^+)_a + (H^+)_b - 4\sqrt{(H^+)_a(H^+)_b} \qquad K_b = \frac{(H^+)_a(H^+)_b}{K_a}$$

Ans. 234. 4.41×10^{-8}, 3.67×10^{-9} mol L^{-1}.

16.15 Alcohol dehydrogenase catalyzes the reaction

$$\text{alcohol} + NAD^+ = \text{aldehyde} + NADH + H^+$$

where NAD^+ is nicotinamide adenine dinucleotide in the oxidized form and NADH in the reduced form. For enzyme isolated from yeast the maximum velocity for ethyl alcohol is given by

$$V(\text{mol L}^{-1}\text{ min}^{-1}) = 2.7 \times 10^4(\text{E})_0$$

where $(\text{E})_0$ is the molar concentration of enzyme of molar mass 150,000 g mol^{-1}. The reaction is conveniently followed spectrophotometrically at 340 nm because NADH has a molar absorbancy coefficient of 620 (mol L^{-1})$^{-1}$ cm^{-1}.

$$\text{Absorbancy} = \log \frac{I_0}{I} = 6200(\text{NADH})d$$

where d is the thickness of the spectrophotometer cuvette. How many grams of enzyme must be placed in a 3-mL reaction mixture to produce a rate of change of absorbancy of 0.001 per minute when a 1.0-cm cuvette is used? *Ans.* 2.7×10^{-9} g.

16.16 What current density i will be obtained for the evolution of hydrogen gas on (*a*) Pt, and (*b*) Fe from 1 mol L^{-1} HCl at 25 °C at an overpotential of 0.1 volt?

Ans. (*a*) 0.012 A cm^{-2}, (*b*) 1.4×10^{-6} A cm^{-2}.

16.17 From the fact that the bimolecular rate constant for the reaction of H^+ with NH_3 is 4.3×10^{10} L mol^{-1} s^{-1}, calculate the reaction radius for this reaction using equation 16.4. The diffusion coefficient of H^+ may be calculated from its mobility (Table 18.2) using equation 18.27. The diffusion coefficient of NH_3 may be neglected because it is quite a bit smaller.

16.18 Pure solutions of the α and β chains of hemoglobin $\alpha_2\beta_2$ can be prepared. Assuming that α and β exist only as monomers in these solutions, and that they react on the first collision, estimate the half-life for the reaction

$$\alpha + \beta \rightarrow \alpha\beta$$

in water at 25 °C. The viscosity of water at this temperature is 8.95×10^{-4} kg m^{-1} s^{-1}. Calculate the half-life if equal volumes of 10^{-6} mol L^{-1} solutions of α and β are mixed.

16.19 An imidazole buffer of pH 7 containing 0.05 mol L^{-1} imidazole has a relaxation time of 2.9×10^{-9} s at 25 °C. What are the values of the rate constants for the reaction

$$C_3N_2H_4 + H^+ \underset{k_{-1}}{\overset{k_1}{\rightleftharpoons}} C_3N_2H_5{}^+$$

The pK for the imidazole at this temperature is 7.21.

16.20 Derive the relation between the relaxation time τ and the rate constants for the reaction $A + B \underset{k_2}{\overset{k_1}{\rightleftharpoons}} C + D$, which is subjected to a small displacement from equilibrium.

16.21 Derive the relation between the relaxation time τ and the rate constants for the mechanism

$$\begin{array}{ccc} A & \underset{k_{-1}}{\overset{k_1}{\rightleftharpoons}} & B \\ K_A \Updownarrow & & \Updownarrow K_B \\ A' & \underset{k_{-1}'}{\overset{k_1'}{\rightleftharpoons}} & B' \end{array}$$

which is subjected to a small displacement from equilibrium. It is assumed that the equilibria, $A \rightleftharpoons A'$, $K_A = (A')/(A)$, and $B \rightleftharpoons B'$, $K_B = (B')/(B)$, are adjusted very rapidly so that these steps remain in equilibrium.

16.22 Calculate the first-order rate constants for the following reactions at 25 °C.

H^+ production

$$HOAc \rightarrow H^+ + OAc^-$$
$$ImH^+ \rightarrow H^+ + Im$$
$$NH_4^+ \rightarrow H^+ + NH_3$$

OH^- production

$$OAc^- + H_2O \rightarrow HOAc + OH^-$$
$$Im + H_2O \rightarrow ImH^+ + OH^-$$
$$NH_3 + H_2O \rightarrow NH_4^+ + OH^-$$

where HOAc is acetic acid and Im is imidazole ($C_3N_2H_4$). The reverse reactions given above may all be assumed to be diffusion controlled with $k = 10^{10}$ L mol^{-1} s^{-1}. Acid dissociation constants at 25 °C are

HOAc	1.75×10^{-5}
ImH^+	1.2×10^{-7}
NH_4^+	5.71×10^{-10}

Which conjugate acid-base pair can play both H^+ and OH^- production roles about equally effectively?

16.23 The solution reaction

$$I^- + OCl^- = OI^- + Cl^-$$

is believed to go by the mechanism

$$OCl^- + H_2O \underset{}{\overset{K_1}{\rightleftharpoons}} HOCl + OH^- \qquad \text{(fast)}$$
$$I^- + HOCl \xrightarrow{k} HOI + Cl^- \qquad \text{(slow)}$$
$$HOI + OH^- \underset{}{\overset{K_2}{\rightleftharpoons}} H_2O + OI^- \qquad \text{(fast)}$$

Derive the rate equation for the forward rate of this reaction that shows the effect of the concentration of OH^-.

16.24 The initial rate v of oxidation of sodium succinate to form sodium fumarate by dissolved oxygen in the presence of the enzyme succinoxidase may be represented by equation 16.42. Calculate V and K_S from the following data.

$(S)/10^{-3}$ mol L^{-1}	10	2	1	0.5	0.33
$v/10^{-6}$ mol L^{-1} s^{-1}	1.17	0.99	0.79	0.62	0.50

16.25 For the fumarase reaction

$$\text{fumarate} + H_2O = \text{L-malate}$$

at pH 7, 25 °C, and 0.01 ionic strength, the Michaelis-Menten parameters have the following values.

$$V_F = (1.3 \times 10^3\ s^{-1})(E)_0 \qquad V_M = (0.8 \times 10^3\ s^{-1})(E)_0$$
$$K_F = 4 \times 10^{-6}\ \text{mol L}^{-1} \qquad K_M = 10 \times 10^{-6}\ \text{mol L}^{-1}$$

where $(E)_0$ is the molar concentration of the enzyme, which has four catalytic sites per molecule. Calculate (*a*) the four rate constants in the mechanism

$$E + F \underset{k_{-1}}{\overset{k_1}{\rightleftharpoons}} EX \underset{k_{-2}}{\overset{k_2}{\rightleftharpoons}} E + M$$

and (*b*) $\Delta G^{\circ\prime}$ for the overall reaction.

16.26 At 25 °C and pH 7.8, the following values are obtained for the Michaelis constant and maximum initial velocity for the forward reaction catalyzed by fumarase.

$$F + H_2O = M \qquad K = (M)_{eq}/(F)_{eq} = 4.4.$$

$$V_F = (0.8 \times 10^3\ \mathrm{s^{-1}})(E)_0$$

$$K_F = 7 \times 10^{-6}\ \mathrm{mol\ L^{-1}}$$

where the enzyme concentration is in moles of enzyme per liter. The enzyme has four catalytic sites per molecule. In some experiments L-malate was added and was found to be inhibitory with a constant

$$K_M = 100 \times 10^{-6}\ \mathrm{mol\ L^{-1}}$$

Calculate the values of the four rate constants in the mechanism

$$E + F \underset{k_{-1}}{\overset{k_1}{\rightleftharpoons}} X \underset{k_{-2}}{\overset{k_2}{\rightleftharpoons}} E + M$$

where E represents an enzymatic site.

16.27 Derive the steady-state rate law for the mechanism

$$E + S \underset{k_{-1}}{\overset{k_1}{\rightleftharpoons}} EP_2 + P_1 \underset{k_{-2}}{\overset{k_2}{\rightleftharpoons}} E + P_2 + P_1$$

in the form including both the forward and reverse reactions. Give the steady-state rate equations for the initial velocities of the forward and reverse reactions. Can all four rate constants be determined from steady-state velocities of the forward and reverse reactions?

16.28 Derive the steady-state rate equation for the mechanism

$$\begin{array}{ccc} E + S & \xrightarrow{k_3} & E' + P \\ k_1 \downarrow\uparrow k_2 & & \\ E' & & \end{array}$$

The Michaelis-Menten form is obtained but, nevertheless, why is this an unsuitable mechanism for explaining enzymatic action?

16.29 A certain enzyme E catalyzes an essentially irreversible reaction S → P. Addition of small amounts of E to a solution of pure S results in a rapid initial production of P, but then the rate of production of P gradually slows down and comes to almost a complete halt, *before* all of the S is converted to P. If a large amount of additional S is now added, there is again a rapid initial production of P that again slows down to almost a complete halt before all the S is converted to P. Explain these findings.

16.30 The Michaelis constant for succinate on succinoxidase is 0.5×10^{-3} mol L^{-1} and the competitive inhibition constant for malonate is 10×10^{-3} mol L^{-1}. In an experiment with 10^{-3} mol L^{-1} succinate and 15×10^{-3} mol L^{-1} malonate, what is the percent inhibition?

16.31 Derive the steady-state rate equation for the mechanism

$$\begin{array}{ccccc} E & + S \underset{k_{-1}}{\overset{k_1}{\rightleftharpoons}} & ES & \xrightarrow{k_2} & E + P \\ \updownarrow K_{EH} & & \updownarrow K_{EHS} & & \\ EH & & EHS & & \end{array}$$

Sketch the shape of the plots of V_S and K_S versus pH.

16.32 As a first approximation the kinetics of the alcohol dehydrogenase reaction

$$\text{alcohol} + \text{NAD}^+ = \text{aldehyde} + \text{NADH} + \text{H}^+$$

may be represented by the mechanism

$$\text{E} + \text{NAD}^+ \xrightarrow{k_1} \text{ENAD}^+$$

$$\text{ENAD}^+ + \text{alcohol} \xrightarrow{k_2} \text{E} + \text{NADH} + \text{aldehyde} + \text{H}^+$$

Derive the steady-state rate equation for the initial velocity.

16.33 What overpotentials E_η are required to produce current densities of 0.1 A cm^2 for the evolution of hydrogen gas from 1 mol L^{-1} HCl at 25 °C on (*a*) Pt, and (*b*) Hg?

CHAPTER 17
PHOTOCHEMISTRY

Photochemistry comprises the study of chemical reactions induced directly or indirectly by light. Ordinary thermal reactions proceeding in the dark acquire their activation energy through random, successive collisions between molecules. Photochemical reactions receive their activation energy by the absorption of photons of light by molecules. Therefore they offer the possibility of a high degree of selectivity in that the energy of a quantum of light may be just sufficient for a particular reaction. Thus the activation step of a photochemical reaction is quite different, and more selective, than the activation of an ordinary (thermal) reaction. Excited electronic states of molecules have different energies and different electron distributions than the ground state, so that they behave chemically in different ways.

17.1 LAWS OF PHOTOCHEMISTRY

There are two basic laws of photochemistry. According to the first, which is due to Grotthus (1817) and Draper (1843), only light that is absorbed can produce photochemical change. This law now seems self-evident, but we must also be aware that there is another effect not visualized by Grotthus and Draper according to which radiation that is not absorbed may stimulate an excited molecule to radiate (Section 11.3).

The second law of photochemistry, proposed by Stark and Einstein (1908–1912), is that a molecule absorbs a single quantum of light in becoming activated.

$$A + h\nu \rightarrow A^* \tag{17.1}$$

Avogadro's number of photons is called an "einstein," just as Avogadro's number of electrons is called a "faraday." The calculation of the energy carried by an einstein of photons has been discussed in Section 11.1. Exceptions to this law have recently been found in simultaneous two-quantum absorptions in systems illuminated with the intense and coherent radiation from a laser.

$$A + 2h\nu \rightarrow A^{**} \tag{17.2}$$

The second law of photochemistry provides the basis for the calculation of quantum yield Φ of a particular process.

$$\Phi = \frac{\text{number of molecules undergoing the process}}{\text{number of quanta absorbed}} \tag{17.3}$$

Alternatively, the quantum yield is expressed by

$$\Phi = \frac{\text{rate of process}}{\text{rate of absorption of radiation}} \tag{17.4}$$

A quantum yield may be calculated for each of the physical or chemical processes that is caused by the absorption of light. If a photoexcited molecule can lose energy only through fluorescence, internal conversion, and intersystem crossing, the sum of the quantum yields for these three processes must equal unity. The quantum yield for chemical reaction can range from a small fraction of unity to a very large number. The quantum yield is a large number when the absorption of light produces a radical or other catalyst that starts a chain reaction of a thermodynamically spontaneous reaction.

Example 17.1 In the photobromination of cinnamic acid to dibromocinnamic acid, using blue light of 435.8 nm at 30.6 °C, a light intensity of 1.4×10^{-3} J s^{-1} produced a decrease of 0.075 millimole of Br_2 during an exposure of 1105 s. The solution absorbed 80.1% of the light passing through it. Calculate the quantum yield.

$$E = \frac{N_A hc}{\lambda} = \frac{(6.02 \times 10^{23}\ \text{mol}^{-1})(6.62 \times 10^{-34}\ \text{J s})(3 \times 10^{8}\ \text{m s}^{-1})}{(4.358 \times 10^{-7}\ \text{m})}$$

$$= 2.74 \times 10^{5}\ \text{J mol}^{-1}$$

$$\text{einsteins absorbed} = \frac{(1.4 \times 10^{-3}\ \text{J s}^{-1})(0.801)(1105\ \text{s})}{(2.74 \times 10^{5}\ \text{J mol}^{-1})} = 4.52 \times 10^{-6}\ \text{mol}$$

$$\text{moles of Br}_2\ \text{reacting} = 7.5 \times 10^{-5}\ \text{mol}$$

$$\Phi = \frac{7.5 \times 10^{-5}}{4.52 \times 10^{-6}} = 16.6$$

17.2 INTRAMOLECULAR PROCESSES

The absorption of light by a diamagnetic molecule generally raises it from a singlet ground state to a singlet excited state. To remind you of this terminology that was used in Section 8.18, the excited electron of a singlet excited state has a spin antiparallel to the one with which it is paired. There are many singlet excited states, but usually only several of those of the lowest energy can be produced with visible or near ultraviolet light. The singlet ground state is represented by S_0 and singlet excited states by $S_1, S_2, \ldots$. After a molecule has been excited to some vibrational level of an excited state, it generally loses vibrational energy very rapidly until it reaches the zeroth vibrational level of that electronic state. An excited molecule may also undergo a very rapid radiationless transition from a higher electronic state to a lower electronic state of the same multiplicity. This process, which is called *internal conversion*, is a first-order process and the rate constants for conversion of S_3 to S_2 and S_2 to S_1 are in the 10^{11}–10^{13} s^{-1} range. Since all the excited singlet molecules are converted to S_1 so rapidly, only this singlet state is shown in Fig. 17.1. As shown in this figure, the S_1 state may then (1) be converted to S_0 by internal

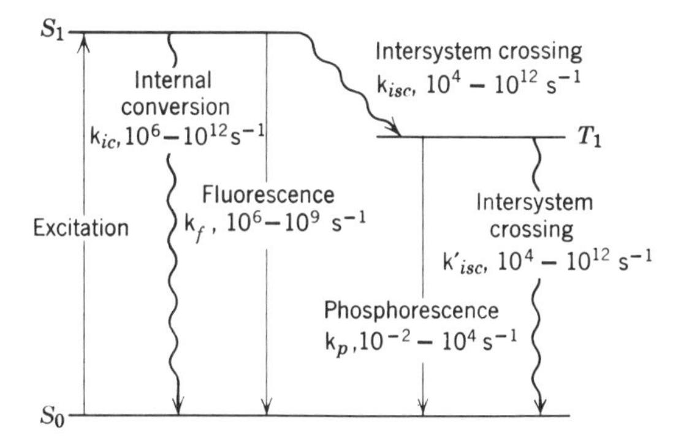

Fig. 17.1 Intramolecular processes resulting from the absorption of radiation and the range of magnitude of the corresponding rate constants. Nonradiative processes are represented by ⇝ and radiative processes are represented by →. (From C. H. J. Wells, *Introduction to Molecular Photochemistry*, Chapman and Hall Ltd., London, 1972. © C. H. J. Wells 1972.)

conversion, (2) fluoresce to the S_0 state, or (3) undergo intersystem crossing to the triplet state T_1. *Intersystem crossing* refers to a radiationless transition between states of different multiplicity (Section 8.18). Because of the spin interchange involved in intersystem crossing, the rate constant for this process is 10^{-2} to 10^{-6} as fast as for internal conversion. The rates of internal conversion and intersystem crossing depend on the energy separation between the states involved; the larger the energy separation, the lower the rate.

For each excited singlet state (S_1, S_2, S_3, etc.) there is a corresponding triplet state (T_1, T_2, T_3, etc.). The difference between the singlet and triplet states is that the electrons are not paired in the triplet states. Since the spin angular momentum of an electron is one-half, the spin for a triplet state is one. The spin angular momentum vector for a triplet state may have one of three orientations in a magnetic field, hence the designation triplet.

A triplet state always has a lower energy than the corresponding singlet state. The explanation is roughly that since the two outer electrons have the same spin, they cannot come very close to each other without violating the Pauli principle. Since the electrons are further apart, there is a decrease in electronic repulsion and the energy of the molecule is lower.

As for the manifold of singlet excited states, the rate constants for internal conversion between excited triplet states are large. Consequently, chemical reactions, whether they occur from a singlet or a triplet state, usually occur from the lowest excited state, S_1 or T_1, respectively.

As shown in Fig. 17.1, the triplet molecules can phosphoresce or undergo intersystem crossing to the S_0 state. The radiation emitted in a transition between states of the same multiplicity (i.e., singlet-singlet transitions or triplet-triplet transitions) is called *fluorescence*, and the radiation emitted in a transition between states of different multiplicity is called *phosphorescence*. Phosphorescence lifetimes are generally considerably longer than fluorescence lifetimes, because transitions between states of different multiplicities are quantum mechanically forbidden to a first approximation.

Figure 17.1 is a simplified representation of the intramolecular processes that may occur if photoexcited molecules do not react chemically and quenchers are not present. This representation is useful for considering organic molecules. The rate processes are all first order, and the usual ranges for the first-order rate constants are given in Fig. 17.1. When a substance is illuminated with a constant intensity, a steady state is reached in which the rates of formation of intermediates are equal to their rates of disappearance. Assuming that all of the radiation is absorbed, the steady-state rate equation for S_1 is

$$I = k_{ic}(S_1) + k_f(S_1) + k_{isc}(S_1) \tag{17.5}$$

where I is expressed in einsteins absorbed per liter second. The steady-state rate equation for T_1 is

$$k_{isc}(S_1) = k'_{isc}(T_1) + k_p(T_1) \tag{17.6}$$

These equations yield the following expressions for the steady-state concentrations of S_1 and T_1.

$$(S_1) = \frac{I}{k_{ic} + k_f + k_{isc}} \tag{17.7}$$

$$(T_1) = \frac{k_{isc}I}{(k'_{isc} + k_p)(k_{ic} + k_f + k_{isc})} \tag{17.8}$$

17.3 FLUORESCENCE

Fluorescence can be regarded as the spontaneous emission of a photon by an excited state concomitant with return to the ground state. The Einstein coefficient A_{mn} for spontaneous emission was discussed in Section 11.3. The coefficient for spontaneous emission may be calculated from the wave functions for the ground and excited states, and it may also be calculated from measured absorption coefficients. If spontaneous emission is the only mode of deactivation of an excited state, the lifetime is given by

$$\tau_0 = \frac{1}{A_{mn}} = \frac{1}{k_f} \tag{17.9}$$

where τ_0 is called the *radiative lifetime* and k_f is the first-order rate constant for fluorescence. We cannot go into the details of the calculations of radiative lifetimes, but simply note the correlation that states that are populated readily are also depopulated readily. In general, it is found that the radiative lifetime is inversely proportional to the molar absorbancy index (Section 11.13).

$$\tau_0 \approx \frac{(10^{-4}\ \mathrm{L\ mol^{-1}\ cm^{-1}\ s})}{\epsilon} \tag{17.10}$$

where the molar absorbancy index ϵ is expressed in L mol^{-1} cm^{-1}. Thus a strongly absorbing compound with $\epsilon = 10^5$ L mol^{-1} cm^{-1} would be expected to have a natural radiative lifetime of about 10^{-9} s, and a weakly absorbing compound with $\epsilon = 10^{-2}$ L mol^{-1} cm^{-1} would be expected to have a natural radiative lifetime of about 10^{-2} s.

The observed lifetime τ of an excited molecule is determined by all the deactivation processes. For the S_1 state of Fig. 17.1 the lifetime of the state is given by

$$\tau_{S_1} = \frac{1}{k_f + k_{ic} + k_{isc}} \tag{17.11}$$

Thus the observed lifetime τ_{S_1} will be less than the radiative lifetime τ_0, which is equal to $1/k_f$.

The quantum yield for fluorescence is given by

$$\Phi_f = \frac{\text{rate of fluorescence emission}}{\text{rate of absorption of radiation}} = \frac{k_f(S_1)}{I} \tag{17.12}$$

Substituting equation 17.7 yields

$$\Phi_f = \frac{k_f}{k_f + k_{ic} + k_{isc}} \tag{17.13}$$

This is the quantum yield for fluorescence in the absence of any quenching or chemical reaction. Under these conditions the natural and observed lifetimes are related by

$$\tau_{S_1} = \Phi_f \tau_0 = \frac{\Phi_f}{k_f} \tag{17.14}$$

The values of k_f determined from τ_{S_1} and Φ_f for benzene and naphthalene are 2×10^6 s^{-1} and 10^6 s^{-1}, respectively.

If fluorescence is simply the spontaneous emission of an excited state referred to in the Einstein theory, it might be expected that the intensity distribution would be the same as that of the absorption spectrum. However, the absorption and fluorescence spectra have an approximate "mirror image" relationship if the spacings of the vibrational levels in the ground state and excited state are similar. This is shown by Fig. 17.2.

The small difference in the 0—0 bands for absorption and fluorescence is due to the difference in the solvation of the initial and final states in the two cases. As shown in Fig. 17.3, the excited state does not become equilibrated with the solvent in the absorption process, and the ground state does not become equilibrated with solvent in the fluorescence process. As indicated in the diagram, the 0—0 fluorescence transition will be of lower energy than the 0—0 absorption transition.

The origin of the fluorescence spectrum may be discussed in greater detail if we restrict our attention to diatomic molecules. Figure 17.4*a* shows the potential energy curves and vibrational levels for the ground state S_0 and first excited singlet state S_1 of a diatomic molecule. Internal conversion rapidly brings the excited molecule to its lowest vibrational level. When a transition occurs to the S_0 level, the

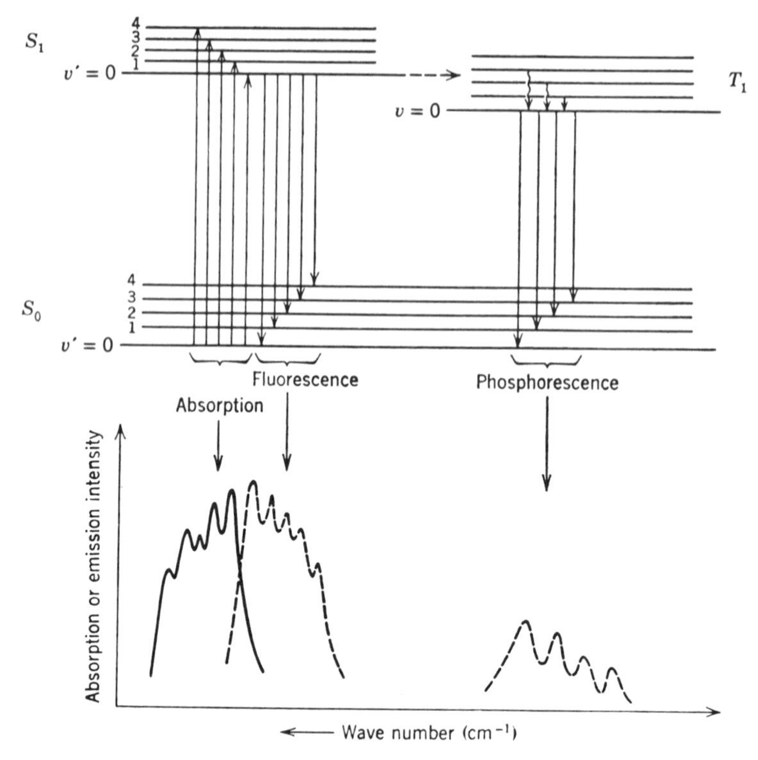

Fig. 17.2 Schematic diagram showing the origin of absorption, fluorescence, and phosphorescence spectra. (From E. F. H. Brittain, W. O. George, and C. H. J. Wells, *Introduction to Molecular Spectroscopy*, Academic Press, New York, 1970.)

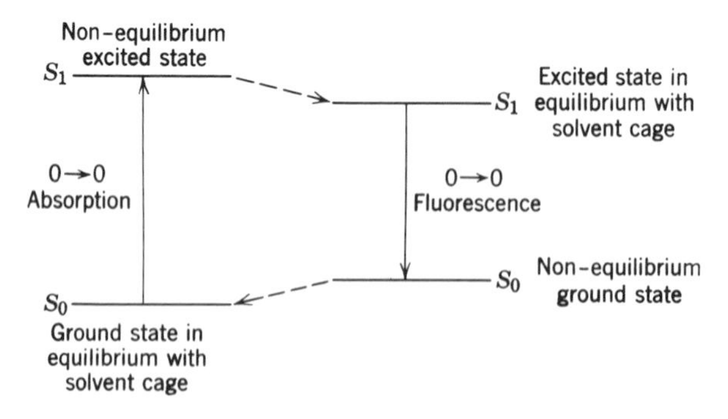

Fig. 17.3 Effect of solute-solvent equilibrium in causing a difference in the energy of the 0—0 absorption transition and the 0—0 fluorescence transition. (From C. H. J. Wells, *Introduction to Molecular Photochemistry*, Chapman and Hall Ltd., London, 1972. © C. H. J. Wells 1972.)

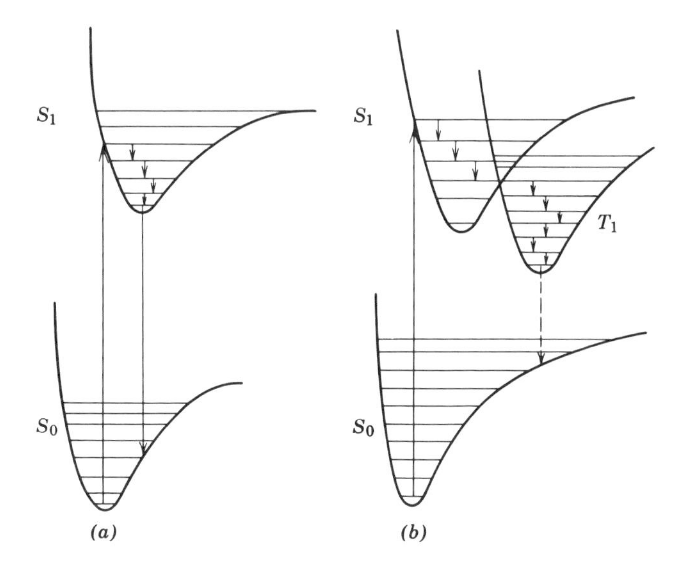

Fig. 17.4 (*a*) Absorption and fluorescence by a diatomic molecule. (*b*) Absorption and phosphorescence by a diatomic molecule. (From J. C. Davis, *Advanced Physical Chemistry*, The Ronald Press Co., New York, 1965.)

molecule is most likely to end up in a higher vibrational level because the potential energy curve for the excited state is displaced to a greater internuclear distance than the ground state curve, and the transition is represented by a vertical line according to the Franck-Condon principle (Section 11.12). In the case illustrated the transition $0 \rightarrow 3$ is the most intense.

Certain fluorescent dyes emit visible light when activated by light in the near ultraviolet or short visible range present in daylight, and this fluorescent light added to the light reflected from cloth gives an appearance of whiter than white.

17.4 PHOSPHORESCENCE

Phosphorescence arises from radiative transitions from the lowest vibrational level of the triplet state T_1 to the various vibrational levels of S_0. Therefore, the phosphorescence spectrum resembles the fluorescence spectrum, as indicated in Fig. 17.2. Since the T_1 state has a lower energy than the S_1 state, the phosphorescence spectrum is observed at longer wavelengths than the fluorescence spectrum. The phosphorescence spectrum is weaker than the fluorescence spectrum because there is more time for energy to be lost by intersystem crossing from T_1 to S_0.

The origin of the phosphorescence spectrum for a diatomic molecule is shown in Fig. 17.4*b*. If, during the loss of vibrational energy from S_1 a molecule happens to have the vibrational energy and internuclear separation that corresponds with the intersection of the potential energy curves for S_1 and T_1, loss of vibrational energy may continue in the T_1 state. At the intersection of the two potential energy curves the nuclei are not moving relative to one another, and the energy is the same in the two states. Under these conditions there is a high probability that a transition will occur in spite of the selection rule that transitions between states of different multiplicity are forbidden. Loss of vibrational energy in the T_1 state continues by internal conversion, as shown. The triplet state has a long lifetime, and so the phosphorescent radiation is emitted slowly.

Returning to Fig. 17.1, we can see that if the rate constant k_{isc} for the intersystem crossing S_1 to T_1 is fast enough, T_1 will be present at an appreciable concentration. If this happens it is very significant for photochemistry, since triplet state molecules may have long lifetimes compared with singlet state molecules, and therefore have a higher probability of undergoing chemical reaction. As in the case of singlet-excited molecules, the natural radiative lifetime for a triplet-excited molecule may be estimated from the molar absorbancy index ϵ for the $S_0 \rightarrow T_1$ transition. Since transitions from the S_0 to the T_1 state are spin forbidden, the absorbancy indices are small. If $\epsilon = 10^{-2}$ L mol^{-1} cm^{-1} s^{-1}, equation 17.10 indicates that a natural radiative lifetime of 10^{-2} s might be expected.

For the T_1 state of Fig. 17.1 the lifetime will be

$$\tau_{T_1} = \frac{1}{k_p + k'_{isc}} \tag{17.15}$$

The quantum yield for phosphorescence is given by

$$\Phi_p = \frac{\text{rate of phosphorescence emission}}{\text{rate of absorption of radiation}} = \frac{k_p(T_1)}{I}$$

$$= \frac{k_p k_{isc}}{(k'_{isc} + k_p)(k_{ic} + k_f + k_{isc})} \tag{17.16}$$

where the second form has been obtained by substituting the expression for the steady-state value of (T_1) given in equation 17.8.

17.5 INTERMOLECULAR PROCESSES

In this section we will emphasize bimolecular processes involving excited molecules. Often, but not always, chemical reaction of one excited molecule with another molecule is accompanied by deactivation of the excited molecule, and the excited state is said to be quenched. Quenching may involve a chemical process such as an addition reaction or an electron transfer process. Quenching an excited molecule (donor D) with a second molecule (acceptor A) may result in the electronic excitation of A with concomitant deactivation of D. This quenching process is often referred to as electronic energy transfer. Since minute traces of impurities (quenchers) can rapidly deactivate excited molecules, substances and solvents used in photochemical studies must be carefully purified. For example, molecular oxygen reacts rapidly with excited molecules ($k \approx 10^9$–10^{10} L mol^{-1} s^{-1}), and it is therefore often important to deoxygenate the solutions under study.

Electronic energy transfer processes may fall into three categories: radiative transfer, short-range (collisional) transfer, and long-range (resonance) transfer. In radiative transfer the donor D emits radiation which is absorbed by the acceptor A.

$$D^* \rightarrow D + h\nu \tag{17.17}$$

$$A + h\nu \rightarrow A^* \tag{17.18}$$

Short-range energy transfer can occur if the distance between donor and acceptor molecules approaches the collision diameter. A collision is not necessarily required, since the energy transfer can occur at distances slightly greater than the collision diameter. According to the Wigner spin conservation rule, the overall spin angular momentum of the interacting pair of molecules must be unchanged in electronic energy transfer. Like spectroscopic selection rules, this rule is not absolutely rigorous, but it is an important guide. Considering only S_0, S_1, and T_1 states, this rule amounts to limiting short-range energy transfer to

$$D^*(S_1) + A(S_0) \rightarrow D(S_0) + A^*(S_1) \tag{17.19}$$

$$D^*(T_1) + A(S_0) \rightarrow D(S_0) + A^*(T_1) \tag{17.20}$$

where D represents the donor molecule and A the acceptor molecule. The efficiencies of these energy transfer processes depend on the relative energies of the states involved. The transfer is relatively efficient if the excited state of D has a greater energy than that of the excited state of A. Energy transfer of this type may be detected by the appearance of fluorescence radiation of $A^*(S_1)$ or phosphorescence radiation of $A^*(T_1)$.

In long-range energy transfer the donor and acceptor molecules are separated by a distance much greater than the collision diameter. The efficiency of the energy transfer in this so-called resonance energy transfer depends on the extent of the overlap of the emission spectrum of the donor and the absorption spectrum of the acceptor. The energy transfer is not efficient unless the decay process $D^* \rightarrow D$ and the excitation process $A \rightarrow A^*$ are allowed electronic transitions. Thus, the most likely long-range energy transfer processes are

$$D^*(S_1) + A(S_0) \rightarrow D(S_0) + A^*(S_1) \tag{17.21a}$$

$$D^*(T_1) + A(S_0) \rightarrow D(S_0) + A^*(S_1) \tag{17.21b}$$

Under favorable circumstances* energy transfer can occur over distances of 5 to 10 nm with rate constants of 10^{10}–10^{11} L mol^{-1} s^{-1}.

The rate constant k_{et} for the energy transfer to an acceptor may be obtained by measuring the triplet lifetime in the presence and in the absence of A. If the triplet state of the donor is deactivated solely by phosphorescence, $T_1 \rightsquigarrow S_0$ intersystem crossing, and bimolecular reaction with an acceptor A, the triplet lifetime is given by

$$\tau_{et} = \frac{1}{k_p + k'_{sic} + k_{et}(\mathrm{A})} \tag{17.22}$$

* N. J. Turro, *Molecular Photochemistry*, W. A. Benjamin, Inc., New York, 1978.

Since the lifetime τ in the absence of the acceptor is given by equation 17.15,

$$\frac{1}{\tau_{et}} = \frac{1}{\tau} + k_{et}(\mathrm{A}) \tag{17.23}$$

The fact that second-order rate constants k_{et} obtained using this relation may equal diffusion-controlled rates (Section 16.2) indicates that energy transfer may take place at every collision of donor and acceptor.

Intermolecular quenching processes are often the key to *photosensitized* reactions. In a photosensitized reaction one molecule (sensitizer) absorbs the light, but another molecule actually undergoes reaction. The sensitizer takes part in the photochemical reaction only in an indirect manner and acts merely as a carrier of the light energy. Electronic energy transfer is one mechanism for photosensitization. For example, 366 nm excitation of benzophenone produces T_1 with virtually unit quantum yield, since k_{isc} is very large.

$$S_0(\text{benzophenone}) \xrightarrow[366\text{ nm}]{h\nu} S_1(\text{benzophenone}) \xrightarrow{k_{isc}} T_1(\text{benzophenone}) \tag{17.24}$$

Norbornadiene is capable of quenching the benzophenone triplet state with a large rate constant, since T_1 for norbornadiene is lower in energy than T_1 for benzophenone. But, unlike benzophenone, norbornadiene is essentially transparent to the 366-nm irradiation. Consequently, the formation of norbornadiene's T_1 state can be photosensitized by benzophenone, as shown in the reaction

$$T_1(\text{benzophenone}) + S_0(\text{norbornadiene}) \xrightarrow{k_{et}} S_0(\text{benzophenone}) + T_1(\text{norbornadiene}) \tag{17.25}$$

and this excited molecule undergoes chemical reaction to produce the valence isomer quadricyclane.

$$T_1\,[\text{norbornadiene}] \xrightarrow{k_r} S_0\,[\text{quadricyclane}] \tag{17.26}$$

This photosensitized valence isomerization illustrates that excited states can relax to nonthermodynamic mixtures, since the norbornadiene-quadricyclane reaction involves an unfavorable $\Delta G°$. Sensitization of such processes to the visible output of the sun may lead to viable approaches to chemical storage of solar energy, provided the reversion reaction can be run cleanly with the release of heat. Numerous photosensitized reactions are known in the gas phase as well as in condensed media involving inorganic and organic sensitizers and reactants. In addition to electronic energy transfer, sensitization mechanisms also involve electron transfer, intermediate adduct, and other primary intermolecular quenching processes in the crucial step of transducing the input light energy.

17.6 CHEMICAL REACTIONS AND THEIR QUANTUM YIELDS

When an excited species undergoes a chemical reaction, we can add another step to the simplified mechanism that we have been discussing. If the excited species (here assumed to be T_1) reacts with reactant R with a bimolecular rate constant k_r, it may be possible to evaluate the rate constant.

$$T_1 + \text{R} \rightarrow \text{stable products} \qquad \text{rate} = k_r(T_1)(\text{R}) \tag{17.27}$$

In the steady state

$$(T_1) = \frac{k_{isc}I}{[k_p + k'_{isc} + k_r(\text{R})][k_{ic} + k_f + k_{isc}]} \tag{17.28}$$

The quantum yield for the production of stable products is

$$\Phi = \frac{k_r(T_1)(\text{R})}{I} \tag{17.29}$$

Substituting equation 17.28 and rearranging yields

$$\frac{1}{\Phi} = \frac{k_f + k_{ic} + k_{isc}}{k_{isc}}\left(1 + \frac{k_p + k'_{isc}}{k_r(\text{R})}\right) \tag{17.30}$$

Even for this simplified mechanism there are so many rate constants to be determined before k_r can be obtained that this is not generally practical and so, in further discussions, we will emphasize quantum yields. Note that from equation 17.3 the quantum yield can be determined by measuring the number of quanta absorbed and the number of molecules reacted. This requires no knowledge of rate constants.

To determine a quantum yield it is necessary to measure the intensity of the light. This may be done by use of a thermopile, which is a series of thermocouples with one set of junctions blackened to absorb all the radiation, which is then converted into heat. The other set of junctions is protected from radiation. The temperature difference between the two sets of junctions is measured by a galvanometer deflection. Galvanometer readings may be converted into joules of radiation per second per square meter striking the thermopile, by calibrating with a standard carbon-filament lamp from the National Bureau of Standards.

The amount of radiation may also be measured with a chemical *actinometer*, in which the amount of chemical change is determined. The yield of the photochemical reaction in the actinometer was determined originally by use of a thermopile. The quantum yields of a few photochemical reactions are summarized in Table 17.1.

Reaction 1 has the same value of Φ from 280 to 300 nm, at low pressures and high pressures, in the liquid state or in solution in hexane. The primary process $HI + h\nu = H + I$ is followed by the reactions $H + HI = H_2 + I$ and $I + I = I_2$, thus giving two molecules of HI decomposed for each photon absorbed. Reaction 2, the dimerization of anthracene, has a quantum yield of unity initially, but the reverse thermal reaction reduces it as the product accumulates. In reaction 3 at 366 nm

Table 17.1 Quantum Yields in Photochemical Reactions at Room Temperature[a]

Reaction	Approximate Wavelength Region, nm	Approximate Φ
1. $2HI \rightarrow H_2 + I_2$	300–280	2
2. $C_{14}H_{10} \rightleftharpoons \frac{1}{2}(C_{14}H_{10})_2$	<360	0–1
3. $2NO_2 \rightarrow 2NO + O_2$	>435	0
	366	2
4. $CH_3CHO \rightarrow CO + CH_4(+C_2H_6 + H_2)$	310	0.5
	253.7	1
5. $(CH_3)_2CO \rightarrow CO + C_2H_6(+CH_4)$	<330	0.2
6. $NH_3 \rightarrow \frac{1}{2}N_2 + \frac{3}{2}H_2$	210	0.2
7. $H_2C_2O_4(+UO_2^{2+}) \rightarrow CO + CO_2 + H_2O(+UO_2^{2+})$	430–250	0.5–0.6
8. $Cl_2 + H_2 \rightarrow 2HCl$	400	10^5
9. $Co(CN)_6^{3-} + H_2O \rightarrow Co(CN)_5(OH_2)^{2-} + CN^-$	313	0.3

[a] W. A. Noyes, Jr., and P. A. Leighton, *Photochemistry of Gases*, Reinhold Publishing Corp., New York, 1941, Appendixes, pp. 415–465; F. Daniels, *J. Phys. Chem.*, **41**, 713 (1938).

the quantum yield is 2 if correction is made for internal screening by the accompanying N_2O_4, which absorbs some light at 366 nm. The reactions are $NO_2 + h\nu = NO_2^*$, $NO_2^* + NO_2 = 2NO + O_2$, where the asterisk indicates an excited molecule. At 435 nm and longer wavelengths no reaction occurs when the radiation is absorbed.

Reaction 4 similarly shows a greater quantum yield at shorter wavelengths. This reaction is interesting because at 300 °C Φ has a value of more than 300, indicating that the free radicals that are first produced by the absorption of light are able to propagate a chain reaction at the higher temperatures. At room temperature the reactions involved in the chain do not go fast enough to be detected. The products given in parentheses are present also but in small amounts.

The experimental determination of the quantum yield constitutes an excellent method for detecting *chain reactions* (Section 15.19). If several molecules of products are formed for each photon of light absorbed, the reaction is obviously a chain reaction in which the products of the reaction are able to promote reaction of other molecules.

Reaction 5 is an example of the fact that the absorption of light in a particular bond does not necessarily cause the rupture of that bond. Acetone, like other aliphatic ketones, absorbs ultraviolet light at about 280 nm. The C=O bond, which we designate as the chromophore, is very strong and does not break to give atomic oxygen. Instead, the absorption energy leads to the cleavage of an adjacent —C— bond that is weaker, thus:

$$(CH_3)_2C{=}O + h\nu \rightarrow CH_3\cdot + CH_3\dot{C}{=}O \qquad (17.31)$$

giving a methyl radical and an acetyl radical. The acetyl radical can then decarbonylate, giving CO and $CH_3\cdot$, or it can react with $CH_3\cdot$ to give back acetone. The methyl radicals can couple to form ethane.

In the photolysis of ammonia, reaction 6, hydrogen atoms are split off, and the low yield is probably due to partial recombination of the fragments. The quantum yield varies with pressure and reaches a maximum at 0.6 to 0.7 Pa.

Reaction 7 illustrates a photosensitized reaction. The photodecomposition of oxalic acid, sensitized by uranyl ion, is so reproducible that it is suitable for use as a chemical actinometer. An actinometer is a photochemical device used to measure the amount of radiation by determining the amount of chemical change. An actinometer is calibrated by use of a thermopile. In the uranyl oxalate actinometer the light is absorbed by the colored uranyl ion, and the energy is transferred to the colorless oxalic acid, which then decomposes. The uranyl ion remains unchanged and can be used indefinitely as a sensitizer. The fact that the molar absorption coefficient of uranyl nitrate is increased by the addition of colorless oxalic acid indicates that a complex is formed. The formation of a loose chemical complex is often necessary for photosensitization.

Reaction 8 is the best known example of a chain reaction. About 1 million molecules react for each quantum absorbed. The molecules of hydrogen chloride formed undergo further reaction with the hydrogen and chlorine atoms produced (Section 15.19). The measurement of the number of molecules per photon gives a measure of the average number of molecules involved in the chain. Initiation of the reaction with a flash of light can result in an explosively fast reaction.

Reaction 9 illustrates the enhanced reactivity of excited state molecules. The inorganic ion $Co(CN)_6^{8-}$ can be recovered unchanged from boiling water as the potassium salt. However, 313 nm irradiation at 25 °C results in aquation to yield $Co(CN)_5(OH_2)^{2-}$ with a quantum yield of 0.30. Given that the excited state has a lifetime of less than 10^{-10} s under the reaction conditions and that a generous rate constant for ground state aquation would be 10^{-6} s^{-1}, the excited state is viewed as being 10^{16} times more reactive than the grand state! Such increases in reactivity are fairly commonly encountered in photochemical work.

17.7 FLASH PHOTOLYSIS

The unstable intermediates in a photochemical reaction are usually present at such low concentrations that they cannot be studied directly. One way to increase their concentrations is to use a very powerful flash of light. The high-energy flash of short duration is obtained by discharging a bank of capacitors through a gas-discharge tube. The flashes are so intense that in some cases practically all the molecules in the reaction tube are dissociated into free radicals and atoms. Powers of 50 MW may be obtained for a few microseconds.

The flash-photolysis method is useful for studying the spectra and decay kinetics of unstable intermediates. The rate of the disappearance of the free radicals can be followed by the rate of increase in the monochromatic transmitted light as measured with an oscilloscope and photomultiplier. The spectrophotometer is set so that the light passing through the illuminated cell is of the wavelength that is absorbed by the free radical. By this technique it has been possible to determine the absorption spectra of radicals such as NH_2, ClO, and CH_3. Another method for obtaining unstable intermediates of photochemical reactions at a concentration at which they can be studied spectroscopically is to form them in a rigid, unreactive medium such as a frozen rare gas at a temperature so low that they have a long lifetime.

17.8 CHEMILUMINESCENCE AND THERMOLUMINESCENCE

Chemiluminescence is the emission of light resulting from certain chemical reactions. For example, the oxidation of ether solutions of magnesium *p*-bromophenyl bromide gives rise to marked chemiluminescence, the greenish blue glow that accompanies the exposure of the solution to air being visible in daylight. The oxidations of decaying wood containing certain forms of bacteria, of luciferin in fireflies, and of yellow phosphorus are further examples.

The effect of radiation on crystals is interesting. Irradiation of alkali halide crystals with X rays give rise to characteristic colors. Sodium chloride becomes yellow and potassium chloride blue, the coloration being due to the absorption of light by electrons that have been released by X rays and are trapped in negative-ion "vacancies" in the crystal lattice. When an irradiated crystal is heated, the trapped electrons are released; in returning to a lower energy level they give off light, a phenomenon known as *thermoluminescence.* If the crystal is heated slowly, a series of light emissions occurs at definite temperatures. The nature of these curves, in which the intensity of light emitted is plotted against the temperature, depends on the extent of the radiation exposure, the impurities present, and other factors. Certain minerals, such as limestones and fluorites, exhibit thermoluminescence even without laboratory exposure to radiation, because they contain traces of uranium in parts per million that have been emitting radiation for geological ages.

17.9 PHOTOGRAPHY

If silver chloride or bromide is mixed with gelatin and exposed very briefly to light, no change is observed; but, when this emulsion is immersed in a solution of a mild reducing agent, such as, for example, pyrogallic acid, the parts that have been exposed to light are reduced to metallic silver much more rapidly than the unexposed parts. The photographic plate consists of a large number of minute grains of crystalline silver halide; some of these grains are completely reduced by the developer to black metallic silver, and others are unaffected. The unaffected grains are dissolved out with sodium thiosulfate ("hypo").

It is necessary for a photon to strike a sensitive spot in the crystal lattice so that the silver halide may be reduced to give a nucleus of silver, which then spreads, on further reduction, to include the whole grain. These sensitive spots seem to be identified with minute traces of silver sulfide in the crystal lattice that are particularly responsive to the action of light. By increasing the number of these sensitive spots and by other means, the speed of photographic films and plates has been greatly increased.

The silver halides respond only to the ultraviolet and to the shorter wavelengths of the visible spectrum but, if certain red dyes, such as dicyanin, are mixed in the emulsion, the plate becomes sensitive also to red. This phenomenon constitutes another example of photosensitization.

17.10 PHOTOSYNTHESIS

In photosynthesis CO_2 and H_2O are converted to the glucose moiety as starch. The overall process can be represented by

$$CO_2 + H_2O \xrightarrow[\text{photons}]{\text{eight}} (CH_2O) + O_2 \tag{17.32}$$

where (CH_2O) represents one-sixth of a glucose moiety. This process, which occurs entirely in the chloroplasts of green cells, involves a large number of steps catalyzed by enzymes.* The chloroplasts contain chlorophyll a, chlorophyll b, carotenes, electron carriers, and enzymes, and have internal membranes that keep reactants separated. In photosynthesis water is oxidized and a low molar mass protein called ferredoxin (Fd) is reduced. One oxygen molecule is produced per eight photons absorbed.

$$2H_2O + 4Fd^{3+} \xrightarrow[\text{photons}]{\text{eight}} 4H^+ + O_2 + 4Fd^{2+} \tag{17.33}$$

Concurrent with this transfer of electrons, 3 molecules of ADP are converted to ATP (see Section 6.5), and so reactions 17.33 should actually be written

$$4Fd^{3+} + 3ADP^{3-} + 3P^{2-} \xrightarrow[\text{photons}]{\text{eight}} 4Fd^{2+} + 3ATP^{4-} + O_2 + H_2O + H^+ \tag{17.34}$$

These are the amounts of reduced ferredoxin and ATP required to reduce 1 mol of CO_2 to glucose in the following "dark" reaction.

$$\begin{aligned} 3ATP^{4-} + 4Fd^{2+} + CO_2 + 2H_2O + H^+ \\ = (CH_2O) + 3ADP^{3-} + 3P^{2-} + 4Fd^{3+} \end{aligned} \tag{17.35}$$

This reaction is carried out in a sequence of steps.

Adding reactions 17.34 and 17.35 yields reaction 17.32. By carrying out the reaction in steps, the chloroplasts are able to use several photons in the visible range to accomplish a reaction that would require a photon of ultraviolet light of wavelength 230 nm if the reaction had to be carried out with a single photon.

---

* J. A. Bassham and M. Calvin, *The Path of Carbon in Photosynthesis*, Prentice-Hall, Englewood Cliffs, N.J., 1957.

Example 17.2 Given that the Gibbs energy change for reaction 17.32 is 477 kJ mol^{-1} of CO_2 reduced to starch and that the energy input between 400 and 700 nm at the earth's surface is equivalent to that at 575 nm, what is the theoretical maximum energy efficiency of photosynthesis by white light?

$$E = N_A h\nu = \frac{(6.02 \times 10^{23}\ \text{mol}^{-1})(6.62 \times 10^{-34}\ \text{J s})(3 \times 10^{8}\ \text{m s}^{-1})}{(575 \times 10^{-9}\ \text{m})(10^{3}\ \text{J kJ}^{-1})}$$

$$= 208\ \text{kJ einstein}^{-1}$$

$$\text{Efficiency} = \frac{477\ \text{kJ mol}^{-1}}{(8)(208\ \text{kJ einstein}^{-1})} = 0.29$$

The efficiency calculated on the basis of total solar radiation at the earth's surface that a plant could actually absorb leads to a maximum efficiency during the maximum growing season of 0.066. Efficiencies of 0.032 have been achieved with corn.

17.11 PHOTOELECTROCHEMICAL CELLS

A photoelectrochemical cell is an electrochemical cell in which the energy of light (rather than an electric power supply) is utilized to produce electrochemical change. In Section 5.11 the standard electromotive force for a hydrogen-oxygen fuel cell was shown to be 1.229 V at 25 °C. Since this calculation assumed reversibility, we also know that a *minimum* applied potential of 1.229 V is required to electrolyze water at 25 °C. Actually, an overvoltage is required and the electrolysis of water does not occur at this reversible potential (Section 16.10). In a photoelectrochemical cell of the type illustrated in Fig. 17.5, the energy of light absorbed by the anode may provide the energy for the electrolysis of water. Whether or not light energy can be converted wholly or partially into stored chemical energy in the form of $H_2(g) + \frac{1}{2}O_2(g)$ or electrical energy that can be dissipated in the load obviously depends on the properties of the anode. If the anode if the *n*-type semiconductor TiO_2 (see Section 19.26) and ultraviolet light is used, electrolysis of water occurs if a potential of 0.25 V is applied across the electrodes. In this cell the TiO_2 is a *photoassistance* agent in that it serves as a receptor for the light and channels excitation energy into the chemical reaction. The optical energy storage efficiency of such a cell may be expressed as

$$\text{Efficiency} = \frac{(\text{energy stored as } H_2) - (\text{energy from power supply})}{(\text{light energy})} \times 100 \qquad (17.36)$$

and efficiencies of several percent have been obtained for the TiO_2 electrode.* The higher efficiencies obtained with $SrTiO_3$ anodes and ultraviolet light rival the light-to-chemical-energy conversion efficiency of any other man-made photochemical conversion device.

* M. S. Wrighton, D. S. Ginley, P. T. Wolczanski, A. B. Ellis, D. L. Morse, and A. Linz, *Proc. Natl. Acad. Sci.*, **72**, 1518 (1975).

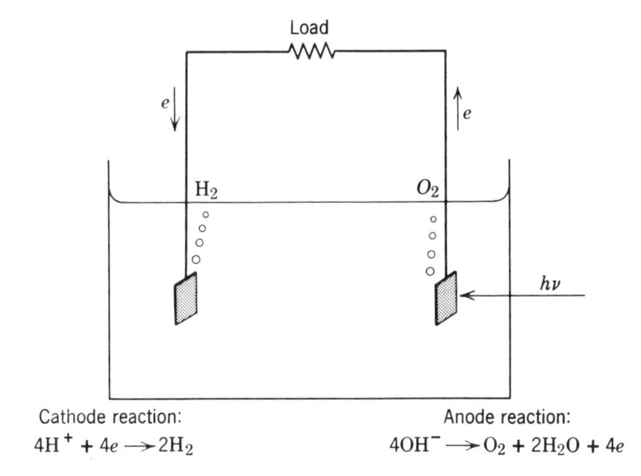

Fig. 17.5 Photoelectrochemical cell for the electrolysis of water.

The energy levels in the semiconductor CdTe make it an even more suitable photoassistance anode, since visible light can be used, but this electrode material decomposes unless a stabilizing electrolyte solution is used. When a stabilizing electrolyte solution is used, the energy of the light absorbed may be converted into electrical energy instead of being stored in $H_2(g) + \frac{1}{2}O_2(g)$.

17.12 LASERS*

Normally a beam of light loses intensity as it passes through an absorbing material. However, if there are molecules in an excited state, stimulated emission (Section 11.3) can occur, and the light beam can gain in intensity. A laser achieves this condition and produces an intense and coherent beam. By coherent we mean that the light waves are in phase. The name "laser" comes from *light amplification* by *stimulated emission* of *radiation.*

In order to get laser action it is necessary to get a population inversion in the absorber. In a population inversion there are more molecules in the *upper* state than in the *lower* state. This cannot be achieved by simply using a high-intensity light source of the proper frequency to excite molecules from a lower level to a higher level. The incident radiation stimulates emission of photons from the upper state at the same rate as they are absorbed by the lower state, and therefore no more than 50% of the molecules can be forced into the upper state by use of an intense source (see equation 11.16). However, if an upper state (e.g., S_1) can be converted to another excited state (e.g., T_1) by a radiationless process, the population of T_1 may become greater than 50%. This process is referred to as optical pumping. If after pumping radiation corresponding with the transition $T_1 \rightarrow S_0$ is passed into the material, stimulated emission occurs.

---

* A. L. Schawlow, *Lasers and Light,* W. H. Freeman Co., San Francisco, 1969.

In a CW (continuous wave) laser the optical pumping is continuous so that a continuous beam is obtained. Energy may be supplied by an electric discharge in a gas laser.

Population inversions may also be achieved in chemical reactions in which products are in excited states. In a chemical laser the energy producing laser radiation comes from the energy of the chemical reaction. For example, a gas stream containing fluorine atoms may be mixed with deuterium (or hydrogen) and carbon dioxide, producing a chain reaction that yields vibrationally excited deuterium (or hydrogen) fluoride. The vibrational-rotational energy from the excited DF pumps the upper CO_2 laser level by intermolecular energy transfer processes. The resulting continuous wave laser action of CO_2 at 10.6 μm can be provided entirely from chemical sources if the fluorine atoms are produced by a reaction such as

$$F_2 + NO = NOF + F \tag{17.37}$$

Lasers have many important uses because of their high peak power, coherence, and high degree of monochromaticity. Flux densities as high as 5×10^{14} W cm^{-2} have been attained. This corresponds with fluctuating electrical fields of the order of 3×10^{8} V cm^{-1} and fluctuating magnetic fields of 10^{6} G.

Lasers are important in chemistry because they can be used to initiate photochemical reactions, and the short pulses permit the study of very fast reactions. Lasers have revolutionized Raman spectroscopy because exposures are greatly reduced, and even weak lines may be detected in experiments of short duration. A laser may be used to ionize one isotopic species of a molecule, owing to the extreme monochromaticity associated with the excitation beam. The ionized molecules may be allowed to react with another substance so that the isotopes may be separated by a chemical method.

References

J. G. Calvert and J. N. Pitts, Jr., *Photochemistry*, Wiley, New York, 1966.
R. B. Cundall and A. Gilbert, *Photochemistry*, Appleton-Century-Crofts, New York, 1970.
R. W. F. Gross and J. F. Bott, *Handbook of Chemical Lasers*, Wiley, New York, 1976.
N. J. Turro, *Molecular Photochemistry*, W. A. Benjamin, Inc., New York, 1967.
C. H. J. Wells, *Introduction to Molecular Photochemistry*, Chapman and Hall, London, 1972.
A. Yariv, *Quantum Electronics*, Wiley, New York, 1975.

Problems

17.1 A certain photochemical reaction requires an activation energy of 126 kJ mol^{-1}. To what values does this correspond in the following units: (*a*) kcal mol^{-1}, (*b*) frequency of light, (*c*) wave number, (*d*) wavelength in nanometers, and (*e*) electron volts?

Ans. (*a*) 30.1 kcal mol^{-1}, (*b*) 3.16×10^{14} s^{-1}, (*c*) 10,500 cm^{-1}, (*d*) 952 nm, (*e*) 1.31 eV.

17.2 A sample of gaseous acetone is irradiated with monochromatic light having a wavelength of 313 nm. Light of this wavelength decomposes the acetone according to the equation

$$(CH_3)_2CO \rightarrow C_2H_6 + CO$$

The reaction cell used has a volume of 59 cm^3. The acetone vapor absorbs 91.5% of the incident energy. During the experiment the following data are obtained.

$$\begin{aligned} \text{Temperature of reaction} &= 56.7\ ^\circ\text{C} \\ \text{Initial pressure} &= 102.16\ \text{kPa} \\ \text{Final pressure} &= 104.42\ \text{kPa} \\ \text{Time of radiation} &= 7\ \text{hr} \\ \text{Incident energy} &= 48.1 \times 10^{-4}\ \text{J s}^{-1} \end{aligned}$$

What is the quantum yield? *Ans.* 0.17.

17.3 A 100-cm^3 vessel containing hydrogen and chlorine was irradiated with light of 400 nm. Measurements with a thermopile showed that 11×10^{-7} J of light energy was absorbed by the chlorine per second. During an irradiation of 1 min the partial pressure of chlorine, as determined by the absorption of light and the application of Beer's law, decreased from 27.3 to 20.8 kPa (corrected to 0 °C). What is the quantum yield?

Ans. 2.6×10^6 mol HCl einstein^{-1}.

17.4 Discuss the economic possibilities of using photochemical reactions to produce valuable products with electricity at 5 cents per milowatt-hour. Assume that 5% of the electric energy consumed by a quartz-mercury-vapor lamp goes into light, and 30% of this is photochemically effective. (*a*) How much will it cost to produce 1 lb (453.6 g) of an organic compound having a molar mass of 100 g mol^{-1}, if the average effective wavelength is assumed to be 400 nm and the reaction has a quantum yield of 0.8 molecule per photon? (*b*) How much will it cost if the reaction involves a chain reaction with a quantum yield of 100? *Ans.* (*a*) \$1.55, (*b*) 1.2 cents.

17.5 For 900 s, light of 436 nm was passed into a carbon tetrachloride solution containing bromine and cinnamic acid. The average power absorbed was 19.2×10^{-4} J s^{-1}. Some of the bromine reacted to give cinnamic acid dibromide, and in this experiment the total bromine content decreased by 3.83×10^{19} molecules. (*a*) What was the quantum yield? (*b*) State whether or not a chain reaction was involved. (*c*) If a chain mechanism was involved, suggest suitable reactions that might explain the observed quantum yield.

Ans. (*a*) 10.1. (*b*) Yes. (*c*)
$$\begin{aligned} \phi CH{=}CHCO_2H + h\nu &= \phi CH{=}CHCO_2H^* \\ \phi CH{=}CHCO_2H^* + Br_2 &= \phi CHBrCHCO_2H^* + Br^* \\ \phi CHBrCHCO_2H^* + Br_2 &= \phi CHBrCHBrCO_2H + Br^* \\ \phi CH{=}CHCO_2H + Br^* &= \phi CHBrCHCO_2H^*. \end{aligned}$$

17.6 Go through the steps of deriving equation 17.30 from the mechanism of Fig. 17.1 plus reaction 17.27.

17.7 A solution of a dye is irradiated with 400 nm light to produce a steady concentration of triplet state molecules. If the triplet state yield is 0.9, and the triplet state lifetime is 20×10^{-6} s, what light intensity, expressed in watts, is required to maintain a steady triplet concentration of 5×10^{-6} mol L^{-1} in a liter of solution. Assume that all of the light is absorbed. *Ans.* 83 kW.

17.8 The quantum yield is 2 for the photolysis of gaseous HI to $H_2 + I_2$ by light of 253.7 nm wavelength. Calculate the number of moles of HI that will be decomposed if 300 J of light of this wavelength is absorbed. *Ans.* 1.27×10^{-3} mol.

17.9 The following calculations are made on a uranyl oxalate actinometer, on the assumption that the energy of all wavelengths between 254 and 435 nm is completely absorbed. The actinometer contains 20 cm^3 of 0.05 mol L^{-1} oxalic acid, which also is 0.01 mol L^{-1} with respect to uranyl sulfate. After 2 hr of exposure to ultraviolet light, the solution required 34 cm^3 of potassium permanganate, $KMnO_4$, solution to titrate the undecomposed oxalic acid. The same volume, 20 cm^3, of unilluminated solution required 40 cm^3 of the $KMnO_4$ solution. If the average energy of the quanta in this range may be taken as corresponding to a wavelength of 350 nm, how many joules were absorbed per second in this experiment? ($\Phi = 0.57$.) *Ans.* 1.25×10^{-2} J.

17.10 A solution absorbs 300 nm radiation at the rate of 1 W. What does this correspond to in einsteins per second? *Ans.* 2.51×10^{-6} einsteins s^{-1}.

17.11 Ketone A dissolved in *t*-butyl alcohol is excited to a triplet state A* by light of 320–380 nm. The triplet may then return to the ground state A or rearrange to give the isomer B, thus

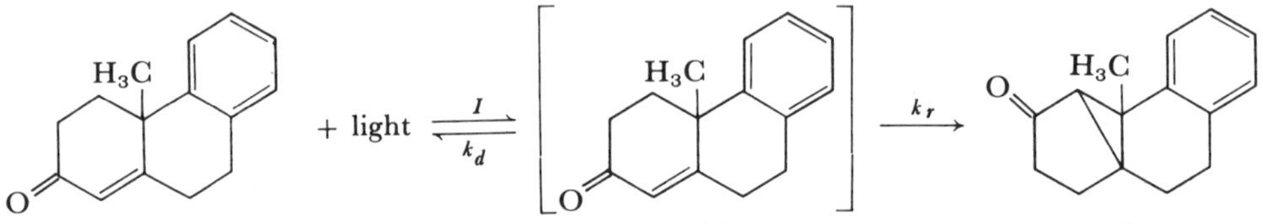

$$\text{A} + \text{light} \underset{k_d}{\overset{I}{\rightleftharpoons}} [\text{A*}] \xrightarrow{k_r} \text{B}$$

The unimolecular rate constant k_d determines the rate of deactivation of excited molecules and the unimolecular rate constant k_r determines the rate of rearrangement. The symbol I represents the number of einsteins per second and it is assumed that each photon absorbed produces one excited molecule of A*. The quantum yield Φ for the formation of B is given by

$$\Phi = \frac{d(\text{B})/dt}{I} = \frac{k_r(\text{A*})}{I}$$

Zimmerman, McCullough, Staley, and Padwa measured the quenching effect of dissolved napthalene and report the following data.

Moles naphthalene L^{-1}	0.0099	0.0330	0.0620	0.0680	0.0775	0.0960
Φ	0.0049	0.0023	0.0020	0.0017	0.0014	0.0012

The quenching rate by naphthalene is controlled by the bimolecular quenching constant of A*, which is equal to the diffusion-controlled constant $k_q = 1.2 \times 10^9$ L mol^{-1} s^{-1}. Assuming a steady state,

$$\frac{d(\text{A*})}{dt} = I - k_r(\text{A*}) - k_d(\text{A*}) - k_q(\text{A*})(\text{N}) = 0$$

where (N) is the concentration of naphthalene.

Calculate k_r and k_d by plotting $1/\Phi$ versus (N) and determining the slope and intercept of the line. *Ans.* $k_r = 1.6 \times 10^5$ s^{-1}, $k_d = 1.9 = 10^7$ s^{-1}.

17.12 The photochemical oxidation of phosgene, sensitized by chlorine, has been studied by G. K. Rollefson and C. W. Montgomery [*J. Am. Chem. Soc.*, **55**, 142, 4025 (1932)]. The overall reaction is

$$2COCl_2 + O_2 = 2CO_2 + 2Cl_2$$

and the rate expression that gives the effect of the several variables is

$$\frac{dc_{CO_2}}{dt} = \frac{kI_0(COCl_2)}{1 + k'(Cl_2)/(O_2)}$$

where I_0 is the intensity of the light. The quantum yield is about two molecules per quantum. Devise a series of chemical equations involving the existence of the free radicals ClO and COCl that will give a mechanism consistent with the rate expression.

Ans.

$$COCl_2 + h\nu \xrightarrow{k_1} COCl + Cl \qquad COCl + Cl_2 \xrightarrow{k_4} COCl_2 + Cl$$

$$COCl + O_2 \xrightarrow{k_2} CO_2 + ClO \qquad Cl + Cl \xrightarrow{k_5} Cl_2$$

$$COCl_2 + ClO \xrightarrow{k_3} CO_2 + Cl_2 + Cl \qquad k = 2,\ k' = k_4/k_2.$$

17.13 Given that solar radiation at noon at a certain place on the earth's surface is 4.2 J cm^{-2} min^{-1}, what is the maximum power output in W m^{-2}? *Ans.* 700 W m^{-2}.

17.14 Calculate the longest wavelength of light that can theoretically decompose water at 25 °C in a one-photon electrochemical process to give $H_2(g)$ and $\frac{1}{2}O_2(g)$ in their standard states. Given: $\Delta G° = 237.2$ kJ mol^{-1} for $H_2O(l) = H_2(g) + \frac{1}{2}O_2(g)$.

Ans. 504 nm.

17.15 If a good agricultural crop yields about 2 tons $acre^{-1}$ of dry organic material per year with a heat of combustion of about 16.7 kJ g^{-1}, what fraction of a year's solar energy is stored in an agricultural crop if the solar energy is about 4184 J min^{-1} ft^{-2} and the sun shines about 500 min day^{-1} on the average? 1 acre = 43,560 ft^2 and 1 ton = 907,000 g. *Ans.* 10^{-3}.

17.16 What intensities of light in J s^{-1} are required to produce a microeinstein s^{-1} at (*a*) 700 nm, and (*b*) 300 nm. (*c*) and (*d*) What are these powers in watts?

17.17 When CH_3I molecules in the vapor state absorb 253.7 nm light, they dissociate into methyl radicals and iodine atoms. Assuming that the energy required to rupture the C—I bond is 209 kJ mol^{-1}, what is the kinetic energy of each of the fragments if they are produced in their ground states?

17.18 Assuming a hypothetical photochemical reaction in which one-tenth of the solar radiation is absorbed and utilized with a quantum yield of unity, how many tons of product can be produced per acre per day, if the molar mass of the product is 100 g mol^{-1}, the average effective light is 510 nm, and the solar radiation is 4.2 J min^{-1} cm^{-2} for 500 min during the day?

17.19 A cold high-voltage mercury lamp is to be used for a certain photochemical reaction that responds to ultraviolet light of 253.7 nm. The chemical analysis of the product is sensitive to only 10^{-4} mol. The lamp consumed 150 W and converts 5% of the electric energy into radiation of which 80% is at 253.7 nm. The amount of the light that gets into the monochromator and passes out the exit slit is only 5% of the total radiation of the lamp. Fifty percent of this 253.7 nm radiation from the monochromator is absorbed in the reacting system. The quantum yield is 0.4 molecule of product per quantum of light absorbed. How long an exposure must be given in this experiment if it is desired to measure the photochemical change with an accuracy of 1%?

17.20 Derive equation 17.8 for the steady-state concentration of T_1 from the mechanism of Fig. 17.1.

17.21 Given that the intensity of solar radiation is 4.2 J cm^{-2} min^{-1}, how much carbon has to be burned to obtain the same amount of heat as the solar radiation on 1 m^2 in an 8-hour day?

17.22 Sunlight between 290 and 313 nm can produce sunburn (erythema) in 30 min. The intensity of radiation between these wavelengths in summer and at 45° latitude is about 50 μW cm^{-2}. Assuming that 1 photon produces chemical change in 1 molecule, how many molecules in a square centimeter of human skin must be photochemically affected to produce evidence of sunburn?

17.23 The quantum yield for the photolysis of acetone

$$(CH_3)_2CO = C_2H_6 + CO$$

at 300 nm is 0.2. How many moles per second of CO are formed if the intensity of the 300-nm radiation absorbed is 10^{-2} J s^{-1}?

17.24 A uranyl oxalate actinometer is exposed to light of wavelength 390 nm for 1980 s, and it is found that 24.6 cm^3 of 0.004 30 mol L^{-1} potassium permanganate is required to titrate an aliquot of the uranyl oxalate solution after illumination, in comparison with 41.8 cm^3 before illumination. Using the known quantum yield of 0.57, calculate the number of joules absorbed per second. The chemical reaction for the titration is

$$2MnO_4^- + 5H_2C_2O_4 + 6H^+ = 2Mn^{2+} + 10CO_2 + 8H_2O$$

17.25 Nitrogen dioxide is decomposed photochemically by light of 366 nm with a quantum yield of 2.0 molecules per photon, according to the reaction

$$2NO_2 \rightarrow 2NO + O_2$$

The thermal reaction runs in the reverse direction. When an enclosed sample of nitrogen dioxide is illuminated for a long period of time, the quantum yield decreases and approaches zero. Suggest a mechanism to explain these facts and write the chemical equations.

17.26 The following reactions describe the photochemical decomposition of hydrogen bromide with light of 253 nm at 25 °C. The primary process dissociates the molecule into atoms of hydrogen and bromine, which can then undergo further reactions. The quantum yield for the primary process is designated by ϕ; the quantum yield for the overall reaction is designated by Φ. The intensity of light absorbed is designated by I.

$$\begin{array}{lll}
(1)\ HBr + h\nu \rightarrow H + Br & \text{Rate} = \phi I & \\
(2)\ H + HBr \rightarrow H_2 + Br & \text{Rate} = k_2 c_H c_{HBr} & \\
(3)\ Br + Br + M \rightarrow Br_2 + M & \text{Rate} = k_3 (c_{Br})^2 c_M & \\
(4)\ H + Br_2 \rightarrow HBr + Br & \text{Rate} = k_4 c_H c_{Br_2} & \\
(5)\ H + H + M \rightarrow H_2 + M & \text{Rate} = k_5 (c_H)^2 c_M & \text{(negligible)}
\end{array}$$

Derive the expression for the quantum yield for the overall reaction and show that early in the reaction $\Phi = 2\phi$.

17.27 Biacetyl triplets have a quantum yield of 0.25 for fluorescence and a measured lifetime of the triplet state of 10^{-3} s. If its phosphorescence is quenched by a compound Q with a diffusion-controlled rate (10^{10} L mol^{-1} s^{-1}), what concentration of Q is required to cut the phosphorescence yield in half?

17.28 Calculate the maximum possible theoretical yield in tons of carbohydrate material $(H_2CO)_n$ that can be produced on an acre of land by green plants or trees during a 100-day growing season. Similar calculations apply to algae growing in a lake or ocean. Assume that the sun's radiation approximately averages 4.2 J cm^{-2} min^{-1} for an 8-hr day and that one-half the area is covered by green leaves. Assume that one-third of the radiation lies between 400 and 650 nm, which is the range of the light absorbed by chlorophyll, and that the average wavelength is 500 nm. Assume that the leaves are thick enough to absorb practically all the light that strikes them. Assume that the quantum yield is 0.12 molecule per photon; that is, 8 photons with chlorophyll can produce 1 H_2CO unit from 1 molecule of CO_2 and 1 molecule of H_2O. Criticize these several assumptions.

17.29 A photochemical reaction of biological importance is the production of vitamin D, which prevents rickets and brings about the normal deposition of calcium in growing bones. Steenbock found that rickets could be prevented by subjecting the food as well as the patient to ultraviolet light below 310 nm. When ergosterol is irradiated with ultraviolet light below 310 nm, vitamin D is produced. When irradiated ergosterol was included in a diet otherwise devoid of vitamin D, it was found that absorbed radiant energy of about 7.5×10^{-5} J was necessary to prevent rickets in a rat when fed over a period of 2 weeks. The light used has a wavelength of 265 nm. (*a*) How many quanta are necessary to give 7.5×10^{-5} J? (*b*) If vitamin D has a molar mass of the same order of magnitude as ergosterol (382), how many grams of vitamin D per day are necessary to prevent rickets in a rat? It is assumed that the quantum yield is unity.

17.30 The electromotive force of the hydrogen-oxygen fuel cell is 1.229 V at 25 °C. To what wavelength of light does this correspond, remembering that this is a two-electron change?

CHAPTER 18

IRREVERSIBLE PROCESSES IN SOLUTION

In this chapter we are concerned with the *rates* with which certain processes occur in solution. The examples to be considered are viscosity, electrical conductivity, and diffusion. Determinations of coefficients of viscosity, diffusion, and sedimentation are useful for calculating the molar mass and shape of macromolecules; these applications are discussed in Chapter 20. The theory for the rates of irreversible processes in solution is handicapped by the lack of an adequate theory for the liquid state, but the theory of irreversible processes provides basic relations for this field.

18.1 VISCOSITY

The coefficient of viscosity was defined in the discussion of the kinetic theory of gases in Section 14.10. This definition applies to laminar flow, that is, flow in which one layer (lamina) slides smoothly relative to another. When the flow velocity is great enough, turbulence develops. The coefficient of viscosity of a liquid may be measured by a number of methods that are illustrated in Fig. 18.1; these methods include the determination of the rate of flow through a capillary, the rate of settling of a sphere in a liquid, and the force required to turn one of two concentric cylinders at a certain angular velocity.

The SI unit of viscosity is the pascal second (Pa s), which is the viscosity of a fluid in which the velocity under a shear stress of 1 Pa has a gradient of 1 m s^{-1} per meter perpendicular to the plane of shear.*

An estimate as to whether the flow will be turbulent may be obtained by calculating a dimensionless quantity referred to as the Reynolds number. The Reynolds number for flow in a tube is defined by $d\bar{v}\rho/\eta$, where d is the diameter of the tube, $\bar{v}$ is the average velocity of the fluid along the tube, ρ is the density of the fluid, and η is its coefficient of viscosity. At flow velocities corresponding with values of the Reynolds number of greater than 2000, turbulence is encountered.

For certain colloidal suspensions and solutions of macromolecules the coefficient of viscosity depends on the rate of shear, and this is referred to as non-Newtonian

* The cgs unit of viscosity is the poise, which is the viscosity of a fluid in which the velocity under a shear stress of 1 dyne cm^{-2} has a gradient of 1 cm s^{-1} per centimeter perpendicular to the plane of shear. A poise is equal to 0.1 Pa s.

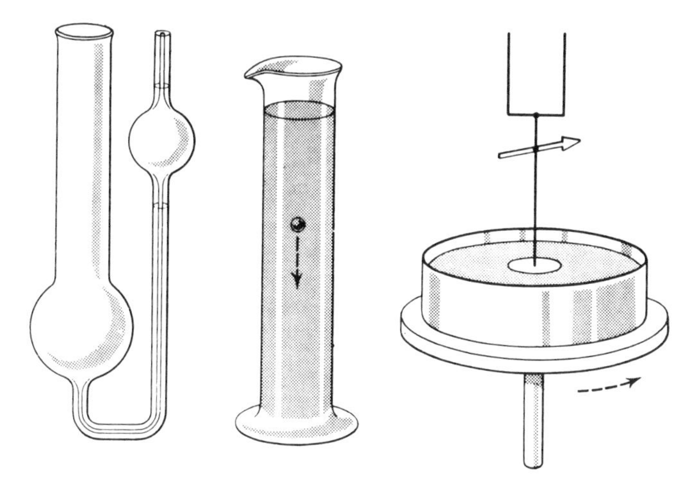

Fig. 18.1 Various types of viscometers.

behavior. If the shear stress orients or distorts the suspended particles, the coefficient of viscosity may decrease as the shear rate is increased.

When a force F is applied to a particle in a solution—as by applying an electric field, if the particle is charged, or a centrifugal field—the particle will be accelerated. As the velocity of the particle increases, it experiences an increasing frictional force. For low velocities the frictional force is given by vf, where v is the velocity and f is the frictional coefficient of the particle. When the velocity is sufficiently high for the frictional force to be equal to the applied force

$$vf = F \tag{18.1}$$

the particle will move with constant velocity.

The frictional coefficient f is of interest because it provides some information about the size and shape of the particle. For spherical particles Stokes showed that for nonturbulent flow,*

$$f = 6\pi\eta r \tag{18.2}$$

where η is the coefficient of viscosity and r is the radius of the spherical particle. The frictional coefficients of prolate and oblate ellipsoids and long rods may be expressed in terms of the radius of a sphere of equal volume and a factor depending on the ratio of the major axis to the minor axis.

The coefficient of viscosity η of a liquid may be determined by measuring the rate of settling of a sphere of known density, as shown in Fig. 18.1. The force causing the sphere to settle in the fluid is equal to its effective mass times the acceleration of gravity; the effective mass is the mass of the sphere minus the mass of the fluid it displaces. If the sphere has a density of ρ, and the density of the

* For a derivation see R. B. Bird, W. E. Stewart, and E. N. Lightfoot, *Transport Phenomena*, Wiley, New York, 1960.

medium is ρ_0, the force causing motion is $\frac{4}{3}\pi r^3(\rho - \rho_0)g$, where g is the acceleration of gravity. When the rate of settling of the sphere in the liquid is constant, the retarding force is equal to the force due to gravity, and so

$$\tfrac{4}{3}\pi r^3(\rho - \rho_0)g = 6\pi\eta r\left(\frac{dx}{dt}\right) \tag{18.3}$$

$$\frac{dx}{dt} = \frac{2r^2(\rho - \rho_0)g}{9\eta} \tag{18.4}$$

Thus, by measuring the velocity dx/dt of settling of a sphere of known r and ρ in a liquid of known density ρ_0, the coefficient of viscosity η may be obtained. This method is especially valuable for solutions of high viscosity, such as concentrated solutions of high polymers. Conversely, a the determination of the rate of settling of colloidal particles of known density in a liquid of known viscosity provides a means for determining the effective particle radius.

The coefficient of viscosity η may also be determined by passing a liquid through a capillary tube and making use of the Poiseuille equation

$$\eta = \frac{P\pi r^4 t}{8Vl} \tag{18.5}$$

where t is the time required for volume V of liquid to flow through a capillary tube of length l and radius r under an applied pressure P.

The viscosities of most liquids decrease with increasing temperature. According to the "hole theory," there are vacancies in a liquid, and molecules are continually moving into these vacancies so that the vacancies move around. This process permits flow, but requires energy because there is an activation energy that a molecule has to have to move into a vacancy. The activation energy is more readily available at higher temperatures, and so the liquid can flow more easily at higher temperatures. The viscosities of several liquids at different temperatures are shown in Table 18.1. The variation of the coefficient of viscosity with temperature may be represented quite well by

$$\frac{1}{\eta} = Ae^{-E_a/RT} \tag{18.6}$$

where E_a is the activation energy for the fluidity, $1/\eta$.

The viscosity of a liquid increases as the pressure is increased because the number of holes is reduced, and it is therefore more difficult for molecules to move around each other.

Table 18.1 Viscosities of Liquids in Pa s (kg m^{-1} s^{-1})

t/°C	0	25	50	75
Water	0.001793	0.000895	0.000549	0.000380
Ethanol	0.00179	0.00109	0.000698	—
Benzene	0.00090	0.00061	0.00044	—
Glycerol	—	0.945	—	—

In contrast with liquids, the viscosity of a gas increases as the temperature increases. The viscosity of a perfect gas is independent of pressure.

18.2 ELECTROLYTIC CONDUCTIVITY

The electric resistance R of a uniform conductor is directly proportional to its length l and inversely proportional to its cross-sectional area A.

$$R = \frac{rl}{A} = \frac{l}{\kappa A} \tag{18.7}$$

where the proportionality constant r is called the *resistivity* and the proportionality constant $\kappa = 1/r$ is called the *conductivity*. In the SI system the conductivity κ has the units $\Omega^{-1}\ m^{-1}$, where the ohm is represented by Ω. Electric conductivities range from $10^8\ \Omega^{-1}\ m^{-1}$ for a metallic conductor at room temperature to $10^{-15}\ \Omega^{-1}\ m^{-1}$ for an insulator like SiO_2.

The electric conductivity κ of an electrolyte solution is made up of contributions from each of the types of ions present.

Resistances may be determined by use of a Wheatstone bridge in which an unknown resistance is balanced against a known resistance. Alternating current is used in measuring the resistances of cells containing electrolytic solutions. When alternating current is used and the electrodes are platinized, the electrolysis that occurs when the current passes in one direction is reversed when the current passes in the other direction. The coating of platinum black, produced by electrolytic deposition, adsorbs gases and catalyzes their reaction. In this way the formation of a nonconducting gas film is prevented.

The conductivity κ of an electrolyte is inversely proportional to the measured resistance of the cell.

$$\kappa = \frac{K_{\text{cell}}}{R} \tag{18.8}$$

The cell constant K_{cell} is determined by measuring the resistance of the cell when it contains a solution of known conductivity. The conductivities of certain standard solutions have been carefully measured.

Example 18.1 When a certain conductance cell was filled with 0.0200 mol L^{-1} potassium chloride, which has a conductivity of $0.2768\ \Omega^{-1}\ m^{-1}$, it has a resistance of $82.40\ \Omega$ at 25 °C, as measured with a Wheatstone bridge; when filled with 0.0025 mol L^{-1} potassium sulfate, it had a resistance of $326.0\ \Omega$.

(*a*) What is the cell constant?

$$K_{\text{cell}} = (0.2768\ \Omega^{-1}\ m^{-1})(82.40\ \Omega) = 22.81\ m^{-1}$$

(*b*) What is the conductivity κ of the K_2SO_4 solution?

$$\kappa = \frac{K_{\text{cell}}}{R} = \frac{22.81\ m^{-1}}{326.0\ \Omega} = 6.997 \times 10^{-2}\ \Omega^{-1}\ m^{-1}$$

18.3 ELECTRIC MOBILITY

The electric mobility u of an ion is its drift velocity in the direction of the electric field, divided by the electric field strength E.

$$u = \frac{dx/dt}{E} \tag{18.9}$$

The drift velocity of an ion is the average velocity in the direction of the field. Because of Brownian motion, an ion undergoes random displacements so that it does not move in a straight line over macroscopic distances.

The electric field E is the resultant electric force per unit positive charge; it therefore has the units N C^{-1}. The electric field is a vector quantity, but we will not use vector notation here because we will be considering only situations where the electric field is in the x direction. The electric potential, which is measured in volts, is the energy per unit charge. Thus 1 V = 1 JC^{-1} = 1 N m C^{-1}. This shows that 1 V m^{-1} = 1 N C^{-1}; therefore the strength of an electric field may also be expressed in V m^{-1}. In the SI system the velocity is expressed in m s^{-1}, and so the electric mobility has the units m s^{-1}/V m^{-1} = m^2 V^{-1} s^{-1}.

The most direct method for determining electric mobilities is to measure the velocity of a boundary between two electrolytic solutions in a tube of uniform cross section through which a current is flowing. For example, if a 0.1 mol L^{-1} solution of potassium chloride is layered over a solution of cadmium chloride in a tube, as illustrated in Fig. 18.2*a*, and a current i is caused to flow through the tube, the potassium ions will move upward toward the negative electrode away from the position of the initial boundary. They will be followed by the slower-moving cadmium ions, so that there will be no gap in the column of electrolyte. Since the concentration of cadmium chloride above the initial boundary position (c'_{CdCl_2}) will, in general, be different from that initially placed below the potassium chloride solution, there will be a change in cadmium chloride concentration at the initial boundary position, which is represented by / / / / / / / in Fig. 18.2*b*.

In order to calculate the electric mobility of potassium ions from the velocity with which they move in the KCl solution, it is necessary to know the electric field strength E in the KCl solution. The electric field strength E is the negative gradient of the electric potential ϕ. When the electric potential varies only in the x direction,

$$E = -\frac{d\phi}{dx} \tag{18.10}$$

For a uniform conductor the difference in the potential per unit distance may be calculated using Ohm's law. For a conductor of unit cross section the difference in potential between two points is equal to the current density I/A, where I is the current and A the area, multiplied by the resistivity $1/\kappa$.

$$E = \frac{I}{A\kappa} \tag{18.11}$$

In the experiment illustrated in Fig. 18.2 the boundary between the KCl and $CdCl_2$ solutions moves with the velocity of the potassium ions in the potassium chloride solution, and so the electric mobility u of the potassium ions may be

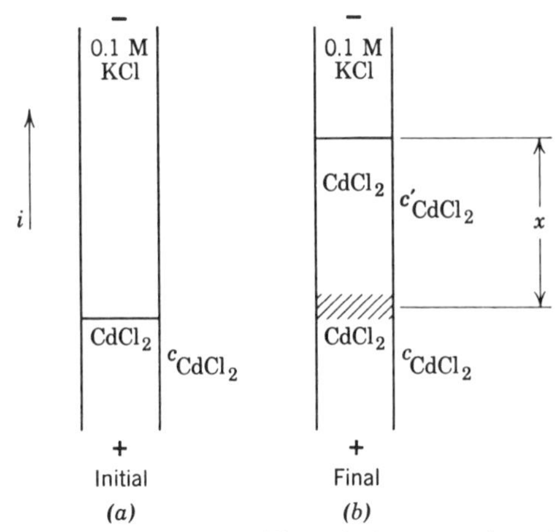

Fig. 18.2 Determining the electric mobility of the potassium ion with the moving-boundary method.

calculated from the distance x moved by the boundary in the time t in an electric field of strength E.

Example 18.2 In the experiment with 0.1 mol L^{-1} potassium chloride illustrated in Fig. 18.2, the boundary moved 4.64 cm during 67 min when a current of 5.21×10^{-3} A was used. The cross-sectional area of the tube was 0.230 cm^2, and $\kappa = 1.29$ $\Omega^{-1}\ m^{-1}$ at 25 °C. Calculate the electric field strength and the electric mobility of the potassium ion.

$$E = \frac{(5.21 \times 10^{-3}\ \text{A})}{(0.230 \times 10^{-4}\ \text{m}^2)(1.29\ \Omega^{-1}\ \text{m}^{-1})} = 176\ \text{V m}^{-1}$$

$$u = \frac{(0.0464\ \text{m})}{(67 \times 60\ \text{s})(176\ \text{V m}^{-1})} = 6.56 \times 10^{-8}\ \text{m}^2\ \text{V}^{-1}\ \text{s}^{-1}$$

To obtain a sharp moving boundary in an experiment such as that illustrated in Fig. 18.2, it is necessary that the leading ion (in this case, potassium) have a higher mobility than the indicator ion (in this case, cadmium). The cadmium chloride solution below the moving boundary (c'_{CdCl_2}) has a lower conductivity than the potassium chloride solution above the boundary. By reference to equation 18.11 it can be seen that the electric field strength is therefore greater in the cadmium chloride solution below the moving boundary than in the potassium chloride solution. Therefore, if potassium ions diffuse down into the cadmium chloride solution, they will be in a stronger electric field and will catch up with the boundary. On the other hand, if cadmium ions diffuse ahead of the boundary, they will have a lower velocity than the potassium ions because of their lower mobility and will soon be overtaken by the boundary. This so-called "adjusting effect" keeps the boundary sharp.

The moving boundary method may be used for the study of mixtures of ions, which may be macromolecular ions such as *proteins*. The study of colloids by this method is referred to as *electrophoresis*. The electric mobilities of a number of small ions at infinite dilution in water at 25 °C are given in Table 18.2.

Table 18.2 Electric Mobilities at 25 °C in Water at Infinite Dilution

	$u/10^{-8}\ m^2\ V^{-1}\ s^{-1}$		$u/10^{-8}\ m^2\ V^{-1}\ s^{-1}$
H^+	36.25	OH^-	20.64
Li^+	4.01	F^-	5.74
Na^+	5.192	Cl^-	7.913
K^+	7.617	NO_3^-	7.406
NH_4^+	7.62	ClO_3^-	6.70
$N(CH_3)_4^+$	4.66	$CH_3CO_2^-$	4.24
Mg^{2+}	5.50	$C_6H_5CO_2^-$	3.36
Ca^{2+}	6.17	SO_4^{2-}	8.29
Pb^{2+}	7.20	CO_3^{2-}	7.18

It is interesting that in the alkali metal and halogen families of the periodic table the lighter ions have lower mobilities. This is a result of the stronger electric field in the neighborhood of the ions with smaller radii, which causes them to be more highly hydrated, by the ion-dipole interaction, than the larger ions. A hydrated ion drags along a shell of water when it moves in a solution between two electrodes, and so it moves more slowly than an unhydrated ion would.

Mobilities increase with the temperature, and the temperature coefficients are very nearly the same for all ions in a given solvent and are approximately equal to the temperature coefficient of the viscosity; for water this is 2% per degree in the neighborhood of 25 °C.

18.4 ELECTRIC MOBILITIES OF HYDROGEN AND HYDROXYL IONS

There is a variety of kinds of evidence* that hydrogen ions in aqueous solutions are hydrated to form $H_9O_4^+$, that is, a trihydrate of hydronium ion H_3O^+ with the following structure.

```
          H   H
           \ /
            O
            :
            H
            |
            O+
           / \
          H   H
        .:     :.
   H—O           O—H
     |           |
     H           H
```

* H. L. Clever, *J. Chem. Ed.*, **40**, 637 (1963).

According to this structure the electric mobility of the hydrogen ion might be expected to be rather low but, as a matter of fact, its mobility is about 10 times that of other ions, except the hydroxyl ion. The high electric mobility of the hydrogen ion is due to the fact that a proton may *in effect* be transferred along a series of hydrogen-bonded (Section 10.8) water molecules by rearrangement of the hydrogen bonds. In the following figure (*a*) shows the initial bonding in a group of oriented water molecules and (*b*) shows the final bonding.

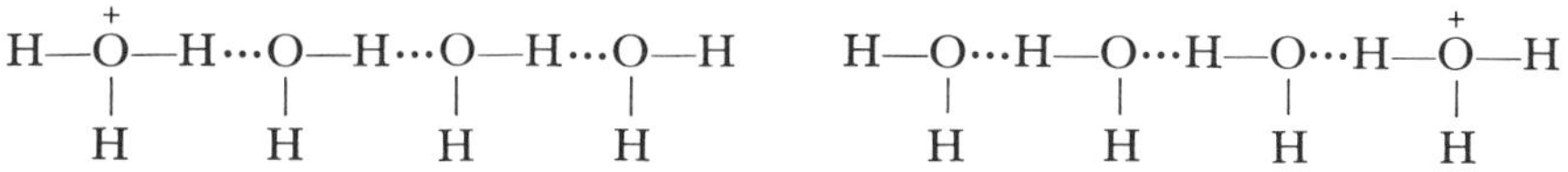

```
   +                                                     +
H—O—H···O—H···O—H···O—H        H—O···H—O···H—O···H—O—H
  |       |      |      |            |       |      |      |
  H       H      H      H            H       H      H      H
          (a)                                (b)
```

In order for another hydrogen ion to be transferred to the right through this group of water molecules, molecular rotations must occur to produce again a favorable orientation for charge transfer.

This model for hydrogen ion mobility helps us to understand the remarkable fact that hydrogen ions move about 50 times more rapidly through ice than through liquid water. In ice each oxygen atom is surrounded by four oxygen atoms at a distance of 0.276 nm in a tetrahedral arrangement. Each hydrogen is near the line through the centers of the oxygen atoms and is about 0.1 nm from one oxygen and 0.176 nm from the other. Hydrogen ions may be conducted rapidly through this structure by the above mechanism because the water molecules are oriented correctly.

The transfer of hydroxyl ion in the opposite direction is illustrated by

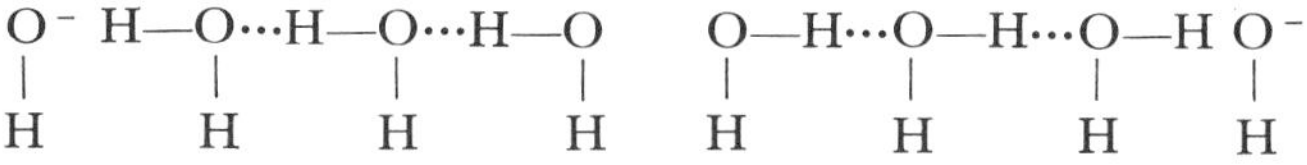

```
O⁻ H—O···H—O···H—O      O—H···O—H···O—H O⁻
|       |      |      |      |       |      |    |
H       H      H      H      H       H      H    H
        (a)                        (b)
```

18.5 RELATION BETWEEN CONDUCTIVITY AND ELECTRIC MOBILITY

The conductivity κ of an electrolyte is the sum of the contributions of all the ionic species in the electrolyte. The electric current contributed by an ion depends on its charge number z_i as well as on its electric mobility. The concentration of ion i expressed in faradays of electric charge is $|z_i|c_i$. If the ions all move with a velocity of u_i the transport of electric charge through a plane perpendicular to the direction of motion is $F|z_i|c_i u_i$. Thus, for the electrolyte as a whole,

$$\kappa = F \sum |z_i| c_i u_i = Fc \sum |z_i| \nu_i u_i \tag{18.12}$$

since equation 18.11 shows that κ is the current density divided by the electric field strength. The number of ions of type i in a molecule of electrolyte is represented by ν_i.

Example 18.3 Calculate the conductivity of 0.100 mol L^{-1} sodium chloride at 25 °C from the mobilities of sodium and chloride ions at this concentration, which are 4.26×10^{-8} and 6.80×10^{-8} $m^2\,V^{-1}\,s^{-1}$, respectively.

$$\begin{aligned}\kappa &= (96{,}485\ \mathrm{C\,mol^{-1}})(0.100 \times 10^{3}\ \mathrm{mol\,m^{-3}})[(4.26 + 6.80) \times 10^{-8}\ \mathrm{m^2\,V^{-1}\,s^{-1}}] \\ &= 1.067\ \Omega^{-1}\,\mathrm{m^{-1}}\end{aligned}$$

18.6 DETERMINATION OF THE ION PRODUCT OF WATER

For pure water or aqueous solutions of low ionic strength, concentrations may be substituted for activities in writing the ion product

$$K_w = (\mathrm{H^+})(\mathrm{OH^-}) \tag{18.13}$$

The concentrations of these ions in pure water may be calculated from its electrical conductivity. At 25 °C the conductivity κ of the purest water* is $5.5 \times 10^{-6}\ \Omega^{-1}\ \mathrm{m^{-1}}$. The concentrations of hydrogen and hydroxyl ions are, of course, equal and may be calculated using equation 18.12 and the values of the limiting ion mobilities in Table 18.2.

$$\kappa = Fc(u_{\mathrm{H^+}} + u_{\mathrm{OH^-}})$$

$$\begin{aligned}c &= \frac{\kappa}{F(u_{\mathrm{H^+}} + u_{\mathrm{OH^-}})} = \frac{5.5 \times 10^{-6}\ \Omega^{-1}\,\mathrm{m^{-1}}}{(96{,}485\ \mathrm{C\,mol^{-1}})(5.689 \times 10^{-7}\ \mathrm{m^2\,V^{-1}\,s^{-1}})} \\ &= 1.00 \times 10^{-4}\ \mathrm{mol\,m^{-3}} = 1.00 \times 10^{-7}\ \mathrm{mol\,L^{-1}}\end{aligned}$$

Thus $K_w = (1.00 \times 10^{-7})^2 = 1.00 \times 10^{-14}$ at 25 °C.

The ion product of water may be determined more precisely through electromotive force measurements. Values of K_w at various temperatures are given in Table 18.3. The ion product for D_2O at 25 °C is 1.54×10^{-15}, showing that the properties of isotopic compounds are very much alike but not identical.

Table 18.3 Ion Product of Pure Water[a]

t/°C	0	10	25	40	50
$K_w/10^{-14}$	0.113	0.292	1.008	2.917	5.474

[a] H. S. Harned and W. J. Hamer, *J. Am. Chem. Soc.*, **55**, 2194 (1933).

18.7 RELATION BETWEEN MOBILITY AND ION FRICTIONAL COEFFICIENT

The mobilities of a number of familiar ions at infinite dilution, given in Table 18.2, may be interpreted in terms of the frictional coefficient f (equation 18.1) of the ion. When an

* F. Kohlrausch and A. Heydweiller, *Z. physik. Chem.*, **14**, 317 (1894).

electric field is applied to an electrolytic solution, an ion is accelerated until it reaches a velocity v where the product of the frictional coefficient f and the velocity v is equal to the force of the field on the charge of the ion. Thus

$$f_i v_i = |z_i| eE \tag{18.14}$$

where z_i is the charge number of ion i. Since the electric mobility is $u_i = v_i/E$, the frictional coefficient of an ion may be calculated from

$$f_i = \frac{|z_i| e}{u_i} \tag{18.15}$$

Apparent radii of the hydrated ions may be calculated from f_i using Stokes' law (equation 18.2).

18.8 DRIVING FORCES IN IRREVERSIBLE PROCESSES

Transport of matter may be caused by a concentration gradient, a centrifugal field or, for ions, by an electric field. In each case the force causing transport can be thought of as the negative gradient of a potential. We have seen that the negative gradient of the electrical potential is the driving force for transport in an electrical field. The negative gradient of the chemical potential is the driving force for diffusion. The negative gradient of the centrifugal potential energy is the driving force for sedimentation.

In discussing transport processes the motion of component i may be expressed in terms of its flux J_i, which is defined as the quantity of component i (expressed in grams or moles) crossing the unit surface area in unit time. If solute molecules at a concentration c_i are all moving with velocity v_i perpendicular to the surface considered, the flux J_i is expressed in terms of the velocity by

$$J_i = v_i c_i \tag{18.16}$$

In the theory for irreversible processes the flux J_i is assumed to be proportional to the gradient of the appropriate potential. If the potential varies only in the x direction,

$$J_i = v_i c_i = -L_i \frac{\partial \phi}{\partial x} \tag{18.17}$$

where ϕ is the potential and L_i is the generalized conductivity.*

Equation 18.17 may be written in terms of the force F_i on a single particle, which is $-(1/N_A)\ \partial\phi/\partial x$.

$$v_i c_i = N_A L_i F_i = N_A L_i v_i f_i \tag{18.18}$$

Thus the conductivity L_i is given by

$$L_i = \frac{c_i}{N_A f_i} \tag{18.19}$$

---

* If two kinds of potential gradient are present, the flux of component i is given by

$$J_i = -L_{i1} \frac{\partial \phi_1}{\partial x} - L_{i2} \frac{\partial \phi_2}{\partial x}$$

Thus fluxes may be coupled. For example, the gradient in potential due to substance 2 may cause a flux of component 1.

Since $N_A f_i$ is the frictional coefficient per mole, the conductivity L_i is proportional to the concentration of the substance being transported and inversely proportional to the frictional resistance.

18.9 DIFFUSION

Fick's first law of diffusion is

$$J_2 = -D_2\left(\frac{dc_2}{dx}\right) \tag{18.20}$$

The diffusion coefficient D_2 of the solute is defined in terms of the flux J_2 of the solute and of its concentration gradient dc_2/dx. If the flux J is expressed in mol m^{-2} s^{-1} and the concentration gradient in mol m^{-4}, it is seen that the diffusion coefficient will have the units of m^2 s^{-1}. The diffusion coefficient for potassium chloride in very dilute aqueous solution at 25 °C is 1.99×10^{-10} m^2 s^{-1}, and for sucrose it is 5.23×10^{-10} m^2 s^{-1}. In this section we want to derive the relationship between D and other properties of the diffusing component.

In diffusion the driving force for the flow of solute is the negative gradient of the chemical potential

$$J_2 = -L_2 \frac{\partial \mu_2}{\partial x} \tag{18.21}$$

The chemical potential μ_i for an ideal solute is given by

$$\mu_2 = \mu_2^\circ + RT \ln c_2 \tag{18.22}$$

Since the temperature and pressure in the cell are constant,

$$\frac{\partial \mu_2}{\partial x} = \frac{\partial \mu_2}{\partial c_2}\frac{\partial c_2}{\partial x} = \frac{RT}{c_2}\frac{\partial c_2}{\partial x} \tag{18.23}$$

Substituting in equation 18.21,

$$J_2 = -\frac{L_2 RT}{c_2}\frac{\partial c_2}{\partial x} \tag{18.24}$$

According to equation 18.19, the frictional coefficient f_2 can be introduced to obtain

$$J_2 = -\frac{RT}{N_A f_2}\frac{\partial c_2}{\partial x} \tag{18.25}$$

Comparison with equation 18.20 shows that for ideal solutions,

$$D_2 = \frac{RT}{N_A f_2} \tag{18.26}$$

which shows that the diffusion coefficient is inversely proportional to the frictional coefficient. This relation was first obtained by Einstein. As we will see in Section 20.4, the radius of spherical particles may be calculated from the measured diffusion coefficient using this relation and Stokes' law (equation 18.2). As another illustration of the use of equation 18.26, equation 18.15 may be inserted to obtain the relationship between the self-diffusion coefficient of an ion and its electric mobility.

$$D_i = \frac{u_i}{|z_i|}\frac{RT}{F} \tag{18.27}$$

The self-diffusion coefficient of an ion may be determined by following the diffusion of radioactivity when a radioactive species of that ion is used.

18.10 THE EQUATION OF CONTINUITY

In studying transport processes the flux J is seldom measured directly. What is measured is the change in concentration with time at various points. In order to relate the flux to the change in concentration, consider the situation illustrated in Fig. 18.3. Flow of solute is occurring in the x direction in a cell of uniform cross section A. We want to calculate the change in concentration in a thin slab of thickness δx. The quantity of material crossing the plane at x in time δt is $J(x)A\ \delta t$, whereas the quantity leaving through the plane at $x + \delta x$ in the same time is $J(x + \delta x)A\ \delta t$, which may be written

$$\left(J(x) + \frac{\partial J}{\partial x}\,\delta x\right)A\ \delta t$$

The net gain in the quantity of material between these hypothetical planes may be expressed in terms of the change of concentration in the volume $A\ \delta x$ or in terms of the difference between these two quantities of material transported.

$$A\ \delta c\ \delta x = JA\ \delta t - \left(J + \frac{\partial J}{\partial x}\,\delta x\right)A\ \delta t = -\frac{\partial J}{\partial x}A\ \delta x\ \delta t \tag{18.28}$$

In the limit, as the distances and times are made smaller,

$$\frac{\partial c}{\partial t} = -\frac{\partial J}{\partial x} \tag{18.29}$$

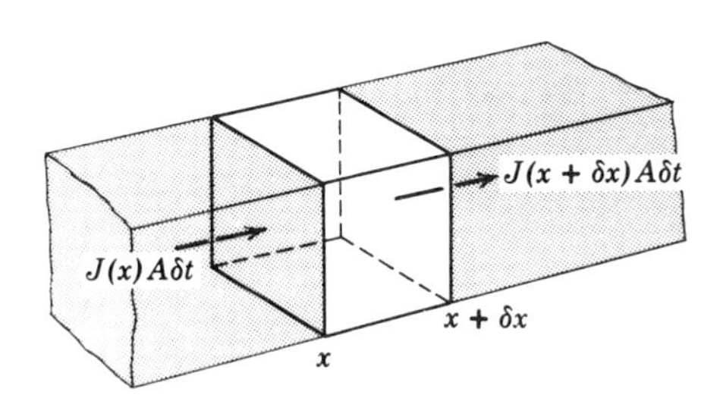

Fig. 18.3 Cell of uniform cross section A in which there is transport by diffusion, sedimentation, or electrical migration.

This is referred to as the equation of continuity. It relates the change in concentration at a given value of x in the cell to the rate of change of flux with distance.

18.11 EXPERIMENTAL MEASUREMENT OF THE DIFFUSION COEFFICIENT

Diffusion coefficients may be measured by a number of different methods.* Probably the most widely used methods are those in which a sharp boundary is formed between solution and solvent, as illustrated in Fig. 18.4. Initially the plot of concentration on the horizontal axis versus height in the cell on the vertical axis has the shape indicated in Fig. 18.4*b*. At a later time this boundary will have become diffuse, and the concentration will vary with height, as illustrated in Fig. 18.4*c*. Instead of an abrupt change in concentration, there is a more gradual one. When the solute is a colored substance, its concentration may be determined photometrically as a function of height. One of the most generally useful methods for determining the diffuseness of the boundary depends on measuring the deflection of light by the refractive-index gradient associated with the concentration gradient.† Since the bending of a light ray in such a boundary is proportional to the refractive index gradient dn/dx, the curve obtained with such a schlieren optical system has the shape indicated in Fig. 18.4*d*. This curve has the shape of a normal probability curve (sometimes referred to as a Gaussian curve) if D is independent of concentration.

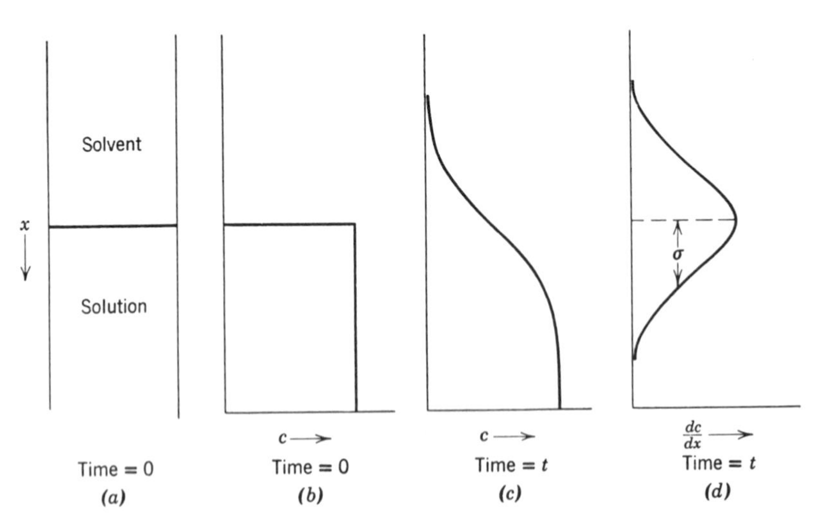

Fig. 18.4 Diffusion of an initially sharp boundary in a cell of uniform cross section.

* L. J. Gosting, *Advances in Protein Chemistry*, Vol. XI, Academic Press, New York, 1956.

† L. G. Longsworth, *Ind. Eng. Chem., Anal. Ed.*, **18**, 219 (1946).

In order to interpret such diffusion experiments, Fick's first law (equation 18.20) is combined with the equation of continuity (equation 18.29) to obtain Fick's second law. Substitution of Fick's first law in the continuity equation yields

$$\frac{\partial c}{\partial t} = \frac{\partial}{\partial x} D \frac{\partial c}{\partial x} \tag{18.30}$$

If the diffusion coefficient D is independent of the concentration and therefore of distance, then

$$\frac{\partial c}{\partial t} = D \frac{\partial^2 c}{\partial x^2} \tag{18.31}$$

which is known as Fick's second law.

To derive the expression for concentration as a function of distance and time for the experiment illustrated in Fig. 18.4, equation 18.31 is integrated with the following boundary conditions: when $t = 0$, $c = c_0$ for $x > 0$, and $c = 0$ for $x < 0$; and when $t > 0$, c approaches c_0 as x approaches ∞ and c approaches 0 as x approaches $-\infty$. The result is

$$c = \frac{c_0}{2}\left(1 + \frac{2}{\sqrt{\pi}} \int_0^{x/2\sqrt{Dt}} e^{-\beta^2}\, d\beta\right) \tag{18.32}$$

where the second term in the parentheses is referred to as the Gaussian error function. The equation for the derivative curve in Fig. 18.4*d* is

$$\frac{\partial c}{\partial x} = \frac{c_0}{2\sqrt{\pi D t}} e^{-x^2/4Dt} \tag{18.33}$$

The function $(2\pi)^{-1/2} e^{-y^2/2}$ is referred to as the normal probability function of y. This bell-shaped probability curve is referred to as a Gaussian curve.

The square of the standard deviation of the experimental bell-shaped curve is

$$\sigma^2 = \frac{\int_{-\infty}^{\infty} x^2 (\partial c/\partial x)\, dx}{\int_{-\infty}^{\infty} (\partial c/\partial x)\, dx} \tag{18.34}$$

Substituting equation 18.33,

$$\sigma^2 = \frac{\int_{-\infty}^{\infty} x^2 e^{-x^2/4Dt}\, dx}{\int_{-\infty}^{\infty} e^{-x^2/4Dt}\, dx} = 2Dt \tag{18.35}$$

where the last form is obtained by using the values of the definite integrals.* Since the standard deviation σ of a Gaussian curve is the half width at the inflection point, and the inflection points are at a height of 0.606 of the maximum ordinate, σ is readily obtained from the experimental curve, and D may be calculated using equation 18.35.

---

*

$$\int_0^{\infty} e^{-a^2x^2}\, dx = \frac{\sqrt{\pi}}{2a}$$

$$\int_0^{\infty} x^{2n} e^{-ax^2}\, dx = \frac{1\cdot 3\cdot 5\cdots(2n-1)}{2^{n+1}a^n} \sqrt{\frac{\pi}{a}}$$

References

R. B. Bird, W. E. Stewart, and E. N. Lightfoot, *Transport Phenomena*, Wiley, New York, 1960.

J. O'M. Bockris and D. M. Drazic, *Electrochemical Science*, Taylor and Francis Ltd., London, 1972.

H. L. Friedman, *Ionic Solution Theory*, Wiley-Interscience, New York, 1962.

R. M. Fuoss and F. Accascina, *Electrolytic Conductance*, Wiley-Interscience, New York, 1959.

S. R. de Groot and P. Mazur, *Nonequilibrium Thermodynamics*, Wiley-Interscience, New York, 1962.

H. S. Harned and B. B. Owen, *The Physical Chemistry of Electrolytic Solutions*, Reinhold Publishing Corp., New York, 1958.

I. Prigogine, *Introduction to the Thermodynamics of Irreversible Processes*, Wiley, New York, 1967.

R. A. Robinson and R. H. Stokes, *Electrolyte Solutions*, Academic Press, New York, 1959.

Problems

18.1 Ten cubic centimeters of water at 25 °C is forced through 20 cm of 2-mm diameter capillary in 4 s. Calculate the pressure required and the Reynolds number.

Ans. 1.14×10^3 N m^{-2}. 1780.

18.2 A steel ball ($\rho = 7.86$ g cm^{-3}) 0.2 cm in diameter falls 10 cm through a viscous liquid ($\rho_0 = 1.50$ g cm^{-3}) in 25 s. What is the viscosity at this temperature?

Ans. 3.46 Pa s.

18.3 How long will it take a spherical air bubble 0.5 mm in diameter to rise 10 cm through water at 25 °C? *Ans.* 0.66 s.

18.4 Using data in Table 18.1 and equation 18.6, estimate the activation energy for water molecules to move into a vacancy at 25 °C. *Ans.* 17.4 kJ mol^{-1}.

18.5 A conductance cell was calibrated by filling it with a 0.02 mol L^{-1} solution of potassium chloride ($\kappa = 0.2768\ \Omega^{-1}$ m^{-1}) and measuring the resistance at 25 °C, which was found to be 457.3 Ω. The cell was then filled with a calcium chloride solution containing 0.555 g of $CaCl_2$ per liter. The measured resistance was 1050 Ω. Calculate (*a*) the cell constant for the cell, and (*b*) the conductivity of the $CaCl_2$ solution.

Ans. (*a*) 126.6 m^{-1}, (*b*) 0.1206 Ω^{-1} m^{-1}.

18.6 A moving boundary experiment is carried out with a 0.1 mol L^{-1} solution of hydrochloric acid at 25 °C ($\kappa = 4.24\ \Omega^{-1}$ m^{-1}). Sodium ions are caused to follow the hydrogen ions. Three milliamperes is passed through the tube of 0.3 cm^2 cross-sectional area, and it is observed that the boundary moves 3.08 cm in 1 hr. Calculate (*a*) the hydrogen ion mobility, (*b*) the chloride ion mobility, and (*c*) the electric field strength.

Ans. (*a*) 3.63×10^{-7} m^2 V^{-1} s^{-1}. (*b*) 7.64×10^{-8} m^2 V^{-1} s^{-1}. (*c*) 23.58 V m^{-1}

18.7 Calculate the conductivity of 0.001 mol L^{-1} HCl at 25 °C. The limiting ion mobilities may be used for this problem. *Ans.* 0.042 61 Ω^{-1} m^{-1}.

18.8 It is desired to use a conductance apparatus to measure the concentration of dilute solutions of sodium chloride. If the electrodes in the cell are each 1 cm^2 in area and are 0.2 cm apart, calculate the resistance that will be obtained for 1, 10, and 100 ppm NaCl at 25 °C. *Ans.* 92,500, 9250, and 925 Ω.

18.9 Calculate the conductivity at 25 °C of a solution containing 0.001 mol L^{-1} hydrochloric acid and 0.005 mol L^{-1} sodium chloride. The limiting ionic mobilities at infinite dilution may be used to obtain a sufficiently good approximation.

Ans. 0.1058 Ω^{-1} m^{-1}.

18.10 Using Stokes' law, calculate the effective radius of a nitrate ion from its mobility (74.0×10^{-9} m^2 V^{-1} s^{-1} at 25 °C). *Ans.* 0.129 nm.

18.11 What is the self-diffusion coefficient of Na^+ in water at 25 °C?

Ans. 1.334×10^{-9} m^2 s^{-1}.

18.12 Using a table of the probability integral, calculate enough points on a plot of c versus x (like Fig. 18.5c) to draw in the smooth curve for diffusion of 0.1 mol L^{-1} sucrose into water at 25 °C after 4 hr and 29.83 min ($D = 5.23 \times 10^{-10}$ m^2 s^{-1}).

18.13 A sharp boundary is formed between a dilute aqueous solution of sucrose and water at 25 °C. After 5 hr the standard deviation of the concentration gradient is 0.434 cm. (a) What is the diffusion coefficient for sucrose under these conditions? (b) What will be the standard deviation after 10 hr? *Ans.* (a) 5.23×10^{-10} m^2 s^{-1}. (b) 0.614 cm.

18.14 Ten cubic centimeters of water at 25 °C is forced through 20 cm of 2-mm diameter capillary in 4 s. Calculate the pressure required and the Reynolds number. Will the flow be laminar?

18.15 Calculate the time necessary for a quartz particle 10 μm in diameter to sediment 50 cm in distilled water at 25 °C. The density of quartz is 2.6 g cm^{-3}. The coefficient of viscosity of water may be taken to be 8.95×10^{-4} kg m^{-1} s^{-1}.

18.16 Estimate the rate of sedimentation of water droplets of 1-μm diameter in air at 20 °C. The viscosity of air at this temperature is 1.808×10^{-5} Pa s.

18.17 Plot log of viscosity of mercury against the reciprocal of the absolute temperature from the following data and calculate the activation energy.

t/°C	0	20	35	98	203
$\eta/10^{-3}$ Pa s	1.661	1.547	1.476	1.263	1.079

18.18 A study of conductivities at high electric field strengths reveals that the conductivity increases slightly with increasing electric field strength. A microsecond pulse at 10 V m^{-1} may be used. Approximately how far will a sodium ion move during such a pulse at room temperature?

18.19 One hundred grams of sodium chloride is dissolved in 10,000 L of water at 25 °C, giving a solution that may be regarded in these calculations as infinitely dilute. (a) What is the conductivity of the solution? (b) This dilute solution is placed in a glass tube of 4-cm diameter provided with electrodes filling the tube and placed 20 cm apart. How much current will flow if the potential drop between the electrodes is 80 V?

18.20 Estimate the conductivity at 25 °C of water that contains 70 ppm by weight of magnesium sulfate.

18.21 In 0.1 mol L^{-1} hydrochloric acid at 0 °C the mobilities of hydrogen and chloride ions are 365×10^{-9} and 79×10^{-9} m^2 V^{-1} s^{-1}, respectively. (a) Calculate the conductivity for this solution at 0 °C. (b) A moving boundary experiment is carried out in a tube with a uniform cross-sectional area of 0.200 cm^2, and sodium ions are caused to follow the hydrogen ions. If a current of 5 mA is passed for 1 hr, how far will the hydrogen ions move? (c) What is the field strength in this experiment?

18.22 Estimate the electric mobility of a $(CH_3)_4N^+$ ion, assuming it is not hydrated in solution at 25 °C. The effective radius may be taken as 0.3 nm.

18.23 Calculate the apparent radii of Li^+ and Na^+ in infinitely dilute aqueous solution from their mobilities in Table 18.2. How do you explain the fact that the ion with the smaller crystal radius (see Table 19.1) has the larger apparent radius in solution?

18.24 If only a small amount of material q is allowed to diffuse through a porous plate from a solution of concentration c'' into a solution of concentration c', Fick's law (equation 18.20) may be written

$$D = \frac{-q}{Kt(c' - c'')}$$

where K is the cell constant that must be determined in an experiment with a substance of known diffusion coefficient. If a 0.10 mol L^{-1} aqueous solution of potassium chloride is allowed to diffuse into water for 12 hr and 38 min at 25 °C, it is found that 1.25×10^{-4} mol of salt diffuses through the porous plate. Calculate D if K has previously been found o be 1.5 cm.

18.25 The self-diffusion coefficient of Ag^+ in dilute aqueous solution at 25 °C is 1.65×10^{-9} m^2 s^{-1}. What is the electric mobility of the silver ion?

18.26 An initially sharp boundary is formed between aqueous solutions containing 0.3 mol L^{-1} glycine and 0.1 mol L^{-1} glycine at 25 °C. $D = 1.022 \times 10^{-9}$ m^2 s^{-1}. Plot c versus x and dc/dx versus x for a diffusion time of 15 hr.

18.27 Using a table of the normal probability function, calculate enough points on a plot of dc/dx versus x (like Fig. 18.4) to draw in the smooth curve for diffusion of 0.01 mol L^{-1} sucrose into water at 25 °C after 3 hr.

18.28 (*a*) Calculate the time required for the half-width of a freely diffusing boundary of dilute potassium chloride in water to become 0.5 cm at 25 °C ($D = 1.77 \times 10^{-9}$ m^2 s^{-1}). (*b*) Calculate the corresponding time for serum albumin ($D = 6.15 \times 10^{-11}$ m^2 s^{-1}).

PART FOUR

STRUCTURES

Structure is so important in chemistry that we have not been able to delay all structural considerations to Part Four; but we have left to this part structural information about crystals and macromolecules. In both cases the relation between structure and function is of special interest.

X-ray diffraction provides a powerful means for the determination of the structures of crystals. This process is simplified by the recognition of the various types of symmetry that the internal structure of a crystal may have. The symmetry in solids is different from the symmetry in individual molecules discussed in Chapter 9. The identification of crystal planes and of the internal structure is illustrated in detail for cubic crystals.

The structures of many solids are determined by the ways spheres are packed, and so this subject is examined in some detail. Even liquids have some regularity in their three-dimensional structure, and liquid crystals have periodicities in one or two dimensions.

The study of the electronic structure of solids introduces new ideas not encountered with isolated molecules. All real crystals are imperfect, and the nature of the imperfections has a big influence on properties. The application of quantum mechanics to crystals helps us understand why some are excellent conductors of electricity and others are insulators; electrical conductivities of solids range over 10^{30}. Semiconductors have intermediate properties, and the modification of these properties by the addition of traces of other substances has provided the basis for a semiconductor industry.

The last chapter in the book is concerned with synthetic high polymers and biological macromolecules. In both cases properties are largely determined by structure, and structure by chemical composition. Linear synthetic polymers tend to have random configurations in solution, but the tightness of coiling is very dependent on the properties of the solvent. Viscosity measurements provide information about average molar mass and shape.

Ultracentrifuge experiments yield information about molar mass and shape of proteins and nucleic acids. Molar masses can be calculated from sedimentation coefficients and diffusion coefficients, or obtained directly from equilibrium ultracentrifuge experiments. Light-scattering measurements also yield molar masses and, for larger macromolecules, information about axial ratio.

CHAPTER 19

CRYSTAL STRUCTURE AND SOLID STATE

In this chapter we will first consider crystal geometry, then X-ray diffraction and the structures of specific crystals, and finally solid-state chemistry. The ideas of point-group symmetry developed in Chapter 9 are basic to the understanding of crystals, but the introduction of translations produces new kinds of symmetry and requires space groups (Section 19.6) in addition to point groups.

In 1912 Laue suggested that the wavelength of X rays might be about the same as the distance between atoms in a crystal so that a crystal could serve as a diffraction grating for X rays. This experiment was carried out by Friederick and Knipping, and they observed the expected diffraction. Almost immediately afterward, W. L. Bragg (1913) improved on the Laue experiment, mainly by substituting monochromatic for polychromatic radiation and by providing a more physical interpretation to the Laue theory of the scattering experiment. Bragg also determined the structures of a number of simple crystals, including those of NaCl, CsCl, and ZnS. Since that time, this tool of structural analysis by single crystal X-ray diffraction has developed into the most powerful method known for obtaining the atomic arrangement in the solid state. Since the 1950s, with the advent of large high-speed computers capable of handling the X-ray data, it has been possible to determine the structures of compounds as complex as proteins.

19.1 LATTICES*

A crystal may be described as a three-dimensional pattern in which a structural motif is repeated in such a way that the environment of every motif is the same throughout the crystal. The motif is often a molecule, but it may also be a group of molecules.

A linear pattern may be described by saying that there is a set of parallel motifs at the end of a vector $\boldsymbol{a}$ and its multiples. These vectors are described by

$$\boldsymbol{T} = u\boldsymbol{a} \tag{19.1}$$

where u is an integer. An example of a linear pattern is shown in Fig. 19.1a. The linear pattern formed in this way can be repeated in a second direction represented

* In discussing crystal geometry we will follow M. J. Buerger, *Introduction to Crystal Geometry*, McGraw-Hill Book Co., New York, 1971, which may be referred to for more detailed information.

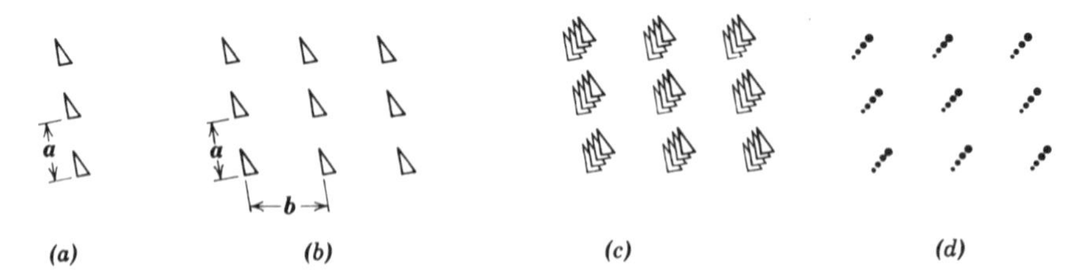

Fig. 19.1 (*a*) One-dimensional pattern with vector ***a***. (*b*) Two-dimensional pattern with vectors ***a*** and ***b***. (*c*) Three-dimensional pattern with vectors ***a***, ***b***, and ***c***. (*d*) Lattice of pattern *c*.

by the vector ***b*** to form a two-dimensional pattern in which every structural pattern has the same environment as every other, as shown in Fig. 19.1*b*. Finally, the whole two-dimensional pattern can be repeated in a third direction ***c*** to produce a translationally ordered pattern in three dimensions, as shown in Fig. 19.1*c*. The structural pattern is repeated at the end of every vector of the form

$$\boldsymbol{T} = u\boldsymbol{a} + v\boldsymbol{b} + w\boldsymbol{c} \qquad u, v, w \text{ are integers} \tag{19.2}$$

For many purposes it is convenient to concentrate on the geometry of the repetition and replace the structural pattern by a point, as shown in Fig. 19.1*d*, to obtain a *lattice*. The lattice may be generated from a single starting point by the infinite repetition of a set of fundamental translations that characterize the lattice. The lattice is described by equation 19.2, but it has no true origin and can be shifted around parallel with itself. Any three noncoplanar vectors ***a***, ***b***, and ***c*** describe a lattice, but a given lattice can be described by an infinite number of sets of three vectors. This is illustrated in two dimensions in Fig. 19.2. The ***a*** vector together with any one of the ***b*** vectors may be chosen to generate the pattern.

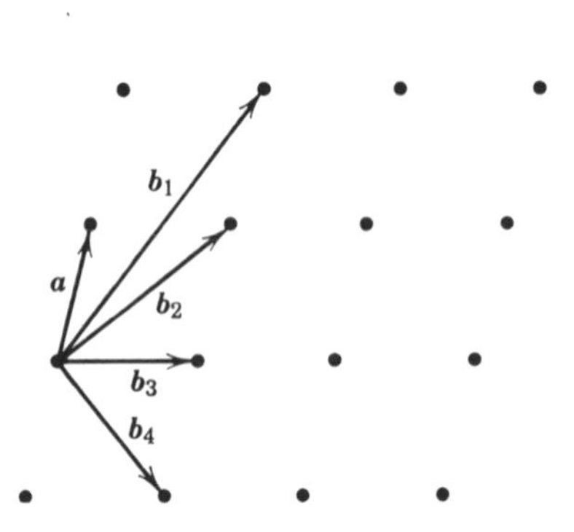

Fig. 19.2 Alternate choices for the second translational vector in a two-dimensional lattice.

19.2 UNIT CELLS

The space occupied by a lattice may be divided into unit cells. The repetition of a cell (with everything in it) in three dimensions generates the entire pattern of a crystal. A given lattice can be blocked out in cells in different ways, as shown in Fig. 19.3. If the corners of the cells include all of the lattice points in the crystal the cell is called a *primitive unit cell*. Primitive cells have one lattice point per cell because each of the corner lattice points is shared by eight cells. A lattice can also be blocked out in cells that do not include all lattice points as corners. This is illustrated by one of the cells in Fig. 19.3. Such cells, referred to as *multiple unit cells*, are useful in simplifying the geometry of crystals for which the primitive cell is oblique, but the multiple cell has two or more edges that are at 90°.

19.3 ROTATIONAL SYMMETRY IN CRYSTALS

A lattice may have various types of rotational symmetry of the type discussed for molecules in Section 9.3; in other words, rotation through an angle of $360°/n$ may bring the lattice into an equivalent position. However, in contrast with individual molecules where rotational axes of order unity to infinity are in principle possible, in crystals only axes of one-, two-, three-, four-, and six-fold rotational symmetry are possible. This statement can be proved by use of Fig. 19.4. The lattice shown in this figure has an axis of n-fold symmetry. The lattice points A_1, A_2, A_3, and A_4 are each separated by distance a. Because of the assumed symmetry, rotation of the lattice about any lattice point through an angle $\alpha = 2\pi/n$ will produce a lattice indistinguishable from the original. Therefore clockwise rotation by α about A_3 and clockwise rotation by α about A_2 requires that there be lattice points at B_1 and

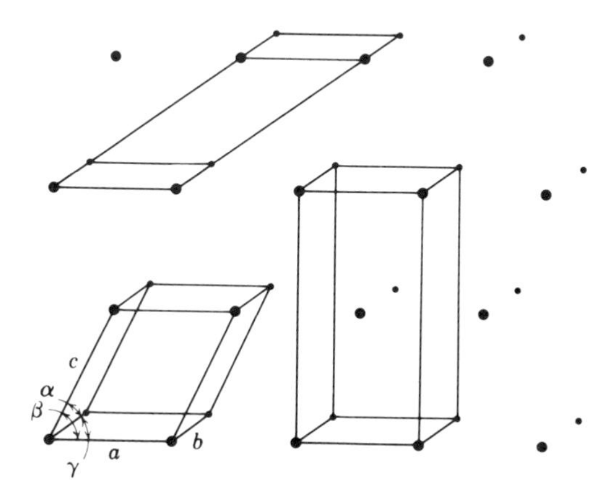

Fig. 19.3 Cells in a lattice.

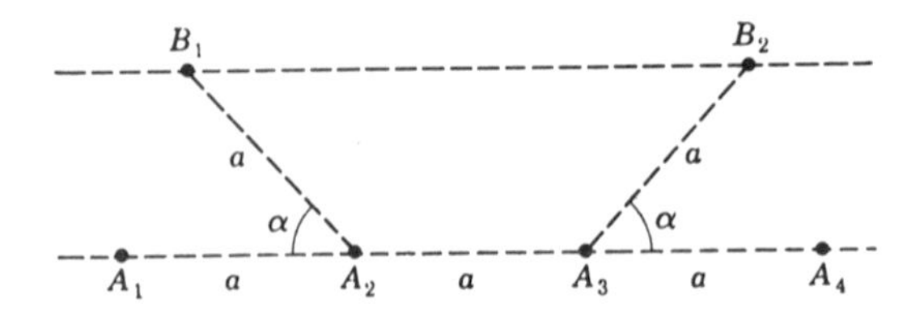

Fig. 19.4 Restriction on rotational order in a crystal. There are n-fold rotational axes at lattice points A_1, A_2, A_3, A_4, B_1, and B_2.

B_2. Since the line B_1B_2 is parallel to A_1A_4, B_1 and B_2 must be separated by an integral multiple of a, represented by ma. Thus

$$a + 2a \cos \alpha = ma \qquad \cos \alpha = \frac{N}{2}$$

where $N = m - 1$. The only values of α that satisfy this equation are 0°, 60°, 90°, 120°, 180°, and 360°, which means that the rotational symmetry of the lattice can be only one-, two-, three-, four-, or six-fold. In Chapter 9, on point-group symmetry, these axes were represented by C_1, C_2, C_3, C_4, and C_6, but crystallographers use the Hermann-Mauguin or international symbols* and in this system the rotational axes are simply referred to by the numbers 1, 2, 3, 4, and 6.

In discussing improper rotations in Chapter 9 we used rotary reflections, but in crystallography the same purpose is accomplished with a rotatory inversion. In a rotary inversion a rotation by 360°/n is followed by inversion through the center of symmetry. The crystallographic rotary inversion axes are represented by $\bar{1}$, $\bar{2}$, $\bar{3}$, $\bar{4}$, and $\bar{6}$, where the number represents the number of equivalent positions in a 360° rotation. The $\bar{1}$ axis is equivalent to an inversion i. A $\bar{2}$ axis is equivalent to a mirror plane. A $\bar{3}$ axis is equivalent to a threefold rotation plus an inversion. A $\bar{6}$ axis is equivalent to a threefold axis and a mirror plane. It is important to note that a rotatory-inversion operation converts an object into its mirror image. Therefore an object that cannot be superimposed on its mirror image cannot possess any element of rotatory-inversion symmetry. In the Hermann-Mauguin system mirror planes are presented by m. A mirror plane perpendicular to an n-fold axis is represented by n/m.

19.4 32 CRYSTALLOGRAPHIC POINT GROUPS

In discussing the symmetry of molecules in Chapter 9 it was remarked that there is in principle an infinite number of point groups. Perfect crystals (crystals grown in a symmetrical environment) can be classified according to the point groups but,

---

* N. F. M. Henry and K. Lonsdale, ed., *International Tables for X-Ray Crystallography*, Vol. I, Symmetry Groups, Kynoch Press, Birmingham, England, 1952.

because of the limitation of crystal lattices to rotational axes 1, 2, 3, 4, and 6 discussed in the preceding section, a crystal must belong to one of 32 crystallographic point groups. In other words, only 32 point groups result from combinations of proper and improper rotations of one-, two-, three-, four-, and sixfold. Although, as we will see, the symmetries of the arrangements of atoms in crystals are more complicated that the 32 crystallographic point groups, the symmetry of crystals that is visible to the eye is the point-group symmetry. Crystallographers have recognized this since the nineteenth century, and so the 32 crystallographic point groups are often referred to as the 32 crystal classes. Certain crystal forms can arise from a single one of the crystal classes, and so these characteristic shapes immediately identify the point group.

19.5 SIX CRYSTAL SYSTEMS

Symmetry determines the types of periodic structures in space that are possible and is used to determine the classification of crystals. For example, if a crystal has a twofold axis of rotation, two of the unit cell axes are perpendicular to each other. If a unit cell has only a twofold axis and there are no other conditions on the dimensions and shape of the cell, the cell is monoclinic and $a \neq b \neq c$, $\alpha = 90°$, $\beta \neq 90°$, $\gamma = 90°$. These latter conditions are a result of the symmetry. The point groups in the monoclinic crystal system are C_2, C_s, and C_{2h}.

If a unit cell belongs to the D_2, C_{2v}, or D_{2h} point groups, there must be three perpendicular lattice vectors; $\alpha = \beta = \gamma = 90°$. Unit cells of this type are said to be orthorhombic. The 32 crystallographic point groups require 14 space lattices (often called Bravais lattices), and their geometries can be referred to six crystal systems shown in Fig. 19.5. The angles α, β, and γ are defined in Fig. 19.3. Symmetry imposes certain restrictions on the lengths of the cell edges (a, b, and c) and the angles between them, and these restrictions are shown. However, it is important to remember that the crystal classes are defined by symmetry and not by the restrictions shown in the last column. The sign $\neq$ is used to mean "not generally equal to," but the angles or edges indicated may be equal accidentally within the error of measurement.

The symmetry of a crystal may be very different from that of the molecule in a crystal. If there are two unsymmetrical molecules in a unit cell, the crystal may have one or more twofold axes. With more molecules in a unit cell, higher symmetries may also be obtained, even though the molecule itself is unsymmetrical. Conversely, symmetrical molecules may crystallize in lattices of lower symmetry.

The position of an atom in a unit cell is designated by giving its coordinates as fractions x, y, z of the unit cell edges a, b, c. The point at (xyz) is located by starting at the origin (0, 0, 0) and moving a distance xa along the a axis, then a distance yb parallel to the b axis, and finally a distance zc parallel to the c axis. The fractional coordinates for the lattice points of a face-centered cubic structure are 0, 0, 0; $\frac{1}{2}, \frac{1}{2}, 0$; $\frac{1}{2}, 0, \frac{1}{2}$; and $0, \frac{1}{2}, \frac{1}{2}$. The remaining lattice points may be obtained by adding unity to each of these coordinates. In expressing the locations of atoms in a

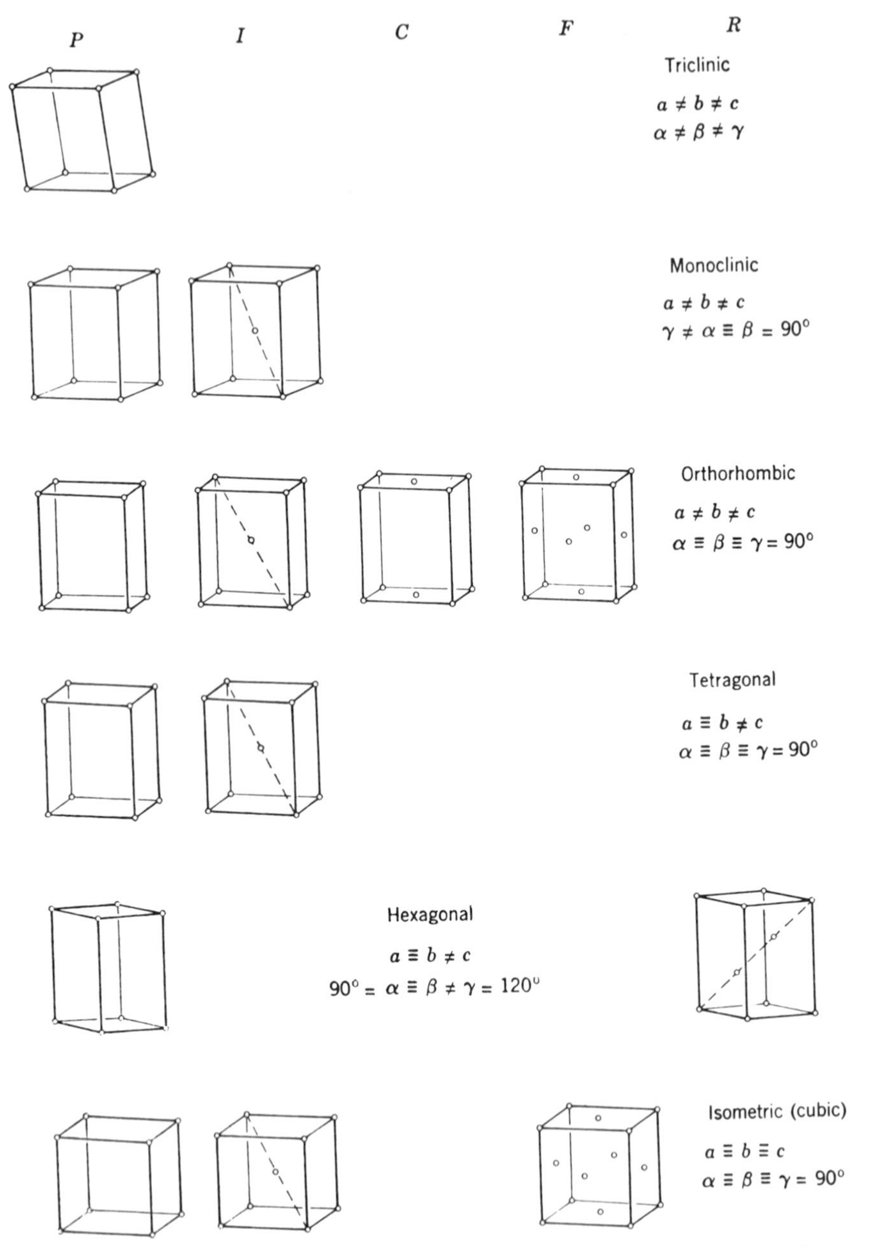

Fig. 19.5 The 14 space lattices (often called Bravais lattices) and the six crystal systems.

unit cell the set of coordinates (000) stands for the locations of all eight corners, that is (100), (111), (101), (110), (001), (011), (010), and (000). In a crystal a lattice point is not necessarily occupied by an atom or molecule, but the lattice point represents a repeating unit.

In a cubic crystal the distance between two points with fractional coordinates x_1, y_1, z_1 and x_2, y_2, z_2 is given by

$$l = a[(x_2 - x_1)^2 + (y_2 - y_1)^2 + (z_2 - z_1)^2]^{1/2} \tag{19.3}$$

The volume of a unit cell is given by

$$V = abc(1 - \cos^2 \alpha - \cos^2 \beta - \cos^2 \gamma + 2 \cos \alpha \cos \beta \cos \gamma)^{1/2} \tag{19.4}$$

where the angles are defined in Fig. 19.3. If the unit cell is cubic, orthorhombic, or tetragonal, the three angles are 90° and the equation reduces to $V = abc$.

19.6 230 SPACE GROUPS

The introduction of the operation of translation in crystals leads to more symmetry operations and their combinations than in the 32 crystallographic point groups. To describe the patterns of crystals, two new kinds of symmetry operations are required; glide planes and screw axes. A glide plane is the combination of a reflection in a plane with a translation by one half of a lattice translation. A screw axis is a combination of a rotation and a fraction of a lattice translation parallel to the axis. The translation accompanying the screw motion must be n/p times the unit translation, where p is an integer and the angle d between successive motifs is $360°/n$. There are 11 possible screw axes represented by the symbol n_p: 2_1, 3_1, 3_2, 4_1, 4_2, 4_3, 6_1, 6_2, 6_3, 6_4, 6_5.

The fact that there are 230 ways in which these symmetry operations may be combined in the three-dimensional patterns of crystals was derived independently by three men: Fedorow, a Russian crystallographer, in 1890; Schoenflies, a German mathematician, in 1891; and Barlow, a British amateur, in 1895. The actual determination of space groups of crystals did not become possible until diffraction techniques were utilized to determine the internal symmetry of crystals. Knowledge of the space group of a crystal simplifies the determination of the structure, because only the asymmetric portion of the unit cell needs to be studied; the rest of the contents may be obtained from symmetry operations.

19.7 DESIGNATION OF CRYSTAL PLANES

The planes through the lattice points of crystals are important because they represent possible crystal faces and because they help us understand X-ray diffraction phenomena. Figure 19.6 shows the lattice points in one plane of a crystal. The c axis is taken as perpendicular to the page. Various sets of planes which are

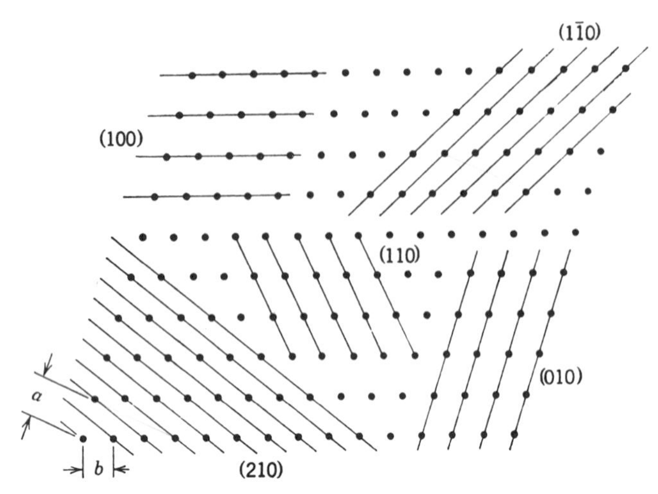

Fig. 19.6 Sets of planes through lattice points as seen along the c axis of a crystal.

parallel to the c axis are indicated in the figure. Actually, there is an infinite number of such sets of planes.

The orientation of a set of planes of a crystal lattice may be specified by means of the intercepts of one of the planes on the three axes a, b, and c of the unit cell. Suppose that a plane intercepts the a axis at a/h, measured from the origin, the b axis at b/k, and the c axis at c/l. This plane is referred to by the indices hkl.* The indices of a set of planes through a lattice may be obtained by counting the number of planes crossed in moving one lattice spacing in the $\boldsymbol{a}$, $\boldsymbol{b}$, and $\boldsymbol{c}$ directions, respectively. For the set of planes in the lower left-hand corner of Fig. 19.6, two planes are crossed in going one lattice distance in the direction of the a axis, and one plane is crossed in going one lattice distance horizontally in the direction of the b axis, whereas no plane would be crossed in going one lattice distance into the paper, since the planes are parallel to the c axis. Thus, this set of planes is designated by the numbers 210. The indices of the set of planes in the upper right-hand corner are $1\bar{1}0$. The negative sign indicates that, if a particular plane is intercepted by going in the positive direction along a, it is necessary to go in the negative direction along b in order to intercept the same plane. This representation for the exterior faces of a crystal and for the internal planes within the crystal will specify the orientation, but not the spatial position of a plane.

The faces of crystals are usually planes with high densities of atoms or molecules, and so they are planes with low indices. Another result of the fact that the external crystal faces are planes with high densities of atoms or molecules is that a given crystalline form of a substance has a constancy of angle between two given faces at

* Often called "Miller indices" because this designation, invented by Whewell in 1825 and Grossman in 1829, was popularized by Miller's textbook of crystallography in 1829. They showed that faces of crystals could be designated by three integers, although nothing was known about the internal structures of crystals.

a given temperature, as discovered by Steno in 1669. Crystal faces with the closest packing of atoms tend to have the lowest Gibbs energy and are therefore the most stable at constant temperature and pressure.

The perpendicular distance d between adjacent planes of a set is given by

$$\frac{1}{d} = \left(\frac{h^2}{a^2} + \frac{k^2}{b^2} + \frac{l^2}{c^2}\right)^{1/2} \tag{19.5}$$

when the unit cell axes are mutually perpendicular (i.e., for orthorhombic, tetragonal, and cubic unit cells). For the general case,

$$\begin{aligned} d = V[&h^2b^2c^2 \sin^2 \alpha + k^2a^2c^2 \sin^2 \beta + l^2a^2b^2 \sin^2 \gamma \\ &+ 2hlab^2c(\cos \alpha \cos \gamma - \cos \beta) + 2hkabc^2(\cos \alpha \cos \beta - \cos \gamma) \\ &+ 2kla^2bc(\cos \beta \cos \gamma - \cos \alpha)]^{-1/2} \end{aligned} \tag{19.6}$$

where V is the unit cell volume given by equation 19.4.

19.8 DIFFRACTION METHODS

In a crystal it is the electrons that scatter X rays. Bragg pointed out that it is convenient to consider that the X rays are "reflected" from a stack of planes in the crystal. For a given stack of planes (hkl) the reflected beam of monochromatic radiation occurs only at a certain angle that is determined by the wavelength of the X rays and the interplanar spacing in the crystal. The relationship correlating these variables is the *Bragg equation*, which may be derived by reference to Fig. 19.7, in which the horizontal lines represent layers in the crystal separated by the distance d. The plane ABC is perpendicular to the incident beam of parallel monochromatic X rays, and the plane LMN is perpendicular to the reflected beam. As the angle of incidence θ is changed, a reflection will be obtained only when the waves are in phase at plane LMN, that is, when the difference in distance between planes ABC and LMN, measured along rays reflected from different planes, is a whole-number multiple of the wavelength. This occurs when

$$FS + SG = n\lambda \tag{19.7}$$

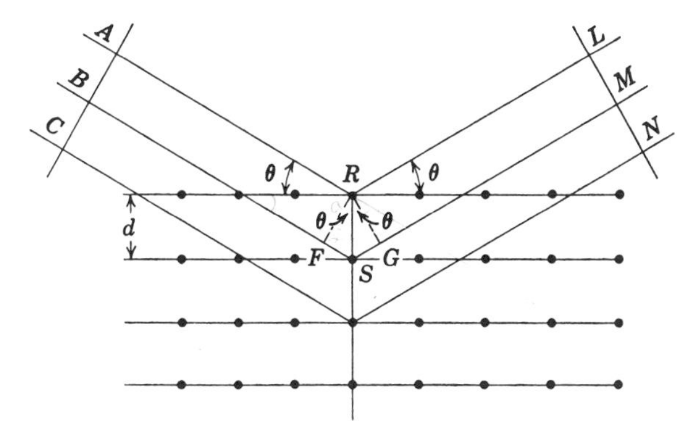

Fig. 19.7 Diagram used in proving the $n\lambda = 2d \sin \theta$.

Since $\sin\theta = FS/d = SG/d$,

$$2d\sin\theta = n\lambda \tag{19.8}$$

This important equation gives the relationship of the distance between planes in a crystal and the angle at which the reflected radiation has a maximum intensity for a given wavelength λ; that is, all the X-ray waves are in phase. If λ is longer than $2d$, there is no solution for n and no diffraction. Thus light waves pass through crystals without being diffracted by the atomic planes. If $\lambda \ll d$ the X rays are diffracted through inconveniently small angles. The Bragg equation does not indicate the intensities of the various diffracted beams. The intensities depend on the nature and arrangement of the atoms within each unit cell.

The reflection corresponding to $n = 1$ for a given family of planes is called the first-order reflection; the reflection corresponding to $n = 2$ is the second-order reflection; and so on. Each successive order exhibits a wider angle. In discussing X-ray reflections it is customary to set $n = 1$ in equation 19.8 and consider that the second-order reflection is from a different stack of planes separated by half the lattice distance, etc. Equation 19.8 may be written

$$\lambda = 2\left(\frac{d}{n}\right)\sin\theta = 2d_{nh,nk,nl}\sin\theta \tag{19.9}$$

where $d_{nh,nk,nl}$ is the perpendicular distance between adjacent planes having the indices nh, nk, nl. The planes nh, nk, nl are parallel to the hkl planes, and the perpendicular interplanar distance is $d_{nh,nk,nl} = d/n$.

To determine the angles at which X rays are diffracted, an oriented single crystal may be rotated in an X-ray beam and the intensity of X rays at the reflection angle determined with a counter. Various types of X-ray cameras have been developed in which the photographic film is moved as the crystal is rotated.

Instead of scattering X rays from a single large crystal, it is convenient (and sometimes necessary when suitable single crystals are not available) in some types of work to pass a collimated beam of X rays through a powdered sample containing microcrystals presumably oriented in random directions. This method was discovered originally by Debye and Scherrer and later by Hull. The reflections may be recorded on a circular photographic film, as illustrated in Fig. 19.8. If coarse crystals are used, the powder pattern is seen to be made up of rings of spots, each spot being produced by a suitably oriented small crystal. If the crystals are very fine, a large number of spots of reflected beams are produced by the different crystal planes, and continuous arcs are obtained on the film.

For cubic crystals, X-ray powder patterns are all that is required to differentiate between primitive, body-centered, and face-centered cubic crystals.

19.9 CUBIC LATTICES

Since the cubic system is the simplest, it is explored in some detail here. There are three independent Bravais lattices that have all the symmetry of a cube: primitive,

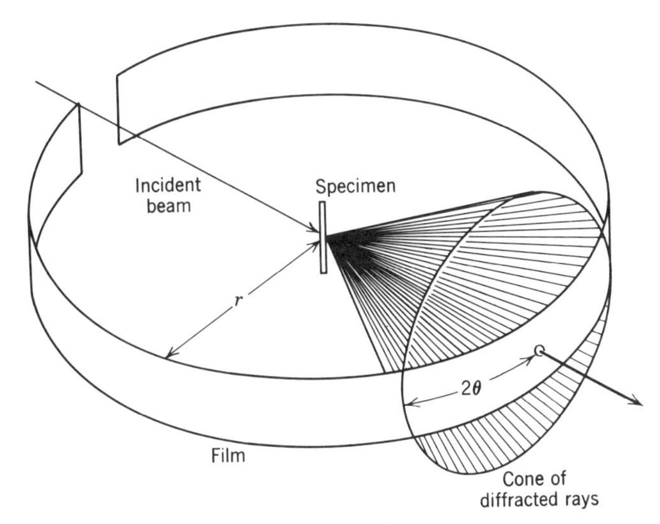

Fig. 19.8 X-ray powder camera.

body-centered, and face-centered. These are illustrated in Fig. 19.9. Since these lattices are based on microscopic translations, they cannot be distinguished by macroscopic examination of the crystals.

In the *primitive cubic lattice* in Fig. 19.9*a* the lattice points shown as black dots at the corners of a cube may represent atoms, ions, molecules, or any repeated structure unit. Three types of reflecting planes are illustrated: 100, 110, and 111. The reflections from these planes occur at the smallest angles of all the possible planes that can be imagined in a cubic crystal, since planes with higher indices are closer together and θ will be larger, according to equation 19.8.

In a cubic crystal the perpendicular distance between a stack of planes with indices h, k, and l is given by

$$d_{hkl} = \frac{a}{\sqrt{h^2 + k^2 + l^2}} \tag{19.10}$$

where a is the length of the side of the unit cell. The spacings in a cubic crystal are obtained by substituting 0, 1, 2, 3, . . . for h, k, and l in this equation.

Exercise I Prove equation 19.10, making use of the fact that the sum of the squares of the sides of a right triangle is equal to the square of the hypotenuse. Test this relation for planes with indices (100), (110), and (111).

For *primitive cubic crystals* d_{hkl} may have the following values: a, $a/\sqrt{2}$, $a/\sqrt{3}$, $a/\sqrt{4}$, $a/\sqrt{5}$, $a/\sqrt{6}$, $a/\sqrt{8}$, etc., where $a/\sqrt{7}$ is missing because 7 cannot be obtained from $h^2 + k^2 + l^2$, where h, k, and l have values of 0, 1, 2, 3, ...

In the *face-centered cubic lattice* illustrated in Fig. 19.9*b* there are lattice points in the center of each face of the unit cell in addition to the lattice points at the

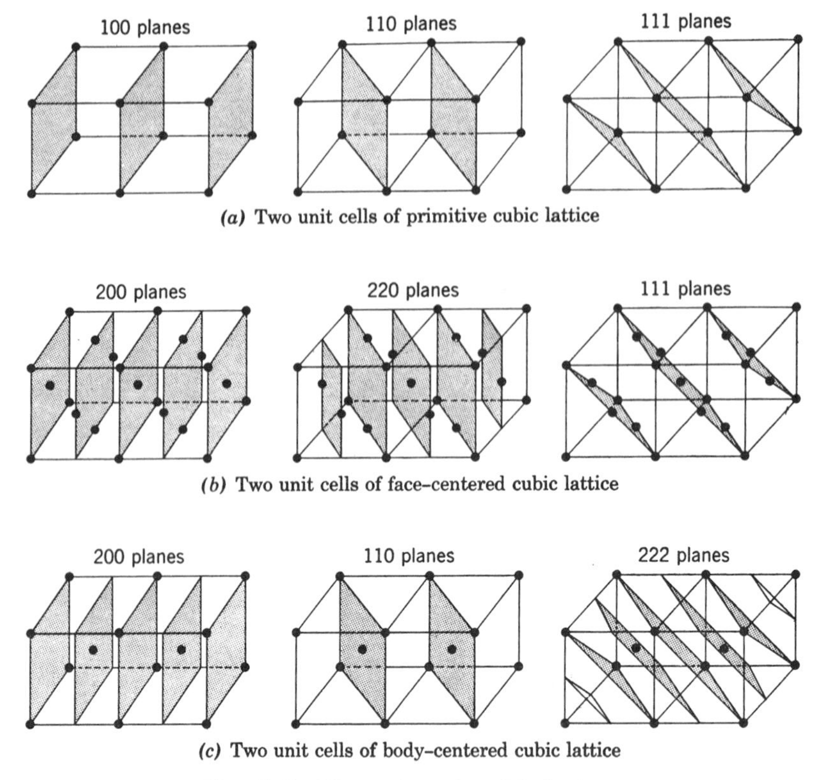

(*a*) Two unit cells of primitive cubic lattice

(*b*) Two unit cells of face-centered cubic lattice

(*c*) Two unit cells of body-centered cubic lattice

Fig. 19.9 Planes through cubic lattices.

corners. One-half of the face-centered lattice points and one-eighth of the corner lattice points belong to the unit cell, making a total of four per cell. Hence, the equivalent positions are xyz; $\frac{1}{2} + x, \frac{1}{2} + y, z$; $\frac{1}{2} + x, y, \frac{1}{2} + z$; $x, \frac{1}{2} + y, \frac{1}{2} + z$. These four equivalent positions also can be given as $000 + xyz$; $\frac{1}{2}\frac{1}{2}0 + xyz$; $\frac{1}{2}0\frac{1}{2} + xyz$; $0\frac{1}{2}\frac{1}{2} + xyz$. NaCl has a face-centered cubic lattice (Fig. 19.12). Since all of the lattice points can be considered to be occupied by Na^+, there are four Na^+ per unit cell. Since the Cl^- ions on the 12 cell edges (in which each edge is shared by four unit cells) and at the center of the unit cell also are related to one another by face-centering, there are four Cl^- per unit cell. The positions of the four Na^+ and four Cl^- ions are as follows.

Four Na^+ 000; $\frac{1}{2}\frac{1}{2}0$; $\frac{1}{2}0\frac{1}{2}$; $0\frac{1}{2}\frac{1}{2}$

Four Cl^- $\frac{1}{2}00$; $0\frac{1}{2}0$; $00\frac{1}{2}$; $\frac{1}{2}\frac{1}{2}\frac{1}{2}$

The diffraction patterns of NaCl and any other face-centered crystal (regardless of the crystal system) show an *absence* of all reflections for which the indices *hkl* are *not all even* or *all odd*. Hence, only the reflections 111, 200, 220, 311, 222, 400, 331 420, etc., are observed. It can be shown mathematically (Section 19.17) that the X-ray scattering from the face-centered, symmetry-related atoms due to the equivalent points at 000; $\frac{1}{2}\frac{1}{2}0$; $\frac{1}{2}0\frac{1}{2}$; $0\frac{1}{2}\frac{1}{2}$ is completely in phase for reflections with indices all even or all odd but completely out of phase for all other *hkl* reflections. Consequently, the spacings corresponding to the *hkl* reflections that may appear on a powder diagram of a face-centered cubic crystal are (from equation 19.10) $a/\sqrt{3}$, $a/\sqrt{4}$, $a/\sqrt{8}$, $a/\sqrt{11}$, $a/\sqrt{12}$, $a/\sqrt{16}$, $a/\sqrt{19}$, $a/\sqrt{20}$, etc.

A body-centered lattice has two positions per unit cell related by translational symmetry, namely xyz; $\frac{1}{2} + x, \frac{1}{2} + y, \frac{1}{2} + z$. Hence there are two lattice points per unit cell at 000 and $\frac{1}{2}, \frac{1}{2}, \frac{1}{2}$ that have identical environments, as shown in Fig. 19.9*c*. An examination of the diffraction data for body-centered crystals shows that *hkl* reflections for which the sum $h + k + l$ is odd are not observed. This means that the scattering of each atom in a body-centered unit cell is completely in phase with that of its corresponding body-centered equivalent atom if the sum of the indices is even, but 180° out of phase if the sum is odd. Accordingly, the interplanar spacings found for a *body-centered cubic lattice* are $a/\sqrt{2}$, $a/\sqrt{4}$, $a/\sqrt{6}$, $a/\sqrt{8}$, etc., which are the distances between the (110), (200), (211), and (220) planes.

The reflections for the three types of cubic crystals are summarized in Fig. 19.10, in which the presence of a reflection is indicated by a line at the corresponding angle of incidence. For the purposes of this illustration, the ratio λ/a is arbitrarily taken as 0.500 for primitive cubic, 0.353 for body-centered cubic, and 0.289 for face-centered cubic. It may be seen that the various types of cubic crystals may be distinguished by their diffraction patterns, since the patterns are qualitatively different. In the powder pattern for primitive cubic the successive lines are closer together, and there is a gap after the sixth line. In the powder pattern for body-

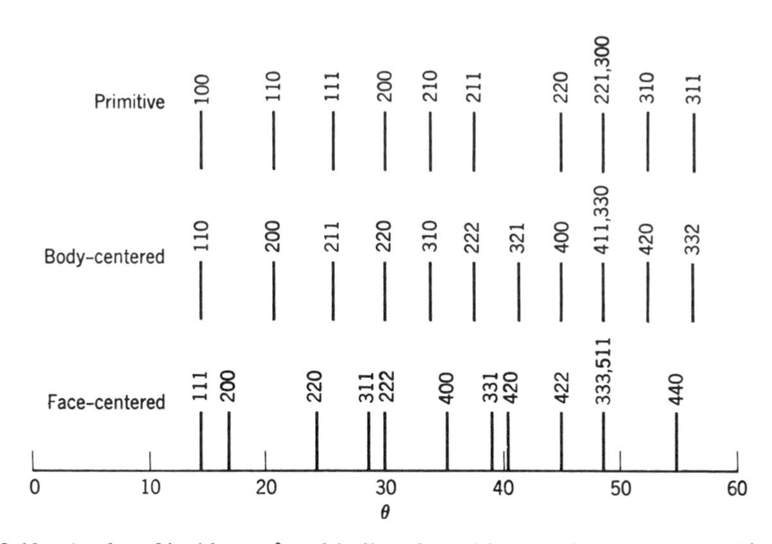

Fig. 19.10 Angles of incidence θ and indices for cubic crystals. The values of λ/a have been chosen arbitrarily to cause the first reflection to fall at the same angle for each type of crystal. For primitive cubic, $\lambda/a = 0.500$; for body-centered cubic, $\lambda/a = 0.353$; for face-centered cubic, $\lambda/a = 0.289$.

centered cubic this gap is filled in. In the powder pattern for face-centered cubic the first two lines are relatively close together, the second line is by itself, and the next two lines are close together. Thus, by use of powder patterns, it is possible to determine whether a cubic crystal is primitive, body centered, or face centered.

19.10 POWDER PATTERNS FOR CUBIC CRYSTALS

The powder patterns of three substances forming cubic crystals are shown in Fig. 19.11. As may be seen by comparison with Fig. 19.10, the reflections for sodium chloride are found to correspond to those expected for a face-centered cubic lattice. The reflection indices have been assigned on this basis. The 100 reflection is missing, and so none of the spacings calculated using the Bragg equation is equal to the length of the side of the unit cell *a*. However, *a* may be calculated from the angle of any reflection by use of equations 19.8 and 19.9. The value of *a* for sodium chloride is 564 pm.

The structural units of the face-centered lattice that have thus been found could be sodium chloride molecules, or there could be a lattice of equal numbers of sodium and chloride ions. If the lattice were made up of sodium chloride molecules, all units of the lattice would be the same, and a more detailed consideration of the

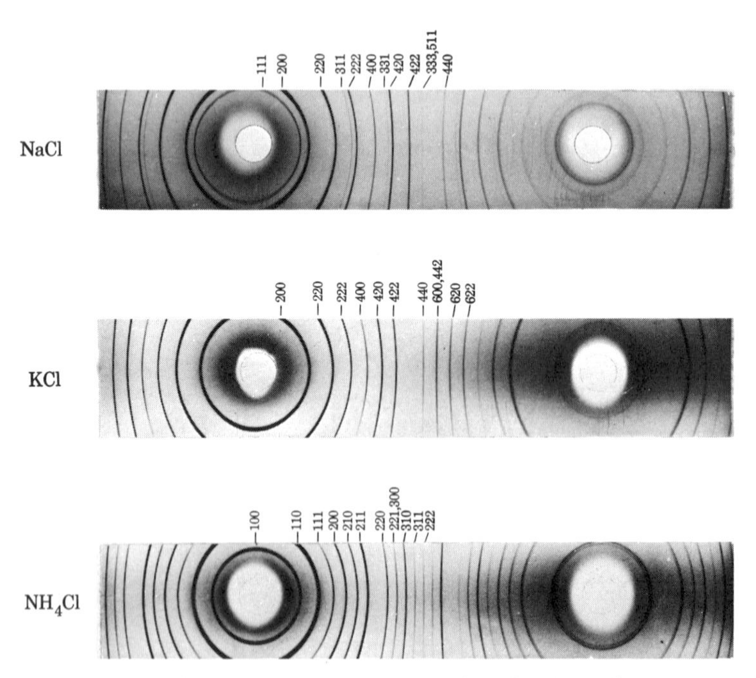

Fig. 19.11 X-ray powder patterns for cubic crystals. The X-ray beam enters through the hole at the right and leaves through the hole at the left (see Fig. 19.8). (Courtesy Prof. S. Bailey of the University of Wisconsin.)

theory shows that the intensities of the X-ray reflections would decrease progressively from first- to second- to third-order reflections. This is true for the 200, 400, and 600 reflections, but not for the 111, 222, 333, ... reflections. It may be noted in Fig. 19.11 that the 111 reflection is weak, whereas 222 is strong and 333 is apparently missing. This fact leads to the requirement of a lattice with ions at the lattice points.

The square connecting sodium ions is drawn in Fig. 19.12, showing that there is a face-centered sodium ion. Similar squares could be drawn on any part of the faces. It is evident that the chloride ions are displaced half a cell edge from the sodium ions.

A further examination of Fig. 19.12 shows that the (111) planes that cut diagonally through the sodium chloride crystal include only sodium ions or only chloride ions. Thus the (222) planes are alternately planes of sodium and chloride ions. It may be remembered that the maximum in X-ray reflections occurs when the angle is such that the paths between successive layers of ions are equal to one wavelength of the reflected radiation. If the rays are reflected from these planes at such an angle that the rays from successive planes of chloride ions differ in path length by one wavelength, the rays coming from successive sodium-ion planes, which are spaced equally between them, will then differ by half a wavelength and cause interference. The interference would be complete except for the fact that the chloride ions have more electrons and scatter X rays more efficiently than the sodium ions. The reflections from the (222) planes, however, have a difference of a whole wavelength between the reflections from the chloride and from the sodium planes, so that there is no interference, and the 222 reflection is intense. The 333 reflection again corresponds to a difference of one-half wavelength between the two sets of reflecting planes, and this interference, combined with the fact that the third-order spectrum is naturally weaker, leads to a very weak reflection.

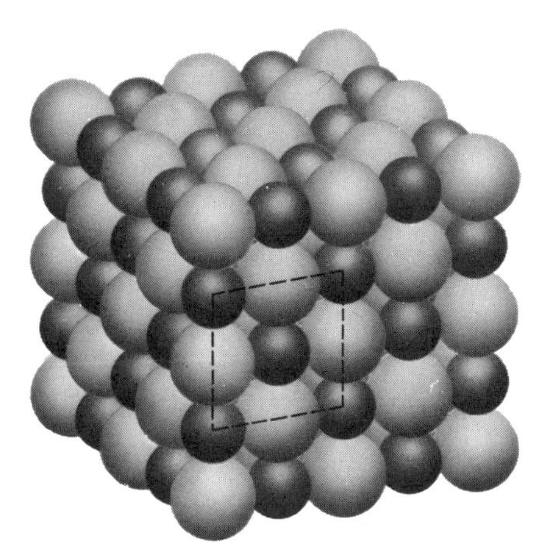

Fig. 19.12 Face-centered lattice of sodium chloride; small spheres, sodium ions; large spheres, chloride ions. Each Na^+ is octahedrally surrounded by six Cl^- and each Cl^- octahedrally surrounded by six Na^+.

The powder pattern for potassium chloride given in Fig. 19.11 is superficially that of a primitive cubic lattice. This seems surprising, since the structure would be expected to be face centered like that for sodium chloride. The reflections look very much like those from a simple cubic structure because the scattering power of the potassium ion is almost exactly equal to that of the chloride ion, since both ions have the argon electronic structure. A more accurate determination of the intensities of reflection shows that potassium chloride does indeed form a face-centered lattice.

The powder pattern for ammonium chloride given in Fig. 19.11 shows that it has a primitive cubic lattice. If the center of a chloride ion is taken as the corner of the unit cell, the ammonium ion lies at the center of the cell, but the crystal is not body-centered cubic, since the ions are not equivalent.

Although the powder camera technique has been widely utilized for fingerprinting compounds, its use to determine the arrangement of atoms is mainly limited to the simple crystal structures comprising the cubic, hexagonal, and tetragonal systems. In general, it is much more convenient to utilize single crystal X-ray techniques to determine the arrangement of atoms. The main advantage of X-ray diffraction over other structural methods is that X-ray diffraction in practically all cases provides a *direct, unique* solution of the structure.

19.11 UNIT-CELL DIMENSIONS

After it is known whether a cubic crystal is of the primitive, face-centered, or body-centered type, the size of the unit cell may be calculated by use of Bragg's law from the angles at which X rays are reflected. For example, when X rays from a palladium target having a wavelength of 58.1 pm are used, the 200 reflection of sodium chloride occurs at an angle of 5.91°. According to Bragg's law,

$$d_{200} = \frac{\lambda}{2 \sin \theta} = \frac{58.1 \text{ pm}}{2 \sin 5.9} = 282 \text{ pm} \tag{19.11}$$

Since the distance between (200) planes is one-half the length of the side of this unit cell, $a = 564$ pm.

If the density and the size of the unit cell are known, the number of atoms, ions, or molecules per unit cell may be calculated. Since the density of sodium chloride is 2.163×10^3 kg m^{-3} at 25 °C, the molar mass is 58.443 g mol^{-1}, and the length of the side of the unit cell is 564 pm.

$$2.163 \times 10^3 \text{ kg}^{-3} = \frac{n(58.443 \times 10^{-3} \text{ kg mol}^{-1})}{(6.022\,05 \times 10^{23} \text{ mol}^{-1})(564 \times 10^{-12} \text{ m})^3} \tag{19.12}$$

$$n = 3.999$$

Thus, as expected, the unit cell contains four sodium ions and four chloride ions.

Example 19.1 Potassium crystallizes with a body-centered cubic lattice and has a density of 0.856×10^3 kg m^{-3}. What is the length of the side of the unit cell a and the distance between (200), (110), and (222) planes? What is the closest distance between atoms and what is the potassium atom radius r?

$$0.856 \times 10^3 \text{ kg m}^{-3} = \frac{2(39.102 \times 10^{-3} \text{ kg mol}^{-1})}{(6.022\,045 \times 10^{23} \text{ mol}^{-1})a^3}$$

$$a = 533.3 \times 10^{-12} \text{ m} = 533.3 \text{ pm}$$

Using equation 19.10,

$$\text{For (200) planes, } d_{200} = 533.3/\sqrt{4} = 266.7 \text{ pm}$$

$$\text{For (110) planes, } d_{110} = 533.3/\sqrt{2} = 377.1 \text{ pm}$$

$$\text{For (222) planes, } d_{222} = 533.3/\sqrt{12} = 154.0 \text{ pm}$$

$$(2r)^2 = \left(\frac{a}{2}\right)^2 + \left(\frac{a}{2}\right)^2 + \left(\frac{a}{2}\right)^2 \qquad 2r = 461.9 \text{ pm} \qquad r = 231.0 \text{ pm}$$

This type of calculation can be used to calculate the Avogadro constant N_A if the density of a crystal, the relative atomic mass, and the unit cell length are determined very accurately. This has been done at the National Bureau of Standards with silicon crystals.* The value of the Avogadro constant determined in this way [$6.022\,097\,6(63) \times 10^{23}$ mol^{-1}] is tied to measurements of other fundamental constants through a least squares adjustment to obtain the best values given in the Appendix.

There is a condition on the ratio of ion sizes that must be satisfied in order for an ionic substance MX to have the NaCl structure. Since the ions are in contact along a cell edge,

$$a = 2(R_+ + R_-) \tag{19.13}$$

Ions cannot overlap along the diagonal of the face of the unit cell. Therefore

$$(4R_-)^2 \leqslant 2a^2 \tag{19.14}$$

Thus

$$a \geqslant 2\sqrt{2}R_- \qquad \text{or} \qquad 2\sqrt{2}\,R_+ \tag{19.15}$$

and

$$2(R_+ + R_-) \geqslant 2\sqrt{2}\,R_- \tag{19.16}$$

$$R_+/R_- \geqslant \sqrt{2} - 1 = 0.414 \tag{19.17}$$

For example, since $R_{Li^+}/R_{Cl^-} = 0.331$ (Table 19.1), it is evident that LiCl cannot have the NaCl structure.

Ionic radii in crystals for a number of anions and cations are summarized in Table 19.1. It is seen that in each column of the periodic table the ionic radius increases with the number of orbital electrons. The radius of an ion is nearly the

* R. D. Deslattes, A. Hemins, R. M. Schoonover, C. L. Carroll, and H. A. Bowman, *Phys. Rev. Letters*, **36**, 898 (1976).

Table 19.1 Ionic Radii in Crystals in pm [a]

Li^+	60	Be^{2+}	31	O^{2-}	140	F^-	136
Na^+	95	Mg^{2+}	65	S^{2-}	184	Cl^-	181
K^+	133	Ca^{2+}	99	Se^{2-}	198	Br^-	195
Rb^+	148	Sr^{2+}	113	Te^{2-}	221	I^-	216
Cs^+	169	Ba^{2+}	135				

[a] L. Pauling, *The Nature of the Chemical Bond*, Cornell University Press, Ithaca, 1960.

same in different crystals because the repulsive force increases very sharply as the internuclear distance becomes smaller than a certain value.

Table 19.2 Covalent Radii for Atoms[a] (radii in pm)

	H	C	N	O	F
Single-bond radius	30	77.2	70	66	64
Double-bond radius		66.7	60	56	
Triple-bond radius		60.3			
		Si	P	S	Cl
Single-bond radius		117	110	104	99
Double-bond radius		107	100	94	89
Triple-bond radius		100	93	87	
		Ge	As	Se	Br
Single-bond radius		122	121	117	114
Double-bond radius		112	111	107	104
		Sn	Sb	Te	I
Single-bond radius		140	141	137	133
Double-bond radius		130	131	127	123

[a] Taken from L. Pauling, *The Nature of the Chemical Bond*, Cornell University Press, Ithaca, N.Y., 1960, which should be consulted for details concerning the source and constancy of these radii.

19.12 INTERATOMIC DISTANCES

It has been found that the distance between two kinds of atoms connected by a covalent bond of a given type (single, double, etc.) is nearly the same in different molecules. The distance between two atoms is taken to be equal to the sum of the bond radii of the two atoms. Since the C—C bond distance is 154 pm in many compounds, the radius for a carbon single bond is taken to be 77 pm. Since the C≡C distance in acetylene is 120 pm, the radius for a carbon triple bond is taken to be 60 pm. By consideration of the bond distances in many compounds it has been possible to build up tables of bond radii, such as Table 19.1, which are useful in describing the structure of molecules. It must be realized, however, that the effective radius of an atom depends in part also on its structure and environment and on the nature of bonds that it forms with other atoms in the molecule.

The interatomic distances and configurations of many molecules and ions have been summarized.*

19.13 CLOSE PACKING OF SPHERES

When the bonding of atoms is not highly directional, it is often found that the lowest energy structure is that in which each atom is surrounded by the greatest possible number of neighbors. It is, therefore, of interest to consider the ways in which uniform spheres can be stacked to form close-packed structures. When spheres are packed in a plane, they arrange themselves so that each sphere is surrounded hexagonally by six others. When the second layer is formed by placing spheres in the hollows on top of the first layer, it is evident that all of the hollows in the first layer are not occupied, as may be seen from Fig. 19.13. When a third layer is added, there is a choice as to whether the spheres in this layer are stacked so that they are not above the spheres in the first layer, as in Fig. 19.13*a*, or are, as in Fig. 19.13*b*. If the spheres in the third layer are not directly above the spheres in the first layer, as shown in Fig. 19.13*a*, the structure has cubic symmetry and the cubic unit cell is face centered. The fact that cubic close packing is really face-centered cubic may be seen from Fig. 19.14. Since this is a close-packed structure, it is of interest to calculate how efficient the packing is. Since the length of the diagonal of the face of a unit cell is $\sqrt{2}a$, the radii of the spheres that just touch are given by $(\sqrt{2}/4)a$. Since there are four spheres per unit cell, the fraction of the volume occupied by spheres is

$$\frac{4(\frac{4}{3}\pi)\left(\frac{\sqrt{2}}{4}\,a\right)^3}{a^3} = 0.7405 \tag{19.18}$$

In cubic close packing each sphere has twelve nearest neighbors, six within its own layer, three in the layer above, and three in the layer below. Metals and inert gases often form cubic close-packed structures.

If the spheres in the third layer are placed over the spheres in the first layer, as shown in Fig. 19.13*b*, and the spheres in the fourth layer are placed over those in the second layer, and so on, the unit cell is hexagonal and the packing is referred to as hexagonal close packing. As in the case of cubic close packing, each sphere has twelve nearest neighbors, and the fraction of the volume occupied by spheres is again 0.7405. The coordinates of the atoms in the unit cell are (000) and $(\frac{1}{3}\frac{1}{3}\frac{1}{2})$. There are two atoms in the unit cell.

The unit cell dimensions in terms of the radius of a sphere are $a = b = 2r$, $c = 4\sqrt{2}r/\sqrt{3}$, and $c/a = 2\sqrt{2}/\sqrt{3} = 1.633$. A number of metals have hexagonal close-packed structures, but c/a usually deviates a little from the ideal ratio of 1.633. This indicates that the atoms are not exactly spherical in shape.

---

* *Tables of Interatomic Distances and Configurations in Molecules and Ions*, The Chemical Society, London, 1958.

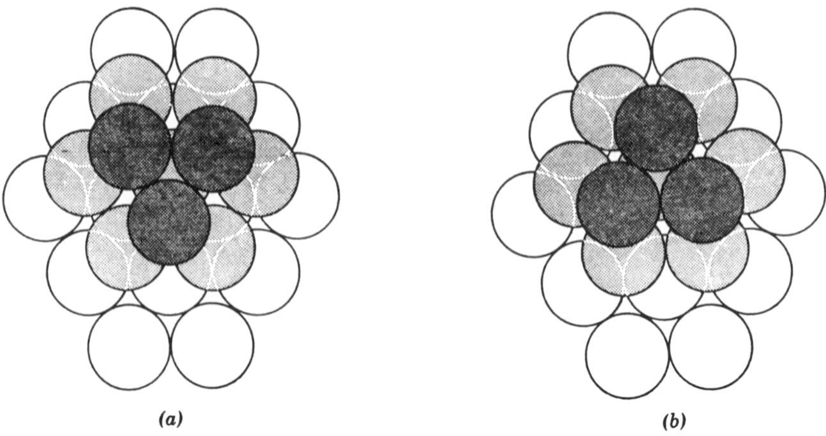

Fig. 19.13 (*a*) Cubic close packing. (*b*) Hexagonal close packing.

The layers of hexagonal close packing may be described as ABABAB···. The layers of cubic close packing may be described by ABCABC···.

Hexagonal close packing and cubic close packing are the only two ways of close packing identical spheres so that the environment of each sphere is identical with the environment of all the other spheres, but there are other ways of close packing spheres so that the environment of each sphere is not identical; for example, ABCABABCAB···. In principle there is an infinity of these other ways.

In a number of crystal structures containing two types of atoms, one type of atoms forms a close-packed structure and the other type occupies interstices between the close-packed spheres. There are two types of interstices between close-packed spheres, tetrahedral sites and octahedral sites. When one sphere rests on three others, the centers of the four spheres lie at the apices of a regular tetrahedron, and the space at the center of this tetrahedron is called a tetrahedral site. Since in any

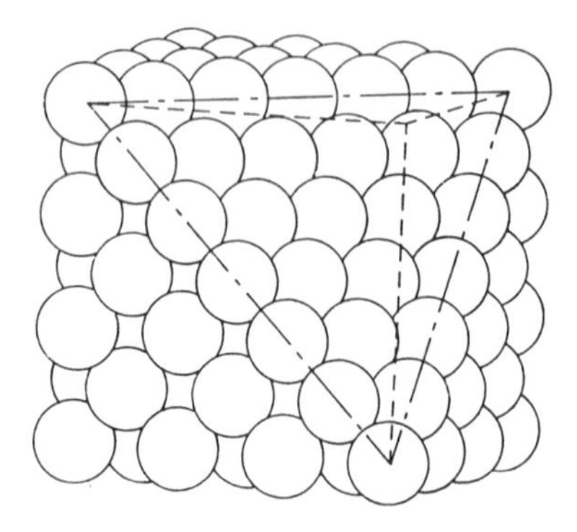

Fig. 19.14 Cubic close packing (face-centered cubic packing). Some atoms have been omitted to show that the close packed planes are (111) planes.

close-packed structure each sphere is in contact with three spheres in the layer below it and three spheres in the layer above it, there are two tetrahedral sites per sphere. Thus if A atoms form a close-packed structure, and B atoms are small enough to fit in the tetrahedral sites, this might be a convenient structure for a compound AB_2. Half of the sites would be occupied in a compound AB. In order for the smaller sphere to occupy a tetrahedral site without disturbing the closest-packed lattice, the radius of the smaller spheres should be no greater than 0.225 of that of the larger spheres.

The other type of site in a closely packed structure is the octahedral site that is surrounded by six spheres whose centers lie at the apices of a regular octahedron. There is one octahedral site for every sphere in the structure so that a compound of the type XY can be accommodated. In order for Y atoms to fit into octahedral sites their radii must be less than 0.414 of the radii of the larger spheres.

Although the majority of metallic elements crystallize with hexagonal close packing or cubic close packing, some crystallize with the body-centered cubic arrangement, which is not a close-packed structure.

Example 19.2 Magnesium forms hexagonal close-packed crystals with $a = 320.9$ pm at 25 °C. What is the density of the metal and the magnesium ion radius?

$$V = a^2c(1 - \cos^2\gamma)^{1/2} = a^2c\sin\gamma$$

Since $c = 1.633a$,

$$V = 1.633(320.9 \times 10^{-12}\ \text{m})^3 \sin 120° = 4.673 \times 10^{-29}\ \text{m}^3$$

$$d = \frac{2(24.305 \times 10^{-3}\ \text{kg mol}^{-1})}{(6.022\,045 \times 10^{23}\ \text{mol}^{-1})(4.673 \times 10^{-29}\ \text{m}^3)} = 1.727 \times 10^3\ \text{kg m}^{-3}$$

$$r = \tfrac{1}{2}a = \tfrac{1}{2}(320.9\ \text{pm}) = 160.5\ \text{pm}$$

19.14 BODY-CENTERED CUBIC STRUCTURE OF SPHERES

In this structure, which is illustrated in Fig. 19.15, each atom has eight nearest neighbors and six other next nearest neighbors slightly further away at the body-centered positions of neighboring cells. By use of the Pythagorean theorem it is readily shown that the distance from the body-centered point to one of the corners of the cubic unit cell is $(\sqrt{3}/2)a$. If the structure is made up of spheres that touch they must have a radius of $(\sqrt{3}/4)a$. The fraction of the volume of the unit cell (and hence of the entire crystal) occupied by spheres is

$$\frac{2(\frac{4}{3}\pi)\left(\frac{\sqrt{3}}{4}a\right)^3}{a^3} = 0.6802 \tag{19.19}$$

A number of metallic elements crystallize in the body-centered cubic structure even though it is not a close-packed structure. The alkali metals and tungsten crystallize in a body-centered cubic structure.

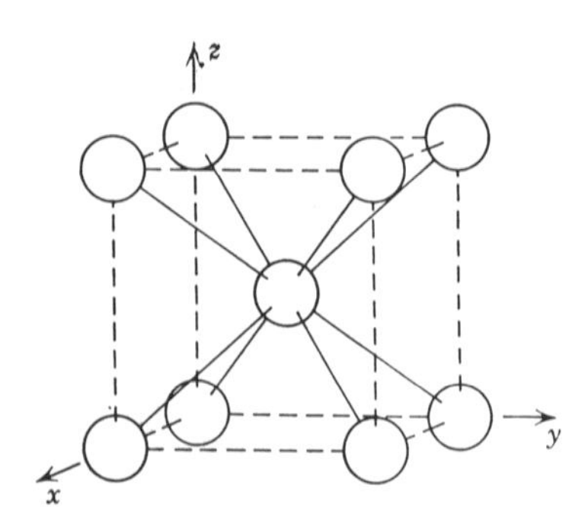

Fig. 19.15 Body-centered cubic structure.

19.15 DIAMOND STRUCTURE

Diamond has a face-centered cubic lattice with atoms at 000 and $\frac{1}{4}\frac{1}{4}\frac{1}{4}$ associated with each lattice point. This structure is represented in two different ways in Fig. 19.16. Since there are two atoms per lattice point, there are eight atoms per unit cell. The unit cell distance for diamond is 356.7 pm. Silicon, germanium, and gray tin also have this structure with unit cell distances of 543.1, 565.7, and 649.1 pm.

Example 19.3 Calculate the C—C bond distance in diamond and the C—C—C angle.

Using equation 19.3 for the points 000 and $\frac{1}{4}\frac{1}{4}\frac{1}{4}$,

$$l = a[(x_2 - x_1)^2 + (y_2 - y_1)^2 + (z_2 - z_1)^2]^{1/2}$$
$$= (356.7 \text{ pm})(\tfrac{3}{16})^{1/2} = 154.5 \text{ pm}$$

which is the C—C distance.

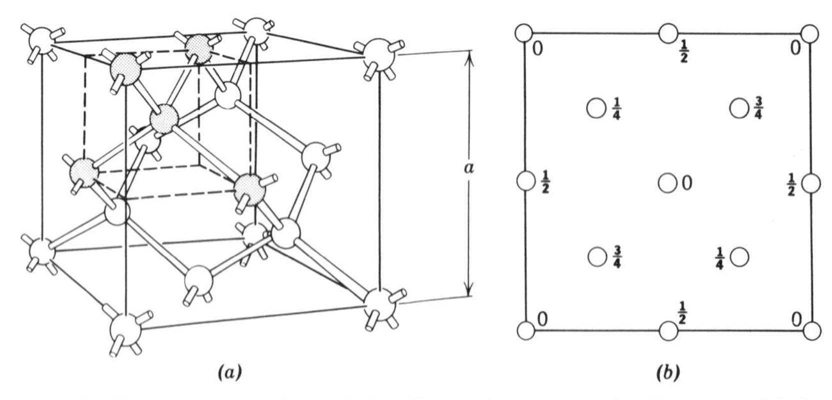

Fig. 19.16 Two representations of the diamond structure. (*a*) Space model showing tetrahedral bonds. (*b*) Projection showing fractional coordinates. It is instructive to draw lines showing the bonds in this projection.

The distance l between a carbon atom at a corner of the unit cell and at a face-centered position is given by

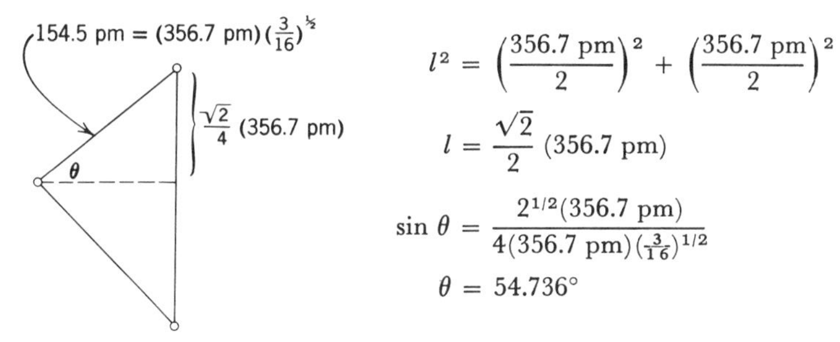

$$l^2 = \left(\frac{356.7\ \text{pm}}{2}\right)^2 + \left(\frac{356.7\ \text{pm}}{2}\right)^2$$

$$l = \frac{\sqrt{2}}{2}(356.7\ \text{pm})$$

$$\sin\theta = \frac{2^{1/2}(356.7\ \text{pm})}{4(356.7\ \text{pm})(\frac{3}{16})^{1/2}}$$

$$\theta = 54.736°$$

The C—C—C bond angle is 2θ or 109.471°, which is referred to as the tetrahedral angle.

19.16 ELECTRON DENSITY FUNCTION

Since X rays are scattered by the electrons of the atoms, the objective of X-ray diffraction experiments is to obtain the electron density function $\rho(xyz)$. The electron density function is the number of electrons per unit volume as a function of the coordinates x, y, and z. Since the electron density $\rho(xyz)$ is a periodic function, it is convenient to represent it by a Fourier series. A periodic function in one direction x may be represented by

$$f(x) = \sum_n A_n \cos n\pi x + \sum_n B_n \sin n\pi x \tag{19.20}$$

where A_n and B_n are coefficients chosen to give $f(x)$ the desired shape. The summations are over all positive and negative integers n. A Fourier series may also be represented by

$$f(x) = \sum_n C_n e^{-in\pi x} \tag{19.21}$$

where $i = \sqrt{-1}$. Series 19.20 and 19.21 are equivalent because

$$e^{-in\pi x} = \cos n\pi x - i \sin n\pi x \tag{19.22}$$

Since the electron density is a function of three variables, we may write

$$\rho(xyz) = \sum_h \sum_k \sum_l F(hkl) \exp\left[-2\pi i(hx + ky + lz)\right] \tag{19.23}$$

where the $F(hkl)$, referred to as the structure factors, are the coefficients to be determined and h, k, and l are the integers over which the series is summed. The coefficients $F(hkl)$ are related to the intensities $I(hkl)$ of radiation reflected from the planes (hkl) by

$$I(hkl) \propto |F(hkl)|^2 \tag{19.24}$$

Measurements of the densities of spots on photographic film or of counts recorded by a Geiger counter for a certain reflection may be subjected to routine corrections to obtain $I(hkl)$ values. Thus a set of $|F(hkl)|^2$ values may be obtained. Unfortunately, what is needed to calculate $\rho(xyz)$ using equation 19.23 are values of $F(hkl)$ instead of $|F(hkl)|^2$. Since $F(hkl)$ is a complex number, we can write

$$F(hkl) = A(hkl) + iB(hkl) \tag{19.25}$$

so that

$$\begin{aligned}|F(hkl)|^2 &= [A(hkl) + iB(hkl)][A(hkl) - iB(hkl)] \\ &= [A(hkl)]^2 + [B(hkl)]^2\end{aligned} \tag{19.26}$$

Since values of $A(hkl)$ and $B(hkl)$ are not obtained directly, indirect methods must be used to obtain these quantities and the electron density function $\rho(xyz)$. This is the phase problem. Fortunately, the number of parameters needed to describe a crystal structure is far smaller than the number of reflections, so that the problem is greatly overdetermined. Several methods are used to get around the phase problem. If heavy atoms are present in the unit cell, they may by themselves determine enough phases so that a Fourier map of the electron density may reveal the position of some of the lighter atoms. Another method that has been used is referred to as isomorphous replacement. In this method, which has been especially useful in determining the structures of protein crystals, crystals are prepared with different heavy atoms and information about phases is obtained by comparing the intensities from the different crystals.

19.17 STRUCTURE FACTOR

The intensity of a beam of X rays scattered by a crystal in a particular direction is proportional to the square of the structure factor $F(hkl)$, introduced in equation 19.23. The structure factor may be related to atomic scattering factors f_i, which depend on the number and distribution of atomic electrons.

$$F(hkl) = \sum_j f_j e^{-2\pi i(hx_j + ky_j + lz_j)} \tag{19.27}$$

where x_j, y_j, and z_j are the atomic coordinates (Section 19.5) of atom j in the unit cell. The atomic scattering factors f_j are functions of $s = (4\pi \sin \theta)/\lambda$. At small s, f_j is equal to the number of electrons on the atom or ion. Thus heavy atoms scatter X rays more effectively than light atoms.

The form of the structure factor expression may be considered to arise as follows. When Bragg's law is satisfied for a given reflection, the amplitude of the wavelet scattered from an atom in one unit cell of the crystal is in phase with the amplitudes of the scattered wavelets from the corresponding atoms in the millions of other unit cells of the crystal. The wavelet scattered by one atom, however, generally will not be in phase with the wavelet scattering by a different atom within the same unit cell, and hence the intensity of the reflection will depend on the extent to which the amplitudes of the different atoms (denoted by the scattering power f_j of the jth atom) are in phase with one another. Since the phase difference between a wavelet scattered by an atom at the *origin* (or corners) of the unit cell and that scattered by an atom whose fractional coordinates are x, y, z is

$$2\pi(hx + ky + lz)$$

the resultant contribution of each atom to the total scattering power of all the atoms in the unit cell (as represented by the structure factor expression) is $f_j e^{2\pi i(hx_j + ky_j + lz_j)}$.

To illustrate the usefulness of the structure factor, let us consider its dependence on the reflection indices for P, C, I, and F lattices (see Fig. 19.5). In P lattices with atoms at the lattice sites, x_j, y_j, and z_j are zero, and so $F(hkl) = \sum f_j$. Thus the structure factor $F(hkl)$ has the same value for all values of h, k, and l, and there will be reflections for all integer values of h, k, and l.

In a C lattice the values of x_j, y_j, and z_j are 000 and $\frac{1}{2}\frac{1}{2}0$, and so the structure factor is given by

$$\begin{aligned} F(hkl) &= fe^{-2\pi i(0+0+0)} + fe^{-2\pi i[(h/2)+(k/2)+0]} = f[1 + e^{-\pi i(h+k)}] \\ &= f[1 + \cos(h + k)\pi - i \sin(h + k)\pi] \end{aligned} \tag{19.28}$$

The structure factor $F(hkl)$ is zero when $h + k = 2m + 1$ with $m = 0, \pm 1, \pm 2, \ldots$. Thus, when $h + k$ is odd, the structure factor $F(hkl)$ is zero and the reflection is extinguished.

In an I lattice the values of x_j, y_j, and z_j are 000 and $\frac{1}{2}\frac{1}{2}\frac{1}{2}$, and so the structure factor is given by

$$F(hkl) = f[1 + e^{-\pi i(h+k+l)}] = f[1 + \cos(h + k + l)\pi - i \sin(h + k + l)\pi] \tag{19.29}$$

The structure factor $F(hkl)$ is zero when $h + k + l$ is odd. Thus 111 is an extinguished reflection for an I lattice.

In an F lattice the values of x_j, y_j and z_j are 000, $\frac{1}{2}\frac{1}{2}0$, $\frac{1}{2}0\frac{1}{2}$, and $0\frac{1}{2}\frac{1}{2}$. Following the procedure described above, it may be shown that if the indices h, k, l are all even or all odd, the reflection is allowed for an F lattice.

19.18 RESULTS FROM X-RAY DIFFRACTION STUDIES

X-ray analyses of thousands of crystal structures have led to detailed knowledge of the geometrical properties of different groups of atoms, including well-established values of bond lengths and angles. The resulting stereochemical principles have been of great help in the determination of new crystal structures, particularly for large molecules of biological origin that are composed of small basic units. Modern crystallographic analyses using data-collecting diffractometers and high-speed computers have enabled the molecular architecture of proteins to be determined. X-ray diffraction data were used in the determination of the structure of deoxyribonucleic acid and in learning about the hydrogen bonding that makes that structure stable.

In certain cases X-ray diffraction may be used to determine the absolute configuration of an optically active substance. In 1951 Bijroet, Peerdeman, and van Bommel studied sodium rubidium (+)-tartaric acid by X-ray diffraction and found that the absolute configuration was the one arbitrarily chosen from the two possible enantiomorphic structures by Fischer 60 years earlier. X-ray diffraction also has been widely used in inorganic chemistry to determine *both* the *structure* and correct *formulas* of many boron hydride and metal carbonyl complexes where erroneous assignments of formulas have been previously made. In the study of the synthetic elements neptunium, plutonium, curium, and americium it was possible to establish quickly the purity of compounds and the chemical composition with exceedingly small amounts of material and without destroying the samples.

The X-ray pattern for an unknown substance may be used to identify it, or the presence of impurities may be detected, and even measured quantitatively, in a known substance. This application is greatly facilitated by such tables as those of the American Society for Testing Materials, which give the spacings calculated from the three strongest powder lines for a very large number of substances.

19.19 NEUTRON DIFFRACTION

The average de Broglie wavelength (Section 8.3) of thermal neutrons is 140 pm at room temperature. An essentially monochromatic beam may be obtained by diffraction from a crystal monochromator that selects a small band of wavelengths from the incident beam obtained from a nuclear reactor. Neutron diffraction also can be used to study the structures of crystals in the form of powders or single crystals. Although the principles of neutron diffraction are similar to those of X-ray diffraction, several fundamental differences between them result in neutron diffraction being a complementary technique to that of X-ray diffraction. Whereas X rays are scattered by electrons, neutrons are scattered primarily by the nuclei in a crystal. Hence, the atomic scattering factors for neutrons do not vary directly with atomic number as do the scattering factors for X rays but, instead, have roughly the same scattering factors (with no dependence on the Bragg scattering angle). This means in contrast to X-ray diffraction that neutron diffraction is especially useful for accurately locating hydrogen atoms in a crystalline structure. For example, in a compound such as uranium hydride, X-ray diffraction was utilized to determine the uranium coordinates and neutron diffraction the hydrogen coordinates.

Since neutrons possess a magnetic moment by virtue of having a spin of $\frac{1}{2}$, there is an additional scattering if the compound contains paramagnetic atoms or ions with unpaired electrons. Thus, neutron diffraction has been widely utilized to investigate structures of magnetic materials such as MnO and Fe_3O_4 in order to determine the arrangement of the atomic magnetic moments in the solids.

19.20 STRUCTURE OF LIQUIDS*

In a perfect crystal the atoms, ions, or molecules occur at definite distances from any individual atom, ion, or molecule that is taken as the center of a coordinate system. In a gas the molecules have random positions at a given time. Liquids are intermediate between crystals and gases in that the molecules are not arranged in a definite lattice, but there is some order. By a detailed analysis of the intensity of the scattered X-rays, it is possible to calculate the distribution of atoms or mole-

---

* Y. Marcus, *Introduction to Liquid State Chemistry*, Wiley, New York, 1977.

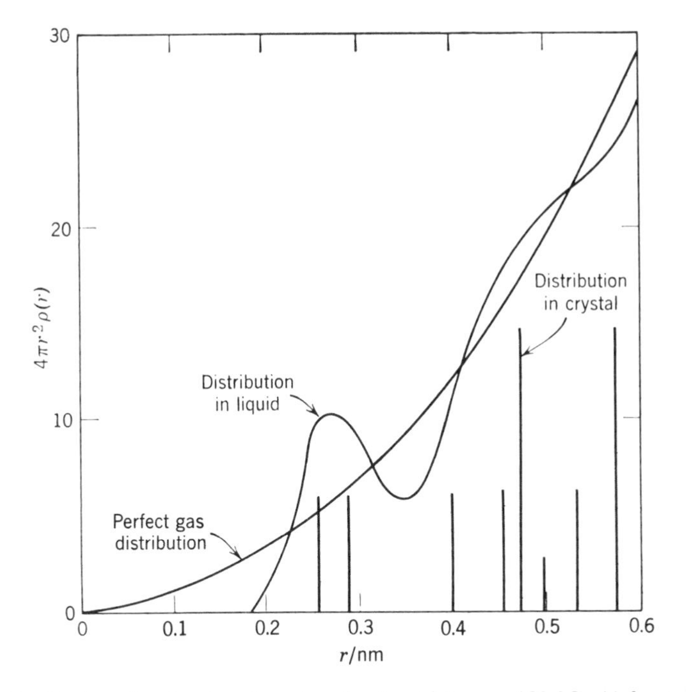

Fig. 19.17 Radial distribution function for liquid Zn at 460 °C. (Adapted from L. S. Darken and R. W. Gurry, *Physical Chemistry of Metals*. Copyright © 1953 by McGraw-Hill, Inc. Used with permission of McGraw-Hill Book Co.)

cules in the liquid and obtain a plot such as Fig. 19.17. The ordinate of this plot gives the probability of finding other atoms at a distance r from a certain atom. This probability is given by $4\pi r^2\rho$, where ρ is the local density of atoms (number of atoms per unit volume). The area under a plot of the radial distribution function $4\pi r^2\rho$ against r between two values of r is equal to the number of atoms contained in the corresponding spherical shell. In Fig. 19.17 the smooth parabolic curve represents the purely random distribution of a perfect mon-atomic gas, and the vertical lines represent the positions and numbers of the atoms in the crystal. As the temperature of a liquid is raised, the maxima and minima in the distribution curve become less pronounced and the distribution function becomes more like that for a gas. The X-ray diffraction method of investigating liquids is useful for determining the nature of the molecules in the liquid state. For example, studies of liquid phosphorus show the existence of P_4 molecules.

The theory of liquids is in a much less satisfactory state than the theories of gases and crystals, but important progress is being made in our understanding of the structure of liquids. The thermodynamic properties of a liquid may be expressed in terms of the radial distribution function.

19.21 LIQUID CRYSTALS

In certain liquids new phases, which resemble both liquid and solid phases, appear when the liquid is cooled. These phases have a translucent or cloudy appearance and are called liquid crystals.

In a liquid of asymmetric molecules the molecular axes are arranged at random. But in liquid crystals there is some kind of alignment. As shown in Fig. 19.18, there are three types of liquid crystals. In *nematic* liquid crystals the long axes of the molecules are lined up. The molecular axes are parallel to each other, but the molecules are not arranged in layers. The word nematic was coined from the Greek root for thread to describe the appearance of this particular type of liquid crystal under a microscope. Nematic liquid crystals have a translucent appearance because they scatter light strongly.

In *cholesteric* liquid crystals the molecular axes are aligned, and the molecules are arranged in layers in which the orientation of the axes shifts in a regular way in going from one layer to the next, as shown in Fig. 19.18. The distance measured perpendicular to the layers through which the direction of alignments shifts 360° is of the order of the wavelength of visible light. As a result of the strong Bragg

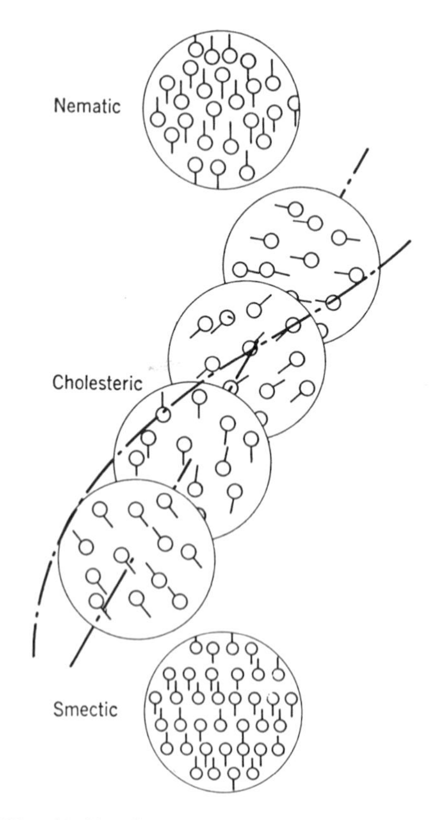

Fig. 19.18 Structures of liquid crystals.

reflection of light, cholesteric liquid crystals have vivid iridescent colors. The pitch of the spiral and the reflected color depends sensitively on the temperature, and so these liquid crystals have been used to measure skin and other surface temperatures. The name cholesteric comes from the fact that many derivatives of cholesterol (but not cholesterol itself) form this type of liquid crystal.

The third type of liquid crystals, *smectic*, are formed by certain molecules with chemically dissimilar parts. The chemically similar parts attract each other, and there is a tendency to form layers as well as to have the molecules aligned in one direction, as illustrated in Fig. 19.18. Smectic phases are soaplike in feel and structure and may have some relationships with cell membranes.

19.22 BINDING FORCES IN CRYSTALS

A number of different types of binding forces are involved in holding crystals together. The physical properties of a crystal are very dependent on the type of bonding.

Ionic crystals are held together by the electrostatic forces between ions. The lattice energy determined from heat of formation and heat of vaporization measurements agrees with that calculated on the assumption that the units of the crystal are ions held together by electrostatic forces.

In ionic crystals there is no fixed directed force of attraction. Although the ionic crystals are strong, they are likely to be brittle. They have very little elasticity and cannot be easily bent or worked. The melting points of ionic crystals are generally high (NaCl, 800 °C; KCl, 790 °C). In ionic crystals some of the atoms may be held together by covalent bonds to form ions having definite positions and orientations in the crystal lattice. For example, in calcium carbonate a carbonate ion does not "belong" to a given calcium ion, but three particular oxygen atoms are bonded to a given carbon atom.

The electric conductivity is low for reasons we will see in the next section.

Covalent crystals, which are held together by covalent bonds in three dimensions, are strong and hard and have high melting points. An example of this type is the diamond structure.

The great difference between graphite and diamond can be understood in terms of the crystal lattice. Graphite has hexagonal networks in sheets like benzene rings. The distance between atoms in the plane is 142 pm, but the distance between these atomic layer planes is 335 pm. In two directions, then, the carbon atoms are tightly held as in the diamond, but in the third direction the force of attraction is much less. As a result, one layer can slip over another. The crystals are flaky, and yet the material is not wholly disintegrated by a shearing action. This planar structure is part of the explanation of the lubricating action of graphite, but this action also depends on absorbed gases, and the coefficient of friction is much higher in a vacuum.

Van der Waals crystals are held together by the same forces (Section 10.14) that cause deviation from the perfect gas law and produce condensation at sufficiently low temperatures. Examples are provided by crystals of neutral organic compounds and rare gases. Since van der Waals forces are weak, such molecular crystals have low melting points and low cohesive strengths.

Hydrogen-bonded crystals are held together by the sharing of protons between electronegative atoms (Section 10.7). Hydrogen bonds are involved in many organic and inorganic crystals and in the structure of ice and water. They are comparatively weak bonds, but they play an extremely important role in determining the atomic arrangement in hydrogen-bonded substances such as proteins and polynucleotides.

Metallic bonds exist only between large aggregates of atoms. This type of bonding gives metals their characteristic properties, opaque, lustrous, malleable, and good conduction of electricity and heat. Metallic bonding is due to the outer, or valence, electrons. The wave functions for these electrons are sufficiently distributed over space that they have appreciable probability densities at distances equal to the interatomic distances in metals. The overlapping of the wave functions for the valence electrons in metals results in orbitals that extend over the entire crystal. The electrons pass throughout the volume of the crystal and, for certain purposes, we may consider that there is an electron gas—except that we will see that it is fundamentally different from other gases.

There is a gradual transition between metallic and nonmetallic properties. Atoms with fewer and more loosely held electrons form metals with the most prominent metallic properties. Examples are sodium, copper, and gold. As the number of valence electrons increases and they are held more tightly, there is a transition to covalent properties.

The close-packed structures are often found in metals because the binding energy per unit volume is maximized. In both of the close-packed structures for atoms of equal size, 12 spheres touch the central sphere, 6 in the midplane, 3 above, and 3 below.

19.23 ELECTRONIC STRUCTURE OF SOLIDS

Theories of the electronic structures of solids have the task of explaining why the properties of solids extend over such a wide range. For example, the electric conductivities range from about 10^8 to 10^{-18} Ω^{-1} m^{-1}.

We have seen in Section 10.3 that as atoms are brought together, their wave functions combine to form bonding and antibonding orbitals. As more atoms are brought together, this process continues. For example, Fig. 19.19 illustrates the splitting that occurs when six hydrogen atoms are brought together in a linear array. The combination of six 1*s* wave functions produces six orbitals, three bonding and three antibonding. As electrons are fed into this energy level scheme, they will first fill the lower energy bonding orbitals, two at a time. As the number of interacting atoms is increased the number of energy levels increases, and they become

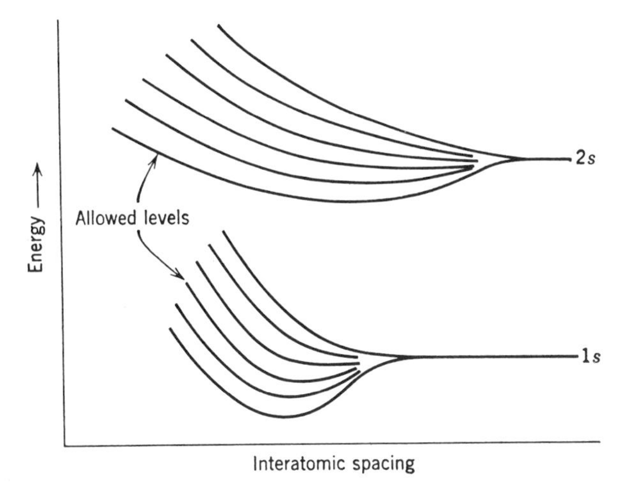

Fig. 19.19 Energy levels for a linear array of six hydrogen atoms as a function of internuclear distance.

more and more closely spaced within a band, but the width of the band at a given internuclear separation does not increase substantially. We use the term *band* to distinguish the groups of levels arising from different atomic orbitals. In Fig. 19.19 there is the 1*s* band and the 2*s* band. Thus, in contrast to molecules, we find in a solid bands of energy levels containing very large numbers of discrete levels, but with relatively large separations in energy between bands. These separations between allowed bands are called energy gaps.

We have been discussing the so-called "tight-binding approximation" for the electronic energy levels of solids. This theory works better for the inner electrons than for the valence electrons. In order to explain the electrical conductivity of solids, we go back to the free electron approximation of Drude, Lorentz, and others. According to this theory, electrons within metals have relatively large mean-free paths and are free to move when an electric field is applied. The electric conductivity κ is given by

$$\kappa = Nqu \tag{19.30}$$

where N is the number of current carriers per unit volume, q is the charge carried by each carrier, and u is its mobility (Section 18.3). Although relatively successful in explaining electric and thermal conductivities, this theory does not lead to correct contributions by the conduction electrons to the heat capacity of a metal. Electrons contribute a very small amount in comparison with the classical expectation (Section 1.12) of $\frac{3}{2}R$.

In order to understand the small contribution of conduction electrons to the heat capacity, we need to remember that electrons are put into energy levels in accord with the Pauli principle, beginning at the lowest energies. At absolute zero the energy levels are filled to a sharply defined energy, referred to as the Fermi energy

E_F. In contrast to the translational energies of gas molecules, where Maxwell-Boltzmann statistics allow any number of particles to have exactly the same energy, electrons follow Fermi-Dirac statistics, which means that only one particle is allowed in each state of the system. According to Fermi-Dirac statistics, the probability $P(E)$ of a state of energy E is

$$P(E) = \frac{1}{e^{(E-E_F)/kT} + 1} \tag{19.31}$$

where E is the energy of the state and E_F is the Fermi energy. As illustrated in Fig. 19.20, $P(E) = 1$ at absolute zero for all states below the Fermi energy, and $P(E) = 0$ for all states above the Fermi energy. At temperatures above absolute zero, some electrons just below the Fermi energy will be promoted to energies above E_F. When $(E - E_F) \gg kT$, the Fermi-Dirac distribution function reduces to

$$P(E) = e^{-(E-E_F)/kT} \tag{19.32}$$

which is the classical Boltzmann distribution. We can see from Fig. 19.20 why valence electrons make such a small contribution to the heat capacity. Only a small fraction of the electrons near the Fermi energy pick up more energy when the temperature is raised one degree.

The Fermi energy is the electrochemical potential of the electrons and determines their tendency to move at an interface, just like the chemical potential does for a substance.

The differences between insulators, metals, and semiconductors may be understood in terms of the extent of filling of energy bands. In an insulator all the bands that contain electrons are completely filled, and the gap to the next band is large. In a metal the highest occupied band is approximately half filled. As the temperature is reduced, metals become better conductors of electricity, because the thermal vibrations of the atoms of the lattice are diminished and so the atoms interfere less with the motion of the conduction electrons. Semiconductors have electrical conductivities in an intermediate range, and their conductivities increase as temperature increases. Pure substances that are semiconductors are called *intrinsic* semiconductors. Semiconductors whose properties are largely determined by impurities are called extrinsic semiconductors (see Section 19.26).

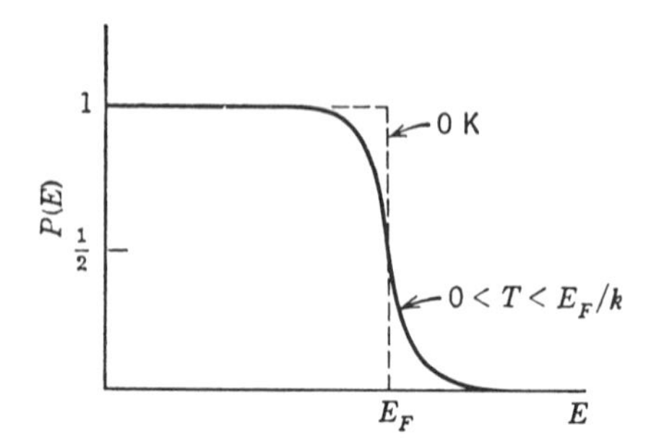

Fig. 19.20 Fermi-Dirac distribution of electron energies in a metal.

19.24 POINT IMPERFECTIONS

Real crystals have imperfections or defects. These imperfections may be at a point, along a line, or over a surface. Properties of crystals such as electrical conductivity, color, rate of diffusion, and mechanical properties may be determined as much by imperfections as by the nature of the host crystal.

A point imperfection may result from the absence of an atom (vacancy), presence of an impurity atom at a lattice site (substitutional impurity atom), presence of an impurity atom at an interstice (interstitial impurity atom), or displacement of an atom to an interstitial site (self-interstitial), as shown in Fig. 19.21.

If the repeating units in the crystal lattice are electrically neutral, vacancies produce no particular problem with respect to the overall balance of electric charge. But in an ionic crystal vacancies must be balanced so that the crystal as a whole is electrically neutral. In a *Frenkel defect* the vacancies are compensated for by an interstitial atom of the same type. In a *Schottky defect* anion and cation vacancies occur in equal numbers. Schottky defects lower the density of the crystal, but Frenkel defects do not change the density significantly.

At thermal equilibrium an otherwise perfect crystal will have a certain number of lattice vacancies, because the Gibbs energy of a crystal is decreased by the presence of disorder in the structure. The probability that a given lattice site is vacant is proportional to the Boltzmann factor $\exp(-E_v/kT)$, where E_v is the energy required to move an atom from a lattice site in the crystal to a lattice site on the surface. For a crystal with N lattice sites and n vacancies,

$$\frac{n}{N-n} = e^{-E_v/kT} \tag{19.33}$$

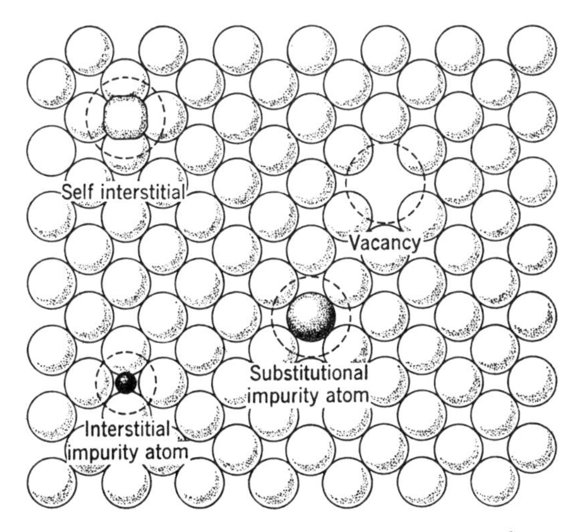

Fig. 19.21 Point imperfections on crystals.

Since E_v is of the order of 1 eV, the number of vacancies is very small at ordinary temperatures.

19.25 LINE DEFECTS AND PLANE DEFECTS

Pure single crystals deform permanently at shear forces orders of magnitude lower than predicted theoretically for perfect crystals. In 1934 line defects were postulated to explain the ease with which single crystals deform permanently. The two principal-line defects are called *edge dislocations* and *screw dislocations*. An edge dislocation corresponds to the edge of an atomic plane that terminates within the crystal instead of passing all the way through. The way in which an edge dislocation facilitates shear in a crystal is illustrated in Fig. 19.22. During shear the dislocation moves across the crystal with the net effect that the top half of the crystal is displaced one lattice distance with respect to the lower half of the crystal. The number of dislocation lines passing through a unit area within an ordinary crystal is of the order of 10^6 cm^{-2} or more.

Near an edge dislocation, the atoms are pushed together above the edge and pulled apart below the edge. Thus impurity atoms with larger diameters than the solvent atoms tend to concentrate below the edge, and impurity atoms with smaller diameters tend to concentrate above the edge. This binding of impurity at the dislocation tends to make it more difficult to move a dislocation in an impure material than in a pure material. Therefore alloys require greater shear forces for permanent deformation than do pure crystals.

A screw dislocation forms a continuous helical ramp of one set of atomic planes about the dislocation line, as shown in Fig. 19.23. A screw dislocation provides for easy crystal growth because atoms can be added at the step. Screw dislocations can move in a crystal subjected to appropriate shear forces, and they give rise to the same type of permanent deformation as an edge dislocation. Real dislocations are often mixtures of edge dislocations and screw dislocations.

Plane defects include the surfaces around the tiny grains that make up crystalline metals. Neighboring grains have unrelated crystallographic orientations, and the

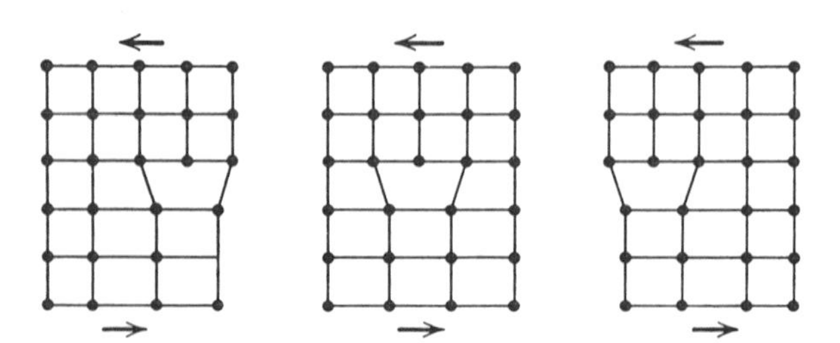

Fig. 19.22 Motion of an edge dislocation under shear. (From N. B. Hannay, *Solid-State Chemistry*. Copyright © 1967. Reprinted by permission of Prentice-Hall, Inc., Englewood Cliffs, N.J.)

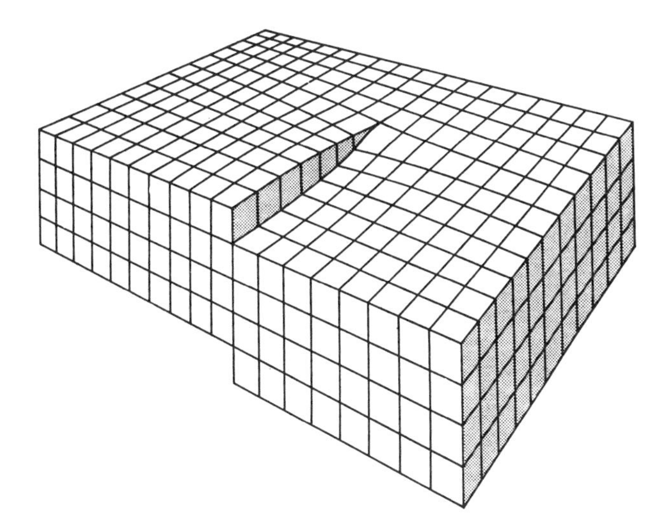

Fig. 19.23 Screw dislocation. (From N. B. Hannay, *Solid-State Chemistry*. Copyright © 1967. Reprinted by permission of Prentice-Hall, Inc., Englewood Cliffs, N.J.)

boundaries between grains are regions of strain in which impurities tend to concentrate. Grain boundaries may be detected by etching a polished metal surface.

Another type of plane defect has already been referred to in connection with the close packing of spheres, that is, irregular stacking of atomic planes (Section 19.13). For example, in face-centered cubic packing the normal stacking sequence is ABCABC· · · . If a B layer accidentally forms on top of a C layer and the reverse stacking sequence proceeds, the sequence is ABCBACBA· · · . The two parts of the crystal on either side of the plane defect are mirror images and are referred to as twins.

19.26 INTRINSIC AND EXTRINSIC SEMICONDUCTORS

A semiconductor is an electronic conductor with electrical conductivity in the range 10^{-7} to 10^4 Ω^{-1} m^{-1}. The conductivities of four semiconductors are summarized in Table 19.3. These solids all have the same crystal structure; the atoms are bonded covalently with four adjacent atoms in a tetrahedral arrangement. Cuprous oxide (Cu_2O), silicon carbide (SiC), and gallium arsenide (GaAs) crystals are also semiconductors.

The electrical conductivity of a pure semiconductor crystal is referred to as the intrinsic conductivity. The intrinsic conductivity is due to the existence of a vacant conduction band that is separated by an energy gap E_g from the filled valence band, as shown in Fig. 19.24. The smaller the energy gap, the higher the concentration of electrons in the conduction band at a given temperature. For every

Table 19.3 Electric Conductivity and Band Gaps for Elements with the Same Crystal Structure

	$\kappa/\Omega^{-1}\ m^{-1}$	E_g/eV
C (diamond)	$<10^{-4}$	6.0
Si	1.5×10^{-3}	1.14
Ge	2	0.67
Sn (gray)	>100	0.08

electron in the conduction band there is a hole in the valence band. Both the electrons and the holes contribute to the electrical conductivity. In an applied electric or magnetic field a hole acts as if it had a positive charge $+e$.

Pure germaniun and pure silicon are weakly ionized solids, just as water is a weakly ionized liquid. At room temperature the intrinsic carrier concentrations ($(e^-) = (h^+)$) in Si and Ge are of the same order of magnitude as the ion concentration ($(H^+) = (OH^-)$) for water.

The addition of impurities to a semiconductor is referred to as doping. Impurities may have a very large effect on electrical conductivity; for example, the addition of one boron atom per 10^5 silicon atoms increases the conductivity of silicon by 10^3 at room temperature. Since the product of the concentrations of electrons and holes is independent of the impurity concentration at a given temperature, the introduction of a small amount of impurity to increase (e^-) must decrease (h^+). Impurities are referred to as *donors* if they can give up an electron to the valence band and as *acceptors* if they can accept electrons from the valence band, leaving holes in the band.

Phosphorus, arsenic, and antimony are donor impurities in silicon or germanium. Silicon and germanium crystallize in the diamond structure (Section 19.15) in which each atom forms four covalent bonds, one with each of its nearest neighbors. Since P, As, and Sb have five valence electrons instead of four, there will be one valence electron from the impurity atom left over after the four covalent bonds are formed. As shown in Fig. 19.25*a*, a phosphorous atom in a silicon lattice has a positive charge and the excess electron remains in the crystal.

For silicon the donor ionization energy E_d is 0.03 eV. Thus the addition of small amounts of phosphorus to silicon produces donor bound levels E_d below the con-

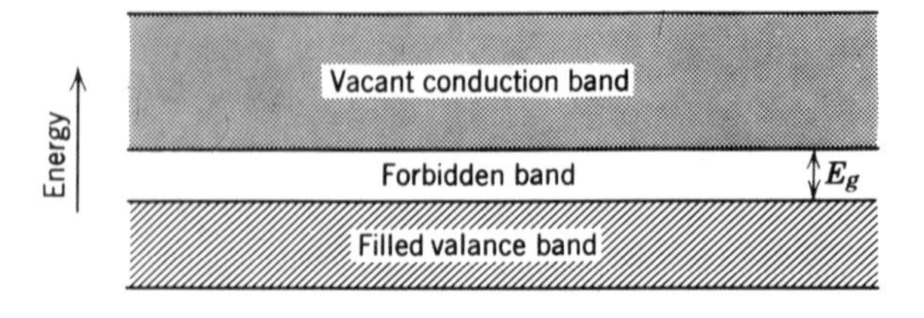

Fig. 19.24 Energy bands for intrinsic conductivity in a semiconductor.

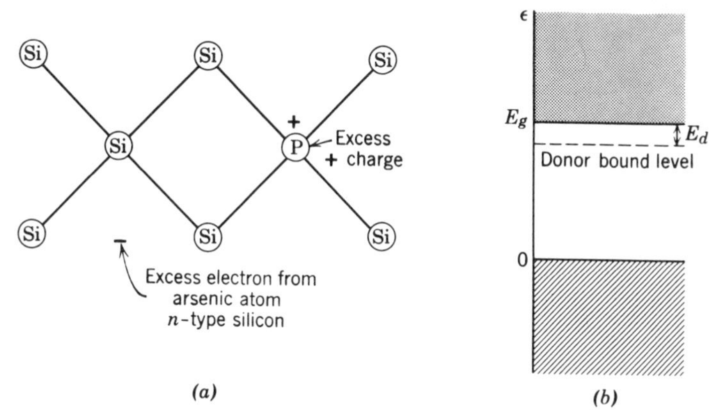

Fig. 19.25 (*a*) Charges resulting from a phosphorus impurity in silicon. (*b*) Energy level for a donor, like phosphorus in silicon. (From C. Kittel, *Introduction to Solid State Physics*. Copyright © 1976. Reprinted by permission of John Wiley & Sons, New York.)

duction band, as shown in Fig. 19.25*b*. Such a material is referred to as *n* type because the conductivity is determined by negative electrons.

Boron, aluminum, and gallium are acceptor impurities in silicon or germanium. Since B, Al, and Ga have three valence electrons instead of four, they can accept an electron from the valence band, as shown in Fig. 19.26*a*. This produces a positive hole, which is then available for conduction. The acceptor ionization energy E_a is 0.045 eV for B in Si. The acceptor bound level is E_a above the top of the valence band, as shown in Fig. 19.26*b*. Such a material is referred to as *p* type because the holes have positive charges.

Added atoms are not necessarily ionized. If the energy required to produce this

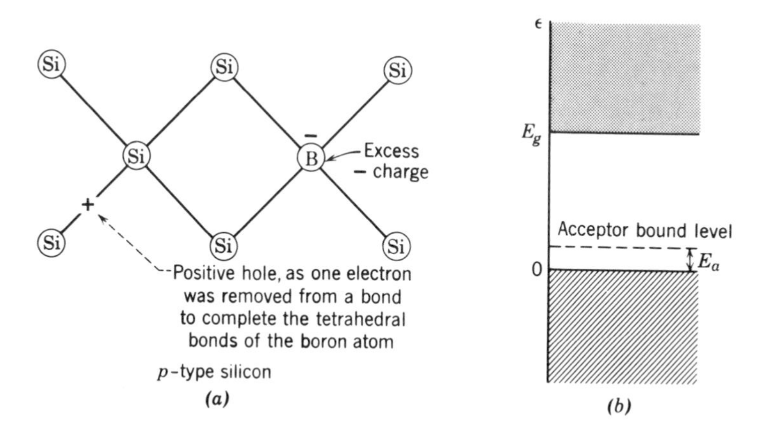

Fig. 19.26 (*a*) Charges resulting from a boron impurity in silicon. (*b*) Energy level for an acceptor, like boron in silicon. (From C. Kittel, *Introduction to Solid State Physics*. Copyright © 1976. Reprinted by permission of John Wiley & Sons, New York.)

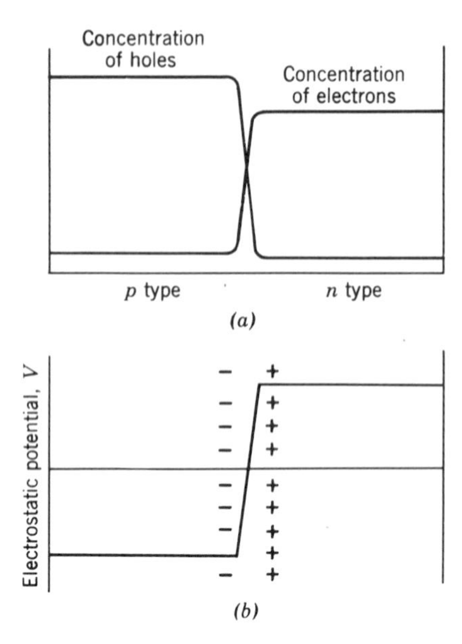

Fig. 19.27 (*a*) Variations of concentrations of holes and electrons across a *p-n* junction. (*b*) Electrostatic potential *V* from acceptor (−) and donor (+) ions near the junction. (From C. Kittel, *Introduction to Solid State Physics*. Copyright © 1976. Reprinted by permission of John Wiley & Sons, New York.)

ionization is quite small, the ionization is essentially complete at room temperature. However, as the temperature is progressively lowered, charge carriers are progressively "frozen out" on the impurity atoms. Thus the ionization of added atoms can be treated as a problem in chemical equilibrium. The donor D and acceptor A ionizations are

$$\mathrm{D} = \mathrm{D}^+ + e^- \tag{19.34}$$

$$\mathrm{A} = \mathrm{A}^- + h^+ \tag{19.35}$$

The concentrations of the species on the right-hand side increase with increasing temperature. The slope of the plot of the logarithm of the concentration of e^- or h^+ versus $1/T$ gives information about the ionization energy of the donor or acceptor.

When a *p*-type semiconductor is brought into contact with an *n*-type semiconductor, a double layer of charge is set up at the junction, as shown in Fig. 19.27. The reason for the double layer of charge can be seen from Fig. 19.27*a*. On the *p* side of the junction there is an excess of free holes and a very low concentration of electrons. (The relative concentration of electrons is exaggerated in this figure.) On the *n* side of the junction there is an excess of electrons relative to holes. The holes on the *p* side have a tendency to diffuse to the *n* side, and the electrons on the *n* side have a tendency to diffuse to the *p* side. This diffusion occurs to a small extent but, when the electrons diffuse into the *p*-type semiconductor, positive charges do not diffuse with them and a layer of negative charge develops. Similarly, the diffusion of positive holes into the n-type conductor is not accompanied by the diffusion of negative charges and a double layer of charge is formed, as shown in Fig. 19.27*b*. Because of the double layer, the electrostatic potential changes at the junction, and there is a built-in electric field.

These p–n junctions are useful because they can be used as rectifiers, solar cells, and other electronic devices. When an electron and a hole recombine, light may be emitted, and this is the mechanism for luminescent p–n diodes.

References

H. J. M. Bowen (et al.) and E. L. Sutton (et al.), suppl. ed., *Tables of Interatomic Distances in Molecules and Ions*, Chemical Society, London, 1965.

M. J. Buerger, *Elementary Crystallography*, Wiley, New York, 1956, repr. 1963.

M. J. Buerger, *Introduction to Crystal Geometry*, McGraw-Hill Book Co., New York, 1971.

J. P. Glusker and K. N. Trueblood, *Crystal Structure Analysis: A Primer*, Oxford University Press, London, 1972.

N. B. Hannay, *Solid State Chemistry*, Prentice-Hall, Englewood Cliffs, N.J., 1967.

International Tables for X-ray Crystallography; Vol. 1, *Symmetry Groups*, 1952; Vol. 2, *Mathematical Tables*, 1959; Vol. 3, *Physical and Chemical Tables*, 1962. The Kynoch Press, Birmingham, England.

C. Kittel, *Introduction to Solid State Physics*, Wiley, New York, 1976.

M. F. C. Ladd and R. A. Palmer, *Structure Determination by X-Ray Crystallography*, Plenum Publ. Corp., New York, 1977.

W. J. Moore, *Seven Solid States*, W. A. Benjamin, Inc., New York, 1967.

L. Pauling, *The Nature of the Chemical Bond*, Cornell University Press, Ithaca, 1960.

D. E. Sands, *Introduction to Crystallography*, W. A. Benjamin, Inc., New York, 1969.

G. A. Somorjai, *The Structure and Chemistry of Solid Surfaces*, Wiley-Interscience, New York, 1969.

M. M. Woolfson, *An Introduction to X-Ray Crystallography*, Cambridge University Press, Cambridge, 1970.

R. W. G. Wycoff, *Crystal Structure* (in five sections with supplements), Wiley-Interscience, New York, 1959.

Problems

19.1 Calculate the angles at which the first-, second-, and third-order reflections are obtained from planes 500 pm apart, using X rays with a wavelength of 100 pm.

Ans. 5.74°, 11.54°, 17.46°.

19.2 The crystal unit cell of magnesium oxide is a cube 420 pm on an edge. The structure is interpenetrating face centered. What is the density of crystalline MgO?

Ans. 3.62 g cm^{-3}.

19.3 Tungsten forms body-centered cubic crystals. From the fact that the density of tungsten is 19.3 g cm^{-3}, calculate (*a*) the length of the side of this unit cell, and (*b*) d_{200}, d_{110}, and d_{222}.

Ans. (*a*) 316 pm, (*b*) $d_{200} = 158$, $d_{110} = 223$, $d_{222} = 91.2$ pm.

19.4 Copper forms cubic crystals. When an X-ray powder pattern of crystalline copper is taken using X rays from a copper target (the wavelength of the $K\alpha$ line is 154.05 pm), reflections are found at $\theta = 21.65°, 25.21°, 37.06°, 44.96°, 47.58°$, and other larger angles. (*a*) What type of lattice is formed by copper? (*b*) What is the length of a side of the unit cell at this temperature? (*c*) What is the density of copper?

Ans. (*a*) Face centered, (*b*) 361.6 pm, (*c*) 8.93×10^3 kg m^{-3}.

19.5 Molybdenum forms body-centered cubic crystals and, at 20 °C, the density is 10.3 g cm^{-3}. Calculate the distance between the centers of the nearest molybdenum atoms. *Ans.* 273 pm.

19.6 Cesium chloride, bromide, and iodide form interpenetrating simple cubic crystals instead of interpenetrating face-centered cubic crystals like the other alkali halides. The length of the side of the unit cell of CsCl is 412.1 pm. (*a*) What is the density? (*b*) Calculate the ion radius of Cs^+, assuming that the ions touch along a diagonal through the unit cell and that the ion radius of Cl^- is 181 pm. *Ans.* (*a*) 3.995×10^3 kg m^{-3}, (*b*) 176 pm.

19.7 The density of potassium chloride at 18 °C is 1.9893 g cm^{-3}, and the length of a side of the unit cell is 629.082 pm, as determined by X-ray diffraction. Calculate the Avogadro constant using the values of the relative atomic masses given in the front cover.

Ans. 6.0213×10^{23} mol^{-1}.

19.8 Calculate the density of diamond from the fact that it has a face-centered cubic structure with two atoms per lattice point and a unit cell edge of 356.7 pm.

Ans. 3.516×10^3 kg m^{-3}.

19.9 (*a*) Metallic iron at 20 °C is studied by the Bragg method, in which the crystal is oriented so that a reflection is obtained from the planes parallel to the sides of the cubic crystal, then from planes cutting diagonally through opposite edges, and finally from planes cutting diagonally through opposite corners. Reflections are first obtained at $\theta = 11° 36', 8° 3'$, and $20° 26'$, respectively. What type of cubic lattice does iron have at 20 °C? (*b*) Metallic iron also forms cubic crystals at 1100 °C, but the reflections determined as described in (*a*) occur at $\theta = 9° 8', 12° 57'$, and $7° 55'$, respectively. What type of cubic lattice does iron have at 1100 °C? (*c*) The density of iron at 20 °C is 7.86 g cm^{-3}. What is the length of a side of the unit cell at 20 °C? (*d*) What is the wavelength of the X rays used? (*e*) What is the density of iron at 1100 °C?

Ans. (*a*) Body centered. (*b*) Face centered. (*c*) 286.8 pm.
(*d*) 57.4 pm. (*e*) 7.84×10^3 kg m^{-3}.

19.10 Insulin forms crystals of the orthorhombic type with unit-cell dimensions of $13.0 \times 7.48 \times 3.09$ nm. If the density of the crystal is 1.315 g cm^{-3} and there are six insulin molecules per unit cell, what is the molar mass of the protein insulin?

Ans. 39,700 g mol^{-1}.

19.11 Aluminum forms face-centered cubic crystals, and the length of the side of the unit cell is 405 pm at 25 °C. Calculate (*a*) the density of aluminum at this temperature, and (*b*) the distances between (200), (220), and (111) planes.

Ans. (*a*) 2.698 g cm^{-3}. (*b*) 202.5, 143.2, 233.8 pm.

19.12 Rutile (TiO_2) forms a primitive tetragonal lattice with $a = 459.4$ pm and $c = 296.2$ pm. There are two Ti atoms per unit cell, one at (000) and the other at $(\frac{1}{2}\frac{1}{2}\frac{1}{2})$. The four oxygen atoms are located at $\pm(uu0)$ and $\pm(\frac{1}{2} + u, \frac{1}{2} - u, \frac{1}{2})$ with $u = 0.305$. What is the density of the crystal? *Ans.* 4.245×10^3 kg m^{-3}.

19.13 A solution of carbon in face-centered cubic iron has a density of 8.105 g cm^{-3} and a unit cell edge of 358.3 pm. Are the carbon atoms interstitial, or do they substitute for iron atoms in the lattice? What is the weight percent carbon? *Ans.* Interstitial, 0.51%.

19.14 A close-packed structure of uniform spheres has a cubic unit cell with a side of 800 pm. What is the radius of the spherical molecule? *Ans.* 283 pm.

19.15 If spherical molecules of 500 pm radius are packed in cubic close packing, and body-centered cubic, what are the lengths of the sides of the cubic unit cells in the two cases? *Ans.* 1414.2, 1154.7 pm.

19.16 Titanium forms hexagonal close-packed crystals. Given the atomic radius of 146 pm, what are the unit cell dimensions and what is the density of the crystal? *Ans.* $a = b = 292$ pm, $c = 477$ pm, 4.517×10^3 kg m^{-3}.

19.17 Metallic sodium forms a body-centered cubic unit cell with $a = 424$ pm. What is the sodium atom radius? *Ans.* 184 pm.

19.18 The diamond has a face-centered cubic crystal lattice, and there are eight atoms in a unit cell. Its density is 3.51 g cm^{-3}. Calculate the first six angles at which reflections would be obtained using an X-ray beam of wavelength 71.2 pm. *Ans.* 9.95°, 11.50°, 16.38°, 19.31°, 20.21°, 25.51°.

19.19 What neutron energy in electronvolts is required for a wavelength of 100 pm? *Ans.* 0.0818 eV.

19.20 At 550 °C the conductivity of solid NaCl is $2 \times 10^{-4}\ \Omega^{-1}\ m^{-1}$. Since the sodium ions are smaller than the chloride ions (see Table 19.1), they are responsible for most of the electric conductivity. What is the ionic mobility of Na^+ under these conditions? *Ans.* 5.61×10^{-14} m^2 V^{-1} s^{-1}.

19.21 What fraction of the lattice sites of a crystal are vacant at 300 K if the energy required to move an atom from a lattice site in the crystal to a lattice site on the surface is 1 eV? At 100 K? *Ans.* 1.56×10^{-17}, 9.08×10^{-6}.

19.22 A diamond has an electric conductivity of $5.1 \times 10^{-5}\ \Omega^{-1}\ m^{-1}$ at 25 °C. Assuming that the mobilities of the carriers are independent of temperature, calculate its conductivity at 35 °C if the energy gap is 6 V. *Ans.* $2.26 \times 10^{-3}\ \Omega^{-1}\ m^{-1}$.

19.23 Calculate the highest-order diffraction line than can be observed for the 100 planes of NaCl using an X-ray tube with a copper target ($\lambda = 154$ pm).

19.24 What is the number of nearest neighbors in atomic crystals of the (*a*) primitive, (*b*) body-centered, and (*c*) face-centered types?

19.25 Make a table of numbers of nearest neighbors and second nearest neighbors for primitive, body-centered cubic, and face-centered cubic lattices.

19.26 From the fact that the length of the side of the unit cell for lithium is 351 pm, calculate the atomic radius of Li. Lithium forms body-centered cubic crystals.

19.27 A crystal has a body-centered structure with a unit cell side of 800 pm. If the lattice points are occupied by spheres, what is their radius?

19.28 By means of the Bragg method, a cubic crystal may be oriented in different directions to obtain d_{100}, d_{110}, and d_{111}. Show that the ratios of the distances between the three different sets of planes are:

Primitive cubic	$d_{100}:d_{110}:d_{111} = 1:0.707:0.578$
Face-centered cubic	$d_{200}:d_{220}:d_{111} = 1:0.707:1.155$
Boby-centered cubic	$d_{200}:d_{110}:d_{222} = 1:1.414:0.578$

19.29 The X-ray powder pattern for molybdenum has reflections at $\theta = 20.25°$, 29.30°, 36.82°, 43.81°, 50.69°, 58.00°, 66.30°, and other larger angles when Cu $K\alpha$ X rays are used ($\lambda = 154.05$ pm). (*a*) What type of cubic crystal is formed by molybdenum? (*b*) What is the length of a side of the unit cell at this temperature? (*c*) What is the density of molybdenum?

19.30 Using X rays with a wavelength of 154 pm from a copper target, it is found that the first reflection from a face of a crystal of potassium chloride at 25 °C is at $\theta = 14° 12'$.

Calculate (*a*) the length of the side of the unit cell for this interpenetrating face-centered lattice, and (*b*) the density of the crystal.

19.31 The density of platinum is 21.45 g cm^{-3} at 20 °C. Given the fact that the crystal is face-centered cubic, calculate the length of the side of the unit cell.

19.32 A substance forms face-centered cubic crystals. Its density is 1.984 g cm^{-3}, and the length of the edge of the unit is 630 pm. Calculate the molar mass.

19.33 The density of calcium oxide is 3.32 g cm^{-3}. If the length of a side of the cubic unit cell is 481 pm, how many molecules of CaO are there per unit cell?

19.34 Potassium bromide has a face-centered cubic lattice, and the edge of the unit cell is 654 pm. What is the density of the crystal?

19.35 Tantalum crystallizes with a body-centered cubic lattice. Its density is 17.00 g cm^{-3}. (*a*) How many atoms of tantalum are there in a unit cell? (*b*) What is the length of a unit cell? (*c*) What is the distance between (200) planes? (*d*) What is the distance between (110) planes? (*e*) What is the distance between (222) planes?

19.36 Aluminum forms face-centered cubic crystals, and at 20 °C the closest interatomic distance is 286.2 pm. Calculate the density of the crystal.

19.37 Calculate the length of the side of the unit cell of potassium iodide (face-centered cubic) from the ionic radii in Table 19.1. Calculate the density of the crystal.

19.38 For uniform spheres of radius r, calculate the length of the side of the unit cell for (*a*) hexagonal close packing, (*b*) face-centered cubic, and (*c*) primitive cubic.

19.39 Cobalt has a hexagonal close-packed structure with a = 250.7 pm. What is its density?

19.40 Calculate the ratio of the radii of small and large spheres for which the small spheres will just fit into octahedral sites in a close-packed structure of the large spheres.

19.41 Si has a face-centered cubic structure with two atoms per lattice point, just like diamond. At 25 °C, a = 543.1 pm. What is the density of silicon? What is the Si—Si bond length?

19.42 The common form of ice has a tetrahedral structure with protons located on the lines between oxygen atoms. A given proton is closer to one oxygen atom than the other and is said to belong to the closer oxygen atom. How many different orientations of a water molecule in space are possible in this lattice?

19.43 What is the de Broglie wavelength of thermal neutrons at 25 °C?

19.44 Silicon is doped with aluminum to the extent of 10^{-8} of the atoms. The conductivity κ is 4.0 Ω^{-1} m^{-1}. What is the mobility of holes? (The density of silicon is 2.4 g cm^{-3}.)

19.45 What concentration of donor atoms in germanium is required to produce an electric conductivity of 10^2 Ω^{-1} m^{-1}? Assume that all donor atoms are ionized and calculate the fraction of the atoms that must be donor atoms. The mobility of electrons in germanium is 0.39 m^2 V^{-1} s^{-1}.

CHAPTER 20

MACROMOLECULES

This term is generally applied to molecules with molar mass greater than about 10,000 g mol^{-1}. Macromolecules in the form of proteins, polynucleotides, and polysaccharides are necessary for life, and their structures make possible complex functions. Macromolecules in the form of synthetic high polymers are the basis for synthetic fibers, plastics, and synthetic rubber. The relation between the physical properties of these materials and their molecular structures is, of course, of utmost importance.

20.1 TYPES OF MACROMOLECULES

Until about 1930 synthetic and natural high polymers were generally regarded as aggregates of smaller molecules. However, at that time Staudinger showed that proteins, rubber, cellulose, and synthetic polymers are made up of long chains of covalently linked molecules. This recognition made it possible to understand rubber-like elasticity in terms of the capacity of coiled polymer chains to accommodate large deformations by the variety of accessible configurations.

Polymers are made up of repeating units called monomers. A polymer may be made up of a single type of monomer, or it may be a copolymer of more than one type of monomer unit. Proteins are copolymers of 22 different aminoacids.

A sample of a synthetic high polymer generally has a wide range of molar masses. However, for most proteins the molecules are identical and all have the same molar mass. Another difference between synthetic high polymers and proteins is that in solution synthetic high polymers have approximately random coil configurations while proteins have configurations in which the polypeptide chains are folded in a specific compact arrangement. The random coil configurations are thermodynamically favored at higher temperatures, and proteins may be denatured by heating; the resulting unfolding of the polypeptide chain generally reduces the solubility of the protein in aqueous solution.

Nucleic acids may be single stranded or double stranded. Double-stranded deoxyribonucleic acid (DNA) forms a double helix that may be very long. The DNA molecules of bacteria may have molar masses as large as 2×10^9 g mol^{-1}, and DNA molecules of animal cells may have molar masses as large as 10^{11} g mol^{-1}.

Special methods are required for the determination of the molar masses of high polymers. Of the colligative properties only the osmotic pressure is useful for macromolecules. Of the methods that are useful, some are based on irreversible processes in solution (velocity sedimentation, diffusion, viscosity, and gel permeation chromatography) and others are thermodynamic (equilibrium ultracentrifugation, osmotic pressure, and light scattering).

20.2 VELOCITY SEDIMENTATION

Particles of finely divided solid sediment in a fluid if their density is higher than that of the suspension medium. Their equivalent radius can be calculated from the rate of sedimentation using the expression for the frictional coefficient, $f = 6\pi\eta r$, given in Section 18.1.

Dissolved molecules tend to sediment in the earth's gravitational field or to float upward, depending on their density relative to that of the solvent, but this tendency is counteracted by the random translational motion of the molecules. However, sufficiently powerful ultracentrifuges have been built to cause even molecules as small as those of sucrose to sediment at measurable rates. Svedberg* was the leader in the development of ultracentrifuges, which he defined as centrifuges adapted for quantitative measurements of convection-free and vibration-free sedimentation. There are two distinct types of ultracentrifuge experiments: (1) those where the velocity of sedimentation of a component of the solution is measured (sedimentation velocity), and (2) those where the redistribution of molecules is determined at equilibrium (sedimentation equilibrium).

The acceleration of a particle in a centrifugal field is equal to $\omega^2 r$, where ω is the angular velocity of the centrifuge in radians per second (i.e., 2π times the number or revolutions per second) and r is the distance of the particle from the axis of rotation. Ultracentrifuges in which r is about 6 cm are commonly operated at 60,000 rpm or 1000 rps, and so the acceleration is

$$\omega^2 r = (2\pi 1000\ \mathrm{s}^{-1})^2(0.06\ \mathrm{m}) = 2.36 \times 10^6\ \mathrm{m\ s}^{-2}$$

Since the acceleration of the earth's field is 9.80 m s^{-2}, the acceleration is 240,000 times greater than in the earth's field.

A solution to be studied in the velocity ultracentrifuge is placed in a cell with thick quartz windows. The cell has a sector shape when viewed at right angles to the plane of rotation of the centrifuge rotor, since the sedimentation takes place radially. As the high-molar-mass component throughout the solution sediments, a moving boundary is formed behind which there is only solvent. The movement of such boundaries in the cell may be followed with the schlieren optical system, as was mentioned in connection with diffusion (Section 18.11).

Figure 20.1 shows the schlieren patterns for an ultracentrifuge experiment with the enzyme fumarase at 50,400 rpm. The second and third photographs were taken 35 and 70 min later than the top photograph. If additional components with different rates of sedimentation had been present, additional peaks would be

---

* T. Svedberg and K. O. Pedersen, *The Ultracentrifuge*, Oxford University Press, Oxford, 1940.

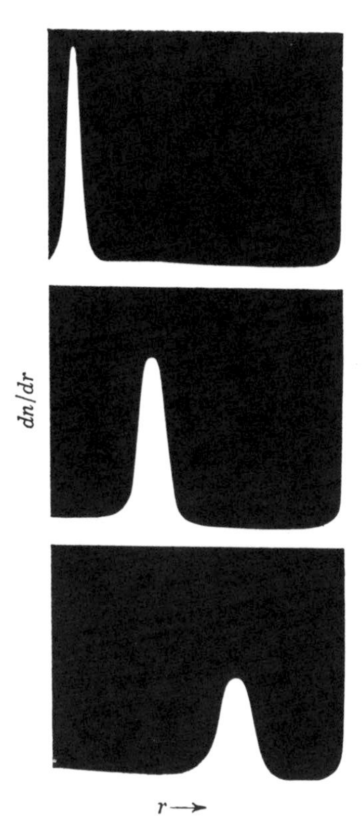

Fig. 20.1 Schlieren patterns for an ultracentrifuge experiment with fumarase at 50,400 rpm. The lower photographs were taken 35 and 70 min later than the top photograph. The protein is dissolved in pH 6.8 phosphate buffer.

evident in the schlieren photograph. Thus, the ultracentrifuge is useful in analyzing complex mixtures such as blood plasma.

When a solution is subjected to a centrifugal field, the solute molecules move away from the axis of rotation if the reciprocal of their partial specific volume $\bar{v}$ is greater than the density ρ of the solution, and otherwise they move toward the axis of rotation. In other words, the direction of "sedimentation" depends on the sign of the Archimedes factor $(1 - \bar{v}\rho)$.

In a centrifugal field a solute molecule is accelerated until the frictional force resisting its motion is equal to the acceleration of the centrifugal field times the effective mass $m(1 - \bar{v}\rho)$, where m is the mass of the molecule. The frictional force is the product of the velocity dr/dt and the frictional coefficient f (Section 18.1). Thus, when the steady-state velocity dr/dt is reached,

$$f\frac{dr}{dt} = m(1 - \bar{v}\rho)\omega^2 r = \frac{M(1 - \bar{v}\rho)\omega^2 r}{N_A} \tag{20.1}$$

where M is the molar mass. The ratio of the velocity to the centrifugal acceleration is called the sedimentation coefficient S.

$$S = \frac{1}{\omega^2 r}\frac{dr}{dt} \tag{20.2}$$

Since ω^2 has the units (seconds)$^{-2}$, the sedimentation coefficient has the unit seconds. The sedimentation coefficients of proteins fall in the range 10^{-13} s to 200×10^{-13} s, and the unit 10^{-13} s is called a *svedberg*.

If a boundary is r_1 centimeters from the axis of the centrifuge at time t_1 and r_2 centimeters from the axis at time t_2, the sedimentation coefficient may be calculated from

$$S = \frac{1}{\omega^2(t_2 - t_1)} \ln \frac{r_2}{r_1} \tag{20.3}$$

which is obtained by integrating equation 20.2.

Example 20.1 In the ultracentrifuge experiment illustrated in Fig. 20.1 the distance from the axis of the ultracentrifuge to the boundary was 5.949 cm in the top photograph and 6.731 cm in the bottom photograph taken 70 min later. Since the speed of the rotor was 50,400 rpm, $\omega^2 = 2.79 \times 10^7\ \text{s}^{-2}$. Using equation 20.3, we see that

$$S = \frac{1}{\omega^2(t_2 - t_1)} \ln \frac{r_2}{r_1} = \frac{2.303 \log (6.731/5.949)}{(2.79 \times 10^7\ \text{s}^{-2})(60 \times 70\ \text{s})} = 10.5 \times 10^{-13}\ \text{s}$$

This is the sedimentation coefficient at 28.2 °C, the temperature of the experiment. Making a correction to 20 °C in water, taking into account the change of viscosity and density, a value of 8.90 svedbergs is obtained.

Substituting equation 20.1 into equation 20.2 yields

$$S = \frac{M(1 - \bar{v}\rho)}{N_A f} \tag{20.4}$$

The sedimentation coefficient by itself cannot be used to determine the molar mass of the sedimenting component unless the molecules are spherical. If the molecules are spherical, $f = 6\pi\eta r$, and equation 20.4 may be used to calculate the molar mass, since

$$M = \frac{4\pi r^3 N_A}{3\bar{v}} \tag{20.5}$$

Since the velocity of sedimentation is so low that there is no appreciable orientation of the molecules, the frictional coefficient involved in sedimentation is taken to be the same as that involved in diffusion. Introduction of equation 18.26 into equation 20.4 yields

$$M = \frac{RTS}{D(1 - \bar{v}\rho)} \tag{20.6}$$

To calculate the molar mass from measured values of S and D, it is necessary to correct sedimentation and diffusion coefficients to the same temperature, usually

20 °C, and if S and D depend appreciably on concentration, to zero concentration. Equation 20.6 has probably been the most widely used in the calculation of molar masses of proteins, and the wide range of molar masses that can be obtained by this method is indicated by Table 20.1.

Table 20.1 Physical Constants of Proteins at 20 °C in Water and Molar Mass

Protein	$S/10^{-13}$ s	$D/10^{-11}$ m^2 s^{-1}	$\bar{v}$/cm^3 g^{-1}	M/g mol^{-1}
Beef insulin	1.7	15	0.72	12,000
Lactalbumin	1.9	10.6	0.75	17,400
Myoglobin	2.06	12.4	0.749	16,000
Ovalbumin	3.6	7.8	0.75	44,000
Serum albumin	4.3	6.15	0.735	64,000
Hemoglobin	4.6	6.9	0.749	64,400
Serum globulin	7.1	4.0	0.75	167,000
Urease	18.6	3.4	0.73	490,000
Tobacco mosaic virus	185	0.53	0.72	40,000,000

Example 20.2 Using the data of Table 20.1, the molar mass of hemoglobin may be calculated as follows (the density of water at 20 °C is 0.9982×10^3 kg m^{-3}).

$$M = \frac{RTS}{D(1 - \bar{v}\rho)}$$

$$= \frac{(8.31\text{ J K}^{-1}\text{ mol}^{-1})(293\text{ K})(4.6 \times 10^{-13}\text{ s})}{(6.9 \times 10^{-11}\text{ m}^2\text{ s}^{-1})[1 - (0.749 \times 10^{-3}\text{ m}^3\text{ kg}^{-1})(0.9982 \times 10^3\text{ kg m}^{-3})]}$$

$$= 64.4\text{ kg mol}^{-1} = 64{,}400\text{ g mol}^{-1}$$

In measuring the sedimentation coefficient of DNA significant concentration effects are encountered at concentrations as low as 10 mg per liter. Taking advantage of the strong nucleotide absorption around 260 nm, the velocity of the sedimenting boundary may be measured even at this low concentration using an ultraviolet absorption optical system. In another method for studying such dilute solutions a thin layer of DNA solution may be placed on a preformed density gradient of CsCl solution created in an ultracentrifuge cell by the centrifugal field causing the sedimentation. In these experiments the density gradient stabilizes the system from convection.

20.3 EQUILIBRIUM ULTRACENTRIFUGATION

Particles or molecules that are denser than the solvent sediment in the earth's gravitational field to an equilibrium distribution in which concentration decreases with vertical height. Particles that are less dense than the solvent have an equilibrium distribution in which concentration increases with vertical height. The increase in gravitational potential energy when a molecule of mass m and partial specific volume $\bar{v}$ is moved from height h_1 to height h_2 in a medium of density ρ is $\Delta\epsilon = m(1 - \bar{v}\rho)g(h_2 - h_1)$. The ratio of the equilibrium concentrations at h_2 and h_1 is given by the Boltzmann distribution $c_2/c_1 = \exp(-\Delta\epsilon/kT)$, or

$$c_2 = c_1 e^{-m(1-\bar{v}\rho)g(h_2-h_1)/kT} \tag{20.7}$$

This equation was used by Perrin to obtain an early value of Avogadro's constant. By measuring c as a function of height for particles of gum gamboge of known mass, he obtained a value of the Boltzmann constant k and calculated N_A from $R = N_A k$, where R is the gas constant.

An ultracentrifuge may be used to observe the sedimentation equilibrium of molecules that do not sediment appreciably in the earth's gravitational field. In order to obtain the equation for equilibrium sedimentation from equation 20.7, we replace the work $m(1-\bar{v}\rho)g(h_2-h_1)$ required to raise a molecule in the earth's gravitational field with the work required to carry a molecule from one value of r to another in the ultracentrifuge cell. The work against the centrifugal field to move one molecule of mass m from r_2 to r_1 is

$$w = \int_{r_1}^{r_2} \omega^2 r m(1-\bar{v}\rho)\,dr = \tfrac{1}{2}\omega^2 m(1-\bar{v}\rho)(r_2^2 - r_1^2) \tag{20.8}$$

Substituting this for the work term in equation 20.7 yields

$$c_1 = c_2 e^{-\omega^2 m(1-\bar{v}\rho)(r_2^2-r_1^2)/2kT} \tag{20.9}$$

or

$$M = \frac{2RT \ln(c_2/c_1)}{(1-\bar{v}\rho)\omega^2(r_2^2 - r_1^2)} \tag{20.10}$$

where c_1 is the concentration at distance r_1 from the axis, c_2 is the concentration at distance r_2, ω is the angular velocity of the rotor (radians per second), and ρ is the density of the solution. Optical methods are used to determine c_2 and c_1. In contrast with velocity sedimentation, this is a thermodynamic measurement that can yield a value for M all by itself. However, the establishment of sedimentation equilibrium is slow, and so this method of determining molar mass is time consuming.

Sedimentation equilibrium may also be used to set up a density gradient due to a low molar mass substance (for example, CsCl) so that a mixture of macromolecules may be separated into bands according to their density. In a density gradient a substance will tend to come to rest at a level where its density is the same as the solution. In 1957 Meselson, Stahl, and Vinograd used CsCl gradients (formed from about 7 mol L^{-1} solutions by sedimentation equilibrium) to separate different types of DNA. By this method it is also possible to separate DNA labeled with ^{15}N from unlabeled DNA. By use of this technique and isotopic nitrogen Meselson and Stahl were able to show in 1958 that when DNA is replicated, each daughter double helix consists of one strand from the parent DNA and one newly synthesized strand.

Example 20.3 Calculate the molar mass of hemoglobin from the fact that in an equilibrium ultracentrifuge experiment at 20 °C, $c_2/c_1 = 9.40$, where $r_1 = 5.5$ cm and $r_2 = 6.5$ cm. The ultracentrifuge is operated at 120 rps, $\bar{v} = 0.749 \times 10^{-3}$ m^3 kg^{-1}, and $\rho = 0.9982 \times 10^3$ kg m^{-3}.

$$M = \frac{2RT}{(1-\bar{v}\rho)\omega^2(r_2^2 - r_1^2)} \ln\frac{c_2}{c_1}$$

$$= \frac{2(8.314\ \text{J K}^{-1}\ \text{mol}^{-1})(293\ \text{K})}{(1 - 0.749 \times 0.9982)(2\pi 120\ \text{s}^{-1})^2[(0.065^2 - 0.055^2)\text{m}^2]} \ln 9.40$$

$$= 63.4\ \text{kg mol}^{-1} = 63{,}400\ \text{g mol}^{-1}$$

20.4 DIFFUSION

Methods for measuring the diffusion coefficient of gases (Chapter 14) and solutes in liquids (Chapter 18) were discussed earlier. The diffusion coefficient of a homogeneous high polymer is of special importance because it may be combined with the sedimentation coefficient to obtain the molar mass. The diffusion coefficient D is related to the frictional coefficient f of a molecule by equation 18.26, which is

$$D = \frac{RT}{N_A f} \tag{20.11}$$

For the special case of spherical particles, molar masses may be estimated from diffusion coefficients alone. Combining equations 18.2 and 20.11 yields

$$D = \frac{RT}{N_A 6\pi\eta r} \tag{20.12}$$

Thus, for spherical particles the radius may be calculated from the measured diffusion coefficient. Since it is more familiar to think of size in terms of molecular mass, the relation between D and the molar mass of the spherical particle may be derived from equation 20.12 by introducing equation 20.5 to obtain

$$D = \frac{RT}{N_A 6\pi\eta}\left(\frac{4\pi N_A}{3M\bar{v}}\right)^{1/3} \tag{20.13}$$

For spherical particles, then, the diffusion coefficient is inversely proportional to the cube root of the molar mass. Of course, if the particles or molecules are not spherical, the value of the molar mass calculated from equation 20.13 will not be correct. This equation, however, does give the maximum molar mass that is consistent with a given D and $\bar{v}$. For a nonspherical particle the frictional coefficient is larger than $6\pi\eta r$ and the molar mass is smaller than that calculated from equation 20.13.

Example 20.4 Using the diffusion coefficient for myoglobin from Table 20.1, calculate the maximum molar mass that this protein can have. The coefficient of viscosity of water at 20 °C is 0.001 005 Pa s. Rearranging equation 20.13,

$$M = \frac{4\pi N_A}{3\bar{v}}\left(\frac{RT}{DN_A 6\pi\eta}\right)^3 = \frac{4\pi(6.022\times10^{23}\ \text{mol}^{-1})}{3(0.749\times10^{-3}\ \text{m}^3\ \text{kg}^{-1})}$$

$$\times\left[\frac{(8.314\ \text{J K}^{-1}\ \text{mol}^{-1})(293\ \text{K})}{(12.4\times10^{-11}\ \text{m}^2\ \text{s}^{-1})(6.022\times10^{23}\ \text{mol}^{-1})6\pi(1.005\times10^{-3}\ \text{Pa s})}\right]^3$$

$$= 17.2\ \text{kg mol}^{-1} = 17{,}200\ \text{g mol}^{-1}$$

The actual molar mass is 16,000 g mol^{-1}.

20.5 INTRINSIC VISCOSITY

The experimental determination of the coefficient of viscosity has been discussed earlier (Section 18.1) in connection with other irreversible processes. Large molecules make considerable contributions to the viscosities of solutions and, since viscosity is readily measured, this provides an important method for studying macromolecules.

The ratio of the viscosity η of a suspension of particles to the viscosity of the solvent η_0 may be represented by an equation of the type

$$\frac{\eta}{\eta_0} = 1 + \nu\phi + \kappa\phi^2 + \cdots \tag{20.14}$$

where ϕ is the volume fraction occupied by particles and ν and κ are constants that depend on the shape of the particles. The quantity η/η_0 is referred to as the relative viscosity, and it is convenient to use the term specific viscosity η_{sp} for $(\eta/\eta_0) - 1$, although neither quantity is, strictly speaking, a viscosity. Thus equation 20.14 may be written

$$\frac{\eta_{sp}}{\phi} = \nu + \kappa\phi + \cdots \tag{20.15}$$

Einstein showed in 1906 that for spheres $\nu = 2.5$. This coefficient increases with axial ratio for prolate and oblate ellipsoids of revolution. It is interesting to note that ν is independent of the size of the particles.

The volume fraction ϕ of a solute may be written $c\bar{v}_2$, where c is the concentration of the solute in mass per unit volume of solution and $\bar{v}_2$ is the partial specific volume. Thus equation 20.15 may be written

$$\frac{\eta_{sp}}{c} = \nu\bar{v}_2 + \kappa c\bar{v}_2{}^2 + \cdots \tag{20.16}$$

The limit of η_{sp}/c as c approaches zero is called the intrinsic viscosity $[\eta]$. Thus, from equation 20.16,

$$\lim_{c\to 0} \frac{\eta_{sp}}{c} = [\eta] = \nu\bar{v}_2 \tag{20.17}$$

Intrinsic viscosities are expressed in deciliters per gram or in $cm^3\ g^{-1}$.

For a solvated macromolecule the volume occupied by a molecule is greater than that calculated using the particle specific volume. If this effect is taken into account, equation 20.17 becomes

$$[\eta] = \nu(\bar{v}_2 + \delta\bar{v}_1) \tag{20.18}$$

where δ is the mass of solvent bound by a gram of solute and $\bar{v}_1$ is the partial specific volume of the solvent. If the particle is not spherical, ν and ρ are both unknown, but values of ν may be calculated for ellipsoids of revolution.

Intrinsic viscosities give strong indications of the axial ratios of molecules in solution. For a spherical, unhydrated protein with $\bar{v} = 0.75\ cm^3\ g^{-1}$, we would expect $[\eta] = 1.9\ cm^3\ g^{-1}$. As shown in Table 20.2, the intrinsic viscosity of ribonuclease is not very much greater than this value. Bushy stunt virus is very nearly spherical, even though it has a molar mass of 10,700,000 g mol^{-1}. The elongated molecules of fibrinogen provide the structure for blood clots, and the even more elongated molecules of myosin are the contractile part of muscle. The very high

Table 20.2 Intrinsic Viscosities in Water at 25 °C[a]

	M/g mol^{-1}	$[\eta]$/cm^3 g^{-1}
Ribonuclease	13,683	2.3
Bovine serum albumin	66,500	3.7
Bushy stunt virus	10,700,000	3.4
Fibrinogen	330,000	27
Tobacco mosaic virus	40,000,000	37
Myosin	493,000	217
DNA	6,000,000	5000

[a] C. R. Cantor and P. R. Schimmel, *Biophysical Chemistry*, W. H. Freeman & Co., San Francisco, 1979.

intrinsic viscosity of double stranded deoxyribonucleic acid (DNA) shows that the molecules are very long and thin. High molar mass DNA may be degraded (reduced in molar mass) by the shear gradients encountered in pipetting solutions. In order to measure the viscosities of DNA solutions at very low shear it has been necessary to use rotating cylinder, or Couette, viscometers.

In solution certain polymer molecules have a randomly kinked and coiled configuration that is constantly changing; in other cases the molecules are not so flexible. The "size" of the molecule, measured, for example, by the average end-to-end distance of the chain, is not directly proportional to the molar mass. The intrinsic viscosity depends, however, on the molar mass. For a series of samples of the same polymer in a given solvent and at a constant temperature, the following empirical relation is obeyed quite well.

$$[\eta] = KM^a \tag{20.19}$$

Here K and a are constants that may be determined by measuring the intrinsic viscosities of a series of samples of a polymer for which the molar masses have been measured by another method. Since values of K and a have been determined for a large number of polymers and solvents, viscosity measurements are widely used to obtain average molar masses for synthetic high polymers.

The value of a is about 0.5 when a theta solvent (Section 20.10) is used. For a good solvent, a is about 0.8 because the molecule is more extended than a random coil.

Example 20.5 Given that the intrinsic viscosity of a sample of DNA ($M = 6 \times 10^6$ g mol^{-1}) is 5000 cm^3 g^{-1}, approximately what concentration of DNA in water would have a relative viscosity of 1.1?

$$\frac{(\eta/\eta_0) - 1}{c} = \frac{0.1}{c} = 5000 \text{ cm}^3 \text{ g}^{-1}$$

$$c = 2 \times 10^{-5} \text{ g cm}^{-3} = 2 \times 10^{-2} \text{ g L}^{-1}$$

Since the molar mass of DNA per nm measured along the axis of the helix is 1920 g mol^{-1}, the length of a molecule of DNA with $M = 6 \times 10^6$ g mol^{-1} is 3100 nm or 3.1 μm.

20.6 OSMOTIC PRESSURE

Measurements of osmotic pressure are useful in the molar mass range up to about 500,000 g mol^{-1}, provided good enough semipermeable membranes may be obtained. The theory of this method has been described in Section 3.13. If the solute is polydisperse the number average molar mass is obtained (Section 20.9).

In solutions of proteins or other colloidal electrolytes it is necessary to distinguish between the *total osmotic pressure*, which would be obtained with a membrane impermeable to both salt and protein, and the *colloid osmotic pressure*, which is obtained with a membrane permeable to salt ions but not to protein. The latter type of membrane is always used when it is desired to obtain the molar mass of the protein or other colloidal electrolyte.

For colloidal electrolytes in solutions with low concentrations of electrolytes, the measured osmotic pressure is greater than that expected for the colloidal ions alone. This is a result of the fact that, although the salt ions may pass through the membrane, they will not be distributed equally at equilibrium. Donnan showed that because of the high molar mass ion on one side of the membrane, the concentration of the small ion of the same sign as the macroion is lower on that side of the membrane than in the salt solution and that this is compensated by an increased concentration of the small ion of opposite charge. The Donnan effect may be reduced by increasing the salt concentration and, if possible, adjusting the pH to the isoelectric point (Section 6.4) of the colloidal electrolyte.

It is convenient to add a term to the simple osmotic pressure equation so that it can be used to represent data over a wider range of concentration.

$$\frac{\Pi}{c} = \frac{RT}{M} + Bc \tag{20.20}$$

The value of B is dependent on polymer-polymer interactions and, for a theta solvent (Section 20.10) is equal to zero. In using equation 20.20 a plot of Π/c versus c is extrapolated to zero concentration. Then M is calculated from

$$\lim_{c \to 0} \frac{\Pi}{c} = \frac{RT}{M} \tag{20.21}$$

20.7 LIGHT SCATTERING

Rayleigh developed the theory for the scattering of light from noninteracting particles and explained the blue color of the sky and the red color of sunsets as being due to the preferential scattering of blue light by the molecules of the atmosphere.

When a molecule is exposed to the oscillating electromagnetic field of a light wave, it is polarized at the frequency of the incident light. The oscillating electric charges in the molecule cause the radiation of light in other directions. As illustrated in Fig. 20.2, the electric field of a plane polarized light wave induces an

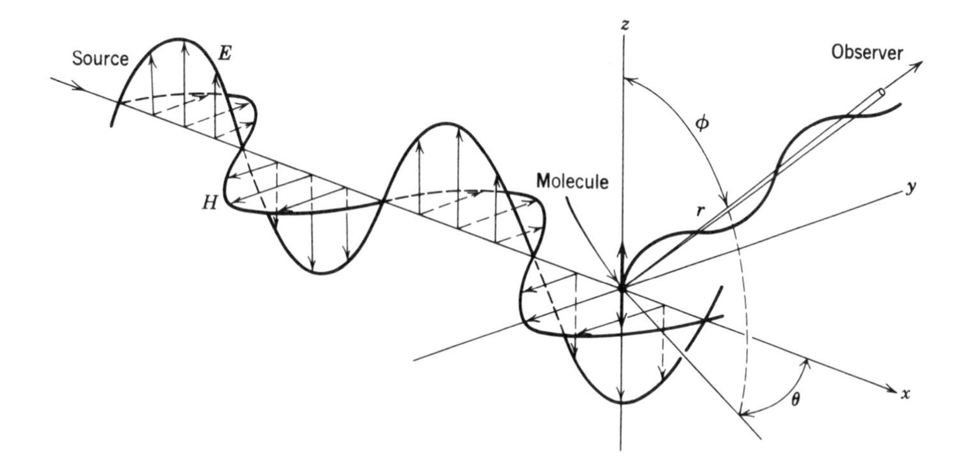

Fig. 20.2 Scattering of plane polarized light by a single molecule. For an isotropic molecule the induced dipole moment is along the z axis. Therefore no light is scattered in the z direction.

oscillating dipole moment in the direction of the electric vector (i.e., along the z axis) if the molecule is isotropic. According to electromagnetic theory, the electric field E_r produced by an oscillating dipole at a distance r from the dipole and an angle ϕ with respect to the direction of polarization (the z axis) is given by

$$E_r = -\left(\frac{\alpha E_0 \sin \phi}{4\epsilon_0 r\lambda^4}\right) \cos 2\pi\nu\left(t - \frac{r}{c}\right) \tag{20.22}$$

where α is the polarizability (Section 14.12) of the molecule, E_0 is the magnitude of the electric vector of the incident light wave, ν is the frequency of the light, and c is the velocity of light. The intensity i of the radiation scattered at angle ϕ is proportional to the square of the amplitude (i.e., the square of the coefficient in equation 20.22). In order to compare i with the intensity I_0 of the incident light, we divide the square of the amplitude by $E_0{}^2$ to obtain

$$\frac{i}{I_0} = \frac{\pi^2\alpha^2 \sin^2 \phi}{\epsilon_0{}^2 r^2 \lambda^4} \tag{20.23}$$

The dependence of the scattering of incident light polarized in the z direction on ϕ is shown in Fig. 20.3a. The scattering is zero in the z-direction.

Usually unpolarized light is used in scattering experiments. The angular dependence of the intensity of scattering of a small (compared with the wavelength of the light) isotropic molecule may be obtained by adding the scattered intensities from two perpendicularly polarized light waves. This yields a scattered intensity of

$$\frac{i}{I_0} = \frac{\pi^2\alpha^2}{2\epsilon_0{}^2 r^2 \lambda^4}(1 + \cos^2 \theta) \tag{20.24}$$

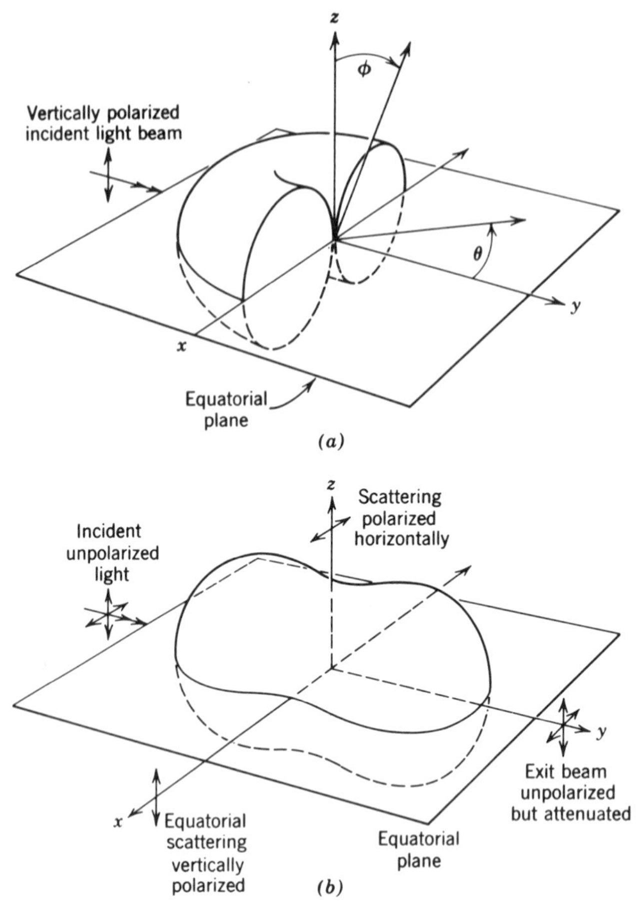

Fig. 20.3 (*a*) Intensity of light scattered by a single molecule from a light wave with its electric vector in the z-direction. The length of the radius vector from the origin indicates the scattered amplitude. (The front half of the diagram is omitted for clarity.) (*b*) Intensity of light scattered by a single molecule from an unpolarized light wave. (From D. F. Eggers et al., *Physical Chemistry*, Wiley, New York, 1964.)

where θ is the angle between the incident beam and the direction of observation. This angular dependence is shown in Fig. 20.3*b*. This shows that the scattered intensities in the forward and reverse directions are equal. The scattered intensity is inversely proportional to the fourth power of the wavelength λ of the light used. Thus, blue light is scattered to a greater extent than red light, and therefore the light transmitted through a suspension of particles is reddish. We are familiar with these effects in the blue sky and red sunset.

Equation 20.24 gives the scattering from a single isolated molecule, but we need to consider the scattering from a large number of molecules. When there is more than one molecule, interference between the light scattered by different molecules has to be considered. If the scattering molecules are arranged in a crystal lattice, it can be shown that the scattering of electromagnetic radiation with a wavelength that is long with respect to the lattice spacing will be zero, except in the forward direction. In other words, for a perfect crystal there will be no scattering, and the incident beam will be transmitted without loss, if there is no absorption. If identical scattering molecules form a liquid, we can consider the scattering to come from small cells in space, each cell being smaller than a wavelength of light, but each containing many molecules of the liquid. If each cell contained the same number of molecules, we would expect that there would be no net scattering, except in the forward direction, as for a crystal. However, fluctuations in the density of a liquid occur at distances of the order of the wavelength of visible light. Therefore, a liquid does scatter light, but it scatters much less than would be expected for that many molecules. When a solute is added, the light scattering increases because there are fluctuations in the concentration of the solute in addition to the fluctuations in the density of the liquid.

The equation for the scattering by a solution may be derived by relating the excess polarizability α of a solute particle over the solvent to the specific refractive index increment of the solute. As shown in Section 10.12, the refractive index n of a solution is related to the excess polarizability α of a solute particle over the solvent by

$$n^2 = n_0^2 = N\alpha/\epsilon_0 \tag{20.25}$$

where n_0 is the refractive index of the solvent and N is the number of solute particles per unit volume. This equation may be rearranged as follows to express the polarizability in terms of the specific refractive index increment $dn/dc = (n - n_0)/c$, where c is the concentration in mass per unit volume, and the molar mass M of the solute.

$$\alpha = \frac{(n^2 - n_0^2)\epsilon_0}{N} = (n + n_0)\frac{(n - n_0)}{c}\frac{c\epsilon_0}{N} = (n + n_0)\frac{dn}{dc}\frac{c\epsilon_0}{N} = 2n_0\frac{dn}{dc}\frac{M}{N_A} \tag{20.26}$$

where the last form has been obtained by replacing $n + n_0$ by $2n_0$, since dilute solutions are considered, and c/N is replaced by M/N_A. Substituting this expression into equation 20.24 yields

$$\frac{i}{I_0} = \frac{2\pi^2 n_0^2 (dn/dc)^2 M^2 (1 + \cos^2\theta)}{\lambda^4 r^2 N_A^2} \tag{20.27}$$

for the scattering of one solute particle. Multiplying by $N = cN_A/M$ to obtain the scattering from N particles per unit volume yields

$$\frac{i}{I_0} = \frac{2\pi^2 n_0^2 (1 + \cos^2\theta)(dn/dc)^2 Mc}{N_A \lambda^4 r^2} \tag{20.28}$$

This equation shows that the excess scattering of the solution over that of the solvent is proportional to the concentration of the solute on a mass per unit volume basis and to the molar mass of the solute. If the molar mass of the solute is unknown, it may be obtained by measuring the other quantities. If this expression is integrated over all angles to obtain the total scattered intensity, it is found that the *transmitted* intensity I is given by

$$I = I_0 e^{-\tau x} \tag{20.29}$$

where τ is the turbidity and x is the thickness of the cell in the direction of the incident beam. This is of the form of Beers' law for light absorption (Section 11.14). The result of this calculation is that

$$\tau = \frac{32\pi^3 n_0^2 (dn/dc)^2 Mc}{3N_A \lambda^2} = HMc \tag{20.30}$$

where

$$H = \frac{32\pi^3 n_0^2 (dn/dc)^2}{3N_A \lambda^4} \tag{20.31}$$

is a constant for a particular solvent, solute, and wavelength. Since the turbidity τ is proportional to molar mass, this method is more useful as the molar mass increases, in contrast to osmotic pressure, which is inversely proportional to molar mass.

Instead of measuring the transmitted intensity and calculating τ using equation 20.29, it is much more sensitive experimentally to determine the scattered intensity using an apparatus such as that shown in Fig. 20.4.

The above theory is for ideal solutions and, in practice, it is necessary to plot Hc/τ versus c and extrapolate to zero concentration to obtain the value of Hc/τ that will yield the correct molar mass from equation 20.30. In fact, it can be shown that for a nonideal polymer solution for which the osmotic pressure (Section 20.6) is given by

$$\frac{\Pi}{cRT} = \frac{1}{M} + Bc + \cdots \tag{20.32}$$

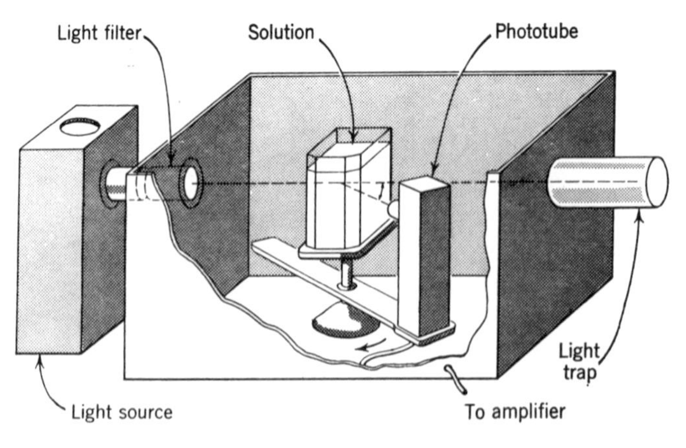

Fig. 20.4 Apparatus for measuring light scattered by a solution.

the turbidity is given by

$$\frac{Hc}{\tau} = \frac{1}{M} + 2Bc + \cdots \tag{20.33}$$

The polarizability α is proportional to the mass of the particle. The intensity of the light scattered by a single particle is proportional to the square of its polarizability and therefore to the square of its mass. If the sample contains a distribution of molar masses, the larger particles make a proportionately larger contribution to the scattering than an equal mass of smaller ones. It can be shown that for a heterogeneous polymer the mass-average molar mass (equation 20.42) is obtained from equation 20.33. The osmotic pressure yields the number-average molar mass.

If one of the dimensions of a molecule is comparable to the wavelength of light, the molecule no longer acts as a single point in scattering light, and there is interference between light waves scattered from different parts of the molecule. It is found that large molecules scatter more light forward, in the direction of the beam, than the small molecules. Thus, for larger molecules or particles, the angular dependence of the light scattering gives information as to shape. The development of lasers (Section 17.12) has made it possible to study the line broadening of light scattered by solutions of macromolecules. Measurement of the spectrum of scattered light permits the determination of the diffusion coefficient.*

20.8 GEL PERMEATION CHROMATOGRAPHY

In gel permeation chromatography a solution of a polymer having a distribution of molar masses is passed through a column packed with a rigid porous gel. The polymer network is swollen by the solvent and has pores of various sizes. As the liquid phase containing the polymer passes through the column, the polymer molecules diffuse into all parts of the gel not mechanically barred to them. The smaller molecules diffuse further into the gel and therefore spend more time in the solid phase of the column. The smaller polymer molecules therefore travel through the column more slowly than the larger polymer molecules, which cannot diffuse into the gel because of their size. Thus a chromatographic separation by size is obtained. Information about molar mass distributions is important because the properties of a polymeric material depend on the range of molar masses as well as on the average molar mass. Molar mass distributions may also be obtained by use of the ultracentrifuge.

---

* B. J. Berne and R. Pecora, *Dynamic Light Scattering*, Wiley, New York, 1976.

20.9 MOLAR MASS DISTRIBUTIONS OF CONDENSATION POLYMERS

The simplest type of polymerization is linear polymerization, in which monomer units are hooked end to end to form a chain without branches. For example, a hydroxyacid, HO—R—CO_2H, can be polymerized to form a polyester

$$\mathrm{HO{-}R{-}\overset{\overset{\displaystyle O}{\|}}{C}{-}O{-}C{-}\overset{\overset{\displaystyle O}{\|}}{C}{-}\cdots{-}O{-}R{-}\overset{\overset{\displaystyle O}{\|}}{C}{-}OH}$$

with the elimination of water. The number of monomer units in a chain is called the *degree of polymerization*, i. Thus $i = M/M_0$, where M is the molar mass of the polymer, and M_0 is the residue mass of the monomer.

A given sample of polyester contains a mixture of molecules, and we can describe this distribution by specifying the fraction of molecules of each different degree of polymerization. The molar mass distribution may be calculated* for the idealized case that the reactivity of the functional groups (carboxyl and hydroxyl for a polyester) is independent of the size of the molecule.

Suppose we start with n_0 monomer molecules AB, for example, $HORCO_2H$, and imagine that the condensation polymerization continues until the number of unreacted A (or B) groups remaining is n. The extent of reaction p is the fraction of A (or B) groups that have reacted.

$$p = \frac{n_0 - n}{n_0} = 1 - \frac{n}{n_0} \tag{20.34}$$

The average number of monomer units in the polymer molecules, $\bar{X}_n$ (number-average degree of polymerization), is equal to the initial number of molecules n_0 divided by the number of molecules remaining. Since each independent molecule has a free A (or B) group,

$$\bar{X}_n = \frac{n_0}{n} = \frac{1}{1 - p} \tag{20.35}$$

where the second form has been obtained by use of equation 20.34. For the average number of monomer units in the polymer molecules to be 100, it is evident that the extent of reaction p will have to be equal to 0.99.

To obtain the distribution of molar masses, consider the probabilities π_i of finding molecules of various degrees of polymerization i in a partially polymerized sample. The probability π_1 of finding an unreacted A (or B) group in the mixture is $1 - p$. The probability π_2 that an AB has reacted with another AB to form ABAB is $p(1 - p)$, since the probability of two independent events is the product of the two independent probabilities for having a bond p and not having a bond $1 - p$. The probability π_3 of finding ABABAB, that is, two successive bonds, is $p^2(1 - p)$. To generalize, consider

$$\pi_1 = 1 - p \qquad \pi_2 = p(1 - p) \qquad \pi_3 = p^2(1 - p) \qquad \pi_i = p^{i-1}(1 - p) \tag{20.36}$$

* P. J. Flory, *J. Am. Chem. Soc.*, **58**, 1877 (1936); P. J. Flory, *Principles of Polymer Chemistry*, Cornell University Press, Ithaca, 1953.

A necessary property of these probabilities is that the sum of the probabilities of all of the different chain lengths in a sample ($i = 1$ to ∞) must equal unity.

$$\sum_{i=1}^{\infty} \pi_i = \sum_{i=1}^{\infty} p^{i-1}(1-p) = (1-p)\sum_{i=1}^{\infty} p^{i-1}$$
$$= (1-p)(1+p+p^2+\cdots) = \frac{1-p}{1-p} = 1 \tag{20.37}$$

Since $(1-p)^{-1} = 1 + p + p^2 + p^3 + \cdots$. The quantity π_i may be interpreted as the mole fraction of i-mer. As may be seen from Fig. 20.5a, the mole fraction decreases steadily with increase in degree of polymerization i.

It is of perhaps greater interest to know the fraction of monomer units that have been incorporated into polymer of a particular degree of polymerization. The probability w_i of finding a monomer unit in an i-mer is equal to the fraction of monomer units in i-mers.

$$w_i = \frac{in_i}{n_0} \tag{20.38}$$

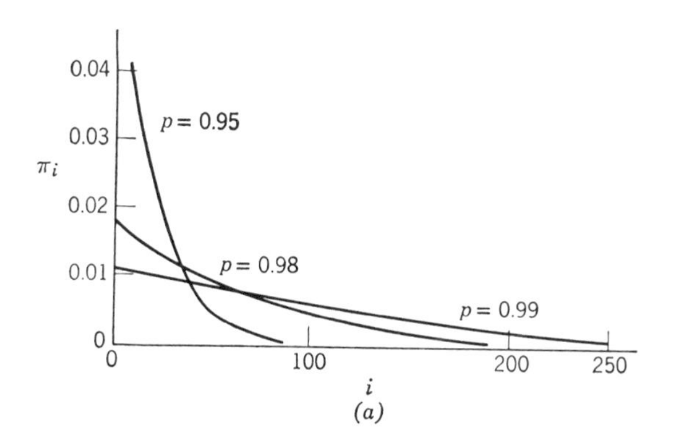

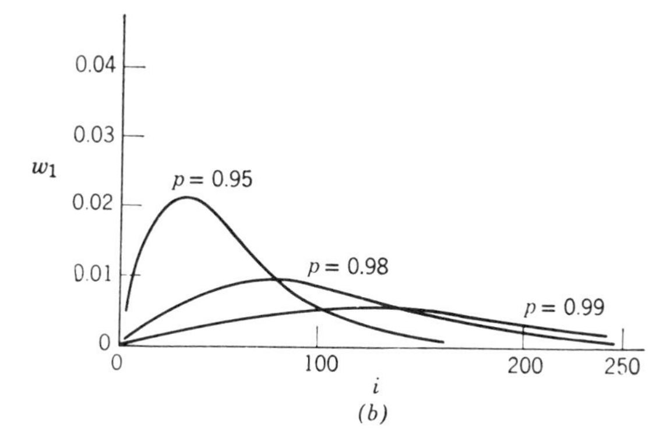

Fig. 20.5 (a) Mole fraction distribution of condensation polymer for extents of reaction p of 0.95, 0.98, and 0.99. (b) Weight fraction distribution of condensation polymer for extends of reaction p of 0.95, 0.98, and 0.99. Note that the extent of reaction must be very close to unity to obtain a high polymer.

where n_i is the number of *i*-mer molecules in the sample and n_0 is the number of monomer units initially present. The weight contributed by each monomer unit is the same, excepting the units that form the head and tail of the chain, and so w_i represents the weight fraction of *i*-mer. Since $n_i/n = p^{i-1}(1 - p)$,

$$w_i = \frac{ip^{i-1}(1 - p)n}{n_0} \tag{20.39}$$

Introducing $n/n_0 = 1 - p$ from equation 20.1, we have

$$w_i = ip^{i-1}(1 - p)^2 \tag{20.40}$$

In contrast with the mole fraction of *i*-mer, π_i, the weight fraction *i*-mer, w_i, goes through a maximum. The weight fractions are plotted for several extents of reaction in Fig. 20.5*b*. (These plots correspond with those in Fig. 20.5*a*.) The essential correctness of these distributions for condensation polymers has been confirmed by fractionating polymer samples by solubility methods into narrow ranges of degree of polymerization and determining the amounts and molar mass of the various fractions.

Synthetic high polymers, and some naturally occurring polymers, have a distribution of molar masses that may be represented by a graph such as Fig. 20.6. The molar mass distribution is, of course, not continuous, but if the molar masses are high the distribution can be considered to be continuous. The ordinate of Fig. 20.6 is the *probability density* P for molar mass M. The quantity $P\,dM$ is the probability of finding molecules having molar masses in the range M to $M + dM$ (see the discussion of the probability density function for molecular velocities). The probability of finding molecules in the range M to $M + dM$ is simply the fraction of the molecules having molar masses in this range.

When there is a range of molar masses, different experimental methods yield different types of average molar mass. The *number-average molar mass* M_n is equal to the mass of the whole sample divided by the number of molecules in it.

$$M_n = \frac{\sum_i n_i M_i}{\sum_i n_i} \tag{20.41}$$

Here n_i is the number of molecules of molar mass M_i per gram of dry polymer. This type of average is obtained from the measurement of the osmotic pressure of solutions containing the polymer, since this property is dependent on the number of molecules of solute per unit volume.

The *mass-average molar mass* weights molecules proportionally to their molar masses in the averaging process; that is, the molar mass M_i is multiplied by the mass n_iM_i of material of that molar mass instead of by the number of molecules. The *mass-average molar mass* M_m is defined by

$$M_m = \sum_i w_i M_i = \frac{\sum_i n_i M_i^2}{\sum_i n_i M_i} \tag{20.42}$$

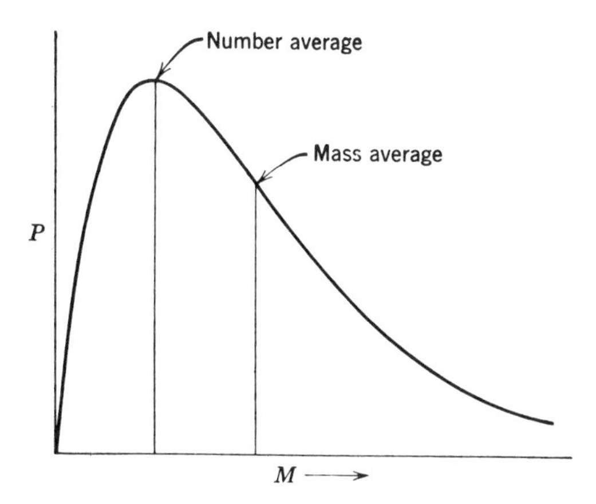

Fig. 20.6 Distribution of molar masses and types of average molar mass. The ordinate gives the probability of occurrence of molecules with a molar mass of M.

This type of average is obtained by the study of light scattering (Section 20.7) because the amount of light scattered depends on the size as well as the number of particles.

The mass-average molar mass must always be greater than the number-average molar mass, except for a sample in which all the molecules have the same weight, so that $M_n = M_m$. The ratio M_m/M_n is a useful measure of the breadth of the molar mass distribution.

Some information about the spread of molar masses in a sample may be obtained by comparing the number-average and mass-average molar masses. The number-average molar mass is simply the number-average degree of polymerization (equation 20.35) times the molar mass of a monomer unit in the polymer M_0.

$$M_n = \sum \pi_i i M_0 = \frac{M_0}{1 - p} \tag{20.43}$$

The mass-average molar mass M_m may be calculated as follows:

$$M_m = \sum iM_i = \sum_i iw_i M_0 \tag{20.44}$$

where w_i is the mass fraction of i-mer. Substituting equation 20.40, we see that

$$M_m = M_0(1 - p)^2 \sum i^2 p^{i-1} = M_0 \frac{1 + p}{1 - p} \tag{20.45}$$

since

$$\sum_i i^2 p^{i-1} = \frac{1 + p}{(1 - p)^3} \tag{20.46}$$

The ratio of these average molar masses is $M_m/M_n = 1 + p$. As we have seen from Fig. 20.5*a* and 20.5*b*, it is necessary for p to be in the range 0.99 to 1 for high polymer to be formed; thus the mass-average molar mass M_m is twice the number-average molar mass M_n, for a high polymer produced by condensation. To obtain high molar mass material, it is important to eliminate traces of monofunctional reactants (acid or alcohol) that cause termination at one end of the chain or the other.

20.10 CONFIGURATIONS OF POLYMER CHAINS

A long polymer chain may assume an enormous number of configurations of essentially identical energy because of the possibility of rotation about single bonds. The average configuration may be calculated for various models of the polymer chain (i.e., certain bond lengths and bond angles) using statistical methods. These theoretical results are of great importance for an understanding of rubberlike elasticity (Section 20.11) and of hydrodynamic and thermodynamic properties of dilute polymer solutions.

If in a polymer chain containing N bonds, each of length b, the bonds are connected by universal joints, the conformation of the polymer can be described in terms of a three-dimensional random walk in which successive steps are completely uncorrelated in direction. The probability $W(L, N)\,dL$ that after N steps the end of the chain will be at a distance between L and $L + dL$ from the origin is given by

$$W(L, N)\,dL = 4\pi\left(\frac{3}{2\pi Nb^2}\right)^{3/2} \exp\left(-\frac{3L^2}{2Nb^2}\right) L^2\,dL \qquad (20.47)^*$$

The plot of $W(L, N)$ goes through a maximum just like the plot for the probability density of molecular speed (Fig. 14.4).

The root-mean-square end-to-end distance $\langle L^2\rangle^{1/2}$ for such a completely flexible chain may be calculated according to the general procedure for calculating averages using the distribution function.

$$\langle L^2\rangle^{1/2} = \left[\int_0^\infty L^2 W(L, N)\,dL\right]^{1/2} = N^{1/2}b \qquad (20.48)$$

Thus it is seen that the root-mean-square end-to-end distance for this simple model of a polymer molecule is proportional to the square root of the number of bonds, and the proportionality factor is the bond length b.

The root-mean-square end-to-end length is greater for real molecules for three reasons.

1. There is a fixed bond angle θ between the bonds forming the backbone of the chain.
2. There are restrictions to free rotations about bonds.
3. Chain segments occupy volume, and this volume is excluded to other segments of the chain.

---

* D. A. McQuarrie, *Statistical Thermodynamics*, Harper and Row, New York, 1973, Chapt. 14.

If the bond angles are all θ it may be shown* that

$$\langle L^2 \rangle^{1/2} = N^{1/2} b \left(\frac{1 + \cos \theta}{1 - \cos \theta} \right)^{1/2} \tag{20.49}$$

In this equation θ is 180° minus the angle between the bonds. Thus, for tetrahedral bonds, 0 = 180° − 109.471° = 70.529° and $\cos \theta = \frac{1}{3}$, so that

$$(1 + \cos \theta)/(1 - \cos \theta)^{1/2} = \sqrt{2}$$

If there are in addition restrictions on free rotations about bonds, the root-mean-square end-to-end distance will be further increased. Recently methods have been developed for calculating configurations of polymer chains in which there are restrictions on internal rotations.†

Actually, the extension of a real polymer chain depends very much on the solvent. When a polymer is dissolved in a good solvent (e.g., when polystyrene is dissolved in toluene), the segment-solvent contacts are energetically more favorable than segment-segment contacts. This causes the molecule to stretch out in solution so that segment-segment contacts are minimized, and the Gibbs energy is a minimum. On the other hand, when a polymer is dissolved in a poor solvent, the segment-segment contacts are favored over segment-solvent contacts. This causes the molecule to tend to ball up in solution in opposition to the thermal motions that tend to extend it. If the solvent is too poor, the polymer will not dissolve.

Intermediate between good and poor solvents is the theta (θ) solvent in which the dimensional contraction resulting from the poorer solvency exactly cancels the increase in size due to the finite volumes of the segments. Cyclohexane is a theta solvent for polystyrene at 34 °C. For a theta solvent the value of B in the osmotic pressure equation 20.32 is equal to zero.

On the basis of equations 20.48 and 20.49 it is possible to understand why the intrinsic viscosity of a high polymer in a theta solvent depends on $M^{1/2}$ (equation 20.19). The polymer molecule in solution may be visualized as a spherical cloud of segments, with the cloud getting thinner as we go out from the center. The size of this spherical cloud is measured by the root-mean-square end-to-end distance $\langle L^2 \rangle^{1/2}$. As an approximation of the spherical cloud, we may replace it with a rigid sphere with a radius proportional to $\langle L^2 \rangle^{1/2}$ and a volume proportional to $\langle L^2 \rangle^{3/2}$. Since, according to equation 20.17, the intrinsic viscosity of a sphere is proportional to the volume per unit mass of the macromolecule, the intrinsic viscosity of a random coil is expected to be proportional to $\langle L^2 \rangle^{3/2}/M$. Since $\langle L^2 \rangle^{1/2}$ is proportional to the number of segments in the chain, or to M, the intrinsic viscosity of a random coil is expected to be proportional to $M^{1/2}$.

---

* V. A. Bloomfield, D. M. Crothers, and I. Tinoco, *Physical Chemistry of Nucleic Acids*, Harper and Row, New York, 1974, p. 158.

† P. J. Flory, *Science*, **188**, 1268 (1975).

20.11 MECHANICAL PROPERTIES OF POLYMERS

Polymeric materials display a very wide range of physical properties: they are hard or soft, leathery or rubbery, brittle or tough, and meltable and nonmeltable. These properties depend on the molecular structure of the polymer and the various properties are obtained by the choice of the polymer and its treatment. The physical properties depend to quite an extent on whether the polymer is (1) completely amorphous, or (2) partly crystalline. In an amorphous polymer the molecular chains are all tangled up in a disordered fashion. Long-chain polymers exist primarily in the amorphous state. Since the long chains are mostly in the randomly coiled configuration, the solid is elastic; when the material is stretched, the chains are extended, and when the stress is removed, the chains take on a more random configuration. Some crystallinity may be present, and to the extent that it is present the material is stiffer.

Amorphous polymers may be glassy, leathery, or rubbery, depending on the temperature. At low temperatures amorphous polymers are in the glassy state, which is like that of a supercooled liquid. As the temperature is raised, there is a transition at the "glass-transition temperature" from the glassy to the leathery state. There is a marked change in physical properties, but no discontinuity in the density. Below the glass-transition temperature even amorphous polymers are rigid and brittle. The atoms and small groups of atoms vibrate about mean positions, but it is not possible for molecular segments to slide over each other. Above the glass-transition temperature an amorphous polymer becomes leathery or elastic and a crystalline polymer becomes more flexible and less brittle. In amorphous polymers large molecular segments are able to slide over each other, and the characteristic plastic properties are obtained. For both amorphous and crystalline polymers the rate of change of density with temperature is greater above T_g that below it because of the greater increase in molecular motion. The transition from the glassy to the rubberlike state generally extends over a range of about 50°, but the temperature range in which it occurs depends on the type of polymer. If there are rather long sections of the molecular chains between cross links and centers of entanglement, long segments will be in Brownian motion, and rubberlike properties will result.

The glass-transition temperature is raised by vulcanization, which introduces cross linkages; that is, bonds are formed between adjacent threadlike molecules, thus restricting the molecular motions.

Many polymeric solids have properties that are intermediate between those of ideal solids and ideal liquids. In a perfectly elastic solid the stress is directly proportional to the strain but independent of the rate of strain. In a perfectly viscous fluid the stress is directly proportional to the rate of strain but independent of the strain itself. If the stress depends on both the strain and the rate of strain, the material is said to be *viscoelastic*. Polymers in particular show such behavior because of the complicated way in which such long chainlike molecules interact with each other.*

* J. D. Ferry, *Viscoelastic Properties of Polymers*, Wiley, New York, 1970.

The mechanical properties of polymers are profoundly dependent on the stereoregularity of their chains. Polymers of monomers like styrene have asymmetric carbon atoms. By using certain solid catalysts it is possible to prepare polymers in which all the adjacent asymmetric atoms, at least for long periods of the chain, have the same steric configuration.

There are two types of stereoregular vinyl polymers, *isotactic* and *syndiotactic*, depending on whether successive pseudoasymmetric carbon atoms have the same or opposite enantiomorphic configurations. These two types of structures are illustrated in Fig. 20.7, along with a structure without order. Structures with a lesser degree of order than isotactic or syndiotactic are referred to as *atactic*. Some polymers (stereoblock polymers) have alternating isotactic and syndiotactic sections. K. Ziegler and G. Natta, who first prepared such *isotactic* polymers, shared the 1963 Nobel prize for chemistry. The regularity in structure of isotactic polymers makes it possible for the polymer chains to fit into a crystal lattice. This increased orderliness increases the melting point, rigidity, and toughness above those of corresponding atactic polymers.

Some polymers contain regions in which the chains are arranged in an orderly, three-dimensional array. These crystalline regions typically have dimensions of the

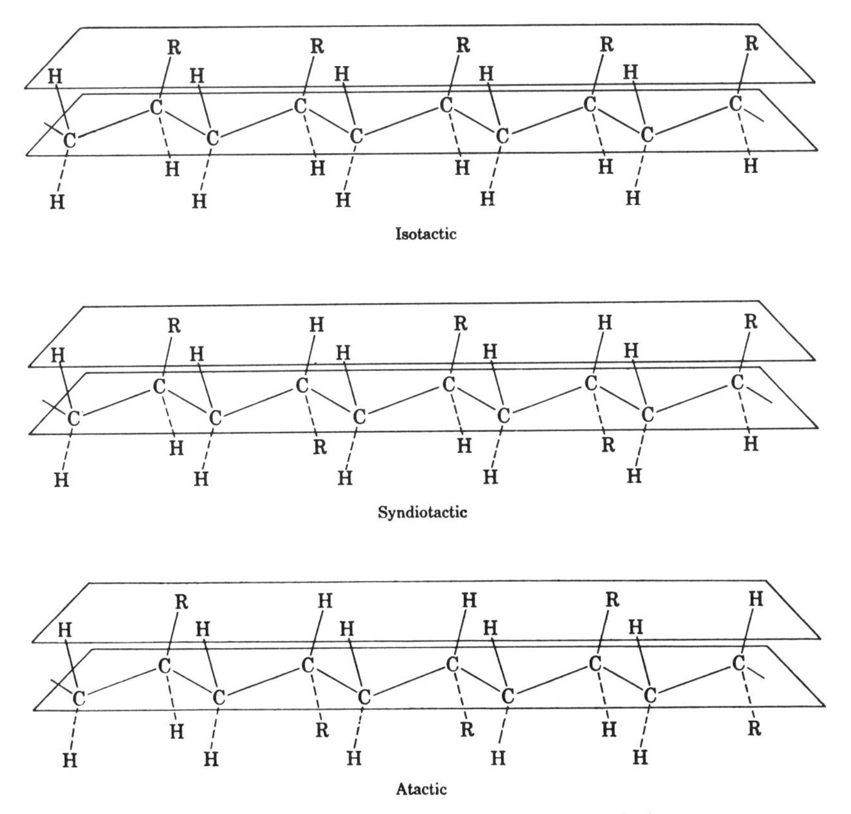

Fig. 20.7 Steric configuration of polymer chains.

order of 10 nm, and they have an important influence on the physical properties of the polymer. The extent of crystallinity increases when the polymer is stretched because this draws the chains together and reduces the random thermal motions. For example, natural rubber is ordinarily amorphous at room temperature but becomes oriented and crystalline when stretched. The amount of crystallinity may be investigated by X-ray diffraction and by studying the volume of the polymer as a function of temperature.

Crystallites occur most frequently in polymers having a comparatively simple and symmetrical structure or polymer chains that tend to associate by the formation of hydrogen bonds. There are crystalline regions in nylon, polyethylene, and other polymers in which the chains have sufficient chemical regularity. The individual crystals in a polymer sample are usually too small to be seen by visible-light microscopy. Larger crystals of long chain polymers may be obtained from solution.

20.12 HELICAL STRUCTURES IN POLYPEPTIDES AND PROTEINS

Proteins are polymers of amino acids ($H_2NCHRCO_2H$) that have molar masses greater than about 10,000 g mol^{-1}. About 22 amino acids occur in proteins, and they are all of the L configuration. In proteins and polypeptides, aminoacid residues are connected through amide bonds.

```
  O
   \\     /
    C—N
   /     \
          H
```

The nuclei of H, N, C, and O lie in a plane, and the connecting bonds are *trans*. Smaller polymers of amino acids called oligopeptides form random coils in solution, but proteins have a more or less fixed three-dimensional structure held together by hydrogen bonds (Section 10.8), disulfide bonds (—S—S—) between cystine residues, and ionic and van der Waals interactions. In the amide group the bond lengths and angles can be considered to be fixed. Only two angles have to be given per alpha carbon to specify the configuration of the chain. If certain angles, which are not prohibited by steric hindrance, are repeated indefinitely, a helical structure results. A peptide helix is characterized by the number n of amino-acid residues per turn of helix and by the distance d traversed parallel to the helix axis per amino-acid residue.

The α helix is sufficiently important to show the structures of the right-handed and left-handed varieties in Fig. 20.8. It is primarily the right-handed helix that exists in proteins. The α helix is held together by hydrogen bonds between a carbonyl oxygen and the N—H of the fourth residue along the chain. This produces

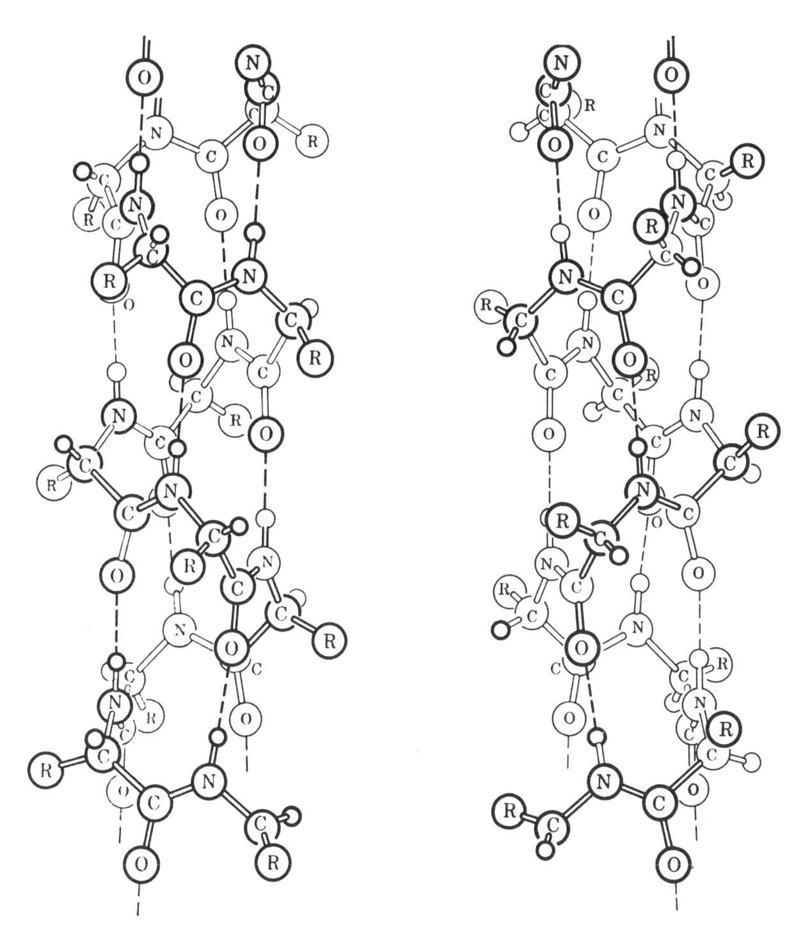

Fig. 20.8 Alpha helix of a polypeptide. A left-handed helix is shown on the left and a right-handed helix on the right.

an especially stable structure with $n = 3.6$ residues per turn. In the direction of the axis of the helix there is a residue every 0.15 nm.

20.13 HELIX-RANDOM COIL TRANSITIONS IN POLYPEPTIDES

A number of synthetic polypeptides show structural transitions in solution when there is a change in temperature or pH. A single polypeptide chain can exist in three states in solution, as illustrated in Fig. 20.9.

If it were possible to have a polypeptide chain in a vacuum, the helix (*a*) would be the stable form at low temperature and the random coil form *c* would be the stable form at high temperatures. For the transition from helix to random coil with

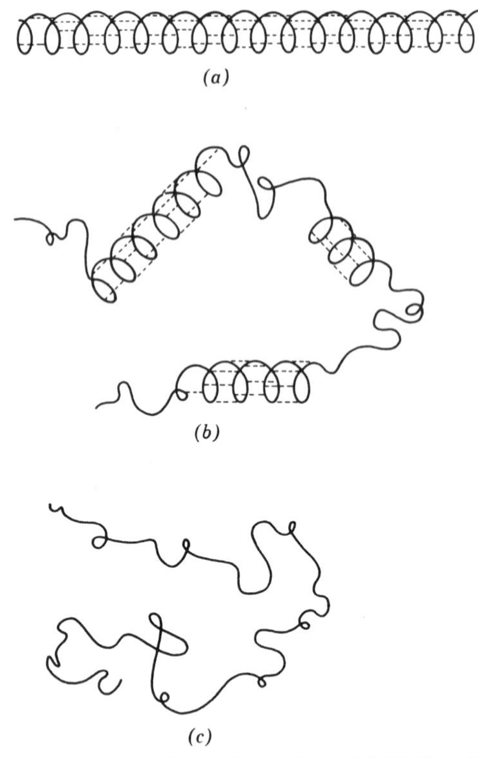

(a)

(b)

(c)

Fig. 20.9 Configurational states of a polypeptide. (*a*) Helix. (*b*) Intermediate form. (*c*) Random coil. [Reprinted with permission from L. Peller, *J. Phys. Chem.*, **63**, 1194 (1959). Copyright by the American Chemical Society.]

ΔH positive and ΔS positive, ΔG would be positive at low temperatures and negative at a sufficiently high temperature. However, the polypeptide interacts strongly with solvent and, as a result, the random coil may have a smaller entropy that the helix in some solvents and a lower enthalpy. In such a solvent, raising the temperature would cause the polypeptide to undergo a transition from random coil to helix. Such a transition for poly-γ-benzyl-L-glutamate in a mixture of dichloroacetic acid and dichloroethane is shown in Fig. 20.10.

At an intermediate temperature the configurational state would be that represented in Fig. 20.9*b*, in which there are alternating random coil and helical regions. The transition from helix to random coil may be much sharper than for a simple chemical equilibrium $A = B$, which occurs in homogeneous solution. In other words, large changes in molecular properties may occur as a result of small changes in temperature, pressure, pH, or other environmental conditions. The change takes place over a narrow range of the environmental variable because of its cooperative nature. In a polypeptide the formation of the first helical section is the least probable. Once the first loop has formed further helix forms more easily. This may be understood in molecular terms as follows: to form the first loop of a helix the

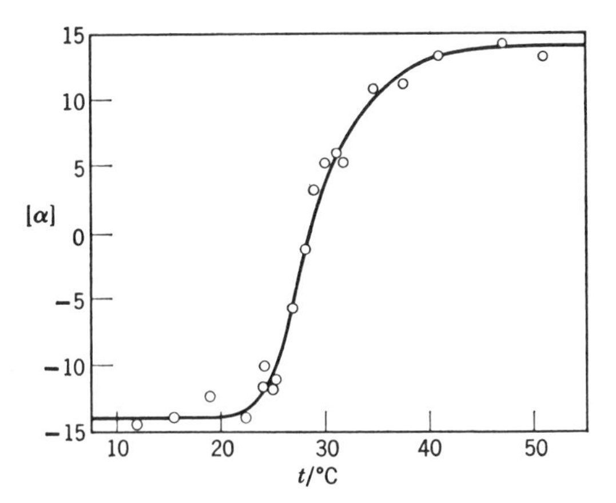

Fig. 20.10 Temperature dependence of the optical rotation of poly-γ-benzyl-L-glutamate (M = 350,000 g mol^{-1}) in dichloroacetic acid-dichloroethane (80:20). [Reprinted with permission from P. Doty and J. T. Yang, *J. Am. Chem. Soc.*, **78**, 499 (1956). Copyright by the American Chemical Society.]

carbonyl oxygen of the ith residue must hydrogen bond with the NH of the $(i + 4)$th residue. This requires that twelve angles have the values corresponding with an alpha helix. Thus the formation of the first loop is unfavorable from an entropy standpoint and is energetically favorable because a single hydrogen bond is formed. However, the addition of helix to a helical sequence requires that only two angles correspond to the needed values per hydrogen bond formed. Thus an additional helix tends to form on existing helical sections instead of in the midst of a random coil. As a consequence, transitions from, for example, helix to coil configuration occur over narrower ranges of temperature or compositions than would otherwise be the case. Ising developed the theory for such transitions and applied it to ferromagnetism.

References

T. M. Birshtein and O. B. Ptitsyn, translated by S. N. Timasheff and M. J. Timasheff, *Conformations of Macromolecules*, Wiley-Interscience, New York, 1966.

V. A. Bloomfield, D. A. Crothers, and I. Tinoco, Jr., *Physical Chemistry of Nucleic Acids*, Harper and Row, New York, 1974.

C. R. Cantor and P. R. Schimmel, *Biophysical Chemistry*, W. H. Freeman & Co., San Francisco, 1979.

R. E. Dickerson and I. Geis, *The Structure and Action of Proteins*, Harper and Row, New York, 1969.

P. J. Flory, *Principles of Polymer Chemistry*, Cornell University Press, Ithaca, 1953.

H. Morawetz, *Macromolecules in Solution*, Wiley-Interscience, New York, 1965.

K. J. Mysels, *Introduction to Colloid Chemistry*, Wiley-Interscience, New York, 1959.

D. Poland, *Cooperative Equilibria in Physical Biochemistry*, Clarendon Press, Oxford, England, 1978.

D. Poland and H. A. Scheraga, *Theory of Helix-Coil Transitions in Biopolymers*, Academic Press, New York, 1970.

C. Tanford, *Physical Chemistry of Macromolecules*, Wiley, New York, 1961.

K. E. Van Holde, *Physical Biochemistry*, Prentice-Hall, Englewood Cliffs, N.J., 1971.

J. W. Williams, *Ultracentrifugation of Macromolecules: Modern Topics*, Academic Press, New York, 1973.

Problems

20.1 Calculate the sedimentation coefficient of tobacco mosaic virus from the fact that the boundary moves with a velocity of 0.454 cm hr^{-1} in an ultracentrifuge at a speed of 10,000 rpm at a distance of 6.5 cm from the axis of the centrifuge rotor.

Ans. 177×10^{-13} s.

20.2 The sedimentation coefficient of myoglobin at 20 °C is 2.06×10^{-13} s. What molar mass would it have if the molecules were spherical? Given: $\bar{v} = 0.749 \times 10^{-3}$ m^3 kg^{-1}, $\rho = 0.9982 \times 10^3$ kg m^{-3}, and $\eta = 0.001\,005$ Pa s. *Ans.* 15,500 g mol^{-1}.

20.3 A sedimentation equilibrium experiment is to be carried out with myoglobin ($M = 16{,}000$ g mol^{-1}) in an ultracentrifuge operating at 15,000 rpm. The bottom of the cell is 6.93 cm from the axis of rotation and the meniscus is 6.67 cm from the axis of rotation. What ratio of concentrations is expected at 20 °C if $\bar{v} = 0.75 \times 10^{-3}$ m^3 kg^{-1} and $\rho = 1.00 \times 10^3$ kg m^{-3}? *Ans.* 2.05.

20.4 The sedimentation and diffusion coefficients for hemoglobin corrected to 20 °C in water are 4.41×10^{-13} s and 6.3×10^{-11} m^2 s^{-1}, respectively. If $\bar{v} = 0.749$ cm^3 g^{-1} and $\rho_{H_2O} = 0.998$ g cm^{-3} at this temperature, calculate the molar mass of the protein. If there is 1 mol of iron per 17,000 g of protein, how many atoms of iron are there per hemoglobin molecule? *Ans.* 68,000 g mol^{-1}, 4.

20.5 The diffusion coefficient for serum globulin at 20 °C in a dilute aqueous salt solution is 4.0×10^{-11} m^2 s^{-1}. If the molecules are assumed to be spherical, calculate their molar mass. Given: $\eta_{H_2O} = 0.001\,005$ Pa s at 20 °C and $\bar{v} = 0.75$ cm^3 g^{-1} for the protein. *Ans.* 512,000 g mol^{-1}.

20.6 A sharp boundary is formed between a dilute buffered solution of hemoglobin ($D = 6.9 \times 10^{-11}$ m^2 s^{-1}) and the buffer at 20 °C. What is the half width of the boundary after 1 hr and 4 hr? *Ans.* 0.0705, 0.1405 cm.

20.7 The diffusion coefficient of hemoglobin at 20 °C is 6.9×10^{-11} m^2 s^{-1}. Assuming its molecules are spherical, what is the molar mass? Given: $\bar{v} = 0.749 \times 10^{-3}$ m^3 kg^{-1} and $\eta = 0.001\,005$ J m^{-3} s. *Ans.* 100,000 g mol^{-1}.

20.8 The protein human plasma albumin has a molar mass of 69,000 g mol^{-1}. Calculate the osmotic pressure of a solution of this protein containing 2 g per 100 cm^3 at 25 °C in (*a*) torr, and (*b*) millimeters of water. The experiment is carried out using a salt solution for solvent and a membrane permeable to salt. *Ans.* (*a*) 719 Pa, (*b*) 73.4 mm H_2O.

20.9 The following osmotic pressures were measured for solutions of a sample of polyisobutylene in benzene at 25 °C.

$c/10^{-2}$ g cm^{-3}	0.500	1.00	1.50	2.00
Π/g/cm^2	0.505	1.03	1.58	2.15

Calculate the number average molar mass from the value of Π/c extrapolated to zero concentration of the polymer. [The pressures may be converted into atmospheres dividing by $(76 \text{ cm atm}^{-1})(13.53 \text{ g cm}^{-3}) = 1028 \text{ g cm}^{-2} \text{ atm}^{-1}$.] *Ans.* $255{,}000 \text{ g mol}^{-1}$.

20.10 A beam of sodium D light (589 nm) is passed through 100 cm of an aqueous solution of sucrose containing 10 g sucrose per 100 cm^3. Calculate I/I_0, where I_0 is the intensity that would have been obtained with pure water, given that $M = 342.30 \text{ g mol}^{-1}$ and $dn/dc = 0.15 \text{ g}^{-1} \text{ cm}^3$ for sucrose. The refractive index of water at 20 °C is 1.333 for the sodium D line. *Ans.* 0.9938.

20.11 The relative viscosities of a series of solutions of a sample of polystyrene in toluene were determined with an Ostwald viscometer at 25 °C.

$c/10^{-2} \text{ g cm}^{-3}$	0.249	0.499	0.999	1.998
η/η_0	1.355	1.782	2.879	6.090

The ratio η_{sp}/c is plotted against c and extrapolated to zero concentration to obtain the intrinsic viscosity. If the constants in equation 20.19 are $K = 3.7 \times 10^{-2}$ and $a = 0.62$ for this polymer, when concentrations are expressed in g/1 cm^3, calculate the molar mass. *Ans.* $500{,}000 \text{ g mol}^{-1}$.

20.12 At 34 °C the intrinsic viscosity of a sample of polystyrene in toluene is 84 $\text{cm}^3 \text{ g}^{-1}$. The empirical relation between the intrinsic viscosity of polystyrene in toluene and molar mass is

$$[\eta] = 1.15 \times 10^{-2} M^{0.72}$$

What is the molar mass of this sample? *Ans.* $296{,}000 \text{ g mol}^{-1}$.

20.13 Given that the intrinsic viscosity of myosin is 217 $\text{cm}^3 \text{ g}^{-1}$, approximately what concentration of myosin in water would have a relative viscosity of 1.5?

Ans. $2.30 \times 10^{-3} \text{ g cm}^{-3}$.

20.14 A sample of polymer contains 0.50 mole fraction with molar mass $100{,}000 \text{ g mol}^{-1}$ and 0.50 mole fraction with molas mass $200{,}000 \text{ g mol}^{-1}$. Calculate (*a*) M_n, and (*b*) M_m.

Ans. (*a*) 150,000, (*b*) $167{,}000 \text{ g mol}^{-1}$.

20.15 For a condensation polymerization of a hydroxyacid in which 99% of the acid groups are used up, calculate (*a*) the average number of monomer units in the polymer molecules, (*b*) the probability that a given molecule will have the number of residues given by this value, and (*c*) the weight fraction having this particular number of monomer units. *Ans.* (*a*) 100, (*b*) 3.7×10^{-3}, (*c*) 3.7×10^{-3}.

20.16 In the condensation polymerization of a hydroxyacid with a residue mass of 200 it is found that 99% of the acid groups are used up. Calculate (*a*) the number-average molar mass, and (*b*) the mass-average molar mass. *Ans.* (*a*) 20,000, (*b*) $39{,}800 \text{ g mol}^{-1}$.

20.17 When rubber is allowed to contract, the mechanical work obtained is equal to the force times the displacement in the direction of the force and is given by $dw = -f\,dL$, where f is force and L is the length of the piece of rubber. If the contraction is carried out reversibly, the first law may be written

$$dU = \left(\frac{\partial U}{\partial T}\right)_L dT + \left(\frac{\partial U}{\partial L}\right)_T dL = T\,dS + f\,dL$$

if pressure-volume work is neglected. (*a*) Show that

$$\left(\frac{\partial S}{\partial T}\right)_L = \frac{1}{T}\left(\frac{\partial U}{\partial T}\right)_L \qquad \left(\frac{\partial S}{\partial L}\right)_T = \frac{1}{T}\left[\left(\frac{\partial U}{\partial L}\right)_T - f\right]$$

(*b*) By using the fact that the order of differentiation used to obtain $\partial^2 S/\partial L\,\partial T$ is immaterial, show that

$$\left(\frac{\partial U}{\partial L}\right)_T = f - T\left(\frac{\partial f}{\partial T}\right)$$

20.18 In polyethene $H(CH_2—CH_2)_nH$ the bond length b is 0.15 nm. What is the root-mean-square end-to-end distance for a molecule with universal joints with a molar mass of 10^5 g mol^{-1}? Taking into account the fact that carbon forms tetrahedral bonds, what is $\langle L^2 \rangle^{1/2}$? *Ans.* 12.7, 17.8 nm.

20.19 A turn of α helix (3.6 residues) is 0.541 nm in length measured parallel to the helix axis. If your hair grows 6 in. per year, how many amino-acid residues must be added to each α helix in the keratin fiber per second? *Ans.* 32 residues per second.

20.20 The sedimentation coefficient of gamma globulin at 20 °C is 7.1×10^{-13} s. Calculate how far the protein boundary will sediment in $\frac{1}{2}$ hr if the speed of the centrifuge is 60,000 rpm and the initial boundary is 6.50 cm from the axis of rotation.

20.21 Using data from Table 20.1, calculate the molar mass of serum albumin.

20.22 Given the diffusion coefficient for sucrose at 20 °C in water ($D = 45.5 \times 10^{-11}$ m^2 s^{-1}), calculate its sedimentation coefficient. The partial specific volume $\bar{v}$ is 0.630 cm^3 g^{-1}.

20.23 A sharp boundary is formed between a solution of hemoglobin in a buffer and the buffer solution at 25 °C. After 10 hr the half-width of the concentration-gradient curve at the inflection point is 0.226 cm. What is the diffusion coefficient of hemoglobin under these conditions?

20.24 The diffusion coefficient of a certain virus having spherical particles is 0.50×10^{-11} m^2 s^{-1} at 0 °C in a solution with a viscosity of 0.001 80 Pa s. Calculate the molar mass of this virus, assuming that the density of the virus is 1 g cm^{-3}.

20.25 The following osmotic pressures of polyvinyl acetate in dioxane were measured by G. V. Browning and J. D. Ferry at 25 °C.

$c/10^{-2}$ g cm^{-3}	0.292	0.579	0.810	1.140
Π/cm of solvent	0.73	1.76	2.73	4.68

Calculate the number-average molar mass. The density of dioxane is 1.035 g cm^{-3}.

20.26 Human blood plasma contains approximately 40 g of albumin (M = 69,000 g mol^{-1}) and 20 g of globulin (M = 160,000 g mol^{-1} per liter. Calculate the colloid osmotic pressure at 37 °C, ignoring the Donnan effect.

20.27 A solution of high polymer in benzene has a concentration of 1 g/100 cm^3 and a refractive index for the sodium D line (589 nm) of 1.5021. The refractive index of benzene under these conditions is 1.5011. The turbidity τ is 2×10^{-4} cm^{-1}. What is the molar mass of the polymer? What is I/I_0 for a 10-cm cell?

20.28 The relation between M and $[\eta]$ or double-stranded linear DNA is $0.665 \ln M = 1.987 + \log([\eta] + 500)$ when $[\eta]$ is expressed in cm^3 g^{-1}. What is the molar mass of DNA that has an intrinsic viscosity of 5000 cm^3 g^{-1}?

20.29 A sample of polystyrene was dissolved in toluene, and the following flow times in an Ostwald viscometer at 25 °C were obtained for different concentrations.

$c/10^{-2}$ g cm^{-3}	0	0.1	0.3	0.6	0.9
Time/s	86.0	99.5	132	194	301

If the constants in equation 20.19 are $K = 3.7 \times 10^{-2}$ and $a = 0.62$ for this polymer, calculate the molar mass.

20.30 Calculate the (a) number, and (b) mass-average molar masses for the following mixture of high-polymer fractions: 1 g of M = 20,000 g mol^{-1}, 2 g of M = 50,000 g mol^{-1}, and 0.5 g of M = 100,000 g mol^{-1}.

20.31 For a condensation polymerization of a hydroxyacid in which 95% of the acid groups is used up, calculate (a) the average number of monomer units in the polymer molecules, (*b*) the probability that a molecule chosen at random will have this number of residues, and (*c*) the weight fraction having this particular number of monomer units.

20.32 For the polymer described in Problem 20.31 calculate the number-average and mass-average molar mass.

20.33 For a polymethylene chain, $H(CH_2)_nH$, of molar mass 100,000 g mol^{-1}, (*a*) calculate the end-to-end distance assuming an exactly linear chain and a C—C distance of 0.154 nm. (*b*) Calculate the root-mean-square end-to-end distance assuming chain is freely jointed. (*c*) Calculate the root-mean-square end-to-end distance assuming that the bond angle is 109° 28′.

20.34 For a free jointed polymethylene chain of molar mass 100,000 g mol^{-1}, plot the probability density for end-to-end distance r versus r.

APPENDIX

UNITS AND PHYSICAL CONSTANTS

SI Units

The International System of Units (abbreviated SI for "Systeme International d'Unites") provides the basis for all types of physical and chemical calculations. This particular system was defined and given official status by the 11th Conference Generale des Poids et Mesures in 1960.* The SI system is founded on the seven base units defined in the following table.

Physical Quantity	Symbol for Quantity	Name of Unit	Symbol for SI Unit	Definition
Length	l	meter	m	1,650,763.73 wavelengths in vacuum of the orange-red line of the spectrum of krypton-86
Mass	m	kilogram	kg	A cylinder of platinum-iridium alloy kept by the International Bureau of Weights and Measures in Paris
Time	t	second	s	The duration of 9,192,631,770 cycles of the radiation associated with a specific transition of the cesium atom
Electric current	I	ampere	A	The magnitude of the current that, when flowing through each of two long parallel wires separated by 1 m in free space, results in a force between the two wires of 2×10^{-7} N for each meter of length
Thermodynamic temperature	T	kelvin	K	Origin is at absolute zero and the triple point of water is 273.16 K
Amount of substance	n	mole	mol	Amount of substance that contains as many elementary entities as there are carbon atoms in 0.012 kg of carbon-12
Luminous intensity	I_v	candela	cd	The luminous intensity, in the perpendicular direction, of a surface of 1/600,000 sq m of a black body at the temperature of freezing platinum under a pressure of 101,325 N m^{-2}

* See *J. Chem. Ed.*, **48**, 569 (1971). M. A. Paul, *J. Chem. Documentation*, **11**, 3 (1971). M. L. McGlashan, *Pure and Applied Chem.*, **21**, No. 1 (1970). M. L. McGlashan, *Physiochemical Quantities and Units*, The Royal Institute of Chemistry, London, 1971. M. L. McGlashan, *Annual Review of Physical Chemistry*, **24**, 51 (1974).

All quantities may be expressed in these units or in terms of derived units obtained algebraically by multiplication and division. The principal derived units used in physical chemistry are given in the following table.

Quantity	Unit	Symbol	Definition
Force	newton	N	$kg\ m\ s^{-2}$
Work, energy, quantity of heat	joule	J	$N\ m\ (= kg\ m^2\ s^{-2})$
Power	watt	W	$J\ s^{-1}$
Pressure	pascal	Pa	$N\ m^{-2}$
Electric charge	coulomb	C	$A\ s$
Electric potential difference	volt	V	$kg\ m^2\ s^{-3}\ A^{-1}\ (= J\ A^{-1}\ s^{-1} = JC^{-1})$
Electric resistance	ohm	Ω	$kg\ m^2\ s^{-3}\ A^{-2}\ (= V\ A^{-1})$
Electric capacitance	farad	F	$A\ s\ V^{-1}\ (= m^{-2}\ kg^{-1}\ s^4\ A^2)$
Frequency	hertz	Hz	s^{-1} (cycle per second)
Magnetic flux density	tesla	T	$kg\ s^{-2}\ A^{-1}\ (= N\ A^{-1}\ m^{-1})$

It is important to be able to convert from other systems of units to SI. Some of these conversion factors are as follows.

Physical Quantity	Name of Unit	Symbol	Equivalent in SI Units
Length	Ångstrom	Å	10^{-10} m (10^{-1} nm)
Energy	electronvolt	eV	$1.602\ 189\ 2 \times 10^{-19}$ J
	wave number	cm^{-1}	1.986×10^{-23} J
	calorie (thermochemical)		4.184 J
	erg		10^{-7} J
Force	dyne		10^{-5} N
Pressure	atmosphere		101.325 kN m^{-2}
	torr		133.322 N m^{-2}
Electric charge	e.s.u.		3.334×10^{-10} C
dipole moment	debye (10^{-18} e.s.u. cm)		3.334×10^{-30} C m
magnetic flux density	gauss	G	10^{-4} T

Values of Physical Constants[a]

Constant	Symbol	Value (with uncertainty)[b]
Speed of light in vacuum	c	$2.997\,924\,58(1) \times 10^{8}$ m s^{-1}
Permittivity of vacuum	$\epsilon_0 = \mu_0^{-1}c^{-2}$	$8.854\,187\,82(5) \times 10^{-12}$ C^2 N^{-1} m^{-2}
Permeability of vacuum	μ_0	$4\pi \times 10^{-7}$ J s^2 C^{-2} m^{-1} (exactly)
Elementary charge	e	$1.602\,189\,2(46) \times 10^{-19}$ C
Planck constant	h	$6.626\,176(36) \times 10^{-34}$ J s
Avogadro constant	N_A	$6.022\,045(31) \times 10^{23}$ mol^{-1}
Atomic mass unit	$u = 10^{-3}$ kg mol$^{-1}/N_A$	$1.660\,565\,5(86) \times 10^{-27}$ kg
Rest mass of electron	m_e	$9.109\,534(47) \times 10^{-31}$ kg
Rest mass of proton	m_p	$1.672\,648\,5(86) \times 10^{-27}$ kg
Rest mass of neutron	m_n	$1.674\,954\,3(86) \times 10^{-27}$ kg
Faraday constant	$F = N_A e$	$9.648\,456(27) \times 10^{4}$ C mol^{-1}
Rydberg constant	R_∞	$1.097\,373\,177(83) \times 10^{7}$ m^{-1}
Hartree energy	H	$4.359\,814(24) \times 10^{-18}$ J
Bohr radius	a_0	$5.291\,770\,6(44) \times 10^{-11}$ m
Bohr magneton	μ_B	$9.274\,078(36) \times 10^{-24}$ J T^{-1}
Nuclear magneton	μ_N	$5.050\,824(20) \times 10^{-27}$ J T^{-1}
Gas constant	R	$8.314\,41(26)$ J K^{-1} mol^{-1}
		$1.987\,192$ cal K^{-1} mol^{-1}
		$0.082\,056\,9$ L atm K^{-1} mol^{-1}
Boltzmann constant	$k = R/N_A$	$1.380\,662(44) \times 10^{-23}$ J K^{-1}

[a] *Manual of Symbols and Terminology for Physiochemical Quantities and Units, International Union of Pure and Applied Chemistry*, Butterworths, London, 1973.

[b] The digits in parentheses following a numerical value represent the standard deviation of that value in terms of the final listed digits.

Greek Alphabet

A	α	Alpha	N	ν	Nu
B	β	Beta	Ξ	ξ	Xi
Γ	γ	Gamma	O	o	Omicron
Δ	δ	Delta	Π	π	Pi
E	ϵ	Epsilon	P	ρ	Rho
Z	ζ	Zeta	Σ	σ	Sigma
H	η	Eta	T	τ	Tau
Θ	θ	Theta	Υ	υ	Upsilon
I	ι	Iota	Φ	ϕ	Phi
K	κ	Kappa	X	χ	Chi
Λ	λ	Lambda	Ψ	ψ	Psi
M	μ	Mu	Ω	ω	Omega

Table A.1 Chemical Thermodynamic Properties at 298.15 K[a]

Substance	ΔH_f° kJ mol^{-1}	ΔG_f° kJ mol^{-1}	S° J K^{-1} mol^{-1}	C_P° J K^{-1} mol^{-1}
$O(g)$	249.170	231.748	160.946	21.912
$O_2(g)$	0	0	205.029	29.355
$O_3(g)$	142.7	163.2	238.82	39.20
$H(g)$	217.965	203.263	114.604	20.786
$H^+(g)$	1536.202			
$H^+(aq)$	0	0	0	0
$H_2(g)$	0	0	130.574	28.824
$OH(g)$	38.95	34.23	183.636	29.886
$OH^-(aq)$	−229.994	−157.293	−10.75	−148.5
$H_2O(l)$	−285.830	−237.178	69.92	75.291
$H_2O(g)$	−241.818	−228.589	188.715	33.577
$H_2O_2(l)$	−187.78	−120.42	109.6	89.1
$He(g)$	0	0	126.040	20.786
$F(g)$	78.99	61.92	158.645	22.744
$F^-(aq)$	−332.63	−278.82	−13.8	−106.7
$F_2(g)$	0	0	202.67	31.30
$HF(g)$	−271.1	−273.2	173.669	29.133
$Cl(g)$	121.679	105.696	165.088	21.840
$Cl^-(aq)$	−167.159	−131.260	56.5	−136.4
$Cl_2(g)$	0	0	222.957	33.907
$ClO_4^-(aq)$	−129.33	−8.62	182.0	
$HCl(g)$	−92.307	−95.299	186.799	29.12
HCl in $100H_2O$	−165.925			
HCl in $200H_2O$	−166.272			
HCl in ∞H_2O	−167.456	−131.168	55.06	−125.5
$Na^+(aq)$	−239.655	−261.872	60.3	
$K^+(aq)$	−251.21	−282.278	102.5	
$NaOH(c)$	−426.73			80.3
NaOH in $100H_2O$	−469.060			
NaOH in $200H_2O$	−469.0			
NaOH in ∞H_2O	−469.595			
$NaCl(c)$	−411.003	−384.028	72.38	49.71
NaCl in $100H_2O$	−406.894			
NaCl in $200H_2O$	−406.752			
NaCl in ∞H_2O	−407.112	−393.041	115.5	
$Br(g)$	111.884	82.429	174.912	20.786
$Br^-(aq)$	−121.55	−103.97	82.4	−141.8

[a] The values in Table A.1 are from D. D. Wagman, W. H. Evans, V. B. Parker, I. Halow, S. M. Bailey, and R. H. Schumm, *Selected Values of Chemical Thermodynamic Properties*, NBS Technical Note 270-3, 1968, and following Technical Notes. The compounds are in the order of elements used in these tables. For the elements represented in Table A.1, this order is O, H, He, F, Cl, Br, I, S, N, P, C, Pb, Al, Zn, Cd, Hg, Cu, Ag, Fe, Pt, Ti, Mg, and Ca. Within a given element section all of the compounds of that element with elements occurring earlier in the order will be found.

Table A.1—*(Continued)*

Substance	ΔH_f° kJ mol^{-1}	ΔG_f° kJ mol^{-1}	S° J K^{-1} mol^{-1}	C_P° J K^{-1} mol^{-1}
$Br_2(l)$	0	0	152.231	75.689
$Br_2(g)$	30.907	3.142	245.354	36.02
$HBr(g)$	−36.40	−53.43	198.585	29.142
$I(g)$	106.838	70.283	180.682	20.786
$I^-(aq)$	−55.19	−51.59	111.3	−142.3
$I_2(c)$	0	0	116.135	54.438
$I_2(g)$	62.438	19.359	260.58	36.90
$HI(g)$	26.49	1.72	206.485	29.158
S(rhombic)	0	0	31.80	22.64
S(monoclinic)	0.34			
$S^{2-}(aq)$	33.1	85.8	−14.6	
$SO_2(g)$	−296.830	−300.194	248.11	39.87
$SO_3(g)$	−395.723	−371.08	256.65	50.67
$SO_4^-(aq)$	−909.267	−744.626	20.083	−29.3
$HS^-(aq)$	−17.6	12.1	62.8	
$H_2S(g)$	−20.63	−33.56	205.69	34.23
$H_2SO_4(l)$	−813.989	−690.101	156.904	138.91
H_2SO_4 in ∞H_2O	−909.27			
$N(g)$	472.704	455.579	153.189	20.786
$N_2(g)$	0	0	191.50	29.125
$NO(g)$	90.25	86.57	210.652	29.844
$NO_2(g)$	33.18	51.30	239.95	37.20
$NO_3^-(aq)$	−207.36	−111.34	146.4	−86.6
$N_2O_4(g)$	9.16	97.82	304.18	77.28
$NH_3(g)$	−46.11	−16.49	192.34	35.06
$NH_4^+(aq)$	−132.51	−79.37	113.4	79.9
$HNO_3(l)$	−174.10	−80.79	155.60	109.87
HNO_3 in ∞H_2O	−207.36			
$P(c)$	0	0	41.09	23.840
$P(g)$	314.64	278.28	163.084	20.786
$PCl_3(g)$	−287.0	−267.8	311.67	71.84
$PCl_5(g)$	−374.9	−305.0	364.47	112.80
C(graphite)	0	0	5.740	8.527
C(diamond)	1.897	2.900	2.377	6.115
$C(g)$	716.682	671.289	157.987	20.838
$CO(g)$	−110.525	−137.152	197.564	29.116
$CO_2(g)$	−393.509	−394.359	213.64	37.11
$CO_2(aq)$	−413.80	−386.02	117.6	
$CO_3^{2-}(aq)$	−677.14	−527.90	−56.9	
$CH(g)$	595.8			
$CH_3(g)$	138.9			
$CH_4(g)$	−74.81	−50.75	186.155	35.309
$CH_4(l)$	−135.44	−65.27	216.40	131.75
$HCO_3^-(aq)$	−69.20	−586.85	91.2	

Table A.1—*(Continued)*

Substance	ΔH_f° kJ mol^{-1}	ΔG_f° kJ mol^{-1}	S° J K^{-1} mol^{-1}	C_P° J K^{-1} mol^{-1}
$HCHO(g)$	−117.2	−113.0	218.66	35.40
$HCO_2H(l)$	−424.72	−361.41	128.95	99.04
$H_2CO_3(aq)$	−699.65	−623.17	187.4	
$CH_3OH(l)$	−238.66	−166.36	126.8	81.6
$CH_3OH(g)$	−200.67	−162.00	239.70	43.89
$CO(NH_2)_2(c)$	−332.88	−196.82	104.60	93.14
$C_2H_2(g)$	226.73	209.20	200.83	43.93
$C_2H_4(g)$	52.26	68.12	219.45	43.56
$C_2H_6(g)$	−84.68	−32.89	229.49	52.64
$CH_3CO_2^-(aq)$	−486.01	−369.41	86.6	−6.3
$CH_3CHO(l)$	−192.30	−128.20	160.3	
$CH_3CO_2H(l)$	−484.5	−390.0	159.8	124.3
$CH_3CO_2H(aq)$	−485.76	−396.56	178.7	
$C_2H_5OH(l)$	−277.69	−174.89	160.7	111.46
$C_2H_5OH(g)$	−235.10	−168.57	282.59	65.44
$NH_2CH_2CO_2H(c)$	−528.10	−368.57	103.51	99.20
$Pb(c)$	0	0	64.81	26.44
$PbO(c, yellow)$	−217.32	−187.90	68.70	45.77
$PbO_2(c)$	−277.4	−217.36	68.6	64.64
$PbSO_4(c)$	−919.94	−813.20	148.57	103.207
$Al(c)$	0	0	28.33	24.35
$Al_2O_3(c)$	−1675.7	−1582.4	50.92	79.04
$Zn(c)$	0	0	41.63	25.50
$Zn^{2+}(aq)$	−153.89	−147.03	−112.1	46.0
$Cd(c)$	0	0	51.76	25.98
$Cd^{2+}(aq)$	−75.90	−77.580	−73.2	
$Hg(l)$	0	0	76.02	27.983
$Hg(g)$	61.317	31.853	174.85	20.786
$Cu(c)$	0	0	33.150	24.435
$Cu^{2+}(aq)$	64.77	65.52	−99.6	
$Ag(c)$	0	0	42.55	25.351
$Ag^+(aq)$	105.579	77.124	72.68	21.8
$Ag_2O(c)$	−31.05	−11.21	121.3	65.86
$AgCl(c)$	−127.068	−109.805	96.2	50.79
$Fe(c)$	0	0	27.28	25.10
$Fe^{2+}(aq)$	−89.1	−78.87	−137.7	
$Fe^{3+}(aq)$	−48.5	−4.6	−315.9	
$Fe_2O_3(c)$	−824.3	−742.2	87.40	103.85
$Fe_3O_4(c)$	−1118.4	−1015.5	146.4	143.43
$FeCl_2(c)$	−341.79	−302.34	117.95	76.65
$FeCl_2$ ($m = 1$)	−423.4	−341.37	−24.7	
$FeCl_3(c)$	−399.49	−334.05	142.3	96.65
$FeCl_3$ ($m = 1$)	−550.2	−398.3	−146.4	
$Pt(c)$	0	0	41.63	25.86

Table A.1—(*Continued*)

Substance	ΔH_f° kJ mol^{-1}	ΔG_f° kJ mol^{-1}	S° J K^{-1} mol^{-1}	C_P° J K^{-1} mol^{-1}
$Ti(c)$	0	0	30.63	25.02
$TiO_2(c)$	−939.7	−884.5	49.92	55.48
$Mg(c)$	0	0	32.68	24.90
Mg^{2+} (m = 1)	−466.85	−454.8	−138.1	
$MgO(c)$	−601.70	−569.44	26.95	37.15
$MgCl_2$ (m = 1)	−801.15	−717.1	−25.1	
$Ca(c)$	0	0	41.42	25.31
Ca^{2+} (m = 1)	−543.83	−553.54	−53.1	
$CaO(c)$	−635.09	−604.04	39.75	42.80
$CaCl_2$ (m = 1)	−877.13	−816.05	59.8	
$CaCO_3$(c, calcite)	−1206.92	−1128.84	92.9	81.88
$CaCO_3$(c, aragonite)	−1207.13	−1127.80	88.7	81.75

Table A.2 Excerpt from JANAF Thermochemical Tables[a]

T/K	C_P° J K^{-1} mol^{-1}	S° J K^{-1} mol^{-1}	$\frac{-(G^\circ - H_{298}^\circ)}{T}$ J K^{-1} mol^{-1}	$H^\circ - H_{298}^\circ$ kJ mol^{-1}	ΔH_f° kJ mol^{-1}	ΔG_f° kJ mol^{-1}
			C(graphite)			
0	0.000	0.000	∞	−1.054	0.000	0.000
298	8.527	5.686	5.686	0.000	0.000	0.000
500	14.627	11.648	6.887	2.381	0.000	0.000
1000	21.543	24.451	12.636	11.816	0.000	0.000
2000	24.539	40.631	22.970	35.321	0.000	0.000
3000	25.342	50.748	30.348	60.300	0.000	0.000
			C(g)			
0	0.000	0.000	∞	−6.535	709.506	709.506
298	20.841	157.992	157.992	0.000	714.987	669.578
500	20.807	168.757	160.352	4.201	716.807	638.252
1000	20.790	183.171	168.573	14.602	717.774	559.049
2000	20.953	197.606	179.891	35.434	715.100	401.145
3000	21.623	206.217	187.322	56.693	711.380	244.969

[a] This table contains excerpts from D. R. Stull and H. Prophet, *JANAF Thermochemical Tables*, 2nd ed., NSRDS-NBS 37, 1971, and supplements: M. W. Chase, J. L. Curnutt, A. T. Hu, H. Prophet, A. N. Syverud, and L. C. Walker, *J. Phys. Chem. Ref. Data*, **3**, 311 (1974); M. W. Chase, J. L. Curnutt, H. Prophet, R. A. McDonald, and A. N. Syverud, *ibid.*, **4**, 1 (1975); *JANAF Thermochemical Tables*, The Dow Chemical Co., Midland, Mich.: $H(g)$, $H^+(g)$, $H^-(g)$, $H_2(g)$, $N_2(g)$, $O(g)$, $O^-(g)$, $O_2(g)$, $e^-(g)$ 3-31-77 and $NH_3(g)$ 6-30-77. In these tables thermodynamic quantities are given beyond the range of phase stability.

Table A.2—*(Continued)*

T/K	C_P° J K^{-1} mol^{-1}	S° J K^{-1} mol^{-1}	$\frac{-(G^\circ - H^\circ_{298})}{T}$ J K^{-1} mol^{-1}	$H^\circ - H^\circ_{298}$ kJ mol^{-1}	ΔH_f° kJ mol^{-1}	ΔG_f° kJ mol^{-1}
			$CH_4(g)$			
0	0.000	0.000	∞	−10.025	−66.906	−66.906
298	35.639	186.146	186.146	0.000	−74.873	−50.815
500	46.342	306.911	190.510	8.201	−80.818	−32.823
1000	71.797	247.446	209.267	38.179	−89.881	19.351
2000	94.399	305.750	243.952	123.595	−92.462	130.486
3000	101.391	345.586	271.563	222.083	−90.579	241.567
			$CO(g)$			
0	0.000	0.000	∞	−8.669	−113.805	−113.805
298	29.142	197.543	197.543	0.000	−110.529	−137.164
500	39.794	212.719	200.857	5.929	−110.022	−155.410
1000	33.183	234.421	212.735	21.686	−112.010	−200.242
2000	36.250	258.600	230.229	56.739	−118.708	−285.989
3000	37.217	273.508	242.325	93.542	−126.332	−366.506
			$CO_2(g)$			
0	0.000	0.000	∞	−9.364	−393.150	−393.150
298	37.129	213.685	213.685	0.000	−393.522	−394.505
500	44.627	234.814	218.187	8.314	−393.677	−394.965
1000	54.308	269.215	235.806	33.405	−394.639	−395.924
2000	60.350	309.210	263.483	91.450	−396.639	−396.442
3000	62.229	334.084	283.131	152.862	−399.057	−395.869
			$Cl(g)$			
0	0.000	0.000	∞	−6.272	119.608	119.608
298	21.836	165.076	165.076	0.000	121.290	105.311
500	22.744	176.640	167.594	4.523	122.261	94.219
1000	22.234	192.318	176.502	15.816	124.311	65.333
2000	21.343	207.393	188.636	37.510	127.047	5.197
3000	21.062	215.982	196.418	58.689	128.876	−56.158
			$HCl(g)$			
0	0.000	0.000	∞	−8.640	−92.127	−92.127
298	29.137	186.795	186.795	0.000	−92.312	−95.302
500	29.305	201.886	190.100	5.891	−92.914	−97.169
1000	31.627	222.798	201.752	21.046	−94.399	−100.805
2000	35.602	246.145	218.664	54.957	−95.575	−106.633
3000	37.246	260.931	230.438	91.479	−96.307	−112.018
			$Cl_2(g)$			
0	0.000	0.000	∞	−9.180	0.000	0.000
298	33.936	222.961	222.961	0.000	0.000	0.000

T/K	C_P° J K^{-1} mol^{-1}	S° J K^{-1} mol^{-1}	$\frac{-(G^\circ - H^\circ_{298})}{T}$ J K^{-1} mol^{-1}	$H^\circ - H^\circ_{298}$ kJ mol^{-1}	ΔH_f° kJ mol^{-1}	ΔG_f° kJ mol^{-1}
500	36.083	241.116	226.903	7.104	0.000	0.000
1000	37.472	266.676	241.095	25.585	0.000	0.000
2000	38.279	292.934	261.178	63.509	0.000	0.000
3000	39.221	308.612	274.542	102.211	0.000	0.000
			$H(g)$			
0	0.000	0.000	∞	−6.197	216.037	216.037
298	20.786	114.604	114.604	0.000	217.999	203.296
500	20.786	125.353	116.960	4.197	219.254	192.983
1000	20.786	139.758	125.173	14.590	222.246	165.540
2000	20.786	154.168	136.478	35.376	226.907	106.872
3000	20.786	162.594	143.875	56.162	229.789	46.170
			$H^+(g)$			
0	0.000	0.000	∞	−6.197	1528.085	
298	20.786	108.834	108.834	0.000	1536.306	1517.102
500	20.786	119.583	111.190	4.197	1541.758	1502.562
1000	20.786	133.988	119.399	14.590	1555.143	1458.183
2000	20.786	148.398	130.708	35.376	1580.590	1351.223
3000	20.786	156.825	138.105	56.162	1604.259	1232.355
			$H^-(g)$			
0	0.000	0.000	∞	−6.197	143.264	
298	20.786	108.847	108.847	0.000	138.909	132.143
500	20.786	119.591	111.202	4.197	135.967	128.386
1000	20.786	134.001	119.411	14.590	128.566	123.650
2000	20.786	148.411	130.721	32.376	112.445	124.788
3000	20.786	156.837	138.118	56.162	94.537	134.788
			$HI(g)$			
0	0.000	0.000	∞	−8.657	28.535	28.535
298	29.158	206.485	206.485	0.000	26.359	1.573
500	29.736	221.656	209.798	5.929	−5.632	−10.096
1000	33.137	243.300	221.656	21.640	−6.745	−14.021
2000	36.623	267.575	239.145	56.865	−6.757	−21.213
3000	37.915	282.700	251.295	94.211	−6.720	−28.464
			$H_2(g)$			
0	0.000	0.000	∞	−8.468	0.000	0.000
298	28.836	130.574	130.574	0.000	0.000	0.000
500	29.259	145.628	133.867	5.883	0.000	0.000
1000	30.204	166.113	145.427	20.686	0.000	0.000

Table A.2—*(Continued)*

T/K	C_P° J K^{-1} mol^{-1}	S° J K^{-1} mol^{-1}	$\frac{-(G^\circ - H_{298}^\circ)}{T}$ J K^{-1} mol^{-1}	$H^\circ - H_{289}^\circ$ kJ mol^{-1}	ΔH_f° kJ mol^{-1}	ΔG_f° kJ mol^{-1}
2000	34.288	188.297	161.829	52.932	0.000	0.000
3000	37.066	202.778	173.197	88.743	0.000	0.000
			$H_2O(g)$			
0	0.000	0.000	∞	−9.904	−238.919	−238.919
298	33.577	188.724	188.724	0.000	−241.827	−228.597
500	35.208	206.413	192.573	6.920	−243.831	−219.078
1000	41.217	232.597	206.614	25.978	−247.885	−192.631
2000	51.103	264.571	228.229	72.689	−251.668	−135.566
3000	55.664	286.273	244.153	126.361	−253.258	−77.145
			$I(g)$			
0	0.000	0.000	∞	−6.197	107.248	107.248
298	20.786	180.682	180.682	0.000	106.847	70.291
500	20.786	191.426	183.037	4.197	76.065	50.304
1000	20.794	205.836	191.246	14.590	77.036	24.163
2000	21.309	220.359	202.573	35.568	78.898	−29.418
3000	22.192	229.170	210.058	57.333	81.262	−84.061
			$I_2(g)$			
0	0.000	0.000	∞	−10.117	65.521	65.521
298	36.878	260.584	260.584	0.000	62.442	19.376
500	37.438	279.814	264.789	7.510	0.000	0.000
1000	37.911	305.930	279.567	26.363	0.000	0.000
2000	38.522	332.398	300.106	64.588	0.000	0.000
3000	39.091	348.126	313.662	103.395	0.000	0.000
			$N(g)$			
0	0.000	0.000	∞	−6.197	470.784	470.784
298	20.786	153.193	153.193	0.000	472.646	455.512
500	20.786	163.942	155.549	4.197	473.884	443.563
1000	20.786	178.351	163.762	14.590	476.503	412.166
2000	20.790	192.757	175.071	35.376	479.955	346.368
3000	20.966	201.209	182.468	56.220	482.516	279.002
			$NO(g)$			
0	0.000	0.000	∞	−9.192	89.772	89.772
298	29.844	210.652	210.652	0.000	90.291	86.596
500	30.489	226.158	214.041	6.058	90.349	84.077
1000	33.987	248.433	226.204	22.230	90.437	77.772
2000	36.648	273.027	244.095	57.861	90.483	65.053
3000	37.468	288.064	256.404	94.977	89.847	52.446

T/K	C_P° J K^{-1} mol^{-1}	S° J K^{-1} mol^{-1}	$\frac{-(G^\circ - H^\circ_{298})}{T}$ J K^{-1} mol^{-1}	$H^\circ - H^\circ_{298}$ kJ mol^{-1}	ΔH_f° kJ mol^{-1}	ΔG_f° kJ mol^{-1}
			$NO_2(g)$			
0	0.000	0.000	∞	−10.188	35.924	35.924
298	36.974	239.923	239.923	0.000	33.095	51.241
500	43.208	260.529	244.333	8.100	32.150	63.839
1000	52.166	293.780	261.437	32.342	32.003	95.726
2000	56.442	331.682	288.052	87.262	33.087	159.000
3000	57.396	354.782	306.691	144.273	32.899	221.940
			$N_2(g)$			
0	0.000	0.000	∞	−8.669	0.000	0.000
298	29.125	191.502	191.502	0.000	0.000	0.000
500	29.577	206.631	194.811	5.912	0.000	0.000
1000	32.698	228.057	206.598	21.460	0.000	0.000
2000	35.987	251.699	223.898	56.141	0.000	0.000
3000	37.049	266.793	235.877	92.738	0.000	0.000
			$N_2O_4(g)$			
0	0.000	0.000	∞	−16.397	18.715	18.715
298	77.258	304.277	304.277	0.000	9.079	97.717
500	97.207	349.234	313.808	17.769	8.761	157.992
1000	119.211	425.007	352.029	72.982	15.188	305.189
2000	129.035	511.649	412.388	198.522	33.062	588.321
3000	131.202	564.459	454.843	328.850	48.995	862.389
			$NH_3(g)$			
0	0.000	0.000	∞	−10.058	−38.920	−38.920
298	35.627	192.602	192.602	0.000	−45.898	−16.380
500	41.991	212.464	196.845	7.812	−49.869	4.778
1000	56.346	246.233	213.648	32.581	−55.074	61.890
2000	72.048	291.062	241.977	98.178	−55.191	179.540
3000	79.496	321.783	263.718	174.184	−51.195	296.144
			$O(g)$			
0	0.000	0.000	∞	−6.724	246.785	246.785
298	21.912	160.946	160.946	0.000	249.170	231.748
500	21.259	172.084	163.398	4.343	250.471	219.572
1000	20.916	186.678	171.816	14.862	252.676	187.736
2000	20.828	201.133	183.280	35.710	255.283	121.667
3000	20.937	209.602	190.740	56.576	256.722	54.522
			$O^-(g)$			
0	0.000	0.000	∞	−6.573	105.730	105.730
298	21.543	157.787	157.787	0.000	101.671	91.416

Table A.2—(*Continued*)

T/K	C_P° J K^{-1} mol^{-1}	S° J K^{-1} mol^{-1}	$\frac{-(G^\circ - H^\circ_{298})}{T}$ J K^{-1} mol^{-1}	$H^\circ - H^\circ_{298}$ kJ mol^{-1}	ΔH_f° kJ mol^{-1}	ΔG_f° kJ mol^{-1}
500	21.108	168.799	160.205	4.297	98.726	85.282
1000	20.878	183.335	168.557	14.774	90.504	74.935
2000	20.811	197.773	179.971	35.606	72.304	66.279
3000	20.799	206.208	187.406	56.409	52.869	67.438
			$O_2(g)$			
0	0.000	0.000	∞	−8.682	0.000	0.000
298	29.372	205.033	205.033	0.000	0.000	0.000
500	31.091	220.589	208.413	6.088	0.000	0.000
1000	34.878	243.475	220.769	22.707	0.000	0.000
2000	37.777	268.655	239.057	59.199	0.000	0.000
3000	39.961	284.399	251.697	98.098	0.000	0.000
			$e^-(g)$			
0	0.000	0.000	∞	−6.197	0.000	0.000
298	20.786	20.874	20.874	0.000	0.000	0.000
500	20.786	31.618	23.230	4.197	0.000	0.000
1000	20.786	46.028	31.439	14.590	0.000	0.000
2000	20.786	60.434	42.748	35.376	0.000	0.000
3000	20.786	68.860	50.141	56.162	0.000	0.000

Table A.3 Relative Atomic Masses and Natural Abundances of Stable Isotopes of Some Elements

Atomic Number	Element	Mass Number	Atomic Mass	Abundance, %
1	H	1	1.007 825	99.99
		2	2.014 10	0.01
3	Li	6	6.015 13	7.5
		7	7.016 01	92.5
6	C	12	12.000 00	98.89
		13	13.003 35	1.11
7	N	14	14.003 07	99.6
		15	15.000 11	0.4
8	O	16	15.994 91	99.76
		17	16.999 14	0.04
		18	17.999 16	0.20
9	F	19	18.998 40	100.0
11	Na	23	22.989 77	100.0
16	S	32	31.972 07	95.0
		33	32.971 46	0.8
		34	33.967 86	4.2
17	Cl	35	34.968 85	75.4
		37	36.965 90	24.6
35	Br	79	78.918 3	50.5
		81	80.916 3	49.5
53	I	127	126.904 4	100.0

INDEX